Wind Farm Noise:
Measurement, Assessment and Control

Wiley Series in Acoustics, Noise and Vibration:

Wind Farm Noise	**Hansen**	**February 2017**
The Effects of Sound on People	Cowan	May 2016
Engineering Vibroacoustic Analysis: Methods and Applications	Hambric et al	April 2016
Formulas for Dynamics, Vibration and Acoustics	Blevins	November 2015

Wind Farm Noise:
Measurement, Assessment and Control

Colin H. Hansen
School of Mechanical Engineering
University of Adelaide
Australia

Con J. Doolan
School of Mechanical and Manufacturing Engineering
University of New South Wales
Australia

Kristy L. Hansen
School of Computer Science, Engineering and Mathematics
Flinders University
Australia

This edition first published 2017

Registered Office
John Wiley & Sons Ltd, The Atrium, Southern Gate, Chichester, West Sussex, PO19 8SQ, United Kingdom

For details of our global editorial offices, for customer services and for information about how to apply for permission to reuse the copyright material in this book please see our website at www.wiley.com.

Library of Congress Cataloging-in-Publication Data

Names: Hansen, Colin H., 1951- author. | Doolan, Con J., author. | Hansen, Kristy L., author.
Title: Wind farm noise : measurement, assessment and control / Colin H. Hansen, Con J. Doolan, Kristy L. Hansen.
Description: Hoboken : John Wiley & Sons Inc., [2017] | Includes bibliographical references and index.
Identifiers: LCCN 2016036077| ISBN 9781118826065 (cloth) | ISBN 9781118826126 (epub) | ISBN 9781118826119 (Adobe PDF)
Subjects: LCSH: Wind power plants–Noise.
Classification: LCC TK1541 .H26 2017 | DDC 621.31/2136–dc23 LC record available at https://lccn.loc.gov/2016036077

A catalogue record for this book is available from the British Library.

Cover Image: Gettyimages/Dazzo

Set in 10/12pt, WarnockPro by SPi Global, Chennai, India

10 9 8 7 6 5 4 3 2 1

This book is dedicated to our families without whose patience it may not have been completed.

There are three sides to every story: your side, my side and the truth. And no-one is lying.

Robert Evans, an American film producer born June 29, 1930.

If we knew what it was we were doing, it wouldn't be called research, would it?

Albert Einstein commenting on research.

Clever is the person who believes half of what he hears. Brilliant is the person who chooses the right half to believe....

Contents

Wiley Series in Acoustics, Noise and Vibration

This book series will embrace a wide spectrum of acoustics, noise and vibration topics from theoretical foundations to real world applications. Individual volumes will range from specialist works of science to advanced undergraduate and graduate student texts. Books in the series will review scientific principles of acoustics, describe special research studies and discuss solutions for noise and vibration problems in communities, industry and transportation.

The first books in the series include those on Biomedical Ultrasound; Effects of Sound on People, Engineering Acoustics, Noise and Vibration Control, Environmental Noise Management; Sound Intensity and Windfarm Noise. Books on a wide variety of related topics.

The books I edited for Wiley, the Encyclopedia of Acoustics (1997), the Handbook of Acoustics (1998) and the Handbook of Noise and Vibration Control (2007) included over 400 chapters written by different authors. Each author had to restrict their chapter length on their special topics to no more than about 10 pages. The books in the current series will allow authors to provide much more in-depth coverage of their topic.

The series will be of interest to senior undergraduate and graduate students, consultants, and researchers in acoustics, noise and vibration and in particular those involved in engineering and scientific fields, including, aerospace, automotive, biomedical, civil/structural, electrical, environmental, industrial, materials, naval architecture and mechanical systems. In addition the books will be of interest to practitioners and researchers in fields such as audiology, architecture, the environment, physics, signal processing and speech.

Malcolm J. Crocker
Series Editor

Preface

Wind farm noise has polarised communities and is featured on numerous web sites that either dismiss its effects on people as a *nocebo* effect or as something in their imagination. There are just as many other web sites that claim wind farm noise has led to serious medical problems in some people and that infrasound generated by wind farms can have far-reaching consequences for the health of people who are exposed. These web sites can be found easily by typing 'wind farm noise' into any internet search engine.

Our intention when writing this book has been to cover all aspects of wind farm noise, including how it is generated, how it propagates, how it is assessed, how it is regulated and what effects it has on people living in the vicinity of wind turbines. Where aspects of wind farm noise are controversial, we have presented what we believe to be an unbiased assessment of the facts. None of the three authors have ever worked for the wind farm industry nor have they been members of any anti-wind-farm organisation. Only the first author has appeared as an expert witness, in a 2010 court proceedings concerned with a wind farm development. This was his only involvement in court proceedings and it was in the capacity of being asked to critique a report prepared by an acoustical consultant for a wind farm operator.

The first two authors have been chief investigators on a number of research projects, funded by the Australian Research Council, on aerodynamic noise generation and the impact of wind farm noise on rural communities. The first author has also spent over 40 years teaching, researching and consulting in acoustics and noise control. The second author has spent nearly 20 years working in the area of aerospace engineering, with a strong focus on aeroacoustics: the science of how objects like rotor blades create sound. Following completion of a PhD in fluid mechanics, the third author has spend the past four years measuring and analysing wind farm noise.

Wind farm noise is a very controversial subject, in that it has been used as a reason for delaying many wind farm projects that together are worth billions of dollars. Most court cases find in favour of the wind farm developer and very few wind farms are prevented from being constructed as a result of court proceedings based on excessive noise, although sometimes the turbine layout has had to be modified to minimise noise impacts on the surrounding communities. Nevertheless, even after wind farms have been constructed, many people complain of the noise keeping them awake at night and causing them to feel ill. In spite of the many reported cases of adverse effects of wind farm noise on people, wind farm proponents insist that wind farm noise is so low in level that it could not possibly be a problem. They often imply that affected people must be developing symptoms as a result of feelings of jealousy over payments received by wind turbine hosts or as a result of anti-wind-farm publicity telling them that wind farms produce such symptoms. Although we are neither pro- nor anti-wind-farm campaigners, we do believe that some people in the vicinity of some wind farms are badly affected by the noise and that further research into this phenomenon is absolutely essential.

We hope that you the reader find the material in this book useful and, where it strays into areas that are controversial, that you find that we have achieved our aim of presenting a balanced point of view.

Colin Hansen
Con Doolan
Kristy Hansen
Adelaide, Australia

1

Wind Energy and Noise

1.1 Introduction

Why write this book about noise generated by wind farms? Many people believe that wind farm noise is a non-issue and that people complain about it because they are unhappy with the lack of financial compensation they receive compared to their neighbours who are hosting the turbines. Other reasons that we often see on pro-wind-farm web sites are that the anti-wind-farm lobby has suggested a range of symptoms are caused by wind farms and that this suggestion has made some people living near wind farms develop these symptoms as a result: the 'so-called' *nocebo* effect. Although the authors of this book would consider themselves neither pro- nor anti-wind-farms, they have taken a sufficient number of their own measurements and spoken to a sufficient number of residents living in the vicinity of wind farms (including wind farm hosts) to appreciate that the character and level of wind farm noise is a problem for a significant number of people, even those who reside at distances of 3 km or more from the nearest turbine.

Although one chapter in this book is concerned with the effects of wind farm noise on people, the main focus of this book is on how wind farm noise is generated and propagated, the characteristics of the noise arriving at residences in the vicinity of wind farms, and measurement procedures and instrumentation, as well as assessment criteria that are necessary for properly quantifying the noise. As many people living in the vicinity of wind farms report 'feeling' vibration when they lie down, vibration generation, propagation and measurement are also discussed in sections in Chapters 4, 5 and 6.

To lay the foundation for the remaining chapters, the rest of this chapter is concerned with a description of how the wind industry has developed in various countries, followed by a brief history of noise studies (including a summary of noise levels generated by large wind turbines), a summary of some public inquiries and wind farm noise regulations, and finally a discussion of the current consensus on wind farm noise and its effects on people.

It is not possible to usefully take part in the wind farm noise debate without having some understanding of acoustics. This is the reason for writing Chapter 2 to follow. First, basic concepts in acoustics necessary for understanding the legislation are discussed. This is followed by a discussion of the fundamentals of frequency analysis, which is an important tool for analysing wind farm noise. Chapter 2 concludes with a discussion of

Wind Farm Noise: Measurement, Assessment and Control, First Edition.
Colin H. Hansen, Con J. Doolan, Kristy L. Hansen.

some advanced concepts of frequency analysis, an understanding of which is essential for practitioners wishing to undertake more advanced analyses of wind farm noise.

Chapter 3 contains an overview of how wind turbines generate noise, while Chapter 4 is about estimating wind turbine sound power levels. Chapter 5 is concerned with using turbine sound power levels and sound propagation models to estimate noise levels in the community. Several propagation models are considered, beginning with the simplest and progressing to the more complex and supposedly more accurate models. Chapter 6 is devoted to a detailed description of procedures and instrumentation for the measurement of wind farm noise and vibration, and includes a discussion of potential errors associated with such measurements. The chapter also includes a discussion on wind tunnel measurements for testing turbine models. Chapter 7 is about the effects of wind farm noise on people, Chapter 8 contains a discussion of various options that can reduce wind turbine noise, both outside of and inside residences, and Chapter 9 contains some suggestions of where we should be heading in terms of wind farm noise research and the reduction of its effects on people.

1.2 Development of the Wind Energy Industry

1.2.1 Early Development Prior to 2000

Mankind has harvested energy from the wind for over a thousand years. The first device designed for this purpose was a vertical-axis, sail-type windmill developed in Persia between 500 and 900 AD. This design appears to have been inspired by boats that used their sails to harness the wind for propulsion. Windmills have been primarily used for water pumping and grain grinding, with the mechanical power developed in the rotating shaft used directly to drive a pump or turn a grindstone. Wind turbines differ from windmills in that they convert the mechanical power into electrical power through use of a generator. They also have a smaller number of blades, since windmills require high torque at low rotor speeds (Manwell et al. 2009); for optimal electrical power generation higher rotor speeds and thus fewer blades are desirable. This is because high rotor speeds result in increased loading and reducing the number of blades reduces stresses on the rotor (Manwell et al. 2009). Another factor to consider is that wind turbine blades are very costly and therefore it is beneficial to minimise their number.

The most common wind turbine configuration that is used today is a horizontal-axis wind turbine (HAWT) and this book will concentrate on aspects of noise associated with this particular design, with a focus on large, industrial-scale wind turbines. The major components of a HAWT are shown in Figure 1.1. The basic principle of operation is that wind causes the blades to rotate and the rotor drives a shaft that is connected, generally via a gearbox, to a generator, which converts the rotational energy into electrical energy.

The power output and rotational speed of a HAWT can be controlled either by designing the blades such that they begin to stall at a certain wind speed (stall control) or by having a mechanism and control system that is able to vary the blade pitch (pitch control, which involves rotation of the blades about the blade axis as opposed to the rotor axis). In a pitch-controlled turbine, the controller will continually adjust the blade pitch to ensure that the power output is optimised for the wind speed being experienced by the blade. A pitch controlled machine can also be easily 'turned off' to protect the turbine when

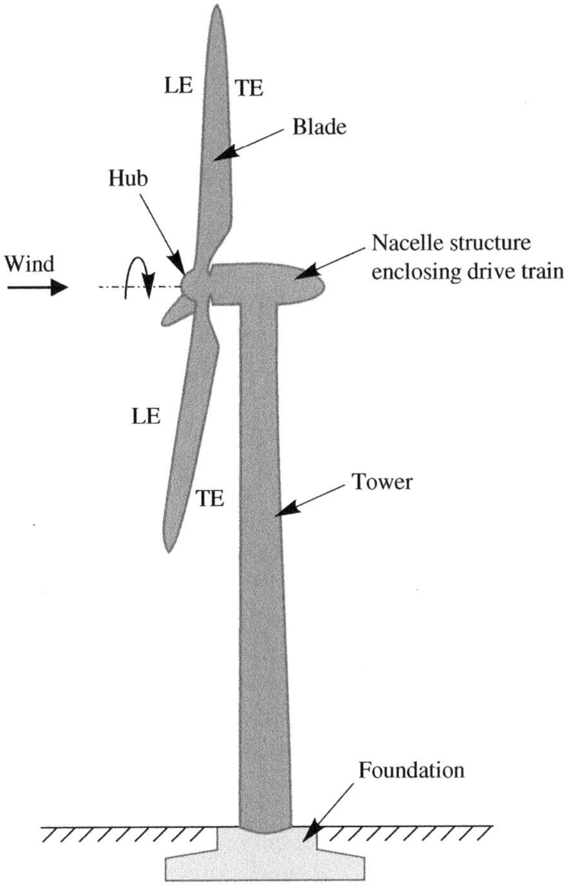

Figure 1.1 Schematic of typical wind turbine: LE, leading edge; TE, trailing edge.

the wind speed becomes too great. This is done by adjusting the pitch of the blades so that they no longer generate appreciable lift. A stall-controlled turbine blade is designed with some twist to ensure the blade stalls gradually along its length. The blade profile also has to be designed so that it stalls just as the wind speed becomes too high, thus reducing the lift force acting on the blade, which in turn limits the blade speed and power. An active stall-controlled turbine is similar to a pitch-controlled turbine in that the pitch is continually adjusted to optimise the power output. However, when the wind speed becomes too great, the stall-controlled turbine will rotate the blades so that they stall, as opposed to a pitch-controlled turbine, which rotates the blades in the opposite direction so that the lift is minimised. In some cases, turbines are also controlled using yaw control. This involves turning the rotor so the blades no longer face directly into the wind. However, this is only used on small turbines and is not relevant to the turbines that are the subject of this book.

Development of large HAWTs for incorporation into electric utilities first began in the early 1930s with the construction of the Balaklava wind turbine in Russia, which was 30 m in diameter, two-bladed and rated to a power of 100 kW. This turbine operated for around two years and generated 200 MWh (Sektorov 1934). In the late 1930s, development of the first megawatt-scale wind turbine began in the USA in a collaborative project between an engineer named Palmer C. Putnam and the Smith company, which was experienced in the construction of hydroelectric turbines and electrical power equipment. The Smith–Putnam HAWT consisted of a two-bladed rotor of diameter 53.3 m, mounted on a truss-type tower at a rotor-axis height of 33.5 m (Putnam 1948). This wind turbine was rated at 1.25 MW and included a number of technological innovations such as blade-pitch control, flapping hinges on the blades to reduce dynamic loading on the shaft, and active yaw control (Spera 2009). Several weeks of continuous operation yielded excellent power production and it was demonstrated that the wind turbine was capable of being inserted into the grid. Unfortunately, development was discontinued in 1945 when a faulty blade spar separated at the repair weld and there were insufficient funds to continue the project.

Over the next 25 years, development proceeded at a modest rate, taking place predominantly in Western Europe, where there was a temporary post-war shortage of fossil fuels that led to increased energy prices. Two HAWT designs emerged from Denmark and Germany during this time, and these would form the basis of future wind turbine development in the 1970s. The 24-m diameter, 200 kW Gedser Mill wind turbine was constructed in Denmark and was designed by Johannes Juul. The rotor consisted of three fixed-pitch blades that were connected with a support frame to improve structural integrity. This frame was removed in later years when the metal blades were replaced with fibreglass ones (Dodge 2006). The rotor was located upwind of the concrete tower and the design was notable for its simplicity, ruggedness and reliability. This wind turbine supplied AC power to the local utility from 1958 until 1967, achieving annual capacity factors of 20% in some years (Spera 2009). The annual *capacity factor* is defined as the ratio of the energy generated in one year to the amount that could be generated if the turbines operated continuously at their maximum power output. In 1967, a mechanical failure resulted in discontinued use of the wind turbine and the machine remained idle for the next 10 years (Auer 2013).

Considerable research effort, with a focus on improved rotor technology, led to the development of the Hütter–Allgaier wind turbine in Germany in the early 1960s. With a diameter of 34 m and rated at 100 kW, it was technologically advanced for its time and included an important design feature of a bearing at the rotor hub that allowed the rotor to 'teeter', in order to minimise the dynamic loading that results from the changes in gyroscopic inertia about the tower axis that arise when the blades of a two-bladed rotor move between the horizontal and vertical positions. A teetering rotor is illustrated schematically in Figure 1.2, which shows the bearing that facilitates the teetering motion. Despite its technological proficiency, the Hütter–Allgaier wind turbine encountered flutter in its long, slender blades, which slowed research progress.

Wind turbines were successfully connected to the grid in France in the period from 1958 to 1964 and the largest such turbine was called the Type Neyrpic, which was 35 m in diameter and rated at 1.1 MW. While this wind turbine demonstrated good performance, its operation was terminated abruptly when the turbine shaft broke.

In the UK, a number of unique 100 kW wind turbine designs were conceived and built in the 1950s with the intention of local grid connection. These turbines operated successfully for a few years, but technical and environmental factors led to the cessation of operations by 1963. Many projects were discontinued during this 25-year period due

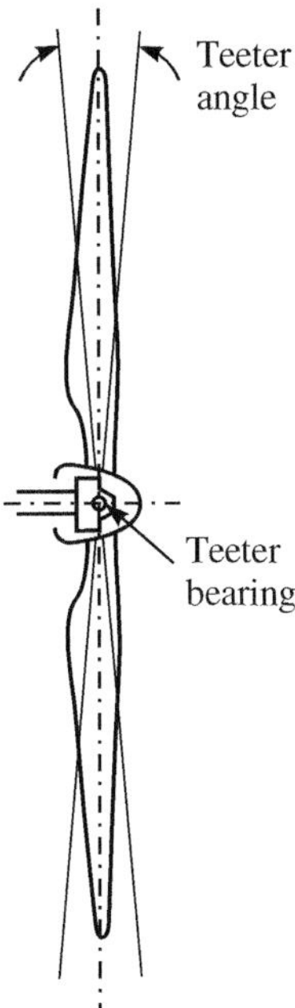

Figure 1.2 Schematic of rotor showing ability to teeter.

to technical issues and adverse weather conditions that resulted in expensive failures. These issues were not investigated further at this time due to a lack of interest in funding research into alternative energy sources, which was directly related to the availability of inexpensive fossil fuels and nuclear resources. Therefore, the Smith–Putnam wind turbine remained the largest in the world until the oil crisis of the 1970s prompted further development in the wind industry.

In the late 1970s, centres were established in Denmark, Germany and the Netherlands for testing of experimental and commercial wind turbines. These centres were also responsible for certification programs for tax or subsidy benefits to ensure that wind turbines met defined standards before entering the market. The International Energy Agency (IEA) was also established in the mid-1970s to encourage cooperation between Western countries on research, policy and development on wind energy. By the early 1990s, several countries had developed wind turbines with power ratings in the megawatt range, including Canada, Denmark, Germany, Italy, the Netherlands, Spain, Sweden, the UK and the USA (International Energy Agency 1989).

Canada pursued a different approach to most countries in the design and construction of megawatt-scale wind turbines, electing to focus on a Darrieus-type vertical-axis wind turbine. The Eolé Darrieus wind turbine was 64 m in diameter, 96 m in height, rated at 4 MW and was completed in 1987 (Richards 1987). Despite being rated at 4 MW, the power was limited to 2.5 MW to increase the lifespan of the turbine. The Eolé was connected to the Hydro-Quebec grid and it operated for over 30 000 h until 1993, generating over 12 GWh of electricity during its lifetime (Tong 2010). It was stopped due to damage to its expensive lower bearing.

Wind turbine development in Denmark proceeded at a modest rate and the size of turbines increased incrementally (Gipe 1995). Two upwind prototypes with a rated power of 630 kW, called Nibe A and B, were constructed in the early 1980s based on a similar design concept to the Gedser wind turbine. Nibe A was stall-controlled whereas Nibe B was pitch-controlled (International Energy Agency 1989), thus enabling a performance comparison to be made between these control mechanisms. The prototypes operated for 15 years, providing a wealth of information that contributed to later development of the wind industry in Denmark (Spera 2009).

The 54-m diameter, 2 MW Tvind wind turbine was a three-bladed downwind machine built by teachers and students from the Tvind school who collaborated with consultants, sub-contractors, volunteers and experts such as Professor Ulrich Hütter. Hütter's influence was evident in the choice of a downwind design (turbine blades downwind of the support tower) and the advanced blade technology. It was later discovered that the downwind configuration resulted in excessive low-frequency noise generation, as will be discussed in Section 1.3. The combination of design principles incorporated into the Gedser-type wind turbines such as the upwind, heavy, three-bladed, asynchronous generator and Hütter/Tvind advanced blade design and root assembly led to a successful combination that influenced future wind turbine development (Maegaard 2013).

In 1982, Germany embarked on an ambitious project to build the 100-m diameter Growian wind turbine, rated at 3 MW, which was the largest HAWT at the time. This wind turbine was a two-bladed downwind machine, which incorporated some of the latest technological innovations including full-span pitch control, carbon filament blades, a tubular steel tower and variable-speed operation. These features would later prove to be successful, but the overall design was over-ambitious for its time and was soon dismantled. In 1991, the 3 MW Aeolus II wind turbine, with a diameter of 80 m, was developed as a collaborative project between Germany and Sweden. In this two-bladed upwind design, advanced blade technology enabled a reduction in weight from 22 tons (existing Swedish Aeolus) to 6 tons (International Energy Agency 1989).

The two-bladed, 1.5 MW Gamma 60 upwind turbine constructed in Italy in 1991 was distinctive for its active yaw control, which provided a means of power regulation above rated speed. The 66-m diameter Gamma 60 design also incorporated a teetered hub as well as a direct current link between the synchronous generator and the step-up transformer. The innovative features of this wind turbine contributed to increased annual energy production, as well as eliminating control components on rotating parts to reduce complexity, thus decreasing manufacturing and maintenance costs (International Energy Agency 1989).

In the Netherlands, development was focussed on the 0.2–1.0 MW range and a number of demonstration projects were initiated by utility companies. The 1 MW NEWECS-45 wind turbine was developed in 1986 and consisted of a 45-m diameter rotor in an upwind configuration. The rotor design consisted of two-blades on which full-span pitch control was implemented and the tower was constructed from tubular steel.

Collaboration between Germany and Spain led to the successful development of the AWEC-60 wind turbine in 1989, which was a 1.2 MW, 60-m diameter, three-bladed upwind machine based in Spain. Whilst the design was based on the German 1.2 MW, WKA-60 wind turbine, further development was undertaken on the electrical system, glass-fibre reinforced blades and the control system in order to reduce the cost (International Energy Agency 1989).

Development of large-scale wind turbines was launched rapidly and successfully in Sweden in the 1980s. The WTS-75 was a two-bladed upwind turbine rated at 2 MW, which possessed some unique design features including a drive-train system with bevelled gears that eliminated the need for power slip rings and a mechanism for raising and lowering all major components. Sweden also constructed the WTS-3 downwind turbine, rated at 3 MW, which produced a relatively large amount of energy compared with other large-scale wind turbines of the 1980s. This design incorporated a teetered hub and a spring-mounted gearbox to reduce the impact of dynamic loading associated with the two-bladed, downwind configuration.

After a number of design iterations, the UK produced the 3 MW LS-1 in 1987, which was a two-bladed upwind turbine of 60 m diameter. The rotor consisted of a teetered hub mounted on elastomeric bearings and the outer 30% of the blades were mounted on rolling element bearings, which enabled variation of the blade pitch angle (Hau 2013). The drive-train design provided control of the rotor speed to within ±5% (Hau 2013).

Following the Arab oil embargo of 1973, the US government invested significant funds into a federal research plan directed towards wind energy development. The first large wind turbine that was developed as part of this program was the MOD-0 configuration in 1975. Over the next decade, this wind turbine was used extensively for testing to identify possible improvements that could be made to the design. It was 38.1 m in diameter, rated at 100 kW and was mounted atop a truss-type tower at a hub height of 30.5 m. The design was similar to the Smith–Putman and Hütter–Allgaier wind turbine designs, where the rotor was two-bladed and located downwind of the tower. Several modifications were made to the MOD-0 over its twelve-year lifetime, including replacement of the truss-type tower with a slender shell tower to reduce wake-induced fatigue loads and incorporation of a teetered hub. The extensive testing that was undertaken also resulted in a large volume of documentation, computer models and control algorithms that form the basis of modern wind turbine technology. During the early stages of the MOD-0 program, an upgraded version of this turbine, rated at 200 kW, was integrated into the grid at four separate locations and designated the MOD-0A. The locations were chosen to ensure that wind power would make up a significant proportion of the input power to the grid, enabling grid connectivity issues to be identified. The wind turbines collectively fed 3.6 GWh into their respective grids during their operating lives (Shaltens and Birchenough 1983) and achieved capacity factors as high as 0.48 (Spera 2009).

The first megawatt-scale wind turbine to be developed as part of the US federal research plan was the MOD-1 configuration in 1979. This model was designed before the problems with the MOD-0 had been identified and understood and therefore was dismantled after only two years of operation. The MOD-1 was 61 m in diameter, rated at 2 MW and resembled the MOD-0 configuration in that it had a downwind, two-bladed rotor, rigidly mounted on a truss-type tower. While this turbine was successfully integrated into the local grid, impulsive loading caused by the substantial wake deficit behind the truss tower resulted in a severe risk of early fatigue as well as environmental problems such as excessive low-frequency noise and electromagnetic interference. The next turbine in the series was the 91.4-m diameter, 2.5 MW MOD-2, which was an upwind design and represented a large technical leap from the earlier models in the US federal plan. The design employed partial-span pitch control on its two blades, which simplified the use of a teetered hub, as only the outer portion of the blades needed to rotate about the blade axis to enable control. A comprehensive testing program was carried out on the MOD-2 design, including investigation of wake-interaction effects, operation strategies and control algorithms. Successful integration into the local grid was also realised, with a group of three MOD-2 wind turbines installed at Goodnoe Hills in 1981, and contributing over 10 GWh to the local grid during 16 000 h of operational time. This group of three formed the first 'wind farm' in the world (Boeing 2015) proving conclusively that groups of wind turbines could operate in a completely automated mode.

The MOD-5B was the next wind turbine in the series to be developed under the federal wind energy program. The design was similar to the MOD-2 wind turbine, in that it consisted of a two-bladed upwind rotor, teetered hub, partial-span pitch control and a tubular steel tower. However, this wind turbine was larger, with a rotor diameter of 97.5 m and a rated power of 3.2 MW. Built in 1987, it was the first large-scale wind turbine to operate at variable speed, which led to improved efficiency and reduced structural loading (Spera 2009). From 1988, the MOD-5B wind turbine was connected to the grid on the island of Oahu in Hawaii and was fully automatic, with software changes made using the local public telephone system (Boeing 2015). The MOD-5B demonstrated excellent performance for such a large and advanced design, and during its lifetime of six years, it ran for 20 561 h and produced 26.8 GWh of electricity.

The largest wind turbine to be built before the year 2000 was the WTS-4, which was a two-bladed downwind machine, rated at 4 MW, with a hub height of 80.4 m and a diameter of 78 m. The support tower was a single 12-sided cylindrical structure of shell construction. Despite its large power rating, this wind turbine produced a relatively small amount of energy during its lifetime (Spera 2009) and it ceased operating after only four years due to a generator failure. The wind turbine was bought for a fraction of its original cost by a local engineer and wind energy enthusiast, who later watched the machine fly to pieces in a storm (Righter 1996).

1.2.2 Development since 2000

Wind power has expanded rapidly since the beginning of the 21st century to the point where there are so many different models of megawatt-rated wind turbines that further consideration of individual models is beyond the scope of this book. The rapid expansion is a result of increased awareness of global warming and eventual fossil fuel depletion, as well as rising concerns over energy security. The amount of global energy generated since the year 2000, as plotted in Figure 1.3, has consistently increased at an average annual rate of approximately 25% and was 17 times higher in 2013 than in 2000.

When describing the relative contribution of wind energy, it is common to refer to the *installed power*, which is the product of the manufacturer's power rating and the number of turbines. This is also referred to as the *installed capacity*. However, this measure does not take into account such factors as wind variability, interactions between wind turbines, lack of grid connectivity and wind turbine malfunctions. Therefore, a more conservative measure is the actual energy generated, which is measured in TWh (terawatt-hours) for large-scale turbines. The annual capacity factor is the ratio of the energy generated over one year to the amount that would be generated if the wind farm

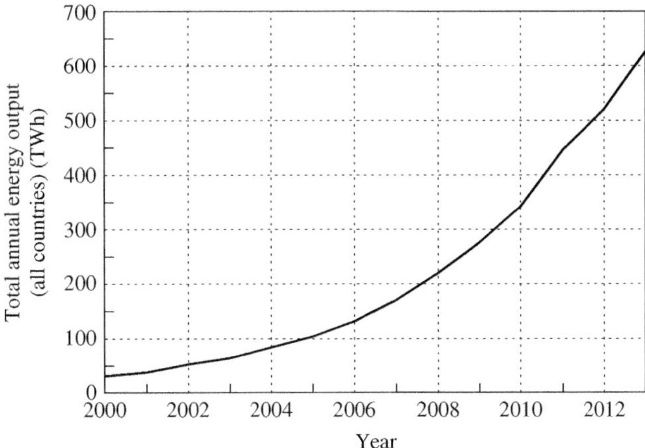

Figure 1.3 Global annual energy output (TWh).

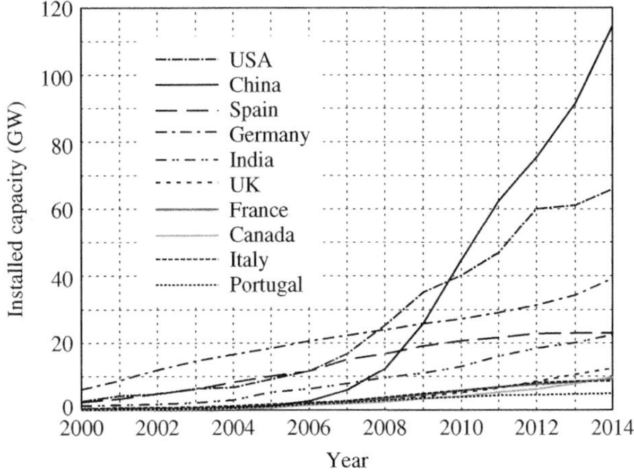

Figure 1.4 Installed capacity of wind power (2000–2014).

operated continuously at its maximum power output. This gives an indication of the overall efficiency of wind energy as a whole.

Figures 1.4–1.6 show the installed power capacity (the rated turbine power in megawatts multiplied by the number of turbines in the wind farm), generated energy (TWh) and annual capacity factor, respectively, for the top ten countries in terms of wind energy generated in 2013. The data in these figures have been compiled from information provided by the IEA, Global Wind Energy Council and the US Energy Information Administration. Germany was leading the world with installed power and generated wind energy in the year 2000 and since then has been increasing its installed power at a rate close to linear. On the other hand, the USA and China's installed power and generated wind energy has been increasing exponentially since the year 2000. As a result, these two countries have emerged as leaders in available and generated

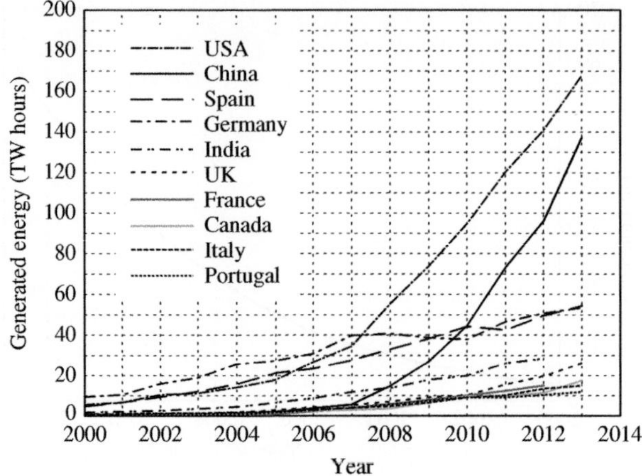

Figure 1.5 Generated wind energy (2000–2014).

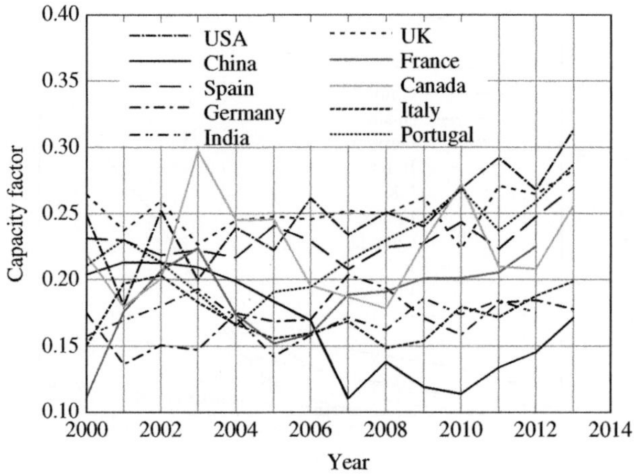

Figure 1.6 Annual capacity factor (2000–2014).

wind power as of 2013/2014. Comparison of the figures for the USA and China reveals that although China has almost double the installed power capacity of the USA, the latter still generates a larger amount of energy. It is therefore not surprising that the annual capacity factor for China is lowest for the ten countries compared in Figure 1.6. Conversely, the USA has the highest annual capacity factor, and has been the world leader for energy generation since 2007.

Technological development of wind turbines has focussed on reduction of costs, increased energy capture and greater reliability. To this end, wind turbines have become progressively larger to take advantage of the high-energy winds that occur at greater altitudes. Several advances have also been made in blade design. These include optimising the blade profile to increase efficiency in low winds, limiting aerodynamic loads in high winds and minimising blade fouling. Advanced composite materials have also been specified in blade designs in place of steel and wood, to improve the strength-to-weight ratio. Most large wind turbines today are variable-speed, pitch-regulated machines, which allows operation at near-optimum ratios between the blade-tip speed and wind speed, thus maximising output power. Modern designs are also predominantly three-bladed, as this number provides the best compromise between aerodynamic efficiency, cost, rotational mass, structural integrity, inertial stability, relatively low tip-speed ratios and aesthetics. The use of fewer blades results in increased aerodynamic efficiency since each blade disturbs the air for the following one. The cost and weight of each blade is substantial, so from this perspective fewer blades are also preferred. Also, the strength and stiffness of each blade is greater when there are fewer blades for a given rotor solidity (total blade planform area divided by swept area). On the other hand, wind turbines with less than three blades experience unbalanced loading during yaw, which can be overcome by using a teetered hub, although this is an extra complication that most manufacturers prefer to avoid. The rotational speed of a three-bladed design is lower than a one- or two-bladed design, resulting in lower tip-speed ratios and hence reduced trailing-edge noise. Many people prefer to look at turbines with three blades rather than one or two blades and since community acceptance is important for wind farm developers, this point is also taken into account.

The drive trains of wind turbines have become lighter and more reliable in recent times, which is an important development, since failure of drive-train components such as the gearbox is costly and the associated downtime is high (Ekwaro-Osire et al. 2011). In the late 1990s, direct-drive generators were introduced as an alternative to gearboxes, but despite their numerous advantages, their size and weight are issues that have prevented widespread use (Spera 2009). These days, the generator components of wind turbines are required to synchronise with electricity grids and they are therefore capable of producing AC electricity, in contrast to early wind turbines, which were developed as stand-alone units and employed DC generators. Control systems for wind turbines have become more sophisticated in recent times as well, with high-speed digital controls enabling processing of data from a number of sensors for optimised power generation. Advanced control algorithms have been developed to facilitate more efficient data processing and optimal actuator responses to sensor inputs. While a number of early wind turbine designs integrated steel truss-type or concrete monopole (single support cylinder or partial-cone) towers into their designs, modern wind turbines consist of a steel, monopole structure with a reinforced concrete foundation.

1.2.3 Support Received by the Wind Industry

Wherever wind energy has been developed successfully, it has been with the aid of government intervention in the form of financial, technical or regulatory support. The reason for this is that, at the time of writing, wind energy is more expensive than energy derived from coal or gas and the industry would be non-viable without financial incentives from governments. However, with many renewable energy targets in place around the globe, it seems that wind power is the least expensive way of achieving them. Of course, it appears that wind turbines are a very clean and environmentally friendly power source, as power is generated without producing greenhouse gases. But are they? To answer this question, one must consider the greenhouse gases that are produced during the construction of wind farms, from transportation of materials to the construction site, and during their maintenance and decommissioning. It is also important to consider the intermittency of wind power and the current lack of energy storage facilities, resulting in significant security and reliability concerns for electrical grids worldwide (Miskelly 2012). Through analysis of power output data provided by the Australian Energy Market Operator, Miskelly (2012) demonstrated that during the full calendar year of 2010, there were over 100 incidences where the entire Eastern Australian grid generated less than 2% of installed capacity. A consequence of these common-mode failures is the need for a rapid response from fossil-fuel-driven power stations, resulting in inefficient operation of these facilities and production of excessive greenhouse emissions at these times.

Recently Weißbach et al. (2013) compared wind energy with other energy generation facilities in terms of its energy return on investment (EROI) value and the number of years to achieve payback on the energy invested in construction, and his results are presented in Table 1.1. The EROI value is the ratio of the usable energy that the energy facility returns during its lifetime to all the invested energy needed to generate this energy. Weißbach further analysed the EROI value in terms of the cost of buffering needed to maintain a continuous power supply, considering the unreliability of the energy source. These values are also included, but must be considered in light of the economic threshold for the EROI value being about 7 (Weißbach et al. 2013), indicating that wind power produces considerably more energy than needed to construct and run the turbines, but that intermittency of supply makes it economically non-viable.

In Table 1.1, CSP is concentrated solar power, such as achieved by an array of mirrors directed at the apex of a tower or a large array of flat or parabolic reflectors. In the case of the more expensive but more efficient parabolic reflectors, sunlight is focussed onto a receiver tube at the focal point of the reflectors, thus heating molten salt as it flows through the tube. In all cases the heat energy generated is used to boil water to drive a steam turbine, which in turn drives a generator to produce electricity. CCGT refers to a combined-cycle gas turbine facility, in which waste heat from the gas turbine is used

Table 1.1 EROI and energy payback times for various energy generation facilities.

Energy generation type	EROI	EROI (buffered)	Energy payback time unbuffered (years)	Energy payback time buffered (years)
Solar PV (roof)	3.9	2.3	6	16
Solar PV (field)	3.8	2.3	6	16
Biomass (corn)	3.5	3.5	0.033	0.033
Wind	16	4	1	5
Solar CSP (parabolic mirror, desert)	21	9.6	1	3.5
CCGT	28	28	0.025	0.025
Coal	30	30	0.167	0.167
Hydro	50	35	2	3
Nuclear	75	75	0.167	0.167

Data from Weißbach et al. (2013).

to generate steam to power a steam turbine, with both turbines driving an electrical generator.

1.3 History of Wind Turbine Noise Studies

Here, some of the earliest reported studies concerning wind farm noise are explored. A large proportion of the work reported here was carried out in the 1980s in response to a noise issue associated with operation of the MOD-1 wind turbine. This was the first well-documented case of acoustic disturbance from a wind turbine that was significant enough to provoke complaints from neighbours. While the noise issue was exacerbated by the fact that the rotor of the MOD-1 was located downwind from the tower, it was shown that a similar mechanism was at play for an upwind rotor (Spencer 1981). The main difference between the two rotor configurations was shown to be the magnitude of the flow deficit encountered by the blades, and consequently the level of noise generated, which is much greater for the downwind configuration (Spencer 1981). Another difference, described by Kelley et al. (1985), is that the blades of a downwind rotor experience transient lift fluctuations due to the periodic vortex shedding that occurs behind the support tower, although the effect is smaller than the flow deficit effect. These differences in the blade inflow conditions cause the noise levels associated with a downwind configuration to be much higher. On the other hand, since upwind turbines also experience a flow deficit as well as inflow turbulence, many of the findings from the studies on downwind turbines are still relevant to modern upwind wind turbine designs.

As mentioned in Section 1.2.1, the MOD-1 was a downwind machine with a two-bladed rotor that was rigidly mounted on a truss-type tower. Detailed investigations carried out by Kelley and his colleagues culminated in a comprehensive report, which identified the issue as unsteady loading imparted to the rotor blades as they passed through the tower wake (Kelley et al. 1985). This phenomenon resulted in high levels of low-frequency impulsive noise that excited structural resonances and interior air volume modes of nearby houses, sometimes causing loose objects to vibrate (Kelley et al. 1985). Measurements indicated that the impulsive character of the noise was directly related to the presence of blade-pass frequency components. Noise propagation was found to be governed by a combination of atmospheric refraction and terrain reflection, which were responsible for focussing the noise towards locations occupied by residences. Due to noise complaints received from about a dozen families living within a 3-km radius of the wind turbine, MOD-1 was slowed down from 35 to 23 RPM and it was found that an 11-dB reduction in sound pressure levels could be achieved (Viterna 1981). On the other hand, there was a corresponding increase in the level of impulsive noise in the 8 and 16 Hz octave bands and while annoyance was reduced, it was not eliminated (Kelley et al. 1988). Noise issues with the MOD-1 turbine prompted a number of investigations on this specific configuration, including field measurements, modelling and wind tunnel experiments.

A predictive model for determining the amplitude of the blade-pass harmonics was developed by Viterna (1981) and implemented in computer software called WTSOUND. The approach was based on theory for aircraft propellers first developed by Gutin in 1937 (Gutin 1948). In summary, the process developed by Viterna (1981) involved the following steps:

1. Calculating the steady aerodynamic blade forces.
2. Determining the variation in these forces due to unsteady aerodynamics.
3. Carrying out a Fourier analysis of the force variation.
4. Calculating sound pressure levels in the acoustic field, by assuming the aerodynamic source to be compact with an effective radius of 75% of the blade span.

The calculated results were in good agreement with the MOD-1 data in the vicinity of two rotor diameters from the wind turbine. However, in the far field, the model underestimated the actual levels of the MOD-1 by 6 dB or more due to propagation effects related to the terrain and atmospheric conditions. Nonetheless, the model accurately recreated the $\sin(f)/f$ spectrum shape characteristic of a pulse of finite width in the time domain, where f is the frequency. For the MOD-1 wind turbine, this pulse of finite length and relatively steep edges resulted from the blade lifting surface passing through a flow deficit. Metzger and Klatte (1981) found that the spectrum envelope was very sensitive to the shape of the flow deficit and that harmonics of the blade-pass frequency in the higher frequency range could be avoided by ensuring that the shape followed a Gaussian profile. One possibility for achieving this was by modifying the tower shape in the vicinity of the rotor blades. However, this would only be effective for one wind direction unless the tower had a lightweight external shell that could rotate as the wind direction changed (Tocci and Marcus 1982).

A model of the MOD-1 wind turbine was constructed and tested in the anechoic wind tunnel at the NASA Langley Research Centre. Researchers carefully scaled the tower details to ensure that the wake would be recreated as accurately as possible (Greene 1981). Results from these experimental studies indicated that the impulsive noise associated with the MOD-1 wind turbine could be significantly reduced by using an upwind configuration. Therefore, one of the primary motives for using an upwind configuration in the MOD-2 wind turbine design was to avoid the impulsive noise issues that were associated with its predecessor (Kelley et al. 1988). The acoustic emissions of the MOD-2 wind turbine were investigated extensively by Kelley et al. (1988) and it was found that the impulsive noise was significantly reduced. Further reduction in the levels and impulsiveness of the low-frequency noise emitted by the MOD-2 machine was achieved by incorporating vortex generators and pitch schedule changes. It is worth noting that the impulsive noise was not eliminated entirely and that the degree of impulsiveness was strongly correlated with the vertical atmospheric stability, the vertical or upwash turbulence length scale and the blade loading (Kelley et al. 1988). Spencer (1981) found that the spectrum shape of a MOD-2 wind turbine with a two-bladed upwind rotor was very similar to the MOD-2 with a downwind rotor in the frequency range from 0 to 45 Hz. The main difference between these spectra was the relative amplitude of the blade-pass harmonics, since the flow deficit associated with the downwind case was much larger than the flow deficit for the upwind turbine.

Apart from impulsive noise generated by blade–tower interaction, a number of other aeroacoustic noise sources associated with wind turbine operation were identified and modelled in the late 1980s and early 1990s. The investigated sources were mainly broadband in nature and resulted from inflow turbulence and airfoil self-noise. Grosveld (1985) found good agreement between predictions of broadband noise and far-field measurements in the vicinity of the two-bladed MOD-OA, MOD-2 and WTS-4 wind turbines and the three-bladed US Windpower Inc. wind turbine. The prediction model considered contributions from inflow turbulence to the rotor, trailing-edge effects and the wake due to a blunt trailing edge and it was found that at low frequencies the dominant source was inflow turbulence noise (Grosveld 1985). Glegg et al. (1987) developed a prediction method for wind turbines that included the source mechanisms of unsteady lift noise, unsteady thickness noise, trailing-edge noise and the noise from separated flow. To determine the inflow turbulence, which is a required input for the unsteady lift and thickness calculations, a detailed model of the atmospheric boundary layer was implemented. Good agreement was obtained between the atmospheric boundary layer model and anemometer measurements, but a 10-dB discrepancy was noted between the measured and calculated acoustic results. Improved correspondence between measurements and predictions was achieved by assuming a turbulence length scale equal to the blade chord. The authors also observed that the presence of the tower on an upwind turbine caused significant acoustic scattering when the rotor blades were close to the tower and hence this effect was also incorporated into their theoretical model (Glegg et al. 1987). However, due to the short duration of this effect, it was found to have a negligible contribution to the average level. On the other hand, the authors noted that it would increase the detectability of the signal.

A review of the aeroacoustic noise generated by large wind turbines was presented by Hubbard and Shepherd (1991) and an additional mechanism of impulsive noise generation was attributed to rotor inflow velocity gradients. Various wind velocity profiles were assumed as inputs to the model developed by Viterna (1981) and the results were compared to measurements recorded up to 80 m from the two-bladed WWG-0600 upwind turbine. There was good agreement between the results, for a specific assumed atmospheric wind velocity profile resulting from the atmospheric boundary layer. On the other hand, the actual wind velocity profile was not measured and therefore it is not known if the assumed velocity profile was accurate.

Propagation of noise from the MOD-1 wind turbine was investigated in detail by Thompson (1982) through analysis of field measurements and development of a computational model. The results indicated that the primary mechanism responsible for enhanced far-field noise levels was atmospheric refraction of acoustic waves caused by vertical wind shear. The influence of ground and surface wave propagation on the enhanced noise levels was found to be negligible in comparison to this effect. Based on the collected data, conditions of adverse noise propagation were predicted to occur about 30% of the time at complex terrain sites. A similar investigation was carried out by Willshire (1985) and Willshire Jr and Zorumski (1987) on the WTS-4 wind turbine. It was shown that low-frequency sound was refracted in the downwind direction, resulting in an attenuation rate of 3 dB per doubling of distance for frequencies below 20 Hz. Predictions of both ray tracing and normal-mode theoretical models supported this observation. In the upwind direction, the absence of a shadow zone was noted for these infrasonic frequencies and the propagating signals indicated a spherical spreading characteristic, resulting in an attenuation of 6 dB/doubling of distance.

The acoustic and vibratory response of buildings to wind farm noise was explored by Stephens et al. (1982) and it was shown that in some circumstances, low-frequency wind turbine noise could be perceived more readily indoors than outdoors. A number of reasons for this phenomenon were presented, including selective attenuation of higher frequencies by the building structure, room modes, structural resonances and noise-induced vibrations. Other complicating factors mentioned included the role of stiffness and air leaks at low frequencies (Stephens et al. 1982). Enhanced perception of indoor low-frequency noise was attributed to an increase in the indoor noise level relative to the outside level at specific frequencies and large variations in sound pressure level as a function of room position (Hubbard and Shepherd 1991).

Thresholds of perception were determined by Stephens et al. (1981) by exposing subjects to a range of impulsive stimuli of the type associated with blade–tower interaction. The test stimuli were synthesised based on MOD-1 data and blade–tower interaction calculations and presented to the listening subjects in an anechoic chamber. The resulting spectra consisted of harmonics of the fundamental frequencies of 0.5 Hz and 1 Hz that were dominated by specific frequencies. The resulting perception thresholds were found to be lower than the pure-tone threshold and it was observed that the chosen fundamental frequency influenced the results. A lower fundamental frequency gave rise to a lower perception threshold. In a later publication (Stephens et al. 1981), the authors presented additional perception curves for use when various levels of background noise were present in addition to the impulsive wind turbine signal. These curves indicated that the perception threshold increased with the level of background noise but that this increase was relatively less for lower frequencies. For a background level of 35 dBA, the perception threshold was still below the pure-tone threshold at all frequencies, according to the ISO389-7 (2005). Comparison of various metrics used in the assessment of low-frequency noise was carried out by Kelley (1987). Evaluators were exposed to simulated signals characteristic of wind turbine noise emissions and they subsequently recorded their perception of the noise according to specified categories and rankings. The researchers then determined the correlation between the stimulus sequences and the evaluators' responses. It was found that people reacted to a low-frequency noise environment and that the A-weighting (see Section 2.2.11) is not an adequate measure of annoyance when low frequencies are dominant (Kelley 1987). The results also indicated that the low-frequency sound level (LSL) (Tokita et al. 1984) and C-weighting (see Section 2.2.11) metrics were the most 'efficient' descriptors of low-frequency noise annoyance.

A method to control the impulsive noise associated with the flow deficit encountered by the wind turbine blades was proposed by Tocci and Marcus (1982). The method involved reduction of the flow deficit through use of airfoil-shaped fairings for the tower, which could be rotated into the appropriate direction to minimise the wake deficit. This technique would also reduce the flow deficit in front of the tower and is therefore relevant for reducing aerodynamic noise associated with blade–tower interaction for modern upwind turbines.

1.3.1 Modern Wind Turbine Sound Power Levels

Readers unfamiliar with acoustics terminology should consult Chapter 2 prior to reading further. More detail on the estimation of turbine sound power is provided in Chapter 4.

The noise output of many modern wind turbines has been reported by Søndergaard (2013) in the form of overall A-weighted sound power levels in dB re 10^{-12} Watts. He showed that the total A-weighted sound power level of turbines with a rated power greater than 2 MW could be described by Eq. (1.1) (within ±5 dB).

$$L_{wA} = 8.8\ \log_{10}(kW) + 59.6 \qquad (\text{dB re } 10^{-12}\text{W}) \tag{1.1}$$

where kW is the turbine rated power in kilowatts.

Søndergaard (2013) also provided low-frequency data in the range 10–160 Hz and showed that the A-weighted sound power level of turbines with a rated power greater than 2 MW could be described by Eq. (1.2) (within ±5 dB).

$$L_{wA} = 10.3\ \log_{10}(kW) + 74.9 \qquad (\text{dB re } 10^{-12}\text{W}) \tag{1.2}$$

The data provided by Søndergaard (2013) were used to derive the relationships in Eqs. (1.1) and (1.2), which show that as the turbine rated power increases, so too does the noise it produces over the entire frequency spectrum, with low-frequency noise increasing by the same amount as mid- and high-frequency noise.

The relative frequency distribution of the noise for turbines with a rated power greater than 2 MW is shown in Figure 1.7, where the decibel difference between the 1/3-octave band sound power level and the total sound power level is plotted.

An interesting result from Sondergard's analysis is that modern pitch-RPM regulated turbines do not produce any more noise as the hub height wind speed increases above about 8 m/s. However, it is quite a different story for stall- and active-stall-regulated turbines. For these turbines, the noise output increases markedly as the wind speed increases above 8 m/s, by anything from 8 to 12 dBA for a wind speed increase to 12 m/s. The 8 m/s wind speed was chosen because this is the wind speed at which all turbines reach 95% of their rated power.

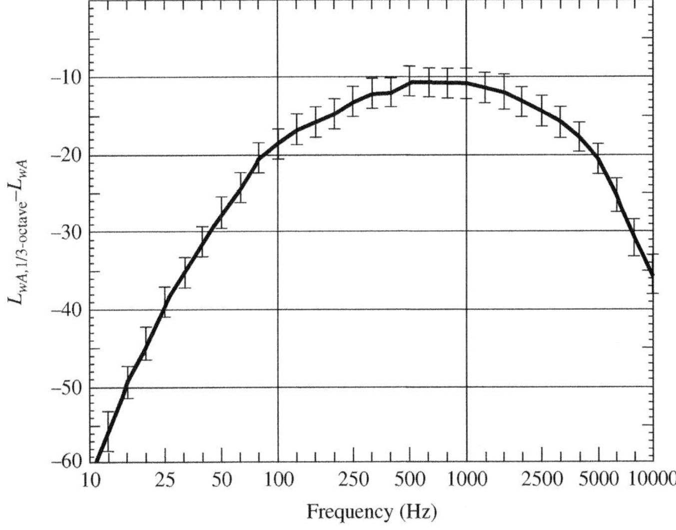

Figure 1.7 Wind turbine 1/3-octave band A-weighted sound power level (dB) normalised to the total A-weighted sound power level, for turbines with a power rating greater than 2 MW. The error bars show the 95% confidence interval around the mean. Data from Søndergaard (2013).

1.4 Current Wind Farm Noise Guidelines and Assessment Procedures

Worldwide, there is a plethora of standards, guidelines and recommended assessment procedures for wind farm noise and its allowable limits. As many as possible are reviewed here but unfortunately it is not possible to capture every single one that has been published. Here we are concerned with assessment procedures and allowable limits – standards that concern sound power measurements or sound propagation predictions are discussed in relevant sections elsewhere in this book and only the parts relevant to allowable limits and assessment guidelines are discussed here.

In the following paragraphs and later in the book, the term *wind shear* is used quite often, so it is useful to discuss its meaning here. Wind shear is a way of describing the variation of wind speed with altitude. For high wind shear conditions, the wind speed increases much more rapidly with altitude than it does under low wind shear conditions. Low wind shear generally occurs during the day, when solar radiation heats the ground and causes mixing in the atmospheric boundary layer within a few hundred metres of the ground, resulting in only a small variation of wind speed with altitude in the boundary layer. At night, in the absence of solar radiation, there is reduced mixing in the boundary layer and it is quite common for wind turbines to be generating a significant amount of power while there is little or no wind close to the ground. One method of quantifying wind shear is with the use of a shear exponent (or coefficient), ξ, which relates wind speed to altitude. The ratio of the wind speed, $U(h)$, at height, h, to the wind speed U_0 at height h_0 is given by,

$$U(h) = U_0\left(\frac{h}{h_0}\right)^{\xi} \tag{1.3}$$

where the wind shear coefficient can vary from 0.1 to 0.6, depending on the ground surface roughness and the atmospheric stability. This is discussed in much more detail in Section 5.2.4.

1.4.1 ETSU-R-97 (used mainly in the UK and Ireland)

ETSU-R-97, a detailed report published in 1996 (ETSU 1996), was perhaps the first government-sponsored work written with the specific purpose of providing detailed planning guidance for the assessment and rating of noise from wind farms. It was prepared by a working group sponsored by the UK Department of Transport and Industry. It follows on from a UK Department of Planning Policy Guidance note (PPG22) published in 1993, which was written at a time when not much was known about wind farm noise and its effects on people, although there was a considerable body of work available from NASA on noise emissions and health effects from the much noisier downwind turbines. ETSU-R-97 is important in that many existing local guidelines for wind farm noise, both in the UK and other countries, have been based on it. However, much more has been discovered about wind farm noise, its characteristics and its effects on people in the past two decades, so the guidance in ETSU-R-97 has become outdated and inappropriate in many areas. The main problems with the recommendations in ETSU-R-97 are listed below.

1. The recommended use of the $L_{A90,10\text{min}}$ metric for measurement of wind farm noise, which is the sound pressure level exceeded 90% of the time in a 10-min period, is based on the assumption that wind farms produce a steady noise at residences, in other words, one that does not vary very much in level over a 10-min period. Many more recent studies, including those of the authors of this book, have shown this to not be the case and although many current guidelines use the more acceptable $L_{Aeq,10\text{min}}$ metric, there are still a number that continue to use the $L_{A90,10\text{min}}$ metric. The $L_{Aeq,10\text{min}}$ metric represents a noise energy average and generally produces wind farm noise levels in the range 1.5 to 3 dB above those obtained using the $L_{A90,10\text{min}}$ metric. However, the $L_{A90,10\text{min}}$ metric is generally considered appropriate for measurement of background noise prior to construction of the wind farm, to minimise contributions from transient noise sources that are not representative of actual background noise levels occurring for the majority of the time. Both of these metrics are defined in Section 2.2.12.
2. ETSU-R-97 recommends that the acceptable nighttime limit for noise should be 43 dBA, based on the outdated WHO guideline that 35 dBA indoors is an acceptable level for people trying to sleep. The WHO has since updated their acceptable indoor noise level for sleep to 30 dBA (with windows open), but most guidelines based on ETSU-R-97 do not account for this change. It is strange that the recommended limit for nighttime is greater than the 35 to 40 dBA limit for daytime and this has received the appropriate criticism in the literature (Cox et al. 2012).
3. Allowable levels (day and night) can be increased to 45 dBA for residents receiving financial benefit as a result of the wind farm.
4. The noise-level limits described above do not account for the low-frequency nature of wind farm noise by the time it reaches the bedrooms in residences more than 1–1.5 km from the nearest turbine.
5. ETSU-R-97 states that noise from a wind farm should be limited to 5 dBA above background noise levels. However, it is stated that this limit should not apply in environments where the daytime background noise levels are less than the limits mentioned in items 2 and 3 above. It was assumed that audible noise above background is not disturbing to people living in low-noise environments, which is why this aspect of ETSU-R-97 has drawn considerable criticism. However, it is clear that an industrial noise source of 35 or 40 dBA superimposed on a typical rural environment of 20 dBA or less would be intrusive to most people.
6. The allowance of 5 dBA above background is based on use of the $L_{A90,10\text{min}}$ metric for measurement of wind farm noise, rather than the usual $L_{Aeq,10\text{min}}$ metric, which means that when comparing with other standards, the allowed increase of $L_{Aeq,10\text{min}}$ over background $L_{A90,10\text{min}}$ is actually at least 7 dBA, and could be more, depending on the difference between $L_{A90,10\text{min}}$ and $L_{Aeq,10\text{min}}$ for wind farm noise. This then exceeds the limit beyond which complaints may be expected.
7. There is no consideration of the common nighttime situation where wind shear is high and there is no (or very little) wind at a residence, while at the same time there is sufficient wind to drive the turbines. This negates the assumption that as the hub height wind speed increases, background noise due to the wind will increase at a greater rate than wind turbine noise.
8. There is no specific adjustment for amplitude modulation (see Section 2.2.7) or maximum allowed modulation depth (see Eq. (2.26)), nor is there any allowance

for random amplitude variation, which is caused by sound from different turbines combining destructively or constructively at any receiver.

9. A curve-fit procedure is used for establishing background $L_{A90,10\text{min}}$ noise levels as a function of measured or estimated wind speed at a height of 10 m above the ground. The curve-fit procedure involves collecting over 2000 data points of noise levels as a function of hub-height wind speed (converted to a standardised 10 m height). At least 500 data points must be for the condition of the wind blowing towards the receiver (±45° from directly downwind). Each data point represents a 10-min average and there is no requirement to separate out nighttime from daytime data. The data points are plotted on a graph of $L_{A90,10\text{min}}$ vs hub-height wind speed and a line of best fit (regression line) is drawn through the data points. Generally, the line is a second- or third-order polynomial (see item 10). Using the line of best fit to establish background noise levels prior to construction of the wind farm means that half the data represent noise levels that are less than the established levels. In establishing background noise levels for the purpose of assessing the intrusiveness of wind farm noise, it would be more reasonable to use levels that are at least one standard deviation (if not two standard deviations) below the average line of best fit. Using one standard deviation would still result in 16% of the data points being less than the specified level, whereas two standard deviations would result in 2.5% of the data being less than the specified level.
10. In the procedure for establishing background noise levels, there is no mention of the electronic noise floor problem of instrumentation, which currently varies between 16 and 18 dBA for most monitoring systems. This means that all measured levels below 26–28 dBA (depending on the actual noise floor) must be corrected to obtain the true noise level. The amount to be subtracted from the measured level can range from 0.5 dB for a measured level 10 dBA above the instrument electronic noise floor to 7 dBA for a measurement 1 dBA above the noise floor. This has a large effect on the order of the curve fit used, which generally should be a straight line or a second-order polynomial rather than the third-order polynomial suggested in the Institute of Acoustics (IOA) good practice guide (IOA 2013) for implementation of ETSU-R-97. In fact, the order of polynomial chosen has a large influence on the resulting 'average' background noise levels at both low and high wind speeds.
11. ETSU-R-97 states that noise levels should be measured using a microphone on a tripod 1.2–1.5 m above the ground in 10-m wind speeds of up to 12 m/s. It is now well known that wind speeds at 10 m of 12 m/s will often result in excessive wind noise on a microphone at a height of 1.2 m, particularly if only a 90-mm diameter primary wind screen is used, with no large secondary wind screen. It is far more practical to measure background noise, as well as wind farm noise, using a microphone mounted on the ground and with a large secondary wind screen in addition to the standard 90-mm wind screen commonly used. This minimises the effect of wind noise on the microphone that results in erroneous measurements at higher wind speeds. Unfortunately, it is difficult to accurately relate the ground-level measurement to what would be measured at a height of 1.5 m, as the relationship between the two measurements is frequency- and ground-type-dependent and is also dependent on interaction between all the contributing noise sources (see Section 6.2.4).

Institute of Acoustics *Good Practice Guidelines*

In 2013, the (British) Institute of Acoustics released a *Good Practice Guide* to the use of ESTU-R-97, in response to a number of instances where it had been applied incorrectly. The *Good Practice Guide* contains a total of 21 recommendations directed at ensuring consistent application of ETSU-R-97 when calculating noise levels in the community that might be expected from a proposed wind farm. The more important recommendations are summarised below.

1. Engage all relevant parties as soon as possible.
2. Background noise surveys should include all areas where the L_{A90} levels due to the wind farm are expected to exceed 35 dBA.
3. Exclude noise levels from existing wind farms in the background noise assessment.
4. Although background noise measurements can be carried out at any time of the year, seasonal effects leading to raised levels should be excluded.
5. Enhanced microphone wind screens should be used. However, these are not defined.
6. Measurement locations for background noise should be 3.5–20 m from a dwelling and noise from local sources should be excluded as much as possible by choice of measurement position. Remaining noise from local sources should be removed from the data. The morning chorus of bird calls should also be removed. If rush-hour traffic at night does not occur regularly, then data including these events should also be removed.
7. Noise measurements should be correlated with wind speed values corresponding to the standardised 10-m height. They should be calculated from from hub-height wind speed, which can either be measured directly or calculated from measurements made at two heights, with the higher of the two being no lower than 60% of hub height. The hub-height wind speed, U_H, is determined from wind speed measurements, U_1 and U_2, at two heights, h_1 and h_2, respectively, (where h_2 is the upper height) using the following two equations.

$$U_H = U_2\left(\frac{h_H}{h_2}\right)^{\xi} \tag{1.4}$$

where h_H is the hub height and the shear coefficient ξ is calculated using

$$\xi = \frac{\log_{10}(U_2/U_1)}{\log_{10}(h_2/h_1)} \tag{1.5}$$

For cases where $U_1 > U_2$, the hub-height wind speed is set equal to U_2.
The standardised wind speed at a height of 10 m is derived from the hub-height wind speed using Eq. (1.6).

$$U_{10\text{m}} = U_H\left[\frac{\log_e\left(\frac{10}{z_0}\right)}{\log_e\left(\frac{h_H}{z_0}\right)}\right] \tag{1.6}$$

where $z_0 = 0.05$ m is the standard ground roughness length used for all ground surfaces, $U_{10\text{m}}$ is the standardised wind speed at 10 m and U_H is the wind speed determined at hub height, either by direct measurement or by use of Eqs. (1.4) and (1.5). Because Eq. (1.6) is only valid for a neutral atmosphere, the standardised wind speed at 10 m will rarely be the same as the actual wind speed measured at 10 m. Also, to be strictly correct, Eq. (1.6) should have a '1+' in the numerator and denominator as shown in Eq. (5.17).

8. Rain-affected data must be excluded from background noise measurements.
9. For background noise measurements, no fewer than 200 valid data points should be recorded in each of the amenity hours (defined below) and nighttime periods, with no fewer than 5 valid data points in any 1-m/s wind-speed bin.
10. Amenity hours are defined as 18:00 to 23:00 every day plus 13:00 to 18:00 Saturday and 07:00 to 18:00 Sunday. Night hours are from 23:00 to 07:00 every night.
11. Directional background noise measurements may be necessary in cases where an existing noise source such as a busy road is on the opposite side of the residence to the proposed wind farm, which would mean that the residence would be in the upwind direction from the road (least background noise) while it would be in the downwind direction from the turbine (most background noise).
12. The noise propagation model described in ISO9613-2 is appropriate for calculating wind farm noise levels at residences and other sensitive receivers, provided that the following considerations are taken into account.
    a) Predictions should be based on octave band analysis.
    b) Topographic screening effects of the terrain should be limited to a reduction of no more than 2 dB, and then only if there is no direct line of sight between the highest point on the turbine rotor and the receiver location.
    c) A ground factor, G, of 0.5 should be used for propagation over any ground except water, paved surfaces or concrete. A soft ground factor of 1.0 should never be used, even if the ground is soft and a ground factor of 0.0 will tend to over-predict noise levels when the ground is soft, although it is appropriate when the ground is hard.
    d) Atmospheric conditions of 10°C and 70% humidity are recommended to represent a reasonably low level of air absorption.
    e) A receiver height of 4 m should be used to reduce the sensitivity of the results to the receiver region ground factor.
    f) A correction of +3 dB (or +1.5 dB if using $G = 0.0$) should be added to the calculated overall A-weighted noise level (see Section 2.2.11) for propagation 'across a valley', if 1.5 times the average of the source and receiver heights is less than the mean height above the ground of the line, R, joining the source and receiver. That is, if $0.75(h_S + h_R) < h_{mean}$. The reason for the increase is the reduced ground absorption effect and the possible presence of more than one reflected path from the turbine to the receiver (see Figure 1.8, which shows two possible reflection paths in a valley (R_1 and R_2).
    g) A measurement uncertainty of 2 dB should be added to measured turbine sound power levels before using them to calculate community noise levels.
    h) In converting L_{A90} noise levels to L_{Aeq} levels, 2 dB should be added to the L_{A90} levels for comparison with the L_{Aeq} predicted levels.

To further expand on the IOA good practice guide, a number of supplementary guidance notes were issued in 2014. The first (IOA 2014a) provides guidance on the following aspects of data collection.

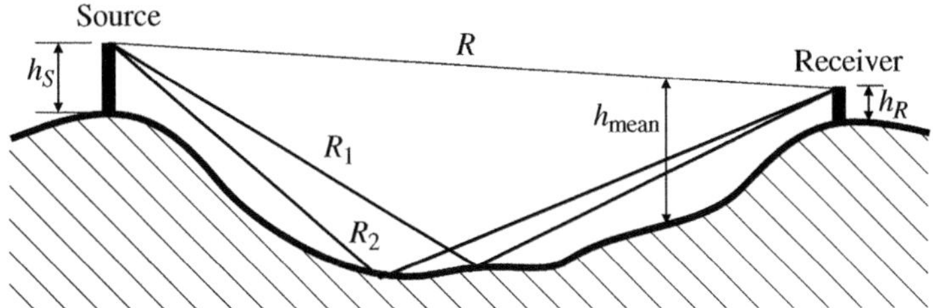

Figure 1.8 Valley between source and receiver

- Secondary wind screens should be used to reduce wind-induced noise on the microphones.
- Background noise measurements should be taken at least 3.5 m from a reflective surface.
- Measurement equipment should not be sited where noise levels are influenced by local noise sources specific to the measurement location.
- Care should be taken in the interpretation of wind speed data obtained using SODAR or LIDAR equipment, which measure wind speed as a function of altitude from a ground-based sonar or laser system, respectively (see Section 6.4.1).

The second note (IOA 2014b) provides guidance on the derivation of curves used to define existing background noise levels. Details are provided on how to analyse the data, including how to account for wind direction effects.

The third guidance note (IOA 2014c) provides more details on how to interpret sound power level data supplied by manufacturers, and how to include uncertainty in the sound power data.

The fourth guidance note (IOA 2014d) explains how to calculate the wind speed at hub height from measurement data at other heights, such as those obtained from an anemometer on a mast or SODAR data. The guidance note also explains how to adjust the wind turbine sound power output in cases where the measured wind speed at 10 m height is used instead of the standardised wind speed calculated as discussed in the previous section. The adjustment used is based on the plotted curve of turbine sound power output vs actual wind speed at a height of 10 m. In this case, the *Good Practice Guidelines* recommend a conservative correction based on shifting the turbine sound power level prediction curve to the 'left' along the wind speed axis (i.e. moving the curve to lower wind speeds) to account for the effects of wind shear. The recommended shift is 1 m/s for turbine hub heights of up to 30 m, 2 m/s for hub heights of up to 60 m and 3 m/s for hub heights of more than 60 m. However, the fourth guidance note provides a more accurate method that can be used if wind shear data are available. This involves either:

- shifting the turbine sound power level prediction curve to the 'left' along the wind speed axis as described above, but using a shift amount based on the measured wind shear rather than a fixed amount;
- correcting the background noise data regression curve (based on measured wind shear values) by shifting the background noise data to the 'right' along the wind speed axis (increased wind speeds for a given background level).

The fifth guidance note (IOA 2014e) is concerned with compliance measurement, mentioning the need for tonal analysis as well as recommending that background noise measurements be repeated with the wind farm shut down. A series of measurements are then taken in downwind conditions, with the turbines operating and shut down, and two sets of data corresponding to the two conditions are then plotted as a function of the standardised 10-m height wind speed. Curves of best fit are then derived for each data set and the turbine noise contribution is then determined by subtracting the level with the turbines off from the level with the turbines on, as described in Section 2.2.9, for each integer wind speed using the noise levels corresponding to the curves of best fit to the measured data.

The sixth guidance note (IOA 2014f) provides guidance for the calculation of noise propagation over water. The guidance note recommends that the ISO9613-2 noise propagation model be used with the hard ground correction, $G = 0$, provided that propagation over water is 50% or more of the total distance from the wind farm to the receiver. Where the body of water is at least 700 m in extent or the turbines are offshore, the expression in Eq. (1.7) should be used to calculate the noise level L_p at a receiver located at a distance r from the turbine rotor. If none of these conditions are met, the ISO9613-2 propagation model should be used.

$$L_p = L_w - 20 \log_{10}(r) - 11 + 3 - A_{\text{atm}} + 10 \log_{10}\left(\frac{r}{700}\right) \tag{1.7}$$

where L_w is the sound power level of the turbine and A_{atm} is the attenuation due to atmospheric absorption (see Section 5.2.2). Propagation modelling is discussed in more detail in Chapter 5.

Problems with the British Institute of Acoustics *Good Practice Guidelines*

Some of the problems associated with implementing ETSU-R-97 according to approach in the IOA *Good Practice Guidelines* and its six accompanying supplements are listed below.

1. There is no mention of accounting for the electronic noise floor of the instrumentation as described on page 20.
2. The simple wind speed measure at 10 m height recommended by ETSU-R-97 is now replaced by an *estimated* wind speed at 10 m height, where the least preferred method of obtaining the 10 m wind speed is to actually measure it at 10 m height, and even then the actual measurement should be adjusted. This is because the 10-m wind speed to be used is a reference wind speed based on the wind speed measured at hub height, h_H. That is, the standardised 10-m wind speed, $U_{10\text{m}}$, is obtained from the turbine hub-height wind speed, U_H, by correcting it to 10-m height, $h_{10\text{m}}$, using a ground roughness factor of $z_0 = 0.05$, without taking into account any atmospheric effects, using the following equation.

$$U_{10\text{m}} = U_H \frac{\log_e(h_{10\text{m}}/z_0)}{\log_e(h_H/z_0)} \tag{1.8}$$

   Clearly, this will produce large errors in the 10-m wind speed in conditions of high wind shear. This explains the large scatter in data obtained when background noise levels are plotted as a function of the hub-height wind speed referenced to 10 m.

3. There is no requirement for turbine manufacturer data to include or exclude an allowance for uncertainty in the sound power measurement.
4. The step change in the ground effect of 3 dB that can occur when there is concave ground between the turbine and receiver (see item 12f above) can result in a 3-dB variation in noise predictions at the margin when the ground just satisfies the concave criterion, as it could depend on the accuracy of the height assumed for the receiver location and the methodology used to calculate the mean propagation height.
5. The allowable noise limits (especially at night) seem too high when compared to WHO recommendations and other jurisdictions. The anomaly of higher allowed noise levels at night than during the day needs to be addressed.

More recent standards such as IEC 61400-11 Ed. 3.0 (IEC 61400-11 2012) use the wind speed at hub height rather than at the 10-m reference height and this is reflected in more recent regulations as well. In this case, the wind speed is what is actually measured at hub height, or derived from the turbine power output, using the power curve supplied by the manufacturer for the particular turbine. This is discussed in more detail in Section 1.5.2.

1.4.2 National Planning Policy Framework for England

This policy framework sets out the UK government's planning policies for England and how these are expected to be applied. It was published in 2012 and updated in 2014 (UK Government 2015). The reason for introducing this guidance note here is that it is a contemporary view of what is acceptable and what is unacceptable in terms of community disturbance from noise. Although the policy provides clear guidelines regarding what action needs to be taken if noise affects a community, it does not provide any guidance regarding what percentage of people or number of people in a community have to be affected before action is taken. The guidance provided by the English National Planning Policy Framework, regarding the types of action that should be taken for various effect levels, is summarised in Table 1.2.

1.4.3 World Health Organisation Guidelines

Although not specifically targeted at wind farm noise, the World Health Organisation (WHO) has issued guidelines for acceptable environmental noise limits for Europe in 1999 (Berglund et al. 1999) and 2009 (Hurtley 2009), which in summary recommend an allowable nighttime indoors noise level of 30 dBA and an allowable outside level of 45 dBA, based on the assumption that the residential structure will result in 15 dBA noise reduction when the windows are open for ventilation. However, this is based on the annoying noise having a spectral distribution similar to traffic noise, which is dominated by mid-frequencies. The allowed levels were derived from surveys in urban and suburban environments where background noise levels are much higher than in rural environments. This results in an industrial noise source introduced into a rural environment being much more annoying than if it were introduced into an urban environment, so this alone is justification for requiring lower outside noise levels than recommended by the WHO documents.

Unlike traffic noise, wind farm noise is dominated by low frequencies at large distances (over 1.0–1.5 km) from the nearest turbine in a wind farm, and as was written in the 1999 WHO document:

Table 1.2 Action to be taken for different levels of noise.

Perception	Outcomes	Effect level	Action
Not noticeable	None	No observed effect	No measures required
Noticeable and not intrusive	Noise can be heard but doesn't cause any change in behaviour or attitude. Can slightly affect the acoustic character of the area but not such that there is a perceived change in the quality of life.	No observed adverse effect	No measures required
Noticeable and intrusive	Noise can be heard and causes small changes in behaviour and/or attitude, e.g. turning up volume of television; speaking more loudly; where there is no alternative ventilation, having to close windows for some of the time because of the noise. Potential for some reported sleep disturbance. Affects the acoustic character of the area such that there is a perceived change in the quality of life.	Observed adverse effect	Mitigate and reduce to a minimum, taking account of the economic and social benefits being derived from the activity causing the noise
Noticeable and disruptive	The noise causes a material change in behaviour and/or attitude, e.g. avoiding certain activities during periods of intrusion; where there is no alternative ventilation, having to keep windows closed most of the time because of the noise. Potential for sleep disturbance resulting in difficulty in getting to sleep, premature awakening and difficulty in getting back to sleep. Quality of life diminished due to change in acoustic character of the area.	Significant observed adverse effect	Avoid, as it is undesirable for such exposure to be caused
Noticeable and very disruptive	Extensive and regular changes in behaviour and/or an inability to mitigate effect of noise leading to psychological stress or physiological effects, e.g. regular sleep deprivation/awakening; loss of appetite, significant, medically definable harm, e.g. auditory and non-auditory.	Unaccept-able adverse effect	Prevent. The impacts on health and quality of life are such that regardless of the benefits of the activity causing the noise, this situation should be prevented from occurring.

Source: English National Planning Framework, Planning Practice Guidance for Noise recommendations, 2014.

> ...if the noise includes a large proportion of low-frequency components, values even lower than the guideline values will be needed, because low-frequency components in noise may increase the adverse effects considerably.

The WHO document also states that when prominent low-frequency components are present, as indicated when the difference between the C-weighted and A-weighted noise levels (see Section 2.2.11) exceeds 10 dB, then measures based on A-weighting are inappropriate. In addition, the residential-structure noise reduction assumed for European houses is much greater than measured in rural properties in more temperate climates such as in Australia. Also, the reduction for noise (such as wind farm noise) that is dominated by low frequencies is much less than for noise characterised by the spectrum typical of traffic noise, which is dominated by mid-range frequencies. It seems that most guidelines written specifically to assess and control the impact of wind farm noise have conveniently omitted the requirements listed below, which follow directly from the WHO guidelines:

- consideration of poorer noise-reducing characteristics of housing structures for low-frequency noise, which may result in higher than 30 dBA noise levels indoors when the noise is dominated by low frequencies such as wind farm noise;
- the greater potential for noise dominated by low frequencies to be more annoying than a more balanced noise spectrum;
- the fact that the WHO document applies to European houses, which are better noise insulators than houses in more temperate environments, such as rural Australia.
- the suggestion of a need for an alternative to the dBA metric for assessment when the noise spectrum is dominated by low frequencies.
- any recognition that background noise levels in rural environments is much less than background noise levels in suburban and urban environments in Europe, on which the WHO guidelines are based.

1.4.4 DEFRA Guidelines

In 2005, the University of Salford prepared guidelines for the assessment of low-frequency noise disturbance (Moorhouse et al. 2005). These guidelines have been applied to wind farm noise due to its dominance by low frequencies at distances exceeding approximately 1.0–1.5 km from the nearest turbine. After a very extensive literature review, the authors proposed that the noise will be annoying if the unweighted L_{eq} exceeds the DEFRA limit values in Table 1.3 in any 1/3-octave band.

Table 1.3 1/3-octave band unweighted noise limits (L_{eq}) recommended by DEFRA to avoid annoyance.

	1/3-octave band centre frequency (Hz)												
	10	12.5	16	20	25	31.5	40	50	63	80	100	125	160
DEFRA limit	92	87	83	74	64	56	49	43	42	40	38	36	34
Pure-tone annoyance threshold (night)	92	88	84	75	62	55	46	39	33	33	32	30	26

If the noise is steady, the limits in the table are increased by 5 dB. A steady noise is one that satisfies either of the following conditions, where the parameters are evaluated for each 1/3-octave band for which the criteria are exceeded:

- $L_{10} - L_{90} < 5$ dB
- The rate of change of sound pressure level ('fast' time weighting) is less than 10 dB/sec. This can be determined using a sound level meter capable of storing short time (<0.1 s) values of L_{eq}.

The DEFRA guidelines may be too high, as they are above the annoyance limits for pure-tone sound derived from the median threshold levels listed in the ISO standard, (ISO1996-1 2003) and the amount of exceedance of the threshold levels for the noise to be annoying Gottlob (1998). Of course, there will be 50% of the population who will be more annoyed than the levels based on median thresholds and to account for more than 97.5% of the population the threshold should be reduced by two standard deviations or 6 dB (see Section 7.3.4). However, if the noise is not tonal, it is likely to be less annoying, especially at the higher frequencies.

1.4.5 Noise Perception Index

Hunt and Hannah (2009) suggest that regulations should be based on the noise perception index (NPI), as first introduced by Hessler (2008) for general environmental noise. The NPI is calculated as follows.

$$\mathrm{NPI} = \frac{1}{N}\sum_{i=1}^{N}\left\{10\log_{10}\left[10^{L_{Aeq}/10} + 10^{(L_{A90})_i/10}\right] - (L_{A90})_i\right\} \tag{1.9}$$

where L_{Aeq} is the calculated noise level at the residence due to the wind farm, L_{A90} is the measured background noise level at the residence prior to construction of the wind farm and N is the number of 10-min periods of ambient noise measurement.

The expected community reaction is listed in Table 1.4 and some examples of NPI values that may be expected for wind farms (as determined by Hunt and Hannah (2009)) are listed in Table 1.5.

Although the NPI is rather over-simplified for wind farm noise, it could be used as a regulatory tool in addition to the other tools discussed in Section 1.6.

Table 1.4 NPI values for a range of receiver sites.

Calculated wind farm L_{Aeq}	Ambient sound level measured prior to wind farm, L_{A90}		
	High	Medium	Low
Low	NPI < 5	5 < NPI < 6	6 < NPI < 8
Medium	5 < NPI < 6	6 < NPI < 8	8 < NPI < 9
High	6 < NPI < 8	8 < NPI < 9	NPI > 9

After Hunt and Hannah (2009)

Table 1.5 Community perception of noise and community response as a function of the NPI value of the noise.

NPI (dBA)	Perception	Predicted community response
≤ 3	Generally imperceptible	No response
3–5	Barely perceptible to perceptible	No response to potentially adverse response
5–10	Perceptible to noticeable	Potentially adverse response to adverse response
> 10	Readily noticeable	Adverse response

1.5 Wind Farm Noise Standards

Environmental noise standards are not very widespread and those that exist generally, with a few exceptions, do not provide adequate guidance regarding acceptable environmental noise levels. In the following subsections, general environmental noise standards are discussed first, followed by a detailed discussion of three standards that are specifically directed at wind farm noise.

1.5.1 General Environmental Noise Standards

The five most well known general standards are discussed below.

ANSI S12.9-4 (2005)

This standard does not specify recommended maximum allowable A-weighted (see Section 2.2.11) levels but does state that the maximum allowable level specified in regulations should be reduced by 5 dB if the noise is impulsive and by a further 5 dB if the noise contains identifiable tones. The sound is considered to be dominated by low frequencies if the C-weighted level exceeds the A-weighted level by more than 10 dB. In that case, to minimise annoyance, the levels in the 16-, 31.5- and 63-Hz octave bands must all be less than 65 dB. There are also adjustments for sounds that increase rapidly in level at a rate greater than 15 dB/s.

ISO 1996

Although the 1971 version of this standard recommends maximum acceptable outdoor noise levels as listed in Table 1.6, the most recent version of the standard, published in 2007, does not recommend any maximum levels.

Other General Environmental Noise Standards

Other environmental noise standards such as BS7445-3 (1991), AS1055.2 (1997) and ASTM E1686-10 (2010) do not recommend acceptable noise limits and provide only vague guidance.

1.5.2 IEC 61400-11

This international standard does not provide advice on acceptable community noise levels, but is focussed on a description of procedures and instrumentation to be used for

Table 1.6 Recommendations for community noise limits according to ISO1996-1971.

District type	Day 7am–7pm (dBA)	Evening 7pm–11pm (dBA)	Night 11pm–7am (dBA)
Rural	35	30	25
Suburban	40	35	30
Urban residential	45	40	35
Urban mixed	50	45	40

the measurement and analysis of acoustic emissions from wind turbines. It includes a description of appropriate measurement positions for a microphone on the ground to measure the sound power of a wind turbine (see Section 4.6), determination of average sound pressure levels and corresponding wind speeds. Of particular interest is the procedure for calculating the uncertainty in the sound pressure level corresponding to a particular wind speed 'bin'. A wind-speed bin is 0.5 m/s wide and centred around integer and half-integer wind speeds. The uncertainty analysis for both sound pressure level and wind speed measurements is discussed in Section 4.6.

The standard uses hub height as the height of the reference wind speed. The wind speed is determined from the electric power output of the turbine for each wind-speed bin. The 'allowed range' of the power curve is defined as the parts of the power output vs wind speed curve for which the following relationship is satisfied:

$$(P_{k+1} - P_{\text{tol}}) - (P_k + P_{\text{tol}}) > 0 \tag{1.10}$$

where k is the wind speed bin number, P_k is the turbine electrical power output corresponding to wind speed bin, k, and P_{tol} is the tolerance on the power reading, which is the same for all bins, at typically 1% to 5% of the maximum value.

For the 'allowed range' of the power curve, the normalised hub-height wind speed, U_H is equal to the wind speed, U_P, corresponding to the turbine electric power curve. For all wind-speed bin numbers not included in the 'allowed range' of the power curve above, the hub-height wind speed, U_H, must be determined from the nacelle anemometer output, U_{nac}, as follows:

$$U_H = \kappa_{\text{nac}} U_{\text{nac}} \tag{1.11}$$

where the coefficient, κ_{nac}, is the value of U_P/U_{nac}, averaged over all wind speed bins included in the 'allowed range' of the power curve.

In cases where background noise levels are to be determined prior to construction of the wind farm, the corresponding wind speeds, U_z, must be measured at a height of at least 10 m above the ground at a typical turbine location, and these measurements are then used to derive corresponding hub-height wind speeds. The hub-height wind speed, U_H for the background noise measurements is then calculated as

$$U_H = \kappa_b U_z \tag{1.12}$$

where the coefficient, κ_b, is the value of U_P/U_z, averaged over all wind speed bins included in the 'allowed range' of the power curve.

The IEC standard also provides a procedure for determining whether or not audible tones are present. This is discussed in more detail in Section 6.2.12.

1.5.3 NZS6808

This standard is the only recent environmental noise standard that suggests acceptable community noise levels. In addition it outlines suitable methods for the prediction and measurement of wind farm noise. Although the standard specifies a generally acceptable noise level of 40 dBA (L_{A90}), the recommended allowed level is 35 dBA in areas of high amenity where existing background noise levels are very low, especially at night, and the average difference between wind farm predicted noise levels and ambient L_{A90} noise levels exceeds 8 dBA. The standard also allows for adjustment of the measured noise level if the noise contains special characteristics such as tonality, impulsiveness, amplitude modulation (see Section 2.2.7) or beating (see Section 2.2.6). The maximum adjustment allowed to account for the more annoying nature of the special characteristic is a 6-dBA increase to the measured noise level.

NZS 6808 (2010) also specifies two methods for calculating community noise levels; one based on ISO9613-2 (see Section 5.6) and the second based on a simplified version of ISO9613-2 (1996).

The standard covers the measurement of background noise levels as a function of wind speed at hub height prior to construction of the wind farm and compliance assessment measurements after construction. One interesting compliance checking technique suggested in the standard is to measure noise levels with and without turbines operating to obtain the contribution due to the turbines. However, the standard states that it is not necessary to turn off turbines that do not contribute significantly to noise levels at the measurement location. These turbines are those that collectively are calculated to contribute 10 dB less than the highest contribution from a single turbine.

Details of recommendations for compliance testing according to this standard are outlined in Section 6.2.14.

1.5.4 AS4959

AS4959 provides advice on choosing analytical models to use for estimating community noise levels using the turbine sound power levels as a basis. Advice is also provided on how to measure $L_{A90,\,10\text{min}}$ background noise levels and how to find the average as a function of wind speed at turbine hub height by regression analysis. For compliance testing, this standard suggests that $L_{A90,\,10\text{min}}$ be measured with the wind farm operational, thus representing a background plus wind farm $L_{A90,\,10\text{min}}$ measurement. The standard suggests that this would be a good approximation to the L_{Aeq} level produced by the wind farm only and thus should be used as a measurement to compare with L_{Aeq} criteria. The justification for this is that for wind farm noise in the absence of background noise, the $L_{A90,\,10\text{min}}$ measurements are lower than the $L_{Aeq,\,10\text{min}}$ measurements by between 1.5 and 2.5 dBA and this is the amount by which background noise may be expected to increase the $L_{A90,\,10\text{min}}$ measurement. This may be a reasonably accurate assumption in some situations, but one would not expect it to apply in all or even the majority of situations, especially at night when rural noise levels are very much lower than implied by the above equivalence. In this case, the L_{Aeq} level would be underestimated by 1.5 to 2 dBA, or possibly more.

An alternative compliance testing method is also provided and involves measuring $L_{Aeq,10min}$ levels directly with the turbines operational and also with them shut down. The turbine contribution is then found by subtracting the $L_{Aeq,10min}$ level without the turbines operating from the $L_{Aeq,10min}$ level with the turbines operating, as described in Section 2.2.9. No advice is provided regarding what community noise levels would be acceptable.

Details of recommendations for compliance testing according to AS 4959 (2010) are outlined in Section 6.2.14.

1.6 Regulations

Most regulations and governmental (or county) ordinances are based on standards and guidelines available at the time the ordinance was drafted. Many state government and local regulations are generated within the framework of a more general act of Parliament, with names such as The Noise Abatement Act, the Environmental Protection Act and the Control of Pollution Act. Under these acts it is unlawful to create a private or public nuisance, with the remedy being an abatement notice served by a local authority such as a council, or a magistrates' court order. Private nuisance is a civil wrong and recognises the right of someone to use their land without unreasonable disturbances from neighbouring property. In practice, the bar is usually set fairly high so that minor problems such as annoyance are not considered to be a nuisance in law.

Statutory nuisances are nuisances that have been declared as such in an Act of Parliament. For example, the 1990 Environmental Protection Act in the UK provides that 'noise emitted from premises and being prejudicial to health or a nuisance' shall constitute a statutory nuisance. However, whether a noise is a nuisance or not is not defined by its character, level or duration, but by the subjective assessment of a court, which makes it very difficult to obtain a ruling against such a huge operation as a wind farm, especially if the court believes that wind farms are of benefit to society as a whole. Thus it is prudent for local government to rely on their own regulations to specify allowable noise levels and any penalties to compensate for the annoying character of the noise.

1.6.1 What Should be Included in a Wind Farm Noise Regulation?

Suggestions for a general wind farm regulation may be found in a paper by the AWED (2016). However, based on our experience with wind farm noise data collection and analysis and residents, we believe that to minimise the impact of wind farm noise on local communities, a local wind farm noise regulation should also include the features outlined in the following subsections.

Minimum Setback Distance to Residences

The setback distance is the distance from the nearest turbine in a wind farm to a residence, provided that the resident is not a turbine host. Shepherd et al. (2011) recommend that the setback distance should be more than 2 km in hilly terrain. However, noise measurements by the authors of this book indicate that for modern wind farms with turbine powers of 3 MW and greater, the setback distance should be much larger than 2 km if intrusive noise at noise-sensitive locations is to be at acceptable levels at night when people are trying to sleep. In addition there should be a minimum setback distance of 4 to 5 times the total turbine height (nacelle height plus blade length) to the nearest neighbour's property line.

Maximum Allowed Calculated External Noise Levels

Maximum allowed calculated external noise levels 30 m from the residential structure should be as listed in Table 1.7 at the exterior of homes of any residents not hosting turbines. These levels are those recommended by the international standard, ISO 1996 (1971).

However, as wind turbines are mostly in rural areas, a level of 25 dBA is very difficult to achieve, which is why the most stringent regulations currently in force recommend a limit of 35 dBA in sensitive noise areas. However, this limit is too high for many residents so 30 dBA may be a reasonable compromise for the maximum allowed noise at the exterior of rural residences.

Kamperman and James (2008) have studied a number of wind farms and their effects on residences and have made the following recommendations.

- An allowed A-weighted (see Section 2.2.11) nighttime limit of 35 dBA or L_{A90}+5 dBA, whichever is *lower*, where L_{A90} is determined using several 10-min measurements during the quiet night hours between 10pm and 4am when the wind is blowing slightly (<2 m/s at 1.5 m above the ground) from the wind farm to the residence and the wind farm is not operating. Daytime levels can be 5 dBA higher.
- A maximum C-weighted noise level of 50 dBC for properties more than 1.6 km from a major highway and 55 dBC for properties less than 1.6 km from a major highway.
- A maximum allowed difference of 20 dB between the measured L_{A90} prior to construction of the wind farm and the C-weighted L_{Ceq} with the wind farm operating.
- A maximum allowable difference in the A-weighted noise level with and without turbines operating. That is, $L_{Aeq} \leq L_{A90} + 5$, where L_{Aeq} is the A-weighted equivalent noise level with the turbines operating and L_{A90} is the A-weighted level exceeded 90% of the time when the turbines are not operating. As turbine L_{Aeq} levels are between 1.5 and 3 dB above turbine L_{A90}, an increase of 5 dBA in the L_{A90} background noise approximately represents a turbine noise source generating the same noise level as the existing background noise, thus raising the background noise level by 3 dBA (see Section 2.2.8).

Table 1.7 ISO1996-1971 recommendations for allowable community L_{Aeq} noise limits.

District type	Day (7am–7pm) (dBA)	Evening (7pm–11pm) (dBA)	Night (11pm–7am) (dBA)
Rural	35	30	25
Suburban	40	35	30
Urban residential	45	40	35
Urban mixed	50	45	40

- To minimise the influence of wind noise on background noise measurements, L_{A90} results are valid when L_{A10} results are no more than 10 dBA above L_{A90} for the same time period.
- A 5 dB penalty is to be applied if any tones are audible.

Adjustment for Uncertainties

Any noise level calculated using a generally accepted propagation model with downwind conditions included, should have 4 dBA added to the calculated level to account for uncertainties in turbine sound powers and uncertainties in the propagation model parameters. A fine example of the under-prediction of wind farm noise levels is presented in Stigwood et al. (2015, Fig. 1). This figure clearly shows that the ISO9613-2 predictions are consistently 2–4 dB above the average of the measured data over the entire range of standardised wind speeds at 10 m. The difference between the predictions and the level exceeded 10% of the time is approximately twice that.

Maximum Allowed Calculated Low-frequency Noise Level

Maximum allowed calculated low-frequency noise level in the range 10–160 Hz should be less than 25 dBA at the exterior of homes of any residents not hosting turbines, with a 4 dBA allowance for calculation uncertainty, so the calculated level is actually less than 21 dBA. The reason for using an A-weighted level is that it is comparable to the existing Danish low-frequency noise requirement for calculating interior noise levels. The DEFRA criteria for external low-frequency noise in Table 1.3 should also be satisfied and perhaps revisited to ensure that the recommended levels are sufficiently low.

Maximum Allowed Amplitude Modulation

A maximum allowed amplitude modulation (AM) amount should be defined and the means by which this should be measured is discussed in Section 6.2.10. However, the metric to be used and the recommended allowable level is still under discussion (AMWG 2015). Alternatively, a penalty should be added to the L_{Aeq} or L_{90} measurements to account for amplitude modulation. Renewable UK (2013) recommend the following:

> ...for AM with a peak to trough level of less than 3 dB there should be no AM penalty; for AM with a peak to trough level of 3–10 dB there is a sliding scale of penalties ranging from 3–5 dB; and for AM with a peak to trough level of ≥ 10 dB there is 5 dB penalty.

However, the alternative of only a penalty on the allowed total A-weighted level has not been generally accepted and is not preferred by many researchers AM is a cause of annoyance even for low-level A-weighted noise, as excessive AM is mainly a low-frequency noise problem (Renewable Energy Foundation 2014). Amplitude modulation is defined and discussed in Section 2.2.7.

Compliance measurements typically require the accumulation of a large number of data points of L_{Aeq} or L_{90} vs standardised wind speed at 10 m or wind speed at hub height. A regression line is then drawn through the data and this is defined as the turbine noise level (after background noise has been accounted for). However, the following points need to be taken into account when deciding how to include a penalty for amplitude modulation.

- Amplitude modulation is most apparent when the L_{Aeq} or L_{90} levels are at their lower end (Stigwood et al. 2015).
- As a result, if the penalty is applied to the individual measurements during which AM was detected, the effect of the penalty on the regression line will be negligible in almost all cases.
- Thus it is recommended that if amplitude modulation exceeds the allowable level for a greater percentage of 10-minute measurements than is considered acceptable, then the AM penalty should be applied to the average data for each wind speed at which AM was detected.

Minimum Turbine Spacings

Allowed minimum turbine spacings should be at least 10 blade lengths (5 rotor diameters) for turbines that are downwind of any others. This applies to all turbine spacings for wind farms where there is no dominant prevailing wind. When turbines are not downwind of any others, the spacing should be at least 7 blade lengths (3.5 rotor diameters). This could apply to turbines arranged along a ridge in a line normal to the prevailing wind direction. If these spacings (or alternative spacings recommended by the turbine manufacturer) are not maintained, turbine sound power levels that are higher than those specified by the manufacturer will result, due to inflow turbulence as a result of wakes from upstream turbines.

Allowance for Increased Noise in Hilly Terrain

An allowance should be made for increased turbine sound power levels for turbines located in hilly terrain where the in flow will be more turbulent and off-axis.

Penalty for Tonal or Impulsive Noise

There should be a 5 dBA penalty applied to the allowable limit if tonal noise is present or if the noise is impulsive. If both are present the penalty should be 8 dBA.

Accounting for Multiple Wind Farms

There should be a requirement that takes into account the development of more than one wind farm in the same general area. The existing ambient sound level should be the one that existed prior to construction of the first wind farm.

Determination of a Compliance Testing Method

A compliance testing method should be defined. Some suggestions regarding how this may be done, based on what has been published in the literature as well as the authors' own experience, are outlined below.

- Noise levels should be measured between the hours of 11pm and 5am when there is high wind shear so that the turbines are rotating and producing more than 70% of their rated power, while the wind speed at the residence where compliance testing is being performed is less than 2 m/s.
- Adequate wind screening of the microphones should be specified for all noise measurements. This means that secondary wind screens should be used and microphones should be mounted on the ground to minimise the wind speed over the microphone.
- As wind farm noise levels are generally quite low, it is difficult to accurately measure L_{Aeq} levels due to the influence of extraneous noise. It is therefore preferable to measure L_{A90} levels and add 2 dB to obtain the approximate L_{Aeq} level. The L_{Aeq} level due to

the turbines is the level that should be compared with the measured background L_{A90} level when determining how many decibells the turbine noise is above background noise levels.

- Measurements should be done with and without the wind farm operating for the same weather conditions. The determination of the wind-farm-only contribution to the noise levels when the wind farm is operating may be determined using one or both of the following methods, although the second one may be a little complex to include in a regulation.
  1. The measured data should be divided into nighttime (11pm to 5am) and the remainder (5am to 11pm). Of particular importance are the nighttime levels. The data should be further divided into turbine OFF and turbine ON levels and only downwind conditions should be considered (with a ±45° angle allowed from direct downwind). Data containing local noise source events should be discarded, as should data recorded when it is raining. In addition, data that includes the early morning bird chorus should be discarded. The remaining A-weighted (see Section 2.2.11) noise-level data should be plotted as a function of wind speed at hub height and a curve of best fit plotted through the data points for the nighttime data and then separately for the remaining (daytime and evening) data. To obtain the wind farm contribution, the level with the turbines OFF should be subtracted logarithmically from the level with the turbines ON, as described in Section 2.2.9, for each integer wind speed, using the noise levels corresponding to the curves of best fit to the measured data. It is well known that turbine noise is most annoying at night at times of high wind shear. This situation often corresponds to there being low or no wind at the residence but significant wind at turbine height, an occurrence that is relatively common in some regions. This results in lower background noise levels at times of high wind shear, so it is suggested that nighttime data should be further divided into low and high wind shear conditions. Long-term measurements by Bigot et al. (2015) showed that month-long average noise levels for an average wind shear exponent of 0.42 (high wind shear) were between 3 and 8 dB higher (depending on the hub-height wind speed) than they were for a shear exponent of 0.25 (low wind shear). This is discussed in more detail in Section 5.2.4.
  2. Alternatively, the data in each grouping just described can be presented as a probability density function plot, as described by Ashtiani (2013, 2015). This method calculates the probability of the existence of a particular dBA level that is due only to the turbine operation. Clearly, the probability of the existence of a particular dBA level will depend on the chosen dBA level and there will be some dBA levels with a much higher probability of occurrence than others, although implementation of the method is quite complex. Its advantage for compliance checking is that it allows regulations to be set along the lines of 'wind farm noise levels should not exceed x dBA for more than 10% of the time'. It also allows the percentage of time that a wind farm is non-compliant to be determined. This method is discussed in more detail in Section 6.2.14.
- In the non-operating condition, all cooling fans and power supplies in the turbine nacelles must be switched off.

- Data recorded prior to construction of the wind farm should also be considered, as even a non-operating wind farm can contribute to background noise levels as a result of wind blowing over the support tower and stationary blades.
- The level of amplitude modulation should be measured using one of the methods described in Section 6.2.10. Perhaps the simplest method is that used by Tachibana (2014), which is described in Section 6.2.10.
- Wind farm operators must be required to make nacelle and mast wind data and turbine output data publicly available in an accessible format to enable compliance checking by any party.
- Compliance management procedures should be required as part of any development approval and should include continuous real-time measurement of noise over the frequency range from 0.5 Hz to 2000 Hz and include quantification of undesirable characteristics such as modulation and tonality.

Instrument Specifications

Desirable instrumentation specifications for background noise and compliance measurements include:

- an electronic noise floor at less than 10 dBA (to allow accurate background noise level measurements down to 20 dBA or corrected background noise measurements to 13 dBA)
- the ability to measure linearly down to 0.5 Hz or linearly down to 10 Hz and then down to 0.5 Hz with a calibrated 'roll-off'.
- secondary wind screens in addition to primary wind screens to protect measurement microphones from wind-induced noise, as discussed in Section 6.2.3.

No Change to Turbine Model

The wind farm developer should not be permitted to change the turbine model after development approval has been given without undertaking another noise prediction study with the new turbines. If the resulting sound levels at any residence are higher than originally calculated, an amendment to the development approval should be sought.

Acoustic Study Funding

The government body responsible for compliance assessment should be funded by the wind farm developer to choose and hire an acoustic consultant to do the testing independent of any influence from the developer.

Decommissioning Costs

Although not related to noise, it is important that any regulation specifies a required upfront payment for decommissioning (of the order of $350 000 per turbine in 2011 US dollars, put into a long-term investment account or bonds), as it is likely that the company installing the wind farm will not be around when its design life expires (usually 25 years).

Property Value Assessment

Another item not related to noise but important nevertheless, is that the value of properties within 5 km of a wind farm should be assessed by an independent commercial property assessor appointed by the local council prior to construction, and the difference between that valuation and any sale price achieved within one year post construction should be paid to the resident by the wind farm operator at the time of sale. Alternatively, if the property cannot be sold for a reasonable price, the wind farm operator should be required to purchase the property at the market value prior to construction of the wind farm.

1.6.2 Existing Noise Ordinances and Regulations

There are a multitude of noise ordinances and regulations in existence, so only a summary of some of these are listed in Tables 1.8–1.10 below (see Fowler et al. (2013) and Batasch (2005)). In the tables, allowed evening levels are similar to daytime levels except where indicated.

It is abundantly clear from Tables 1.8, 1.9 and 1.10 that there is very little commonality between different jurisdictions, both in terms of the allowed noise levels and the metric used to quantify them. Notes that should be read in conjunction with Tables 1.8–1.10 are listed below. The notes that are relevant to the various jurisdictions in the tables are indicated by superscript numbers adjacent to the name of the applicable country, state or county.

1. The quantities $L_{A90,10\text{min}}$, L_{Aeq}, L_{dn} and L_{den} are defined in Section 2.2.12.
2. In many jurisdictions, a 5-dB penalty is applied if the turbine noise has readily identifiable tonal components. In Germany the penalty is 3 dB or 6 dB, depending on the distinctiveness of the tones. However, there is a relatively stringent definition of what constitutes a tone and this is discussed in more detail in Section 6.2.12,
3. In Norway, France, the Netherlands and many places in the USA, no tonal penalty is applied.
4. Daytime usually refers to the hours between 7am and 7pm, evening between 7pm and 10pm and nighttime is between 10pm and 7am. Exceptions are:
   - the UK, where 10pm is replaced with 11pm and evening is defined as between 6pm and 11pm
   - Arapahoe County in Colorado, where the 11pm is replaced with 7pm
   - the Netherlands, where 10pm is replaced with 11pm and evening is defined as between 7pm and 11pm
   - Italy, where daytime is 7am to 9pm, evening is 9pm to 11pm and nighttime is 11pm to 7am.
5. In Italy, there is an additional requirement that a noise source cannot exceed existing L_{A90} background noise levels in residential areas by more than 3 dBA at night and 5 dBA during the day, except for industrial areas and residences near major roads.
6. In some of the cases listed in the tables, the levels are not legislated but are merely guidelines or standards. However, local authorities usually follow the guidelines in terms of setting allowable levels for each development.
7. In some of the cases listed in the tables, different allowable levels are specified for locations where the background noise is low. These are the values that are listed in the tables, as almost all wind farms are located away from densely populated areas so that background noise levels around typical wind farm developments are almost always low.
8. In Flanders, Belgium, the evening limits (between 7pm and 10pm) are the same as the nighttime limits.

Table 1.8 Noise metrics and threshold limits for Western USA.

State or county	Setback distance	Noise metric	Rural		Residential		Residential, near industry	
			Day	Night	Day	Night	Day	Night
California (Alameda)	N/A	L_{50}	–	45	–	45	–	45
California (Contra Costa)	N/A	L_{dn}	60	60	60	60	60	60
California (Fairfield)	N/A	L_{Aeq}	–	45	–	45	–	45
California (Fresno)	N/A	L_{50}	–	45	–	45	–	45
California (Kern)	N/A	L_{dn}	50	50	50	50	50	50
California (Kern)	N/A	$L_{A8.3}$	45	45	45	45	45	45
California (Monterey)	N/A	L_{den}	45–55	45–55	45–55	45–55	45–55	45–55
California (Morro Bay)	N/A	L_{Aeq}	–	45	–	45	–	45
California (Riverside)	2–3000 ft	L_{Aeq}	60	60	60	60	60	60
California (Sacramento)	N/A	L_{50}	–	50	–	50	–	50
California (San Bernadino)	N/A	L_{Aeq}	–	45	–	45	–	45
California (San Francisco)	N/A	L_{Aeq}	55	50	55	50	55	50
California (San Joaquin)	N/A	L_{Aeq}	–	45	–	45	–	45
California (Santa Cruz)	N/A	L_{Aeq}	–	40	–	45	–	45
California (Solano)	2000 ft	L_{Aeq}	44	44	44	44	44	44
California (Solano)	2000 ft	L_{den}	50	50	50	50	50	50
Colorado	N/A	L_{Aeq}	50	50	50	50	50	50
Colorado (Arapahoe)[3]	$H + B$	L_{Aeq}	55	50	55	50	55	50
Nevada (Lyon)[3]	$2(H + B)$	L_{Aeq}	55	55	55	55	55	55
New Mexico (San Miguel)[3]	0.5 miles	L_{Aeq}	<bk	<bk	<bk	<bk	<bk	<bk
Oregon[3,13,15]	N/A	L_{A50}	36	36	36	36	36	36
Washington	N/A	L_{Aeq}	60	50	60	50	70	70
Washington	N/A	L_{A25}	65	55	65	55	75	75
Washington	N/A	$L_{A16.7}$	70	60	70	60	80	80
Washington	N/A	$L_{A2.5}$	75	65	75	65	85	85
Wyoming (Laramie)[3]	$5.5H$	L_{Aeq}	50	50	50	50	50	50
Wyoming (Plympton)[3,12]	N/A	$L_{Ceq} - L_{Aeq}$	15	15	15	15	15	15

Background noise is usually measured using the $L_{A90,\,10\text{min}}$ metric. However, the wind farm noise is measured using the metric indicated in column 3 of the table. The term <bk, means that introduced wind farm noise must be less than existing background noise levels, which implies that the introduced noise can increase existing noise levels by a maximum of 3 dB; H is the hub height of the turbine; B is blade length.

Table 1.9 Noise metrics and threshold limits for Eastern USA.

State or county	Setback distance	Noise metric	Rural Day	Rural Night	Residential Day	Residential Night	Residential, near industry Day	Residential, near industry Night
Georgia[3]	Various	L_{Aeq}	55	55	55	55	55	55
Illinois[3,13]	N/A	L_{eq}	Table 1.11					
Indiana (Tipton)[3,13]	N/A	L_{eq}	Table 1.11					
Maine (Freedom)[21]	13H	L_{Aeq}	40	35	40	35	40	35
Michigan[3,10]	N/A	L_{Aeq}	55	55	55	55	55	55
Michigan (Huron)[2,10]	N/A	L_{A90}	55	55	55	55	55	55
Minnesota[3]	N/A	L_{Aeq}	50	50	50	50	50	50
Minnesota (Lincoln)[3]	750 ft	L_{Aeq}	50	50	50	50	50	50
New York (Jefferson)[2,10]	1000 ft / 5 H	L_{A10}	50	50	50	50	50	50
North Carolina[3]	$2.5(H + B)$	L_{Aeq}	55	55	55	55	55	55
North Carolina (Carteret)[16]	$2.5(H + B)$	L_{Aeq}	35	35	35	35	35	35
Pennsylvania (Potter)[3]	1750 ft / $5(H + B)$	L_{Aeq}	bk+5	bk+5	bk+5	bk+5	bk+5	bk+5
Wisconsin	N/A	L_{Aeq}	50	45	50	45	50	45
Wisconsin (Shawano)[2,14]	1250 ft / $3.1(H + B)$	L_{Aeq} Table 1.12	bk+5	bk+5	bk+5	bk+5	bk+5	bk+5

Background noise is usually measured using the $L_{A90,\,10\text{min}}$ metric. However, the wind farm noise is measured using the metric indicated in column 3 of the table. The term, bk+5, means that introduced wind farm noise must be less than 5 dBA above the existing background noise levels; H is the hub height of the turbine; B is the blade length.

9. The South Australian guidelines distinguish between rural living and rural industry, with the latter defined as any farming residence on a farm producing any rural produce for sale. This means that farmers can be subjected to 5 dBA more noise than non-farming residents, irrespective of how noisy their farming operation may be and irrespective of whether it produces noise at night.
10. In the UK, Michigan in the USA, Jefferson County in New York USA, New Zealand and all states of Australia, the allowed level is the greater of the level specified in the table and existing background noise plus 5 dB, where the background noise levels are determined from 10-min $L_{A90,\,10\text{min}}$ data points (that is, the level exceeded 90% of the time during the 10-min period), plotted as a function of wind speed at hub height, which can be measured at hub height or calculated from a measurement at a lower height. Daytime background noise level data are combined with nighttime noise level data and plotted as a function of 10-min averaged hub-height wind speed, where the 10-min wind speed average is the same period as the 10-min noise level. A total of 2000 data points are needed, with at least 500 representing the condition

Table 1.10 Noise metrics and threshold limits for countries outside the USA. *H* is the total height of the turbine (hub height plus blade length) and *B* is the blade length. Wind speeds are at 10 m height.

State or county	Setback distance	Noise metric	Rural		Residential		Residential, near industry	
			Day	Night	Day	Night	Day	Night
Australia (SA)[2,9,10]	1000 m	$L_{A90,10min}$	40	40	40	40	40	40
Australia (VIC)[2,10]	2000 m	$L_{A90,10min}$	40	40	40	40	40	40
Australia (QLD)[2,10,20]	1500 m	L_{Aeq}	37	35	37	35	37	35
Australia (NSW)[2,10]	2000 m	L_{Aeq}	35	35	35	35	40	40
Australia (WA)[2,10]	2000 m	L_{Aeq}	35	35	35	35	40	40
Belgium (Flanders)[3,8]	6*B*	L_{Aeq}	48	43	44	39	44	39
Belgium (Wallonia)[3]	3*H*	L_{Aeq}	45	45	45	45	45	45
Canada (Alberta)[3,18]	N/A	L_{Aeq}	–	40	–	43–46	–	43–46
Canada (British Columbia)[3]	N/A	L_{Aeq}	40	40	40	40	40	40
Canada (Manitoba)[3,19]	N/A	L_{Aeq}	40	40	40	40	40	40
Canada (New Brunswick)[3,19]	N/A	L_{Aeq}	40	40	40	40	40	40
Canada (Ontario)[3,19]	N/A	L_{Aeq}	40	40	45	45	45	45
Canada (Prince Edward Is.)[3]	3*H*	–	–	–	–	–	–	–
Denmark[2,17]	4*H*	L_{Aeq}	42–44	42–44	37–39	37–39	37–39	37–39
Denmark[2,17]	4*H*	$L_{Aeq,LF}$	25(in)	20(in)	25(in)	20(in)	25(in)	20(in)
France[3,11]	N/A	L_{Aeq}	35	35	35	35	35	35
Germany[2,11]	N/A	L_{Aeq}	50	35	50	35	55	40
Ireland	N/A	L_{Aeq}	40	40	40	40	40	40
Italy[4,5]	N/A	L_{Aeq}	50	40	55	45	60	50
New Zealand[2,10]	N/A	$L_{A90,10min}$	40	35	40	40	40	40
Norway[3]	N/A	L_{den}	45	45	45	45	45	45
Sweden[2]	N/A	L_{Aeq} at 8 m/s	35	35	40	40	40	40
Switzerland	N/A	L_{Aeq}	50	40	50	40	50	40
The Netherlands[3,4]	N/A	L_{den}	47	47	47	47	47	47
The Netherlands[3,4]	N/A	L_{night}	41	41	41	41	41	41
United Kingdom[2,4,10]	N/A	$L_{A90,10min}$	35–40	43	35–40	43	35–40	43

The label, (in) means indoor noise level.

where the wind is blowing from the wind farm to the measurement location (±45° from the direct downwind direction).

11. For France and Germany the allowed level is the greater of the level specified in the table and existing background noise plus 5 dB during the day, and plus 3 dB at night. In France, an adjustment is allowed so that allowable noise limits can be raised by 1, 2 or 3 dBA when the exceedance occurs for periods of 4–8 h, 2–4 h or 20–120 min, respectively.
12. The town of Plympton in Wyoming, USA has focussed its regulations entirely on the low-frequency and infrasonic part of the noise spectrum. No tones, in the frequency range 0–20 Hz and with a sound pressure level exceeding 50 dB

when energy-averaged over several minutes, are allowed inside of a residence if the peaks exceed the RMS sound pressure level by more than 10 dB. In addition, the maximum allowed difference between the unweighted sound pressure level in the range 0.1 Hz and above and the A-weighted level (see Section 2.2.11) cannot exceed 20 dB.

13. Tipton County in Indiana and Oregon specify allowable octave band limits in the frequency range from 63 Hz to 8 kHz (see Table 1.11).

Table 1.11 Octave band noise limits in Illinois (L_{eq}) and Oregon (L_{50}).

	Octave band centre frequency (Hz)								
	31.5	**63**	**125**	**250**	**500**	**1000**	**2000**	**4000**	**8000**
Illinois, day	75	74	69	64	58	52	47	43	40
Illinois, night	69	67	62	54	47	41	36	32	32
Tipton County (IN)	–	75	70	65	59	53	48	44	41
Oregon, day	68	65	61	55	52	49	46	43	40
Oregon, night	65	62	56	50	46	43	40	37	34

14. Shawano County in Wisconsin, USA specifies the allowable 1/3-octave band noise limits (L_{eq}) listed in Table 1.12.

Table 1.12 1/3-octave band noise limits in Shawano County (WI) (L_{eq}). The 70 dB in bands 2–12.5 Hz is for each band in that range.

1/3-octave band centre frequency (Hz)																
2–12.5	**16**	**20**	**25**	**31.5**	**40**	**50**	**63**	**80**	**100**	**125**	**250**	**500**	**1k**	**2k**	**4k**	**8k**
70	68	68	67	65	62	60	57	55	52	50	47	45	42	40	37	35

15. In Oregon USA, the allowed increase of any noise source above the existing ambient is 10 dBA (L_{A50}) and the minimum allowed ambient noise level for this consideration is 26 dBA, so that the maximum allowed noise level is never less than 36 dBA, even in very quiet environments.
16. In Carteret County, North Carolina, the allowed level can be exceeded any number of times for a maximum of 5 consecutive minutes at any property line.
17. In Denmark, the lower limits apply to a wind speed of 6 m/s and the upper limits apply to a wind speed of 8 m/s. Also the low-frequency limit ($L_{Aeq,LF}$) is a calculated indoor level as described in Section 5.4.
18. In Alberta, Canada, the levels are for residences that are more than 500 m from a heavily travelled road or railway and which are not subjected to frequent aircraft overflights. For residences between 30 and 500 m from a heavily travelled road or railway, 5 dB is added to the levels in the table and for residences less than 30 m

from a heavily travelled road or railway or which are subjected to frequent aircraft overflights, 5 dB is added to the levels in the table.

19. In Manitoba, New Brunswick and Ontario, Canada the allowable levels in rural and residential areas vary as a function of wind speed at 10 m height, as shown in Table 1.13.

Table 1.13 Allowed wind farm L_{Aeq} noise levels in three Canadian provinces

10-m height wind speed (m/s)	4	5	6	7	8	9	10	11
Allowed everywhere, Manitoba	40	40	40	43	45	49	51	53
Allowed everywhere, New Brunswick	40	40	40	43	45	49	51	53
Allowed rural, Ontario	40	40	40	43	45	49	51	
Allowed residential, Ontario	45	45	45	45	45	49	51	

20. In Queensland, Australia, nighttime (10pm to 6am) background noise levels are separated from daytime levels when determining how much greater wind farm noise is than background noise levels. In addition, maximum allowed C-weighted (L_{Ceq}) levels of 60 dBA at night and 65 dBA during the day are specified.
21. The limits in the town of Freedom, Maine, USA are 'not to exceed' limits and in addition there are C-weighted limits that are not to be exceeded. The post-construction C-weighted measurement cannot be more than 20 dBC above the pre-construction measured A-weighted ambient noise level. In addition, 50 dBC is not to be exceeded at any time.

1.7 Inquiries and Government Investigations

There have been numerous court cases in which residents have opposed planned wind farm developments as well as existing developments, and these are too numerous to list here. There are very few examples of successful litigation by residents, although in some cases wind farm operations have been limited to daytime and evening hours.

There have been a number of government-related inquiries, investigations and action plans directed at wind farms in a number of countries. Many of these include wind farm noise as a priority issue. Those mentioned below are those for which information is publicly accessible.[1]

1.7.1 Australia 2010–2014

In Australia, there have been four inquiries into the effects of wind farms on health: two by the National Health and Medical Research Council (NHMRC) and two by the Australian Senate.

1 The conclusions of the various reports are mostly reproduced verbatim, but with occasional minor edits.

NHMRC 2010, 'Rapid Review of the Literature on the Effects of Wind Farms on Human Health'

This study, undertaken by the NHMRC, was a 'rapid review' of the literature and concluded the following:

- There are no direct pathological effects from wind farms and any potential impact on humans can be minimised by following existing planning guidelines.
- There is currently no published scientific evidence to positively link wind turbines with adverse health effects.
- Noise levels from wind turbines have been assessed as negligible; that is, they appear to be no different to levels found in other everyday situations.
- A survey of all known published results of infrasound from wind turbines found that wind turbines of contemporary design, where rotor blades are in front of the tower, produce very low levels of infrasound.

This study attracted considerable criticism as it ignored much evidence that acoustical and medical experts considered should have been included. It has now been rescinded in light of the more recent study described below and reported in February 2015.

NHMRC 2013/14/15, 'Comprehensive Review of the Effects of Wind Farms on Human Health'

The NHMRC undertook a comprehensive review of the literature in 2013 and 2014 to determine whether or not wind farms produced any adverse health effects. All of the reviewed studies were considered irrelevant or of poor quality and thus were not able to be used to arrive at a conclusion. The review then used parallel studies of the effects of other noise sources with similar A-weighted noise levels (see Section 2.2.11). In addition, no studies were identified that specifically looked at possible effects on human health of infrasound. The findings published by the NHMRC are listed below and these are followed by a detailed critique of the entire approach:

- There is currently no consistent evidence that wind farms cause adverse health effects in humans.
- There is no direct evidence that exposure to wind farm noise affects physical or mental health. There are unlikely to be any significant effects on physical or mental health at distances greater than 1500 m from wind farms.
- There is consistent but poor-quality direct evidence that wind farm noise is associated with annoyance. Bias of different kinds and confounding factors are possible explanations for the associations observed.
- There is less consistent poor-quality direct evidence of an association between sleep disturbance and wind farm noise. However, sleep disturbance was not objectively measured in the studies and bias of different kinds and confounding factors are possible explanations for the associations observed. While chronic sleep disturbance is known to affect health, the parallel evidence suggests that wind farm noise is unlikely to disturb sleep at distances of more than 1500 m.
- There is no direct evidence that considered possible effects on health of infrasound or low-frequency noise from wind farms.
- Background evidence indicates that wind farm noise is generally in the range of 30–45 A-weighted decibels (dBA) (see Section 2.2.11) at a distance of 500–1500 m from a

wind farm and below 30–35 dBA beyond 1500 m. Although individuals may perceive aspects of wind farm noise at greater distances, it is unlikely that it will be disturbing at distances of more than 1500 m.

In summary, the NHMRC has labelled all evidence as poor, bringing them to the first conclusion stated above. Of course, there is ample evidence in the literature that does not support this conclusion, nor the conclusion that wind farms have no effect on sleep for residents at distances greater than 1500 m. The study seems to be biased towards the wind farm industry, as the first statement could equally have been written as 'There is currently no consistent evidence that wind farms do *not* cause adverse health effects in humans', which would have resulted in an entirely different perception by the community. The emphasis in the report on 'confounding factors' and the many statements that findings of health effects or sleep problems could be due to causes other than wind farms also negates the findings of many worthwhile studies published in peer-reviewed journals.

As one may expect, the NHMRC review findings have attracted a substantial amount of criticism. As stated by Hanning (2015),

> ...in considering the evidence, NHMRC adopted inappropriately strict evidential criteria. This is the reactionary approach to public health risks and is clearly not in the public interest. Action in defence of the public health does not require certainty. In addition, they have turned the burden of proof on its head. It is the wind industry's duty to prove the safety of its activities not that of the public.

However, the review does recommend that more research should be done.

South Australian Parliamentary Committee on Wind Farm Developments in South Australia

This committee was formed in 2012, received over 150 submissions and produced no publicly available report or recommendations.

Australian Senate, Community Affairs References Committee, 2011

The Australian Senate Community Affairs References Committee investigated the social and economic impact of rural wind farms in 2011 and published a report with the following recommendations related to wind farm noise.

- Noise standards adopted by the states and territories for the planning and operation of rural wind farms should include appropriate measures to calculate the impact of low-frequency noise and vibrations indoors at impacted dwellings.
- Further consideration should be given to the development of policy on separation criteria between residences and wind farm facilities.
- The Commonwealth Government should initiate as a matter of priority thorough, adequately resourced epidemiological and laboratory studies of the possible effects of wind farms on human health. This research must engage across industry and the community, and include an advisory process representing the range of interests and concerns.
- The National Acoustics Laboratories should conduct a study and assessment of noise impacts of wind farms, including the impacts of infrasound.

- The draft National Wind Farm Development Guidelines should be redrafted to include discussion of any adverse health effects and comments made by NHMRC regarding the revision of its 2010 public statement.

None of these recommendations have been implemented by 2015, four years after their publication.

Australian Senate, Select Committee on Wind Turbines, 2015

The Select Committee on Wind Turbines was established in 2014 with the following terms of reference relevant to wind turbine noise:

- the role and capacity of the National Health and Medical Research Council in providing guidance to state and territory authorities;
- implementation of planning processes in relation to wind farms, including the level of information available to prospective wind farm hosts;
- adequacy of monitoring and compliance governance of wind farms;
- application and integrity of national wind farm guidelines.

In August, 2015, the committee produced a final report containing the following recommendations related to wind farm noise:

1. An Independent Expert Scientific Committee on Industrial Sound (IESC) should be established, with funding from a levy on wind turbine operators accredited to receive renewable energy certificates, for the purpose of:
    - conducting independent, multidisciplinary research into the adverse impacts and risks to individual and community health and wellbeing associated with wind turbine projects and any other industrial projects which emit sound and vibration energy;
    - establishing a formal channel to communicate its advice and research priorities and findings to the Environmental Health Standing Committee;
    - developing a single national acoustic standard on audible noise from wind turbines that is cognisant of the existing standards, Australian conditions and the signature of new turbine technologies;
    - developing a national acoustic standard on infrasound, low-frequency sound and vibration from industrial projects;
    - developing a national standard on minimum buffer zones;
    - developing a template for State Environment Protection Agencies to adopt a fee-for-service licensing system (see recommendation 9, below);
    - developing a guidance note proposing that State Environment Protection Authorities be responsible for monitoring and compliance of wind turbines and suggesting an appropriate process to conduct these tasks;
    - providing scientific and technical advice to State Environment Protection Authorities, the Clean Energy Regulator and the Federal Health Minister to assist in assessing whether a proposed or existing wind farm project poses risks to individual and community health;
    - publishing information relating to its research findings and providing the Federal Health Minister with research priorities and research projects to improve scientific understanding of the impacts of wind turbines on the health and quality of life of affected individuals and communities;

- providing guidance, advice and oversight for research projects commissioned by agencies such as the National Health and Medical Research Council and the Commonwealth Scientific and Industrial Research Organisation relating to sound emissions from industrial projects.

This committee was established in October, 2015.

2. The National Environment Protection Council should establish a National Environment Protection (Wind Turbine Infrasound and Low Frequency Noise) Measure (NEPM), developed from the findings of the IESC.
3. If the Regulator of the Renewable Energy (Electricity) Act 2000 receives an application from a wind power station that is properly made under section 13, the Regulator must:
   - seek the advice of the IESC on whether the proposed project, over its lifetime, poses risks to individual and community health;
   - confer with the Federal Minister for Health and the Commonwealth Chief Medical Officer to ascertain the level of risk that the proposed project poses to individual and community health; and
   - not accredit the power station until such time as the Federal Minister for Health is satisfied that these risks have been mitigated.
4. Provision should be made in the Renewable Energy (Electricity) Act 2000, that compliance with the National Environment Protection (Wind Turbine Infrasound and Low-Frequency Noise) Measure (NEPM) is a pre-requisite for wind energy generators to receive Renewable Energy Certificates and that wind energy generators operating in states that do not require compliance with the NEPM are ineligible to receive Renewable Energy Certificates. Existing and approved wind farms should be given a period of no more than five years in which to comply.
5. The Commonwealth Government should introduce National Wind Farm Guidelines which each Australian State and Territory Government should reflect in their relevant planning and environmental statutes.
6. The Renewable Energy (Electricity) Act 2000 should be amended to enable partial suspension and point in time suspension of renewable energy certificates for wind farm operators that are found to have breached the conditions of their planning approval.
7. A National Wind Farm Ombudsman should be established, with funding from a levy on wind turbine operators accredited to receive renewable energy certificates, to handle complaints from concerned community residents about the operations of wind turbine facilities. This position was filled in October, 2015.
8. Data collected by wind turbine operators relating to wind speed, basic operation statistics including operating hours and noise monitoring should be made freely and publicly available on a regular basis.
9. All State Governments consider shifting responsibility for monitoring wind farms in their jurisdiction from local councils to the State Environment Protection Authority.
10. The Federal Department of the Environment prepare a quarterly report collating the wind farm monitoring and compliance activities of the State Environment Protection Authorities.
11. The National Health and Medical Research Council (NHMRC) continue to monitor and publicise Australian and international research relating to wind farms and

health. The NHMRC should fund and commission primary research that the IESC identifies as necessary.
12. A national regulatory body be established under Commonwealth legislation for the purpose of monitoring and enforcing wind farm operations.

1.7.2 Canada

In Canada, there have been two enquiries into the adverse health effects of wind farms, one by Health Canada begun in 2012 and reported on in 2014 and one by The Council of Canadian Academies, reported on in 2015, which was sponsored but not influenced by Health Canada. This latter study was undertaken by an expert panel that included medical experts and acoustics experts.

Health Canada, 2014

In 2014, Health Canada reported the results of a wind farm noise study that had the following objectives:

- Investigation of the prevalence of health effects or health indicators among a sample of Canadians exposed to WTN (wind turbine noise) using both self-reported and objectively measured health outcomes.
- Application of statistical modelling in order to derive exposure response relationships between WTN levels and self-reported and objectively measured health outcomes.
- Investigation of LFN and infrasound from wind turbines as potential contributing factors towards adverse community reaction.

The conclusions of the study, which have been criticised by a number of medical professionals (for example, Krogh and McMurtry (2014)), are listed below.

- Self-reports of diagnosis of a number of health conditions were not found to be associated with exposure to WTN levels.
- Self-reported stress, as measured by scores on the Perceived Stress Scale, was not found to be related to exposure to WTN levels.
- Exposure to WTN was not found to be associated with any significant changes in reported quality of life for any of the four domains, nor with overall quality of life and satisfaction with health.
- Annoyance towards noise, shadow flicker, blinking lights, vibrations, and visual impacts were found to be statistically associated with increasing levels of WTN. The relationship between noise and community annoyance is stronger than any other self-reported measure, including sleep disturbance. Additional findings related to WTN are listed below.
    - A statistically significant increase in annoyance was found when WTN levels exterior to residences exceeded 35 dBA.
    - Reported WTN annoyance was statistically higher in the summer, outdoors and during evening and nighttime.
    - WTN annoyance significantly dropped in areas where calculated nighttime background noise exceeded WTN by 10dB or more.
    - Annoyance was significantly lower among the participants who received personal benefit.

  - WTN annoyance was found to be statistically related to several self-reported health effects including, but not limited to, blood pressure, migraines, tinnitus, dizziness and perceived stress.
  - WTN annoyance was found to be statistically related to measured hair cortisol, systolic and diastolic blood pressure.
  - The findings support a potential link between long term high annoyance and health.
  - Health and well-being effects may be partially related to activities that influence community annoyance, over and above exposure to wind turbines.
  - Self-reported and measured health effects were not dependent on the particular levels of noise, or particular distances from the turbines, and were also observed in many cases for road traffic noise annoyance.
- Calculated outdoor WTN levels near the participants' homes were not found to be associated with sleep efficiency, the rate of awakenings, duration of awakenings, total sleep time, or how long it took to fall asleep.

The most significant shortcoming of the above study was that there was an assumption that increasing wind turbine noise levels should be correlated to increasing numbers of people suffering symptoms. Perhaps a more reasonable assumption would be that a fixed percentage of people are susceptible to wind farm noise and that the range of wind farm noise levels that can produce a reaction or symptom in all of the sensitive people is quite large.

Another shortcoming is that the noise levels used to correlate health effects were based on calculated average levels. There was no consideration of the spectrum shape of noise arriving at different residences, how it varied with time, nor whether pulsating or rumbling noise was present. Thus any attempt to just correlate A-weighted average predicted noise levels (see Section 2.2.11) using a relatively simple noise propagation model cannot yield reliable results.

The results of the Health Canada study were reported in detail in a number of publications in 2016 (Michaud et al. 2016a,b,c).

Council of Canadian Academies, 2015

The Council of Canadian Academies is an independent not-for-profit organisation that supports independent, science-based, authoritative expert assessments to inform public policy development in Canada. The Minister of Health (responsible for Health Canada) asked the Council to find an answer to the question, 'Is there evidence to support a causal association between exposure to wind turbine noise and the development of adverse health effects?' The following sub-questions were also included in the request.

- Are there knowledge gaps in the scientific and technological areas that need to be addressed in order to fully assess possible health impacts from wind turbine noise?
- Is the potential risk to human health sufficiently plausible to justify further research into the association between wind turbine noise exposure and the development of adverse health effects?
- How does Canada compare internationally with respect to the prevalence and nature of reported adverse health effects among populations living in the vicinity of commercial wind turbine establishments?

- Are there engineering technologies and/or other best practices in other jurisdictions that might be contemplated in Canada as measures that may minimise adverse community response towards wind turbine noise?

The expert panel identified about 30 symptoms and adverse health outcomes that have been attributed to exposure to wind turbine noise, based on a broad survey of the internet and peer-reviewed literature. Empirical evidence related to any associations between these health outcomes and exposure to wind turbine noise was then collected, resulting in an examination of approximately 300 publications, of which 38 were found to be relevant to the health effects of wind turbine noise. The study concluded the following (see the full report, CCA (2015)):

1. Standard methods of measuring sound may not properly identify nor quantify low-frequency noise nor the amplitude modulation of that noise.
2. The evidence is sufficient to establish a causal relationship between exposure to wind turbine noise and annoyance.
3. There is limited evidence to establish a causal relationship between exposure to wind turbine noise and sleep disturbance.
4. The evidence suggests a lack of causality between exposure to wind turbine noise and hearing loss.
5. There is inadequate evidence of a direct causal relationship between exposure to wind turbine noise and stress, although stress has been linked to other sources of community noise.
6. For all other health effects considered (fatigue, tinnitus, vertigo, nausea, dizziness, cardiovascular diseases, diabetes, and so on), the evidence was inadequate to come to any conclusion about the presence or absence of a causal relationship with exposure to wind turbine noise.
7. Knowledge gaps prevent a full assessment of public health effects of wind turbine noise.
8. Research on long-term exposure to wind turbine noise would provide a better understanding of the causal associations between wind turbine noise exposure and certain adverse health effects.
9. Technological development is unlikely to resolve, in the short term, the current issues related to perceived adverse health effects of wind turbine noise.
10. Impact assessments and community engagement provide communities with greater knowledge and control over wind energy projects and therefore help limit annoyance.

1.7.3 Denmark 2013

The Danish Ministry for Health and Prevention has commissioned a study into the health effects of wind turbines, which was announced by the Minister for Health and Prevention on 2 July 2013. The announcement stated:

> In the absence of epidemiological studies of wind turbine noise, effects on the cardiovascular system by long-term exposure to wind turbine noise can not be completely excluded at the present time. I am therefore pleased today to announce that the Environmental Protection Agency will ask the Danish Cancer Society

to prepare a detailed description of a study of possible effects of wind turbine noise.

However, the results were not available at the time of writing.

1.7.4 Northern Ireland 2013

In November, 2013, the Committee for the Environment of the Northern Ireland Assembly agreed to carry out a full inquiry into wind energy. One of the terms of reference was 'To compare the perceived impact of wind turbine noise and separation distances with other jurisdictions and other forms of renewable energy development'. The recommendations, published in January, 2015 (Northern Ireland Assembly: Committee for the Environment 2015), that are related to noise are as follows.

- The standard conditions (see IOA (2013), Appendix B) which were developed by the Institute of Acoustics, and which have been endorsed in Scotland, England and Wales, should be routinely attached to planning consents in Northern Ireland.
- The Department should review the use of the ETSU-97 guidelines on an urgent basis, with a view to adopting more modern and robust guidance for measurement of wind turbine noise, with particular reference to current guidelines from the World Health Organisation.
- Arrangements should be put in place for on-going long-term monitoring of wind turbine noise.
- The Department, working with local universities, should commission independent research to measure and determine the impact of low-frequency noise on those residents living in close proximity to individual turbines and wind farms in Northern Ireland.
- The Department, taking into account constraints on the availability and suitability of land for the generation of wind energy, should specify a minimum separation distance between wind turbines and dwellings.

1.7.5 Scotland

In early 2015, the Scottish Conservatives, then a minor opposition party in the Scottish Parliament, launched an action plan for rural Scotland that included a pledge for local councils to be able to place a moratorium on new wind farms and facilitate compensation for those whose property prices had dropped. This was partly a result of the 6000 public objections to large-scale wind farms that had been lodged in the preceding year.

1.7.6 Wales

In 2011 the Petitions Committee was presented with a petition requesting that the Welsh Government pass a statute controlling the noise from wind turbines during the hours between 6pm and 6am. After an extensive investigation the committee made the following recommendations to the Welsh Government.

1. The Statutory Planning Guidance should be amended to introduce buffer zones that maintain the current 500 metres minimum distance between dwellings and turbines, and increase the separation distance in specified appropriate circumstances up to 1500 metres.

2. ETSU-R-97 guidelines should be revised to take into account the lower ambient noise levels in rural areas and the latest research and World Health Organisation evidence on the effects of noise on sleep disturbance.
3. Faulty turbines should be switched off at specified times overnight as soon as a fault affects its noise emissions
4. The Institute of Acoustics Working Group should carry out meaningful consultation with people living close to wind turbines so that their experiences can help to shape the conclusions and recommendations of the Group that are expected to be published in September 2012.

1.8 Current Consensus on Wind Farm Noise

The only consensus related to wind farm noise is that there is no consensus on wind farm noise. Wind farms are a political 'hot potato'. Governments and environmentalists see them as an easy way of achieving renewable energy targets and saving the planet from greenhouse gas overload. Wind farms are seen by many as a great economic benefit, so one can understand the lack of sympathy in the general population for wind farm neighbours who complain about the noise and/or attempt to halt wind farm developments or shut down existing wind farms. There are two main views on wind farm noise. One view is that wind farms do not produce any significant noise and certainly not any noise at a level that should annoy or affect the health of anyone. As may be expected, this view is held universally by wind turbine manufacturers and wind farm developers and most (but not all) other people who benefit financially from the wind farm industry. Another view that is held by many people living in the vicinity of wind farms (including some wind farm hosts) is that wind farm noise is annoying, it results in sleep deprivation and in some cases results in serious health problems. Whether health problems are a result of sleep deprivation and stress associated with continually experiencing an intrusive noise or whether they are a direct physiological result of the character of wind farm noise is not clear at the time of writing and a considerable amount of research is needed to clarify this issue. There are emotional outbursts on the web from people supporting both sides of the argument and unfortunately some of the personal insults and attacks that can be found on some pro-wind farm web sites are unhelpful at best.

References

AWED 2016 Sample wind farm ordinance. Alliance for Wise Energy Decisions. http://wiseenergy.org/Energy/Model_Wind_Law.pdf.

AMWG 2015 Methods for rating amplitude modulation in wind turbine noise. Technical report, Institute of Acoustics Noise Working Group (Wind Turbine Noise): Amplitude Modulation Working Group.

AS 4959 2010 Acoustics – measurement, prediction and assessment of noise from wind turbine generators. Standards Australia.

Ashtiani P 2013 Generating a better picture of noise immissions in post construction monitoring using statistical analysis. In: *5th International Meeting on Wind Turbine Noise*, Denver, Colorado.

Ashtiani P 2015 Spectral discrete probability density function of measured wind turbine noise in the far field. *6th International Meeting on Wind Turbine Noise*, Glasgow, UK.

Auer P 2013 *Advances in Energy Systems and Technology*, vol. 1. Academic Press.

Batasch M 2005 Regulation of wind turbine noise in the Western United States. *First International Meeting on Wind Turbine Noise: Perspectives for Control*, Berlin, Germany.

Berglund B, Lindvall T and Schwela D 1999 Guidelines for community noise. WHO.

Bigot A, Slaviero D, Mirabel C and Dutilleaux P 2015 Influence of vertical temperature gradient on background noise and on long-range noise propagation from wind turbines. *6th International Meeting on Wind Turbine Noise*, Glasgow, UK.

Boeing 2015 MOD-2/MOD-5B Wind turbines. http://www.boeing.com/history/products/mod-2-mod-5b-wind-turbine.page.

CAA 2015 Understanding the evidence: wind turbine noise. Technical report, Council of Canadian Academies, Expert Panel on Wind Turbine Noise and Human Health.

Cox R, Unwin D and Sherman T 2012 Wind turbine noise impact assessment: where ETSU is silent. Technical report, National Wind Watch.

Dodge D 2006 Illustrated history of wind power development. http://www.telosnet.com/wind/.

Ekwaro-Osire S, Jang TH, Stroud A, Durukan I, Alemayehu F, Swift A and Chapman J 2011 Gear with asymmetric teeth for use in wind turbines. In: Proulx, T (ed), *Experimental Mechanics on Emerging Energy Systems and Materials*, vol. 5. Springer.

ETSU 1996 ETSU-R-97: Assessment and rating of noise from wind farms. Technical report, UK Department of Trade and Industry, Energy Technology Support Unit.

Fowler K, Koppen E and Mathis K 2013 International legislation and regulations for wind turbine noise. *5th International Meeting on Wind Turbine Noise*, Denver, Colorado.

Gipe P 1995 *Wind Energy Comes of Age*. John Wiley & Sons.

Glegg S, Baxter S and Glendinning A 1987 The prediction of broadband noise from wind turbines. *Journal of Sound and Vibration*, **118**(2), 217–239.

Gottlob D 1998 German standard for rating low-frequency noise immissions. *Proceedings of Internoise98*, Christchurch, New Zealand, pp. 812–817.

Greene G 1981 Measured and calculated characteristics of wind turbine noise. Technical report CP2185, NASA.

Grosveld F 1985 Prediction of broadband noise from horizontal axis wind turbines. *Journal of Propulsion and Power*, **1**(4), 292–299.

Gutin L 1948 On the sound field of a rotating propeller (From Phys. Zeitschr. der Sowjetunion, Bd. 9, Heft 1, 1936, pp 57–71). Technical report TM1193, NASA.

Hanning C 2015 Submission to the Australian Senate Select Committee on Wind Turbines. Technical report, University Hospitals of Leicester.

Hau E 2013 *Wind Turbines – Fundamentals, Technologies, Application, Economics*. Springer.

Hessler G 2008 The noise perception index (NPI) for assessing noise impact from major industrial facilities and power plants in the US. *Noise Control Engineering Journal*, **56**(5), 374–385.

Hubbard H and Shepherd K 1991 Aeroacoustics of large wind turbines. *Journal of the Acoustical Society of America*, **89**(6), 2495–2508.

Hunt M and Hannah L 2009 The use of Noise Perception Index (NPI) for setting wind farm noise limits. *Third International Meeting on Wind Turbine Noise*, Aalborg, Denmark.

Hurtley C 2009 *Night noise guidelines for Europe*. WHO Regional Office Europe.

IEC 61400-11 2012 Wind turbines – part 11: Acoustic noise measurement techniques, edn 3.0. International Electrotechnical Commission.

IOA 2013 A good practice guide to the application of ETSU-R-97 for the assessment and rating of wind turbine noise. Technical report, IOA.

IOA 2014a A good practice guide to the application of ETSU-R-97 for the assessment and rating of wind turbine noise. Supplementary guidance note 1: data collection. Technical report, Institute of Acoustics.

IOA 2014b A good practice guide to the application of ETSU-R-97 for the assessment and rating of wind turbine noise. Supplementary guidance note 2: data processing and derivation of ETSU-R-97 background curves. Technical report, Institute of Acoustics.

IOA 2014c A good practice guide to the application of ETSU-R-97 for the assessment and rating of wind turbine noise. Supplementary guidance note 3: sound power level data. Technical report, Institute of Acoustics.

IOA 2014d A good practice guide to the application of ETSU-R-97 for the assessment and rating of wind turbine noise. Supplementary guidance note 4: wind shear. Technical report, Institute of Acoustics.

IOA 2014e A good practice guide to the application of ETSU-R-97 for the assessment and rating of wind turbine noise. Supplementary guidance note 5: post completion measurements. Technical report, Institute of Acoustics.

IOA 2014f A good practice guide to the application of ETSU-R-97 for the assessment and rating of wind turbine noise. Supplementary guidance note 6: noise propagation over water for on-shore wind turbines. Technical report, Institute of Acoustics.

International Energy Agency 1989 R&D Wind Energy Annual Report. WECS Executive Committee, National Energy Administration.

ISO1996 1971 Acoustics: Assessment of noise with respect to community response. International Organization for Standardization

ISO9613-2 1996 Acoustics: Attenuation of sound during propagation outdoors. International Organization for Standardization

ISO1996-1 2003 Acoustics: Description, measurement and assessment of environmental noise–part 1: Basic quantities and assessment procedures (reviewed 2012). International Organization for Standardization

ISO389-7 2005 Acoustics: Reference zero for the calibration of audiometric equipment – part 7: Reference threshold of hearing under free-field and diffuse-field listening conditions. International Organization for Standardization

Kamperman G and James R 2008 The 'How to' guide to siting wind turbines to prevent health risks from sound. http://www.windturbinesyndrome.com/wp-content/uploads/2008/11/kamperman-james-10-28-08.pdf.

Kelley N 1987 A proposed metric for assessing the potential of community annoyance from wind turbine low-frequency noise emissions. Technical report, Solar Energy Research Inst.

Kelley N, McKenna H, Hemphill R, Etter C, Garrelts R and Linn N 1985 Acoustic noise associated with the MOD-1 wind turbine: its source, impact, and control. Technical report, US Department of Energy.

Kelley N, McKenna H, Jacobs E, Hemphill R and Birkenheuer N 1988 Technical Report TR-3036. the MOD-2 wind turbine: Aeroacoustical noise sources, emissions, and potential impact. Technical report, NASA.

Krogh C and McMurtry R 2014 Health Canada and wind turbines: Too little too late? Candadian Medical Association blog. http://cmajblogs.com/health-canada-and-wind-turbines-too-little-too-late/.

Maegaard P 2013 *Wind Power for the World: The Rise of Modern Wind Energy*. Stanford Publishing.

Manwell J, McGowan J and Rogers, AL 2009 *Wind energy explained: theory, design and application*, 2nd edn. John Wiley & Sons.

Metzger F and Klatte R 1981 Downwind rotor horizontal axis wind turbine noise prediction. *Proceedings of NASA Workshop on Wind Turbine Dynamics*, Cleveland, Ohio, vol. **1**, pp. 425–430.

Michaud D, Feder K, Keith S, Voicescue S, Marro L, Than J, Guay M, Denning A, Bower T, Villeneuve P, Russell E, Koren G and van den Berg F 2016a Self-reported and measured stress related responses associated with exposure to wind turbine noise. *Journal of the Acoustical Society of America*, **139**(3), 1467–1479.

Michaud D, Feder K, Keith S, Voicescue S, Marro L, Than J, Guay M, Denning A, D'Arcy M, Bower T, Lavigne E, Murray B, Weiss S and van den Berg F 2016b Exposure to wind turbine noise: Perceptual responses and reported health effects. *Journal of the Acoustical Society of America*, **139**(3), 1443–1454.

Michaud D, Keith S, Feder K, Voicescue S, Marro L, Than J, Guay M, Bower T, Denning A, Lavigne E, Chantal W, Janssen S, Leroux T and van den Berg F 2016c Personal and situational variables associated with wind turbine noise annoyance. *Journal of the Acoustical Society of America* **139**(3), 1455–1466.

Miskelly P 2012 Wind farms in Eastern Australia – recent lessons. *Energy & Environment*, **23**(8), 1233–1260.

Moorhouse A, Waddington D and Adams M 2005 Proposed criteria for the assessment of low frequency noise disturbance. Technical report, University of Salford.

Northern Ireland Assembly: Committee for the Environment 2015 *Report on the Committee's Inquiry into Wind Energy*, Vol. 1–6. Northern Ireland Stationery Office.

NZS 6808 2010 Acoustics – wind farm noise. Standards New Zealand.

Putnam P 1948 *Power from the Wind*. Van Nostrand.

Renewable Energy Foundation 2014 The efficacy of the RUK AM condition. http://www.ref.org.uk/publications/310-the-efficacy-of-the-ruk-am-condition.

Renewable UK 2013 The development of a penalty scheme for amplitude modulated wind farm noise: description and justification. Technical report, Renewable UK.

Richards B 1987 Initial operation of project EOLE 4MW vertical-axis wind turbine generator. *Proceedings of Windpower '87*, San Francisco, USA.

Righter R 1996 *Wind Energy in America: A History*. University of Oklahoma Press.

Sektorov V 1934 The first aerodynamic three-phase electric power plant in Balaclava. *L'Elettrotecnica*, **21**(23–24), 538–542.

Shaltens R and Birchenough A 1983 Operational results for the experimental DOE/NASA Mod-OA wind turbine project. Technical report TM-83S17, NASA.

Shepherd D, McBride D, Welch D, Dirks K and Hill E 2011 Evaluating the impact of wind turbine noise on health-related quality of life. *Noise and Health*, **13**(54), 333.

Søndergaard B 2013 Low frequency noise from wind turbines: Do the Danish regulations have any impact? *5th International Meeting on Wind Turbine Noise*, Denver, Colorado, pp. 28–30,

Spencer R 1981 Noise generation of upwind rotor wind turbine generators. *Proceedings of NASA Workshop on Wind Turbine Dynamics*, Cleveland, Ohio, pp. 419–423.

Spera D 2009 *Wind Turbine Technology: Fundamental Concepts in Wind Turbine Engineering*, 2nd edn. ASME.

Stephens D, Shepherd K and Grosveld F 1981 Establishment of noise acceptance criteria for wind turbines. *Intersociety Energy Conversion Engineering Conference*, Atlanta, USA, pp. 2033–2036.

Stephens D, Shepherd K, Hubbard H and Grosveld F 1982 Guide to the evaluation of human exposure to noise from large wind turbines. Technical report, NASA.

Stigwood M, Large S and Stigwood D 2015 Cotton Farm wind farm long term community noise monitoring project – 2 years on. *6th International Meeting on Wind Turbine Noise*, Glasgow, UK.

Tachibana H 2014 Outcome of systematic research on wind turbine noise in Japan. *Proceedings of Internoise 2014*, Melbourne, Australia.

Thompson D 1982 Noise propagation in the atmosphere's surface and planetary boundary layers. *Proceedings of Internoise 82*, San Francisco, USA.

Tocci G and Marcus E 1982 A parametric evaluation of wind turbine noise. *Proceedings of Internoise 82*, San Francisco, USA.

Tokita Y, Oda A and Shimizu K 1984 On the frequency weighting characteristics for evaluation of infra and low frequency noise. *Proceedings of Internoise 84 and Noise-Con 84*, Honolulu, Hawaii, USA, pp. 917–920.

Tong W 2010 *Wind Power Generation and Wind Turbine Design*. WIT Press.

UK Government 2015 English planning practice guidance: noise. http://planningguidance.communities.gov.uk/blog/guidance/noise/noise-guidance/.

Viterna L 1981 The NASA LeRC wind turbine sound prediction code. Technical report CP-2185, NASA.

Weißbach D, Ruprecht G, Huke A, Czerski K, Gottlieb S and Hussein A 2013 Energy intensities, EROIs (energy returned on investment), and energy payback times of electricity generating power plants. *Energy*, **52**, 210–221.

Willshire Jr W and Zorumski W 1987 Low-frequency acoustic propagation in high winds. *Noise-Con 87: Proceedings of the National Conference on Noise Control Engineering*, State College, PA, USA, vol. **1**, pp. 275–280.

Willshire W 1985 Long range downwind propagation of low-frequency sound. NASA Langley Research Center.

2

Fundamentals of Acoustics and Frequency Analysis

2.1 Introduction

The purpose of this chapter is to describe the concepts in acoustics that are necessary to properly understand wind turbine noise generation, propagation, assessment and regulation.

In Section 2.2 we introduce the basic concepts and definitions related to the physics of sound and noise, which will allow the reader to understand the meaning of sound pressure and sound pressure level (and decibels), sound intensity level, sound power level and A-weighting. Most definitions have been internationally standardised and are listed in standards publications such as IEC60050-801 (1994). In Section 2.3 we discuss the concept of converting the time-varying sound pressure signal from a microphone into its component frequencies. Section 2.4 is concerned with some more advanced concepts associated with frequency analysis that are useful for understanding the character of wind farm noise and perhaps most useful to practitioners who are responsible for measuring and analysing noise produced by wind farms at receiver locations.

2.2 Basic Acoustics Concepts

Noise can be defined as a 'a disagreeable or undesired sound'. From the acoustics point of view, sound and noise constitute the same phenomenon of atmospheric pressure fluctuations about the mean atmospheric pressure; the differentiation between sound and noise is greatly subjective. What is sound to one person can very well be noise to somebody else. Wind turbines can a sound that is sweet and gentle to wind farm developers and to many wind farm hosts but they can sound excruciatingly annoying to many people forced to live in their near vicinity.

Sound (or noise) is the result of pressure variations or oscillations in an elastic medium (e.g. air, water, solids), generated by a vibrating surface or turbulent fluid flow. Sound propagates in the form of longitudinal (as opposed to transverse) waves, involving a succession of compressions and rarefactions in the elastic medium, as illustrated by Figure 2.1a. When a sound wave propagates in air (which is the medium considered in this book), the oscillations in pressure are above and below the ambient atmospheric

Wind Farm Noise: Measurement, Assessment and Control, First Edition.
Colin H. Hansen, Con J. Doolan, Kristy L. Hansen.

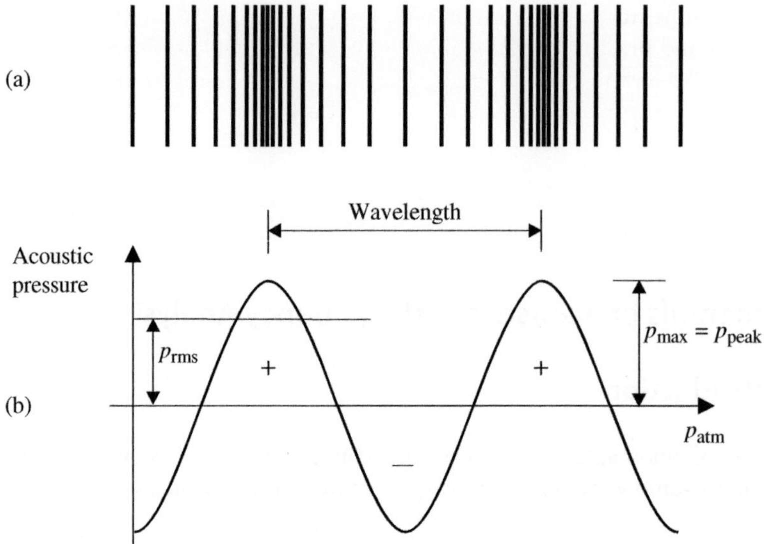

Figure 2.1 Representation of a sound wave: (a) compressions and rarefactions caused in air by the sound wave; (b) graphic representation of pressure variations above and below atmospheric pressure.

pressure and are very much smaller than atmospheric pressure (an amount of approximately 1/50 000 of atmospheric pressure for a loud noise of 100 dB).

2.2.1 Root Mean Square Sound Pressure

The root mean square (RMS), referred to in the previous section, is a way of quantifying a sound pressure that is varying with time. First of all, the atmospheric pressure is subtracted from the total pressure measurement. This is done in practice by the microphone, which has a pressure equalisation mechanism so that both sides of its diaphragm are subjected to atmospheric pressure, but only one side is subjected to the varying pressure associated with the sound wave. This is achieved by exposing the rear of the diaphragm to the atmosphere via a small-diameter hole or tube. As a result, it is only possible for very slowly varying sound pressures to influence the pressure in the cavity backing the microphone diaphragm. The larger the hole diameter, the higher will be the frequency at which pressure equalisation will occur, and thus the lowest frequency that the microphone can measure.

The microphone output provides a sound pressure that varies positively and negatively about a zero mean pressure and the instrumentation attached to the microphone, such as a sound level meter, squares all instantaneous values of the signal and averages the results over some arbitrary time interval. The square root is then taken of this average value to produce an RMS value. For a single-frequency sound wave, the peak value is 1.41 times the RMS value (as illustrated in Figure 2.1b) but for sound containing many frequencies such as random noise, the peak value may be many times (up to 8 or 10 times) the RMS value and for this type of sound a crest factor is defined as the ratio of the maximum absolute value (or peak) sound pressure divided by the RMS sound pressure.

Modern instrumentation is invariably digital in nature, which means that the time-varying sound pressure signal from the microphone is sampled with a specified time interval between the instantaneous samples. The time interval between samples defines the sample rate and for most audio instrumentation that measures in the audio range, this is 48 kHz (or 48 000 samples per second). Thus the squaring operation in the process of obtaining RMS levels is applied to each sample in the specified time interval. The squared values are then summed and the result divided by the number of samples in the specified time interval before taking the square root to find the RMS sound pressure, p_{RMS}. In mathematical terms, for a time-varying sound pressure signal of k samples, the RMS value is

$$p_{\text{RMS}} = \sqrt{\frac{1}{N}\sum_{k=0}^{N-1} p^2(k)} \tag{2.1}$$

The square of the RMS sound pressure is often written as $\langle p^2 \rangle_t$, where the subscript t represents a time average as opposed to a spatial average of the sound pressure. In the remainder of the book, the subscript t will be omitted for brevity. Beginning the k index at zero rather than 1 is a convention that has been in use for a long time. The reason is that it is mathematically convenient, especially for more complex analyses, although this may not be apparent here. Another reason is that the first sample is taken at time $t = 0$, so it follows that the first subscript should be zero.

2.2.2 Statistical Descriptors and Their Use

There are a number of statistical descriptors for time-varying signals that can prove useful when analysing data. The most important ones for our applications are listed below.

Mean

The mean of a time-domain sampled signal, which is N samples in duration, is

$$\overline{x} = \frac{1}{N}\sum_{k=0}^{N-1} x(k) \tag{2.2}$$

Variance and Standard Deviation

The variance is defined as the square of the standard deviation, as follows

$$\sigma_x^2 = \frac{1}{N-1}\sum_{k=0}^{N-1} (x(k) - \overline{x})^2 \tag{2.3}$$

where the $(N - 1)$ in the denominator instead of N results in an unbiased estimate.

Central Moments

Central moments are used to define higher-order statistical measures that can be used to evaluate and assess the quality of a recorded signal. The ith central moment of a time-varying signal, $x(t)$, is given by

$$M_i = \text{E}[(x - \overline{x})^i] = \frac{1}{N-1}\sum_{k=0}^{N-1} (x(k) - \overline{x})^i \tag{2.4}$$

where, if $i = 1$, the $(N - 1)$ in the denominator is replaced with N and $\mathrm{E}(X)$ is the expected value of a quantity, X, averaged over a long time period.

Comparing Eqs. (2.2) and (2.4), it can be seen that $M_1 = 0$; that is,

$$M_1 = \frac{1}{N}\left[\left(\sum_{k=0}^{N-1} x(k)\right) - N\overline{x}\right] = \overline{x} - \overline{x} = 0 \tag{2.5}$$

Comparing Eqs. (2.3) and (2.4), it can be seen that $M_2 = \sigma_x^2$.

Skewness

Skewness of a time-varying signal is a dimensionless measure of the degree of asymmetry of a signal about its mean, and is defined in terms of the central moment of the signal as

$$S_x = \frac{M_3}{\sigma_x^3} \tag{2.6}$$

A skewness value that differs from a value of zero indicates a deviation of the data set from a Gaussian distribution, which may be a property of the noise source or may indicate a fault with the instrumentation.

Kurtosis

Kurtosis is similar to variance, except data values are raised to the fourth power rather than the second power. This results in emphasis on the large values in the signal and suppression of the influence of the small values. For a normal distribution the kurtosis value is 3. If the kurtosis value is larger than 3, then the tails in the distribution are higher than for a normal distribution and vice versa. The kurtosis value for a pure tone signal is 1.5.

When comparing different records in a signal, a higher than usual kurtosis value could indicate a 'spike' in the signal, such as may be measured by a microphone when someone slams a door.

Kurtosis is a dimensionless parameter defined as

$$K_x = \frac{M_4}{\sigma_x^4} \tag{2.7}$$

As the kurtosis of a normal distribution is 3, some prefer to define kurtosis in terms of excess kurtosis, K_{ex} as

$$K_{\mathrm{ex}} = \frac{M_4}{\sigma_x^4} - 3 \tag{2.8}$$

2.2.3 Amplitude, Frequency, Wavelength, Wavenumber and Speed for Single-frequency Sound

Sound waves that consist of a pure tone (single frequency) only are characterised by:

- the amplitude of pressure changes, which can be described by the maximum pressure amplitude, $p_{\max}$, or the RMS amplitude, p, and is expressed in pascals (Pa);
- the wavelength, λ, which is the distance travelled by the pressure wave during one cycle.

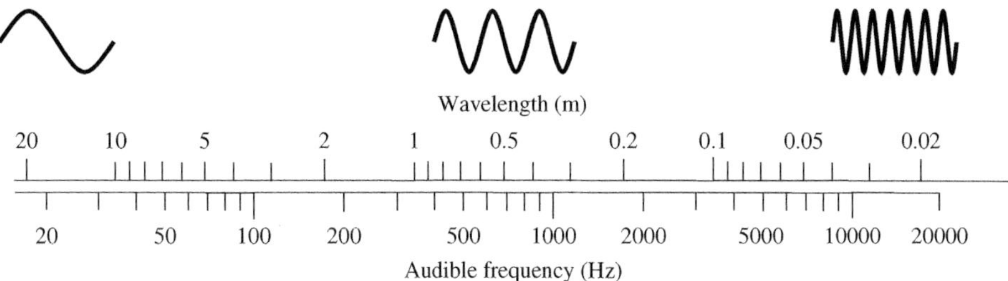

Figure 2.2 Wavelength in air versus frequency at 20°C.

- the frequency, f, which is the number of pressure variation cycles in the medium per unit time or, simply, the number of cycles per second, and is expressed in Hertz (Hz). Noise is usually composed of many frequencies combined together. The relation between wavelength and frequency can be seen in Figure 2.2.
- the period (T), which is the time taken for one cycle of a wave to pass a fixed point. It is related to frequency by

$$T = 1/f \tag{2.9}$$

The speed of sound propagation, c, the frequency, f, and the wavelength, λ, are related by the following equation.

$$c = f\lambda \tag{2.10}$$

The speed of propagation, c, of sound in air is 343 m/s at 20°C and 1 atmosphere pressure. At other temperatures (not too different from 20°C), it may be calculated using

$$c = 332 + 0.6T_c \quad \text{(m/s)} \tag{2.11}$$

where T_c is the temperature in degrees Centigrade. Alternatively, making use of the equation of state for gases, the speed of sound may be written as (Bies and Hansen 2009):

$$c = \sqrt{\gamma R T_k / M} = \sqrt{\frac{\gamma P}{\rho}} \quad (\text{m s}^{-1}) \tag{2.12}$$

where T_k is the temperature in Kelvin (273.15 + the temperature in °C), R is the universal gas constant, which has the value 8.314 J/mol/K, and M is the molecular weight, which for air is 0.029 kg/mole. For air, the ratio of specific heats, γ, is 1.402. P is the atmospheric pressure (101.4 kPa at sea level) and ρ is the density of the medium through which the sound is travelling (1.204 kg/m^3 at 20°C and sea level).

The wavenumber, k, of a single-frequency wave is a commonly used term and is related to the wavelength by

$$k = 2\pi/\lambda \tag{2.13}$$

All of the properties just discussed above (except the speed of sound) apply only to a pure tone (single-frequency) sound, which is described by the oscillations in pressure shown in Figure 2.1.

2.2.4 Units for Sound Pressure Measurement

Sound pressure is measured in pascals (Pa), the same units as used for atmospheric pressure. Sound pressure level is measured using the decibel (dB) scale and these are the units that are most commonly used to report sound levels. Converting sound pressure in pascals to a decibel level requires the use of the logarithm operator and the introduction of a reference sound pressure. The reason for the introduction of a reference sound pressure is that a decibel number represents the ratio between two quantities; in this case, between the actual sound pressure in pascals and a reference number. The reason for using decibels rather than pascals is that the ear has a very large dynamic range for hearing, from the smallest of sounds at 20 μPa (normal hearing threshold) to the loudest (gunshot) at 2000 Pa, and the decibel scale is a means of compressing that from 0 dB to 160 dB. Anything above 140 dB is considered to result in permanent hearing damage, even for a short exposure time.

The reference sound pressure used for calculating decibels is equivalent to the normal threshold of hearing ($p_{\text{ref}} = 20\ \mu\text{Pa}$) at 1.8 kHz, at which our hearing sensitivity is close to that at our most sensitive frequency of 3.3 kHz (see Figure 2.3). Of course, this is an average level as all people are different, with some having even greater sensitivity while others have less. The sound pressure level, L_p in decibels is related to the RMS sound pressure, p, in pascals by the following equation.

$$L_p = 10\log_{10}\frac{\langle p^2\rangle}{p_{\text{ref}}^2} = 10\log_{10}\langle p^2\rangle - 10\log_{10}p_{\text{ref}}^2 = 10\log_{10}\langle p^2\rangle + 94\ \text{(dB)} \tag{2.14}$$

In the discipline of underwater acoustics (and any other liquids), the same equations apply, but the reference pressure used is 1 μPa, rather than 20 μPa.

Hearing Sensitivity

The international standard, ISO 226 (2003), provides hearing threshold curves and curves of equal loudness. An equal loudness curve tells us the sound pressure level

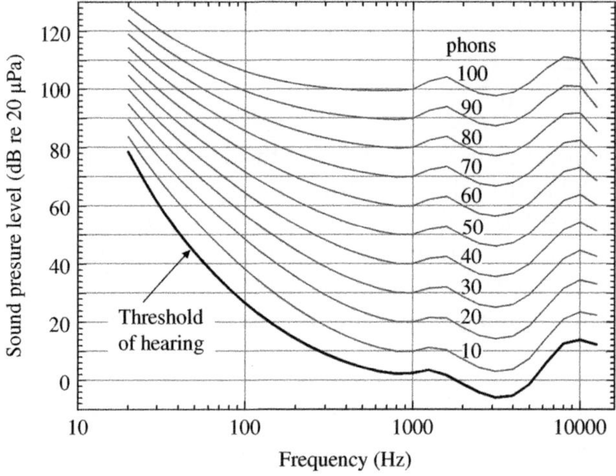

Figure 2.3 Threshold of hearing and equal loudness contours (data from ISO226 (2003)).

needed for a sound at a particular frequency to sound as loud as sound at 1000 Hz. The numerical label on each curve is the sound pressure level at 1000 Hz. For example, a sound at 1000 Hz with a 30-dB sound pressure level will have the same loudness as a 100-Hz sound with a sound pressure level of 56 dB. The curves plotted in Figure 2.3 have been derived from the equations provided in the ISO 226 standard. The equal loudness curves are in phons. The phon value of a sound is the sound pressure level of a pure tone at 1000 Hz that is judged to be equally as loud as the sound being considered.

2.2.5 Sound Power

As discussed in the previous section, sound pressure is a measure of the pressure fluctuations in the atmosphere as a result of the action of a sound source. Sound pressure is the quantity that our ear responds to. In this section we discuss sound power, which is a measure of the total energy radiated by a sound source. This quantity, together with information about the directional nature of sound radiated by the sound source, is used to estimate the sound pressure level that will be experienced by a receiver at some distance from the source. Of course, atmospheric effects as well as effects of obstacles and reflection from the ground must also be taken into account when estimating the sound level that will be experienced. Means for calculating the sound power from an aerodynamic source such as a wind turbine are discussed in Chapter 4, and means for taking into account atmospheric, ground and obstacle effects on sound propagation are discussed in Chapter 5.

The sound power, L_w, in watts may be converted to a decibel quantity using a reference quantity of $W_{\text{ref}} = 10^{-12}$ watts and the following equation

$$L_w = 10 \log \frac{W}{W_{\text{ref}}} = 10 \log W + 120 \quad \text{(dB)} \tag{2.15}$$

For a simple spherical source radiating sound uniformly in all directions and away from the ground and any other reflecting surfaces, the sound power, W, produced by the source can be related to the square of the RMS sound pressure, $\langle p^2 \rangle$, measured at some distance, r, from the source by

$$W = \frac{4\pi r^2 \langle p^2 \rangle}{\rho c} \quad \text{(watts)} \tag{2.16}$$

where ρ is the density of air (1.204 kg/m^3 at 20°C) and c is the speed of sound in air (343 m/s at 20°C).

Alternatively, if it is assumed that $\rho c \approx 400$ (it is actually 413 at room temperature, but the approximation only gives a 0.15 dB error), the Eq. (2.16) can be written in terms of sound power level and sound pressure level as

$$L_w = L_p + 20 \log_{10} r + 11.0 \quad \text{(dB)} \tag{2.17}$$

As noise predictions are generally only quoted to the nearest decibel, the 0.15 dB error is acceptable. However, if desired, 0.15 dB can be subtracted from the calculated sound pressure level, but the final result should never be presented with a greater accuracy than 0.1 dB as it is not possible to measure any more accurately than that.

Sometimes in the non-scientific literature sound intensity is used mistakenly in place of sound pressure. In fact, although sound intensity is related to sound pressure, it is a different quantity. Unlike sound pressure, it is a vector quantity, defined by a level and a direction. For a simple sound wave such as a plane or spherical wave propagating outdoors away from its source and any obstacles, sound intensity, I, is related to the RMS sound pressure, p, the density of the atmosphere, ρ, and the speed of sound, c, by

$$I = \frac{p^2}{\rho c} \tag{2.18}$$

The sound intensity level, L_I, in decibels, uses a reference quantity, $I_{\text{ref}} = 10^{-12}$ watts/m^2, and is defined as follows

$$L_I = 10 \log_{10} \frac{I}{I_{\text{ref}}} = 10 \log_{10} I + 120 = L_p - 0.2 \quad \text{dB} \tag{2.19}$$

The 0.2 dB constant in the above equation only applies for a plane or spherical wave at sea level and at a temperature of 20°C. Other altitudes and pressures result in different values for the speed of sound and density of the atmosphere in Eq. (2.18).

2.2.6 Beating

When two tones of very small frequency difference are presented to the ear, one tone, which varies in amplitude with a modulation frequency equal to the difference in frequency of the two tones, will be heard. When the two tones have exactly the same frequency, the frequency modulation will cease. When the tones are separated by a frequency difference greater than what is called the 'critical bandwidth' (see Section 2.3.3), two tones are heard. When the tones are separated by less than the critical bandwidth, one tone of modulated amplitude is heard, where the frequency of modulation is equal to the difference in frequency of the two tones. The latter phenomenon is known as *beating*.

Let two tonal sounds of amplitudes A_1 and A_2 and of slightly different frequencies, ω and $(\omega + \delta\omega)$ be added together. It will be shown that a third amplitude modulated sound is obtained. The total pressure due to the two tones may be written as:

$$p_1 + p_2 = A_1 \cos \omega t + A_2 \cos(\omega + \Delta\omega)t \tag{2.20}$$

where one tone of amplitude, A_1, is described by the first term and the other tone of amplitude, A_2, is described by the second term in Eq. (2.20).

Assuming that $A_1 \geq A_2$, defining $A = A_1 + A_2$ and $B = A_1 - A_2$, setting the first term on the right-hand side equal to $A_1 \cos(\omega + \Delta\omega/2 - \Delta\omega/2)$ and using well known trigonometric identities, Eq. (2.20) may be rewritten as follows:

$$\begin{aligned} p_1 + p_2 &= A \cos(\Delta\omega/2)t \cos(\omega + \Delta\omega/2)t \\ &\quad + B \cos(\Delta\omega/2 - \pi/2)t \cos(\omega + \Delta\omega/2 - \pi/2)t \end{aligned} \tag{2.21}$$

When $A_1 = A_2$, $B = 0$ and the second term in Eq. (2.21) is zero. The first term is a cosine wave of frequency $(\omega + \Delta\omega)$ modulated by a frequency $\Delta\omega/2$. At certain values of time, t, the amplitude of the wave is zero; thus the wave is described as *fully modulated*. If B is non-zero as a result of the two waves having different amplitudes, a modulated wave is still obtained, but the depth of the modulation decreases with increasing B and the wave is described as *partially modulated*. If $\Delta\omega$ is small, the familiar beating phenomenon is obtained (see Figure 2.4). The figure shows a beating phenomenon for a number of cases involving different frequency differences and different relative amplitudes of the two waves. Beating is only perceptible if the frequency difference between the two signals (or beat frequency) is less than about 10 Hz.

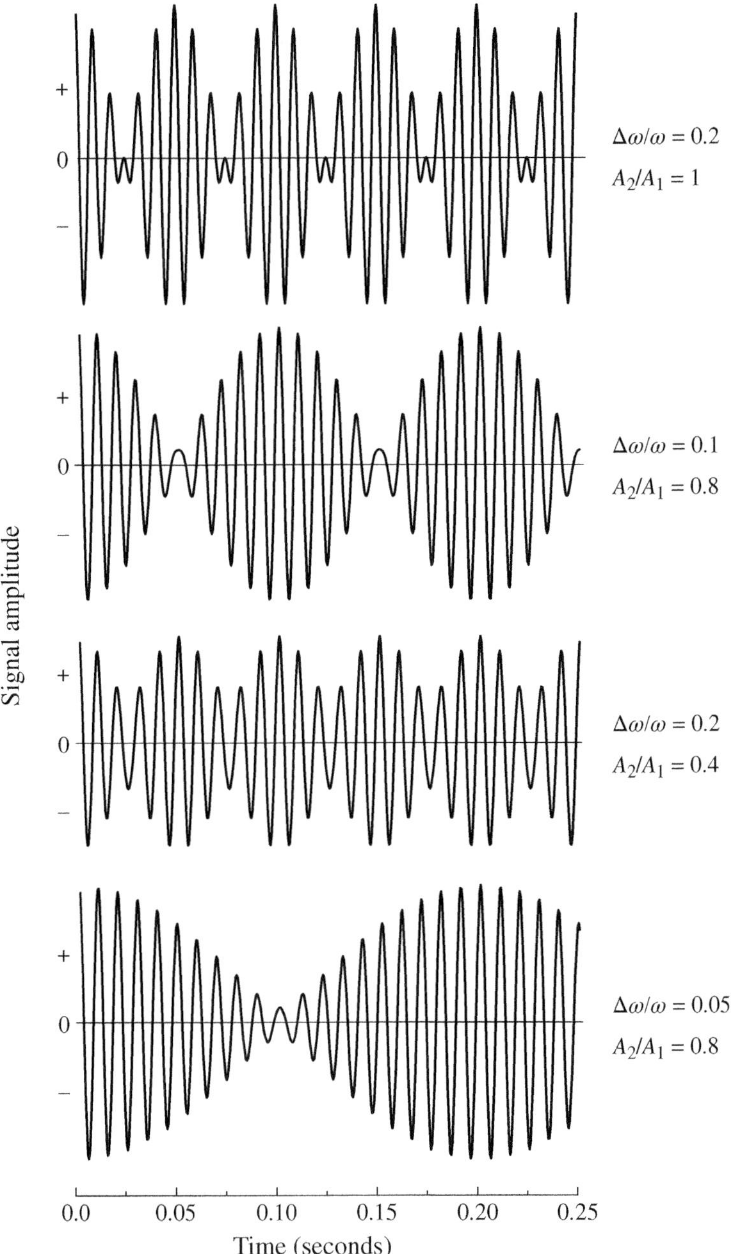

Figure 2.4 Illustration of beating with different relative amplitudes and frequency differences for a primary signal at 100 Hz.

It is interesting to note that if the signal in Figure 2.4a were analysed on a very fine resolution spectrum analyser, only two peaks would be seen; one at each of the two interacting frequencies. There would be no peak seen at the beat frequency as there is no energy at that frequency even though we apparently 'hear' that frequency.

2.2.7 Amplitude Modulation and Amplitude Variation

Amplitude modulation (AM) is defined as the periodic variation in amplitude of a noise or vibration signal. In relation to wind farm noise, it is defined by the (UK) Institute of Acoustics (IOA) as (AMWG 2015),

> ...periodic fluctuations in the level of broadband noise from a wind turbine (or wind turbines), the frequency of the fluctuations being the blade-pass frequency of the turbine rotor, as observed outdoors at residential distances in free-field conditions.

There are two types of AM commonly identified, although the IOA recommend that there is no need to adopt separate definitions for AM (AMWG 2015). However, for the purpose of clarity the two types of AM commonly encountered are described here.

The first is an inherent part of wind turbine operation and is a result of the rotation of the blades causing changes in the trailing-edge noise-radiation direction relative to a stationary receiver. The directivity remains constant relative to the blade but the blade is rotating, which causes noise radiation to vary from an upwards direction when the blades are moving upwards to a downwards direction when the blades are moving down. A stationary receiver will experience sound pressure level variations as a direct result and these variations will be periodic at the blade-pass frequency. This type of AM is referred to as 'normal AM' (AMWG 2015; Renewable UK 2013) and corresponds to an approximate 3-dB modulation depth (see Eq. (2.26)), which in the past has been considered expected and acceptable. It manifests as a swishing sound, reflecting the trailing-edge noise spectrum. This type of AM is more apparent close to a turbine.

However, in many cases, enhanced AM (EAM) or other AM (OAM) is generated by turbines and is heard at large distances. This type of AM is generated by a different mechanism and manifests as a 'thumping' or 'whooming' noise, which is characterised by relatively low frequencies (Van den Berg 2005). The generation mechanism for this type of AM could be transient stall, which can occur when turbines lie in the wakes of upstream turbines or when there is high wind shear, so that blades entering the higher wind speed near and at the top of their path, stall due to their angle of attack being too high for the corresponding wind speed. Another mechanism could be a result of the periodic unloading of the blades as they pass the tower, generating relatively high level higher-order harmonics in the frequency range 20–80 Hz.

The extent of AM is often characterised in terms of a modulation depth. This concept can be quite difficult to define for a realistic noise spectrum so we will begin here by defining it for a modulated single-frequency tone.

The sound pressure as a function of time that will be experienced when a tonal sound of frequency, f, is modulated with a frequency, f_m, is given by,

$$p(t) = A(1 + \mu m(t)) \cos(2\pi f t + \phi) \tag{2.22}$$

where μ is defined as the modulation index (Renewable UK 2013); ϕ is an arbitrary phase angle and may be set equal to zero for our purposes here. The modulation function, $m(t)$, can be represented as a simple cosine function as,

$$m(t) = \cos(2\pi f_m t + \phi_m) \tag{2.23}$$

where ϕ_m is an arbitrary phase angle that may be set equal to zero for our discussion here. The quantity, A is the time-averaged value of the amplitude of the signal being modulated. The modulation signal, $m(t)$, of Eq. (2.23) can be derived directly from $p(t)$ of Eq. (2.22) using a Hilbert transform (see Section 2.4.10) to find an expression for the envelope signal, $e(t)$, which is equal to $A(1 + \mu m(t))$.

The maximum and minimum amplitudes, $A_{\max}$ and $A_{\min}$ respectively, of the modulated wave form are given by

$$A_{\max} = A(1 + \mu) \quad \text{and} \quad A_{\min} = A(1 - \mu) \tag{2.24}$$

and

$$\mu = \frac{A_{\max} - A_{\min}}{A_{\max} + A_{\min}} \tag{2.25}$$

where $A_{\max}$ and $A_{\min}$ are defined in Figure 2.5b. As mentioned previously, A is the mean amplitude of the signal being modulated (the 10-Hz signal in Figure 2.5b).

The modulation depth, R_{mm}, in decibels is defined as (Renewable UK 2013):

$$R_{mm} = 20 \log_{10}\left(\frac{A_{\max}}{A_{\min}}\right) = 20 \log_{10}\left(\frac{1+\mu}{1-\mu}\right) = (L_p)_{\max} - (L_p)_{\min} \tag{2.26}$$

where $(L_p)_{\max}$ is the maximum sound pressure level if the y-axis scale is in decibels and $(L_p)_{\min}$ is the minimum sound pressure level.

The problem of quantifying wind farm AM is a more complex than discussed for a single-frequency sound being modulated. Although modulation of single-frequency sound such as gearbox noise may occur for a wind farm, it is more common to find broadband sound modulated by one or more frequencies related to the turbine blade-pass frequency.

Amplitude variation of a signal includes both periodic AM and random amplitude variations. These latter variations are generally the result of completely different mechanisms to those that are responsible for AM, and may include amplitude variation as a result of beating. Another cause of amplitude variation in wind farm noise is variation in the interference between noise coming from different turbines, as a result of varying atmospheric conditions affecting the relative phase of the sounds arriving at a residence.

Beating, AM and random amplitude variation are all illustrated in Figure 2.5. In part (a) for the beating case, the two signals are 10 Hz and 11 Hz, resulting in a difference frequency (modulation frequency) of 1 Hz. The amplitude of the 11-Hz signal is 0.8 times the amplitude of the 10-Hz signal. This could represent beating between two different harmonics of the blade-pass frequency or harmonics from two different turbines. In part (b), a signal of frequency, $f = 10$ Hz is amplitude modulated at a frequency of $f_m = 1$ Hz, with a modulation index of 0.8. This could represent turbine tonal noise at 10 Hz being modulated by the blade-pass frequency at 1 Hz. Part (c) represents a time series measurement at a distance of 3 km from a wind farm. The figure represents an example of AM that may be measured in practice for a wind farm. Part (d) shows a typical time-varying signal representing wind farm noise that is not amplitude modulated

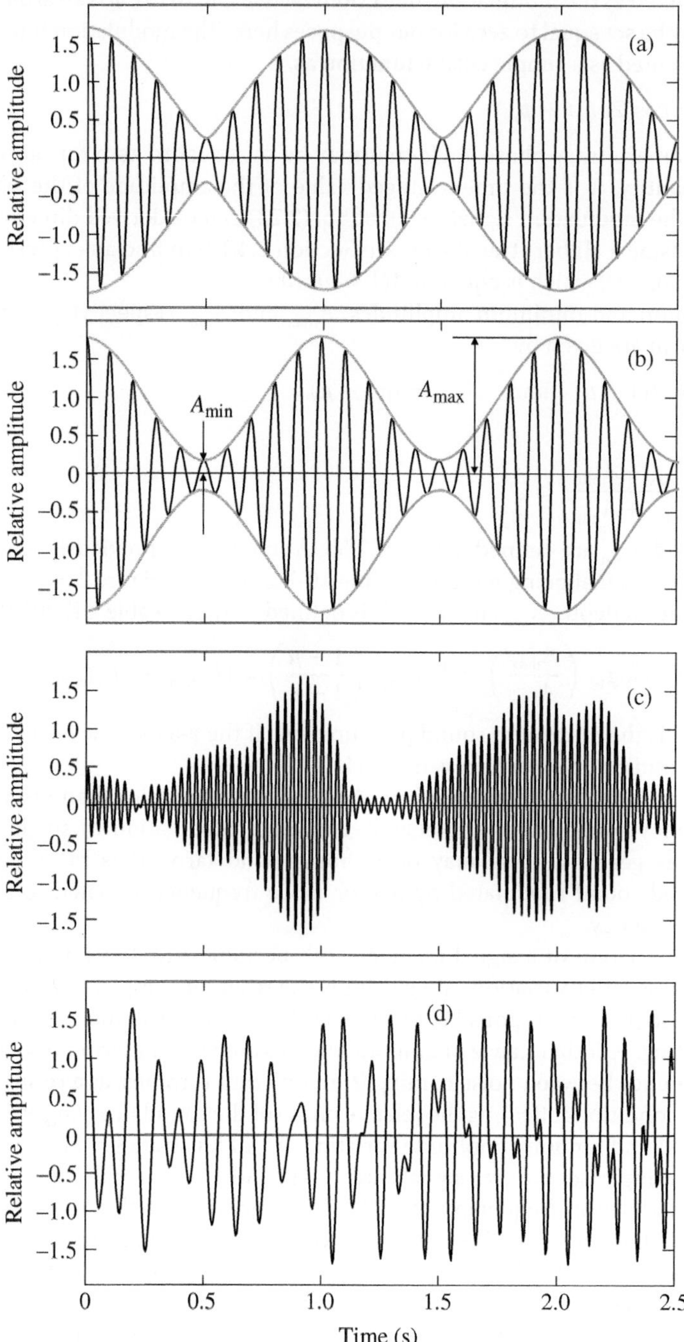

Figure 2.5 Examples illustrating the difference between beating, AM and random amplitude variation: (a) beating; (b) pure AM; (c) typical AM signal from a wind turbine; (d) random amplitude variation.

in a periodic way. Thus it is referred to as an amplitude-varying noise but not an AM noise.

Although the time-domain representations of beating and AM look to be very similar in Figure 2.5, they are very different in the frequency domain. The 'beating' signal will show only two peaks in the frequency spectrum; one at each of the two frequencies that are combining to form the beating wave form: 10 Hz and 11 Hz in this case. The 'amplitude-modulated' signal will show three peaks: one at the frequency of the tone being modulated and one at each side of this frequency and separated by the modulation frequency. For the example in the figure, these three peaks would be at 9 Hz, 10 Hz and 11 Hz.

2.2.8 Decibel Addition

If one knows the sound level in dB produced at a particular location by each of two or more noise sources, then the total sound level is obtained by converting each sound level in dB to a squared sound pressure, adding the squared sound pressures and then converting the result back to decibels. There is a shorthand way of implementing the above steps in a single equation as follows

$$L_{pt} = 10 \log_{10} \sum_{i=1}^{N} 10^{L_{pi}/10} \tag{2.27}$$

where L_{pt} is the total sound pressure level as a result of summing N sound pressure levels, L_{pi}, $i = 1, 2 \ldots N$. This is best illustrated by example. Let us assume that Turbine 1 produces a sound level of 32 dBA at Residence 1, Turbine 2 produces a level of 34 dBA at Residence 1 and Turbine 3 produces a level of 35 dBA at Residence 1. What is the total sound level at residence 1 in dB? This can be calculated as follows

$$L_{pt} = 10 \log_{10}(10^{32/10} + 10^{34/10} + 10^{35/10}) = 38.6 \text{ (dB)} \tag{2.28}$$

As an alternative, Table 2.1 may be used.

Referring to the previous example, we can use Table 2.1 for the addition (even though there are three levels to be added). First we add the levels from Turbines 1 and 2 to get a total of $34 + 2.1 = 36.1$ dB. Then we add this total to the level from Turbine 3 to obtain a new total of $36.1 + 2.5 = 38.6$ dB, which is the same result as obtained previously.

In practice, the sound pressure level measurements are averaged over time and A-weighted (see Section 2.2.11) to give $L_{pt} = L_{Aeq,t}$. The averaging time specified in various regulations is usually 10 min for wind farm noise. Equation (2.27) can be used to add together 1/3-octave or octave band data to obtain total sound pressure levels. It can also be used to add together three 1/3-octave bands to obtain octave band data. Octave and 1/3-octave bands are defined in Section 2.3.2.

It is important to note that this sort of addition relies on the added noises being incoherent; that is, the noises cannot have a defined phase relationship, which implies noises of different frequencies or combinations of randomly varying amplitudes and frequencies that are unrelated to one another (random noise).

If the noises to be added are related via a fixed phase, then they are coherent and they are also both tones of the same frequency. In this case, the sound pressure level due to two tones of the same frequency combined together must be found by taking the phase relationship into account (Bies and Hansen 2009). If the tones are in-phase at

Table 2.1 Table for combining decibel levels

Difference between the two dB levels to be added (dB)	Amount to be added to the higher level in order to get the total level (dB)
0	3.0
1	2.5
2	2.1
3	1.8
4	1.5
5	1.2
6	1.0
7	0.8
8	0.6
9	0.5
10	0.4

the measurement point, then two tones of the same sound pressure level will produce a total sound pressure level of 6 dB greater than that produced by one tone alone. This principle underpins the measurement of wind turbine sound power levels. The measurement microphone is placed on a hard reflecting surface on the ground and 6 dB is subtracted from the measured sound pressure level to obtain the free-field sound pressure level to be used in the sound power calculation. This approach is justified because the direct sound and sound reflected from the hard surface will be in-phase, because the hard reflecting surface does not induce a phase shift and the path length of the direct and reflected sound is essentially the same, at least at frequencies in the range of interest for wind farm noise.

2.2.9 Decibel Subtraction

One may well ask why we would wish to subtract one decibel level from another. In the assessment of wind farm noise this procedure is used to determine the noise level at a residence due to a wind farm, by subtracting the noise levels measured before the wind farm was constructed from the total noise levels measured afterwards to obtain the contribution just from the wind farm. Of course, in the case of wind farms, the large variation in noise levels due to atmospheric conditions and the presence and absence of other noise sources at various times means that this calculation is very approximate at best. Subtraction of one noise level from another is similar to the addition of two noise levels. The decibel levels are converted to sound pressure squared levels and one is subtracted from the other before converting back to decibel levels. A shorthand way of implementing this procedure can be written as

$$L_{ps} = 10 \log_{10}(10^{L_{pt}/10} - 10^{L_{pb}/10}) \tag{2.29}$$

where L_{ps} is the sound pressure level due to the source only, L_{pt} is the total measured sound pressure level and L_{pb} is the background sound pressure level.

This is best illustrated by the following example. Suppose the noise level measured at a residence before a wind farm was constructed was 35 dB. After construction, the measured noise level was 39 dB. The contribution due to the wind farm only is calculated as follows.

$$L_p \text{ (wind farm only)} = 10 \log_{10} (10^{39/10} - 10^{35/10}) = 36.8 \text{ (dB)} \tag{2.30}$$

2.2.10 Noise Source Directivity

The previous discussion has been based on the assumption that the noise source under consideration was radiating sound uniformly in all directions, referred to as an *omni-directional* source. However, in practice the presence of reflecting surfaces near the source, such as the ground, and the directional sound radiation properties of the source itself will affect the sound pressure experienced by the receiver.

The presence of a reflecting plane, such as the ground, restricts the space into which a source can radiate and if the source power is assumed to be unaffected, the sound pressure level must increase as a result of sound reflected from the ground combining with sound travelling directly from the sound source. This is discussed in detail in Chapter 5. The directivity as a result of the source properties is difficult to evaluate for a wind farm so the sound power is estimated using a sound pressure measurement in the downwind direction, which is considered to be the worst case in terms of sound levels experienced by a distant receiver. This sort of measurement does go some way towards taking into account the wind farm directivity.

Measurements Taken Close to the Ground

If the microphone measuring a sound level is located close to a reflecting surface (less than about one-tenth of a wavelength), the sound pressure of the source image (or reflected wave) is added to that at the receiver due to the direct sound wave; in other words, pressure doubling occurs, with an apparent increase in sound pressure level of 6 dB. For infrasound, a microphone at the usual measurement height of 1.5 m should record very close to the same sound level as a microphone on the ground. This is because the 1.5-m height will be less than approximately 1/10 of a wavelength of sound for all frequencies below about 20 Hz.

For most practical sound sources radiating non-tonal noise, when the receiver is located further than a half wavelength from a reflecting surface, the path difference between the direct and reflected waves is usually sufficiently large for the two waves to combine with random phase; the squared pressures, $\langle p^2 \rangle$, will add with an apparent increase in sound pressure level of 3 dB. For frequencies at which 1.5 m represents more than 1/10 of a wavelength and less than 1/2 of a wavelength (20–115 Hz approximately), the apparent increase in sound pressure level as a result of reflection from an acoustically hard ground surface will decrease gradually from 6 dB to 3 dB.

If the noise is tonal in nature then sound pressures of the direct and reflected waves must be added using coherent addition, taking into account the phase shift on reflection and relative phase shift due to the differences in lengths of the two propagation paths.

2.2.11 Weighting Networks

Attempts to present a single decibel number to describe the annoyance of environmental noise has led to the use of weighting networks, whereby the level of noise is adjusted as a function of frequency in an attempt to replicate how an average normal ear would hear. These weighting networks are more suitable for some types of noise than others. The most commonly used weighting network in environmental noise measurement and assessment is the A-weighting network, which, when applied to a noise measurement, results in a number followed by 'dBA'.

The C-weighting network does not apply as much attenuation of low-frequency noise as does the A-weighting network and is meant for louder noises such as from aircraft flyover near airports. However, it is also used to evaluate low-frequency noise by comparing the C-weighted level to the A-weighted level. In 1987, Kelley (1987) published proposed metrics for assessing community annoyance from wind turbine noise and these used the C-weighted metric as well as another metric known as the *low-frequency sound level* (LSL), first proposed by Tokita et al. (1984). Some regulations specify an allowed difference of 15 or 20 dB between the C-weighted level and the A-weighted level before a low-frequency penalty is applied to the allowed A-weighted (or dBA) level. Low-frequency noise is commonly defined as noise having a frequency between 20 and 200 Hz, although some researchers have defined it as having a frequency between 20 and 160 Hz.

To assess infrasound (defined as noise below 20 Hz), the G-weighting has been developed. This only covers the frequency range up to 315 Hz but puts most emphasis on noise below 20 Hz. In an attempt to standardise instrumentation for unweighted measurements, the Z-weighting has been developed, which is essentially a zero-weighting defined down to 10 Hz.

Most sound-measuring instruments have the option to apply various weighting networks electronically. Alternatively, it is sometimes convenient to measure 1/3-octave or octave band sound levels and apply the weighting correction corresponding to the centre frequency of the band to the entire band. This is done by simply adding the correction in decibels (usually negative) to the measured unweighted level in the frequency band of interest, as illustrated in the example below. However, this latter method is not as accurate as applying the weighting network electronically to the signal being measured.

The characteristics of the four weighting networks described above are illustrated in Figure 2.6, with 1/3-octave band values tabulated in Table 2.2. For example, if the linear reading at 125 Hz were 90.0 dB re 20 μPa, then the A-weighted reading would be 90.0 − 16.1 = 73.9 dB(A).

Example

1 Given the sound spectrum shown in line 1 of the table below, find the overall unweighted (linear) sound level in decibels and the A-weighted sound level in dB(A).

Octave band centre frequency (Hz)	31.5	63	125	250	500	1000	2000	4000	8000
Linear level (dB)	55	55	50	45	40	41	35	30	25
A-weighting correction (dB)	−39.4	−26.2	−16.1	−8.6	−3.2	0	1.2	1.0	−1.1
A-weighted level (dBA)	15.6	28.6	33.9	36.4	36.8	41.0	36.2	31.0	23.9

Solution

The linear level is calculated using:

$$L_{pt} = 10\log_{10}\sum_{i=1}^{9} 10^{L_{pi}/10} = 59.0 \text{ (dB)}$$

where L_{pi} are the levels shown in line 2 of the table. The A-weighted overall level is found by adding the A-weighting corrections (see Table 2.2) to line 2 to obtain line 4 and then adding the levels in line 4 using the above expression to give: $L_{pt} = 44.9$ dB(A).

Apparent Loudness

It is well known that the apparent loudness of a sound is dependent on its frequency as well as the sound pressure, and that the variation of loudness with frequency also depends to some extent on the sound pressure. In other words, the difference between the threshold of perception and the level at which sound is described as 'very loud' is much less at low frequencies than it is at mid and high frequencies. This means that the difference in level between 'just perceptible' and 'annoying' is much smaller at very low frequencies. Also, subjective differences in perception are also very much more apparent for low-frequency noise so that a low-frequency noise that is extremely annoying to one person may not be perceptible to another, even though both people's hearing is in the normal range. These ideas are especially relevant to wind farm noise, which is often dominated by low frequencies at residential locations.

2.2.12 Noise Level Measures

When reading regulations concerning wind farm noise, the quantity to be measured and compared to an allowed level is not necessarily the same from one regulation or country

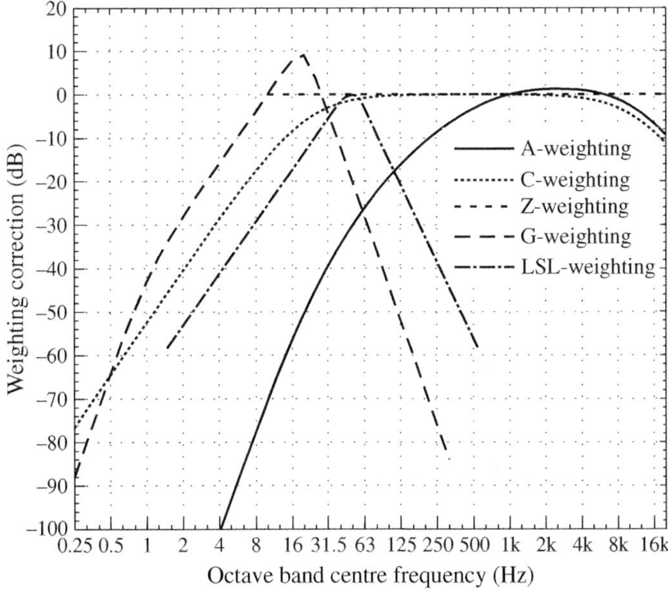

Figure 2.6 Various weighting curves used in the assessment of noise.

Table 2.2 Weighting corrections (dB) at 1/3-octave band centre frequencies to be added to unweighted signal.

Frequency (Hz)	A	C	G	Z	LSL	Frequency (Hz)	A	C	G	Z	LSL
0.25	−197.6	−77.0	−88.0	–	–	80	−22.5	−0.5	−36.0	0.0	−9.0
0.315	−189.5	−73.0	−80.0	–	–	100	−19.1	−0.3	−44.0	0.0	−15.0
0.4	−181.5	−69.0	−72.1	–	–	125	−16.1	−0.2	−52.0	0.0	−21.0
0.5	−173.5	−65.0	−64.3	–	–	160	−13.4	−0.1	−60.0	0.0	−27.0
0.63	−165.5	−60.9	−56.6	–	–	200	−10.9	0.0	−68.0	0.0	−33.0
0.8	−157.4	−56.9	−49.5	–	–	250	−8.6	0.0	−76.0	0.0	−39.0
1.0	−148.6	−52.5	−43.0	–	–	315	−6.6	0.0	−84.0	0.0	−45.0
1.25	−140.6	−48.5	−37.5	–	–	400	−4.8	0.0	–	0.0	−51.0
1.6	−132.6	−44.5	−32.6	–	−57.0	500	−3.2	0.0	–	0.0	−57.0
2.0	−124.6	−40.6	−28.3	–	−53.0	630	−1.9	0.0	–	0.0	–
2.5	−116.7	−36.6	−24.1	–	−49.0	800	−0.8	0.0	–	0.0	–
3.15	−108.8	−32.7	−20.0	–	−45.0	1 000	0.0	0.0	–	0.0	–
4.0	−100.9	−28.8	−16.0	–	−41.0	1 250	0.6	0.0	–	0.0	–
5.0	−93.1	−25.0	−12.0	–	−37.0	1 600	1.0	−0.1	–	0.0	–
6.3	−85.4	−21.3	−8.0	–	−33.0	2 000	1.2	−0.2	–	0.0	–
8.0	−77.8	−17.7	−4.0	–	−29.0	2 500	1.3	−0.3	–	0.0	–
10.0	−70.4	−14.3	0.0	0.0	−25.0	3 150	1.2	−0.5	–	0.0	–
12.5	−63.4	−11.2	4.0	0.0	−21.0	4 000	1.0	−0.8	–	0.0	–
16.0	−56.7	−8.5	7.7	0.0	−17.0	5 000	0.5	−1.3	–	0.0	–
20.0	−50.5	−6.2	9.0	0.0	−13.0	6 300	−0.1	−2.0	–	0.0	–
25.0	−44.7	−4.4	3.7	0.0	−9.0	8 000	−1.1	−3.0	–	0.0	–
31.5	−39.4	−3.0	−4.0	0.0	−5.0	10 000	−2.5	−4.4	–	0.0	–
40	−34.6	−2.0	−12.0	0.0	−1.0	12 500	−4.3	−6.2	–	0.0	–
50	−30.2	−1.3	−20.0	0.0	0.0	16 000	−6.6	−8.5	–	0.0	–
63	−26.2	−0.8	−28.0	0.0	−3.0	20 000	−9.3	−11.2	–	0.0	–

to the next. Here, the more commonly used quantities in noise assessment are defined and explained.

Equivalent Continuous Sound Pressure Level

The equivalent continuous sound pressure level, L_{eq}, the equivalent continuous sound pressure level, is defined in terms of the time-varying sound pressure level, $L_p(t)$, as follows:

$$L_{eq} = 10 \log_{10} \left[\frac{1}{T} \int_0^T 10^{L_p(t)/10} \mathrm{d}t \right] \quad \text{(dB)} \tag{2.31}$$

which is essentially the time-averaged squared sound pressure converted to decibels. The averaging of pressure squared rather than pressure means that higher instantaneous levels have more effect on the average than lower levels.

Equivalent Continuous A-weighted Sound Pressure Level

The equivalent continuous A-weighted sound pressure level, L_{Aeq}, characterises fluctuating noise as an equivalent steady-state A-weighted level. It is found by replacing the unweighted sound pressure level in Eq. (2.31) with the A-weighted sound pressure level, L_{pA}, and is defined as:

$$L_{Aeq} = 10 \log_{10} \left[\frac{1}{T} \int_0^T 10^{L_{pA}(t)/10} \mathrm{d}t \right] \quad \text{(dBA)} \tag{2.32}$$

where T is a specified period of time over which the sound level is averaged and $L_{pA}(t)$ is the A-weighted sound pressure level at time t.

Of course, we can use an equation similar to Eq. (2.32) to define equivalent continuous C-weighted, G-weighted and Z-weighted sound levels, just by substituting G, C or Z for A in two places in the equation.

Statistical Sound Levels

The quantity L_x is the sound level (in dB) that is exceeded x percent of the time and the quantity, L_{Ax} is the A-weighted sound level (in dBA) that is exceeded x percent of the time. When written as $L_{x,10\text{min}}$, it means that the time period of the measurement is 10 min. Thus, $L_{A90,\,10\text{min}}$ means the sound level that was exceeded 90% of the time during a 10-min measurement period.

Many regulations for wind farm noise are written in terms of an L_{A90} level, which is the A-weighted sound level that is exceeded 90% of the time. The idea of this approach is that wind farm noise is supposedly relatively steady, whereas background noise fluctuates considerably. So the idea is that an L_{A90} measurement will remove most of the background noise and just measure the wind farm noise. Of course this is an erroneous assumption because wind farm noise fluctuates wildly at residential locations more than a few hundred meters distant as a result of varying atmospheric conditions affecting both the generation and propagation of the noise. Use of an L_{A90} measurement effectively ignores the worst 90% of wind farm noise. A more useful descriptor would be an L_{A50} measurement, which is the noise level that is exceeded 50% of the time. It would be even more useful to concentrate wind farm measurements in the period from midnight to 5am, when the influence from other sound sources is at a minimum and then use L_{Aeq} measurements. The best way of separating wind farm noise from background noise is to take measurements with the wind farm shut down and with it operational for similar atmospheric conditions (temperature and wind profiles) and in downwind conditions (when the wind is blowing from the wind farm to the receiver).

The measurements should also include data that represent temperature-inversion conditions. These measurements should also be done at night and the wind farm power output should be checked to make sure it is producing the design amount for the particular wind conditions (which should be 8–14 m/s at hub height). Of course, none of the above descriptors are useful for characterising low-frequency noise and its possible effects on the health and wellbeing of residents in the vicinity of wind farms. Also, the measurements with the wind farm turned off do not fully account for infrasound generated by a wind farm, as it has been shown that wind blowing past stationary blades and support towers can generate infrasound tones at similar levels, but different frequencies to those measured at blade-pass frequencies and harmonics when turbines are operating (Huson 2015).

Day–night Level

The day–night noise level, L_{dn} or DNL, is the noise level averaged over day and night but with a weighting applied to the night level, as follows.

$$L_{dn} = 10\log_{10}\left[\frac{15\times 10^{0.1L_{\text{day}}} + 9\times 10\times 10^{0.1L_{\text{night}}}}{24}\right] \tag{2.33}$$

where L_{day} is the L_{Aeq} value averaged over the 15 daytime hours from 7am to 10pm and L_{night} is the L_{Aeq} value averaged over the 9 nighttime hours from 10pm to 7am.

Day–evening–night Level

The day-evening-night noise level, L_{den} or CNEL, is the noise level averaged over day, evening and night but with weightings applied to the evening and night levels, as follows.

$$L_{den} = 10\log_{10}\left[\frac{12\times 10^{0.1L_{\text{day}}} + 3\sqrt{10}\times 10^{0.1L_{\text{evening}}} + 9\times 10\times 10^{0.1L_{\text{night}}}}{24}\right] \tag{2.34}$$

where L_{day} is the L_{Aeq} value averaged over the 12 daytime hours from 7am to 7pm, L_{evening} is the L_{Aeq} value averaged over the 3 evening hours from 7pm to 10 pm and L_{night} is the L_{Aeq} value averaged over the 9 nighttime hours from 10pm to 7am. Unfortunately, there are several variations for the definition of day, evening and night hours: in Italy daytime extends from 7am to 9pm and evening is from 9pm to 11pm, so the constants 12, 3 and 9 in the above equation become 14, 2 and 8, respectively. In some jurisdictions, the evening extends from 7pm to 11pm and in these cases the constants 3 and 9 in the above equation become 4 and 8, respectively.

Characterisation and Assessment of Low-frequency Noise

As will be seen in the discussion in Chapter 7, low-frequency noise can have a severe adverse effect on some people and a reliable method is needed to quantify it and assess its potential adverse effect. At the time of writing, only two suggested methods for characterising low-frequency noise have been published (Jakobsen 2012; Moorhouse et al. 2005). These methods are discussed in detail in Sections 5.4 and 1.4.4, respectively.

2.2.13 Sound in Rooms

As most complaints about wind farm noise are about excessive levels indoors, it is useful to have an understanding of sound fields in rooms or enclosed spaces, especially low-frequency sound fields.

Sound in rooms is strongly affected by the reflective properties of the walls, floor and ceiling and any other objects within the room. At low frequencies (less than 200 Hz) most surfaces are very reflective and this results in the establishment of quite strong room acoustic resonances (or standing waves). Such resonances are characterised by regions of much higher than average sound pressure levels (antinodes) and other areas of much lower than average sound pressure levels (nodes). As the frequency of the incoming sound increases above 200 Hz, the room surfaces become more absorbent and the room resonances less apparent, so there is not as great a difference in sound level between areas of high sound pressure level and areas of low sound pressure level. Also, as the frequency increases, the room resonances become more closely spaced in frequency and this results in the perception that the sound field is more uniformly distributed throughout the room.

At low frequencies, where the room resonances are particularly strong and well separated in frequency, it is possible that even broadband low-frequency noise entering the room will be selectively filtered by the room response and sound like tonal noise. In addition, as a result of the house structure having a larger attenuating effect on mid- and high-frequency noise transmission than it does on low-frequency noise transmission, low-frequency noise will be even more apparent than would be expected from analysis of data external to the dwelling.

When low-frequency sound enters a room, the resulting sound waves spread out in all directions. When the advancing sound waves reach the walls, floor or ceiling of the enclosure they are reflected, generally with a small loss of energy, eventually resulting in waves travelling around the enclosure in all directions. If each path that a wave takes is traced around the enclosure, there will be certain paths of travel that repeat upon themselves to form normal modes (or standing waves). At certain frequencies, waves travelling around such paths will arrive back at any point along the path in phase. Amplification of the wave disturbance will result and the normal mode will be resonant. When the incoming sound contains energy at frequencies corresponding to the resonance frequencies of one or more room modes, the interior space will respond strongly, being only limited by the absorption present in the enclosure.

In an enclosure at low frequencies, the number of resonance frequencies within a specified frequency range will be small. Thus, at low frequencies, the response of a room as a function of frequency and location will be quite irregular; that is, the spatial distribution in the reverberant field will be characterised by pressure nodes and anti-nodes. As an example, a sound source is introduced into the corner of a room as shown in Figure 2.7. The source produces a single frequency, which is slowly increased without changing the source output level. The sound level at another corner is measured with a microphone. Room corners are chosen for excitation and measurement, as all room modes have antinodes in these locations, so none will be missed. Results for a large room of volume 180 m^3 and dimensions 6.840 × 5.565 × 4.720 m are shown in Figure 2.7.

Modal response is by no means peculiar to rectangular or even regular-shaped rooms. Modal response characterises enclosures of all shapes. Splayed, irregular or odd numbers of walls will not prevent resonances and accompanying pressure nodes and antinodes in an enclosure constructed of reasonably reflective walls.

The resonance frequency, f_n (Hz), of mode, n, of a rectangular shaped room may be calculated using the following simple expression.

$$f_n = \frac{c}{2}\sqrt{\left[\frac{n_x}{L_x}\right]^2 + \left[\frac{n_y}{L_y}\right]^2 + \left[\frac{n_z}{L_z}\right]^2} \quad \text{(Hz)} \tag{2.35}$$

where c is the speed of sound (typically 343 m/s), n_x, n_y and n_z are the number of nodal planes in the room in the x, y and z directions, respectively, and can have any integer value from zero upwards.

If we use the dimensions of a typical bedroom of 4.2 × 3.5 × 2.4 m in Eq. (2.35), the lowest resonance frequency is about 41 Hz. Below this frequency the sound pressure

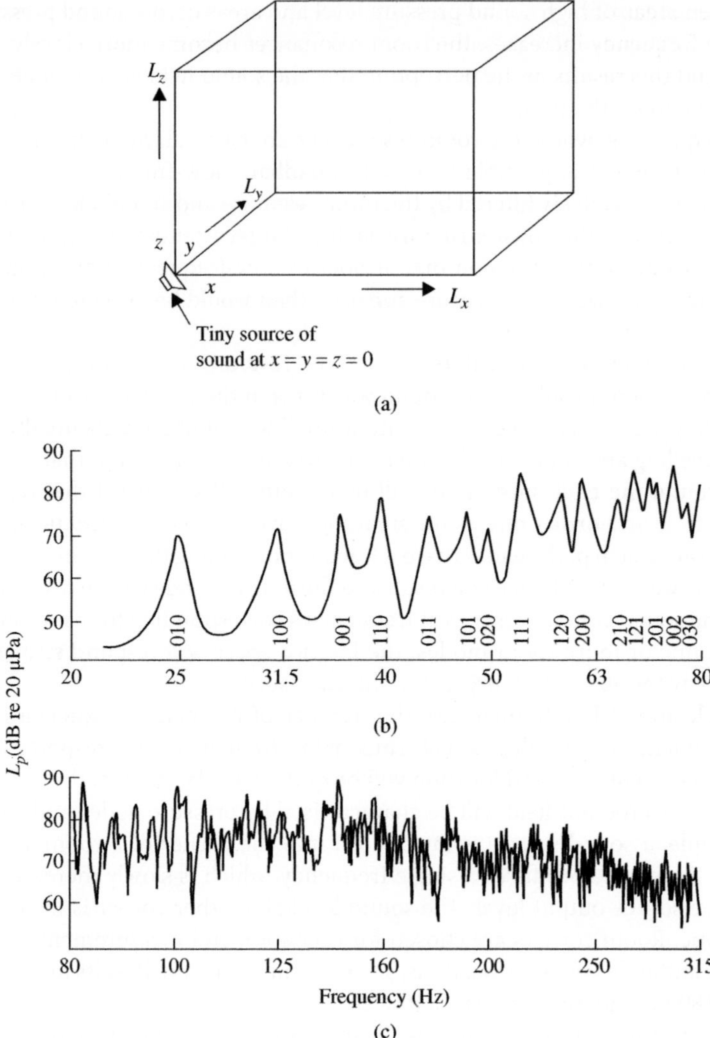

Figure 2.7 Sound level in the corner of a rectangular room as function of frequency, generated by a sound source with an output that is the same at all frequencies: (a) rectangular room of dimensions 6.84 × 5.56 × 4.72 m with a loudspeaker in the corner; (b) room response from 20–80 Hz, with mode orders in the *x*, *y* and *z* directions indicated on the figure (e.g. 010); (c) room response from 80–315 Hz.

level will be relatively uniform throughout the room. However, room resonances will be fairly well defined all the way up to the low-frequency upper limit of 200 Hz.

The approximate number of modes, N, which may be excited in the frequency range from zero up to f hertz, is given by the following expression for a rectangular room (Morse and Bolt 1944):

$$N = \frac{4\pi f^3 V}{3c^3} + \frac{\pi f^2 S}{4c^2} + \frac{fL}{8c} \tag{2.36}$$

where c is the speed of sound, V is the room volume, S is the room total surface area and L is the total perimeter of the room, which is the sum of lengths of all edges.

At very low frequencies, another resonance effect can occur in a room. This resonance is a result of the interaction between the room volume and an opening into the room, such as a partly open window or door. It is known as a Helmholtz resonance and it can be particularly important for wind farm noise as it can amplify indoor infrasonic noise in the 3–5 Hz range and above. This resonance effect was measured and reported by Hansen et al. (2015), along with the means for its calculation. The Helmholtz resonance frequency, f_0, can be calculated using

$$f_0 = \frac{c}{2\pi}\sqrt{\frac{\rho S_w}{mV}} \tag{2.37}$$

where c is the speed of sound in air (343 m/s), ρ is the density of air (1.2 kg/m^3), V is the air volume in the house, m is the average mass per unit area of the walls, roof and ceiling, and S_w is the surface area of the walls and roof.

A discussion of the sound field in rooms in the mid- to high-frequency range, where the sound field is more uniform, is provided in Bies and Hansen (2009), so only the important results are provided here. The most useful result is the relationship between sound pressure level, L_p in a room and the sound power, L_w of a sound source injecting energy into the room (which could be through a window, wall or roof). Thus,

$$L_p = L_w + 10\log_{10}\left[\frac{D_\theta}{4\pi r^2} + \frac{4(1-\alpha)}{S\alpha}\right] + 10\log_{10}\left[\frac{\rho c}{400}\right] \tag{2.38}$$

where r is the distance between the sound source and the measurement point, ρ is the density of air, c is the speed of sound in the air in the room (typically 343 m/s), D_θ is the directivity factor of the sound source (typically 2) and α is the average sound absorption coefficient (between 0 and 1.0) of the room surfaces. The quantity, α, can be determined by measuring the room reverberation time, T_{60}, in 1/3-octave or octave bands. The reverberation time is the time taken (in seconds) for the sound field in a room to decay by 60 dB once the sound source is switched off. In practice, the 30-dB decay time is measured and the result doubled. To measure the decay time in a room of volume V and room surface area (walls, floor and ceiling) S, 1/3-octave or octave bands of noise are introduced into the room using a random noise generator, amplifier and loudspeaker and the average absorption coefficient α is related to the reverberation time by

$$\alpha = \frac{55.25V}{ScT_{60}} \tag{2.39}$$

2.3 Basic Frequency Analysis

Sounds usually encountered in practice and especially those emitted by wind farms are not pure tones. In general, sounds are complex mixtures of pressure variations that vary with respect to phase, frequency, and amplitude. For such complex sounds, there is no simple mathematical relation between the different characteristics. However, any signal may be considered as a combination of a certain, possibly infinite, number of sinusoidal waves, each of which may be described as outlined above. These sinusoidal components constitute the frequency spectrum of the signal. The conversion of a time series representation of a signal to a frequency spectrum is referred to as Fourier analysis. In modern instrumentation, where the signal produced by the microphone is sampled using analogue-to-digital converters, the transformation of the signal from the time domain to the frequency domain is done using what is known as the discrete Fourier transform (DFT). The DFT is implemented in instrumentation using the fast Fourier transform (FFT), which is a much faster way of calculating the result than using the classical DFT formulae (Cooley and Tukey 1965). More details and explanation of the DFT and FFT may be found in the excellent book written by Brandt (2010). FFT analysis is discussed in more detail in Section 2.4.

This section will be mainly concerned with frequency analysis in the form of 1/3-octave and octave bands. In this type of analysis, the frequency spectrum is divided into standard frequency bands that include a specified range of frequencies, so that a particular frequency band represents the sum of all the energies associated with each frequency in the frequency band. This sort of analysis is a compromise in terms of detail between the single number A-weighted sound level (dBA) and the large amount of frequency spectrum data associated with FFT analysis (see Section 2.4 for more detail).

To illustrate longitudinal wave generation, as well as to provide a model for the discussion of sound spectra, the example of a vibrating piston at the end of a very long tube filled with air will be used, as illustrated in Figure 2.8.

Suppose the piston in Figure 2.8 moves forward. Since the air has inertia, only the air immediately next to the face of the piston moves at first; the pressure in the element of air next to the piston increases. The element of air under compression next to the piston will expand forward, displacing the next layer of air and compressing the next elemental volume. A pressure pulse is formed and this travels down the tube with the speed of sound, c. Suppose the piston stops and subsequently moves backward; a rarefaction is formed next to the surface of the piston and this follows the previously formed compression down the tube. If the piston again moves forward, the process is repeated, with the net result being a "wave" of positive and negative pressure transmitted along the tube.

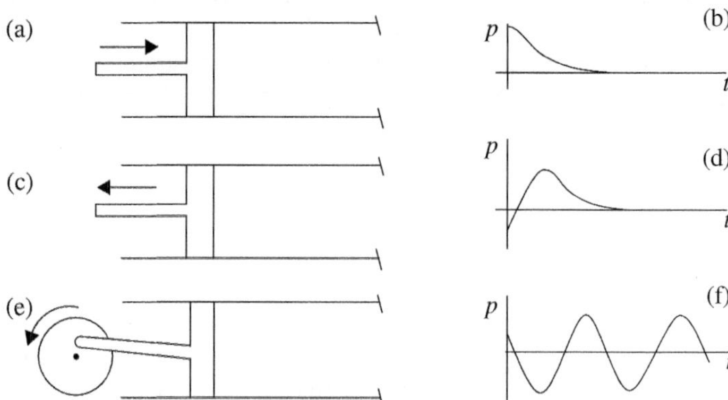

Figure 2.8 Sound generation illustrated: (a) the piston moves right, compressing air as in (b); (c) the piston stops and reverses direction, moving left and decompressing air in front of the piston, as in (d); (e) the piston moves cyclically back and forth, producing alternating compressions and rarefactions, as in (f). In all cases disturbances move to the right with the speed of sound.

If the piston moves with simple harmonic motion, a sine wave is produced; that is, at any instant the pressure distribution along the tube will have the form of a sine wave, or at any fixed point in the tube the pressure disturbance, displayed as a function of time, will have a sine wave appearance. Such a disturbance is characterised by a single frequency. The pressure variations at a fixed location in the tube and the corresponding spectra are illustrated in Figure 2.9a,b for various types of motion of the piston.

If the piston moves irregularly but cyclically, for example, so that it produces the waveform shown in Figure 2.9c, the resulting sound field will consist of a combination of sinusoids of several frequencies. The spectral (or frequency) distribution of the energy in this particular sound wave is represented by the frequency spectrum of Figure 2.9d. As the motion is cyclic, the spectrum consists of a set of discrete frequencies.

Although some sound sources have single-frequency components, most sound sources (including wind turbines) produce a very disordered and random waveform of pressure versus time, as illustrated in Figure 2.9e. Such a wave may or may not have one or more periodic components in addition to the random noise. However, Fourier analysis will enable the representation of the waveform as a collection of waves of all frequencies and the relative amplitude (in dB) of all frequencies making up the waveform will be shown. For a random type of wave, the sound pressure squared in a band of frequencies is plotted as shown, for example, in the frequency spectrum of Figure 2.9f.

Two special kinds of spectra are commonly referred to as *white random noise* and *pink random noise*. White random noise contains equal energy per hertz and thus has a constant spectral density level. Pink random noise contains equal energy per measurement band and thus has an octave and 1/3-octave band level that is constant with frequency.

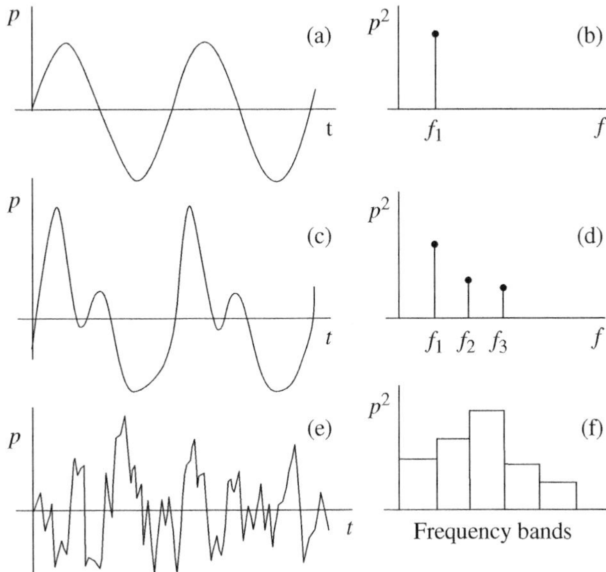

Figure 2.9 Spectral analysis illustrated: (a) disturbance p varies sinusoidally with time t at a single frequency f_1, as in (b); (c) disturbance p varies cyclically with time t as a combination of three sinusoidal disturbances of fixed relative amplitudes and phases; the associated spectrum has three single-frequency components f_1, f_2 and f_3, as in (d); (e) disturbance p varies erratically with time t, with a frequency band spectrum as in (f).

2.3.1 Digital Filtering

The signal output by a microphone or accelerometer is a time-varying analogue voltage. Modern instrumentation takes this signal and converts it to digital form by sampling its value a set number of times per second (or hertz). This is called the sample rate or sampling frequency. The Nyquist frequency is half the sampling frequency and represents the highest sound or vibration frequency that can be determined from the sampled signal. If frequencies higher than the Nyquist frequency exist in the signal, then they will 'fold back' at lower frequencies, causing errors in the amplitudes of the frequencies of interest (those below the Nyquist frequency). This effect is known as *aliasing* and is discussed in more detail in Section 2.4.5. An analog filter (called an anti-aliasing filter, see Section 2.4.5) is used to severely reduce the amplitude of frequencies higher than the highest frequency of interest, so as to minimise the effects of aliasing. Some instrumentation samples the signal at a very high frequency so that a simple analog anti-aliasing filter can be used. Then the resulting signal is digitally filtered and 'downsampled' to obtain the desired sampling rate. Although in theory the sampling frequency only needs to be twice the highest frequency of interest for analysis in the frequency domain, it is common to choose a sample rate 2.5 times higher, to allow for the finite steepness of the roll-off of the anti-aliasing filters. However, for time-domain filtering, such as implementation of an A-weighting filter, the sample rate should be at least 10 times the highest frequency of interest in the time-domain signal.

As well as providing anti-aliasing filters for downsampled signals in instrumentation with a very high sampling rate, digital filters are used to provide A-weighting and C-weighting, as well as octave and 1/3-octave band filtering. In addition, digital filters with continually adjusting coefficients are used to generate the cancelling signal in an active noise-control system. This is discussed in Chapter 8.

The basic form of a digital filter is a tapped delay line, whereby the output sample at a particular time consists of the weighted sum of a fixed number of preceding samples. This fixed number is referred to as the number of filter taps. There are two types of basic digital filter:

- the finite impulse response filter (FIR), which uses only input data samples to derive the output
- the infinite impulse response filter (IIR), which uses previous output samples as well as input samples to derive the output.

The output, $y(k)$ at time sample k for both filters can be represented mathematically as

$$\begin{aligned} y(k) = {} & b_0 x(k) + b_1 x(k-1) + \cdots + b_N x(k-N) \\ & + a_1 y(k-1) + a_2 y(k-2) + \cdots + a_M y(k-M) \end{aligned} \tag{2.40}$$

where the a and b filter coefficients represent the weightings that are applied to the various samples prior to them being summed, and the integers in brackets represent the sample numbers. The number of taps on the filter based on input data samples is N and the number of taps on the filter based on output data samples is M. For the FIR filter, all the a coefficients are set equal to zero. Eq. (2.40) is more easily understood by referring to the diagrammatic representations of the filters in Figures 2.10 and 2.11. In the figures, z^{-1} represents a single sample time delay.

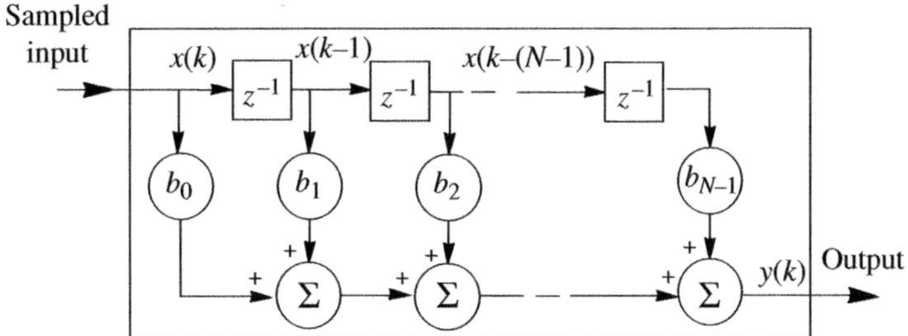

Figure 2.10 Finite impulse response (FIR) filter structure.

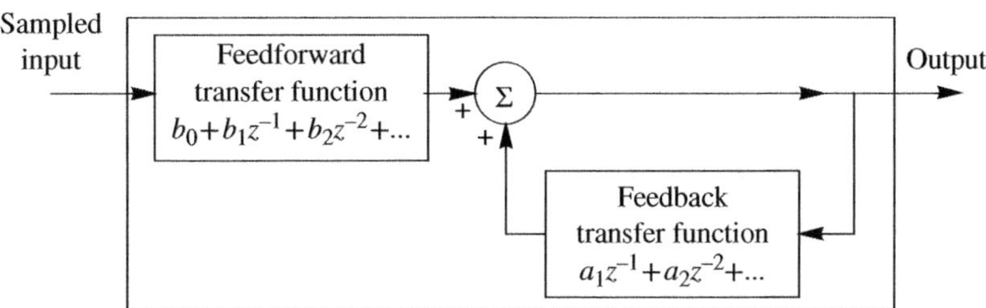

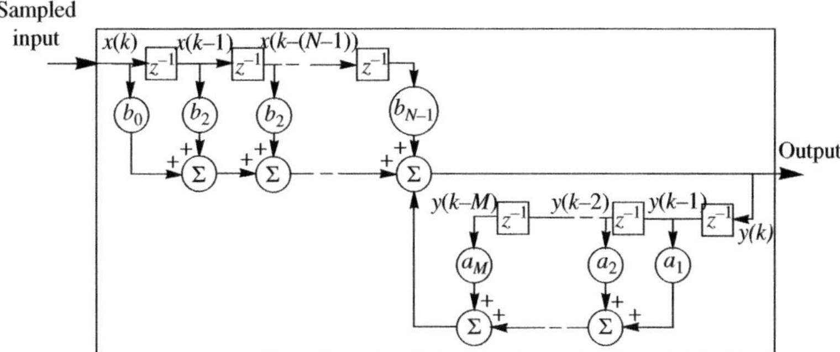

Figure 2.11 Infinite impulse response (IIR) filter structure.

2.3.2 Octave Band and 1/3-Octave Band Analysis

As mentioned in Section 2.2.11, one simple form of frequency analysis is to divide the frequency spectrum into octave or 1/3-octave bands so that the level in each band represents the combined sound levels for all frequencies contained in that band. In practice, this is done either by using digital filters, one for each band, or by FFT analysis, a process discussed in detail in Section 2.4. The analysis is done by first passing the analogue time series signal through an octave or 1/3-octave filter and the output of the filter is then passed through a squaring circuit followed by an averaging circuit. Of course, if the signal is a train of digital samples, the squaring and averaging is done digitally.

Each octave or 1/3-octave band is defined by a lower frequency limit, an upper frequency limit and a centre frequency. Standard values for these three frequencies are summarised in Table 2.3. The band number (BN) is related to the band centre frequency,

f_C and the band centre frequency is related to the upper, f_u, and lower limiting frequencies, f_ℓ, by

$$BN = 10 \log_{10} f_C \quad \text{and} \quad f_C = \sqrt{f_\ell \ f_u} \tag{2.41}$$

The band limits in Table 2.3 are calculated using

$$f_u / f_\ell = 2^{1/N} \tag{2.42}$$

where $N = 1$ for octave bands and $N = 3$ for 1/3-octave bands. The bandwidth, Δf of each band may be calculated using

$$\Delta f = f_C \frac{2^{1/N} - 1}{2^{1/2N}} \tag{2.43}$$

which is equal to $0.2316 f_C$ for 1/3-octave bands and $0.7071 f_C$ for octave bands. The above equations result in values very close to those in the table.

By small adjustments in the calculated values recorded in the table, it has been possible to arrange the 1/3-octave centre frequencies so that ten times their logarithms are the band numbers of column one on the left of the table. Consequently, as may be observed, the 1/3-octave centre frequencies repeat every decade in the table. Octave band results can also be obtained from the narrowband power spectrum obtained using FFT analysis. This is not as straightforward as adding together logarithmically the levels corresponding to all frequencies in the particular 1/3-octave or octave band, and is discussed in Section 2.

Of course, it is possible to analyse a sound signal into narrower bands, for example 1/12 octave bands. This possibility is provided in some FFT instrumentation.

2.3.3 Octave and 1/3-Octave Filter Rise and Settling times

One important aspect of analysis of wind farm sound signals using octave or 1/3-octave bands is the filter rise time, which is the time it takes for the filter to measure the true value of a continuous signal. So if the signal is varying rapidly, especially at low frequencies, it is not possible for the output of the filter to track the rapidly varying input, resulting in a considerable error in the RMS output, especially if the signal has high crest factors (the ratio of the peak value of a signal to its RMS value). From Table 2.4, which was assembled with data provided by Bray and James (2011), it can be seen that the impulse response (see Section 2.4.13) duration to achieve 90% (or 1 dB error) of the true magnitude of the signal for a 1/3-octave band filter centred at 1 Hz is approximately 5 s or five full cycles. It is clear that use of such a 1/3-octave filter will not correctly measure the energy associated with rapidly varying sound having relatively high crest factors, as may be experienced near wind farms. As the centre frequency of the octave or 1/3-octave band becomes lower, the required sampling time to obtain an accurate measure of the signal becomes larger. For higher frequencies not listed in Table 2.4, the rise time decreases by a factor of ten for each decade increase in frequency. Octave band filters would have rise times of one third of the rise time of a 1/3-octave filter with the same centre frequency.

One way of increasing the ability of the measurement system to track and measure accurately sound signals with high crest factors is to increase the bandwidth of the filter used for the measurement so that we no longer use 1/3-octave or octave band filters. A suggestion by Bray and James (2011) is to use filter bandwidths that are equal

Table 2.3 Preferred octave and 1/3-octave frequency bands.

Band number	Octave band centre frequency	1/3-octave band centre frequency	Band limits	
			Lower	Upper
−1		0.8	0.7	0.9
0	1	1	0.9	1.1
1		1.25	1.1	1.4
2		1.6	1.4	1.8
3	2	2	1.8	2.2
4		2.5	2.2	2.8
5		3.15	2.8	3.5
6	4	4	3.5	4.4
7		5	4.4	5.6
8		6.3	5.6	7
9	8	8	7	9
10		10	9	11
11		12.5	11	14
12	16	16	14	18
13		20	18	22
14		25	22	28
15	31.5	31.5	28	35
16		40	35	44
17		5	44	57
18	63	63	57	71
19		80	71	88
20		100	88	113
21	125	125	113	141
22		160	141	176
23		200	176	225
24	250	250	225	283
25		315	283	353
26		400	353	440
27	500	500	440	565
28		630	565	707
29		800	707	880
30	1000	1000	880	1130
31		1250	1130	1414
32		1600	1414	1760
33	2000	2000	1760	2250
34		2500	2250	2825
35		3150	2825	3530

(*Continued*)

Table 2.3 (Continued)

Band number	Octave band centre frequency	1/3-octave band centre frequency	Band limits	
			Lower	Upper
36	4000	4000	3530	4400
37		5000	4400	5650
38		6300	5650	7070
39	8000	8000	7070	8800
40		10 000	8800	11 300
41		12 500	11 300	14 140
42	16 000	16 000	14 140	17 600
43		20 000	17 600	22 500

Table 2.4 1/3-octave filter rise times for a 1-dB error.[a)]

1/3-octave centre frequency (Hz)	Rise time (ms)	1/3-octave centre frequency (Hz)	Rise time (ms)	1/3-octave centre frequency (Hz)	Rise time (ms)
1.00	4989	10.0	499	100	49.9
1.25	3963	12.5	396	125	39.6
1.60	3148	16.0	315	160	31.5
2.00	2500	20.0	250	200	25
2.50	1986	25.0	199	250	19.9
3.15	1578	31.5	158	315	15.8
4.00	1253	40.0	125	400	12.5
5.00	995	50.0	99.5	500	9.95
6.30	791	63.0	79.1	639	7.91
8.00	628	80.0	62.8	800	6.28

a) Standard ANSI S1.11 sixth-order filter (ANSI/ASA S1.11 2004).

to the critical bandwidth of our hearing mechanism, which is approximately 100 Hz at our lowest hearing frequencies. Thus the filter bandwidth for measuring low-frequency noise and infrasound from wind farms should be 100 Hz (actually from 0.5–100 Hz) if we are to measure crest factors in a way similar to that in which our hearing mechanism experiences them. Bray and James (2011) recommend using a fourth order Butterworth 'Bark 0.5' (see Section 6.2.7) band-pass filter centred on 50 Hz with a rise time of approximately 8.8 ms, which according to Bray and James (2011) simulates the approximately 10-ms response of our hearing mechanism. This response is ten times faster than the 'fast' response on a sound level meter, which implies that the 'fast' response underestimates the true rise time of our hearing mechanism. Thus we hear much higher peaks than are measured by the 'fast' sound level meter measurements as illustrated by comparing parts (a) and (b) and also parts (c) and (d) in Figures 2.12 and 2.13.

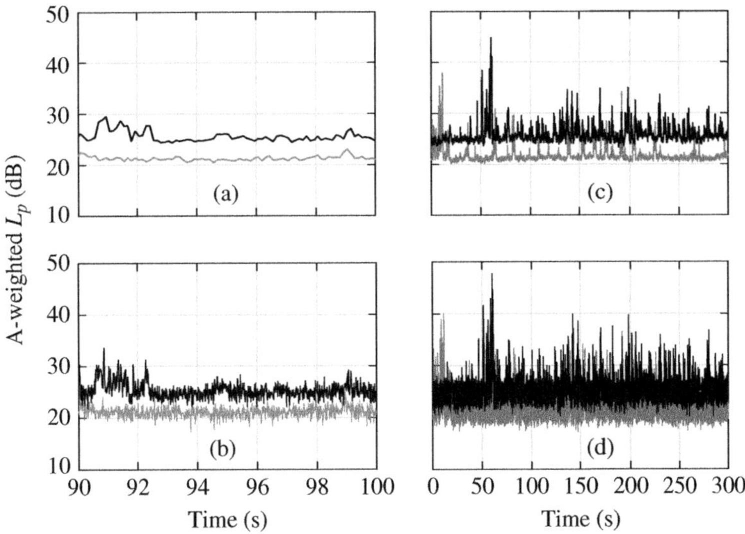

Figure 2.12 A-weighted sound pressure levels (SPL) measured at 3:30am outside a residence 3.3 km from a 37 × 3.3 MW turbine wind farm, with the turbines on (black) and off (grey), for a wind speed of 10.4 m/s at the hub of the nearest turbine and 0 m/s at the measurement point. (a) A-weighted level outdoors (10-s period, 125-ms averaging or 'fast' response, L_{Aeq}[OFF] = 21 dBA, L_{Aeq}[ON] = 26 dBA; (b) the same, but with 10-ms averaging; (c) A-weighted level outdoors (5-min period, 125-ms averaging, L_{Aeq}[OFF] = 27 dBA, L_{Aeq}[ON] = 22 dBA); (d) the same but with 10-ms averaging.

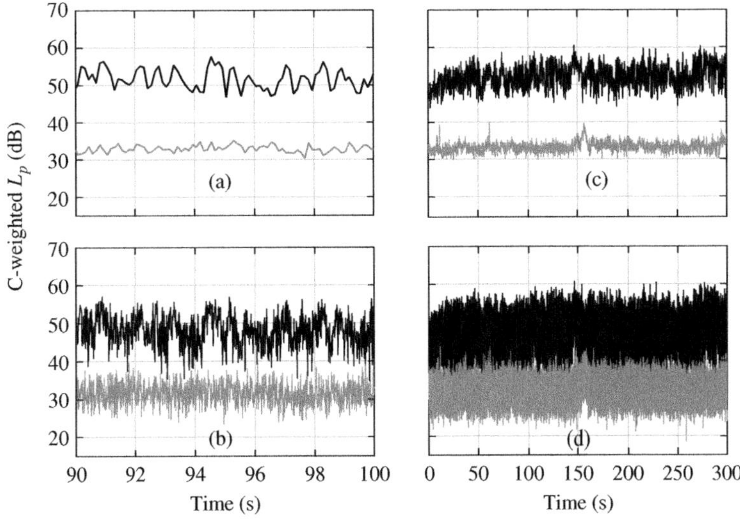

Figure 2.13 C-weighted sound pressure levels (SPL) measured at 3:30am outside an unoccupied residence 3.3 km from a 37 × 3.3 MW turbine wind farm with the turbines on (black) and with them of (grey) for a wind speed of 10.4 m/s at the hub of the nearest turbine and 0 m/s at the measurement point. (a) C-weighted level outdoors (10-s period, 125-ms averaging or 'fast' response, L_{Ceq}[OFF] = 32 dBC, L_{Ceq}[*ON*] = 52 dBC); (b) the same but with 10-ms averaging; (c) C-weighted level outdoors (5-min period, 125–ms averaging, L_{Ceq}[OFF] = 33 dBC; L_{Ceq}[*ON*] = 53 dBC); (d) the same but with 10-ms averaging.

In addition to filter rise time, it is also important to take into account the settling time of the RMS detector in order to achieve 0.5-dB accuracy, which is in addition to the rise time discussed above. For a single-frequency signal of frequency, f, the settling time for 0.5-dB accuracy will be approximately $3/f$.

For a random signal, the calculated statistical error, ϵ, is given by

$$\epsilon = \frac{4.34}{\sqrt{BT_A}} \quad \text{dB} \tag{2.44}$$

where T_A is the averaging time in seconds and B is the bandwidth of the filter in Hz. There is a 68.3% chance of any estimate being within $\pm\epsilon$, 95.5% chance of it being within $\pm 2\epsilon$ and a 99.7% chance of it being within $\pm 3\epsilon$.

2.4 Advanced Frequency Analysis

In this section, various aspects of FFT analysis will be discussed. We have already discussed octave and 1/3-octave band analysis. FFT analysis usually allows much more frequency resolution (each component in the frequency spectrum representing smaller frequency spans) to be obtained, which is an advantage when trying to identify wind turbine sound sources or sound field characteristics such as AM.

FFT analysis is the process of transforming a time-varying signal into its frequency components so that we obtain a plot or table of signal amplitude as a function of frequency. A general Fourier representation of a periodic time-varying signal of period T consisting of:

- a fundamental frequency $f_1 = 1/T$, represented by $x_1(t) = x(t + T)$
- various harmonics, n, of frequency f_n, represented by $x_n(t) = x(t + nT)$ where $n = 2, 3, \ldots$

takes the following form:

$$x(t) = \sum_{n=1}^{\infty} A_n \cos(2\pi n f_1 t) + B_n \sin(2\pi n f_1 t) \tag{2.45}$$

As an example, we can examine the Fourier representation of a square wave, as shown in Figure 2.14. The first four harmonics in part (b) are described by the first four terms in Eq. (2.45), where for n odd, $A_n = 4/(\pi n)$, and for n even, $A_n = 0$ and $B_n = 0$. The component characterised by frequency, nf_1, is usually referred to as the nth harmonic of the fundamental frequency, f_1, although some call it the $(n-1)$th harmonic. Use of Euler's well known equation (Abramowitz and Stegun 1964) allows Eq. (2.45) to be rewritten in the following alternative form

$$x(t) = \frac{1}{2}\sum_{n=0}^{\infty}[(A_n - \mathrm{j}B_n)\mathrm{e}^{\mathrm{j}2\pi n f_1 t} + (A_n + \mathrm{j}B_n)\mathrm{e}^{-\mathrm{j}2\pi n f_1 t}] \tag{2.46}$$

where the $n = 0$ term has been added for mathematical convenience. However, it represents the zero-frequency (DC) component of the signal and is usually considered to be zero.

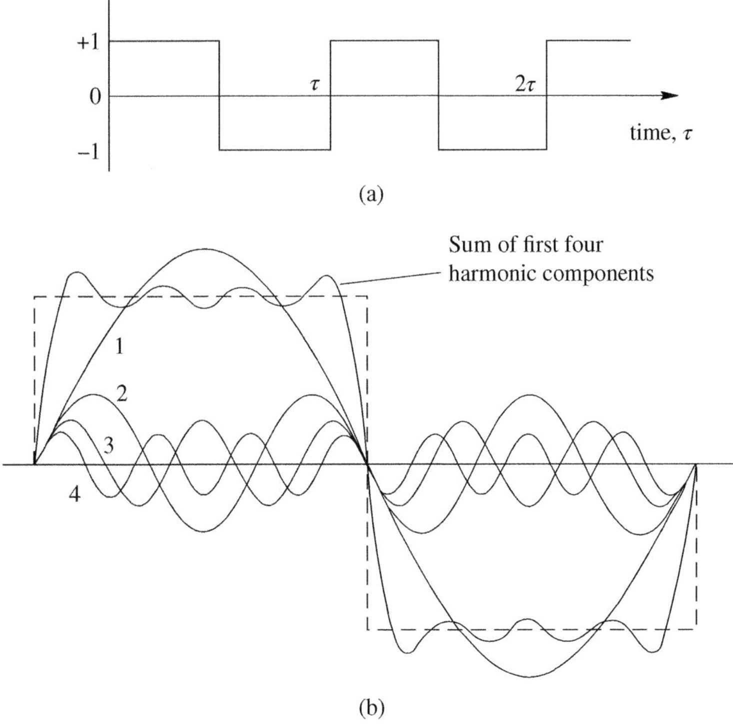

Figure 2.14 An example of Fourier analysis of a square wave: (a) periodic square wave in the time domain; (b) the first four harmonic components of the square wave in (a).

A further reduction is possible by defining the complex spectral amplitude components $X_n = (A_n - \mathrm{j}B_n)/2$ and $X_{-n} = (A_n + \mathrm{j}B_n)/2$. Denoting the complex conjugate by *, the following relation may be written

$$X_n = X^*_{-n} \tag{2.47}$$

It can be seen that Eq. (2.47) is satisfied, thus ensuring that the right-hand side of Eq. (2.46) is real. The introduction of Eq. (2.47) into Eq. (2.46) allows the following more compact expression to be written

$$x(t) = \sum_{n=-\infty}^{\infty} X_n \mathrm{e}^{\mathrm{j}2\pi n f_1 t} \tag{2.48}$$

The spectrum of Eq. (2.48) now includes negative as well as positive values of n, giving rise to components $-nf_1$. The spectrum is said to be two-sided and the amplitudes of the negative frequency components mirror their positive counter-parts; that is, the negative frequency spectrum is the mirror image of the positive spectrum. The spectral amplitude components, X_n, may be calculated using the following expression

$$X_n = \frac{1}{T} \int_{-T/2}^{T/2} x(t) \mathrm{e}^{-\mathrm{j}2\pi n f_1 t} \mathrm{d}t \tag{2.49}$$

The spectrum of squared amplitudes is known as the power spectrum. The mean of the instantaneous power of the time-varying signal, $[x(t)]^2$, averaged over the period T is

$$W_{\mathrm{mean}} = \frac{1}{T}\int_0^T [x(t)]^2 \mathrm{d}t \tag{2.50}$$

Substitution of Eq. (2.45) in Eq. (2.50) and integrating gives

$$W_{\mathrm{mean}} = \frac{1}{2}\sum_{n=1}^{\infty}[A_n^2 + B_n^2] \tag{2.51}$$

Equation (2.51) shows that the total power is the sum of the powers of each spectral component, provided that no windowing of the time-domain signal was done prior to taking the FFT (see Section 2.4.4).

The previous analysis may be extended to the more general case of random noise by allowing the period, T, to become indefinitely large. In this case, X_n becomes $X_D(f)$, a continuous function of frequency, f. It is to be noted that whereas the units of X_n are the same as those of $x(t)$, the units of $X_D(f)$ are those of $x(t)$ per hertz. With the proposed changes, Eq. (2.49) takes the following form:

$$X_D(f) = \int_{-\infty}^{\infty} x(t)\mathrm{e}^{-\mathrm{j}2\pi ft}\mathrm{d}t \tag{2.52}$$

The spectral density function, $X_D(f)$, is complex, characterised by a real and an imaginary part (or amplitude and phase). Equation (2.48) becomes

$$x(t) = \int_{-\infty}^{\infty} X_D(f)\mathrm{e}^{\mathrm{j}2\pi ft}\mathrm{d}f \tag{2.53}$$

Equations (2.52) and (2.53) form a Fourier transform pair, with the former referred to as the forward transform and the latter as the inverse transform. In practice, a finite sample time T is always used to acquire data and the spectral representation of Eq. (2.48) is the result calculated by spectrum analysis equipment. This latter result is referred to as the spectrum and the spectral density is obtained by multiplying by the sample period, T. Where a time function is represented as a sequence of samples taken at regular intervals, an alternative form of Fourier transform pair is as follows. The forward transform is

$$X(f) = \sum_{k=-\infty}^{\infty} x(t_k)\mathrm{e}^{-\mathrm{j}2\pi f t_k} \tag{2.54}$$

The quantity $X(f)$ represents the spectrum and the inverse transform is

$$x(t_k) = \frac{1}{f_s}\int_{-f_s/2}^{f_s/2} X(f)\mathrm{e}^{\mathrm{j}2\pi f t_k}\mathrm{d}f \tag{2.55}$$

where f_s is the sampling frequency and t_k is the time corresponding to sample k in the time series data.

The form of Fourier transform pair used in spectrum analysis instrumentation is the DFT, for which the functions are sampled in both the time and frequency domains. Thus

$$x(t_k) = \sum_{n=0}^{N-1} X(f_n)\mathrm{e}^{\mathrm{j}2\pi nk/N} \quad k = 0, 1, \ldots (N-1) \tag{2.56}$$

$$X(f_n) = \frac{1}{N}\sum_{k=0}^{N-1} x(t_k)\mathrm{e}^{-\mathrm{j}2\pi nk/N} \quad n = 0, 1, \ldots (N-1) \tag{2.57}$$

where k and n represent discrete sample numbers in the time and frequency domains, respectively, and $X(f_n)$ represents the *amplitude* of the nth component in the frequency spectrum. In Eq. (2.56), the spacing between frequency components (or frequency resolution) in hertz is dependent on the time, T, to acquire the N samples of data in the time domain and is equal to $1/T$ or f_s/N. Thus the effective filter bandwidth, B, is equal to $1/T$. The four Fourier transform pairs are shown graphically in Figure 2.15. In Eqs. (2.56) and (2.57), the functions have not been made symmetrical about the origin, but because of the periodicity of each, the second half of each sum also represents the negative half period to the left of the origin, as can be seen by inspection of the figure.

The frequency components above $f_s/2$ in Figure 2.15 can be more easily visualised as negative frequency components and in practice the frequency content of the final spectrum must be restricted to less than $f_s/2$. This is explained in Section 2.4.5, where aliasing is discussed.

The DFT is well suited to the digital computations performed in instrumentation or by frequency analysis software on a personal computer. Nevertheless, it can be seen by referring to Eq. (2.56) that to obtain N frequency components from N time samples, N^2 complex multiplications are required. Fortunately, this figure is reduced by the use of the FFT algorithm to $N \ \log_2 N$, which for a typical case of $N = 1024$, speeds up computations by a factor of 100. This algorithm is discussed in detail by Randall (1987).

2.4.1 Auto Power Spectrum and Power Spectral Density

The auto power spectrum (sometimes called the power spectrum) is the most common form of spectral representation used in acoustics and vibration. The auto power spectrum is the spectrum of the squared RMS values of each frequency component, whereas the frequency spectrum discussed previously was a spectrum of the *amplitudes* of each frequency component. The two-sided auto power spectrum, $S_{xx}(f_n)$ may be estimated by averaging a large number of squared amplitude spectra, $X(f_n)$, and applying a scaling factor to account for conversion from an amplitude spectrum to an RMS spectrum ($\sqrt{2}$) and to account for the application of a windowing function to the sampled data (see Section 2.4.4). Estimation of the scaling factor, S_A, is discussed in Section 2.4.4. Thus,

$$S_{xx}(f_n) \approx \frac{S_A}{Q}\sum_{i=1}^{Q} X_i^*(f_n)X_i(f_n) = \frac{S_A}{Q}\sum_{i=1}^{Q} |X_i(f_n)|^2 \quad n = 0, 1, \ldots (N-1) \tag{2.58}$$

where i is the spectrum number and Q is the number of spectra over which the average is taken. The larger the value of Q, the more closely will the estimate of $S_{xx}(f_n)$ approach its true value.

The *power spectral density*, $S_{Dxx}(f_n)$ (or PSD(f_n)) can be obtained from the power spectrum by dividing the amplitudes of each frequency component by the frequency spacing, Δf, between adjacent components in the frequency spectrum or by multiplying by the time, T, to acquire one record of data. Thus, the two-sided power spectral density is:

$$S_{Dxx}(f_n) \approx \frac{TS_A}{Q}\sum_{i=1}^{Q}|X_i(f_n)|^2 = \frac{S_A}{Q\Delta f}\sum_{i=1}^{Q}|X_i(f_n)|^2 \quad n = 0, 1, \ldots (N-1) \tag{2.59}$$

where 'two-sided' indicates that the spectrum extends to negative as well as positive frequencies. The time blocks are usually overlapped by up to 50% to decrease the random error in the PSD estimate (Brandt 2010), as explained in Section 2.4.6.

The auto power spectrum is useful for evaluating tonal components in a spectrum, although for random noise it is more appropriate to use the PSD function of Eq. (2.59). The auto power spectrum is used to evaluate spectra that contain tonal

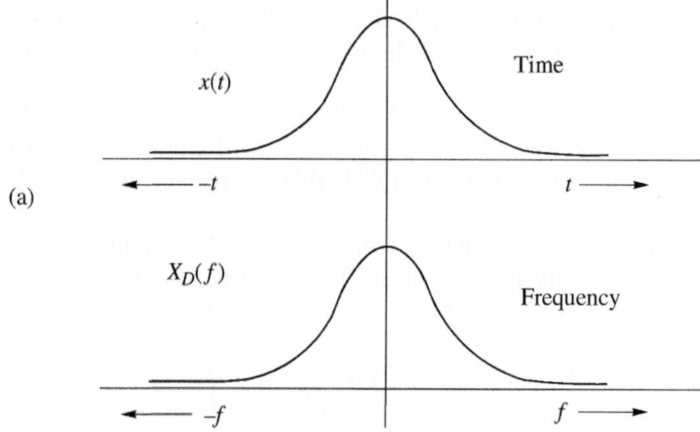

Figure 2.15 Various Fourier transform pairs (after Randall (1987)). The dashed lines indicate a periodically repeating sequence:

a) Integral transform; signal infinitely long and continuous in both the time and frequency domains

$$X_D(f) = \int_{-\infty}^{\infty} x(t)e^{-j2\pi ft}dt \quad \text{and} \quad x(t) = \int_{-\infty}^{\infty} X_D(f)e^{j2\pi ft}df;$$

b) Fourier series; signal periodic in the time domain and discrete in the frequency domain

$$X(f_n) = \frac{1}{T}\int_{-T/2}^{T/2} x(t)e^{-j2\pi f_n t}dt \quad \text{and} \quad x(t) = \sum_{n=-\infty}^{\infty} X(f_n)e^{j2\pi f_n t};$$

c) Sampled function; signal discrete in the time domain and periodic in the frequency domain

$$X(f) = \sum_{k=-\infty}^{\infty} x(t_k)e^{-j2\pi ft_k} \quad \text{and} \quad x(t_k) = \frac{1}{f_s}\int_{-f_s/2}^{f_s/2} X(f)e^{j2\pi ft_k}df;$$

d) Discrete Fourier transform; signal discrete and periodic in both the time and frequency domains

$$X(f_n) = \frac{1}{N}\sum_{k=0}^{N-1} x(t_k)e^{-j2\pi nk/N} \quad \text{and} \quad x(t_k) = \sum_{n=0}^{N-1} X(f_n)e^{j2\pi nk/N}$$

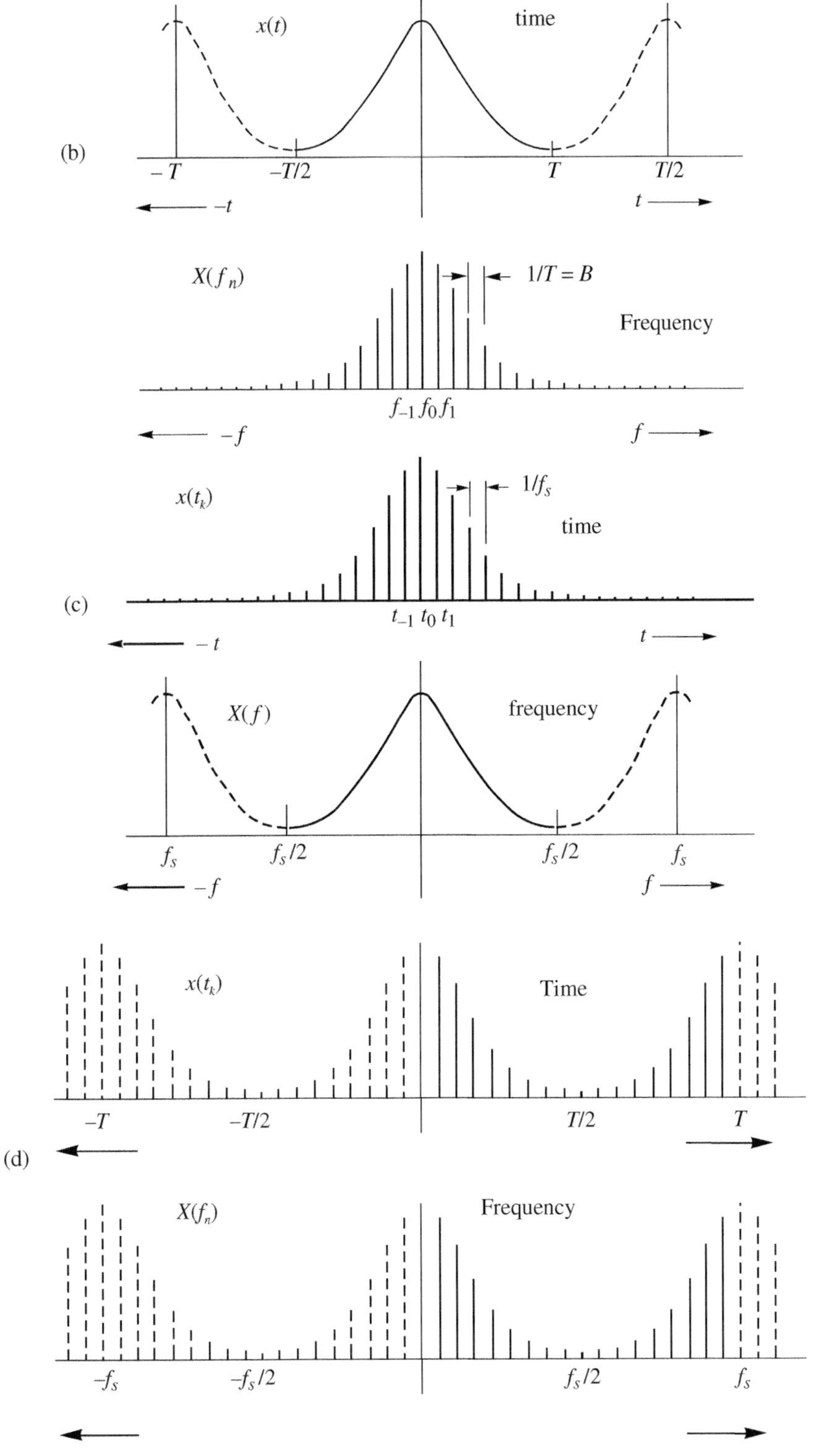

Figure 2.15 (*Continued*)

components because, unlike the PSD, it is able to give the true energy content of a tonal component. This is because the bandwidth of a tone is not the same as the frequency spacing in the spectrum, and is often much smaller. This results in the spectral amplitude of a tone being independent of the frequency resolution of the FFT analysis, provided that the tonal frequency corresponds to the frequency of one of the spectral lines (see Sections 2.4.3 and 2.4.4). Thus dividing a tonal amplitude by the spectral resolution to obtain the PSD will result in a significant error in the tonal amplitude. In real systems, the frequency of a tone may vary slightly during the time it takes to acquire a sufficient number of samples for an FFT and also from one FFT to another during the averaging process to obtain power spectra. In this case the tone may be spread out in frequency so that its amplitude will depend on the frequency resolution; a better estimate of the amplitude will be obtained with a coarse frequency resolution. A sufficiently coarse resolution would enable the range of frequency variation to be captured in a single-frequency bin in the spectrum.

The problems of errors in tonal amplitudes can be avoided by calculating the PSD from the auto-correlation function, $R_{xx}(\tau)$, which is the covariance of the time series signal $x(t)$ with itself, time shifted by τ seconds. It is defined as

$$R_{xx}(\tau) = \langle x(t)x(t+\tau)\rangle_t = \mathrm{E}[x(t)\cdot x(t+\tau)] = \lim_{T\to\infty}\frac{1}{T}\int_{-T/2}^{T/2} x(t)\cdot x(t+\tau)\,\mathrm{d}t \quad (2.60)$$

where E[x] is the expected value of x and $\langle x\rangle_t$ is the time-averaged value of x. The auto-correlation function is discussed in more detail in Section 2.4.16. The PSD, $S_{Dxx}(f_n)$ is obtained from the auto-correlation function by substituting $R_{xx}(\tau)$ for $x(t)$, τ for t, $S_{Dxx}(f_n)$ for $X(f_n)$ and $S_{Dxx}(f)$ for $X_D(f)$ in the equations in the caption of Figure 2.15. For example, for the caption for part (c) in the figure,

$$S_{Dxx}(f) = \sum_{k=-\infty}^{\infty} R_{xx}(\tau_k)\mathrm{e}^{-\mathrm{j}2\pi f\tau_k} \quad (2.61)$$

and

$$R_{xx}(\tau_k) = \frac{1}{f_s}\int_{-f_s/2}^{f_s/2} S_{Dxx}(f)\mathrm{e}^{\mathrm{j}2\pi f\tau_k}\mathrm{d}f \quad (2.62)$$

However, the auto-correlation function is very computationally intensive to calculate, so this method is not in common use. In fact, the reverse is more often the case; the auto-correlation of a data set is found by taking the inverse FFT of the PSD.

For random noise, the frequency resolution affects the spectrum amplitude; the finer the resolution the smaller will be the amplitude. For this reason we use PSDs for random noise for which the effective frequency resolution is 1 Hz.

In practice, the single-sided power spectrum, $G_{xx}(f_n)$, (positive frequencies only) is the one of interest and this is expressed in terms of the two-sided auto power spectrum

$S_{xx}(f_n)$ as follows.

$$G_{xx}(f_n) = \begin{cases} 0 & f_n < 0 \\ S_{xx}(f_n) & f_n = 0 \\ 2S_{xx}(f_n) & f_n > 0 \end{cases} \tag{2.63}$$

A similar expression may be written for the single-sided PSD, $G_{Dxx}(f_n)$, as follows

$$G_{Dxx}(f_n) = \begin{cases} 0 & f_n < 0 \\ S_{Dxx}(f_n) & f_n = 0 \\ 2S_{Dxx}(f_n) & f_n > 0 \end{cases} \tag{2.64}$$

If successive spectra, $X_i(f_n)$, are averaged, the result will be zero, as the phases of each spectral component vary randomly from one record to the next. Thus in practice, auto power spectra are more commonly used, as they can be averaged together to give a more accurate result. This is because auto power spectra are only represented by an amplitude; phase information is lost when the spectra are calculated (see Eq. (2.58)). The same reasoning applies to the PSD (power per hertz), which is obtained from the auto power spectrum by dividing the amplitude of each frequency component by the frequency spacing, Δf, between adjacent components.

2.4.2 Linear Spectrum

Sometimes one finds results of a spectral analysis presented in terms of linear rather than the squared values of an auto power spectrum. Each frequency component in the linear spectrum is calculated by taking the square root of each frequency component in the auto power spectrum.

2.4.3 Leakage

Leakage is the phenomenon that occurs when we take a DFT using a finite time window; that is, by taking a finite number of data samples. This results in a spectrum containing discrete frequency components separated by a frequency interval. The number of frequency components and the frequency separation interval, Δf are set by the sampling frequency, f_s and the total sampling time (or measurement time), T. The frequency separation between adjacent components in the spectrum is given by

$$\Delta f = \frac{1}{T} \tag{2.65}$$

and the number of discrete frequency components, N, in the spectrum is given by

$$N = Tf_s \tag{2.66}$$

Each discrete frequency component (or spectral line) in the spectrum is like a band-pass filter with a characteristic response, $W(f_n)$, defined by a sinc function; that is

$$W(f_n) = T\frac{\sin(\pi f_n T)}{\pi f_n T} = T\mathrm{sinc}(f_n T) \quad n = 0, 1, \ldots (N-1) \tag{2.67}$$

This means that a sinusoidal signal equal to the exact frequency of a spectral component will be given the correct amplitude value in the frequency spectrum, as the sinc function has a value unity at frequency, f_n. However, for a sinusoidal signal with a frequency half way between two spectral lines, the energy of the signal will be split between the two adjacent lines (or frequency bins) and neither line will give the correct result. In fact, for this case the error will result in a value that is 36% (or 1.95 dB) too small.

2.4.4 Windowing

As mentioned in Section 2.4.3, leakage occurs when calculating the DFT of a sinusoidal signal with a non-integer number of periods in the time window used for sampling. The error is caused by the truncation of the continuous signal by using a finite time window, which causes a discontinuity when the two ends of the record are effectively joined in a loop as a result of the DFT. Leakage can be reduced by using a windowing function applied to the time window such that all samples in the time record are not given equal weighting when calculating the DFT. In fact the window may be configured so that samples near the beginning and end of the time window are weighted much less than samples in the centre of the window. This minimises the effect of the signal being discontinuous at the beginning and end of the time window. The discontinuity without weighting causes side lobes to appear in the spectrum for a single frequency, as shown by the solid curve in Figure 2.16, which is effectively the same as applying a rectangular window weighting function. In this case, all signal samples before sampling begins and after it ends are multiplied by zero and all values in between are multiplied by one. In the figure the normalised frequency of $1 \times 1/T$ represents the frequency resolution or number of hertz between adjacent frequency bins in the spectrum.

A better choice of window is one that places less weight on the signal at either end of the window and maximum weight on the middle of the window. One such weighting, called a Hanning window, is illustrated in Figure 2.16. The result of weighting the input signal in this way is shown by the dashed curve in the figure. Even though the main lobe is wider, representing poorer frequency resolution, the side lobe amplitudes fall away more rapidly, resulting in less contamination of adjacent frequency bins.

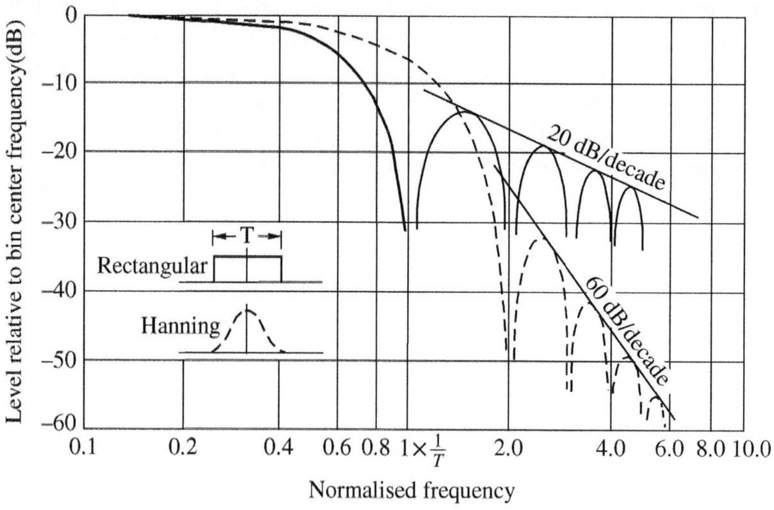

Figure 2.16 Comparison of the filter characteristics of the rectangular and Hanning time-weighting functions for a power spectrum (after Randall (1987).

Table 2.5 Properties of the various time weighting functions.

Window type	Highest side lobe (dB)	Side lobe fall off (dB/decade)	Normalised noise bandwidth (bins) B_{en}	Maximum power spectrum amplitude error (dB)	$\rho(1)$[a]	Window total energy (dB)
Rectangular	−13	−20	1.00	3.92	0.5	1.0
Triangular	−27	−40	1.33	1.83	0.25	
Hanning	−32	−60	1.50	1.43	0.167	−4.26
Hamming	−43	−20	1.36	1.75	0.235	−4.01
Blackman	−58	−20	1.73	1.10	0.09	
Blackman–Harris (3-term)	−67	−20	1.70	1.13	0.096	−5.13
Blackman–Harris (4-term)	−92	−20	2.00	0.83	0.038	−5.88
Kaiser–Bessel ($\alpha = 1.74$)	−40	−20	1.41	1.62	0.208	
Kaiser–Bessel ($\alpha = 2.59$)	−60	−20	1.68	1.16	0.103	
Kaiser–Bessel ($\alpha = 3.39$)	−80	−20	1.90	0.9	0.053	
Kaiser–Bessel ($\alpha = 3.5$)	−82	−20	1.93	0.9		
Gaussian ($\alpha = 3.0$)	−55	−20	1.64	1.25	0.106	−4.51
Poisson ($\alpha = 3.0$)	−24	−20	1.65	1.46	0.151	
Poisson ($\alpha = 4.0$)	−31	−20	2.08	1.03	0.074	
Flat-top	−93	−20	3.75	0.01		−7.56
Welch	−21	−36.5	1.20	2.22	0.345	−2.73

a) See Eq. (2.95).

The properties of various weighting functions are summarised in Table 2.5. In the table, the highest side lobe is the number of decibels (in the auto power spectrum) by which the signal corresponding to the highest side lobe will be attenuated compared to a signal at the filter centre frequency. The 'side lobe' fall off is illustrated in Figure 2.16, where the side lobes are the peaks to the right of the normalised frequency of 1.0.

The noise bandwidth in Table 2.5 is an important quantity in spectrum analysis. It is defined as the bandwidth of a rectangular filter that would let pass the same amount of broadband noise. It is especially useful in calculating the RMS level of power in a certain bandwidth in a spectrum such as a 1/3-octave band. This is discussed in more detail in Section 2.4.4.

The maximum amplitude error is the amount that a signal will be attenuated when it has a frequency that lies exactly mid-way between the centre frequencies of two adjacent filters (corresponding to normalised frequency of 0.5 in Figure 2.16).

The best weighting function for amplitude accuracy in the frequency domain is the flat top (the name refers to the weighting in the frequency domain, whereas the window shape refers to the weighting in the time domain). This is often used for calibration of instrumentation because of its uniform flat frequency response over the bandwidth, $B(= 1/T)$, which results in the measured amplitude of the spectral component being independent of small variations in signal frequency around the band centre frequency, thus making this window suitable for instrument calibration with a tonal signal.

However, the flat top window provides poor frequency resolution (distinction between adjacent frequency lines). Maximum frequency resolution (and minimum amplitude accuracy) is achieved with the rectangular window, so this is sometimes used to separate two spectral peaks that have a similar amplitude and a small frequency spacing. Good compromises that are commonly used are the Hanning window and the Kaiser–Bessel window, the former used when amplitude accuracy is more important and the latter for separation of closely spaced frequencies.

When transient signals (signals that occur for a time shorter than the sampling interval) are to be analysed, the best window is a rectangular one. However, if the transients are repetitive and several occur during a data sampling period, then a Hanning weighting function may be used.

When a Fourier analysis is performed in practice using the DFT algorithm, the resulting frequency spectrum is divided into a number of bands of finite width. Each band may be considered as a filter, the shape of which is dependent on the weighting function used. If the frequency of a signal falls in the middle of a band, its amplitude will be measured accurately. However, if it falls midway between two bands, the error in power spectrum amplitude varies from 0.0 dB for the flat top window to 3.9 dB for the rectangular window. At the same time, the frequency bands obtained using the flat top window are 3.77 times wider so the frequency resolution is 3.77 times poorer than for the rectangular window. In addition, a signal at a particular frequency will also contribute to the energy in other nearby bands, as can be seen by the shape of the filter curve in Figure 2.16. This effect is known as spectral leakage and it is minimised by having a high negative value for the side lobe fall off in Table 2.5.

Amplitude Scaling to Compensate for Window Effects

The effect of a non-rectangular window is to remove information and energy from the signal, resulting in an amplitude error. This must be compensated for by using an amplitude correction factor, A_f. This amplitude correction factor is used to calculate a scaled *amplitude* spectrum $X_s(f_n)$ from Eq. (2.57) as follows

$$X_s(f_n) = \frac{A_f}{N} \sum_{k=0}^{N-1} x(t_k)\, w(k)\, \mathrm{e}^{\mathrm{j}2\pi kn/N} \qquad n = 0, 1, \ldots (N-1) \tag{2.68}$$

where N is the number of discrete frequency components in the spectrum and

$$A_f = \frac{N}{\sum\limits_{k=0}^{N-1} w(k)} \tag{2.69}$$

In the above equation, $w(k)$ is the window weighting function for each sample, k, in the time domain used to calculate the frequency spectrum. If $w(k) = 0$, then the kth sample value is set equal to 0 and if $w(k) = 1$, the sample value is unchanged.

The scaling factor, S_A for the amplitude spectrum in Eq. (2.68) is thus

$$S_A = \begin{cases} A_f^2/N^2 & n > 0 \\ A_f^2/(2N^2) & n = 0 \end{cases} \tag{2.70}$$

For an auto power spectrum or PSD, the scaling factor to be used in Eq. (2.59) has an additional term to that of Eq. (2.70) to account for the leakage of energy into adjacent bins as a result of application of a windowing function. Thus,

$$S_A = \begin{cases} A_f^2/(N^2 B_{\text{en}}) & n > 0 \\ A_f^2/(2N^2 B_{\text{en}}) & n = 0 \end{cases} \tag{2.71}$$

where B_{en} is defined by

$$B_{\text{en}} = \frac{N \sum_{k=0}^{N-1} w^2(k)}{\left(\sum_{k=0}^{N-1} w(k) \right)^2} \tag{2.72}$$

Eqs. (2.59), (2.71) and (2.72) are described as a Welch estimate of the PSD and represent the most commonly used method of spectral analysis in instrumentation and computer software. One example is the **pwelch** function in MATLAB®. However, this method of obtaining the PSD has associated bias errors that decrease as the frequency resolution in the original frequency spectrum, $X(f_n)$, becomes finer (that is, smaller). An estimate of the bias error, ϵ_b, for the particular case of a Hanning window with a frequency resolution of Δf is (Schmidt 1985a,1985b),

$$\epsilon_b \approx \frac{(\Delta f)^2 G''_{xx}(f_n)}{6 G_{xx}(f_n)} + \frac{(\Delta f)^4 G''''_{xx}(f_n)}{72 G_{xx}(f_n)} \tag{2.73}$$

where the prime represents differentiation with respect to frequency and four primes represent the fourth derivative. It can be seen from the above that where there are tones (which produce large values of the second derivative in particular), the error will be large.

Each windowing function identified in Table 2.5 requires different equations to calculate the coefficients, $w(k)$. The coefficients represent the quantity that data sample k, in the time series data, is multiplied by before being included in the data set used for taking the FFT. These equations are listed below for each of the windows identified in Table 2.5.

Rectangular Window For a record in the time domain that is a total of N samples in length (producing N discrete frequency components in the frequency domain), the window coefficients corresponding to each sample k in the time series record, are given by

$$w(k) = 1 \quad 1 \le k \le N \tag{2.74}$$

Triangular Window For a record that is a total of N samples in length, the window coefficients corresponding to each sample k in the record for N odd are given by

$$w(k) = \begin{cases} 2k/(N+1); & 1 \le k \le (N+1)/2 \\ 2 - 2k/(N+1); & \dfrac{(N+1)}{2} + 1 \le k \le N \end{cases} \tag{2.75}$$

and for N even, the coefficients are given by

$$w(k) = \begin{cases} (2k-1)/N; & 1 \le k \le N/2 \\ 2-(2k-1)/N; & \dfrac{N}{2}+1 \le k \le N \end{cases} \tag{2.76}$$

The coefficients of another version of a triangular window are given by

$$w(k) = 1 - \left|\frac{2k-N+1}{L}\right| \tag{2.77}$$

where L can be $N, N+1$, or $N-1$. All alternatives converge for large N.

Hanning Window The Hanning window is the one most commonly used in spectrum analysis and is the one recommended for PSD analysis. For a record that is a total of N samples in length, the window coefficients corresponding to each sample k in the record are given by

$$w(k) = 0.5\left[1-\cos\left(\frac{2\pi k}{N-1}\right)\right] \quad 0 \le k \le N-1 \tag{2.78}$$

Hamming Window For a record that is a total of N samples in length, the window coefficients corresponding to each sample k in the record are given by

$$w(k) = \alpha - \beta\cos\left(\frac{2\pi k}{N-1}\right) \quad 0 \le k \le N-1 \tag{2.79}$$

where the optimum values for α and β are 0.54 and 0.46 respectively.

Blackman Window For a record that is a total of N samples in length, the window coefficients corresponding to each sample k in the time series record are given by

$$w(k) = a_0 - a_1\cos\left(\frac{2\pi k}{N-1}\right) + a_2\cos\left(\frac{4\pi k}{N-1}\right) \quad 0 \le k \le N-1 \tag{2.80}$$

where $a_0 = (1-\alpha)/2$, $a_1 = 1/2$, and $a_2 = \alpha/2$. For an unqualified Blackman window, $\alpha = 0.16$.

Blackman–Harris Window For a record that is a total of N samples in length, the window coefficients corresponding to each sample k in the time series record for a three-term window are given by

$$w(k) = a_0 - a_1\cos\left(\frac{2\pi k}{N-1}\right) + a_2\cos\left(\frac{4\pi k}{N-1}\right) \quad 0 \le k \le N-1 \tag{2.81}$$

where $a_0 = 0.42323$, $a_1 = 0.49755$ and $a_2 = 0.07922$.

For a four-term window, the equation for the coefficients is given by

$$w(k) = a_0 - a_1\cos\left(\frac{2\pi k}{N-1}\right) + a_2\cos\left(\frac{4\pi k}{N-1}\right) - a_3\cos\left(\frac{6\pi k}{N-1}\right) \quad 0 \le k \le N-1 \tag{2.82}$$

where $a_0 = 0.35875$, $a_1 = 0.48829$, $a_2 = 0.14128$ and $a_3 = 0.01168$.

Kaiser–Bessel Window For a record that is a total of N samples in length, the window coefficients corresponding to each sample k in the record are given by

$$w(k) = \frac{I_0\left(\pi\alpha\sqrt{1-\left(\frac{2k}{N-1}-1\right)^2}\right)}{I_0(\pi\alpha)} \qquad 0 \le k \le N-1 \tag{2.83}$$

where I_0 is the zero-th order modified Bessel function of the first kind. The parameter α determines the tradeoff between main lobe width and side lobe levels. Increasing α widens the main lobe and increases the attenuation of the sidelobes. To obtain a Kaiser window that provides an attenuation of β dB for the first sidelobe, the following values of $\pi\alpha$ are used

$$\pi\alpha = \begin{cases} 0.1102(\beta - 8.7); & \beta > 50 \\ 0.5842(\beta - 21)^{0.4} + 0.07886(\beta - 21); & 21 \le \beta \le 50 \\ 0; & \beta < 21 \end{cases} \tag{2.84}$$

A typical value of α is 3, and the main lobe width between the nulls is given by $2\sqrt{1+\alpha^2}$.

Gaussian Window For a record that is a total of N samples in length, the window coefficients corresponding to each sample k in the record are given by

$$w(k) = \exp\left[-\frac{1}{2}\left(\frac{k-(N-1)/2}{\sigma(N-1)/2}\right)^2\right]; \quad 0 \le k \le N-1 \quad \text{and} \quad \sigma \le 0.5 \tag{2.85}$$

where σ is the standard deviation of the Gaussian distribution. An alternative formulation is given by

$$w(k) = \exp\left[-\frac{1}{2}\left(\frac{\alpha k}{(N-1)/2}\right)^2\right] \qquad -(N-1)/2 \le k \le (N-1)/2 \tag{2.86}$$

where $\alpha = (N-1)/2\sigma$ and $\sigma \le 0.5$.

Poisson (or exponential) Window For a record that is a total of N samples in length, the window coefficients corresponding to each sample k in the record are given by

$$w(k) = \exp\left(-\left|k - \frac{N-1}{2}\right|\frac{1}{\tau}\right) \qquad 0 \le k \le N-1 \tag{2.87}$$

For a targeted decay of D dB over half of the window length, the time constant τ is given by

$$\tau = \frac{8.69N}{2D} \tag{2.88}$$

An alternative formulation is given by

$$w(k) = \exp\left(\frac{-\alpha|k|}{(N-1)/2}\right); \quad -(N-1)/2 \le k \le (N-1)/2 \tag{2.89}$$

where $\exp\ \{x\} = e^x$ and $\alpha = (N-1)/2\tau$.

Flat-top Window In frequency analysis, the flat-top window is used mainly for calibration of instrumentation with a tone. The frequency of the calibration tone need not be close to the centre frequency of a bin in the frequency spectrum to obtain an accurate result; in fact an accurate result will be obtained for any calibration frequency.

For a record that is a total of N samples in length, the window coefficients corresponding to each sample k in the record are given by

$$w(k) = a_0 + a_1 \cos\left(\frac{2\pi k}{N-1}\right) + a_2 \cos\left(\frac{4\pi k}{N-1}\right) + a_3 \cos\left(\frac{6\pi k}{N-1}\right) + a_4 \cos\left(\frac{8\pi k}{N-1}\right) \quad 0 \le k \le N-1 \tag{2.90}$$

The coefficients are:

- $a_0 = 0.21557895$
- $a_1 = 0.41663158$
- $a_2 = 0.277263158$
- $a_3 = 0.083578947$
- $a_4 = 0.006947368$.

The coefficients may also be normalized to give $a_0 = 1.0$, $a_1 = 1.93$, $a_2 = 1.29$, $a_3 = 0.388$ and $a_4 = 0.028$.

Welch Window For a record that is a total of N samples in length, the window coefficients corresponding to each sample, k in the record are given by

$$w(k) = 1 - \left(\frac{2k - N + 1}{N-1}\right)^2 \tag{2.91}$$

Power Correction and RMS Calculation

It is often of interest to determine the RMS value or decibel level of an auto power spectrum over a defined frequency range within the spectrum. For example, one may wish to compute 1/3-octave or octave band decibel levels from an auto power spectrum covering the range from 1–10 000 Hz. One may think that all one needs to do is add logarithmically (see Section 2.2.8) the various frequency components contained within the band of interest to obtain the required result. Adding logarithmically the frequency component amplitudes in an auto power spectrum is the same as converting the decibel amplitude levels to RMS^2 quantities, adding them together and then converting back to decibels. Unfortunately, finding the RMS value or decibel level of a band within a spectrum is not as simple as described above, as energy from each frequency bin leaks into adjacent bins, giving a result that is too large. Thus the calculation needs to be divided by a correction factor, B_{en}, defined in Eq. (2.72), which is different for each windowing function. The correction factor, called the *normalised noise bandwidth* is listed in Table 2.5 for various windowing functions. B_{en} is a constant for a particular window and for a Hanning window it is 1.5. The RMS value of an auto spectrum between frequency locations f_{n_1} and f_{n_2} is thus given by

$$x_{\text{RMS}}(n_1, n_2) = \sqrt{\frac{\sum_{n=n_1}^{n_2} G_{xx}(f_n)}{B_{\text{en}}}} \tag{2.92}$$

where $G_{xx}(f_n)$ is defined in Eq. (2.63). The correction factor holds even if the spectrum consists of only a single tone. This is because even for a single tone, energy appears in the two frequency bins adjacent to the one containing the frequency of the tone.

Calculating the RMS value or decibel level of a PSD requires a slightly different process because the PSD is already scaled such that the area under the curve corresponds to the mean square of the signal. In this case the RMS value of a signal between two spectral lines, n_1 and n_2, is given by

$$x_{\mathrm{RMS}}(n_1, n_2) = \sqrt{\Delta f \sum_{n=n_1}^{n_2} G_{Dxx}(f_n)} \tag{2.93}$$

where Δf is the frequency resolution used to obtain the original auto power spectrum that is converted to a PSD (which is the same as the PSD frequency resolution) and $G_{Dxx}(f_n)$ is the single-sided PSD, which is defined in Eq. (2.64).

2.4.5 Sampling Frequency and Aliasing

The sampling frequency is the frequency at which the input signal is digitally sampled. If the signal contains frequencies greater than half the sampling frequency, then these will be "folded back" and appear as frequencies less than half the sampling frequency. For example, if the sampling frequency is 20 000 Hz and the signal contains a frequency of 25 000 Hz, then the signal will appear in the spectrum as 5000 Hz. Similarly, if the signal contains a frequency of 15 000 Hz, this signal also will appear as 5000 Hz. This phenomenon is known as 'aliasing' and in a spectrum analyser it is important to have analogue filters that have a sharp roll-off for frequencies above about 0.4 of the sampling frequency. Aliasing is illustrated in Figure 2.17.

2.4.6 Overlap Processing

When a limited amount of time is available for collecting data, more accurate results can be obtained by implementing overlap processing, as this allows more spectra to be averaged. For this case, the time series data are divided into a number of records and then a DFT is performed on each segment. For an example case of a 50% overlap, this means that the first segment analysed is the first time record, the second segment is the second half of the first time record appended to the beginning of the first half of the second record, the third segment is the second record, the fourth segment is the second half of the second time record appended to the beginning of the first half of the third record, and so on. Even though the same data are used in more than one DFT, the effect of overlap analysis is to provide more spectra to average, which results in a smaller error in the final averaged spectrum. However, the *effective number* of averages is slightly less than the actual number of averages when overlap processing is used. The effective number of averages is window dependent and can be calculated using

$$\frac{Q_e}{Q_d} = \frac{Q/Q_d}{1 + 2\sum_{i=1}^{Q} \frac{Q-i}{Q}\rho(i)} \tag{2.94}$$

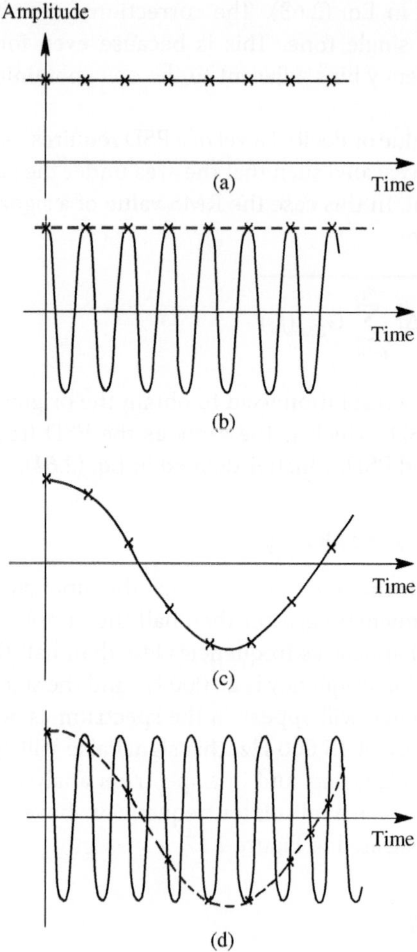

Figure 2.17 Illustration of aliasing: (a) zero-frequency or DC component; (b) spectrum component at sampling frequency f_s interpreted as DC; (c) spectrum component at $(1/N)f_s$; (d) spectrum component at $[(N+1)/N]f_s$ interpreted as $(1/N)f_s$. After Randall (1987).

where Q is the number of overlapping segments used, Q_d is the number of non-overlapping segments in the original data set, Q_e is the effective or equivalent number of averages, which give the same variance or uncertainty in the averaged DFT as the same number of averages using independent data. The quantity, $\rho(i)$ is defined as

$$\rho(i) = \frac{\left[\sum_{k=0}^{N-1} w(k)w(k+iD)\right]^2}{\left[\sum_{k=0}^{N-1} w^2(k)\right]^2} \tag{2.95}$$

where N is the number of discrete frequency components in the spectrum, $w(k)$ is dependent on the windowing function used, $D = round[N(1 - P/100)]$ and P is the percentage overlap. For an overlap percentage up to 50%, the only non-zero value of $\rho(i)$ is when $i = 1$. Values of $\rho(1)$ for various windows are included in Table 2.5 for a 50% overlap.

Overlap processing is particularly useful when constructing sonograms (see Section 6.2.13), as overlap processing results in smaller time intervals between adjacent spectra, resulting in better time resolution. For example, with 50% overlap, three spectra are obtained with the same number of samples, thus representing the same time period as two spectra with non-overlap processing.

Overlap processing can also be used to improve frequency resolution, Δf, by using more samples, N, in the FFT ($\Delta f = f_s/N$) than is used with non-overlap processing for the same number of effective averages. This is useful when there is a limited-length data set and the maximum possible accuracy and frequency resolution is needed.

2.4.7 Zero Padding

Zero padding is the process of adding zeros to extend the number of samples in a record in the time domain prior to taking the DFT. This results in a frequency spectrum with frequency components more tightly spaced, which has resulted in some users thinking that they have achieved a finer frequency resolution. In fact, the apparent finer frequency resolution is actually an interpolation between the frequency bins that would exist with no zero padding, so no more information has been gained. Higher frequency resolution can only be achieved with a longer sampling time. Thus zero padding is not considered a useful tool in this context. Nevertheless, when one is analysing a transient that has zero amplitude outside the sampling time window, zero padding can result in finer frequency resolution than if only the length of the transient had been used.

Note than when zero padding is used, an additional scaling factor has to be used to calculate the amplitude of the resulting spectrum. If the spectrum contains n data points and m zeros, the scaling factor with which the resulting power spectral amplitudes have to be multiplied to get the correct values is $(m + n)/n$.

2.4.8 Uncertainty Principle

The uncertainty principle states that the frequency resolution, Δf, (equal to the effective filter bandwidth, B) of a Fourier transformed signal is equal to the reciprocal of the time, T_A, to acquire the sampled record of the signal. Thus, for a single spectrum, $\Delta f T_A = 1$. An effectively higher $\Delta f T_A$ product can be obtained by averaging several spectra together until an acceptable error is obtained according to Eq. (2.44), where B is the filter bandwidth (equal to Δf or frequency resolution) and T_A is the total sample time.

2.4.9 Time Synchronous Averaging and Synchronous Sampling

Time synchronous averaging is a slightly different process to synchronous sampling, although both are intended for use with noise and vibration signals obtained from rotating equipment such as wind turbines. With time synchronous averaging, the aim is to obtain averaged time-domain data by averaging data samples that correspond to the same angular location of a rotor. This is done prior to taking a DFT. For example, if the rotor is rotating at 16 RPM, the idea would be to use a tachometer signal to indicate when each revolution started and then obtain averaged data at fixed angular intervals by interpolating between data samples, if necessary, when the speed changes. This results in samples that do not correspond to the same angular locations as they did prior to the first speed change since sampling began. In this way, each DFT will be the result of a calculation based on the average of a number of synchronised time samples, so the frequency scale will now be replaced with a scale representing multiples of the fundamental frequency. However, the fundamental and its harmonics will be much more clearly visible than if the time samples were separated by fixed time intervals, especially for the case of a wind turbine characterised by slight speed changes as the wind speed varies. This is especially true for variable speed rotors that characterise more modern turbines. It is still desirable to obtain several records of averaged data so that the resulting auto power spectra can be averaged. As this method of analysis is for the purpose of identifying tonal signals, it is not suitable for PSD calculations.

In the case of wind turbines, it is often better to use the blade-pass frequency instead of using the rotor rotational speed as a basis. For a three-bladed rotor, this would result in an effective rotational speed that was a factor of three greater; 48 RPM for the above example. In this way, the frequency axis will be multiples of the blade-pass frequency rather than multiples of the rotor rotational speed.

Synchronous sampling is slightly different to the process described above. Rather than taking samples at fixed time intervals and then interpolating and averaging in the time domain, synchronous sampling involves sampling the signal at fixed angular increments of the rotating rotor so that when the rotor speed changes the interval between time samples changes. A DFT is then taken of each record in the time domain and the resulting frequency spectrum has multiples of the fundamental rotational frequency along its axis. This method is often referred to as 'order tracking'. Again, it is only used for tonal analysis with the auto power spectrum and, as it is intended for tonal analysis, it is not suitable for obtaining a PSD.

2.4.10 Hilbert Transform

The Hilbert transform is often referred to as envelope analysis, because it involves finding a curve that envelopes the peaks in a signal. The signal may be a time-domain signal or a frequency-domain signal. Any regular harmonic variation in the envelope signal represents an amplitude modulation (AM) of the original signal, which can be quantified by taking an FFT of the envelope signal. In wind farm noise applications, the Hilbert transform is applied to a time series signal as one means of quantifying AM (see Section 2.2.7).

The Hilbert transform applied to a time-domain (or time series) signal, $x(t)$, can be represented mathematically as

$$\mathcal{H}\{x(t)\} = \tilde{x}(t) = \frac{1}{\pi}\int_{-\infty}^{\infty} x(\tau)\left(\frac{1}{t-\tau}\right)\mathrm{d}\tau = \frac{1}{\pi}x(t) * \left(\frac{1}{t}\right) \tag{2.96}$$

where $*$ represents the convolution operator (see Section 2.4.15).

The Fourier transform of $\tilde{x}(t)$ is given by

$$\tilde{X}(f) = -\mathrm{j}\ \mathrm{sign}(f) X(f) \tag{2.97}$$

or alternatively,

$$\tilde{X}(f) = \begin{cases} \mathrm{e}^{-\mathrm{j}\pi/2} X(f) & f > 0 \\ 0 & f = 0 \\ \mathrm{e}^{\mathrm{j}\pi/2} X(f) & f < 0 \end{cases} \tag{2.98}$$

An amplitude modulated time-domain signal, $m(t)$, (see Section 2.2.7), can be calculated from the Hilbert transform of the total time-domain signal, $x(t)$, which for our purposes can represent sound pressure as a function of time. To calculate the envelope function, $e(t)$, which is the envelope of the original signal, the following equation is used (Brandt 2010):

$$e(t) = \sqrt{x^2(t) + \tilde{x}^2(t)} \tag{2.99}$$

An envelope for an amplitude modulated signal is shown in Figure 2.18.

The Hilbert transform, $\tilde{x}(t)$ can be calculated from the ordinary DFT, $X(f_n)$, of the original signal as follows (Brandt 2010):

$$\tilde{x}(k) = \frac{2}{N} \mathrm{Im}\left[\sum_{n=0}^{N/2} X(f_n) \mathrm{e}^{\mathrm{j}2\pi nk/N}\right] \tag{2.100}$$

where N is the total number of samples used to calculate the Fourier transform and Im[ ] represents the imaginary part of the complex number in brackets (see Appendix A).

2.4.11 Cross-spectrum

The cross-spectrum is a measure of how much one noise represented in the time domain as $x(t)$ and frequency domain as $X(f_n)$ may be related to another noise represented in the time domain as $y(t)$ and frequency domain as $Y(f_n)$. For example, it may be used to determine the extent to which indoor noise, $y(t)$, in a residence may be caused by exterior

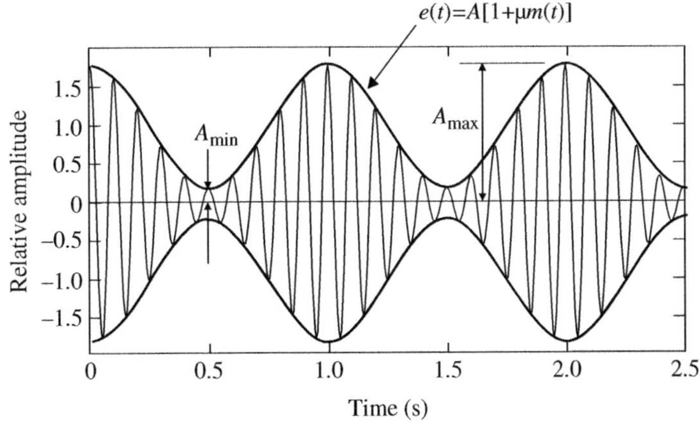

Figure 2.18 Envelope analysis with the Hilbert transform.

noise, $x(t)$, where $y(t)$ is considered to be the system output and $x(t)$ the system input. It is also used to estimate the coherence between two noise signals, which is another way of quantifying by how much one signal is related to another.

The two-sided cross-spectrum and cross-spectral density (CSD) have similar forms to the auto power spectrum and PSD of Eqs. (2.58) and (2.59), respectively, and the single-sided cross-spectrum, $S_{xy}(f_n)$ is calculated from the two-sided spectrum in a similar way to the auto power spectrum. Thus,

$$S_{xy}(f_n) \approx \frac{S_A}{Q} \sum_{i=1}^{Q} X_i^*(f_n) Y_i(f_n) \quad n = 0, 1, \ldots (N-1) \tag{2.101}$$

where i is the spectrum number, $X_i(f_n)$ and $Y_i(f_n)$ are complex spectral components corresponding to frequency f_n and Q is the number of spectra over which the average is taken. The larger the value of Q, the more closely will the estimate of $S_{xy}(f_n)$ approach its true value. In the equation, the asterisk * represents the complex conjugate, $X_i(f_n)$ and $Y_i(f_n)$ are instantaneous spectra and $S_{xy}(f_n)$ is estimated by averaging over a number of instantaneous spectrum products obtained with finite time records of data. In contrast to the auto power spectrum, which is real, the cross-spectrum is complex, characterised by an amplitude and a phase.

In practice, the amplitude of $S_{xy}(f_n)$ is the product of the two amplitudes $|X(f_n)|$ and $|Y(f_n)|$ and its phase is the difference in phase between $X(f_n)$ and $Y(f_n)$ $(= \theta y - \theta x)$. This function can be averaged because, for stationary signals, the relative phase between $x(t)$ and $y(t)$ is fixed and not random.

The *cross-spectral density*, $S_{Dxy}(f_n)$ (or CSD(f_n)) can be obtained from the cross-spectrum by dividing the amplitudes of each frequency component by the frequency spacing, Δf, between adjacent components in the cross-spectrum or by multiplying by the time, T, to acquire one record of data. Thus, the two-sided CSD is:

$$S_{Dxy}(f_n) \approx \frac{TS_A}{Q} \sum_{i=1}^{Q} X_i^*(f_n) Y_i(f_n) \quad n = 0, 1, \ldots (N-1) \tag{2.102}$$

The CSD can also be obtained directly from the cross-correlation function, which is defined as

$$R_{xy}(\tau) = \langle x(t)y(t+\tau) \rangle_t = \mathrm{E}[x(t)y(t+\tau)] = \lim_{T\to\infty} \frac{1}{T} \int_{-T/2}^{T/2} x(t)y(t+\tau)\mathrm{d}t \tag{2.103}$$

where E[x] is the expected value of x. Using the cross-correlation function, the CSD may be written as

$$S_{Dxy}(f) = \int_{-\infty}^{\infty} R_{xy}(\tau) \mathrm{e}^{-\mathrm{j}2\pi f\tau} \mathrm{d}\tau \tag{2.104}$$

and in the sampled domain (corresponding to Figure 2.15d),

$$S_{Dxy}(f_n) = \frac{1}{N} \sum_{k=0}^{N-1} R_{xy}(\tau_k) \mathrm{e}^{-\mathrm{j}2\pi nk/N} \tag{2.105}$$

In practice, the two-sided cross-spectrum, $S_{xy}(f_n)$, is expressed in terms of the one-sided cross-spectrum $G_{xy}(f_n)$, where

$$G_{xy}(f_n) = \begin{cases} 0 & f_n < 0 \\ S_{xy}(f_n) & f_n = 0 \\ 2S_{xy}(f_n) & f_n > 0 \end{cases} \tag{2.106}$$

A similar expression may be written for the single-sided CSD, $G_{Dxy}(f_n)$, as follows

$$G_{Dxy}(f_n) = \begin{cases} 0 & f_n < 0 \\ S_{Dxy}(f_n) & f_n = 0 \\ 2S_{Dxy}(f_n) & f_n > 0 \end{cases} \tag{2.107}$$

Note that $G_{xy}(f_n)$ and $G_{Dxy}(f_n)$ are complex, with real and imaginary parts referred to as the co-spectrum and quad-spectrum respectively. They are usually expressed in terms of an amplitude and a phase or as a real and imaginary part. As for auto power spectra and PSDs, the accuracy of the estimate of the cross-spectrum improves as the number of records over which the averages are taken increases. The statistical error for a stationary Gaussian random signal is given as (Randall 1987):

$$\epsilon = \frac{1}{\sqrt{\gamma_{xy}^2(f_n) Q}} \qquad n = 0, 1, \ldots (N-1) \tag{2.108}$$

where $\gamma_{xy}^2(f_n)$ is the coherence function relating noise signals, $x(t)$ and $y(t)$ (see Section 2.4.12), and Q is the number of averages.

The amplitude of $G_{xy}(f_n)$ gives a measure of how well the two functions $x(t)$ and $y(t)$ correlate as a function of frequency and the phase angle of $G_{xy}(f_n)$ is a measure of the phase shift between the two signals as a function of frequency.

2.4.12 Coherence

The coherence function is a measure of the degree of linear dependence between two signals, as a function of frequency. It is calculated from the two auto power spectra and the cross-spectrum as follows

$$\gamma_{xy}^2(f_n) = \frac{|G_{yx}(f_n)|^2}{G_{xx}(f_n) G_{yy}(f_n)} = \frac{|G_{xy}(f_n)|^2}{G_{xx}(f_n) G_{yy}(f_n)} \qquad n = 0, 1, \ldots (N-1) \tag{2.109}$$

By definition, $\gamma^2(f_n)$ varies between 0 and 1, with 1 indicating a high degree of linear dependence between the two signals, $x(t)$ and $y(t)$. Thus in a physical system where $y(t)$ is the output and $x(t)$ is the input signal that causes the output, the coherence is a measure of the degree to which $y(t)$ is linearly related to $x(t)$. If random noise is present in either $x(t)$ or $y(t)$, then the value of the coherence will diminish. Other causes of a diminished coherence are insufficient frequency resolution in the frequency spectrum or poor choice of window function. A further cause of diminished coherence is a time delay, of the same order as the length of the record, between $x(t)$ and $y(t)$.

The main application of the coherence function is in checking the validity of frequency-response measurements (see Section 2.4.13). Another more direct application is the calculation of the signal, S, to noise, N, ratio as a function of frequency

$$S/N = \frac{\gamma_{xy}^2(f_n)}{1 - \gamma_{xy}^2(f_n)} \quad n = 0, 1, \ldots(N-1) \tag{2.110}$$

In this relatively narrow definition of S/N, the 'signal' is the component in the response, $y(t)$, that is caused by $x(t)$; 'noise' refers to anything else in $y(t)$. The coherence will always be unity by definition if only one spectrum (rather than the average of many spectra) is used to calculate $G_{xx}(f_n)$, $G_{yy}(f_n)$ and $G_{yx}(f_n)$.

For cases in which there are many inputs and one or more outputs, it is of interest to estimate the degree of correlation between one group of selected inputs $X(n) = [x_1(n), \ldots, x_m(n)]$ and one output, $y[n]$. This is the basis of the concept of multiple coherence, defined as (Potter 1977):

$$\gamma_{Xy}^2 = \frac{G_{Xy}^{\mathrm{H}}(f_n) G_{XX}^{-1}(f_n) G_{Xy}(f_n)}{G_{yy}(f_n)} \tag{2.111}$$

where γ_{Xy}^2 is the multiple coherence function between the vector of inputs X and the output, y, G_{Xy} is the m dimensional vector of the cross-spectrum between the inputs, X and the output, y, G_{XX} is the $m \times m$ matrix of the power spectrum and cross-spectrum of the vector of inputs, and G_{yy} is the power spectrum of the output. The power spectrum and cross-spectrum quantities in Eq. (2.111) can be replaced with their equivalent PSD terms with no change in result.

2.4.13 Frequency-response (or Transfer) Function

The frequency-response function $H(f_n)$ (or FRF) is defined as:

$$H(f_n) = \frac{Y(f_n)}{X(f_n)} \quad n = 0, 1, \ldots(N-1) \tag{2.112}$$

The FRF is the Fourier transform of the system impulse-response function, $h(t_k)$. It is a convenient way of quantifying the relative amplitude of and the phase between two signals as a function of frequency. The impulse response of a system is the system output as a function of time following an impulse input (a very short, sudden input).

In practice, it is desirable to average $H(f_n)$ over a number of spectra, but as $Y(f_n)$ and $X(f_n)$, $n = 0, 1, \ldots, (N-1)$ are both instantaneous spectra, it is not possible to average either of them. For this reason it is convenient to modify Eq. (2.112). There are a number of possibilities, one of which is to multiply the numerator and denominator by the complex conjugate of the input spectrum. Thus

$$H_1(f_n) = \frac{Y(f_n)X^*(f_n)}{X(f_n)X^*(f_n)} = \frac{G_{xy}(f_n)}{G_{xx}(f_n)} \quad n = 0, 1, \ldots(N-1) \tag{2.113}$$

A second version is found by multiplying with $Y^*(f)$ instead of $X^*(f)$. Thus

$$H_2(f_n) = \frac{Y(f_n)Y^*(f_n)}{X(f_n)Y^*(f_n)} = \frac{G_{yy}(f_n)}{G_{yx}(f_n)} \quad n = 0, 1, \ldots(N-1) \tag{2.114}$$

Either of the above two forms of FRF are amenable to averaging, but $H_1(f_n)$ is the preferred version if the output signal, $y(t)$, is more contaminated by noise than the input signal $x(t)$, whereas $H_2(f_n)$ is preferred if the input signal, $x(t)$, is more contaminated by noise than the output (Randall 1987).

2.4.14 Coherent Output Power

Coherent output power calculations allow one to determine what contribution a particular sound source may be making to a particular acoustic measurement. This has particular relevance to wind turbine measurements, where it is desirable to estimate the contribution of wind farm noise to the total noise levels measured at a residential location several kilometres away. The coherent output power process can also be used to eliminate extraneous noise from a signal.

Although coherent power calculations for obtaining the contribution of a particular sound source to an acoustic measurement are relatively simple, practical implementation of the procedure requires considerable care. To be able to determine the coherent output power of a sound source, it is necessary to be able to obtain an uncontaminated signal of just the noise source itself. This usually requires a measurement to be made close to the noise source. If the auto power spectrum of the measurement made close to a wind farm is $G_{xx}(f_n), n = 0, 1, \ldots, (N-1)$ and the auto power spectrum of the contaminated measurement at the residence is $G_{yy}(f_n)$, then the auto spectrum, $G_{cc}(f_n)$ of the contribution of the wind farm at the residence is given by

$$G_{cc}(f_n) = |H_1(f_n)|^2 G_{xx}(f_n) = \frac{|G_{xy}(f_n)|^2}{G_{xx}(f_n)} \quad n = 0, 1, \ldots (N-1) \tag{2.115}$$

where N is the number of data points in the auto power spectrum. $G_{cc}(f_n)$ can also be written in terms of the coherence between the two signals, so that

$$G_{cc}(f_n) = \gamma^2(f_n) G_{yy}(f_n) \quad n = 0, 1, \ldots (N-1) \tag{2.116}$$

Similar relationships also apply for power spectral densities with the functions $G(f_n)$ replaced with $G_D(f_n)$ in the above two equations. It is assumed in all relationships that the measurement, $G_{xx}(f_n)$ is uncontaminated by noise not related to the wind farm. In practice, a small amount of contamination leads to the results being slightly less accurate.

It is possible to use Eqs. (2.115) and (2.116) to calculate the relative contributions of a number of sources to a particular sound pressure measurement, provided the signals $G_{xx}(f_n), n = 0, 1, \ldots, (N-1)$ acquired near each sound source do not contaminate one another significantly.

Obtaining a measurement of the wind farm noise at a large distance from the residence at the same time as the measurement is taken close to the wind farm so that a cross-spectrum can be calculated produces some logistical problems, which can be overcome with some attention to detail. First, it is important that the time taken to acquire one record of data should be much longer (by a factor of at least 2) than the time taken for the sound to travel from the wind farm to the residence. For example, the time for sound to travel from a microphone near the wind farm to a residence located 3 km away is 3000/343 s. In this case, the time to acquire a record of data must be longer than about 18 s.

The second major problem with finding the cross-spectrum of signals from two microphones located 3 km apart is that it is not feasible to have a connecting cable that is 3 km or even 1.5 km long. The alternative is to transmit the signal wirelessly using an FM transmitter and receiver circuits at the distant microphone and recording instrument, respectively.

2.4.15 Convolution

Convolution in the time domain is the operation that is equivalent to multiplication in the frequency domain. Multiplication of two functions, $X(f)$ and $H(f)$, in the frequency domain $\{X(f)H(f)\}$ involves multiplying the spectral amplitude of each frequency component in $X(f)$ with the amplitude of the corresponding frequency component in $H(f)$. The equivalent operation of convolution of the two time-domain signals, $x(t) * h(t)$ is a little more complex. That is,

$$y(t) = x(t) * h(t) = \int_{-\infty}^{\infty} x(\tau)h(t - \tau)\mathrm{d}\tau \tag{2.117}$$

The Fourier transform of the convolved signal, $x(t) * h(t)$ is $\{X(f)H(f)\}$.

Deconvolution is the process of determining $h(t)$ of Eq. (2.117) from known signals, $y(t)$ and $x(t)$. For example, $h(t)$ may be a system impulse response that is to be determined from input and output signals. Deconvolution is often performed in the frequency domain so that in the absence of any significant noise in the signals,

$$H(f) = Y(f)/X(f) \tag{2.118}$$

The inverse Fourier transform is then taken of $H(f)$ to obtain $h(t)$. If noise is present in the output signal, $y(t)$, then the estimate for $h(t)$ will be in error. The error may be reduced using Weiner deconvolution, but this is a complex operation and beyond the scope of this book.

If the two time functions are represented by sampled data, $x[n], n = 0, 1, 2, \ldots (N - 1)$ and $h[m], m = 0, 1, 2 \ldots (M - 1)$, such as obtained by a digital data acquisition system, the output of the convolution, $y[k], k = 0, 1, 2, \ldots (N + M - 2)$, of the two signals at sample number k, is given by

$$y[k] = x[k] * h[k] = h[k] * x[k] = \sum_{n=0}^{N-1} x[n]h[k - n] \tag{2.119}$$

where terms in the sum are ignored if $[k - n]$ lies outside the range from 0 to $(M - 1)$.

Convolution is often used to define the relationship between three signals of interest: the input to a system, $x[k]$, the system impulse response, $h[k]$, and the output, $y[k]$, from the system, as indicated in Eq. (2.119). As can be seen from the equation, each sample in the input signal contribute to many samples in the output signal.

With sampled data, the DFT is used to obtain frequency spectra. However, the inverse discrete Fourier transform (IDFT) of the product of two spectra in the frequency domain will no longer represent a linear convolution in the time domain. Instead, the time-domain equivalent is circular convolution, which results from the associated periodicity of the DFT and is not equal to the linear convolution. The circular convolution operation is denoted by ⊛ rather than by $*$ for normal linear convolution. Thus,

$$y(k) = \mathrm{IDFT}\{X(f_n)H(f_n)\} = x[k] \circledast h[k] = \sum_{n=0}^{N-1} x[n]h[((k - n))_N] \tag{2.120}$$

where $[((k-n))_N]$ is calculated as $[k-n]$ mod N, meaning that $[((k-n))_N]$ is the integer remainder after $[k-n]$ has been divided by ℓN, where ℓ is the largest integer possible for $[k-n] \geq \ell N$. If $[k-n]$ is negative, then ℓ will be negative. Note that, for example, -1 is a larger integer than -2. If $\ell = 0$, $[((k-n))_N] = [k-n]$, a positive number.

A circular convolution is obtained when an IDFT is taken of the product of two DFTs that correspond to two time-domain signals. This circular convolution can be converted to a linear convolution over one cycle of the DFT by using zero padding (see Section 2.4.7) of the two original time-domain signals before taking their DFT. This is discussed by (Brandt 2010, Ch. 9).

2.4.16 Auto-correlation and Cross-correlation Functions

The auto-correlation function of a signal, $x(t)$, in the time domain is a measure of how similar a signal is to a time-shifted future or past version of itself. If the time shift is τ seconds, then the auto-correlation function is

$$R_{xx}(\tau) = \mathrm{E}[x(t)x(t+\tau)] \tag{2.121}$$

If $\tau = 0$, the auto-correlation function is equal to the variance of the signal, σ_x^2.

The cross-correlation between two different time signals, $x(t)$ (the input) and $y(t)$ (the output) is

$$R_{xy}(\tau) = \mathrm{E}[x(t)y(t+\tau)] \tag{2.122}$$

Auto-correlation and cross-correlation function definitions are applicable to random noise signals only if the average value of one record of samples is not much different to that for subsequent records or the average of many records. Random noise signals typically satisfy this condition provided that the signal does not vary too much.

The cross-correlation function can be used to find the acoustic delay between two signals originating from the same source. In this case, the delay is the time difference represented by the maximum value in the cross-correlation function. If the speed of sound is known, the delay allows one to determine the distance between the two microphones that are providing the two signals.

The following relationships hold for real signals, $x(t)$ and $y(t)$.

$$R_{xx}(\tau) = R_{xx}(-\tau) \tag{2.123}$$

and

$$R_{yx}(\tau) = R_{xy}(-\tau) \tag{2.124}$$

As direct calculation of correlation functions is difficult and resource intensive, they are usually estimated from spectra. This is done by first estimating a PSD as in Eqs. (2.59) and (2.64), where the function $X_i(f_n)$ has had a rectangular window applied (which is effectively a multiplication by unity, as all $w(k)$ in Eq. (2.68) are unity for a rectangular window). The PSD needs to be zero padded with as many zeros as data points to avoid circular convolution, as explained in detail by (Brandt 2010, Ch. 9).

We first compute a PSD estimate using Q spectrum averages as in Eq. (2.59), and for each spectrum, we use a $2N$ record length with the last half of the samples set equal to zero. So we obtain for the single-sided PSD,

$$G_{Dxx}(f_n) = \frac{4S_A}{Q\Delta f}\sum_{i=1}^{Q}|X_{iz}(f_n)|^2 \tag{2.125}$$

where the subscript, z indicates that the spectrum has been calculated with an equal number of zeros as data samples, added to the data set. $\Delta f = f_s/N, f_s$ is the sample rate, N is the number of samples in each data segment or record, $X_i(f_n)$ is an unscaled spectrum and the single-sided PSD, G_{Dxx}, has been multiplied by 2 to account for the added zeros. As we have N data samples in each record, and another N zeros added to the end, the total length of each record is $2N$. As a rectangular window has been used, the scaling factor, S_A is simply equal to $1/(4N^2)$, as there are $2N$ samples in the record. The ith unscaled, zero-padded spectrum, $X_{iz}(f_n)$, obtained by passing time samples through a rectangular window, is given by

$$X_{iz}(f_n) = \sum_{k=0}^{2N-1} x_i(t_k)\mathrm{e}^{-\mathrm{j}2\pi kn/N}; \quad n = 0, 1, \ldots, (2N-1) \tag{2.126}$$

where x_i represents the ith complete data record in the time domain, which has been doubled in size using zero padding, t_k is the time that the kth sample was acquired and the subscript, z indicates a spectrum with zero padding.

The IDFT is taken of Eq. (2.125) to obtain the quantity, $\hat{R}_{xx}(t_k)$ and this is used to compute an estimate, $\hat{R}_{xx}(i)$, for the auto-correlation function for the ith record, using (Brandt 2010):

$$\hat{R}_{xx}(i) = \begin{cases} \dfrac{N-k}{N\Delta t}\hat{R}_{xx}(k) & k = 0, 1, \ldots, (N-1) \\[2ex] \dfrac{k-N}{N\Delta t}\hat{R}_{xx}(k) & k = N, (N+1), \ldots, (2N-1) \end{cases} \tag{2.127}$$

where Δt is the time interval between samples and the hat over a variable indicates that it is an estimate.

The cross-correlation between two signals, $x(t)$ and $y(t)$ can be computed in a similar way to the auto-correlation. In this case we replace the PSD with the CSD and begin by computing the CSD using 50% zero padding and Q spectrum averages as in Eq. (2.59), and for each spectrum we use a $2N$ record length with the last half of the samples set equal to zero. So we obtain for the single sided CSD,

$$G_{Dxy}(f_n) = \frac{4S_A}{Q\Delta f}\sum_{i=1}^{Q} X_{iz}^*(f_n)Y_{iz}(f_n) \tag{2.128}$$

where $Y_{iz}(f_n)$ and $X_{iz}(f_n)$ are unscaled spectra and the single-sided CSD, G_{Dxy} has been multiplied by 2 to account for the added zeros. As before, the scaling factor, S_A is simply equal to $1/(4N^2)$, as there are $2N$ samples in the record. The ith unscaled, zero-padded spectrum, $Y_{iz}(f_n)$, obtained by passing time samples through a rectangular window, is similar to that for $X_{iz}(f_n)$, which is given in Eq. (2.126), and is given by

$$Y_{iz}(f_n) = \sum_{k=0}^{2N-1} y_i(t_k)\mathrm{e}^{-\mathrm{j}2\pi kn/N}; \quad n = 0, 1, \ldots, (2N-1) \tag{2.129}$$

The IDFT is taken of Eq. (2.129) to obtain the quantity, $\hat{R}_{xy}(t_k)$ and this is used to compute an estimate for the cross-correlation function for the ith record, using (Brandt 2010):

$$\hat{R}_{xy}(i) = \begin{cases} \dfrac{N-k}{N\Delta t}\hat{R}_{xy}(k) & k = 0, 1, \ldots, (N-1) \\[2ex] \dfrac{k-N}{N\Delta t}\hat{R}_{xy}(k) & k = N, (N+1), \ldots, (2N-1) \end{cases} \tag{2.130}$$

2.4.17 Maximum Length Sequence

A maximum length sequence (MLS) excitation signal is sometimes used in acoustics to obtain measurements of transfer functions, which include loudspeaker response functions and the noise reduction from outside to inside of a house. A schematic of a typical measurement system is shown in Figure 2.19. An MLS signal is a digitally synthesised binary series of samples, which is mapped to −1 and 1 respectively, to produce a signal that is symmetric about zero. The clock rate is the number of times a new value is output per second by the shift register generating the binary sequence, and this should be at least 2.5 times the maximum frequency of interest in the transfer function to be measured. However, the sample rate is recommended to be 10 times the maximum frequency of interest Ljung (1999). For the measurement of transfer functions, MLS analysis has the following advantages:

- It can minimise the effects on the measurement of noise not generated by the test system.
- The spectral content of an MLS signal closely resembles white noise with a flat power spectral shape.
- It is deterministic, and this property, together with the previous one, has attracted the description of *pseudo-random noise*.

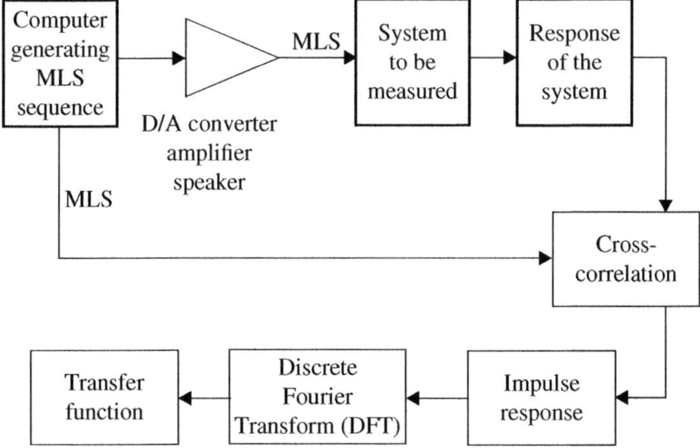

Figure 2.19 Schematic arrangement for the measurement of a system transfer function (such as outside-to-inside noise levels in a house) using an MLS signal.

- It has a pre-determined temporal length before repeating. This length, in terms of the total number of samples in a sequence, is $L = 2^M - 1$ samples, where M is the order of the MLS and this is usually 12 or greater in practical systems. Note that f_s/L, where f_s is the sample rate, must be longer than the impulse response of the system being measured. In the case of measuring noise reductions from outside to inside of a house, it would have to be greater than the propagation time of the noise from the outside source to the inside microphone.
- It has a low crest factor, thus transferring a large amount of energy to the system being excited, thus achieving a very high signal-to-noise ratio. For every doubling in the number of sequences that are averaged, the signal-to-noise ratio is improved by 3 dB.
- The cross-correlation of the input and output of a system using an MLS is equal to the impulse response of the system between the input and output. As discussed above, taking an FFT of the cross-correlation yields the system transfer function. For the case of the house mentioned above, this transfer function is the noise reduction from outside to inside in decibels.
- The use of cross-correlation with MLS to obtain transfer functions rejects all noise not correlated with the MLS signal and so MLS is effective in finding transfer functions in noisy environments such as at low frequencies when measuring house outside-to-inside noise reductions.
- MLS suppresses the DC part of the system response.

The MLS sequence is generated digitally using a maximal linear feedback shift register, which cycles through every possible binary value (except all zeros) before repeating. A shift register of length M contains M bits, each of which can have a value of 1 or 0. The length of an MLS sequence before it repeats itself is related to the number of bits, M, in the shift register used to generate it. If the shift register length is $M = 20$, then the number of samples in the MLS sequence prior to repeating is $2^{20} - 1$ ($N =$ 1 048 575 samples). The −1 is there because the case of all bits in the shift register being 0 is excluded. An MLS generating system with a shift register of length 5 is implemented, as illustrated in Figure 2.20.

The output from the shift register chain at time k is either a 1 or a zero and this is mapped to −1 and +1 respectively before being transmitted to an amplifier.

A modulo-2 sum (or XOR sum, denoted by ⊕) is the process of combining bits in two binary numbers to produce a third number. The two bits in the same position in the two numbers to be summed, are combined to produce a value for the bit in the same position in the new number by following the rules:

- if both of the two bits being added are one, then the result is zero
- if both of the two bits being added are zero, then the result is zero
- if one of the bits is one and the other is zero, then the result is one.

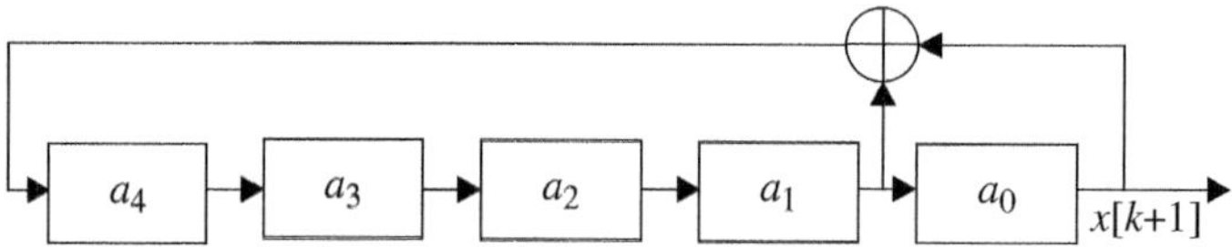

Figure 2.20 The next value for register bit a_4 is the modulo-2 sum of a_0 and a_1, indicated by the ⊕ symbol.

The values of the bits corresponding to sample number $k + 1$ in the MLS being generated with a shift register of length 5 can be calculated recursively from the values corresponding to sample k, using the following relations:

$$\begin{cases} a_4[k+1] = a_0[k] + a_1[k]; \quad \text{(modulo-2 sum)} \\ a_3[k+1] = a_4[k] \\ a_2[k+1] = a_3[k] \\ a_1[k+1] = a_2[k] \\ a_0[k+1] = a_1[k] \end{cases} \tag{2.131}$$

2.5 Summary

The acoustic principles necessary for understanding noise generation, propagation and reception from a wind farm are well established and have been outlined in this chapter. There are many techniques that can be used to analyse noise measurement data to accurately define the noise character and its potential for annoyance (see Chapter 6). These techniques have as their basis, the material discussed here.

References

Abramowitz M and Stegun I 1964 *Handbook of Mathematical Functions.* National Bureau of Standards, USA.

AMWG 2015 Methods for rating amplitude modulation in wind turbine noise. Technical report, Institute of Acoustics Noise Working Group (Wind Turbine Noise): Amplitude Modulation Working Group.

ANSI/ASA S1.11 (R2009) 2004 Specification for octave, half-octave and third-octave band filter sets. American National Standards Institute and Acoustical Society of America.

Bies D and Hansen C 2009 *Engineering Noise Control: Theory and Practice,* 4th edn. Spon Press.

Brandt A 2010 *Noise and Vibration Analysis: Signal Analysis and Experimental Procedures.* John Wiley & Sons.

Bray W and James R 2011 Dynamic measurements of wind turbine acoustic signals, employing sound quality engineering methods considering the time and frequency sensitivities of human perception. *Proceedings of Noise-Con,* Portland, Oregon, pp. 25–7.

Cooley J and Tukey J 1965 An algorithm for the machine calculation of complex Fourier series. *Mathematics of Computation,* **19**(90), 297–301.

Hansen K, Hansen C and Zajamšek B 2015 Outdoor to indoor reduction of wind farm noise for rural residences. *Building and Environment,* **94**, 764–772.

Huson W 2015 Stationary wind turbine infrasound emissions and propagation loss measurements. *6th International Meeting on Wind Turbine Noise,* Glasgow, UK.

IEC60050-801 1994 International electrotechnical vocabulary – Chapter 801: Acoustics and electroacoustics. International Electrotechnical Commission.

ISO 226 2003 Acoustics: Normal equal-loudness-level contours. International Standards Organisation.

Jakobsen J 2012 Danish regulation of low frequency noise from wind turbines. *Journal of Low Frequency Noise, Vibration and Active Control,* **31**(4), 239–246.

Kelley N 1987 A proposed metric for assessing the potential of community annoyance from wind turbine low-frequency noise emissions. Technical report, Solar Energy Research Inst., Golden, CO, USA.

Ljung L 1999 *System Identification – Theory For the User,* 2nd edn. Prentice Hall.

Moorhouse A, Waddington D and Adams M 2005 Proposed criteria for the assessment of low frequency noise disturbance. Technical report, University of Salford.

Morse P and Bolt R 1944 Sound waves in rooms. *Reviews of Modern Physics,* **16**(2), 69–150.

Potter R 1977 Matrix formulation of multiple and partial coherence. *Journal of the Acoustical Society of America,* **61**(3), 776–781.

Randall R 1987 *Frequency Analysis,* 3rd edn. Brüel and Kjær.

Renewable UK 2013 Wind turbine amplitude modulation: research to improve understanding as to its cause and effect. Technical report, Renewable UK.

Schmidt H 1985a Resolution bias errors in spectral density, frequency response and coherence function measurements, I: general theory. *Journal of Sound and Vibration,* **101**(3), 347–362.

Schmidt H 1985b Resolution bias errors in spectral density, frequency response and coherence function measurements, III: application to second-order systems (white noise excitation). *Journal of Sound and Vibration,* **101**(3), 377–404.

Tokita Y, Oda A and Shimizu K 1984 On the frequency weighting characteristics for evaluation of infra and low frequency noise *Proceedings of Inter-noise 84 and Noise-Con 84,* pp. 917–920 Institute of Noise Control Engineering.

Van den Berg G 2005 Mitigation measures for night-time wind turbine noise *First International Meeting on Wind Turbine Noise: Perspectives for Control,* Berlin, Germany.

3

Noise Generation

3.1 Introduction

This chapter provides an overview of how wind turbines create noise. The main emphasis is on aerodynamic noise generation, but other noise sources, such as machinery, are discussed and reviewed at the end of the chapter.

The treatment of aerodynamic noise begins with a physical description of turbulence and how it creates sound. Most wind turbine noise is considered to be *aerodynamic*. In other words, the noise is generated by turbulent fluid flow interacting with the rotor blades or other solid objects. As discussed in this chapter, turbulence itself, in the absence of other objects, will create sound. This is often termed *turbulent sound production* and is the basis for the production of wind turbine noise. For wind turbines, turbulent sound production, in isolation, is weak. However, turbulent sound production becomes more efficient (and more noticeable at large distances) when the turbulent pressure field interacts with or is scattered by the wind turbine blades. Therefore, after the fundamentals of turbulent sound production are presented, this chapter provides a summary of aeroacoustics and the common theories used to understand aerodynamic noise generation. This will introduce the aeroacoustic terminology that describes the noise generated by the wind turbine rotor, known as *rotor noise*, and particular subsets of rotor noise, such as trailing-edge noise. After a description of the unsteady aerodynamic environment experienced by a wind turbine, the individual aerodynamic noise generation mechanisms are reviewed.

Important noise sources that are reviewed include the various airfoil self-noise mechanisms (trailing-edge noise, separation-stall noise and tip noise) that contribute to low and high-frequency broadband noise ('swish') as well as tonal noise. The noise generated by the interaction of turbulence with the leading edges of wind turbine blades, mostly low-frequency, is also described. The effect of wind shear, amplitude modulation and the effect of the tower on rotor noise are explained. The chapter is completed by sections discussing the effects of blade flexibility on aerodynamic noise generation and a review of current information regarding wind turbine machinery noise.

Although parts of this chapter use advanced mathematics to describe the sound generation mechanisms, the less mathematically inclined reader can still gain considerable physical understanding of wind turbine noise generation by skipping the equations and just reading the text.

Wind Farm Noise: Measurement, Assessment and Control, First Edition.
Colin H. Hansen, Con J. Doolan, Kristy L. Hansen.

3.1.1 Definitions

To aid the interpretation of the following chapter, it is important to define some aerodynamic and acoustic terms. The Reynolds number is a special non-dimensional number used to provide an indication of the effect of fluid viscosity on the air flow over the wind turbine blade. It is defined as

$$Re_b = \frac{Ub}{\nu} \tag{3.1}$$

where b is the blade chord (defined as the distance between the leading and trailing edges of the airfoil, measured in a straight line), ν is the kinematic viscosity of the air and U is the relative air velocity incident on the blade (see Section 4.4.2). In the special case of a fixed wing or blade in a wind tunnel test, $U = U_\infty$, where U_∞ is the freestream air velocity. In the case of a wind turbine blade, U_∞ is the wind velocity and U is the vector sum of the rotational velocity of the blade, Ωr (where Ω is the rotational rate of the rotor in radians per second and r is the radial distance from the rotor centre to the blade segment under consideration) and $U_\infty - v_i$ (the difference between the wind velocity and the induced velocity of the rotor, v_i; see Section 4.4.2 for details). Note that we use the subscript b to denote the length scale used to calculate the Reynolds number; in this case it is the blade chord, b.

If the Reynolds number is sufficiently low ($Re_b \sim 100\,000$, for example), the boundary layers on the surfaces of the airfoil at the trailing edge are likely to be laminar.[1] If the Reynolds number is high (say, $Re_b \sim 1\,000\,000$), then the boundary layers at the trailing edge are likely to be turbulent. Note that the process whereby the laminar boundary layers turn into turbulent ones (a process known as *transition*) is affected by many things, including the surface roughness of the airfoil, the airfoil shape and its angle of attack. The boundary layer can also be forced into a turbulent state by using a boundary-layer trip. A trip consists of a region of increased roughness (or height) placed near the leading edge of the airfoil. The increased roughness causes the flow to destabilise and pass directly into a turbulent state. The advantage of using a trip is to increase the airfoil's stall angle, as a turbulent boundary layer is less inclined to separate than a laminar one (see next paragraph). The disadvantages of a trip are higher skin-friction drag and noise, due to the turbulent boundary layer containing more energy than one that transitions naturally.

An airfoil's angle of attack (defined as the angle the airfoil makes with the incoming flow) controls the amount of lift it produces. If the angle of attack becomes too large, the airfoil will stall, which means that the amount of lift it produces will begin to fall as the angle of attack continues to increase. The stall angle is the angle of attack where the lift is a maximum. Physically, stall occurs when the boundary layer on the surface of the airfoil separates from it. Separation is a specific term that signifies the process where the boundary layer appears to lift from the surface of the airfoil, creating a region of recirculating flow beneath it. Importantly, the streamlines no longer follow the curvature of the

1 Laminar flow is defined as smooth and regular flow that is not turbulent. Turbulent flow is defined as irregular and tortuous flow.

airfoil, so the flow accelerations required to produce lift are diminished (or destroyed) resulting in loss of lift. If the angle of attack is increased so that it is much higher than the stall angle, it is sometimes termed as 'deep-stall'. This is a subjective term, which may be considered relevant to the case where the angle of attack is so high that there is negligible lift produced by the airfoil or when the separation point has moved close to the leading edge of the airfoil.

A vortex is defined as a region of rotating fluid. Vortices are useful to describe what we observe on wind turbine blades. These are tip vortices (regions of rotating fluid streaming from the tips of the blades) and small vortices or 'eddies' within the turbulent boundary layer on the wind turbine blades themselves. Vortices are normally rotating, turbulent flow regions where the flow is dissipated via viscous shear (owing to the velocity gradient within the vortex) over time.

The Mach number is the ratio of fluid speed to the speed of sound,

$$M_{\infty} = \frac{U_{\infty}}{c} \tag{3.2}$$

where U_{∞} is, in this particular case, the freestream speed of the flow and c is the speed of sound. Thus in this case, Eq. (3.2) represents the freestream Mach number. The subscript normally identifies the speed used to determine the Mach number. For example, if the tip speed were used instead, the blade-tip Mach number would be $M_{\mathrm{tip}} = U_{\mathrm{tip}}/c$. The Mach number represents the compressibility of the flow field and is often used to relate the flow to the acoustic field.

Aeroacoustic sources of sound are normally described as either monopoles, dipoles or quadrupoles. Monopoles are the simplest acoustic source; they are generated by unsteady volumetric compression and expansion of fluid (for example, by a vibrating surface, fluid injection, combustion or the movement of a turbine blade with respect to a fixed observer; see Section 3.2.3). They radiate sound waves omnidirectionally and are the most efficient form of acoustic source. Dipoles can be considered mathematically as the superposition of two monopoles of alternate phase. In other words, the monopoles are 180° out of phase, so that when one is contracting, the other is expanding. Physically, they represent the sound field created by an unsteady force (or momentum flux) applied to the surrounding air. They are directional, in that the maximum sound pressure is observed when the listener is aligned with the force vector producing the sound. Dipoles are a common source model used to describe the noise generated by wind turbines as it is the unsteady forces generated by the air flow over the blades that creates most of the noise. The final sound source we describe here is the quadrupole, consisting of four monopoles superimposed with varying phase and orientation (there are different types of quadrupole, depending on the orientation of the base monopoles, the lateral and longitudinal). Two of the monopoles are in phase and the other two are 180° out of phase with the first two, as shown in Figure 3.1. The two orientations that describe the lateral and longitudinal quadrupoles are illustrated in the figure. Quadrupoles represent the sound created by turbulence itself (the strength of quadrupoles is the turbulent shear stress). As the Mach number of wind turbine blades is low, quadrupoles are not normally included in the analysis of wind turbine noise, but are included in some of the discussions in this chapter for completeness and to illustrate the relative importance of the unsteady aerodynamics with respect to noise generation.

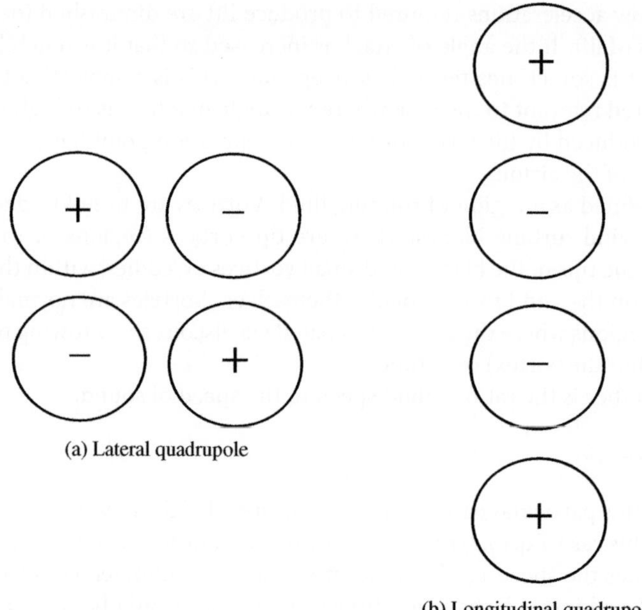

Figure 3.1 Two types of quadrupole: (a) the four monopoles in a lateral quadrupole; (b) the orientation of four monopoles in a longitudinal quadrupole. Each monopole is described as a circle, with its relative phase indicated by a plus or minus sign, as shown.

3.2 Aeroacoustics

3.2.1 Turbulence and Sound

According to the theory of Lighthill (1952), unsteady fluid flow will make sound. When the flow is turbulent, the sound produced will be broadband noise, reflecting the random nature of fluid turbulence. Imagine a turbulent flow far from any boundary. Turbulent flow can be thought of as a mixture of rotational regions of fluid known as *eddies*, of various sizes or scales, from large, energy-containing eddies, to small, dissipative ones. The scales are linked in an 'energy cascade', where large eddies are unstable and break down into smaller ones until they reach a lower limit,[2] where they stabilise, and the turbulent energy is dissipated by viscous heating. Figure 3.2 is a schematic representation of the turbulent energy cascade, showing how an eddy with characteristic velocity u_0 and scale l_0 is dissipated. While this general view is adequate for the following discussion, further reading on turbulent flows can be found in the literature (Pope 2000; Tennekes and Lumley 1972).

There are a number of ways to view the mechanism of turbulent sound production, but the convenient and physically descriptive model of Ribner (1981) is used here, despite its original use for explaining jet noise. Remembering that turbulence is a random process, we can think of the pressure varying throughout turbulent flow in a chaotic manner,

2 Known as the Kolmogorov scale.

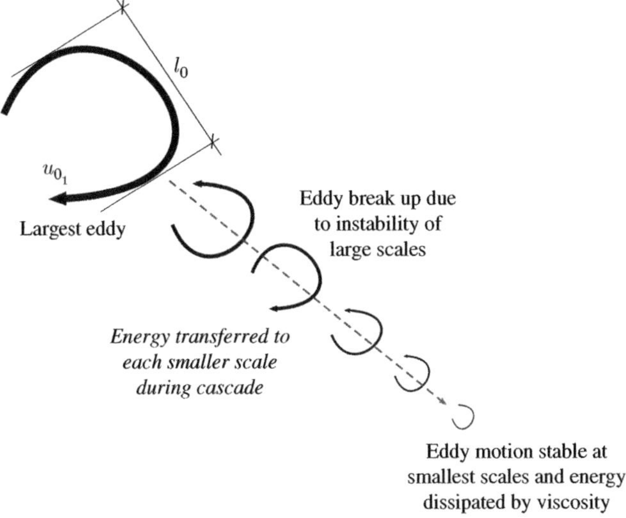

Figure 3.2 The turbulent energy cascade.

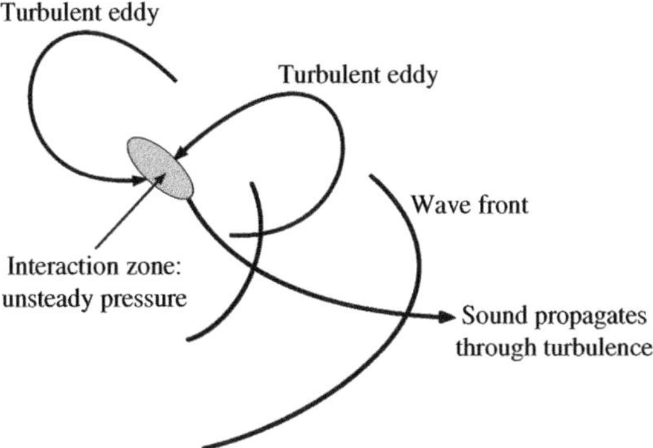

Figure 3.3 Phenomenological model of turbulent sound production. Adapted from Ribner (1981).

with the turbulent eddies interacting with each other, as shown in Figure 3.3. This interaction creates unsteady, fluctuating regions of pressure, which are sources of sound. Most of the turbulent energy will be retained in the flow (and ultimately dissipated or convected away), but a small part will be propagated to the far field as sound, due to the compressibility of the fluid. In effect, the unsteady expansion and compression of fluid, driven by inertial effects (momentum), generates sound (Ribner 1981). As shown in Figure 3.3, sound waves generated in turbulence will refract as they encounter gradients of wind velocity and/or temperature, making their exact location and nature difficult to determine.

Ribner (1981) also provides a straightforward method of mathematically describing turbulent sound production. Using the dilatation model, the source strength per unit volume, $\mathbf{q}$, can be approximated as:

$$\mathbf{q}(\mathbf{y}, t) \sim \frac{\partial^2 [V]}{\partial t^2} \sim -\frac{1}{c^2} \frac{\partial^2 [p(\mathbf{y}, t)]}{\partial t^2} \tag{3.3}$$

where c is the local speed of sound and the source exists at vector location $\mathbf{y}$ in the source region that has volume V (that is, the region of space containing the turbulent eddies). The square brackets [ ] indicate retarded time.[3] The second derivative is needed because it is the acceleration of the fluid that is important for generating sound. Thus, using a free-field Green's function (meaning that the sound propagates away from the source as spherical waves), the solution for pressure p at a point in the acoustic far field is

$$p(\mathbf{x},t) = \frac{1}{4\pi r}\int_V \left[-\frac{1}{c^2}\frac{\partial^2}{\partial t^2}(p(\mathbf{y},t))\right] \mathrm{d}V \tag{3.4}$$

where $\mathbf{x}$ is the vector position of the observer in the far field and $r = |\mathbf{x} - \mathbf{y}|$ is the magnitude or distance to the observer. It has been shown (Ribner 1962) that local pressure in the noise source region, $p(\mathbf{y}, t)$ can be related to local momentum flux so that Eq. (3.4) becomes

$$p(\mathbf{x},t) = \frac{1}{4\pi r}\int_V \left[-\frac{1}{c^2}\frac{\partial^2}{\partial t^2}(\rho v_x^2)\right] \mathrm{d}V \tag{3.5}$$

where v_x is the velocity component in the source pointed to the observer and ρ is the fluid density. Complete and more rigorous mathematical descriptions of the derivation of this law can be found in Lighthill (1952) and Proudman (1952).

What is important about Eq. (3.5) is that it illustrates Lighthill's law; that sound pressure due to turbulence is proportional to the fourth power of velocity. Additionally, mean square pressure (proportional to sound power) is thus of the form,

$$\overline{p}^2 \sim \frac{1}{x^2c^2}\int_V \left(\overline{\frac{\partial^2(\rho v_x^2)}{\partial t^2}}\right)^2 L^3\, \mathrm{d}V \tag{3.6}$$

which illustrates the famous eighth sound power law of Lighthill (1952) ($W \propto U^8$, where W is sound power (watts) and U is a characteristic velocity (m/s); see Section 3.2.2. In Eq. (3.6), L refers to a turbulence correlation length (in metres), over which the turbulence is coherent.

3.2.2 The Effect of Solid Surfaces

While turbulent flow, in isolation, is an important noise source for high-speed jets (Tam et al. 2008), it is a very weak source of noise for wind turbines. Maximum Mach numbers of wind turbine flows are approximately $M = U_0/c \approx 0.1 - 0.2$, whereas for jet noise $M \approx 1$. The eighth power law dependency illustrated in Eq. (3.6) is found through dimensional analysis; that is, the second derivative in time can be considered

$$\frac{\partial^2}{\partial t^2} \propto f^2 \propto \frac{U_0^2}{L^2}$$

where f is the frequency of the noise. Hence, the term in the integral of Eq. (3.6) can be expressed as:

$$\left(\overline{\frac{\partial^2(\rho v_x^2)}{\partial t^2}}\right)^2 \propto U_0^8$$

3 Retarded time is the time at the observer minus the propagation time.

where $U_0 = Mc$, and is the characteristic velocity of the system (tip speed for a wind turbine, or centreline nozzle exit speed for a jet). Hence the sound power radiated from turbulence on a wind turbine will be rather small compared with that of a jet. In fact, it is most likely unnoticeable.

Low Mach number flows can create significant levels of noise via scattering from solid surfaces. Curle (1955) extended Lighthill's theory to show that surfaces support sources of sound whose strength depends upon the unsteady forces they apply to the surrounding fluid. This is intuitive, as unsteady flow (turbulence) will create unsteady forces on a solid object.

Curle (1955) showed that, assuming a low Mach number flow, the sound radiated by a rigid, stationary body can be written as the gradient of a surface pressure integral

$$p(\mathbf{x}, t) = \frac{1}{4\pi}\frac{\partial}{\partial x_i}\int\int_S \frac{\ell_j}{r}[p\delta_{ij}]dS(\mathbf{y}) \tag{3.7}$$

where $\mathbf{y}$ denotes a point on the rigid, stationary surface, separated from the observation point $\mathbf{x}$ by the distance r, and ℓ_j are the components of the unit vector that are normal to the surface. The square brackets denote a value taken at the retarded time $t - r/c$ and δ_{ij} is the Dirac delta function.[4]

When the integration is carried out on a surface of acoustically compact dimensions (see Section 3.3.2 for a definition of compactness), Curle's equation can be simplified to predict the acoustic far field generated at an observation point by an unsteady vector point force, $\mathbf{F}$, applied on a compressible fluid[5] that is initially at rest,

$$p(\mathbf{x}, t) = -\frac{1}{4\pi}\frac{\partial}{\partial x_i}\left[\frac{F_i}{r}\right] = \frac{1}{4\pi c}\frac{x_i}{r^2}\left[\frac{\partial F_i}{\partial t}\right] \tag{3.8}$$

where F_i are the three vector components of the resulting force applied on the fluid. Einstein's summation convention is assumed, which means that when the subscript is used, to sum over each of the vector components (aligned with the chosen coordinate system or, more precisely, aligned with the unit vectors associated with the chosen coordinate system). Equation (3.8), containing vectors $\mathbf{x} = (x_1, x_2, x_3)^T$ and $\mathbf{F} = (F_1, F_2, F_3)^T$ (the superscript T means transpose to a column vector), can be re-written as,

$$p(\mathbf{x}, t) = \frac{1}{4\pi c}\frac{1}{r^2}\left[x_1\frac{\partial F_1}{\partial t} + x_2\frac{\partial F_2}{\partial t} + x_3\frac{\partial F_3}{\partial t}\right] \tag{3.9}$$

This directly links the unsteady force on an object to sound production. Thus an estimation of the unsteady aerodynamic force will also provide a useful approximation of the sound field.

3.2.3 The Effect of Moving Solid Surfaces

The theory of Curle is appropriate for objects that are stationary with respect to an observer, but a wind turbine blade is a moving source. To understand how this affects the production of sound and the nature of the source, we must consider the theory of Ffowcs-Williams and Hawkings (1969), which takes source movement into account.

---

4 $\delta_{ij} = 1$ when $i = j$ and zero otherwise.

5 A compressible fluid is one whose density changes with pressure and/or temperature. Air is a compressible fluid.

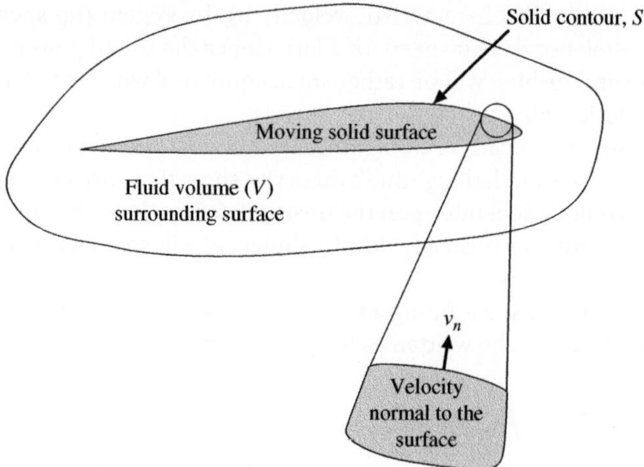

Figure 3.4 A moving, rigid body with surface S, surrounded by a fluid volume V. The velocity normal to a small element of the surface v_n is illustrated.

Consider a volume of fluid, V, enclosing a rigid body,[6] as shown in Figure 3.4. The volume is divided into two regions by the surface contour, S, which has an arbitrary velocity. The Ffowcs-Williams–Hawkings (FW-H) equation for the sound pressure at vector location $\mathbf{x}$ and at time t is stated as follows:

$$p(\mathbf{x}, t) = \frac{1}{4\pi}\frac{\partial^2}{\partial x_i \partial x_j}\int_V \left[\frac{T_{ij}J}{r|1-M_r|}\mathrm{d}^3\eta\right] - \frac{1}{4\pi}\frac{\partial}{\partial x_i}\int_S \left[\frac{p_{ij}n_j A}{r|1-M_r|}\right]\mathrm{d}S(\eta) + \frac{1}{4\pi}\frac{\partial}{\partial t}\int_S \left[\frac{\rho v_n}{r|1-M_r|}\right]\mathrm{d}S(\eta) \tag{3.10}$$

where the square brackets indicate evaluation at retarded time and v_n is the normal velocity of the surface. Lighthill's stress tensor, for the case of wind turbines, has individual terms approximated by

$$T_{ij} = \rho u_i u_j = \rho \begin{pmatrix} u_1^2 & u_1 u_2 & u_1 u_3 \\ u_2 u_1 & u_2^2 & u_2 u_3 \\ u_3 u_1 & u_3 u_2 & u_3^2 \end{pmatrix} \tag{3.11}$$

where the fluid velocity vector components (those aligned with the unit vectors of the chosen coordinate system) are defined as $\mathbf{u} = (u_1, u_2, u_3)^T$. The first differential term in Eq. (3.10) is defined using Einstein's summation convention as,

6 A rigid body will not change shape when a force or pressure is applied to it. It is also impermeable, so that the fluid velocity at the surface is zero.

$$\frac{\partial^2}{\partial x_i \partial x_j} \int_V \left[\frac{T_{ij}J}{r|1-M_r|} \mathrm{d}^3\eta \right] = \frac{\partial^2}{\partial x_i \partial x_j}[Z_{ij}]$$

$$= \begin{array}{l} \dfrac{\partial^2[Z_{11}]}{\partial x_1^2} + \dfrac{\partial^2[Z_{12}]}{\partial x_1 \partial x_2} + \dfrac{\partial^2[Z_{13}]}{\partial x_1 \partial x_3} \\ + \dfrac{\partial^2[Z_{21}]}{\partial x_2 \partial x_1} + \dfrac{\partial^2[Z_{22}]}{\partial x_2^2} + \dfrac{\partial^2[Z_{23}]}{\partial x_2 \partial x_3} \\ + \dfrac{\partial^2[Z_{31}]}{\partial x_3 \partial x_1} + \dfrac{\partial^2[Z_{32}]}{\partial x_3 \partial x_2} + \dfrac{\partial^2[Z_{33}]}{\partial x_3^2} \end{array} \tag{3.12}$$

The term $[Z_{ij}]$ is defined by inspection of the RHS of Eq. (3.12).

The second differential term in Eq. (3.10) is defined as (with $[X_i]$ defined by inspection),

$$\frac{\partial}{\partial x_i} \int_S \left[\frac{p_{ij}n_j A}{r|1-M_r|} \right] \mathrm{d}S(\eta) = \frac{\partial}{\partial x_i}[X_i]$$

$$= \frac{\partial[X_1]}{\partial x_1} + \frac{\partial[X_2]}{\partial x_2} + \frac{\partial[X_3]}{\partial x_3} \tag{3.13}$$

Figure 3.4 illustrates the normal velocity of a small element of the surface. The solution is evaluated using a new coordinate system denoted by η, chosen so that the sources are at rest in the η-frame. The source is assumed to move with a Mach number M_r relative to the observer. The factor J is a volume divergence term that accounts for volume distortions of the body; similarly, A is an area divergence term to take into account area change. As the body is rigid in this case, $A = J = 1$.

From Eq. (3.10), Ffowcs-Williams and Hawkings (1969) show that aerodynamic sound, in general, can be produced by three source distributions:

- a distribution of quadrupoles (usually associated with turbulent flow), described in Eq. (3.10) through the use of the Lighthill stress tensor, T_{ij}
- a distribution of dipoles (usually associated with aerodynamic force)
- monopoles, if the surface is moving (representing a volume-displacement effect).

Noise due to quadrupoles (the first term on the right-hand side of Eq. (3.10)) is often associated with turbulence (see Section 3.2.1). This noise may often be neglected when considering wind turbines due to the low associated Mach number of the air flow. The second term in Eq. (3.10) is associated with unsteady force and is often called *loading noise*. In this term, the quantity denoted as $p_{ij}n_j$ is the force per unit area acting normally from the surface into the fluid. It is a vector quantity, with three components aligned with the unit vectors of the chosen coordinate system. The monopole term (the remaining term in Eq. (3.10)) is due to fluid displacement of the object as it passes through the fluid. It is often called *thickness noise*. These terms were first used to describe helicopter rotor noise (Brentner and Farassat 2003), which has some similarity with wind turbine noise.

3.3 Aerodynamic Noise Generation on Wind Turbines

An illustration of the aerodynamic environment of a wind turbine is shown in Figure 3.5. The turbine extracts energy from the wind using three long blades that rotate due to the aerodynamic forces acting upon them. While the performance of wind turbines is often analysed as a steady system, in reality the aerodynamic environment presented to the wind turbine blades is unsteady.

3.3.1 The Aerodynamic Environment of a Wind Turbine

The incoming flow to a wind turbine is a turbulent atmospheric boundary layer. This presents two unsteady aerodynamic conditions on the rotor. First, the mean velocity profile of the boundary layer (that is, the mean variation of wind speed with height from the ground) presents a region of wind shear (speed gradient) to the rotor, imparting aerodynamic loads on each radial blade section that varies with azimuth (rotational angle). The fundamental frequency of this variation is approximately equal to that of rotor rotation, sometimes referred to as the blade-passing frequency. Second, the turbulent boundary layer consists of turbulent eddies of various scales (see Section 3.2.1) that vary in size across its height. Simply, there are, on average, larger eddies at the top of the boundary layer than near the ground. These eddies interact with the rotor blades to create further unsteady aerodynamic loads on each rotor blade. The frequency of excitation due to the convected atmospheric turbulence will be higher than the blade passing frequency and is related to the turbulent velocity spectrum of the air that the blade encounters (see Section 3.3.5).

The blades will also encounter an unsteady aerodynamic environment as they pass the tower. The tower is a large, symmetric obstacle presented to the oncoming flow. Hence the flow will change upstream of the tower in order to negotiate its passage. Figure 3.6 illustrates an (idealised) aerodynamic environment about the tower. It shows a plan view

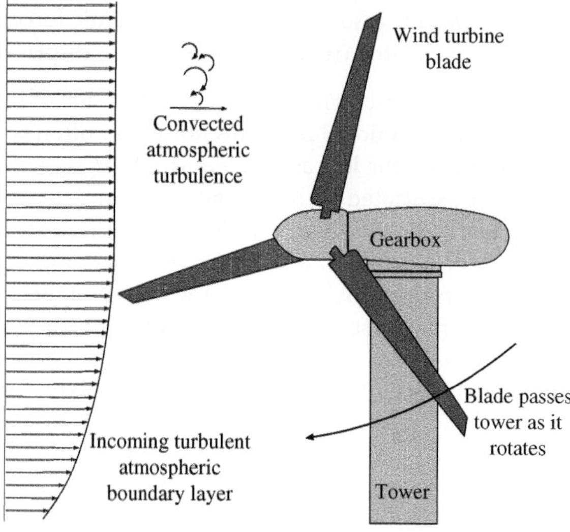

Figure 3.5 The aerodynamic environment of a wind turbine.

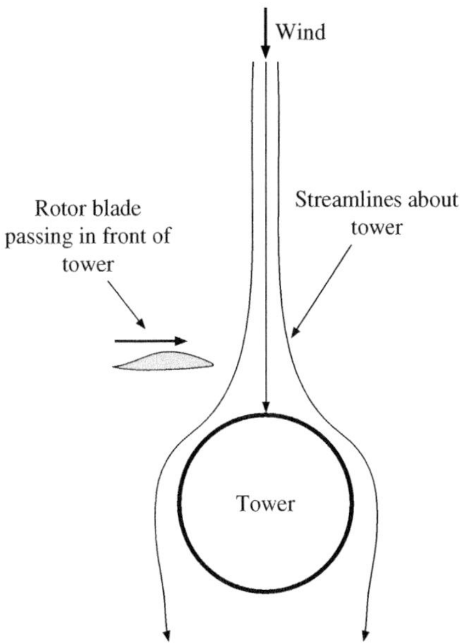

Figure 3.6 The aerodynamic environment around the tower: the figure shows a horizontal sectional view through the tower and rotor plane, illustrating the motion of the blade through the streamlines about the tower.

of a section through the tower and rotor. The wind approaches the tower from the top of the page. Streamlines in the plane of the page are also shown, and these are distorted upstream of the tower because it is an obstacle in the flow.

A section through one of the rotor blades is also shown in Figure 3.6. As the blade rotates, a section of the blade will pass through the distorted upstream streamlines. This will have the effect of changing the angle of attack of the rotor blade as it passes in front of the tower. This is therefore another important unsteady aerodynamic environment about the wind turbine (see also Doolan et al. (2012)). It should be noted that the model shown here can only be considered as approximate, because it assumes a simplified model of the fluid dynamics. In reality, the blade passing through the perturbed streamlines upstream of the tower will cause further distortion. The vorticity in the boundary layer and tip will have an effect on the streamlines after blade passage and may interact with a following blade. Blade–tower interaction (BTI) noise is discussed in Section 3.3.7.

Having considered the aerodynamic environment imposed externally on the blades, our attention is now turned to the aerodynamics of the blade itself. Figure 3.7 shows a sketch of the aerodynamic environment about a blade tip. As the blade tip rotates, it encounters atmospheric turbulence at the leading edge, causing unsteady aerodynamic lift, as already discussed.

The blade also develops a boundary layer on its surface. A boundary layer is created due to viscous shear stress in the fluid. Fluid particles on the blade surface impose shear on neighbouring particles throughout the boundary layer. The boundary-layer height is the distance beyond which the shear-affected zone has little influence on the surrounding flow. Large wind turbines for power generation usually have turbulent boundary

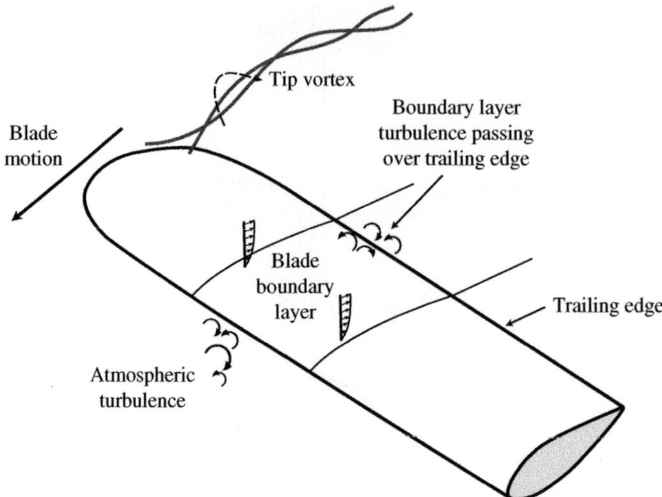

Figure 3.7 The aerodynamic environment of a blade tip.

layers (rather than laminar ones) on their blade surfaces due to the high associated Reynolds numbers, especially at the faster-moving tips.

The boundary layer imparts broadband, random pressure fluctuations to the surface of the blade, which have smaller length-scales and higher intensity levels than the upstream, convecting atmospheric turbulence. It is the interaction of boundary layer turbulence with the trailing edge of the blade that creates noise. Trailing-edge noise has been identified (Oerlemans et al. 2007) as one of the most important sources of noise on a wind turbine. This noise source is described in Section 3.3.2.

Under certain operational conditions, the boundary layer may separate from the surface of the blade, resulting in stall. Such a flow condition may occur:

- when the angle of attack is high
- under yawed operation (operation in a cross-flow)
- under low flow conditions (especially on the inner portions of the rotors)
- under high wind shear conditions
- if the blade shape becomes distorted from ice accretion or dirt and bug deposits.

The effect of stall is to create unsteady loading on the blade and increase the turbulence passing the trailing edge. Separation-stall noise is discussed in Section 3.3.3.

The tip of the blade is an interesting unsteady flow environment that also creates noise. Lift is created on the blade aerodynamically by supporting a pressure difference across it; a higher pressure exists below the blade than on the top. At the tip of the blade, the high-pressure fluid below the blade is drawn towards the lower-pressure zone above the blade. This effect, combined with the movement of the blade, creates what is known as a *trailing tip vortex*. These are often noticed behind jet airliners as they pass overhead through cold skies; water and ice condense in the low-pressure vortex cores of aircraft wing-tip vortices, become visible and are sometimes called contrails. The same trailing vortex system occurs behind wind turbine blades. The creation of the tip vortex is a turbulent process creating unsteady flow about the tip. The noise created by the creation of tip vortices is described in Section 3.3.4.

3.3.2 Trailing-edge Noise

Trailing-edge noise occurs when unsteady flow encounters the sharp trailing edge of an airfoil. For a large energy-generating turbine, the Reynolds numbers are usually large enough to ensure that the boundary layer passing over the trailing edge is turbulent with a wide range of scales. This results in broadband noise generation. At lower Reynolds numbers, more coherent eddies are formed near the trailing edge. As the flow energy is concentrated in fewer eddies in more restricted wavelengths, tonal noise can be generated. Tonal noise like this is sometimes associated with smaller wind turbines, but can be eliminated using boundary-layer trips. Blunt trailing-edge noise is associated with vortex shedding from a blunt trailing edge and can occur at most Reynolds numbers. These three types of trailing-edge noise are discussed in the following paragraphs.

Turbulent Trailing-edge Noise

The sudden impedance change that occurs at the trailing edge of an airfoil scatters the unsteady pressure within the boundary layer, creating acoustic waves that are experienced by an observer in the far field as noise (Chase 1972, 1975). In terms of acoustic sources, the scattering process makes the quadrupole sources in the turbulent boundary layer more efficient radiators of sound, as documented by Ffowcs-Williams and Hall (1970). Moreover, Ffowcs-Williams and Hall (1970) found that the noise power is proportional to the fifth power of the Mach number ($W \propto M^5$, where W is the sound power in watts). For the low Mach number case of a wind turbine (where the tip Mach number is $M_{tip} \sim 0.1 - 0.25$), trailing-edge noise is thus a more efficient radiator than a dipole, which has a sixth-power dependency on Mach number ($W \propto M^6$).

Observers near wind turbines often report a 'swishing' character to the noise (Bakker et al. 2012; Doolan 2013; Van den Berg 2004) and mostly, this can be attributed to the combined effects of trailing-edge noise directivity and convective amplification due to the rotating blades (Oerlemans and Schepers 2009) (see also Section 3.3.6 for a discussion of enhanced amplitude modulation and wind-shear effects). Directivity of the trailing-edge noise source is due to the nature of the scattering process. Amiet (1976) provides an edge-scattering theory for trailing-edge noise and it has been used here to demonstrate its directivity. While this theory is for a flat plate of finite chord b, it may be considered as an appropriate assumption for wind turbine blades, as they are relatively thin compared with the acoustic wavelength (that is, $h/\lambda << 1$, where h is the blade thickness and λ is the wavelength).

Figure 3.8 compares the directivity patterns of trailing-edge noise when the wavelength is either much larger, smaller or of the same order as the chord b.[7] When the wavelength is much larger than the airfoil chord, b (Figure 3.8a), the airfoil is a compact radiator and the resulting radiation pattern is similar to that of a dipole. At the other extreme, when the wavelength is much smaller than the chord (Figure 3.8c), the airfoil appears as a semi-infinite half-plane, and the resulting directivity has a cardioid shape, as first shown by Ffowcs-Williams and Hall (1970). In between these cases, when the wavelength is of the same order as the chord or smaller (Figure 3.8b), the waves generated at the trailing edge are also scattered from the leading edge, setting up an interference pattern. It is this 'forward-looking' nature of trailing-edge noise directivity, coupled with

7 Note that chord is represented as b and speed of sound as c.

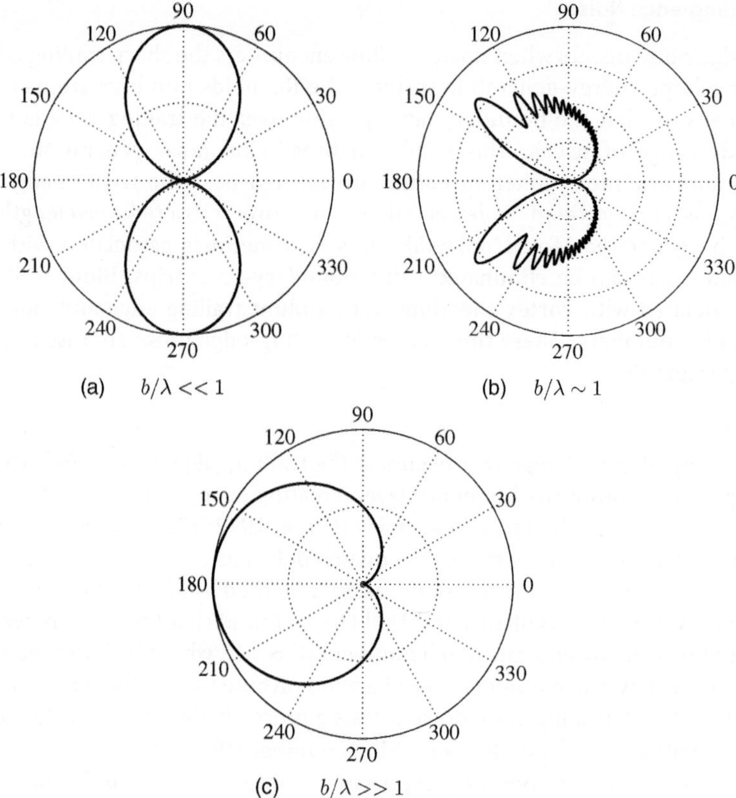

Figure 3.8 Directivity patterns of trailing-edge noise using the theory of Amiet (1976). The origin is at the trailing edge location and the flow is assumed from left to right; b is chord, λ is wavelength.

the rotating motion of the blades, that gives the 'swish-like' qualities of wind turbine noise. For modern turbines b/λ is between 0.4 and 4 (Oerlemans and Schepers 2009).

When a source moves and approaches an observer, its amplitude will rise due to *convective amplification*. According to Oerlemans and Schepers (2009), this effect provides most of the contribution to the asymmetry in a wind turbine's radiation pattern, and can be approximated by

$$D_{CA} \approx \frac{1}{(1 - M\cos\zeta)^4} \tag{3.14}$$

where D_{CA} is an amplification factor that describes the increase in sound level of a source moving at Mach number M at an angle ζ, defined as the angle between the direction of flow velocity over the blade and the source–observer line. A *Doppler* frequency shift must also be accounted for by increasing or decreasing the radiated frequency depending on the relative velocities and positions of the trailing edge and the observer.

A recent review of experimental and theoretical work on trailing-edge noise was performed by Doolan and Moreau (2015) who related the the findings to wind turbines. It was found that there are many aspects of wind turbine noise that are still not well understood, mainly due to a lack of data. For instance, there has been no direct comparison of

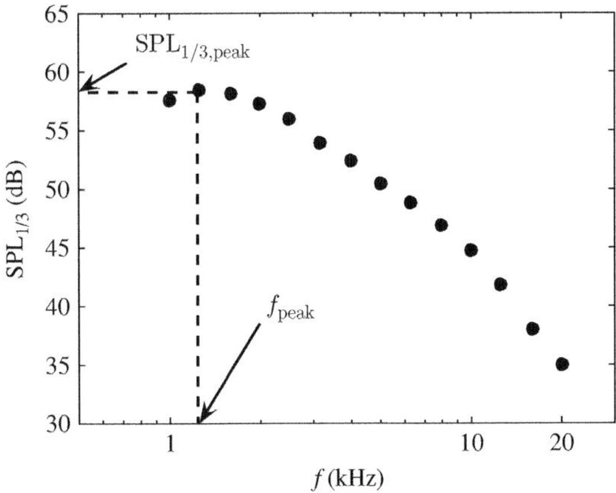

Figure 3.9 Typical 1/3-octave turbulent trailing noise spectrum, $M = 0.174, Re = 1.54 \times 10^6$. SPL: sound pressure level. Source: Herr (2007).

a wide range of integral, peak and spectral data sets. Moreover, it is necessary to see how the various data sets compare at higher Reynolds numbers, at the range where modern and future wind turbines operate.

An important example relates to the so-called 'low-frequency' peak in the 1/3-octave spectrum. A typical example of a 1/3-octave band turbulent trailing-edge noise spectrum (of a stationary blade mounted in a wind tunnel) is shown in Figure 3.9, measured in an anechoic wind tunnel (Herr 2007). The low-frequency peak (defined as f_{peak} with level $\text{SPL}_{1/3,\text{peak}}$, as indicated in the figure) is normally assumed to be a typical part of trailing-edge noise spectra. However, there are some authors who attribute the peak to facility scattering effects and/or data-processing techniques (Devenport et al. 2010a, 2013; Migliore and Oerlemans 2004). In contrast, and as reviewed by Doolan and Moreau (2015), modelling and analysis using boundary-layer wall pressure models (Herr et al. 2010) do predict a low-frequency peak in the noise data. It should also be noted that computational models (Marsden et al. 2008; Wolf et al. 2012) also predict the low-frequency peak in simulated spectra. It is therefore the opinion of the authors that the low-frequency peak is part of a typical trailing-edge noise spectrum.

Considering that most of the acoustic energy of trailing-edge noise is contained in the frequency bands close to the peak, it is important to properly characterise its level and frequency, as lower-frequency noise will not be attenuated with distance as much as high-frequency components, therefore increasing the risk of them being heard at greater distances and being associated with annoyance.

To further illustrate the importance of the low-frequency peak, the peak radiating frequency was calculated for a range of Reynolds numbers for a fixed tip Mach number of $M = 0.175$, as displayed in Figure 3.10. The chord was varied to change Re_b for a fixed Mach number. The peak frequency was calculated assuming a mean peak radiating Strouhal number of $St_{\text{peak}} = \frac{f_{\text{peak}}\delta^*}{U} = 0.069$, which represents the mean of all data compared in the review of trailing-edge noise data by Doolan and Moreau (2015). The boundary-layer displacement thickness at the trailing edge was calculated using the empirical correlations of Brooks et al. (1989). Peak radiating frequencies can be seen to be in the audible range over all Reynolds numbers shown. In particular, as Reynolds

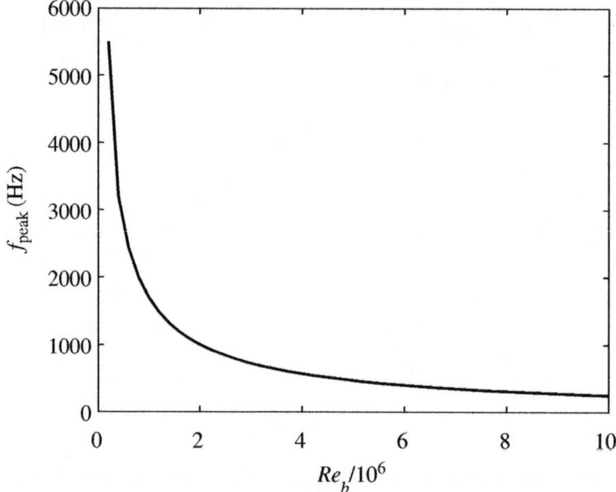

Figure 3.10 Peak radiating frequency vs blade Reynolds number (Re_b) for a fixed tip Mach number of $M = 0.175$.

numbers increase, the frequency gradually decreases, making the noise possibly more annoying and perceptible over longer distances, due to the lower levels of atmospheric and ground absorption at lower frequencies. While this analysis has many simplifying assumptions, it highlights the importance of properly resolving the low-frequency peak. Considering that future wind turbines (Ceyhan 2012) are planning to have blade Reynolds numbers in excess of 10×10^6, such a result needs to be considered when planning expansions in the wind energy industry.

Tonal Trailing-edge Noise: Instability Noise

If the boundary layer on an airfoil is transitional,[8] a particular type of tonal noise can be generated at the airfoil's trailing edge. Original studies of this type of noise (Paterson et al. 1973) suggested that the noise was from vortex shedding, but further investigation (Arbey and Bataille 1983; Tam 1974) indicated that the production of tonal noise at low Reynolds numbers can be explained by an aeroacoustic feedback loop, as shown in Figure 3.11. An excellent review of the topic was recently published by Arcondoulis et al. (2010).

Figure 3.11 summarises many of the aspects considered essential to the formation of a feedback loop. Central to all feedback-loop theories is the generation of instabilities in the boundary layer of a type known as Tollmein–Schlicting waves (or TS waves for short, see Schlichting et al. (2000)). These instabilities occur naturally in boundary layers that are undergoing transition to turbulence. This process can become reasonably prolonged over airfoils at low-to-moderate Reynolds numbers, so that TS waves may exist along most of the chord. TS waves are convecting pressure disturbances and create sound as they pass the trailing edge. When they form in the boundary layer, they consist of many sizes (or wavelengths), and a frequency spectrum shows them as a 'broadband hump', centred about a main frequency determined by instability theory. The feedback loop is

8 A transitional boundary layer is one that is in the process of transitioning from a laminar to a turbulent state. It occurs when the Reynolds number is in a low-to-moderate range; $10^5 \lesssim Re_b \lesssim 10^6$.

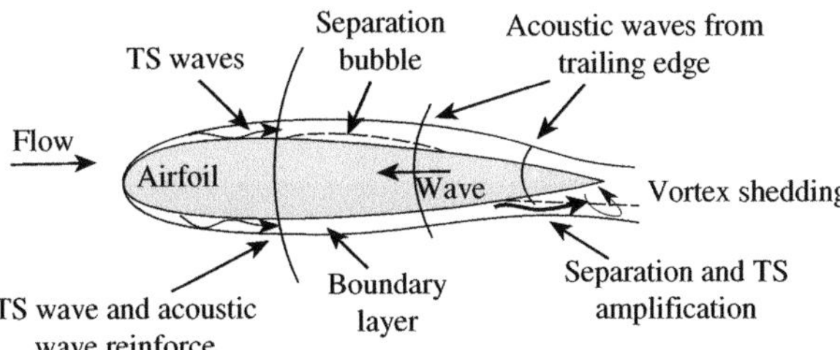

Figure 3.11 The aeroacoustic feedback loop responsible for airfoil tonal noise generation at low Reynolds numbers.

generated when the acoustic wave formed by a TS wave at the trailing edge propagates upstream. If the phase of the acoustic wave and the phase of the TS wave at the point of first instability match (Kingan and Pearse 2009), then the TS wave is reinforced, and the amplitude of this particular wave is amplified above the level of the original broadband hump. The presence of flow separation near the trailing edge will amplify instabilities in the boundary layer (Nash et al. 1999), further enhancing the feedback loop.

When measurements are taken of noise of an airfoil with boundary layer instabilities, a spectrum similar to the illustrated example in Figure 3.12 is obtained. Figure 3.12 shows the estimated broadband hump as a dashed line and a typical measured spectrum as the solid line. Usually, the tones are equally spaced in the frequency domain and the frequencies, when plotted against flow speed, form a 'ladder structure', which can be described by empirical power laws (Paterson et al. 1973). The existence of the ladder structure is usually taken to confirm the existence of a feedback loop.

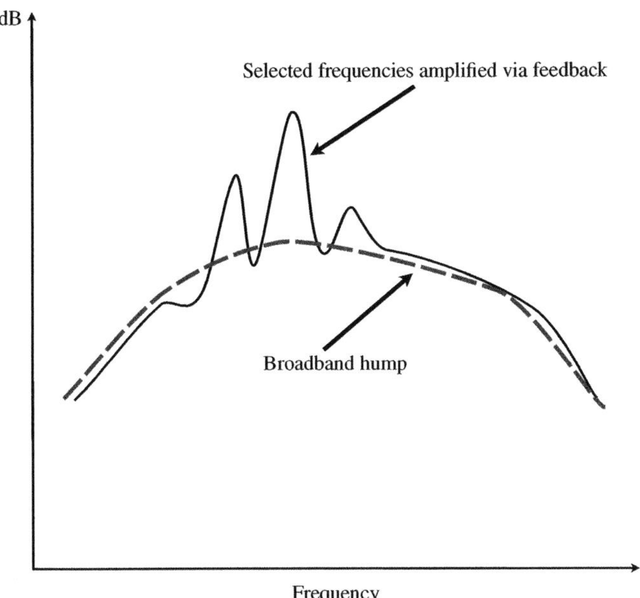

Figure 3.12 Typical spectrum for airfoil instability noise, assuming tones are created by selective amplification by a feedback loop.

It is interesting to consider the role of a separation bubble on a feedback loop (illustrated as a dashed line in Figure 3.11). Separation bubbles often occur on the suction side of airfoils with transitional boundary layers. The separation and/or reattachment points of the bubble may be highly receptive to acoustic waves, as recently found by Schumacher et al. (2014) in their study of feedback loops. However, computer simulations (Jones et al. 2010) show that laminar separation bubbles are not receptive to acoustic waves in this situation and it is the airfoil leading edge that is the point that closes a feedback loop.

It must be noted that agreement on the existence of a feedback loop in the scientific community is far from unanimous. Some researchers (Nash et al. 1999; Tam and Ju 2012) believe that the airfoil, by itself, cannot support the feedback loop and that its existence is due to spurious acoustic wave reflections in the test facility. The debate is open and remains an interesting academic problem. In terms of wind turbines, this type of tonal noise may develop on blades at low Reynolds number, say at inboard locations on a large turbine or on small wind turbines. It is unlikely that tonal noise at inboard locations on a large wind turbine will be at a sufficiently high amplitude to be noticeable, but instability noise is potentially a problem on small wind turbines. Instability noise is easily treated by tripping the boundary layer near the leading edge. This removes the spatial coherence of the instabilities (or forces turbulent transition), thus preventing energy focussing at any hydrodynamic wavelength. Further, wind turbine blades quickly become dirty and contaminated with bugs (and ice in cold climates) that will, in effect, trip the boundary layer. Therefore instability tonal noise is a minor problem compared with other forms of aerodynamic noise on a turbine blade.

Tonal Trailing-edge Noise: Blunt Trailing-edge Noise

When the Reynolds number becomes large ($Re_b \geq 10^5$) or the boundary layers on both sides of the blade are tripped (either intentionally, using trip tape or turbulators, or unintentionally, by dirt, bugs or ice accretion), turbulent boundary layers are created. If the airfoil trailing edge is *blunted*, in other words it has a finite thickness at the trailing edge, the turbulent boundary layers can merge and form a turbulent vortex street (a repeating pattern of swirling vortices), similar to that found behind a cylinder. This occurs because the base pressure is low in the near wake of the blunted blade; the low pressure creates a large-scale flow instability that results in vortex shedding. Blake (1986) presents an excellent overview of turbulent vortex shedding from blunt trailing edges. All practical wind turbine rotor blades will have blunted trailing edges as a consequence of the manufacturing process.

Figure 3.13 illustrates the aerodynamic environment of a blunt airfoil placed in a uniform flow. Turbulent boundary layers exist on each side of the airfoil and pass over the edge. If the edge is of sufficient thickness, vortex shedding occurs, creating unsteady pressure at the trailing-edge region. Given the alternating sign of the vorticity of each shed vortex, the unsteady pressure is exactly out-of-phase at the edge, creating unsteady lift forces, which support a dipole noise source. If the wavelength of sound is much larger than the chord, the blade is acoustically compact and will radiate noise with an intensity proportional to velocity raised to the sixth power. In the non-compact case, where the wavelength of sound is smaller than the chord, sound intensity will be proportional to flow velocity to the fifth power, thus representing a considerable increase in the sound-radiating efficiency.

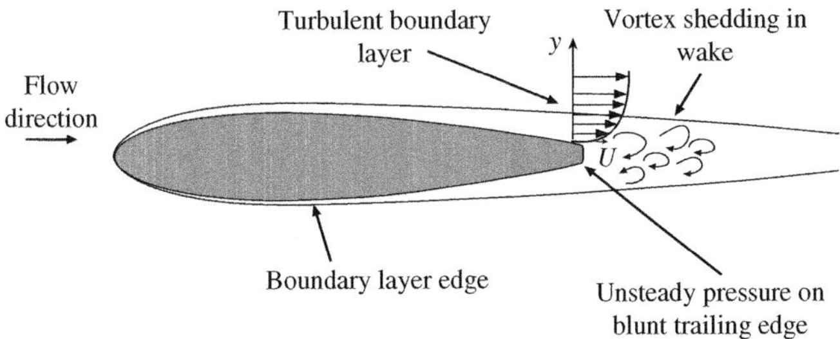

Figure 3.13 Blunt airfoil with vortex shedding.

Blake (1986) provides the following criteria for assessing compactness of airfoils: the airfoil or blade is compact if

$$\frac{\omega b}{c} < 2\pi \tag{3.15}$$

where ω is the frequency in radians per second. The airfoil is non-compact at high frequencies when the leading and trailing edges are at least one wavelength apart, so

$$\frac{\omega b}{c} > 2\pi \tag{3.16}$$

This raises an interesting point that warrants further discussion (see also Blake (1986)). In the non-compact case, the leading edge and trailing-edge noise sources (created by turbulence interaction at the leading edge (Section 3.3.6) and turbulent boundary layer/wake interaction at the trailing edge) will be acoustically separated. This means that they can be considered as separate sources that may alter the radiation pattern and its dependence on Mach number. Further, if both $\omega b/U > 1$ and $\omega b/c > 2\pi$, the leading edge and trailing-edge noise sources can be treated independently of each other (see Section 4.4.2 for a definition of U in terms of U_∞). For the specific case of wind turbines, these criteria are normally achieved. For example, for a tip speed of 60 m/s and frequency of 500 Hz, the chord must be larger than 686 mm to be acoustically compact and larger than 19 mm to achieve aerodynamic independence. The chord of most large wind turbine generators (power output greater than 1 MW) will meet these requirements.

There is some disagreement in the literature as to what amount of trailing-edge bluntness (defined as the ratio h/δ^*, where h is the trailing edge thickness and δ^* is the boundary-layer displacement thickness at the trailing edge) is required to create vortex shedding and tonal noise. Blake (1986) suggests that if $h/\delta^* \lesssim 0.3$, vortex shedding will be suppressed. However, Herr and Dobrzynski (2005) show that strong vortex shedding is noticeable for $h/\delta^* = 0.18$, so some caution needs to be exercised when trying to predict vortex shedding based on edge thickness.

Brooks and Hodgson (1981) indicate that when $h/\delta^* > 4$, the turbulent trailing-edge noise component changes in character from a tone to a broad peak. This is most likely due to the changes in the turbulent flow field when the edge thickness increases. When this occurs, there is more opportunity for shear layer instabilities to reduce the coherence of the vortex roll-up process in the near wake. The reduced spatial coherence manifests itself as temporal de-correlation of the effective noise source at the edge, resulting in spectral broadening of the tonal peak in the noise spectrum.

The frequency of blunt trailing-edge noise has been found to scale according to a Strouhal number, St, based on wake thickness, y_f, (Blake 1986)

$$St = \frac{\omega y_f}{U} \sim 1 \quad (3.17)$$

More specifically, measurements (Brooks and Hodgson 1981) show that a Strouhal number based on thickness is appropriate for NACA 0012 airfoils, where

$$St = \frac{fh}{U} = 0.1 \quad (3.18)$$

Recently, Moreau and Doolan (2013) showed that this Strouhal number relationship also holds for thin, flat-plate models exhibiting blunt trailing-edge noise tones in their noise spectra. In this case $0.7 \leq h/\delta^* \leq 0.84$ and the Reynolds number based on chord was $1.6 \leq Re_b/10^5 \leq 4.2$, where Re_b is defined in Eq. (3.1). Herr and Dobrzynski (2005) were also able to show reasonable agreement for a flat plate at much higher Reynolds numbers based on chord ($2.1 \leq Re_b/10^6 \leq 7.9$). Grosveld (1985) successfully used Eq. (3.18) for predicting full-scale wind turbine turbulent blunt trailing-edge noise.

The Strouhal number relation described above will change with blunt trailing-edge shape. Blake (1986) provides the most comprehensive guide to predicting how blunt trailing-edge shape affects vortex shedding frequency as well as noise amplitude.

3.3.3 Separation-stall Noise

Figure 3.14 illustrates the flow field that develops when an airfoil's angle of attack is high enough to induce stall. Typically, an airfoil has a predefined stall angle, beyond which the boundary layer separates from its surface. Stall occurs because of the adverse pressure gradient that develops on the upper (or suction) surface of the airfoil in the chord-wise direction, and has the effect of reducing the local flow velocity. Eventually, the reduction of flow velocity causes a region of reverse flow in the lower region of the boundary layer. When this happens, the boundary layer is said to have separated, as there exists a wind-speed profile in the upper region of the shear layer (above the separation line) that resembles a boundary layer. Below the separation line exists the separated region, or behind the airfoil, the separated wake. The separated region contains many turbulent eddies. The energy-containing eddies are much larger and coherent than those in the attached boundary layer case. Consequently, the interaction of these eddies and

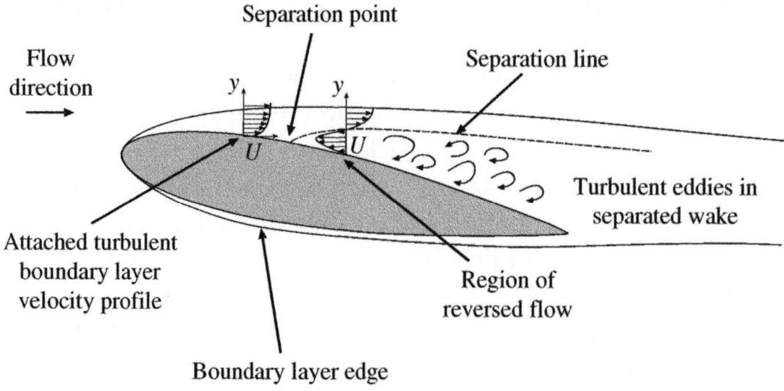

Figure 3.14 Flow field that develops over an airfoil just prior to stall.

the blade creates noise (known as separation-stall noise), which is usually produced at a lower frequency and at a higher amplitude than trailing-edge noise.

There is little previously published information concerning separation-stall noise. Paterson et al. (1975) produced what is most probably the earliest study of separation-stall noise. Here, the noise and flow created when an upstream vortex impinged on a wing tip was experimentally investigated. The object of the study was to investigate broadband helicopter noise sources, but the results are applicable to wind turbines. It was shown that stall noise is created by large eddies in the separated flow interacting with the trailing edge. Further, stall noise, once initiated, was found to be relatively insensitive to external flow conditions, but external flow conditions obviously are the cause of stall to begin with.

Fink and Bailey (1980) is the most cited work in this area. This is a study of aircraft high-lift device noise reduction. The authors show that separation-stall can increase noise levels by over 10 dB, relative to an attached, turbulent trailing-edge noise case. Brooks et al. (1989) have provided the most comprehensive data set of separation-stall noise as well as an empirical model for its prediction. Brooks et al. (1989) show that at high Reynolds number, the peak radiating Strouhal number[9] increases during stall but, because the effective separated boundary layer height increases dramatically during separation-stall, the overall effect is a lowering of peak radiating frequency. Brooks et al. (1989) also show that real radiating noise levels may increase by up to 10 dB.

The most recent work on stall noise was performed by Moreau et al. (2009). Here a combined, experimental, numerical and analytical program was attempted. The experimental results showed the existence of two stall regimes; a light stall regime (beginning at angles of attack $5° \leq \alpha \leq 15°$ for a NACA 0012 profile) that occurs just after the initiation of separation and a deep stall regime at high angles of attack ($\alpha \geq 26°$). Each shows a reduction of peak radiating frequency and an increase in amplitude, as observed in previous studies. Interestingly, the narrowband analyses of their noise spectra show that separation-stall noise is associated with two tonal, low-frequency peaks, that intensify and lower in frequency as the airfoil operates in the deep stall region. These results can only be considered preliminary, as Moreau et al. (2009) state that the results may be influenced by strong facility interaction effects, which were confirmed by numerical simulation. Further work is necessary. The effect of stall on amplitude modulation is discussed in Section 6.2.10.

3.3.4 Tip Noise

As discussed in Section 3.3, lift is created on the rotor blades due to the pressure differential on each surface of the blade. Figure 3.7 shows the formation of a vortex at the tip (the *tip-vortex*), created by the combined effects of pressure difference, viscous shear and forward movement of the blade tip. In fact, the trailing vortex is a direct consequence of the conservation of angular momentum (see the theoretical development in Katz and Plotkin (2006)). As the vortex forms over the tip, the turbulent boundary layer is swept into the vortex itself, creating a complex three-dimensional flow environment where turbulent flow, different in nature to that at the trailing edge for most of the blade, exists adjacent to the tip region and outer-most region of the trailing edge. Hence there

9 Strouhal number is defined as $St = \frac{f\delta^*}{U}$, where δ^* is the boundary layer displacement thickness at the airfoil trailing edge (nominally, the boundary layer on the suction side of the airfoil is used when it is placed at an angle of attack), f is frequency and U is defined in Section 4.4.2.

are two possible sources of tip noise: turbulence near the tip-edge itself and turbulence created by the trailing vortex as it passes over or near the trailing edge at the outermost radius of the blade. The production of noise is believed to be via an edge-scattering process similar to trailing-edge noise (see Section 3.3.2, Ffowcs-Williams and Hall (1970) and Amiet (1976)).

There are few published studies relating to tip noise and most relate to fixed-wing or stationary tests in wind tunnels. Many of the early studies concerned with tip-vortex noise originated in the helicopter industry, and were concerned with impulsive blade–vortex interaction (BVI) noise (Coton et al. 2004; Doolan et al. 2001; Mantay et al. 1977; Paterson et al. 1975; Wang et al. 2002), where a trailing vortex interacts with a rotor blade. This causes the familiar thumping noise associated with helicopters in descent, or in the burbling noise created by ingestion of the main rotor wake into the tail rotor. Vortex interactions such as these are thought not to occur on wind turbines because the tip vortices are swept away from the rotor blades by the oncoming air flow and therefore do not have the opportunity to interact with them. Rather, it is noise generated by *vortex formation* at the rotor tip that is of interest here. Hence, the term *tip noise* refers to vortex formation noise rather than other noise sources linked with trailing vortices.

Some of the first experimental tip-noise measurements were performed by Kendall (1978) on fixed flap and wing models. Using a directional microphone, they found that the noise was concentrated at the tips for installed flaps at an angle of attack, but tip noise diminished when the flaps entered stall and the tip vortex lost strength. Similarly, Fink and Bailey (1980) used an acoustic mirror to show increased noise levels towards the tips of a flap and wing model, which reduced as stall approached and the tip vortex system weakened.

Early models of tip noise (called vortex formation noise here) were published by George et al. (1980), who developed an analytical tip-noise model for helicopter rotors (see George (1978) for some additional references to earlier work). The model assumed tip noise was generated by additional turbulence passing the trailing edge of the blade, and was hence based on the theory initially outlined by Amiet (1976). George et al. (1980) found that the analytical formulation compared reasonably with some experimental results. Also, tip-noise was characterised as a high-frequency phenomenon and was found to be important when inflow turbulence is low and the rotors are highly loaded. George and Chou (1984) presented another model of tip noise based on the formation of vortices at the edges of delta wings, with similar levels of agreement.

More detailed experiments were presented by Brooks and Marcolini (1986), who used better noise measurement techniques than earlier studies. Tip-noise spectra were extracted from the measurements by comparing data obtained from two- and three-dimensional airfoils using a scaling and subtraction process. The data confirmed that tip noise is important at high frequencies. Further, it was shown that flat (or square) tips produce more noise than rounded ones. An empirical model was developed using the data, which forms part of the model described in Brooks et al. (1989), which is still in widespread use. The results indicate that tip noise may increase overall levels by up to 5 dB, but care must be taken when extrapolating to full-size wind turbines.

Wagner et al. (1996) summarises the results from some European Union research programs, the aim of which was to reduce noise from wind turbines by modifying tip shape.

A wide variety of tip shapes have been tested, but most do not alter overall noise levels significantly. It was found that bevelling the trailing edge near the tip (*not* the tip edge) resulted in reductions of 3–5 dB. This suggests that trailing-edge noise is the dominant noise mechanism near the tip (either due to high velocity near the tip and/or higher turbulence due to the tip vortex). Moreover, bevelling the trailing edge was shown theoretically by Howe (1988) to reduce trailing-edge noise.

More recently, beamforming results of full-scale wind turbines (Oerlemans et al. 2007) provide an explanation as to why modifying the tip shape has little effect on overall noise levels, while altering trailing edge shape does. In their beamforming images, Oerlemans et al. (2007) show that the peak noise-radiating regions are just inboard of the blade tip edge, and from the trailing edge region. Hence it is the fast-moving, highly turbulent air passing over the edge near the tip (but not over it) that is the dominant source of sound. Thus modifying the trailing edge shape and properties (for example through serrations, an approach recently implemented by Oerlemans et al. (2009)) is likely the most successful noise mitigation strategy to employ on wind turbines.

While tip noise has been studied for over 30 years, there is surprisingly little quality noise and flow data available to validate models or to investigate them more fully. To properly understand tip noise and its relevance to turbines, more studies are required, especially to understand the differences between fixed-wing wind tunnel tests and rotating-blade turbine tests.

3.3.5 Turbulence–Leading-edge Interaction Noise

When turbulent flow encounters an airfoil or wind turbine blade, it will create unsteady lift. On a wind turbine, unsteady lift is caused by the change in angle of attack that the blade experiences when turbulent eddies within the atmospheric boundary layer interact with its leading edge. According to the theory of (Curle 1955, see also Section 3.2.2), unsteady lift is a source of sound, known as turbulence–leading-edge interaction noise or inflow-turbulence noise. This was confirmed semi-quantitatively by Clark and Ribner (1969) and more recently and precisely by Leclercq and Doolan (2009).

Amiet (1975) provides a useful theory for the prediction of turbulence–leading-edge interaction noise (see also Paterson and Amiet (1977), who apply measurements to theory). The theory assumes that an airfoil can be approximated by a thin rigid plate and that the unsteady lift can be determined (in the frequency domain) using a transfer function between the incoming unsteady turbulent flow and the resulting lift. Using the theories of Curle (1955) and Kirchoff (Lamb 1932), the unsteady lift can be converted into a far-field sound pressure field. It should be noted that the radiation pattern of turbulent leading-edge noise depends on the frequency of the incoming turbulent gust. That is, if the turbulent eddy size approaching the blade is much larger than its chord, then the sound radiates as a dipole source, and the blade sectional lift will set the source strength and be proportional to the sixth power of the relative wind velocity (U^6), where U is as defined in Section 4.4.2. If, however, the frequency becomes high, so that the turbulent eddies are smaller than the chord size, then the lift response will become localised towards the leading edge. The directivity pattern will change to something more similar to trailing-edge noise (Section 3.3.2) and will be proportional to the fifth power of the relative wind velocity (U^5).

Wagner et al. (1996) provide a useful scaling analysis that illustrates typical eddy sizes and frequencies associated with turbulence–leading-edge interaction noise. If a turbulent eddy with a length-scale (size) of Λ convects towards a blade with velocity U_c, the convective wavenumber of the eddy will be $k_c = \omega/U_c = 2\pi/\Lambda$, where ω is the angular frequency in radians, defined as $\omega = 2\pi f$, where $f \sim U_c/\Lambda$ is frequency. Similarly, the acoustic wavenumber will be $k = \omega/c = 2\pi/\lambda$, where c is the speed of sound and λ is the wavelength of sound. The frequency of sound created by a blade passing through turbulence is determined by the size of the eddy in the turbulent fluid (represented by Λ) and the speed with which the blade passes through it, relative to the eddy (approximated by U_c). Given that the ratio of acoustic to convective wavenumbers is the convective Mach number ($M_c = U_c/c$), we obtain the following useful expression:

$$\frac{k}{k_c} = M_c \sim \frac{\Lambda}{\lambda} \tag{3.19}$$

Further manipulation using the blade chord, b, gives a non-dimensional relationship between frequency and eddy size.

$$St_b = \frac{fb}{U_c} = \frac{1}{(\Lambda/b)} \tag{3.20}$$

where St_b is the Strouhal number based on blade chord, b. Equation (3.20) can be used to estimate eddy sizes responsible for noise generation at various frequencies and to estimate when the blade can be considered acoustically compact or not. When $St_b \approx 1$, the eddy size is approximately the same size as the chord; when $\Lambda/b > 1$ the blade is a compact noise source and it is a non-compact noise source for $\Lambda/b < 1$. It is reasonable to assume that the turbine tip speed is much greater than the free-stream wind speed, U_∞. In this case, for a typical tip Mach number of $M_{\text{tip}} \approx M_c \approx 0.25$ and tip chord of 1.5 m, the eddy size to frequency relationship is shown in Table 3.1. Here it is assumed that the tip speed is much larger than the wind speed to simplify the estimation of $M_c = U_c/c$. In reality, U_c is the vector sum of the wind velocity and tip velocity.

For the representative test case used to generate the data in Table 3.1, a frequency of $f = 57$ Hz is the point where the blade chord is about the same size as a turbulent eddy. As the frequency under consideration shifts from values lower than 57 Hz to values higher than 57 Hz, the blade transitions from being considered compact (lift generated at the approximate centre of pressure, radiating like a dipole) to non-compact (lift generated closer to the leading edge, where radiation becomes more complex because it is due to many independent surface dipoles).

Although it is interesting to understand the relevant turbulent length scales, the noise level is determined by the turbulent energy spectrum of the atmospheric boundary layer.

Table 3.1 Turbulent eddies and their relationship to a typical wind turbine blade.

Frequency, f (Hz)	1	10	250	500	1000
Acoustic wavelength, λ (m)	343	34.3	1.372	0.686	0.343
Eddy size, Λ (m)	85.75	8.575	0.343	0.172	0.086
Eddy size to chord, Λ/b	57.17	5.717	0.229	0.114	0.057

Assumes $b = 1.5$ m, $c = 343$ m/s and $M_c = 0.25$. At $f = 57$ Hz, $\Lambda/b = 1$.

Measurements (Van der Hoven 1957) of the turbulent power spectrum in the atmosphere at 100 m altitude show that the majority of the turbulent kinetic energy resides at approximately 0.02 Hz. However, significant energy is available for the production of audible noise in the inflow-turbulence power spectrum up to approximately 2000 Hz (Wagner et al. 1996).

Understanding unsteady blade loading has been a successful way to predict helicopter rotor noise (George 1978; George and Chou 1984; George and Kim 1977) and this methodology has been extended to wind turbines (Glegg et al. 1987; Grosveld 1985). Recently, there have been some further developments that aid in our understanding of turbulence–leading-edge interaction noise.

Migliore and Oerlemans (2004) experimentally investigated inflow-turbulence noise from six wind turbine blade airfoil models in a wind tunnel. They showed that the sharper the leading edge (smaller leading edge radius), the higher the measured noise level. Similarly, Moreau and Roger (2005) showed in their wind tunnel measurements of three airfoils of varying thickness in turbulent flow, that noise is reduced as the airfoil model thickness increases. Further, they show that the angle of attack has little influence on noise level. Note that we refer here to the total airfoil thickness, rather than the thickness of a blunted trailing edge, which was referred to in Section 3.3.2.

Mish and Devenport (2006a, b) investigated the unsteady surface pressure and lift response (directly related to noise) of a NACA 0015 airfoil to incoming turbulence in a wind tunnel, and were able to explain some of the effects seen in earlier studies. Measured surface pressure spectra showed a reduction of up to 5 dB with increasing angles of attack, compared with theory, for low reduced frequencies ($\omega b/(2U_c) < 5$).[10] Conversely, at reduced frequencies greater than 5, significant increases in surface pressure spectral levels were observed. It was determined that the reduction in lift spectral level at low frequencies was due to distortion of the inflow caused by the strain rates imposed on the flow field by the airfoil leading edge, which is not accounted for in many theoretical models.[11] It was also shown that this effect only occurs for eddies that are smaller than the chord (non-compact sound radiation cases); eddies larger than the chord are not significantly affected, presumably because the strain rates in the fluid about the leading edge are not able to significantly alter large eddies of this type. The high-frequency increase in lift spectral levels was shown to be a mean-loading effect, in that the thickening of the boundary layer on the suction side created large turbulent eddies over the airfoil, resulting in higher fluctuating pressure amplitudes. Referring to Table 3.1, wind turbine noise above approximately 57 Hz will be radiated from a non-compact source, where the eddies interacting with the airfoil are significantly smaller than the chord. Hence the results of Mish and Devenport (2006a, b) appear to be particularly relevant to understanding wind turbine noise generation.

Further light is shed on turbulence interaction noise by recent work by Devenport et al. (2010b), who measured leading-edge noise from turbulence interaction with an airfoil in a wind tunnel. Glegg and Devenport (2010) confirmed numerically that the level of high-frequency noise reduces as airfoil thickness increases. The effect of angle of attack

10 The reduced frequency is a standard method to present frequency normalised to the airfoil dimensions and flow. It is defined as $\omega b/(2U_c)$, where $\omega = 2\pi f$, f is frequency and U_c is the convective velocity of eddies approaching or passing over the airfoil.

11 Mish and Devenport (2006a) published an excellent literature review of theoretical and experimental studies relating to unsteady surface pressure and lift generation due to turbulent gust–airfoil interaction.

on lift response was found to be significant, but the effect on noise is lost when the response function is averaged with the isotropic turbulence energy spectrum. Interestingly, calculations for anisotropic turbulence show increases in noise, but different airfoil geometries may reduce it. Thus, the main conclusion of recent laboratory and theoretical work over the last decade is that airfoil geometry and inflow turbulence properties significantly affect turbulence–leading-edge interaction noise, but still, further work is necessary to fully understand the problem, especially for the conditions experienced by wind turbines.

3.3.6 Wind-shear Noise

Consider again the overall aerodynamic environment of a wind turbine, as illustrated in Figure 3.5. There are two noise sources associated with the wind shear or incoming atmospheric boundary-layer velocity profile that must be discussed. It should be noted that these are only postulated noise sources and have not been proven. They are gaining more acceptance in the wind turbine aeroacoustic community, but more research is needed to prove they exist and are important.

The first noise source is loading noise. As the blades pass through the atmospheric boundary layer, the angle of attack and dynamic pressure will change due to the changing free-stream velocity conditions. Thus the aerodynamic forces on the rotor blades will change with time. This temporal variation in aerodynamic force will create noise, as shown by Curle's theory (see Section 3.2.2). If important, this noise source is expected to be infrasonic or low-frequency, as it will vary with the blade-pass frequency (typically 1 Hz).

If the local atmospheric conditions become stable, such as at night (Van den Berg 2004), noise of a 'thumping' nature has been noticed from wind farms. One theory that can explain this phenomenon is also related to the incoming atmospheric boundary-layer velocity profile. While 'swish' (see Section 3.3.2) is the usual explanation for amplitude modulation of wind turbine noise (Oerlemans and Schepers 2009), the more intense 'thumping' noise is considered to be a different type of amplitude modulation, known as 'other' or 'enhanced' amplitude modulation (shortened to OAM or EAM) as recently discussed by Oerlemans (2011).

EAM is thought to be created during periods when the atmospheric boundary layer becomes stable; that is, when the velocity of the wind approaching the top of the rotor disc is significantly higher than that approaching the bottom of the disc. If the free-stream velocity at the top of the disc is high, then there is potential for aerodynamic stall to occur at the tip of each turbine blade. The angle of attack of the blade is defined by the difference between the angle formed by the vector sum of the incoming wind velocity and blade rotational velocity, and the pitch angle of the blade. If the incoming wind velocity increases rapidly from the bottom of the rotor disc to the top, the angle of attack may be pushed past its stall angle. As discussed above (Section 3.3.3), separation-stall noise is associated with increased levels of noise, particularly at low frequency. As the separation will last for only part of the blade rotation, the sudden increase in level due to stall onset may create the 'thumping' noise reported by others.

Oerlemans (2011) provide some analysis and argument for EAM, but it is yet to be comprehensively proven. Other recent measurements of full-scale rotor noise using beamforming (Lee et al. 2012) show that the method of wind turbine operation controls noise production. Lee et al. (2012) show that stall-regulated turbines can expect that inflow-turbulence noise (see Section 3.3.5) will be higher than separation-stall noise for most operating conditions. More modern pitch-regulated turbines were shown not to exhibit separation-stall noise at all and turbulent trailing-edge noise (Section 3.3.2) dominated the noise spectrum. Hence there is no clear consensus in the literature. It is most likely that EAM is due to a complex interaction of operation mode, atmospheric stability, propagation effects and aeroacoustics that we are only really beginning to understand.

3.3.7 Blade–Tower Interaction Noise

Blade–tower interaction (BTI) noise is normally associated with downwind turbines, where the interaction of the rotating blades with the wake of the tower creates impulsive noise. Hubbard et al. (1983) accurately and concisely described the noise generated through BTI for downwind turbines. The time-trace of the acoustic measurement shows impulsive peaks associated with the blade passing the tower. As the blade passes into the wake, unsteady aerodynamic loading occurs, resulting in loading noise (see Section 3.2.3). The narrowband spectrum shows many large, low-frequency harmonics related to blade loading caused by wake interaction. For downwind machines, these low-frequency noise components proved to be very annoying and were one of the reasons for the development of modern upwind turbines.

While strong BTI noise is not observed for upwind turbines, it is likely that low-frequency loading noise will be generated. Doolan et al. (2012) showed that impulsive noise may be generated by the interaction of the blades with the perturbed flow upstream of the tower. Figure 3.6 illustrates the phenomenon. The flow over the tower creates a region of non-uniform flow upstream of the tower, represented by the curved streamlines in Figure 3.6. As the rotor blade passes through this perturbed-flow region, the angle of attack changes on the blade, causing a fluctuation in lift force. This fluctuation in lift force creates radiated sound with a timescale associated with the size of the perturbed-flow region upstream of the tower.

To illustrate the timescales and approximate waveform associated with the upstream blade–tower interaction, an approximate model was created by Doolan et al. (2012). The model uses potential flow theory to estimate the flow field upstream of the tower. This is a valid use of potential flow theory, as no boundary-layer separation occurs in this region and the effects of viscosity are small. Using the flow field estimate, the variation of angle of attack with time is estimated for a blade section passing though the perturbed-flow region. This angle of attack history is then converted into time-varying lift using thin airfoil theory. Using the theory of Curle (1955) and assuming a compact source, the source strength can be estimated by taking the time derivative of the lift. Using this method, an estimate of BTI noise source strength, appropriately non-dimensionalised, is

$$\frac{\dot{L}D_T}{U_{\rm tip}qb\ell} = 2\pi\dot{\alpha}\frac{D_T}{U_{\rm tip}} \tag{3.21}$$

where $\dot{L}$ is the time derivative of lift of the small section of blade under analysis, D_T is the tower diameter, q is the dynamic pressure of the flow approaching the blade tip, $U_{\rm tip}$ is the speed of the blade tip, b is the blade chord, ℓ is the length of the span wise region

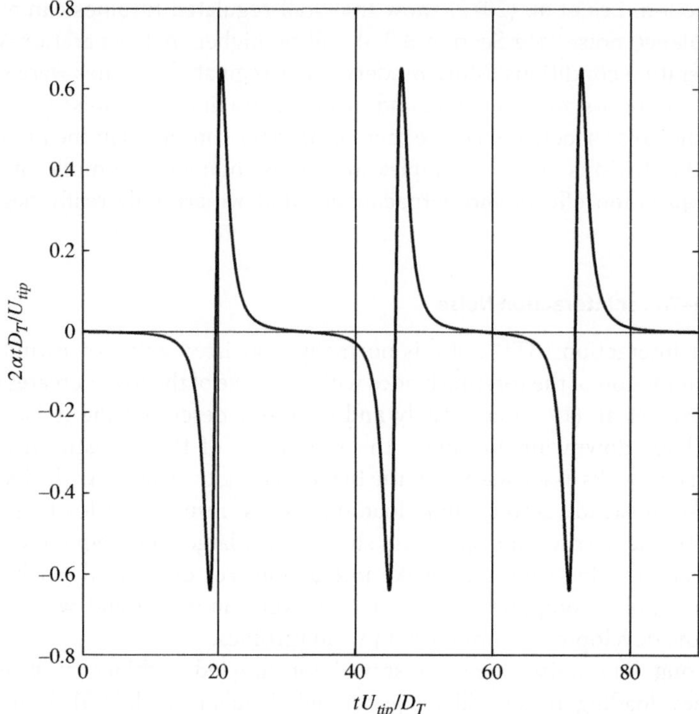

Figure 3.15 Time variation of BTI noise source strength over one revolution of an upwind wind turbine.

of the blade under analysis (assumed to be the outer 20% of the rotor blade) and $\dot{\alpha}$ is the time derivative of the blade angle of attack.

To illustrate the nature of the upstream BTI, Figure 3.15 shows the variation of the strength of the BTI noise source during one complete revolution of a representative upwind turbine. This turbine was assumed to have a rotor diameter of 94 m and a tower height of 100 m. A wind speed of 9.75 m/s, a rotational speed of 14.7 RPM and a chord of 1.5 m was also assumed. Time is shown in a non-dimensional form using the tower diameter and tip speed to determine an appropriate normalising timescale. The noise source calculation assumes the diameter of the tower $D_T = 4$ m and the rotor disc is positioned 4 m upstream of the tower. The calculation was also performed for the blade-tip region of the rotor.

As shown in Figure 3.15, three pulses are generated during each revolution. The creation of each pulse occurs when a blade passes the tower and interacts with the perturbed-flow region. Such a repetitive impulsive noise source will contain a variety of low-frequency components, but the magnitude of this source is at this stage unknown. Hence more work remains to be done to understand the role of BTI noise for upwind turbines. Perhaps the impulsive aerodynamic loading manifests itself through the vibration response of the blades. It may be speculated that the repeated impulsive loading of the blades as they pass though the perturbed flow about the tower will excite their various elastic modes. It may be that vibro-acoustic noise created through impulsive excitation is more noticeable than the direct loading noise. Again, more work needs to be done to understand if these noise sources are indeed an important part of the low-frequency noise spectrum.

Another, speculative, noise source is the interaction between trapped trailing vortices, located between the rotor disc and tower, with a following blade. Consider a blade rotating past the tower. The trailing vortex from the tip is convected downstream by the wind. However, in the presence of the tower, the vortex will become 'trapped' or slowed, stretch about the tower and eventually dissipate. If the vortex slows enough so that its velocity and pressure fields are still large enough to significantly interact with the next blade as it passes the tower, then it is possible that upstream BTI loading noise could be intensified. While plausible, vortex-enhanced BTI remains a hypothesis that needs further investigation.

3.3.8 Thickness Noise

A brief theoretical description of thickness noise was given in Section 3.2.3 and Figure 3.4. In essence, it is the sound produced by the volume displacement of fluid caused a moving object with respect to a fixed observer. It is sometimes referred to as *steady* thickness noise as it is created by the steady motion of a blade surface. Because wind turbines rotate with a blade-passing frequency of approximately 1 Hz, steady thickness noise has in the past been considered to be unimportant (Wagner et al. 1996). This was because the low rotational rate and relatively low tip speed were thought to create sources that were too weak to be audible. There has not been direct experimental confirmation of this idea and further work is necessary to prove or disprove it.

Another form, known as *unsteady* thickness noise, has been shown to be important for wind turbines. Unsteady thickness noise with respect to wind turbines was first discussed by Glegg et al. (1987), who reformulated the original theory of Hawkings (Glegg 1987). Glegg argued that when turbulence interacts with the moving rotor blade, unsteady lift and drag are produced. Unsteady loading noise, associated with the lift component of this interaction, is what is normally considered to be the dominant source of noise. However, when the observer is in the plane of rotation of the blades, the loading noise disappears due to its dipole directionality, and so to does trailing-edge noise, because of the scattering effect of the leading edge (see Section 3.3.2) at moderate frequencies. However, noise is detected in the plane of the rotor (Glegg 1987), and this is attributed to the unsteady drag experienced by the rotor blades as they encounter turbulent gusts.

Similar to loading noise, a turbulent gust will create unsteady drag on the blade, and thus create a dipole source of sound that radiates fore and aft of the blade section on which it is created. The strength of this source is somewhat lower (by about 10 dB) than other sources, but is noticeable. Another source of unsteady thickness noise is flow separation, discussed in Section 3.3.3. Just as separation creates unsteady lift, it will also create unsteady drag and create in-plane noise sources with dipole directivity. While the importance of these sources was shown by Glegg et al. (1987), there have been relatively few discussions in the literature since. One reason may be that the distribution of drag dipoles can be accommodated in the second term of the Ffowcs Willams and Hawkings equation (Eq. (3.10)), as another dipole term. Despite these comments, and considering the importance attributed to these sources by Glegg et al. (1987), care should be taken to include them in rotor-noise calculations. In-plane noise sources may be an important part of the EAM mechanism discussed in Section 3.3.6.

3.4 Aero-elasticity and Noise

Nearly all aeroacoustic analyses of wind turbine blades assume that the blades are rigid. However, wind turbine blades are very long, slender structures, having grown from 30 m to approximately 120 m in length over the last two decades. Therefore the aero-elasticity of wind turbine blades is important and has mainly been assessed from a structural and fatigue life-assessment perspective. Hansen et al. (2006) and Rasmussen et al. (2003) provide excellent reviews of the current state of the art concerning the prediction of aero-elastic response of a wind turbine blade. Hansen (2007) gives an overview of the physics concerning the main types of aero-elastic vibration that occur in wind turbines: *stall-induced vibration* or *stall flutter* is the instability most common on stall-regulated wind turbines while *classical flutter* is the instability associated with pitch-regulated, variable-speed wind turbines.

Stall-induced vibrations (referred to here as stall flutter) occur when the blades experience aerodynamic stall, usually when they are operating near or beyond the stall boundary; that is, when the angle of attack of the blade section is high enough to cause separation of the boundary layer from the surface. Stall-induced vibration can cause large bending moments and other oscillatory loads, affecting the life of the rotor blades. There are no reported noise data directly attributed to stall-induced vibration, but it would seem that the proposed EAM mechanism (Oerlemans 2011) discussed in Section 3.3.6 would be linked. Direct acoustic radiation from stall-induced vibration is low, because the frequencies of the important excited structural modes are small (typically 0.5–4 Hz). Thus it is most likely that stall-induced vibration will modulate the amplitude of separation-stall noise created by the blades.

The 'abruptness' of the blade section stall characteristics will control the onset of stall-induced vibration, or in other words, how quickly lift is lost on the blade after it passes into stall controls the amount of vibration. This is because the rate of lift reduction post-stall sets the amount of negative aerodynamic damping in the system, thus determining the amount of structural damping required to regulate blade vibration. As structural damping is often low, stall-induced vibration is likely during blade stall. It has been found that the introduction of turbulence generators (small triangles or even sand attached along the leading edge of the blade) can reduce the risk of stall-induced vibration by ensuring a turbulent boundary layer on the surface of the blade. A turbulent boundary layer will delay separation and hence stall. The use of turbulence generators also significantly reduces the rate of lift reduction with angle of attack during post-stall operation (that is, they reduce the amount of negative aerodynamic damping). While effective, the introduction of turbulence generators will reduce the power-extraction performance of the wind turbine. Other factors that influence stall-induced vibration include the overall turbine dynamics (including the tower) and the level of structural damping.

Classical flutter is a problem for aeronautical systems; this mechanism has been studied in detail and is included in the design process for aircraft (Fung 1993). Here, a torsional vibration mode of the blade couples with a bending vibration mode through aerodynamic excitation and damping. Classical flutter has highly negative aerodynamic damping (resulting in ever-increasing amplitude once initiated) that cannot be compensated for by structural elements, so designers avoid its occurrence by operating the aircraft away from the so-called *flutter boundary*. To date, there have been no reported cases of classical flutter on a wind turbine, but it is speculated that it might become a problem for pitch-regulated, variable-speed wind turbines as they become larger and have longer, more flexible blades. Hansen (2007) provides the results of a case study for wind turbine blade flutter concerning the virtual 5-MW pitch-regulated turbine designed by NREL, and these illustrate the problems that may occur in future systems. Again, the structural frequencies excited by classical flutter are predicted to be low (< 10 Hz) with corresponding small values of radiation efficiency, so the intensity of direct sound radiation will most likely be low. Flutter may have a modulating effect on aerodynamic noise, but this has not been investigated to date.

Kim et al. (2012) provide the only published study that investigates the effects of blade flexibility (or fluid–structure interaction) on aerodynamic noise from a wind turbine. In this work, the unsteady vortex lattice method was used to predict the aerodynamic loads. Non-linear composite beam theory was coupled with the aerodynamics to calculate the dynamic response of the blades, while semi-empirical models were used to estimate trailing-edge and turbulence–leading-edge interaction noise (see Section 3.3.5), sometimes also referred to as turbulence-ingestion noise. Kim et al. (2012) showed that elastic blades decrease broadband noise because blade flexibility reduces the angle of attack as the load on the blades increases. The effect was more pronounced at moderate speeds when the pitch control system set the mean angle of attack to approximately 5°. At this operating condition, small reductions in angle of attack had a significant effect on noise reduction (1–3 dB). At higher speeds, the angle of attack was near 0° and was less sensitive to small reductions in angle of attack, so there was almost no change in noise due to the flexibility of the blades.

3.5 Other Noise Sources

The other major noise sources from a wind turbine are in the machinery used for conversion of the wind into electrical energy. This is often termed machinery noise and can be grouped into following main sources (Pinder 1992):

- gearbox
- generator
- cooling fans
- auxiliary equipment: oil coolers, hydraulic systems for pitch control.

Machinery noise can be transmitted in one of two ways:

- via airborne paths, with the noise transmitted directly from the source to the receiver
- via structure-borne paths, with the noise transmitted along other structural elements of the turbine before being radiated to the receiver.

Important transmission paths on a wind turbine include the nacelle, hub, tower and rotor blades.

There is very little information regarding machinery noise from wind turbines in the literature. The most informative is a paper by Pinder (1992), who has provided the results of a very useful noise-ranking study. While the results have been developed for wind turbines from the early 1990s, the ranking of the sources is expected to be applicable to modern turbines.

Pinder's study indicates that gearbox noise is the dominant machinery-noise source on a turbine. Further, if the gearbox is mounted to the nacelle housing or machinery raft, structure-borne noise will most likely dominate the machinery-noise spectrum. Gearbox noise escaping from nacelle openings is the next important noise source, followed by structure-borne noise transmitted to the hub, blades, generator and tower (in that order). Cooling fans are the next important source of machinery noise, and this noise is transmitted through nacelle openings or air duct grilles. Generator noise may be transmitted via airborne or structure-borne paths in a similar manner to gearbox noise, but it is ranked lower than cooling-fan noise in terms of importance.

Given the significance of gearbox noise, it is important to understand the construction of wind turbine gearboxes and trends in their development. Dawson and MacKenzie (2014) provide an overview of modern wind turbine gearbox designs and the typical tonal noise frequencies produced by them. Many modern turbines use a multi-stage gearbox, utilising a planetary gear stage that feeds two parallel stages. The aim of the gearbox is to convert the relatively low rotational speed of the turbine rotor shaft to the high-speed needed to run the generator. The speed of the gearbox output shaft needs to turn at the correct speed to generate either 50 Hz (Europe, Australia) or 60 Hz (North America) AC power, and this speed depends on the number of poles that the generator has. Dawson and MacKenzie (2014) provide an analysis of gear mesh frequencies for a typical multi-stage wind turbine gearbox, and this is summarised in Table 3.2. The table shows that typical gear mesh frequencies are low and range from 20–550 Hz.

Transmission errors between gear teeth result in tonal radiated noise at the gear mesh frequencies. These transmission errors may occur due to surface imperfections on the gear teeth from the manufacturing process, misalignment, wear and also elastic deflection from the inter-teeth forces. Wind turbine gearboxes are subjected to very high torque from the main rotor and require rather high gear ratios (22–75) to achieve the high-speed output requirements of the generator. Wind gusts lead to misalignment of the drive train, resulting in transmission errors, noise and gradual failure of the gear components (Ragheb and Ragheb 2010). In addition to gear-based transmission errors, bearings can fail under the random high-amplitude loads imposed by wind gusts. This is especially true for the bearings used at the high-speed output shaft of the gearbox. Bearing noise has not been identified as a noise source in the literature, but practically it can be expected to become noticeable at some stage during the gearbox lifetime.

Table 3.2 Typical wind turbine gearbox gear mesh frequencies

Component	Gear mesh frequency (Hz)
Planet gears	22–25
Low speed shaft sun gear	65–75
Low speed shaft gear	100–120
Intermediate speed shaft pinion	100–120
Intermediate speed shaft gear	500–600
High speed shaft pinion	500–600

Sources: Dawson and MacKenzie (2014) and the authors' own measurements.

The tonal noise created by transmission errors can be significant. Long-term recordings of wind turbine noise (Baath 2013) show tonal components over the 20–550 Hz frequency range that are 5–10 dB above broadband noise levels. In the frequency spectrum, some of these tones are represented by very narrow peaks and are only resolved at narrow frequency resolution and long time averaging (and thus may not be observed in a 1/3-octave representation); others are broad enough to be classed as a tone under some noise standards. If tonal noise is present, a penalty of up to 5 dB can be added to the wind turbine noise output, thus increasing the distance required between the wind farm and dwellings to ensure public amenity.

Amplitude modulation of gearbox noise is also common, resulting in observable tones the frequencies of which are side-bands of the gear mesh frequencies. This is due to the cyclic nature of the torque loading on the gearbox, misalignment of the shafts, irregularities of the gear teeth spacing (either from manufacturing errors, wear or elastic deflection) and damage of gear teeth (Dawson and MacKenzie 2014).

In an attempt to resolve some of the problems associated with wind turbine gearboxes, there have been attempts to remove them completely from the design. The European company Enercon has developed the direct-drive wind turbine, explained in some detail by Ragheb and Ragheb (2010). In this design, the generator has been designed with 50–150 poles, thereby requiring only a slow rotational speed to create 50–60 Hz electrical power, thus eliminating the need for a gearbox. Such a design has been implemented in Europe and Japan, with good results, but is considered to be expensive by some parts of the wind power industry, as it needs special construction techniques and has logistical issues (Cotrell 2002; Oyague 2009).

3.6 Summary and Outlook

This chapter has provided a concise overview of turbulence, aeroacoustic theory and the various mechanisms of noise production on a wind turbine rotor blade. While many of the underlying theories of aeroacoustics are mature, there are still many aspects of wind turbine noise generation that are unresolved and deserve more attention.

In the case of turbulent trailing-edge noise (Section 3.3.2), many aspects of the noise source need further investigation. There is still no consensus on the existence and level of the 'low-frequency' peak. Further, recent reviews (Doolan and Moreau 2015) point to a lack of data that does not allow effective comparison and where it does, there is considerable disagreement in some cases. This is especially true at high Reynolds number, where data are scarce. There is also little noise data for wind turbine airfoil shapes other than the NACA 0012. Given the ever-increasing size of wind turbines, there is a need for more experimental and numerical data to properly understand the role of airfoil shape, Reynolds number, mean flow and turbulence properties (anisotropy, nonhomogeneity, and so on) on radiated noise.

While not a major issue for wind turbines, there is also academic controversy surrounding airfoil tonal noise due to instabilities. While most studies point to the existence of a feedback loop, others suggest otherwise. There is therefore a need for further investigation of tonal noise generation. Turbulent blunt trailing-edge noise, a more common noise source for wind turbines, is reasonably well understood, yet there remains some disagreement as to what thickness of the trailing edge will initiate turbulent vortex shedding of sufficient strength and span-wise coherency to be audible.

Separation-stall noise (Section 3.3.3) is an important noise source, yet there is little data to establish a proper understanding of its generation. While it generally is agreed that noise will increase under this condition, exact mechanisms are not understood and more work is needed. Similarly, there are few recent studies that comprehensively explain tip noise (Section 3.3.4) and more work is required to relate the flow fields at the tip to noise generation and to understand the differences between fixed-wing wind tunnel tests and rotating-blade turbine tests.

Turbulence-leading edge integration noise (or inflow noise, Section 3.3.5) has been studied extensively, with the major noise-generating physics reasonably well understood. Recent important work has revealed the role of blade thickness and loading effects on noise generation, but more work is needed to understand noise generation in the conditions experienced by wind turbines, such as the effects of anisotropic incoming turbulence and blade profile.

Wind-shear is another noise source that is not well understood (Section 3.3.6). It has recently been identified as a part of so-called 'enhanced amplitude modulation' (EAM), the cause of the 'thumping' noise noticed by some people in the vicinity of wind farms. EAM is part of the separation-stall noise phenomenon discussed above. While still the subject of speculation, noise due to wind shear and other meteorological conditions requires further attention. Another speculative noise source that needs to be studied is blade–tower-interaction (BTI), the interaction of a blade with the upstream, curved streamlines of the flow. Such a noise source may also contribute to the 'thumping' character, but it remains a theory until further work is done. Unsteady thickness noise (Section 3.3.6) may also contribute to EAM, but again, more work is needed.

Depending on the operating condition, blade flexibility has been calculated to reduce broadband noise slightly, or not at all.

Machinery noise is dominated by gearbox noise, which has a unique character because it is dominated by amplitude-modulated tones created by transmission errors between gear teeth.

Overall, it can be concluded that we have a reasonable understanding of the major noise generating physics of wind turbines, but many of the finer details that can help us distinguish and explain the peculiarities of wind turbine noise remain unexplored. This is mainly because there still remains a lack of fundamental experimental and numerical data obtained under controlled conditions that can isolate and investigate specific noise mechanisms. Our ability to design quieter turbines for future large wind farms will be limited if this important work is not undertaken.

References

Amiet R 1975 Acoustic radiation from an airfoil in a turbulent stream. *Journal of Sound and Vibration*, **41**(4), 407–420.

Amiet R 1976 Noise due to turbulent flow past a trailing edge. *Journal of Sound and Vibration*, **47**(3), 387–393.

Arbey H and Bataille J 1983 Noise generated by airfoil profiles placed in a uniform laminar flow. *Journal of Fluid Mechanics*, **134**, 33–47.

Arcondoulis E, Doolan C, Zander A and Brooks LA 2010 A review of trailing edge noise generated by airfoils at low to moderate Reynolds number. *Acoustics Australia,* **38**(3), 129–133.

Baath LB 2013 Noise spectra from wind turbines. *Renewable Energy,* **57**(C), 512–519.

Bakker R, Pedersen E, van den Berg G, Stewart R, Lok Wand Bouma J 2012 Impact of wind turbine sound on annoyance, self-reported sleep disturbance and psychological distress. *Science of the Total Environment,* **425**, 42–51.

Blake WK 1986 *Mechanics of Flow-induced Sound and Vibration,* vol. 2. Academic Press.

Brentner K and Farassat F 2003 Modeling aerodynamically generated sound of helicopter rotors. *Progress in Aerospace Sciences,* **39**(2), 83–120.

Brooks T, Pope D and Marcolini M 1989 Airfoil self-noise and prediction. Reference publication 1218, NASA.

Brooks TF and Hodgson T 1981 Trailing edge noise prediction from measured surface pressures. *Journal of Sound and Vibration,* **78**(1), 69–117.

Brooks TF and Marcolini MA 1986 Airfoil tip vortex formation noise. *AIAA Journal,* **24**(2), 246–252.

Ceyhan O 2012 Towards 20 MW wind turbine: high Reynolds number effects on rotor design. *50th AIAA Aerospace Sciences Meeting,* AIAA 2012-1157.

Chase D 1972 Sound radiated by turbulent flow off a rigid half-plane as obtained from a wavevector spectrum of hydrodynamic pressure. *Journal of the Acoustical Society of America,* **52**(3B), 1011–1023.

Chase D 1975 Noise radiated from an edge in turbulent flow. *AIAA Journal,* **13**(8), 1041–1047.

Clark P and Ribner H 1969 Direct correlation of fluctuating lift with radiated sound for an airfoil in turbulent flow. *Journal of the Acoustical Society of America,* **46**(3), 802–805.

Coton FN, Marshall JS, Galbraith RAM and Green RB 2004 Helicopter tail rotor orthogonal blade vortex interaction. *Progress In Aerospace Sciences,* **40**(7), 453–486.

Cotrell JR 2002 A preliminary evaluation of a multiple-generator drivetrain configuration for wind turbines. *ASME 2002 Wind Energy Symposium,* pp. 345–352.

Curle N 1955 The influence of solid boundaries on aerodynamic sound. *Proc. Roy. Soc. London,* **A231**, 505–514.

Dawson B and MacKenzie N 2014 Tonal characteristics of wind turbines drive trains. *Inter-Noise and Noise-Con Congress and Conference Proceedings,* vol. **249**, pp. 2699–2708.

Devenport W, Burdisso R, Camargo H, Crede E, Remillieux M, Rasnick M and Van Seeters P 2010a Aeroacoustic testing of wind turbine airfoils. Subcontract Report NREL/SR-500-43471, National Renewable Energy Laboratory.

Devenport WJ, Staubs JK and Glegg SAL 2010b Sound radiation from real airfoils in turbulence. *Journal of Sound and Vibration,* **329**(17), 3470–3483.

Devenport WJ, Burdisso RA, Borgoltz A, Ravetta PA, Barone MF, Brown KA and Morton MA 2013 The kevlar-walled anechoic wind tunnel. *Journal of Sound and Vibration,***332**(17), 3971–3991.

Doolan C 2013 A review of wind turbine noise perception, annoyance and low frequency emission. *Wind Engineering,* **37**(1), 97–104.

Doolan CJ and Moreau DJ 2015 Review of airfoil trailing edge noise and the implications for wind turbines. *International Journal of Aeroacoustics,* **14**(5–6), 811–832.

Doolan C, Coton Fand Galbraith R 2001 Surface pressure measurements of the orthogonal vortex interaction. *AIAA Journal*, **38**, 88–95.

Doolan C, Moreau D and Brooks L 2012 Wind turbine noise mechanisms and some concepts for its control. *Acoustics Australia*, **40**(1), 7–13.

Ffowcs-Williams J and Hall L 1970 Aerodynamic sound generation by turbulent flow in the vicinity of a scattering half plane. *Journal of Fluid Mechanics*, **40**(4), 657–670.

Ffowcs-Williams J and Hawkings D 1969 Sound generation by turbulence and surfaces in arbitrary motion. *Proceedings of the Royal Society London*, **A264**(1151), 321–342.

Fink M and Bailey D 1980 Airframe noise reduction studies and clean-airframe noise investigation. Final report, United Technologies Research Center, East Hartford, CT.

Fung YC 1993 *An Introduction to the Theory of Aeroelasticity*. Dover Publications.

George AR 1978 Helicopter noise: state-of-the-art. *Journal of Aircraft*, **15**(11), 707–715.

George AR and Chou S 1984 Broadband rotor noise analyses. Contractor report 3797, NASA.

George AR and Kim YN 1977 High-frequency broadband rotor noise. *AIAA Journal*, **15**(4), 538–545.

George AR, Najjar FE and Kim YN 1980 Noise due to tip vortex formation on lifting rotors. *6th AIAA Aeroacoustics Conference* AIAA 80-1010.

Glegg S 1987 Significance of unsteady thickness noise sources. *AIAA Journal*, **25**(6), 839–844.

Glegg SAL and Devenport WJ 2010 Panel methods for airfoils in turbulent flow. *Journal of Sound and Vibration*, **329**(18), 3709–3720.

Glegg S, Baxter S and Glendinning A 1987 The prediction of broadband noise from wind turbines. *Journal of Sound and Vibration*, **118**(2), 217–239.

Grosveld F 1985 Prediction of broadband noise from horizontal axis wind turbines. *Journal of Propulsion and Power*, **1**(4), 292–299.

Hansen MH 2007 Aeroelastic instability problems for wind turbines. *Wind Energy*, **10**(6), 551–577.

Hansen MOL, Sørensen JN, Voutsinas S, Sørensen N and Madsen HA 2006 State of the art in wind turbine aerodynamics and aeroelasticity. *Progress In Aerospace Sciences*, **42**(4), 285–330.

Herr M 2007 Design criteria for low-noise trailing-edges. *13th AIAA/CEAS Aeroacoustics Conference*, AIAA 2007-3470.

Herr M and Dobrzynski W 2005 Experimental investigation in low-noise trailing edge design. *AIAA Journal*, **43**(6), 1167–1175.

Herr M, Appel C, Dierke Jand Ewert R 2010 Trailing-edge noise data quality assessment for CAA validation. *16th AIAA/CEAS Aeroacoustics Conference*, AIAA 2010-3877.

Howe MS 1988 The influence of surface rounding on trailing edge noise. *Journal of Sound and Vibration*, **126**(3), 503–523.

Hubbard H, Grosveld F and Shepherd K 1983 Noise characteristics of large wind turbine generators. *Noise Control Engineering Journal*, **21**(1), 21–29.

Jones LE, Sandberg RD and Sandham ND 2010 Stability and receptivity characteristics of a laminar separation bubble on an aerofoil. *Journal of Fluid Mechanics*, **648**(1), 257–296.

Katz J and Plotkin A 2006 *Low-speed Aerodynamics*, vol. **13**, 2nd edn. Cambridge University Press.

Kendall JM 1978 Measurements of noise produced by flow past lifting surfaces. *AIAA 16th Aerospace Sciences Meeting*, AIAA 78-239.

Kim H, Lee S, Son E, Lee Sand Lee S 2012 Aerodynamic noise analysis of large horizontal axis wind turbines considering fluid–structure interaction. *Renewable Energy*, **42**, 46–53.

Kingan M and Pearse J 2009 Laminar boundary layer instability noise produced by an aerofoil. *Journal of Sound and Vibration*, **322**(4–5), 808–828.

Lamb H 1932 *Hydrodynamics*. Cambridge University Press.

Leclercq D and Doolan C 2009 The interaction of a bluff body with a vortex wake. *Journal of Fluids and Structures*, **25**(5), 867–888.

Lee GS, Cheong C, Shin S and Jung SS 2012 A case study of localization and identification of noise sources from a pitch and a stall regulated wind turbine. *Applied Acoustics*, **73**, 817–827.

Lighthill M 1952 On sound generated aerodynamically: General theory. *Proceedings of the Royal Society London*, **A221**, 564–587.

Mantay WR, Shidler PA and Campbell RL 1977 Some results of the testing of a full-scale Ogee tip helicopter rotor; acoustics, loads, and performance. AIAA 77-1340 in *AIAA 4th Aeroacoustics Conference*, Atlanta, Georgia, USA.

Marsden O, Bogey C and Bailly C 2008 Direct noise computation of the turbulent flow around a zero-incidence airfoil. *AIAA Journal*, **46**(4), 874–883.

Migliore P and Oerlemans S 2004 Wind tunnel aeroacoustic tests of six airfoils for use on small wind turbines. *Journal of Solar Energy Engineering* **126**, 974.

Mish PF and Devenport WJ 2006a An experimental investigation of unsteady surface pressure on an airfoil in turbulence – Part 1: Effects of mean loading. *Journal of Sound and Vibration*, **296**(3), 417–446.

Mish PF and Devenport WJ 2006b An experimental investigation of unsteady surface pressure on an airfoil in turbulence – Part 2: Sources and prediction of mean loading effects. *Journal of Sound and Vibration* **296**(3), 447–460.

Moreau DJ and Doolan CJ 2013 Noise-reduction mechanism of a flat-plate serrated trailing edge. *AIAA Journal*, **51**(10), 2513–2522.

Moreau S and Roger M 2005 Effect of angle of attack and airfoil shape on turbulence-interaction noise, *11th AIAA/CEAS Aeroacoustics Conference*, AIAA 2005–2973.

Moreau S, Roger M and Christophe J 2009 Flow features and self-noise of airfoils near stall or in stall. *15th AIAA/CEAS Aeroacoustics Conference*, AIAA 2009–3198.

Nash E, Lowson M and McAlpine A 1999 Boundary-layer instability noise on aerofoils. *Journal of Fluid Mechanics*, **382**, 27–61.

Oerlemans S 2011 An explanation for enhanced amplitude modulation of wind turbine noise. Contractor report NLR-CR-2011-071, Nationaal Lucht en Ruimtevaartlaboratorium.

Oerlemans S and Schepers J 2009 Prediction of wind turbine noise and validation against experiment. *International Journal of Aeroacoustics*, **8**(6), 555–584.

Oerlemans S, Sijtsma P and Mendez Lopez B 2007 Location and quantification of noise sources on a wind turbine. *Journal of Sound and Vibration*, **299**(4-5), 869–883.

Oerlemans S, Fisher M, Maeder T and Kögler K 2009 Reduction of wind turbine noise using optimized airfoils and trailing-edge serrations. *AIAA Journal*, **47**(6), 1470–1481.

Oyague F 2009 Gearbox modeling and load simulation of a baseline 750-kW wind turbine using state-of-the-art simulation codes. Technical report NREL/TP-500-41160, National Renewable Energy Laboratory.

Paterson R and Amiet R 1977 Noise and surface pressure response of an airfoil to incident turbulence. *Journal of Aircraft*, **14**(8), 729–736.

Paterson R, Vogt P and Munch C 1973 Vortex noise of isolated airfoils. *Journal of Aircraft*, **10**(5), 296–302.

Paterson RW, Amiet RK and Munch CL 1975 Isolated airfoil-tip vortex interaction noise. *Journal of Aircraft*, **12**(1), 34–40.

Pinder JN 1992 Mechanical noise from wind turbines. *Wind Engineering* **16**(2), 158–168.

Pope S 2000 *Turbulent Flows*. Cambridge University Press.

Proudman I 1952 The generation of noise by isotropic turbulence. *Proceedings of the Royal Society of London. Series A, Mathematical and Physical Sciences*, **214**(1116), 119–132.

Ragheb A and Ragheb M 2010 Wind turbine gearbox technologies. *1st International Nuclear and Renewable Energy Conference (INREC), 2010*, pp. 1–8.

Rasmussen F, Hansen MH, Thomsen K, Larsen TJ, Bertagnolio F, Johansen J, Madsen HA, Bak C and Hansen AM 2003 Present status of aeroelasticity of wind turbines. *Wind Energy*, **6**(3), 213–228.

Ribner HS 1962 Aerodynamic sound from fluid dilatations: a theory of the sound from jets and other flows. Technical report, University of Toronto, Institute of Aerophysics.

Ribner HS 1981 Perspectives on jet noise. *AIAA Journal*, **19**(12), 1513–1526.

Schlichting H, Gersten K and Gersten K 2000 *Boundary-layer Theory*. Springer Verlag.

Schumacher KL, Doolan CJ and Kelso RM 2014 The effect of a cavity on airfoil tones. *Journal of Sound and Vibration*, **333**(7), 1913–1931.

Tam C 1974 Discrete tones of isolated airfoils. *The Journal of the Acoustical Society of America*, **55**, 1173.

Tam C, Viswanathan K, Ahuja K and Panda J 2008 The sources of jet noise: experimental evidence. *Journal of Fluid Mechanics*, **615**, 253–292.

Tam CK and Ju H 2012 Aerofoil tones at moderate Reynolds number. *Journal of Fluid Mechanics*, **690**, 536–570.

Tennekes H and Lumley J 1972 *A First Course in Turbulence*. MIT press.

Van den Berg G 2004 Effects of the wind profile at night on wind turbine sound. *Journal of Sound and Vibration*, **277**(4-5), 955–970.

Van der Hoven I 1957 Power spectrum of horizontal wind speed in the frequency range from 0.0007 to 900 cycles per hour. *Journal of Meteorology*, **14**(2), 160–164.

Wagner S, Bareiss R and Guidati G 1996 *Wind Turbine Noise*. Springer Verlag.

Wang T, Doolan C, Coton F and Galbraith R 2002 Experimental study of the three dimensionality of orthogonal blade–vortex interaction. *AIAA Journal* **40**(10), 2037–2046.

Wolf WR, Azevedo JLF and Lele S 2012 Convective effects and the role of quadrupole sources for aerofoil aeroacoustics. *Journal of Fluid Mechanics*, **708**, 502–538.

4

Wind Turbine Sound Power Estimation

4.1 Introduction

This chapter provides means to estimate and measure the sound power generated by wind turbines. Here, the terms 'estimate' and 'predict', are used interchangeably, while acknowledging the strict definition of these terms in statistics. To put our discussion into context, we *predict* the sound power from a wind turbine using empirical models of the various noise sources to *estimate* the individual source strengths. This chapter is limited to methods for wind turbine sound power estimation; methods for estimating the sound pressure level at a large distance from a turbine are detailed in Chapter 5.

We begin by describing various empirical wind turbine sound power prediction methods. As the level of sophistication of the models increase, various aeroacoustic theories are used (see Chapter 3) to obtain estimates of the sound power. The models outlined in this chapter can be used by engineers to predict sound power from wind turbines, either at the design stage or as part of a noise assessment. Blade element momentum theory (BEMT) is also summarised in order to provide a relatively straightforward means of taking into account the effects of blade rotation – without resorting to computational fluid dynamics – in the analyses required to estimate the strengths of the individual noise sources. After the sound power estimation methods are described, a short discussion is provided on recent developments in computational techniques used for wind turbine sound power prediction.

At the end of the chapter, means to measure the sound power output of turbines operating in the field are discussed. A discussion of measurement uncertainty is also included.

4.2 Aerodynamic Noise Prediction

4.2.1 Types of Prediction Methods

Noise prediction methods have many important uses. Incorporation of noise prediction methods into an overall design process will ensure that a wind turbine will produce as much power as possible from the wind while also minimising noise emissions, or at least ensuring that the wind turbine will be compliant with the rules set out by the local regulatory body where the wind farm will be constructed. Other uses include wind

Wind Farm Noise: Measurement, Assessment and Control, First Edition.
Colin H. Hansen, Con J. Doolan, Kristy L. Hansen.

turbine placement or wind farm design. Sophisticated numerical models can be used to understand the noise-generating physics and better understand experimental results obtained from wind tunnels or field testing.

There are a number of different ways of calculating noise emissions from a wind turbine, with varying levels of accuracy and complication. As pointed out by (Wagner et al. 1996, pp. 67–143), Lowson (1993a) has provided a useful classification of wind turbine noise estimation techniques[1]:

Class I These are simple algebraic models that provide estimates of overall sound power level, based on turbine operational parameters. They are calibrated to limited data sets and are considered to give a 'rule-of-thumb' estimate.

Class II These models consider each wind turbine noise source in isolation using a sub-modelling approach. They typically use semi-empirical models that are based on theoretical considerations of the type of noise source. The noise source sub-models are normally calibrated to aeroacoustic wind tunnel data. Class II models are most frequently used to predict noise from wind turbines.

Class III These are sophisticated models of wind turbine noise that incorporate as many details of the turbine operation as possible. Such models include computational techniques where complex, time-dependent fluid dynamics simulations of the entire turbine are combined with a noise model. Computational facilities are only now becoming sufficiently powerful and affordable to begin to attempt these types of simulations.

4.3 Simple Models

The rules-of-thumb for wind turbine sound power estimation are summarised by (Wagner et al. 1996, pp. 67–143) and listed below. The basis for these power estimates is that emitted sound power is proportional to the mechanical or rated generating power of the turbine,

$$W \propto P_{WT} \tag{4.1}$$

where W is the sound power (in watts) and P_{WT} is the rated electrical power output of the wind turbine (also in watts). This leads to the simplest Class I prediction formula (Lowson 1993a),

$$L_{wA} = 10 \log_{10} P_{WT} + 50 \tag{4.2}$$

where L_{wA} is the overall A-weighted sound power level. Models for sound power level based on operational and geometrical parameters are (see (Wagner et al. 1996, pp. 67–143)):

$$L_{wA} = 22 \log_{10} D + 72 \tag{4.3}$$

$$L_{wA} = 50 \log_{10} U_{\text{tip}} + 10 \log_{10} D - 4 \tag{4.4}$$

where D is the rotor diameter in metres and U_{tip} is the rotor tip velocity in metres per second.

---

1 This is a useful classification for all low-Mach-number rotating-blade problems (Carolus and Schneider 2000).

A simple model for sound pressure level at an observer (Hagg 1990, cited in Wagner et al. (1996)) is,

$$L_{pA} = C_1 \log_{10} U_{\text{tip}} + C_2 \log_{10} \left(n_B \frac{A_B}{A_R} \right) + C_3 \log_{10} C_T$$
$$+ C_4 \log_{10} \left(\frac{D}{r} \right) - C_5 \log_{10} D - C_6 \qquad (4.5)$$

where L_{pA} is the overall A-weighted sound pressure level at a distance of $r_0 = \sqrt{r^2 - h^2}$, n_B is the number of blades, A_B is the blade area (m^2), A_R is the rotor area (m^2), C_T is the axial force coefficient, r is the distance from the rotor hub to the observer in metres, r_0 is the distance from the tower base to the observer in metres and h is the hub height above the observer in metres.

The axial force (thrust) coefficient can be estimated using (Frandsen 2007),

$$C_T \approx \frac{3.5(2U - 3.5)}{U^2} \approx \frac{7 \text{ m/s}}{U} \qquad (4.6)$$

where U is the wind velocity incident on the blade and is defined in Section 4.4.2. The constants that should be used in Eq. (4.5) are listed in Table 4.1.

These simple relationships should be used for engineering approximations only. For more accurate calculations, Class II or III models should be used for sound power calculations; the methods in Chapter 5 should be used to calculate the sound pressure level at an observer for a given predicted or measured turbine sound power output, L_w. Class II and III sound power prediction methods will be summarised in the following sections.

4.4 Semi-empirical Methods (Class II Models)

4.4.1 Overall Framework

Semi-empirical methods are a popular means for determining both sound power and level from wind turbines and can be considered as Class II models. Used appropriately, they can provide insight into the dominant noise-producing mechanisms at different frequencies and explain phenomena such as swish. Example publications where semi-empirical models have been used for wind turbines include Boorsma and Schepers (2011); Fuglsang and Madsen (1996); Grosveld (1985); Lowson (1993b); Moriarty and Migliore (2003); Oerlemans and Schepers (2009).

Table 4.1 Constants for use in Eq. (4.5).

Constant	Value (dB)
C_1	63.3
C_2	11.5
C_3	2.5
C_4	20.0
C_5	10.0
C_6	27.5

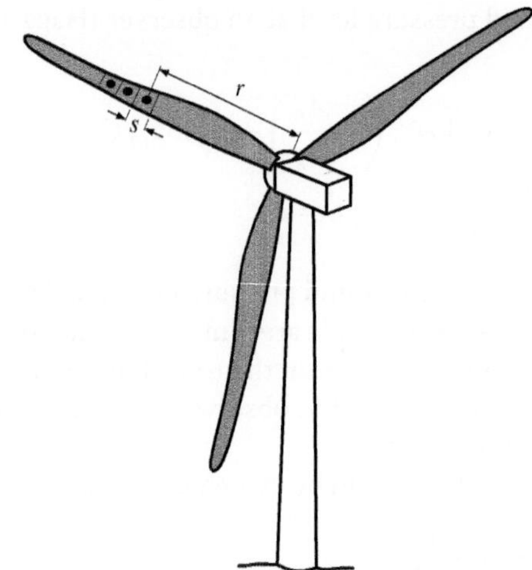

Figure 4.1 Overall framework for semi-empirical modelling of wind turbine noise.

Semi-empirical models are normally applied using a strip-wise technique, as illustrated in Figure 4.1. Each blade is divided into small radial segments with length s, each one a distance r from the centre of the rotor disk. The overall rotor radius is R.

The overall framework of the method can be described as a number of steps:

1. Obtain wind turbine operating conditions, geometrical data and aerodynamic properties.
2. Divide each blade into radial segments.
3. Perform an aerodynamic analysis to determine aerodynamic conditions affecting each segment.
4. Use the aerodynamic analysis results (angle of attack, incoming flow velocity) to determine the boundary layer properties on the blade, especially at the trailing edge.
5. Apply semi-empirical noise prediction models to each segment, for each of the two noise sources – airfoil self-noise and turbulent inflow noise – to obtain the corresponding sound power levels in 1/3-octave bands.
6. Sum logarithmically the sound power levels obtained from each segment and each 1/3-octave band to obtain an overall sound power level for the turbine, assuming each segment is incoherent.

4.4.2 Aerodynamic Analysis

To estimate the aerodynamic noise generated by each rotor segment acting on the air flow, an efficient technique for calculating the flow conditions encountered by each blade segment is required. A method to determine local blade-segment inflow conditions is to use blade-element-momentum theory (BEMT). Leishman (2006) provides an excellent description of BEMT for both helicopters and wind turbines and here, a useful summary will be provided to enable wind turbine sound power calculations to be undertaken.

BEMT is used to solve for the local inflow conditions on each blade segment by performing a momentum balance across successive annuli of the rotor disk. An individual blade segment of width, s (as shown in Figure 4.1), is shown in Figure 4.2. Assuming

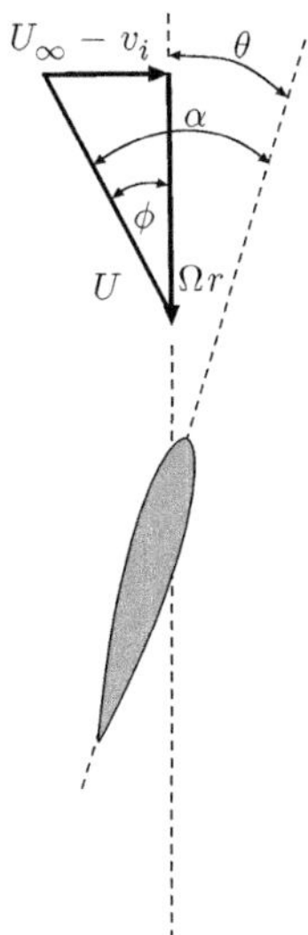

Figure 4.2 Wind turbine blade element model.

that the turbine is operating in an unyawed state,[2] the important flow velocities and angles are shown for a reference frame moving with the blade. In Figure 4.2, U_∞ is the local wind velocity approaching the blade element (averaged over one revolution), v_i is the local induced velocity due to extraction of energy from the wind and ϕ is the local inflow angle, which is a function of the local inflow velocity, $U_\infty - v_i$. The velocity U is the vector sum of Ωr (where Ω is the rotational rate of the rotor in radians per second and r is the radial distance from the rotor centre to the blade segment under consideration) and the local inflow velocity.

The local pitch angle of the blade element is θ (this includes pitch and local twist angles), hence the blade element angle of attack, α, is

$$\alpha = \theta + \phi \tag{4.7}$$

The problem is to calculate the local inflow angle ϕ. This will vary along the rotor radius due to varying geometric parameters, aerodynamic/loading characteristics and local wind velocities (as the turbine operates in a boundary layer). To solve the problem, the local inflow angle can be expressed as (Leishman 2006):

$$\phi = \tan^{-1}\left(\frac{U_\infty - v_i}{\Omega r}\right) \approx \frac{U_\infty - v_i}{\Omega r} = \left(\frac{U_\infty - aU_\infty}{\Omega R}\right)\left(\frac{\Omega R}{\Omega r}\right) = \left(\frac{1-a}{\bar{r}X_{\mathrm{TSR}}}\right) \tag{4.8}$$

Here, the small angle assumption is used and $\bar{r} = r/R$, where R is the rotor radius and r is defined in Figure 4.1. The local induction factor $a = v_i/U_\infty$ and X_{TSR} is the tip-speed ratio,

$$X_{\mathrm{TSR}} = \frac{\Omega R}{U_\infty} \tag{4.9}$$

2 An unyawed state means that the turbine is facing directly into the wind.

The local induction factor can be calculated for each blade using

$$a(\bar{r}, X_{\text{TSR}}) = -\sqrt{\left(\frac{\sigma X_{\text{TSR}} C_{l_\alpha}}{16} + \frac{1}{2}\right)^2 - \frac{\sigma X_{\text{TSR}} C_{l_\alpha} (X_{\text{TSR}} \theta \bar{r} + 1)}{8}} + \left(\frac{\sigma X_{\text{TSR}} C_{l_\alpha}}{16} + \frac{1}{2}\right) \tag{4.10}$$

where $\sigma = N_b b / \pi R$ is the local solidity with N_b equal to the number of blades and b is the chord of the local blade element (defined as the distance between the leading and trailing edges of the airfoil, measured in a straight line). The local aerodynamics are taken into account using the derivative

$$C_{l_\alpha} = \frac{\partial C_l}{\partial \alpha} \tag{4.11}$$

where C_l is the local section lift coefficient.

As C_l is a function of α, the solution of Eq. (4.10) is iterative. For the first iteration, or for an initial approximation, $C_{l_\alpha} = 2\pi$ rad^{-1}, which is the lift-slope for thin airfoils. Look-up tables for the actual section airfoil profiles should be used to obtain accurate results. It must be noted that Eq. (4.10) is valid only for $0 \leq a \leq 0.5$.

Leishman (2006) points out that if an ideal distribution of θ can be obtained across the rotor (through the selection of twist and pitch angles), then a condition of maximum energy extraction can occur and $a = 1/3$. This condition may also be useful for noise calculations during maximum energy harvesting.

Rotor-tip loss effects can be taken into account through modification of BEMT (Leishman 2006). A tip-loss factor F can be derived to take into account the loss in momentum over the rotor plane

$$F = \frac{2}{\pi} \arccos(\exp(-\varphi)) \tag{4.12}$$

where

$$\varphi = \frac{N_b}{2} \left(\frac{\bar{r} - \bar{r}_0}{1 - a}\right) X_{\text{TSR}} \tag{4.13}$$

where $\bar{r}_0 = r_0 / R$ is the non-dimensionalised root-cut-out distance, and r_0 is the distance from the centre of the rotor disk to the root of the rotor blade.

Thus Eq. (4.10) becomes

$$a(\bar{r}, X_{\text{TSR}}, F) = -\sqrt{\left(\frac{\sigma X_{\text{TSR}} C_{l_\alpha}}{16F} + \frac{1}{2}\right)^2 - \frac{\sigma X_{\text{TSR}} C_{l_\alpha} (X_{\text{TSR}} \theta \bar{r} + 1)}{8F}} + \left(\frac{\sigma X_{\text{TSR}} C_{l_\alpha}}{16F} + \frac{1}{2}\right) \tag{4.14}$$

where $F = 1$ is the usual starting guess for calculations. Again Eq. (4.14) is only valid for $0 \leq a \leq 0.5$. A modified equation is provided by Leishman (2006) for cases when $0.5 < a \leq 1$

$$a(\bar{r}, X_{\text{TSR}}, F) = -\sqrt{\left(\frac{\sigma X_{\text{TSR}} C_{l_\alpha}}{16F} - \frac{1}{2}\right)^2 - \left(\frac{1}{2} - \frac{\sigma X_{\text{TSR}} C_{l_\alpha}(X_{\text{TSR}}\theta\bar{r} + 1)}{8F}\right)} - \left(\frac{\sigma X_{\text{TSR}} C_{l_\alpha}}{16F} - \frac{1}{2}\right) \tag{4.15}$$

While BEMT is a desirable and reasonably straightforward method to obtain aerodynamic parameters for a noise analysis, it has its limitations. BEMT cannot properly take into account the wake distortion caused by vertical wind profiles (due to the atmospheric boundary layer), yawed operation or transient effects. To cope with these effects and still keep computational requirements reasonable, vortex wake models are used. Vortex wake models were originally developed for helicopters and have been extended for use on wind turbines. They work by determining the trajectory of vortex filaments from the rotors. In some cases, vortex filaments are determined a priori in space (the 'prescribed' method). Alternatively, a 'free' method is used, in which the filaments are allowed to convect and deform according to the turbine aerodynamic environment. A good summary of vortex wake models can be found in Leishman (2006).

The constant improvement in computational power is now allowing more complicated aerodynamic simulations of entire turbines. Computational fluid dynamics modelling can be performed by solving the unsteady Reynolds-averaged Navier–Stokes (RANS) equations or by solving the more computationally demanding spatially averaged Navier–Stokes equations (large eddy simulation or LES). RANS modelling makes use of temporal averaging and the use of turbulence models in order to cope with the viscous flow physics. LES, on the other hand, resolves eddy motion at scales above the size of the local grid and uses a turbulence model for only the finest scales of turbulence. Each are computationally demanding and need the use of special moving grids and other techniques.

RANS and LES usually need to use wall models to cope with the grid resolution requirements at the blade surface. In some cases, a line-actuator model can be used to replace the blade surface in an LES simulation. The line actuator approach models the blade as a source of momentum flux, the strength of which is set by the aerodynamic parameters of the blade surface, which dramatically reduces the computational requirements. Such techniques are available to engineers but require vast amounts of time and resource to implement properly. A recent review of computational aerodynamic methods for wind turbines was published by Miller et al. (2013).

4.4.3 Boundary-layer Estimates

Once an estimate of angle of attack and local inflow velocity can be made for each blade element, an estimate of the boundary layer height must be made to perform airfoil noise predictions. As with the aerodynamic predictions summarised in the previous sections, there are a number of approaches that can be used.

A straightforward method is to use the BPM empirical model for boundary layer height (Brooks et al. 1989).[3] This model is based on detailed measurements for tripped

3 The BPM model is named after Brooks, Pope and Marcolini, who first proposed the model.

and untripped NACA 0012 airfoils over a Reynolds number (based on chord) range of 40 000–1.5×10^6 and angles of attack of $0° \leq \alpha \leq 25.2°$. An explanation of the BPM model is provided in Appendix B.

The BPM boundary layer model is, however, only completely accurate for NACA 0012 airfoils under the specific conditions used to obtain the data for the empirical model. The BPM data were obtained for both tripped and untripped boundary layers, but the tripped case used a rather heavy trip, causing a large and possibly unrealistic boundary layer height. In order to obtain boundary layer information for arbitrary airfoil shapes, angles of attack and Reynolds numbers, more sophisticated techniques are required.

Panel methods are potential flow solvers widely used in aeronautical design. They can be used with viscous calculation methods to obtain estimates of boundary layer properties at the trailing edge of wind turbine blades for noise estimates. A popular two-dimensional panel method solver is XFOIL (Drela 1989)[4]. The rotation of the wind turbine blade affects the aerodynamics of the airfoil section, particularly near stall. These effects have been incorporated in an updated version of XFOIL, called RFOIL (Timmer and Van Rooij 2003; Van Rooij 1996). Both XFOIL and RFOIL can be used for boundary layer estimates that are suitable for noise predictions using semi-empirical models.

Finally, boundary layer estimates can be made using computational fluid dynamics. While more computationally demanding, continual increases in available computational power are making two-dimensional RANS simulations almost routine. However, it remains to be seen if computational results using RANS are more reliable than those using the panel method codes XFOIL and RFOIL. Large eddy simulation is still too computationally demanding to use for boundary layer estimates for semi-empirical sound power predictions.

4.4.4 Airfoil Noise Models

Trailing-edge Noise

A general form for the sound power level, L_w, for a 1/3-octave band of centre frequency, f, for trailing-edge noise of an individual blade segment can be written, following Boorsma and Schepers (2011); Lowson (1993b), as

$$L_{w,\mathrm{TE}} = 10\log_{10}(4\pi\delta^* M^5 s) + A \tag{4.16}$$

where δ^* is the displacement boundary layer thickness at the trailing edge of the blade segment, s is the length of the blade segment (see Figure 4.1), $M = U/c$, is the Mach number of the flow incident on the blade segment, U is defined in Section 4.4.2 and A is a spectral shape and amplitude function. The function A varies depending on the particular semi-empirical model used; common models will be summarised below.

Lowson (1993b) provides a straightforward model that is only a function of 1/3-octave band centre frequency, f, given by,

$$A = A(f) = 10\log_{10}\left\{\frac{4(f/f_p)^{2.5}}{\left[1+(f/f_p)^{2.5}\right]^2}\right\} + 128.5 \tag{4.17}$$

4 Available for free download at http://web.mit.edu/drela/Public/web/xfoil/.

The peak frequency, f_p, is calculated using (Brooks et al. 1989)

$$f_p = \frac{0.02UM^{-0.6}}{\delta^*} \tag{4.18}$$

where U is the velocity of the flow incident on the blade segment (see Figure 4.2). The Lowson model does not use a full aerodynamic analysis to obtain the boundary layer conditions at the trailing edge of the blade segment. Instead, a modified empirical flat-plate boundary layer model is used

$$\frac{\delta^*}{b} = 0.046 \mathcal{B} Re_b^{-0.2} \tag{4.19}$$

where $\mathcal{B}$ is a boundary layer thickness parameter (chosen to be between 2 and 4), $Re_b = \frac{Ub}{\nu}$ is the Reynolds number based on the chord of the blade segment and ν is the kinematic viscosity of the air.

Grosveld (1985) also provides a reasonably straightforward method for calculating A. He modifies the helicopter noise model of Schlinker and Amiet (1981) to obtain,

$$A = A(f) = 10\log_{10}\left\{\left(\frac{St}{St_p}\right)^4\left[\left(\frac{St}{St_p}\right)^{1.5} + 0.5\right]^{-0.4}\right\} + 65.18 \tag{4.20}$$

where

$$St = \frac{f\delta}{U} \tag{4.21}$$

with the boundary layer thickness, $\delta \approx 8\delta^*$, which can be calculated using the methods outlined in Section 4.4.3, and U is defined in Section 4.4.2. The peak Strouhal number is taken to be $St_p = 0.1$.

Brooks et al. (1989) provide the most popular, and detailed, semi-empirical trailing-edge model, known as the BPM model. This model takes into account boundary layer height, angle of attack, Reynolds number and Mach number. The BPM model also considers separation noise, tip noise, laminar vortex shedding and blunt trailing-edge noise, as part of the spectral function A. A full listing of the BPM model, re-worked so it is in an appropriate form to use in Eq. (4.16), is given in Appendix B.

The BPM model is in wide use, particularly for wind turbines. It has been incorporated into the US National Renewable Energy Laboratory wind turbine prediction code, NAF Noise (Moriarty and Migliore 2003)[5]. NAFNoise incorporates the original BPM boundary layer model as well as improvements using XFOIL (Drela 1989). It also includes a trailing-edge model based on surface pressure measurements and inflow noise models. The BPM model is also used in other analyses, such as Boorsma and Schepers (2011), Fuglsang and Madsen (1996) and Oerlemans and Schepers (2009).

Tip Vortex Formation Noise

The 1/3-octave band sound power level for tip vortex formation noise is

$$L_{w,\text{tip}} = 10\log_{10}[4\pi M^5(1 + 1.036\alpha_{\text{tip}})^3\ell^2] + C \tag{4.22}$$

5 Available at https://wind.nrel.gov/designcodes/simulators/NAFNoise/.

where C is a spectral function and ℓ is the separation over the trailing edge due to the tip vortex (an indication of the tip vortex core size) and $M = U/c$ is the Mach number of the flow incident on the blade segment. A semi-empirical model for this (part of the BPM model) was determined by Brooks et al. (1989). This model specifies

$$C = -30.5[\log_{10}(St'') + 0.3]^2 + 126 \tag{4.23}$$

$$St'' = \frac{f\ell}{U_{\max}} \tag{4.24}$$

where f is the 1/3-octave band centre frequency and $U_{\max}$ is defined as

$$U_{\max} = U(1 + 1.0036\alpha_{\text{tip}}) \tag{4.25}$$

U is defined in Section 4.4.2 and α_{tip} is the geometric angle of attack the tip makes with the oncoming flow (see Figure 4.2 for an illustration of α at any location along the blade). This is suitable for long aspect ratio wings and rotors, such as wind turbines. Some additional correction may be warranted, taking into account the loading at the tip (Brooks et al. 1989), yet there is little to guide those wishing to implement a noise model. At present, the best estimate for tip angle of attack (α_{tip}) for Class II noise models can be obtained from the BEMT model presented earlier in this chapter.

The separation distance, ℓ, for rounded tips is

$$\frac{\ell}{b} \approx 0.008\alpha_{\text{tip}} \tag{4.26}$$

and for square tips

$$\frac{\ell}{b} \approx \begin{cases} 0.0230 + 0.0169\alpha_{\text{tip}} & (0^\circ \le \alpha_{\text{tip}} \le 2^\circ) \\ 0.0378 + 0.0095\alpha_{\text{tip}} & (2^\circ < \alpha_{\text{tip}}) \end{cases} \tag{4.27}$$

where b is the chord at the tip for square tips and for round tips it is the chord at the tip before the rounding process.

Note that these expressions for ℓ/b are provided by Brooks et al. (1989) and are based on limited data. More work is needed to better understand the flow and noise-producing environment at the tips of rotors to improve the accuracy of these models.

Blunt Trailing-edge Noise

An empirical model for sound power generated by rotor blunt trailing edges is

$$L_{w,\text{BTE}} = 10\log_{10}[4\pi M^{5.5}h] + D \tag{4.28}$$

where h is the thickness of the airfoil at the trailing edge, $M = U/c$, is the Mach number of the flow incident on the blade segment and D is a spectral function, developed as part of the BPM model and defined in Appendix B.

4.4.5 Inflow Noise Model

The sound power level in 1/3-octave bands for turbulent inflow noise is

$$L_{w,\text{IN}} = 10\log_{10}\left(4\pi M^5 s\mathcal{L}\frac{u'^2}{U^2}\right) + B \tag{4.29}$$

where s is the length of the blade segment (see Figure 4.1), $\mathcal{L}$ is the integral length scale of the turbulent flow encountering the leading edge of the blade segment and $M = U/c$ is the Mach number based on the inflow air velocity, U. The integral length scale is a dimension that represents, statistically, the eddy size of turbulence. In this case, it is an eddy that represents turbulence at a particular point in the atmospheric boundary layer. Formally, it is defined as the integral of the point velocity auto-correlation function,

$$\mathcal{L} = U_\infty \int_0^\infty R_{uu}(\tau)\,\mathrm{d}\tau \tag{4.30}$$

where R_{uu} is the auto-correlation function of velocity and τ is the time delay (Bendat and Piersol 2011, Ch. 5). Unfortunately, this formal definition is not useful for practical noise calculations. Useful values of $\mathcal{L}$ can be obtained using the empirical methodology presented in the ESDU 85020 standard, which is also conveniently coded into a spreadsheet calculation tool: ESDU 01008.[6] Note that (Wagner et al. 1996, pp. 67–143) suggest that a value $\mathcal{L} \approx 100$ m can be assumed.

To complete the definitions for Eq. (4.29), the variance of the turbulent vertical velocity component is u'^2 (which can also be determined using ESDU 85020) and B is a spectral shape and level function. A convenient and popular inflow noise model for B was developed by Lowson (1993b), based on the work of Amiet (1975), which will be summarised. This model is divided into low- and high-frequency regimes; the high-frequency model should be used if the frequency is above a critical frequency, f_c

$$f_c = \frac{\beta^2 U}{4b} \tag{4.31}$$

where $\beta = \sqrt{1 - M^2}$. The high-frequency model ($f \geq f_c$) for B is

$$B_{hf} = 10\log_{10}\left(\frac{\hat{\omega}^3}{(1+\hat{\omega}^2)^{7/3}}\right) + 110.9 \tag{4.32}$$

where $\hat{\omega}$ is a non-dimensionalised wavenumber,

$$\hat{\omega} = \frac{\pi f b}{U} \tag{4.33}$$

The low-frequency model ($f < f_c$) is

$$B_{lf} = B_{hf} + 10\log_{10}\left(\frac{K_l}{1+K_l}\right) \tag{4.34}$$

where

$$K_l = 10 S^2 M \frac{\hat{\omega}^2}{\beta^2} \tag{4.35}$$

and

$$S^2 = \frac{1}{\frac{2\pi\hat{\omega}}{\beta^2} + \frac{1}{1+2.4\hat{\omega}/\beta^2}} \tag{4.36}$$

6 ESDU standards and computer software can be found at https://www.esdu.com/.

Lowson (1993b) uses the ESDU estimates for turbulent inflow velocity at height, h, from the ground,

$$\frac{U_\infty}{u_\tau} = 2.5 \log_e(h/z_0) \tag{4.37}$$

where z_0 is the ground roughness height, and u_τ is the friction velocity of the atmospheric boundary layer. This can be approximated by

$$u_\tau = 0.4U_{10} \log_e(10/z_0) \tag{4.38}$$

where U_{10} is the velocity at 10 m height. The turbulence level is given by

$$u' = 2.75u_\tau \tag{4.39}$$

4.4.6 Prediction of Total Sound Power

Sound power in each 1/3-octave band is calculated by assuming all sources combine to form a monopole source located at the hub. The method involves the addition of the sound powers from each blade segment and source, assuming they are incoherent, to obtain an overall sound power for the particular 1/3-octave band of centre frequency, f. That is,

$$L_{w,\mathrm{tot}} = 10 \log_{10}\left(N_B \sum_{i=1}^{N} 10^{L_{w,i}/10}\right) \tag{4.40}$$

where $L_{w,\mathrm{i}}$ is the 1/3-octave band sound power level of the ith blade segment, considering all sources. Here, it is assumed there are N blade segments (with $i = 1$ closest to the hub and $i = N$ at the tip) and N_B blades. For the ith blade segment when $i \neq N$(not at the tip), it is defined as,

$$L_{w,\mathrm{i}} = 10 \log_{10}\{10^{(L_{w,\mathrm{TE},i})/10} + 10^{(L_{w,\mathrm{BTE},i})/10} + 10^{(L_{w,\mathrm{IN},i})/10}\} \tag{4.41}$$

where $L_{w,\mathrm{TE},i}$ is the trailing-edge sound power for the ith segment, $L_{w,\mathrm{BTE},i}$ is the blunt trailing-edge sound power for the ith segment and $L_{w,\mathrm{IN},i}$ is the inflow turbulence sound power for the ith segment for the 1/3-octave band of centre frequency, f. For the blade segment at the tip ($i = N$), it is defined as,

$$L_{w,\mathrm{i}} = 10 \log_{10}\{10^{(L_{w,\mathrm{TE},N})/10} + 10^{(L_{w,\mathrm{BTE},N})/10} + 10^{(L_{w,\mathrm{tip}})/10} + 10^{(L_{w,\mathrm{IN},N})/10}\} \tag{4.42}$$

The source is assumed to be located at the hub and to propagate as a monopole. To obtain the sound pressure level at an observer location, the power must be corrected for geometrical spreading, atmospheric attenuation and refraction and ground effects. These corrections are discussed in Chapter 5.

The sound power levels, $L_{w,\,\mathrm{tot}}$, and all other sound power levels in the preceding sections of this chapter are in 1/3-octave bands and f in the equations refers to the 1/3-octave band centre frequency.

4.5 Computational Methods (Class III Models)

As the cost of computational power decreases, it is becoming feasible to perform full-scale wind turbine flow simulations and use these to predict the generated sound power, the so-called Class III models described in Section 4.2. However, there have been very few full rotor *blade-resolved*[7] noise predictions based on a solution of the Navier–Stokes equations. The only recent example is the work of Nelson et al. (2012), who solved the unsteady Reynolds-averaged Navier–Stokes (URANS) equations about a full turbine, including the tower. Acoustic predictions were performed using PSU WOP-WOP, Pennsylvania State University's helicopter noise prediction code that is based on the Ffowcs-Williams–Hawkings!equations (Ffowcs-Williams and Hawkings 1969) modified for use with wind turbines. While an impressive simulation requiring large computational resources, the noise simulation data were not compared with experimental results, so only the feasibility of the technique was demonstrated. Further, the use of the URANS equations restricts the ability of the code to resolve small eddies, hence the prediction of trailing-edge noise will be limited in accuracy. Nevertheless, this study points to the future of wind turbine simulation.

It can be expected that the use of blade-resolved LES will start to become feasible in the near future, as computational tools improve, further enhancing the capability of noise calculation techniques. A recent review of blade-resolved RANS, LES and DES (detached eddy simulation) turbine simulations was performed as part of the study published by Carrión et al. (2014). These authors went on to show that computational fluid dynamics is suitable to predict complex aspects of turbine wakes, such as vortex breakdown.

Another approach recently used to predict full-scale wind turbine noise is to use unsteady vortex lattice methods. Such an approach was used traditionally for the prediction of the low-frequency 'thump' from downwind turbines (Wagner et al. 1996, pp. 67–143). Recently, Kim et al. (2012) combined the unsteady vortex lattice method with beam theory and semi-empirical noise models to examine the effects of aero-elasticity on noise production. It was shown that blade flexibility reduces broadband noise production, due to a reduction in angle of attack as the blades are loaded.

4.6 Estimations of Sound Power From Measurements

The most widely used measurement procedure for determining the sound power emitted from wind turbines is the IEC 61400-11 (2012) standard. The procedure described in this standard is specific to measurements in the vicinity of a wind turbine, where instrumentation is positioned close enough to minimise uncertainties in propagation losses and far enough to assume plane-wave propagation. According to this procedure, a microphone is placed on a measurement board at a specific distance from a wind turbine and the measured sound pressure level is used to estimate the sound power level. The calculated value represents the 'apparent' sound power level of a point source located at the rotor centre that would result in the measured sound pressure level. Sound power data are estimated over a range of wind speeds to enable detailed characterisation of the source under various operating conditions. The source characterisation procedure presented in IEC 61400-11 (2012) also includes an assessment of tonality characteristics of the noise and this is discussed in more detail in Section 6.2.12. The purpose of the IEC

7 That is, the computational mesh is extended to the blade surface so that the boundary layer is resolved.

61400-11 (2012) standard is to provide a consistent and accurate approach to measurement and analysis of wind turbine acoustic emissions for application in various contexts.

4.6.1 Instrumentation

According to IEC 61400-11 (2012), the following requirements relating to instrumentation must be fulfilled for sound power measurements:

1. The sound level meter must conform to Class 1 according to IEC 61672-1 (2013) and must satisfy the requirements of this standard between 20 and 11 200 Hz.
2. The microphone diaphragm diameter must be $\leq$ 13 mm.
3. The microphone - together with its preamp - must be mounted on a ground board, with the diaphragm perpendicular to the board and adjacent to its centre. The ground board is to be made from an acoustically hard material such as plywood or metal, where the thickness is at least 12 mm in the case of wood and 2.5 mm in the case of metal.
4. The microphone must be protected from wind noise with a primary wind screen consisting of half of a standard 90 mm wind screen that covers the microphone, which is placed in the centre of the ground board. To improve the signal-to-noise ratio in high winds and to ensure that masking noise is not overestimated (in the case of tonality assessment – see Section 6.2.12), a secondary wind screen is suggested. This secondary wind screen should have a minimum diameter of 450 mm and should be placed symmetrically over the primary wind screen. The insertion loss of the secondary wind screen must be measured and corrected for, in accordance with the specifications in IEC 61400-11 (2012).
5. The microphone must be calibrated at least at one frequency immediately before and after each set of measurements (or if microphones are disconnected and reconnected) and the calibrator must conform to the requirements in IEC 60942 (2003).
6. The wind speed is determined using the turbine power curve; that is, the plot of produced electrical power against wind speed at hub height. The power curve must refer to the relevant wind turbine model and preferably would have been measured according to IEC 61400-12-1 (2005) or IEC 61400-12-2 (2013). The power curve can only be used over intervals of wind speed where both of the following conditions are satisfied (IEC 60942 2003):
   - there is only a single value of wind speed that corresponds to a specific output power value
   - the slope of the power curve is positive so that for the kth wind-speed bin, the corresponding turbine electrical power, P_k, satisfies the following equation.

   $$(P_{k+1} - P_{\text{tol}}) - (P_k + P_{\text{tol}}) > 0 \tag{4.43}$$

   The first condition above is generally not satisfied at high wind speeds where the output power has reached its maximum value and no longer increases with wind speed. If either of the above conditions are not satisfied, the wind speed from the nacelle-mounted anemometer may be used, but the nacelle speed must be corrected by multiplying it by the average ratio of the power curve wind speed to the nacelle wind speed for all wind speeds that satisfy the two requirements listed above.
7. Other required measurements include temperature and atmospheric pressure at 1.5 m height and wind turbine rotor speed. Measurement of the blade pitch angle is recommended.

4.6.2 Procedure

With regards to measurement procedure, the following criteria must be satisfied:

1. The horizontal distance from the microphone to the wind turbine tower base, R_0, should be $R_0 = H + D/2$, where H is the hub height of the turbine and D is the diameter of the turbine rotor (twice the blade length). The tolerance of this positioning is $\pm$ 20%, maximum $\pm$ 30 m and the measurement accuracy is required to be $\pm$ 2%. The inclination angle, ϕ, which is the angle between the horizontal and the direct line of sight to the turbine hub at the measurement position, is between 25° and 40°. The position of the measurement microphone with respect to the wind turbine is shown in Figure 4.3.
2. Sound-reflecting surfaces should not contribute more than 0.2 dB to the measured levels.
3. The ground board should be positioned flat on the ground and the measurement area made level with the existing ground.
4. All acoustic measurements must be recorded and stored for later analysis.
5. Data corrupted by extraneous noise should be discarded.
6. As a minimum, the analysis must include hub-height wind speeds from 0.8 to 1.3 times the wind speed at 85% of the maximum turbine electrical power, rounded to bin-centre wind speeds. Wind-speed bins are 0.5 m/s in width and centred around integer and half-integer wind speeds. Wind speeds at the lower bound of each wind-speed bin are included; wind speeds at the upper bound are excluded.
7. Background noise must be measured before and after each measurement series with the wind farm shut down, ensuring that the instrumentation set-up is the same and the wind conditions are similar.
8. A broad range of wind speeds should be covered.
9. At least 180 measurements are required overall for both total noise and background noise.
10. At least 10 measurements are required in each wind-speed bin for both total noise and background noise. In some cases, 30 spectra may be required where tones are found to be intermittent.

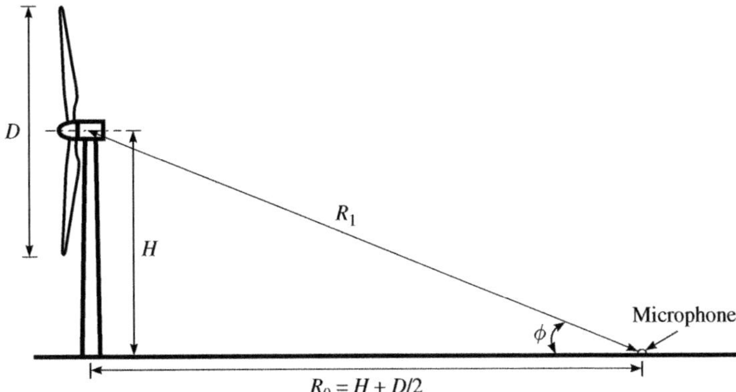

Figure 4.3 Placement of microphone for determining the sound power level of a wind turbine according to IEC 61400-11 (2012).

11. Measurements are valid when the microphone is positioned downwind from the wind turbine, within the range of ±15°. Of course, measurements at other angles can be obtained for use in propagation models for predicting wind farm noise in directions other than downwind.

4.6.3 Data Analysis

The aim of the sound power analysis is to determine, using statistical methods, A-weighted sound power spectra in 1/3-octave bands and overall A-weighted sound power level. This involves measuring 10-s periods of the equivalent A-weighted noise level and the A-weighted 1/3-octave spectrum over frequencies from 20 Hz to 10 kHz. The average wind speed is determined for concurrent 10-s intervals and the 1/3-octave spectra are normalised, divided into wind-speed bins and averaged as discussed below. It is important to note that acoustic data are energy averaged (see Section 2.2.8), whereas wind speed data are arithmetically averaged. Where the average value of the wind speed is not located at the wind-speed bin centre, linear interpolation between the bin centre and an adjacent bin is used to determine the sound pressure level associated with the wind-speed bin centre for each 1/3-octave band. The selection of the adjacent bin (out of the two possible – one on each side) is determined by which side of the bin centre that the average wind speed lies. The chosen adjacent bin is the one that lies furthest from the average wind speed, so that linear interpolation (rather than extrapolation) can be used to find the sound pressure level corresponding to the bin-centre frequency. The same procedure is applied to the background noise. The operational noise spectrum for each wind-speed bin is then corrected by logarithmically subtracting the background noise spectrum.

Results are to be marked with an asterisk if the difference between the sum of the 1/3-octave bands of the operational noise and the sum of the 1/3-octave bands of background noise is between 3 and 6 dB. If this difference is 3 dB or less, the result for the corresponding wind-speed bin is not reported. The uncertainty associated with the analysis is determined according to details provided later in this section. The expected standard deviation (or standard uncertainty) may be taken as ±2 dB in the absence of data (Keith et al. 2016).

Normalised 1/3-Octave Spectra

Third octave spectra are normalised to account for logarithmic summation of the individual 1/3-octave bands, generally resulting in a slightly higher value than the measured overall level. This is because there is leakage of a small amount of energy from each 1/3-octave band into adjacent 1/3-octave bands due to the properties of 1/3-octave filters. The normalisation procedure is as follows:

1. The calculated A-weighted sound pressure level, $L_{Aeq,o,j}$, is determined through logarithmic summation of the levels in individual 1/3-octave bands, $L_{Aeq,i,j}$, using Eq. (4.44).

$$L_{Aeq,o,j} = 10 \log_{10} \sum_{i=1}^{28} 10^{(L_{Aeq,i,j}/10)} \tag{4.44}$$

Here, the subscript i refers to the 1/3-octave band number (where $i = 1$ for the 20-Hz 1/3-octave band, and so on) and the subscript j refers to the jth measurement period.

2. The difference, Δ_j between the measured equivalent A-weighted noise level, $L_{Aeq,j}$ and $L_{Aeq,o,j}$ is found using Eq. (4.45).

$$\Delta_j = L_{Aeq,j} - L_{Aeq,o,j} \tag{4.45}$$

3. The difference, Δ_j is added to each 1/3-octave band to give the normalised 1/3-octave spectrum, $L_{Aeq,n,i,j}$.

$$L_{Aeq,n,i,j} = L_{Aeq,i,j} + \Delta_j \tag{4.46}$$

The normalised 1/3-octave spectra are corrected for the presence of a secondary wind screen, if applicable, using the insertion loss values determined according to Section 6.2.3.

Average Spectra per Wind-speed Bin

The normalised 1/3-octave spectra are sorted into wind-speed bins, k. To reduce complexity, we will use the equivalent notation expressed in Eq. (4.47) from this point onwards, consistent with IEC 61400-11 (2012).

$$L_{Aeq,n,i,j,k} = L_{i,j,k} \tag{4.47}$$

The average sound pressure level, $\overline{L}_{i,k}$ for each 1/3-octave band, i, is determined using Eq. (4.48).

$$\overline{L}_{i,k} = 10 \log_{10}\left(\frac{1}{N}\sum_{j=1}^{N} 10^{(L_{i,j,k}/10)}\right) \tag{4.48}$$

where N is the number of measurements in wind-speed bin, k.

Average Wind Speed

The average wind speed, $\overline{V}_k$ is determined through calculation of the arithmetic average of wind speeds in bin k as expressed in Eq. (4.49).

$$\overline{V}_k = \frac{1}{N}\sum_{j=1}^{N} V_{j,k} \tag{4.49}$$

where $V_{j,k}$ is the 10-s average wind speed value calculated for measurement period j.

Uncertainty Analysis

Uncertainty components of both Type A and Type B are considered as part of the sound power analysis and these quantities are expressed in the form of standard uncertainties, which are combined using the method of combination of variances to give the combined standard uncertainty. Type A uncertainties are determined based on the statistics of a number of measurements of the same quantity, whereas Type B uncertainties are estimates based on the expected variations about a mean due to such factors as (IEC 61400-11 2012):

1. calibration ($\pm$0.3 dB)
2. instrument response (frequency dependent; see Section 6.2.16)
3. measurement board ($\pm$0.5 dB)
4. wind screen insertion loss ($\leq$ 0.4 dB at frequencies below 1 kHz)

5. distance and direction (±0.2 dB)
6. air absorption (for which it is best not to correct per IEC 61400-11 (2012))
7. weather conditions (±0.8 dB)
8. wind speed measured using nacelle anemometer or meteorological mast (±1.2 dB)
9. wind speed derived using power curve (±0.3 dB).

Items 1–7 relate to sound pressure level measurement whereas items 8 and 9 are associated with wind speed measurement.

The listed Type B uncertainty ranges have to be converted to expanded uncertainties as discussed in Section 6.2.16. The standard uncertainty for Type B contributions is found using Eq. (6.42) and assuming a rectangular distribution (IEC 61400-11 2012). Type A and Type B uncertainties that have both been converted to expanded uncertainties are then combined using Eq. (6.43).

The latest IEC standard (IEC 61400-11 2012) outlines a rather complex procedure that takes into account the covariance between the measured sound pressure level and the measured wind speed. The covariance is a measure of the linear relation between random variables and is taken into account in this situation because the sound pressure level and wind speed are correlated. The number of measurements available in each wind-speed bin is also considered in the analysis. The uncertainty associated with the operational and background measurements is then combined. The applied method is slightly different if the operational levels in a given 1/3-octave band are less than 3 dB higher than the background levels and the result is marked with square brackets. For details on this method, the reader is advised to consult IEC 61400-11 (2012).

Apparent Sound Power Levels

The apparent sound power level in each wind-speed bin and 1/3-octave band, $L_{WA,i,k}$ is calculated from the background corrected sound pressure level data that corresponds to the wind-speed bin centres, $L_{V,c,i,k}$. An adjustment of 6 dB is made to account for the pressure doubling that occurs due to the presence of the measurement board and it is assumed that spherical spreading occurs between the source and receiver. Thus,

$$L_{wA,i,k} = L_{V,c,i,k} - 6 + 10 \log_{10} \left(\frac{4\pi R_1^2}{S_0} \right) \tag{4.50}$$

where R_1 is the direct distance, in metres, from the rotor centre to the microphone, S_0 is the reference area and $S_0 = 1\ \text{m}^2$.

The overall A-weighted sound power level in each wind-speed bin is found by logarithmically summing the levels in all 1/3-octave bands. The overall uncertainty for each wind-speed bin is also calculated.

Typical 1/3-octave sound power plots for a 3-MW wind turbine operating at various hub height wind speeds are shown in Figure 4.4 (Kaiser-Wilhelm-Koog GmbH 2005). Error bars have been included to indicate the uncertainty associated with the measurements in each third-octave band.

4.6.4 Comments on Turbine Sound Power Measurements

Typically sound power measurements are undertaken by the wind turbine manufacturer under ideal conditions, so the actual sound power emitted by the wind turbine installed in the field will vary depending on topography, proximity to other wind turbines, surface

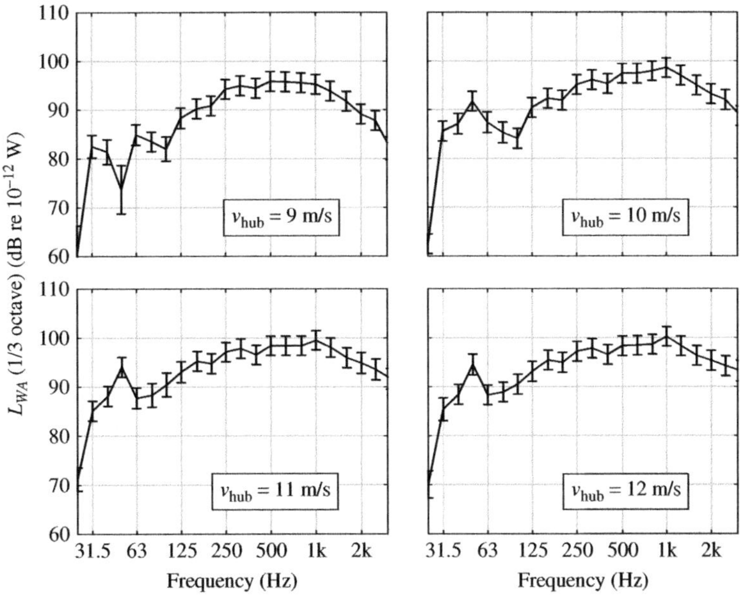

Figure 4.4 Typical 1/3-octave power plot obtained using the measurement procedure outlined in IEC 61400-11 (2012). Data from Kaiser-Wilhelm-Koog GmbH (2005).

roughness and meteorological conditions. In situations where the topography differs significantly from the flat terrain used for sound power measurements, differences in the results are expected. In cases where the terrain is undulating, the resulting increase in the inflow turbulence results in a significant increase in the wind turbine sound power output. Turbulence from the wake of upwind turbines will also result in increased inflow turbulence. Currently the influence of topography and turbine layout on wind turbine noise generation is not well understood and more research in this area is needed. Surface roughness is an important factor when deriving hub height wind speed from measurements made at another height, as roughness length is contained in Eq. (5.17), which is used for this calculation. Therefore, if the roughness at the location where the wind turbine is installed differs from the ideal site used to calculate sound power, there can be variations in the results (Søndergaard 2005). Meteorological conditions also influence the level of inflow turbulence, as well as the vertical inflow angle. The existence of high levels of wind shear can also cause increased turbine noise levels as a result of the blades experiencing stall, as discussed in Sections 3.3.6 and 6.2.10.

4.6.5 Possible Improvements to Procedures for Measuring Turbine Sound Power Levels

An alternative procedure for determining turbine sound power levels was described by Buzduga and Buzduga (2015). The main advantage of this alternative method is that the contribution of background noise to the measurements can be taken into account without shutting down the wind turbine. The method involves measuring at two positions that are in line with the wind turbine hub (in other words, in the downwind direction) and satisfy the positioning requirements outlined in IEC 61400-11 (2012), as described

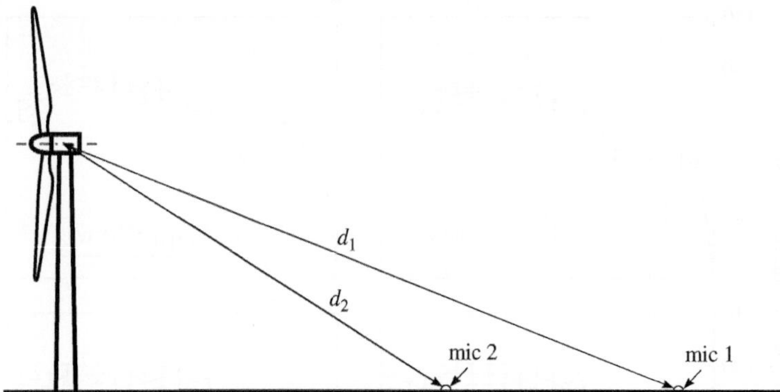

Figure 4.5 Placement of microphones for determining the sound power level of a wind turbine without having to measure backgound noise.

in Section 4.6.2 and shown in Figure 4.5. Assuming that the directivity and excess attenuation are the same for both measurement positions, the difference in sound pressure level at positions 1 and 2 is given by Eq. (4.51).

$$L_{s2} - L_{s1} = 20\log_{10}\left(\frac{d_1}{d_2}\right)^2 \tag{4.51}$$

where L_{s2} and L_{s1} are the sound pressure levels measured at positions 1 and 2, respectively and d_1 and d_2 are the direct line distances between the measurement positions 1 and 2 and the hub centre and $d_1/d_2 > 1$.

The noise level measured at positions 1 and 2, L_{s+n}, can be considered as a logarithmic sum of the noise level associated with the wind turbine, L_s and the background noise level, L_n as shown in Eq. (4.52).

$$10^{L_{s,i}/10} + 10^{L_{n,i}/10} = 10^{L_{s+n,i}/10} \tag{4.52}$$

where the index $i = 1, 2$ refers to measurement positions 1 and 2, respectively.

This results in three simultaneous equations that can be solved to give the sound pressure level at position 1, as follows,

$$L_{s1} = 10\log_{10}(10^{L_{s+n,2}/10} - 10^{L_{s+n,1}/10}) - 10\log_{10}\left(\left(\frac{d_1}{d_2}\right)^2 - 1\right) \tag{4.53}$$

Some additional constraints are imposed by Buzduga and Buzduga (2015) to maintain consistency with IEC 61400-11 (2012):

- The combined wind turbine and background sound pressure level must exceed the wind turbine only sound pressure level; that is, $L_{s+n,i} > L_{s,i}$.
- Data are only reported when the wind turbine noise exceeds the ambient levels by more than 3 dB; that is, when $L_{s,i} - L_{n,i} > 3$ dB.
- The measurement is rejected when the level measured at position 1 is greater than or equal to the measured level at position 2; that is, when $L_{s+n,1} \geq L_{s+n,2}$.

4.7 Summary

This chapter has provided a summary of various calculation and measurement methods to estimate the sound power generated by a wind turbine during operation. The calculation methods are classed as either I, II or III, depending on their level of sophistication. Class I predictions are quick to perform, yet are the least accurate. Class II predictions involve a mix of aerodynamic and aeroacoustic theory and use empirical noise models, and are the most popular and feasible method of estimating sound power. Class III predictions are the most sophisticated as they attempt to model as much of the noise-producing physics and noise propagation as possible. While there are some notable attempts at progressing Class III models, it is not feasible to use them until the necessary computational power becomes more affordable.

The last part of the chapter has outlined procedures for determining the sound power generated by wind turbines and included a discussion of the uncertainty associated with these measurements.

References

Amiet R 1975 Acoustic radiation from an airfoil in a turbulent stream. *Journal of Sound and Vibration*, **41**(4), 407–420.

Bendat J and Piersol A 2011 *Random Data: Analysis and Measurement Procedures*. Wiley.

Boorsma K and Schepers J 2011 Enhanced wind turbine noise prediction tool SILANT. *Proceedings of the Fourth International Meeting on Wind Turbine Noise*, Rome, Italy.

Brooks T, Pope D and Marcolini M 1989 Airfoil self-noise and prediction. Reference publication 1218, NASA.

Buzduga V and Buzduga A 2015 A new method for determining the wind turbine noise based on the constant divergence of the sound pressure level. *6th International Meeting on Wind Turbine Noise*, Glasgow, UK.

Carolus T and Schneider M 2000 Review of noise prediction methods for axial flow fans. *Proceedings of InterNoise 2000*, Nice, France.

Carrión M, Woodgate M, Steijl R, Barakos G, Gomez-Iradi S and Munduate X 2014 Understanding wind-turbine wake breakdown using computational fluid dynamics. *AIAA Journal*, **53**(3), 1–15.

Drela M 1989 *XFOIL: An analysis and design system for low Reynolds number airfoils. Low Reynolds Number Aerodynamics*, vol **54**, pp. 1–12. Springer.

ESDU 01008 2010 Computer program for wind speeds and turbulence properties: flat or hilly sites in terrain with roughness changes.

ESDU 85020 2002 Characteristics of atmospheric turbulence near the ground.

Ffowcs-Williams J and Hawkings D 1969 Sound generation by turbulence and surfaces in arbitrary motion. *Proceedings of the Royal Society London*, **A264**(1151), 321–342.

Frandsen ST 2007 Turbulence and turbulence-generated structural loading in wind turbine clusters. Technical report Risø-R-1188(EN), Risø National Laboratory, Denmark.

Fuglsang P and Madsen H 1996 Implementation and verification of an aeroacoustic noise prediction model for wind turbines. Technical report Riso-R-867(EN), Riso National Laboratory, Denmark.

Grosveld F 1985 Prediction of broadband noise from horizontal axis wind turbines. *Journal of Propulsion and Power*, **1**(4), 292–299.

Hagg F 1990 Aerodynamic noise reduced design of large advanced wind turbines. *Proceedings of the European Community Wind Energy Conference*, Madrid, Spain, pp. 384–388.

IEC 60942 Ed.1.0 2003 Electroacoustics – sound calibrators. International Electrotechnical Commission.

IEC 61400-11 Ed.3.0 2012 Wind turbines – Part 11: Acoustic noise measurement techniques. International Electrotechnical Commission.

IEC 61400-12-1 2005 Wind turbines. Part 12-1: Power performance measurements of electricity producing wind turbines. International Electrotechnical Commission.

IEC 61400-12-2 2013 Wind turbines. Part 12-2: Power performance of electricity-producing wind turbines based on nacelle anemometry. International Electrotechnical Commission.

IEC 61672-1 2013 Electroacoustics – sound level meters. Part 1: Specifications. International Electrotechnical Commission.

Kaiser-Wilhelm-Koog GmbH 2005 Report of acoustical emissions of a wind turbine generator system of the type v90-3mw, mode 0 near Bökingharde (Germany). Technical Report WT 4245/05, Kaiser-Wilhelm-Koog GmbH.

Keith S, Feder K, Voicescue S, Soukhoftsev V, Denning A, Tsang J, Broner N, Richarz W and van den Berg F 2016 Wind turbine sound power measurements. *Journal of the Acoustical Society of America*, **139**(3), 1431–1435.

Kim H, Lee S, Son E, Lee S and Lee S 2012 Aerodynamic noise analysis of large horizontal axis wind turbines considering fluid–structure interaction. *Renewable Energy*, **42**, 46–53.

Leishman J 2006 *Principles of Helicopter Aerodynamics*, 2nd edn. Cambridge University Press.

Lowson MV 1993a Assessment and prediction of wind turbine noise. Technical report ETSU W/13/00284/REP, Flow Solutions Ltd, Bristol, United Kingdom.

Lowson MV 1993b A new prediction model for wind turbine noise. *International Conference on Renewable Energy – Clean Power 2001, 1993*, London, pp. 177–182.

Miller A, Chang B, Issa Rand Chen G 2013 Review of computer-aided numerical simulation in wind energy. *Renewable and Sustainable Energy Reviews*, **25**, 122–134.

Moriarty P and Migliore P 2003 Semi-empirical aeroacoustic noise prediction code for wind turbines. Technical report, NASA Center for Aerospace Information.

Nelson CC, Cain AB, Raman G, Chan T, Saunders M, Noble J, Engeln R, Dougherty R, Brentner KS and Morris PJ 2012 Numerical studies of wind turbine acoustics. *50th AIAA Aerospace Sciences Meeting*, Nashville, TN, vol. 6.

Oerlemans S and Schepers J 2009 Prediction of wind turbine noise and validation against experiment. *International Journal of Aeroacoustics*, **8**(6), 555–584.

Schlinker R and Amiet R 1981 Helicopter rotor trailing edge noise. Contractor report 3470, NASA.

Søndergaard B 2005 Noise measurements according to IEC 61400-11. How to use the results. *First International Meeting on Wind Turbine Noise: Perspectives for Control*, Berlin, Germany.

Timmer W and Van Rooij R 2003 Summary of the Delft University wind turbine dedicated airfoils. *Journal of Solar Energy Engineering*, **125**(4), 488–496.
Van Rooij R 1996 Modification of the boundary layer calculation in RFOIL for improved airfoil stall prediction. Report IW-96087R, TU-Delft, the Netherlands.
Wagner S, Bareiss R and Guidati G 1996 *Wind Turbine Noise*. Springer Verlag.

5

Propagation of Noise and Vibration

5.1 Introduction

Almost all wind farm developments require wind farm noise level predictions to be made for residences that may be affected by noise. Generally, to predict the noise level outdoors at a particular residence due to a wind farm, it is necessary to know:

- the turbine sound power level
- the distance of the residence from each turbine in the wind farm
- the terrain character and type of ground cover between each turbine and the residence
- the meteorological conditions.

As ground surface conditions vary considerably over long distances and as atmospheric wind and temperature conditions vary with time and location, any calculations made using standard propagation models are approximate. Results should therefore always be reported with an expected uncertainty value as well as a description of the meteorological conditions to which the predictions apply. In addition, the sound level arriving at a receiver is affected by the sound power output of the wind farm, which in turn is affected by turbine spacing and associated wake and turbulence effects, wind shear, turbine rotational synchronicity, tower height, blade length and turbine power settings. Wind farm noise generation and sound power estimation are discussed in Chapters 3 and 4. The measurement of sound power level and sound pressure level are discussed in Chapter 6.

Wind farms also produce modulated noise and often noise with multiple tonal components, which may not add significantly to the average noise level but which may add greatly to the annoyance and disturbance experienced by residents. The result is that any wind farm noise level predicted at a residence by a propagation model will not necessarily be representative of the annoyance of the noise. Moreover, due to all of the uncertainties involved, it may not be a very accurate representation of the range of noise levels that will be experienced either. As will be seen in following sections, at best we may expect an uncertainty in the predicted noise level at residences of ±3 to 4 dBA for any specified atmospheric condition.

Wind Farm Noise: Measurement, Assessment and Control, First Edition.
Colin H. Hansen, Con J. Doolan, Kristy L. Hansen.

The prediction of the level of noise that arrives at a sensitive receiver location from a wind farm relies on mathematical models to simulate all of the variables that influence the propagation of the noise, and this is referred to as *propagation modelling*. The treatment of propagation modelling here begins with an introduction to the general physical principles that underpin propagation models and a discussion of the various effects that influence the sound level arriving at the receiver. These effects include:

- atmospheric absorption
- ground reflections and absorption
- terrain and manmade barriers
- scattering due to vegetation and atmospheric turbulence
- meteorological effects such as atmospheric stability and wind and temperature gradients
- reflections from vertical surfaces
- effects of large source heights
- the effect of the spreading out of the sound rays as they move further away from the sound source ('spherical spreading'), which results in a 6-dB reduction in sound pressure level for each doubling of distance from the noise source.

The preceding effects are taken into account separately in most propagation models, although they can sometimes be combined. The effects of the ground cover, terrain profile between the wind farm and the location at which the sound level is to be estimated, atmospheric absorption, atmospheric wind and temperature gradients and vegetation are referred to as 'excess attenuation', as they generally add to the reduction in sound level as a result of distance from the noise source. However, sometimes the excess attenuation can be negative, (especially the attenuation due to ground reflection effects), which means that it adds to the expected noise level, resulting in a reduction of noise of less than 6 dB for each doubling of distance from the source. Excess attenuation effects are calculated separately for each octave or 1/3-octave band, although some propagation models combine two or three excess attenuation effects together. As these excess attenuation effects are frequency dependent, the noise spectrum is divided into octave bands for the simpler propagation models and 1/3-octave bands for the more complex and accurate models. Finer frequency resolution than 1/3-octave cannot be justified due to the lack of accuracy of the model predictions.

Following the discussion of general principles, various propagation models that have been used for wind farm noise predictions are discussed, beginning with some very simple models and ending with two very complex models, with three models in between that are neither very simple nor overly complex. The description of propagation models ends with a simple model for the calculation of noise levels at a receiver from an offshore wind farm.

It is also of interest to be able to determine the effect of wind farm noise on interior noise levels in residences, so that the potential for sleep disturbance and annoyance can be assessed. For this reason, the transmission of noise from outside to inside residences is discussed in Section 5.13.

The chapter finishes with a brief discussion of the generation of ground vibration by a wind farm, how this vibration is propagated to a sensitive receiver and how it is detected at the receiver location.

5.2 Principles Underpinning Noise Propagation Modelling

The various propagation models discussed in the following sections are used to calculate the sound pressure level at a receiver as a function of the sound power level of a single sound source such as a wind turbine. Sound pressure level contributions at a receiver location from individual turbines are combined on an energy basis (see Section 2.2.8), with the relative phase of sound arriving from different turbines not taken into account.

Accurate estimations of noise levels at receiver locations require accurate estimations of the attenuation of sound due to propagation from the turbines to a receiver location as well as accurate estimates of sound power levels radiated by the turbines making up the wind farm. The remainder of this chapter discusses how to calculate the attenuation of sound due to propagation from the turbines to the receiver location. Estimates and measurements of turbine sound power are discussed in Chapter 4. Accounting for uncertainty in these estimates is discussed in Section 5.12. Generally, worst-case turbine sound power levels should be provided by the turbine manufacturer, but this is not always the case.

In propagation modelling, the turbine noise source is modelled as a point source, which is a valid representation provided that the receiver distance from the turbine is greater than approximately two rotor diameters. Although propagation models based on calculations at single frequencies produce 'interference dips' at frequencies above a few hundred hertz, these dips are considerably reduced when frequency averaging over a 1/3-octave band and also as a result of other effects such as atmospheric turbulence and the continual movement of the rotor sound sources responsible for the noise. An 'interference dip' is the reduction in sound pressure level at particular frequencies due to destructive interference between the direct sound ray and the ground-reflected ray. There can also be other interference dips at low and infrasonic frequencies due to the interference between sound rays arriving from different turbines in a wind farm. This can be made worse if turbines are rotating synchronously such that blades from several turbines pass the tower at the same time. All types of sound, including low-frequency sound and infrasound, can vary quite substantially in level at a particular location as a result of continually varying atmospheric wind and temperature gradients. As an example, the sound level at a receiver location 2 km from the nearest turbine in a wind farm and downwind from it, can increase by up to 20 dB over what is calculated for a neutral atmospheric condition (Verheijen et al. 2011). Propagation models are generally used to quantify worst-case noise levels corresponding to downwind propagation in the presence of a temperature inversion (as often occurs at night).

The general equation for calculating the octave band or 1/3-octave band sound pressure level at a receiver, located at a distance, r, from the ith turbine in a wind farm, for a given octave band or 1/3-octave band turbine sound power level, L_{wi}, of the ith turbine noise source is as follows.

$$L_{pi} = L_{wi} - 20 \log_{10} r - 11 - A_{Ei}; \quad \text{(dB)} \tag{5.1}$$

where the turbine noise source is considered to be a point source and where the sound power level, L_{wi}, for a wind turbine is based on the sound pressure level measurement close to the turbine in the downwind direction to simulate worst-case conditions as described in Section 4.6.

The excess attenuation A_{Ei}, for sound travelling to the receiver from sound source, *i*, is calculated using,

$$A_{Ei} = A_{\mathrm{atm}} + A_{\mathrm{gr}} + A_{\mathrm{bar}} + A_{\mathrm{met}} + A_{\mathrm{misc}} \tag{5.2}$$

Sometimes the effect due to spherical spreading is included as A_{div} in the excess attenuation equation instead of the general equation, and in this case Eq. (5.1) becomes

$$L_{pi} = L_{wi} - A_{Ei}; \quad \text{(dB)} \tag{5.3}$$

and Eq. (5.2) becomes

$$A_{Ei} = A_{\mathrm{div}} + A_{\mathrm{atm}} + A_{\mathrm{gr}} + A_{\mathrm{bar}} + A_{\mathrm{met}} + A_{\mathrm{misc}} \tag{5.4}$$

The excess attenuation components, A_{div}, A_{atm}, A_{gr}, A_{bar}, A_{met} and A_{misc} refer respectively to attenuation due to spherical spreading, absorption by the atmosphere, effects due to the ground, effects of barriers and obstacles, meteorological effects such as wind and temperature gradients and miscellaneous effects such as vegetation screening, reflections from vertical surfaces, source height and in-plant screening (screening due to objects in an industrial facility, such as buildings and tanks). In the more advanced propagation models, the ground effect and barrier effect are combined and then further combined with the effects of atmospheric refraction caused by vertical wind and temperature gradients as well as the effects of atmospheric turbulence. The ground effect is usually negative, resulting in an increased sound pressure level at the receiver.

The sound pressure level at a particular receiver location due to a single turbine, calculated using Eq. (5.1), is combined with the sound pressure level calculated for all other turbine sources to give the total sound pressure level, using Eq. (2.27). The A-weighted sound pressure level in each band is then calculated by arithmetically adding the A-weighting to each octave or 1/3-octave band, using Table 2.2. The overall A-weighted sound pressure level generated by each turbine at the receiver location is then calculated by logarithmically adding all of the octave *or* 1/3-octave band A-weighted levels (depending on which propagation model is being used), using Eq. (2.27).

While it is generally recognised that the various components of attenuation are inter-related and not simply additive, only the very complex propagation models attempt to combine two or three effects. However, the simpler models approximate the total excess attenuation as a linear (or arithmetic sum) of the individual effects. That is, if two individual attenuations are 3 dB and 2 dB respectively, the overall excess attenuation would be 5 dB.

Prior to discussing individual propagation models, some general observations will be made about each of the excess attenuation effects just mentioned. Where calculation of the effects is the same in all propagation models, details will be discussed here and then referenced later in the discussion of each propagation model.

5.2.1 Spherical Spreading, A_{div}

The reduction (attenuation) in sound level as a function of distance from the source is known as spherical spreading and is a result of the sound ray spreading out into an ever-increasing volume. The value in decibels that has to be subtracted from the source sound power level, L_w, to account for spherical spreading and consequently to obtain the sound pressure level at a specified distance from the noise source, is a function of the type of source. Most sources can be modelled as point, line or plane sources and the most appropriate model for a wind turbine is a point source for a receiver at distances greater than about 200–300 m from the turbine, which are the normal distances to residences affected by wind farm noise and for which noise impact calculations are done at the planning stage. In this case, the spherical spreading term for use in Eq. (5.4) is given by,

$$A_{\text{div}} = 10 \log_{10} (4\pi r^2) = 20 \log_{10} r + 11 \quad \text{(dB)} \tag{5.5}$$

In some texts, a source on hard ground is modelled as if it were radiating into a hemispherical half-space so the '11' in Eq. (5.5) is replaced with '8'. This is a valid approach provided that the ground-effect term, A_{gr}, in Eq. (5.4) is set equal to zero. However, it is preferable to leave the '11' in Eq. (5.5) and to account for all ground surfaces (including hard ground) with the ground-effect term, A_{gr}, in Eq. (5.4).

When do Wind Farms Behave like Line Sources?

It is well documented (Bies and Hansen 2009) that noise from a line of point sources, such as traffic, will decay at a rate of 3 dB per doubling of distance from the centre of the line of sources, whereas sound from a single point source decays at a rate of 6 dB per doubling of distance from the source, provided the receiver is more than a wavelength from the source. The correct sound level at any receiver location more than $(b/2) \cos \alpha_\ell$ (see Figure 5.1) from the centre of a line of incoherent point sources will be obtained by calculating the contribution of each source to the receiver sound level and then adding the contributions incoherently, as described in Section 2.2.8. However, as the receiver location moves away from the centre of the line of point sources, the sound does not decay at a rate of 6 dB per doubling of distance as would be expected for a point source. In fact, the equation for the spherical spreading part of the excess attenuation in Eq. (5.4) for a line of three or more sources, if the total sound pressure level from all sources is to be calculated, is:

$$A_{\text{div}} = 10 \log_{10} \left(\frac{\alpha_u - \alpha_\ell}{4\pi r_0 b} \right) \tag{5.6}$$

where α_u and α_ℓ are the angles subtended between the lines joining the receiver to each end of the line of sources and the line from the receiver that is perpendicular to the line of sources (or its extension), as illustrated in Figure 5.1. Also, r_0 is the normal distance from

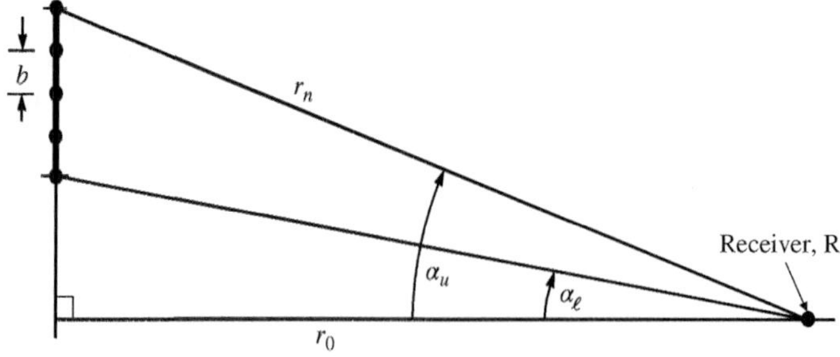

Figure 5.1 Geometry for a line of point sources.

the receiver to the line of sources or its extension and b is the average separation distance between sources, again as shown in Figure 5.1. The dependence on r_0 rather than r_0^2 would indicate a 3-dB attenuation per doubling of distance from the noise source. However, as the receiver becomes further away from the line of sources, the $(\alpha_u - \alpha_\ell)$ quantity (in radians, not degrees, where 1 radian = 57.3 degrees) becomes smaller, so the attenuation rate increases above 3 dB per doubling of distance until it eventually reaches 6 dB per doubling of distance. For a line of 37 wind turbines 400 m apart, the attenuation rate due to spherical spreading from 1 km to 2 km directly out from the centre of the line of turbines in a direction normal to the line joining the turbines is 3.4 dB; between 2 km and 4 km it is 3.9 dB and between 10 km and 20 km it is 5.6 dB. It can therefore be seen that for a real wind farm, sound does not decay anywhere near 6 dB per doubling of distance within a few kilometres of the nearest turbine. The attenuation rate in a direction at a shallow angle from the line joining all turbines, as a function of distance from the end turbine, is slightly less than the values mentioned above. However, as mentioned, the sound level at any particular location (at non-infrasonic frequencies) will be correctly calculated by considering each turbine as a point source and adding the contributions of each source incoherently to obtain the total receiver sound pressure level.

However, as the frequency extends lower than 20 Hz, the sound wavelengths become very long and sound arriving at the receiver from different wind turbines becomes more coherent. In addition, in this frequency range the sound from a wind turbine is dominated by tonal frequencies related to the blade-pass frequency. This means that sound arriving from different turbines can constructively and destructively interfere, which greatly affects the attenuation rate with increasing distance from the wind farm. It is quite probable that at these frequencies, there will be locations, P_i, where the sound from two or more turbines is in phase and produces a level that is greater than at some other location less than half the distance of P_i from the nearest turbine. In addition, the effects of wind and atmospheric turbulence, together with small variations in turbine rotational speeds results in the relative phase between sound arriving at any given location from different turbines changing with time, so that locations of constructive interference can change to ones of destructive interference in tens of seconds or at even shorter time intervals, thus creating a considerable variation in the sound level experienced by a stationary receiver.

For the case of a single wind turbine, the noise is generated by a number of sources and, as an additional complication, the aerodynamic sources associated with blade motion are moving. The added complication of moving sources and their distribution over the plane of the rotor makes analysis of sound propagation very complex, so propagation models usually approximate the turbine as a single stationary point source. The assumption of the noise source being located at hub height can lead to errors in the calculation of the sound pressure level arriving at the receiver location where, in downwind conditions (corresponding to a downward-refracting atmosphere), more than one ground-reflected ray will arrive at the receiver. The higher the assumed height of the source, the greater will be the distance from the turbine at which more than one ground-reflected ray will arrive at the receiver. Infrasound and low-frequency noise (ILFN) are generated at the point where each blade passes the tower as a result of the flow-velocity deficit experienced. In some cases, ILFN may also be generated as the blades approach the top of their arc (Renewable UK 2013), when dynamic stall may occur due to the blade having a stall angle of attack as it enters the faster-moving air above the boundary layer. This latter generation mechanism is yet to be definitively proven. However, it is important to know which mechanism is dominating the low-frequency sound and infrasound if the correct source height is to be used and the location at which multiple ground reflections affect the noise propagation is to be determined. Beyond this location, the attenuation of sound with distance for a single turbine source will decrease at a rate close to 3 dB per doubling of distance. It is important to note that once a receiver is a horizontal distance, d, of more than 1.5 rotor diameters from a turbine, for the purposes of sound propagation calculations, the turbine may be considered as a stationary point source. In fact, Makarewicz (2011) shows analytically that a turbine with a rotor height, h and a blade length, B_L, may be considered as a point source if the following relationship is satisfied.

$$\sqrt{d^2 + h^2} > B_L\sqrt{5} \tag{5.7}$$

5.2.2 Atmospheric Absorption, A_{atm}

The excess attenuation, A_{atm}, due to atmospheric absorption is given by

$$A_{\mathrm{atm}} = \alpha_a d \quad \text{(dB)} \tag{5.8}$$

where d is the distance (in metres) travelled by the sound ray from the source to the receiver.

The attenuation, α_a in decibels per metre of travel through the atmosphere for a pure tone frequency, f, is given by the standard ANSI/ASA S1.26 (2014) as

$$\alpha_a = 8.686 f^2 \left\{ \left[1.84 \times 10^{-11} \left(\frac{P_a}{P_r}\right)^{-1} \left(\frac{T}{T_r}\right)^{1/2} \right] + \left(\frac{T}{T_r}\right)^{-5/2} \times \left[0.01275\, \mathrm{e}^{-2239.1/T} \left(\frac{f_{ro}}{f_{ro}^2 + f^2}\right) + 0.1068\, \mathrm{e}^{-3352/T} \left(\frac{f_{rN}}{f_{rN}^2 + f^2}\right) \right] \right\} \quad \text{(dB/m)} \tag{5.9}$$

where T is the ambient atmospheric temperature in Kelvin (273.15 +°C), T_r is the reference temperature in Kelvin, P_a is the ambient atmospheric pressure, P_r is the reference atmospheric pressure (101 325 Pa),

$$f_{ro} = \left(\frac{P_a}{P_r}\right) \left[24 + \left(\frac{(4.04 \times 10^4 h)(0.02 + h)}{0.391 + h} \right) \right] \quad \text{(Hz)} \tag{5.10}$$

and

$$f_{rN} = \left(\frac{T}{T_r}\right)^{-1/2} \left(\frac{P_a}{P_r}\right) \left\{ 9 + 280h\, \mathrm{e}^{-4.170[(T/T_r)^{-1/3} - 1]} \right\} \quad \text{(Hz)} \tag{5.11}$$

where h is the % molar concentration of water vapour which may be calculated from the % relative humidity, h_{rel} by

$$h = h_{\mathrm{rel}} \left(\frac{P_{\mathrm{sat}}}{P_r}\right) \left(\frac{P_a}{P_r}\right)^{-1} \tag{5.12}$$

where

$$\frac{P_{\mathrm{sat}}}{P_r} = 10^V \tag{5.13}$$

and

$$V = 10.79586\left[1 - \left(\frac{273.16}{T}\right)\right] - 5.02808\log_{10}\left(\frac{T}{273.16}\right) + 1.50474\times10^{-4}\left\{1 - 10^{-8.29692[(T/273.16)-1]}\right\} + 0.42873\times10^{-3}\left\{-1 + 10^{4.76955[1-(273.16/T)]}\right\} - 2.2195983 \tag{5.14}$$

or V can be calculated to a close approximation using

$$V = -6.8346\left(\frac{273.16}{T}\right)^{1.261} + 4.6151 \tag{5.15}$$

5.2.3 Ground Effect, A_{gr}

The effect of the ground on sound arriving at a receiver from a wind turbine is treated differently by the different noise propagation models and will be discussed quantitatively in the sections dealing with each model. However, it is of interest to qualitatively discuss the various rays that arrive at the receiver from the source as a result of interaction with the ground. These should all be included in the ground effect part of the excess attenuation in an accurate noise propagation model by use of the spherical-wave reflection coefficient rather than just the plane-wave reflection coefficient (see Appendix C). There are three waves that should be considered, as listed below. The simplest sound propagation models only include the first reflected wave type, as they only use the plane-wave reflection coefficient and not the spherical-wave reflection coefficient.

- **Reflected wave** is the wave that is reflected from the ground; it is included in all propagation models. It combines at the receiver with the direct wave from the source and can either reinforce or interfere with the direct wave, respectively increasing or decreasing the sound level, depending on the relative phase between the direct and reflected wave and whether or not atmospheric effects affect the coherence between the two waves. The location of specular reflection can be found by assuming an image source exists directly beneath the actual source and as far below the ground surface as the actual source is above it. A straight line is then drawn between the image source and the receiver. The point of specular reflection is where this line intersects the ground surface, as shown by point O in Figure 5.2.
- **Ground wave** is an additional contribution to the reflected wave field arising from diffusion of the image source as a result of spherical wave incidence on the ground. This wave is taken into account by use of the spherical-wave reflection coefficient in the more advanced propagation models.
- **Surface wave** is an entirely separate wave that propagates along the ground surface, with exponentially decreasing amplitude above the ground. This wave is not affected by atmospheric refraction due to wind and temperature gradients and is more likely to exist at low frequencies, which could explain why low-frequency sound is not attenuated as much as mid- and high-frequency sound in upwind directions. On the other hand, in downwind directions it is likely to be insignificant compared to the ground wave and ground-reflected waves. This wave is also taken into account by use of the spherical-wave reflection coefficient in the more advanced propagation models.

At very low frequencies (below about 20 Hz) ground absorption is negligible and, in addition, the direct and ground-reflected waves can combine coherently at the receiver.

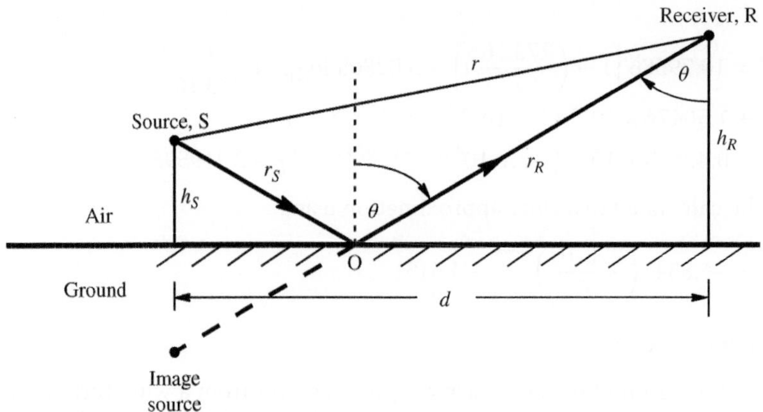

Figure 5.2 Geometry illustrating reflection of a wave incident on the ground surface.

As the difference in distance travelled by the direct and ground-reflected waves from the turbine source to a receiver 1.5 m above the ground, is a small fraction of a wavelength at these frequencies, it is appropriate to allow a 6-dB rather than 3-dB ground correction (Jakobsen 2012). Between 20 and 100 Hz, the ground correction changes gradually from 6 dB to 3 dB for grass-covered ground; that is, an increase over the sound pressure level due to the direct ray only. At frequencies higher than 25 Hz, the ground effect should be calculated more accurately using the models discussed in the remainder of this chapter. For hard ground such as water or concrete, the frequency at which the ground correction begins to deviate from 6 dB is much higher than it is for grass-covered ground. As the distance from the source becomes sufficiently large, rays that have been reflected more than once from the ground will begin to arrive at the receiver and beyond this distance the ground effect can become even greater than 6 dB. This is discussed in more detail in Section 5.2.4.

5.2.4 Meteorological Effects, A_{met}

If the weather conditions are not accurately known, then it is generally assumed that meteorological effects will result in a sound level variation about the predicted level at the receiver, as shown in Table 5.1.

The variations in Table 5.1 are primarily due to variations in vertical wind shear and temperature gradients, which cause variations in the vertical sound-speed gradient. This in turn causes the sound rays to bend towards or away from the ground, depending on whether the resulting vertical sound-speed gradient is positive or negative, and this causes significant variations in the sound level arriving at the receiver.

Wind shear is the common name for a vertical wind speed profile in which the wind speed changes with height above the ground. Due to surface friction effects, the wind is almost always characterised by an increasing speed as distance from the ground increases. Generally wind shear is greater at lower wind speeds and the spread in its value is usually greater at night. Wind shear and its variance are greater for flat terrain than for hilly terrain (Bowdler 2009).

In the direction upwind from the source (that is, the direction into the wind from the source), sound rays are diffracted upwards away from the ground and may result

Table 5.1 Variability in sound level predictions due to meteorological influences, A_{met} (dB), including both upwind and downwind conditions.

Octave band centre frequency (Hz)	Distance from the source (m)			
	100	200	500	1000
	A_{met} (dB)			
63	+1	+4, −2	+7, −2	+8, −2
125	+1	+4, −2	+6, −4	+7, −4
250	+3, −1	+5, −3	+6, −5	+7, −6
500	+3, −1	+6, −3	+7, −5	+9, −7
1000	+7, −1	+11, −3	+12, −5	+12, −5
2000	+2, −3	+5, −4	+7, −5	+7, −5
4000	+2, −1	+6, −4	+8, −6	+9, −7
8000	+2, −1	+6, −4	+8, −6	+9, −7

in a *shadow zone*. In this zone, the only sound in the mid- and high-frequency ranges that reaches a receiver is what is scattered due to atmospheric turbulence or due to other obstacles impacted by the sound ray, although this may not be the case for IFLN (see Section 5.2.3). Thus in the upwind direction, the sound experienced at a receiver is usually less than would be expected in the absence of wind. Although this means that propagation in the upwind direction is not the worst case in terms of the noise level experienced at a receiver, it is still of interest for wind farm noise. This is because the attenuation of low-frequency sound and infrasound in the upwind direction is less than it is for mid- and high-frequency sound for the reasons discussed in Section 5.2.3.

In the direction downwind from the source, sound rays are refracted towards the ground, so that the sound level at large distances from the source is likely to be greater than would be expected in the absence of wind. At a sufficient distance from the source, there will be multiple rays striking the ground, with all except one having been reflected from the ground more than once. At distances greater than the shortest distance from the source that multiple reflections occur, there will be less attenuation of sound with distance than might otherwise be expected and this is the main reason that noise propagation models (except for the Nord2000 model, which has a correction for multiple ground reflections occurring between the source and receiver) are not considered reliable at distances greater than 1 km from the source. However, as wind turbine noise sources are very high above the ground, this distance can sometimes be extended to 3 km (see Section 5.2.4).

In rural areas, where wind farms may be found, atmospheric temperature inversions often occur in addition to wind shear in the early hours of the morning when there is a clear, cloud-free sky. In these cases, the atmospheric temperature increases with altitude close to the ground, instead of the more normal condition where it decreases with altitude. The layer of air in which the temperature is increasing is known as the *inversion layer* and this is usually just a few hundred metres thick. The effect of an atmospheric temperature inversion is to cause sound rays to bend towards the ground in much the same way as they do when propagating downwind. Conversely, the more normal (especially during the daytime) non-inverted atmospheric temperature profile results in sound rays that are refracted upwards, away from the ground, thus resulting in reduced sound pressure levels at the receiver. However, only a light downwind condition needs to exist to counteract the effect of the usual atmospheric temperature gradient in non-inverted conditions.

Atmospheric stability, wind gradients and temperature gradients also affect the sound power output of the turbines as well as the propagation of noise from the turbines to a receiver.

When calculating the appropriate sound power to be used in the modelling, it is sometimes necessary to be able to determine the hub-height wind speed from anemometer measurements at a different height. The wind speed, $U(h)$, at height, h, can be calculated approximately from the wind speed, U_0, at height, h_0 as follows,

$$U(h) = U_0 \left(\frac{h}{h_0} \right)^{\xi} \tag{5.16}$$

where ξ is the wind shear coefficient, which is a function of surface roughness and atmospheric stability (see Figure 5.3).

Values of the wind shear coefficient may also be derived from data provided by Sutton (1953). Values of ξ for various ground surfaces and neutral (adiabatic) atmospheric conditions (Pasquill stability category D; see Table 5.8) are listed in Table 5.2. The last four

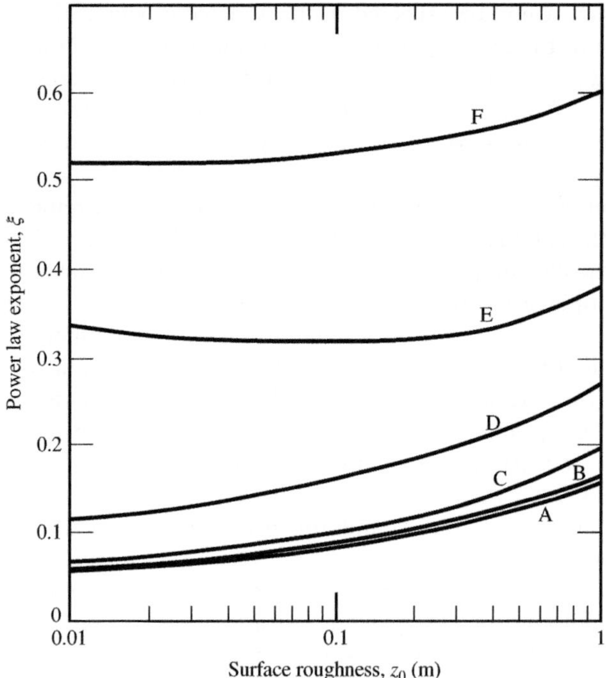

Figure 5.3 Estimates of the wind shear coefficient for wind speed as a function of surface roughness, based on averaged measurements between 10 and 100 m above the ground (after Irwin (1979)). The exponent is a function of Pasquill stability categories, A–F, (see Table 5.3) and surface roughness.

Table 5.2 Values of the empirical constant ξ for a neutral (adiabatic) atmosphere (Pasquill stability category D).

Type of ground surface	ξ
Very smooth (mud flats, ice)	0.08
Snow over short grass	0.11
Swampy plain	0.12
Sea	0.12
Lawn grass, 1 cm high	0.13
Desert	0.14
Snow cover	0.16
Thin grass, 10 cm high	0.19
Air field	0.21
Thick grass, 10 cm high	0.24
Country side with hedges	0.29
Thin grass, 50 cm high	0.36
Beet field	0.42
Thick grass, 50 cm high	0.43
Grain field	0.52

After Sutton (1953).

values in the table are inconsistent with curve 'D' in Figure 5.3, so they may not be valid.

For a neutral atmosphere (Pasquill stability category D), an expression based on surface roughness only is sometimes used; in some cases it is also used without justification for other atmospheric conditions. It is often referred to as the *logarithmic velocity profile*:

$$U(h) = U_0 \left(\frac{\log_e(h/z_0)}{\log_e(h_0/z_0)} \right) \tag{5.17}$$

where:

- z_0 is the roughness length (see Table 5.3, Table 5.4)
- h is the height at which wind speed is to be determined
- h_0 is the anemometer height (usually 10 m above the ground) corresponding to the measured wind speed, U_0.

The surface roughness, z_0 can also be estimated using measurements of the wind speed at two different heights in a stable atmosphere and using Eq. (5.17) as a basis to give (IEC 61400-11 Ed.3.0 2012):

$$z_0 = \exp \left(\frac{U(h)\log_e h_0 - U_0 \log_e h}{U(h) - U_0} \right) \tag{5.18}$$

When the atmosphere is very stable (as often occurs at night) or very unstable (as sometimes occurs during the day), Eq. (5.17) is not a very satisfactory estimator for

Table 5.3 Estimates of roughness length for various ground surface types (Davenport 1960; Wieringa 1980, 1992).

Surface type	Roughness length, z_0 (m)
Still water or calm open sea, unobstructed downwind for 5 km	0.0002
Loose sand and snow	≈ 0.0002
Open terrain with a smooth surface such as concrete runways in airports, mowed grass	0.0024
Tidal flat or flat desert, no vegetation, no obstacles	0.0002–0.0005
Fallow ground	0.001–0.004
Short grass or moss	0.008–0.03
Open flat terrain; grass, few isolated obstacles	0.03
Long grass and heather	0.02–0.06
Low crops	0.04–0.09
Low crops, occasional large obstacles	0.10
Agricultural land with some houses and 8-m tall sheltering hedgerows within a distance of about 500 m	0.10
High mature grain crops	0.12–0.18
High mature grain crops, scattered obstacles	0.25
Continuous bushland	0.35–0.45
Parkland, bushes, numerous obstacles	0.5
Regular large obstacle coverage (suburb, pine forest)	0.4–1.6
Tropical forest	1.7–2.3
City centre with high- and low-rise buildings	≥ 2

Table 5.4 Alternative estimates of roughness length for various ground surface types.

Surface type	Roughness length, z_0 (m)
Water, snow or sand	0.0001
Open flat land, mown grass, bare soil	0.01
Farmland with some vegetation	0.05
suburbs, towns, forests, many trees & bushes	0.3

From IEC 61400-11 Ed.3.0 (2012).

the wind speed profile based on a measurement at a height of 10 m (as discussed in Section 6.4.3). A better estimate will be obtained using Eq. (5.16), together with Figure 5.3.

The noise propagation model CONCAWE takes the vertical sound-speed gradient into account indirectly by using atmospheric stability classes, as discussed in Section 5.5. The ISO model discussed in Section 5.6 also takes the vertical sound-speed gradient into account indirectly, but instead of using atmospheric stability classes, it calculates the noise level at the receiver for worst-case atmospheric conditions (downwards-refracting atmosphere). The NMPB-2008 model provides calculations for two different stability conditions: a 'neutral' atmosphere (Pasquill stability category D) and a downward-refracting atmosphere. The Nord2000 and Harmonoise models can calculate receiver noise levels for any specified logarithmic or linear sound-speed gradient for which the sound speed varies monotonically with height above the ground. The Harmonoise model uses the lin-log profile directly to calculate the radius of curvature of a ray emitted by the source, whereas the Nord2000 model calculates an equivalent linear profile, which would not seem to be as representative of the actual profile. However, the lin-log profile does not produce a circular ray path so the Harmonoise analysis is based on a calculation of the maximum height reached by a sound ray and a circular path that reaches the same height, which is a good approximation of the actual path. Regardless of the method used to obtain it, the radius of the equivalent circular path is then used to calculate the approximate travel times and travel distances for the direct and ground-reflected rays arriving at the receiver.

The Nord2000 and Harmonoise propagation models require that the ray path lengths and ray travel times be calculated to a reasonable degree of accuracy, so that rays from the same source arriving via different paths can be combined coherently or partially coherently. The relative error in the calculated propagation path length, in terms of wavelengths, and the relative phase error between different arrays arriving at the receiver increase with frequency, so it is unlikely that the coherent addition of direct and ground-reflected rays will be very accurate at mid- and high-frequencies. However, the accuracy improves when averaging over octave or 1/3-octave bands.

It has been established (Rudnik 1957) that the radius of curvature of a sound ray is dependent upon the vertical gradient of the speed of sound, which can be caused either by a wind gradient or by a temperature gradient, or by both. It has also been shown (Piercy et al. 1977) that refraction due to either a vertical wind gradient or a vertical temperature gradient produces equivalent acoustic effects, which are essentially additive. A sound ray travelling at an angle, ψ_S above (or below) the line parallel to the ground plane will have a curved path with radius of curvature, R_c at height, h, given by the following equation (De Jong and Stusnik 1976), where c is the speed of sound at height, h. When R_c is positive the sound rays are curved downwards and when R_c is negative the sound rays are curved upwards:

$$R_c = \frac{c}{\left[\frac{\mathrm{d}c}{\mathrm{d}h}\right] \cos \psi_S} \tag{5.19}$$

A non-zero sonic gradient, $\mathrm{d}c/\mathrm{d}h$, implies that the speed of sound varies with altitude, which means that the radius of curvature of a sound ray will vary as the ray height varies. The sonic gradient, $\mathrm{d}c/\mathrm{d}h$, is a direct result of vertical atmospheric wind and temperature gradients and in the following analysis it will be assumed that the wind contribution to the sound-speed gradient adds to the temperature gradient contribution. The total vertical gradient can thus be expressed as:

$$\frac{\mathrm{d}c}{\mathrm{d}h} = \frac{\mathrm{d}\boldsymbol{U}}{\mathrm{d}h} + \left[\frac{\partial c}{\partial h}\right]_T \tag{5.20}$$

where the term, $\left[\frac{\partial c}{\partial h}\right]_T$ on the right-hand side is evaluated by assuming a linear vertical temperature profile. The sonic gradient due to the vertical atmospheric temperature profile can be calculated from a knowledge of the vertical atmospheric temperature profile and the use of Eq. (2.12). The latter equation may be rewritten more conveniently for the present purpose as follows,

$$c = c_{00}\sqrt{T/273} \tag{5.21}$$

where T is the temperature in Kelvin and c_{00} is the speed of sound at sea level, one atmosphere pressure and 0°C (331.3 m/s).

The vertical sound-speed gradient resulting from the atmospheric temperature gradient is found by differentiating Eq. (5.21) with respect to h to give:

$$\left[\frac{\partial c}{\partial h}\right]_T = \frac{\mathrm{d}T}{\mathrm{d}h}\frac{\partial c}{\partial T} = \frac{\mathrm{d}T}{\mathrm{d}h}T^{-1/2}\frac{c_{00}}{2\sqrt{273}} = 10.025\frac{\mathrm{d}T}{\mathrm{d}h}[T_0 + 273]^{-1/2} = A_m \tag{5.22}$$

In Eq. (5.22), $\mathrm{d}T/\mathrm{d}h$ is the vertical temperature gradient (in °C/m) and T_0 is the ambient temperature in °C at 1 m height. Typical values of A_m range from −0.1 to +0.1 s^{-1}.

The component of the sonic gradient due to the wind, $\mathrm{d}\boldsymbol{U}/\mathrm{d}h$, is evaluated either directly using Eq. (5.23) or indirectly by calculating the contribution to the radius of curvature of the sound ray due to the wind gradient, as explained below.

To make the problem of finding the ray path length from source to receiver tractable, an equivalent linear sound-speed profile is determined and used, enabling the radius of curvature to be approximated as a constant value, independent of the altitude of the sound ray. This results in a circular ray path.

Here, procedures are given for calculating the radius of curvature, R_c, of the refracted ray, using two different methods. For both methods, expressions are needed for the wind speed and air temperature as a function of altitude above the ground and these can then be linearised and combined to find an expression for the overall sound-speed gradient as a function of height h above the ground.

It will be assumed that the vertical temperature profileis linear; that is, the verticaltemperature gradient is constant and given by Eq. (5.22), with the height set as midway between the source and receiver heights. It will also be assumed that the gradient due to wind, given by Eq. (5.23), gives rise to an effective speed of sound gradient of equal magnitude.

The wind gradient is a vector. Hence $\boldsymbol{U}$ is the velocity component of the wind in the direction from the source to the receiver and measured at height h above the ground. The velocity component $\boldsymbol{U}$ is positive when the wind is blowing in the direction from the source to the receiver and negative for the opposite direction.

Two methods for calculating the total sonic gradient and consequently the radius of curvature of a sound ray will now be discussed. For both methods, the actual ray path is approximated by a circular path, which is generally close to the true path shape. The circular-path approximation is necessary in order to make the problem tractable and effectively requires the sonic gradient to be approximated by an equivalent linear gradient. In a downward-refracting atmosphere (positive radius of curvature), at a sufficient source–receiver separation distance, one or more ground-reflected rays will arrive at the receiver in addition to the direct ray with no ground reflection. The calculation of the radius of curvature for the direct ray will be discussed first, and this will be followed by a discussion of the ground-reflected rays.

Direct Calculation of the Sonic Gradient

This method is not used in any standard noise propagation models, but is outlined here as it is a somewhat simpler calculation and, as shown in Section 6.4.3, it is more accurate in situations in which atmospheric stability conditions are not neutral (that is, Pasquill stability criterion not equal to 'D'). It begins with Eq. (5.16) and differentiates it with respect to height, giving the following expression for the expected wind gradient at height, h.

$$\frac{\mathrm{d}\boldsymbol{U}}{\mathrm{d}h} = \xi \frac{\boldsymbol{U}(h)}{h} \tag{5.23}$$

Equations (5.23) and (5.22) are substituted into Eq. (5.20) to obtain the total sonic gradient. The height h is fixed to a height mid-way between the source and receiver heights, resulting in a linear sonic gradient. Equation (5.23) is then used with Eqs. (5.19), (5.20) and (5.22) to calculate the radius of curvature due to the combined effects of atmospheric wind and temperature gradients.

The angle, ψ_S, at which the ray actually leaves the source is determined iteratively, starting with $\psi_S = 0$, calculating the corresponding value of R_c using Eqs (5.19), (5.20), (5.22) and (5.23) and then calculating the distance, d_c, from the source at which the ray is at a height equal to the receiver height. Referring to Figure 5.4, we may write the following for d_c:

$$d_c = 2R_c \sin(\psi_S + \varphi)\cos\varphi \tag{5.24}$$

where

$$\varphi = \arctan\frac{(h_S - h_R)}{d_c} \tag{5.25}$$

As d_c appears on the left-hand side of Eq. (5.24) and also in the equation for φ, and we eventually require that $d_c \approx d$, it is more efficient to replace d_c with d in Eq. (5.25).

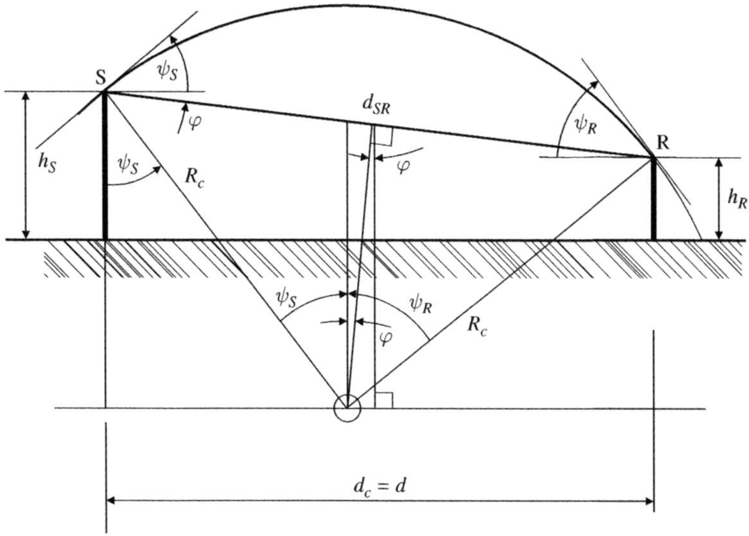

Figure 5.4 Geometry for a sound ray originating at source, S and arriving at receiver, R.

The above equations are also valid for the receiver being lower than the source, in which case φ is negative. The value of ψ_S is incremented (or decremented) by small amounts until $d_c \approx d$ to the required accuracy (usually a few percent).

This method for calculating the radius of curvature of a sound ray is especially useful in cases where the measurements of the sound speed as a function of height above the ground (such as obtained by a SODAR system) can be approximated better by a linear fit than a logarithmic fit.

Indirect Calculation of the Sonic Gradient

Although the general principles of this approach are used in the Harmonoise and Nord2000 models, the detailed calculations of the radius of curvature are different. We will begin here with what is common to both models and then discuss the differences in the detailed calculations. Both models begin with Eq. (5.17) for the wind speed profile and Eq. (5.22) for the temperature profile to obtain the following equation for the sound speed, $c(h)$, at height, h.

$$c(h) = B_m \log_e \left(\frac{h}{z_0} \right) + A_m h + c_0 \tag{5.26}$$

where c_0 is the speed of sound at height, $h = 0$, corresponding to the air temperature adjacent to the ground, with c_0 calculated using Eq. (5.21), with values of c and T corresponding to ground level. This equation is then differentiated to obtain the sonic gradient.

$$\frac{\mathrm{d}c}{\mathrm{d}h} = A_m + \frac{B_m}{h + z_0} \tag{5.27}$$

Substituting Eq. (5.27) into (5.19) gives the following expression for the radius of curvature of the sound ray.

$$\frac{1}{R_c} = \frac{1}{c} \left[A_m + \frac{B_m}{h + z_0} \right] \cos \psi_S \tag{5.28}$$

If A_m and B_m can be evaluated, then it is possible to use Eq. (5.28) to find the radius of curvature of a ray at any height, h and Eq. (2.12) to find the corresponding speed of sound, c.

The coefficient, A_m is the gradient of the speed of sound with height due to the atmospheric temperature profile and is given by

$$A_m = \left[\frac{\partial c}{\partial h} \right]_T \tag{5.29}$$

where $\left[\frac{\partial c}{\partial h} \right]_T$ is given by Eq. (5.22). A_m can also be determined by fitting a straight line to the measured atmospheric temperature profile. Typical values of A_m range from -0.1 to $+0.1$ s^{-1}.

The coefficient, B_m in Eq. (5.26) arises directly from Eq. (5.17) and may be expressed as,

$$B_m = \frac{U_0}{\log_e \left(\frac{h_0}{z_0} \right)} \tag{5.30}$$

Alternatively, the coefficient, B_m can be found by fitting a curve to SODAR data that provides wind speed as a function of altitude. Typical values of B_m range from −1 to +1 ms^{-1}, although higher values are possible at night in the presence of a 'low-level jet'. The above equations for $c(h)$ and B_m are strictly only valid for neutral atmospheric conditions. In cases where the atmosphere is stable, as often occurs at night, it is better to replace the first term in Eq. (5.26) with the right-hand side of Eq. (5.16), with ξ determined from Figure 5.3 or by curve fitting SODAR data, as discussed in Section 6.4.3.

The Nord2000 model uses the expressions for A_m and B_m together with Eq. (5.26) to calculate the speed of sound at the source and receiver heights and from those values, calculates the average sonic gradient between the source and receiver using

$$\frac{\Delta c}{\Delta h} = \frac{c(h_R) - c(h_S)}{h_R - h_S} \tag{5.31}$$

This gradient is then used to calculate the radius of curvature of the sound ray (by substituting $\Delta c/\Delta h$ for $[\mathrm{d}c/\mathrm{d}h]$ in Eq. (5.19)), which is then used to calculate the travel distance for the ray travelling along the path from the source to the receiver using Eq. (5.19). Details of the calculation of the travel distance and associated propagation time are provided in Appendix D, where similar details are also provided for the reflected ray.

The Harmonoise model uses the radius of curvature of the direct ray as a basis for a coordinate transformation, so that the ground is given that radius and all sound ray paths are then straight lines. So the Harmonoise model does not need to calculate radii of curvature for reflected rays.

For calculating the radius of curvature of the direct ray path, the Harmonoise model first expresses the radius of curvature (see Eq. (5.19)) of the sound ray in terms of the contribution, R_A, due to the temperature gradient and the contribution, R_B, due to the wind gradient, as follows:

$$\frac{1}{R_c} = \frac{\cos\psi_S}{c}\left[\left(\frac{\partial c}{\partial h}\right)_T + \frac{\mathrm{d}U}{\mathrm{d}h}\right] = \frac{1}{R_A} + \frac{1}{R_B} \tag{5.32}$$

where c is the speed of sound at height h in stationary air.

The radius of curvature of a sound ray leaving the source due to the atmospheric temperature gradient can be derived using Eq. (5.22) to give:

$$\frac{1}{R_A} = \frac{A_m \cos\psi_S}{c} \tag{5.33}$$

where c is the speed of sound at the source height and A_m is given by Eq. (5.22). The $\cos\psi_S$ term is missing from the 2007 Harmonoise documentation (de Roo et al. 2007). However, it was included in the earlier documentation (Nota and van Leeuwen 2004). Referring to Figure 5.5, for a downward-refracting atmosphere,

$$\cos\psi_S = \frac{\sqrt{R_A^2 - [(d/2) - \sin\varphi\sqrt{R_A^2 - (d/2)^2}]^2}}{R_A} \tag{5.34}$$

Thus Eq. (5.33) may be rewritten as

$$\frac{1}{R_A} = \left[\left(\frac{c}{A_m}\right)^2 + \left(\frac{d}{2} - \sin\varphi\sqrt{R_A^2 - (d/2)^2}\right)^2\right]^{-1/2} \tag{5.35}$$

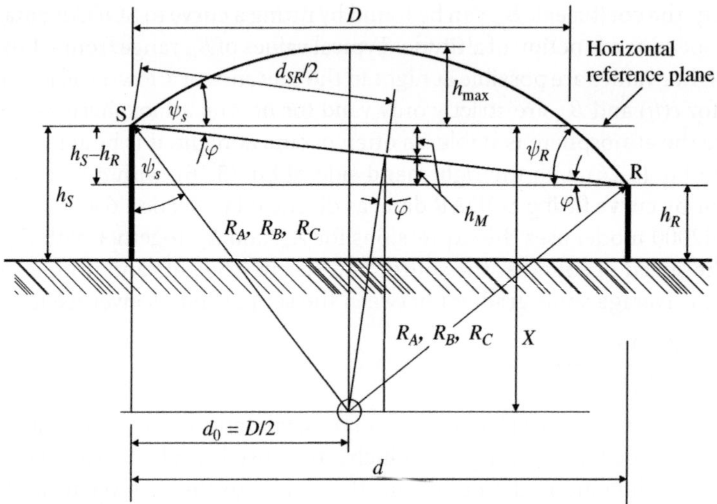

Figure 5.5 Geometry for calculating the radius of curvature of a sound ray originating at source, S, and arriving at receiver, R, for a source higher than the receiver and for the maximum ray height between the source and receiver.

As having R_A on both sides of Eq. (5.35) results in some difficulty in calculating R_A, the equation may be written approximately as follows for small φ, a condition that occurs when the source–receiver separation is large compared to the difference in source and receiver heights:

$$\frac{1}{R_A} \approx \left[\left(\frac{c}{A_m}\right)^2 + \left(\frac{d}{2}\right)^2\right]^{-1/2} \tag{5.36}$$

An expression for R_B, the radius of curvature due to the wind gradient, will now be derived for the case of the source height greater than the receiver height and for the maximum ray height lying between the source and receiver. Reference will be made to Figure 5.5. As a good first approximation, the speed of sound (corresponding to stationary air) used in the preceding and following expressions corresponds to the atmospheric temperature at a height mid-way between the source and receiver. Using a fixed speed of sound results in a circular ray path, which makes the analysis tractable.

Referring to the book by Salomons (2001, Eq. 4.9), the maximum height reached by the sound ray as a result of wind-gradient-induced curvature is given by

$$h_{max} = D\sqrt{\frac{B_m}{2\pi c}} = DK \tag{5.37}$$

where the height is above a horizontal reference plane and the corrected definition of D is the horizontal distance between the two points at which the sound ray arc strikes the horizontal reference plane, as shown in Figure 5.5. For the purposes of the following analysis, the 'reference plane' is defined as a horizontal plane of height equal to the source height when the source is higher than the receiver and a horizontal plane equal to the receiver height when the receiver is higher than the source. The quantity, h_{max} is the maximum height of the sound ray above this plane. This differs from the definitions used in Harmonoise documentation and in the book by Salomons, where h_{max} is defined as the maximum height of the sound ray above the ground and it is assumed that the source and receiver heights are small compared to h_{max}. The latter definition results in the analysis being invalid for very high sound sources such as wind turbines. However, the new definition of the reference plane described above allows for high sources and receivers in the model and will be the one used here to enable the Harmonoise model to be applied to sound propagation from wind turbines.

From Eq. (5.37), the quantity K is given by

$$K = \sqrt{\frac{B_m}{2\pi c}} \tag{5.38}$$

and inspection of Figure 5.5 shows that D is given by

$$D = 2\sqrt{R_B^2 - X^2} \tag{5.39}$$

The quantity h_{max} can also be written in terms of R_B and X as

$$h_{max} = (R_B - X) \tag{5.40}$$

Using Equations (5.37), (5.39) and (5.40), the following expression can be derived:

$$\frac{R_B}{X} = \gamma = \frac{1 + 4K^2}{1 - 4K^2} \tag{5.41}$$

where γ is a constant that only depends on B_m. Rearranging Eq. (5.41) gives,

$$R_B^2 = \gamma^2 X^2 \tag{5.42}$$

Referring to Figure 5.5, the following expression can be written for the radius of curvature, R_B, of the sound ray.

$$R_B^2 = \left(\frac{d_{SR}}{2}\right)^2 + \left(\frac{X - |h_M|}{\cos\varphi}\right)^2 \tag{5.43}$$

where φ is the angle between the horizontal and the line joining the source and receiver, d_{SR} is the straight-line distance between the source and receiver, as shown in Figure 5.5 and h_M is the height above the reference plane of the mid-point of the line joining the source S and receiver R and is given by

$$|h_M| = \frac{|h_S - h_R|}{2} \tag{5.44}$$

Setting equal the right-hand sides of Eqs. (5.42) and (5.43), and substituting $\cos^{-2}\varphi = (1 + \tan^2\varphi)$, gives

$$\gamma^2 X^2 = (X - |h_M|)^2(1 + \tan^2\varphi) + (d_{SR}/2)^2 \tag{5.45}$$

Rearranging gives

$$X^2(1 + \tan^2\varphi - \gamma^2) + X[-2|h_M|(1 + \tan^2\varphi)] + (d_{SR}/2)^2 + h_M^2(1 + \tan^2\varphi) = 0 \tag{5.46}$$

This is a standard quadratic equation ($AX^2 + 2BX + C = 0$), with the solution

$$X = \frac{-B \pm \sqrt{B^2 - AC}}{A} \tag{5.47}$$

where the negative solution to the square root is chosen and

$$\begin{aligned} A &= 1 + \tan^2\varphi - \gamma^2 \\ B &= -|h_M|(1 + \tan^2\varphi) \\ C &= (d_{SR}/2)^2 + h_M^2(1 + \tan^2\varphi) \end{aligned} \tag{5.48}$$

where $d_{SR} = d/\cos\varphi = \sqrt{d^2 + (h_S - h_R)^2}$.

Equation (5.46) may be rewritten by dividing each term by X^2 to produce an equation in $1/X$ with the solution

$$\frac{1}{X} = \frac{-B \pm \sqrt{B^2 - AC}}{C} \tag{5.49}$$

where the positive solution of the square root is chosen.

Using Eqs. (5.37) and (5.40), we obtain the following for R_B, which can be substituted into Eq. (5.32) to obtain the overall radius of curvature:

$$\frac{1}{R_B} = \frac{1}{|X| + h_{\max}} = \frac{1}{|X| + KD} \tag{5.50}$$

The only quantity in Eq. (5.50) which requires additional calculations is D. Inspection of Figure 5.5 and use of Eqs (5.37) and (5.40), we obtain:

$$R_B = \sqrt{\left(\frac{D}{2}\right)^2 + X^2} = |X| + KD \tag{5.51}$$

Rearranging gives,

$$\left(\frac{D}{2}\right)^2 + X^2 = X^2 + K^2D^2 + 2|X|KD \tag{5.52}$$

so,

$$D = \frac{2|X|K}{0.25 - K^2} \tag{5.53}$$

If $AC > B^2$, the sound rays are assumed to not be curved and $(1/X) = (1/R_B) = 0$. The angle, ψ_S, at which the sound ray leaves the source can be derived from inspection of Figure 5.5:

$$\psi_S = -\varphi + \arcsin\left(\frac{d}{2R_c\cos\varphi}\right) \tag{5.54}$$

where

$$\varphi = \arctan\left(\frac{h_S - h_R}{d}\right) \tag{5.55}$$

The preceding equations apply for cases of the receiver higher or lower than the source and also for the cases where the maximum height of the sound ray above the reference plane is not between the source and receiver. These cases are illustrated in Figures 5.6–5.8. When the receiver is higher than the source, the reference plane is set equal to the receiver height. The maximum ray height is then referenced to this plane. If the ray height is referenced to the ground plane as is done in the Harmonoise

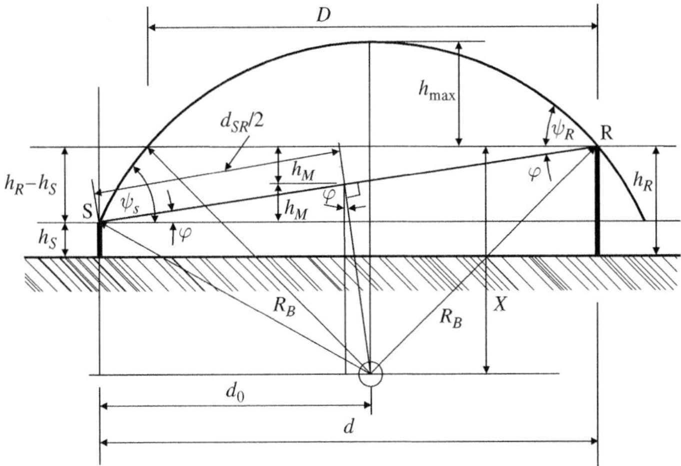

Figure 5.6 Geometry for calculating the radius of curvature of a sound ray originating at source, S and arriving at receiver, R, for a receiver higher than the source and for the maximum ray height between the source and receiver.

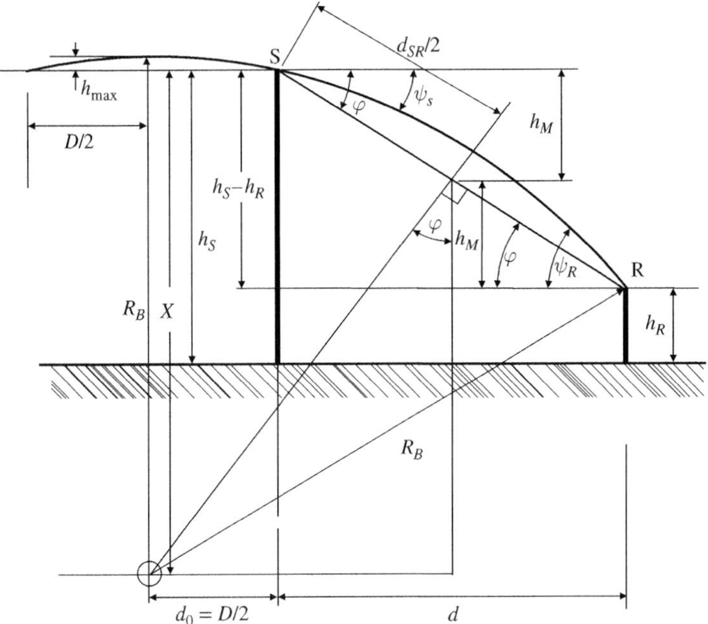

Figure 5.7 Geometry for calculating the radius of curvature of a sound ray originating at source, S and arriving at receiver, R, for a source higher than the receiver and for the maximum ray height *not* between the source and receiver.

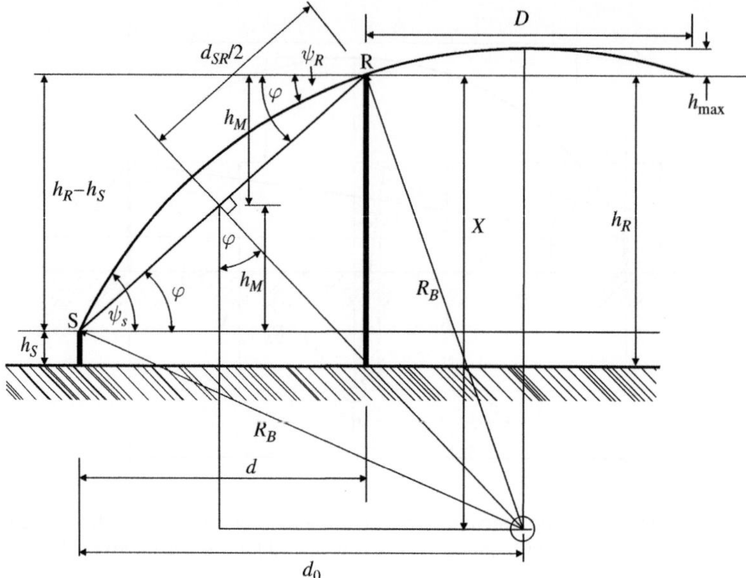

Figure 5.8 Geometry for calculating the radius of curvature of a sound ray originating at source, S and arriving at receiver, R, for a receiver higher than the source and for the maximum ray height *not* between the source and receiver.

documentation, no results can be obtained for high sources or high receivers. However, Eq. 4.9 in Salomons (2001) is valid for the reference plane choices used here.

The angle φ is positive when $h_S > h_R$ and negative when $h_S < h_R$. It represents the angle between the horizontal and the line joining the source and receiver. For cases where the source is high compared to the source–receiver separation distance, the centre of the circular arc may be above ground level, as shown in Figure 5.9. It can be shown that the same equations apply in this case as well.

The previous analysis may be compared to the approximate expression given in earlier Harmonoise documentation (Nota and van Leeuwen 2004) and also used in the Nord2000 model documentation (Plovsing 2014).

$$\frac{1}{R_B} \approx \frac{8}{d_{SR}}\sqrt{\frac{B_m}{2\pi c}} \tag{5.56}$$

The radius of curvature is used in the Nord2000 and Harmonoise propagation models to calculate the path lengths and propagation times for sound travelling from the source to the receiver over a propagation path that has been curved as a result of atmospheric wind gradients and temperature gradients. Taking into account the ray curvature also allows more accurate calculation of barrier attenuation, as discussed in Appendix E.

Calculation of Ray Path Lengths and Propagation Times

The expression for the radius of curvature of the sound ray can be used to calculate the angles at which the sound ray leaves the source and arrives at the receiver (see Figure 5.10), using Eq. (5.54). The included angle θ for the curved path between the

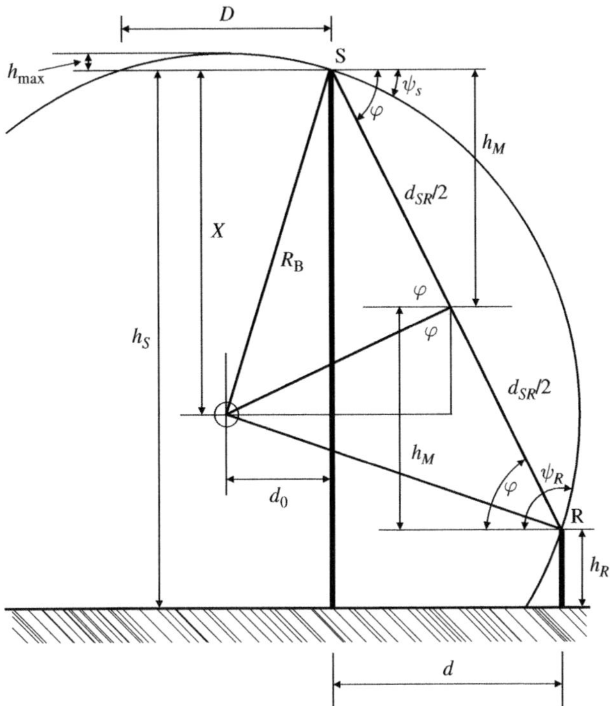

Figure 5.9 Geometry illustrating the arrangement for determining h_{max} and X when the source–receiver distance is comparable to the source height.

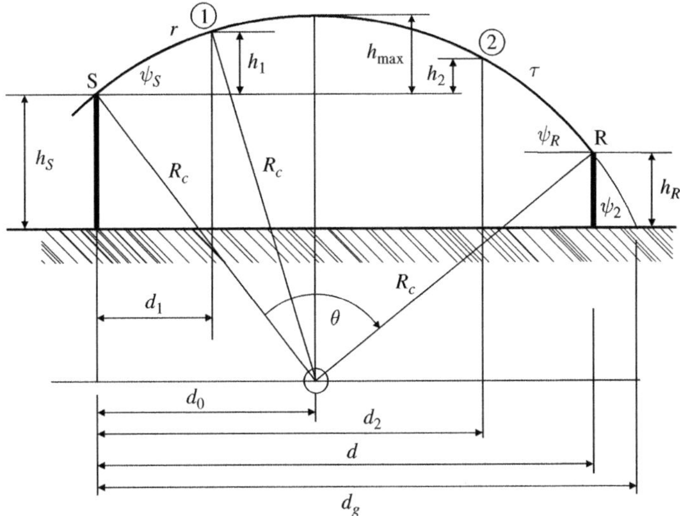

Figure 5.10 Geometrical parameters for calculating the height and centre of curvature of a curved sound ray beginning at point S and ending at point R.

source S and receiver R is given by

$$\theta = \psi_S + \psi_R \tag{5.57}$$

The horizontal distance d_0 of the centre of the circular sound ray path from the source S is given by

$$d_0 = \sqrt{R_c^2 - (R_c - h_{\max})^2} = R_c \sin\psi_S \tag{5.58}$$

and ψ_S is defined by Eq. (5.54). Note that d_0 will be negative if $\psi_S < 0$ (corresponding to the ray leaving the source at an angle below the horizontal). Negative d_0 implies that the centre of the arc is on the opposite side of the source to the receiver.

For distances d_1 less than d_0 from the source, the height, h_1, of the ray at any specified distance, d_1, is given by

$$h_1 = \sqrt{R_c^2 - (d_0 - d_1)^2} - R_c \cos\psi_S \tag{5.59}$$

For distances d_2 greater than d_0 from the source, the height, h_2, of the ray at any specified distance, d_2, is given by

$$h_2 = \sqrt{R_c^2 - (d_2 - d_0)^2} - R_c \cos\psi_S \tag{5.60}$$

The preceding analysis for calculating the radius of curvature of the direct sound ray differs slightly from the approach recommended in the Harmonoise model. The difference is that in the preceding analysis, the 'reference plane' is set at a height equal to the highest of the source and receiver rather than on the ground, and this allows high sources and receivers to be considered. The radius of curvature of the sound ray that is used in the Harmonoise model is very much the same as in the preceding analysis for low sources, because the Harmonoise model uses a reference plane on the ground, which is equivalent to using a reference plane at the source height if the source height is small compared to the maximum height reached by the ray.

The preceding analysis also differs from the approach adopted in the Nord2000 model, which is described in Appendix D. In addition, the Nord2000 model contains a procedure for finding the reflection location for ground-reflected rays.

The ray variables that need to be calculated in order to determine the contribution of the direct and ground-reflected rays from the same source when they arrive at the receiver are listed below. In the following items, the source may be the actual noise source or the top of a barrier and the receiver may be the actual receiver or the top of a barrier:

- the travel time τ along the sound ray from source to receiver
- the travel distance r along the sound ray from source to receiver.

The distance, r, travelled by a direct ray to the receiver over an arc of a circle of radius, R_c, with an included angle of θ radians (see Eq. (5.57)), is given by

$$r = R_c\theta \tag{5.61}$$

There are three different situations for calculating an arc length and all arc lengths are calculated from the originating point of the arc at the source or the top of a barrier. The different situations for calculating arc lengths are listed below.

1. *Arc begins at the source, leaving it at an angle ψ_S above the horizontal, and has an end point after the maximum height of the arc.* Referring to Figure 5.10, and in particular to location 2 on the arc, the angle ψ_R (radians), for use in Eq. (5.57) to calculate θ and hence the arc length, is given by

$$\psi_R = \arcsin \frac{d - d_0}{R_c} \tag{5.62}$$

2. *Arc begins at the source, leaving it at an angle, ψ_S above the horizontal, and has an end point prior to the maximum height that the arc could reach if it continued.* This situation is represented by location 1 on the arc in Figure 5.10. Equation (5.62) applies in this case as well, and the resulting value of ψ_R will be negative.
3. *Arc begins at the source, leaving it at an angle $-\psi_S$ below the horizontal, and thus must have an end point after the maximum height of the arc* (see Figure 5.11). This case occurs if ψ_S of Eq. (5.54) is negative. Equation (5.62) applies in this case as well with d_0 negative, as the centre of the arc is on the opposite side of the source to the receiver, as shown in Figure 5.11, and the resulting value of ψ_R will be positive.

In all cases, Eq. (5.57) applies for calculating the included angle θ for use in Eq. (5.61), provided that the correct signs are used for ψ_S and ψ_R.

The calculation of the arc length from a source point to the receiver point is done by substituting Eq. (5.57) into Eq. (5.61), with ψ_S defined by Eq. (5.54) and ψ_R defined by Eq. (5.62). The radius of curvature for use in Eq. (5.62) is given by Eq. (5.32).

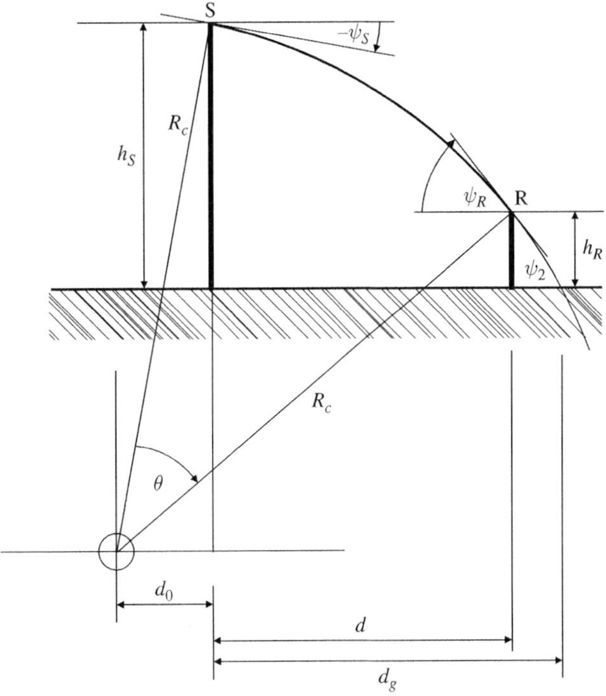

Figure 5.11 Geometrical parameters for a direct ray from source to receiver, leaving the source at an angle $-\psi_S$ below the horizontal.

Alternatively, if the iterative ray tracing method is used to find R_c, then ψ_S is incremented and decremented from $\psi_S = 0$ and the radius of curvature for each value of ψ_S is calculated using Eqs. (5.19), (5.20), (5.22) and (5.23).

To calculate the travel time, τ, from the source to the receiver, it is necessary to take into account the variation in sound speed over different parts of the arc due to them being at differing altitudes. The simplest way to do this is to divide the arc up into a number of segments, each with a particular height above the ground and associated constant sound speed. The sound speed corresponding to a particular height above the ground is given by Eq. (5.26). The arc length, r_i, for each segment, i, with a mid-point at height, h_i and corresponding speed of sound, $c(h_i)$, is calculated as described above and the propagation time, τ_i, is calculated using

$$\tau_i = r_i / c(h_i) \tag{5.63}$$

The approach adopted in the Nord2000 model (see Appendix D) is to integrate the travel time as a function of sound speed over the arc travelled by the ray. This is the same approach as described above if the arc segments are made infinitesimally small.

Ground-reflected Rays – Single Ground Reflection

For a ray reflected from the ground the following additional parameters are needed:

- the location of the point of reflection relative to the source position
- the travel distance r_S along the sound ray from source to reflection point
- the travel distance r_R along the sound ray from reflection point to receiver
- the ground-reflection angle ψ_2 (grazing angle) for a ray.

Depending on the heights of the source and receiver and their separation distance, there may be more than one ground-reflected ray arriving at the receiver for the case of a downward-refracting atmosphere. If the source–receiver separation distance is sufficiently small for specified source and receiver heights, there will only be a single ground-reflected ray arriving at the receiver. However, as the separation distance increases or the source or receiver heights decrease below a certain amount, the number of ground-reflected rays will exceed 1.

The CONCAWE, ISO9613 Part 2 and the NMPB-2008 models only consider a single ground reflection. The Nord2000 model finds a reflection point for a single ground-reflected ray for each ground segment (see Section E.3) and then makes an adjustment to the sound pressure level at the receiver to account for multiple ground-reflected rays, as discussed in Section 5.8.3. When calculating the ray path lengths and travel times for the ground-reflected rays in the Nord2000 model, the radius of curvature used for the ray path is different to that used for the direct path. This apart, the calculation procedure is similar, with the receiver at the ground-reflection point for the path between the source and reflection point and with the source at the ground-reflection point for the path between the ground-reflection point and the receiver.

The Harmonoise model only considers a single ground reflection, which is contributed to in varying amounts by each ground segment. The model uses the radius of curvature calculated in the previous section to perform a coordinate transformation, resulting in a curved ground profile for which straight-line rays can be used. It is relatively straightforward then to find the point of specular reflection of a ground-reflected ray between the source and receiver for each ground segment, as described in Section 5.9. However, the effect of multiple ground reflections is not taken into account. The specular reflection point is calculated for each ground segment (or its extension), then the contribution of the reflection from each ground segment is calculated using a Fresnel weighting factor as described in Appendix G, with the segmentation of a ground profile into straight-line segments discussed in Section E.3. The curved-ground analogy can only be used if the source–receiver separation distance is less than 20% of the radius of curvature of the sound rays. In the Harmonoise model, the same radius of curvature is used for both direct and ground-reflected rays.

Ground-reflected Rays – Multiple Ground Reflections

As shown by Salomons, at reasonably large distances from a source, rays will group into sets of four, which all reach approximately the same height, given by Eq. (5.64) for a logarithmic sound-speed profile.

$$h_{\max,n} = \frac{d}{n}\sqrt{\frac{B_m}{2\pi c_0}} \tag{5.64}$$

The actual number, N_{gr}, of ground-reflected rays that may be expected can be calculated approximately as follows (Salomons 2001),

$$N_{gr} \approx \frac{8h_{\max,1}}{h_S + h_R} \tag{5.65}$$

Equation (5.65) is valid when a ground reference plane is used for the definition of $h_{\max}$. However, when using a higher reference plane, such as one located at the source height, as suggested in this book, there will be a relatively small error in the number N_{gr} that is calculated using Eq. (5.65), as it will affect the calculation of the distance from the source to the first ground reflection. At low and infrasonic frequencies, atmospheric absorption is negligible and there is no loss on reflection from the ground. If we assume (Salomons 2001) that all ground-reflected rays travel approximately the same distance, and that they can be combined together assuming random phase relationships, then the increase in expected sound pressure level at the receiver over what would be experienced in the absence of any ground reflections is given by

$$\Delta L = 10\log_{10}(N_{gr} + 1) \tag{5.66}$$

It is interesting to do an example calculation using Eqs. (5.65) and (5.66). Comparing sound levels for a 40-m high noise source and 2-m high receiver at distances of 1 km and 2 km from the source, and assuming only spherical spreading effects, then for normal spherical spreading, the level at 2 km will be 6 dB lower than the level at 1 km (20 $\log_{10}$ (2/1)). For this configuration, assuming $B_m = 1$ and $c_0 = 343$, Eq. (5.64) gives $h_{\max,1} = 21.5$ m at 1 km and 43 m at 2 km. The number of ground reflections corresponding to the 1-km distance is four and for the 2-km distance it is eight. From Eq. (5.66), the difference in ΔL between the two distances is 2.6 dB, which translates to a decay rate of 2.6 dB per doubling of distance, rather than the usual 6 dB. So it seems that for very low frequencies and infrasound, we can expect a decay rate close to 3 dB per doubling of distance rather than 6 dB, once more than one ground reflection arrives at the receiver. This effect is not so pronounced in the mid- and high-frequency ranges due to the absorbing properties of both the atmosphere and the ground, which explains why most noise prediction models give satisfactory results for noise sources for which the A-weighted noise levels are dominated by mid- and high-frequency noise. However, at large distances from the noise source, especially at night, when a low-level jet or downwind conditions exist or when a strong temperature inversion may be present, one will find that low-frequency noise and infrasound will often dominate the noise spectrum more so than during the day, even if they do not dominate close to the source. This may explain why people living between 1 and 5 km from a wind farm often complain of annoying noise, which would be expected for a noise spectrum dominated by low-frequency noise that varies significantly with time (due to varying atmospheric wind and temperature profiles and turbulence). The distance, d_{mult}, corresponding to the arrival of more than one ground-reflected ray arriving at the receiver can be calculated approximately by rearranging Eqs. (5.64) and (5.65) to give,

$$d_{\text{mult}} = \frac{h_S + h_R}{4}\sqrt{\frac{2\pi c_0}{B_m}} \tag{5.67}$$

Equation (5.67) shows that for wind farms, which represent high sound sources, the distance at which there is a transition from 6-dB sound pressure level decay rate per doubling of distance to approximately 3 dB for very low frequencies and infrasound is quite large. Results calculated using Eq. (5.67) are provided in Table 5.5 for a 1.5 m high receiver, various source heights and various sonic gradients, characterised by varying values of B_m. The presence of a downwind low-level jet (LLJ; see next subsection)

Table 5.5 Source–receiver distance at which more than one reflected ray arrives at the receiver. Receiver height of 1.5 m

Source height (m)	B_m	Distance at which multiple reflections begin (m)
2	0.5	57
40	0.5	681
100	0.5	1666
200	0.5	3307
300	0.5	4949
2	1	41
40	1	482
100	1	1178
200	1	2339
300	1	3499
2	5	18
40	5	215
100	5	527
200	5	1046
300	5	1565

is simulated with a large value of B_m, thus increasing noise levels at the receiver over what would normally be expected. The most common time for this to occur is at night (Nunalee and Basu 2014). Instead of a normal power-law exponent, ξ, between 0.1 and 0.2, an LLJ will result in a power-law exponent closer to 0.8 (Nunalee and Basu 2014), which is equivalent to $B_m \approx 5$ (as $B_m \approx \xi U_0$, where U_0 is the wind speed measured at a height of 10 m). In this case, the location at which multiple reflections begin to occur is much closer to the turbine source.

Low-level Jets

Low-level jets (LLJs) are typically a nighttime phenomenon (Baas et al. 2009), characterised by a maximum in the wind speed profile, which is typically located 100 to 500 m above the ground. They are also characterised by a much higher than expected wind gradient (wind shear) near the ground, which results in much more severe than expected curvature of the sound rays due to atmospheric refraction. This results in the distance from the source at which multiple ground reflections begin being significantly reduced (see Section 5.2.4), and thus reduces the applicability of noise propagation models to shorter propagation distances than one would expect. The presence of an LLJ will result in considerably higher than expected noise levels downwind of a wind farm.

LLJs are more prevalent near flat terrain and near coastal areas and can have a significant effect on noise propagation from offshore turbines. An LLJ is defined by Bass (2011) as the case when the maximum in the vertical atmospheric wind speed profile is at least 2 m/s faster as well as more than 25% faster than the next minimum above. A minimum is neglected if the wind speed decreases by less than 1 m/s before increasing again. Jets are typically characterised by a maximum wind speed of 4–14 m/s at an altitude between 80 and 400 m. LLJs only occur when there is a temperature inversion and thus they are a nighttime phenomenon, reaching their maximum strength in the early hours of the morning, around 5 am to 6 am.

5.2.5 Barrier Effects, A_{bar}

The treatment of the effect of a barrier on sound attenuation is different for the different propagation models and this effect is described (where relevant) within the discussion of the individual models in the following sections.

5.2.6 Miscellaneous Propagation Effects, A_{misc}

Miscellaneous effects on propagation include vegetation screening, reflections from vertical surfaces, source-height effects and in-plant screening. These effects are taken into account differently by different models and details are discussed in the sections concerned with each model. Some additional comments on source-height effects are included in the following subsection.

Source-height Effects

For wind turbines, it is arguable (although not universally agreed) that, as a result of the large height of the noise source, the path difference between direct and ground-reflected rays arriving at the receiver is sufficiently large that the two rays arriving at the receiver can be combined incoherently. In this case, it is sufficient to calculate the reflection loss for the ground-reflected ray using the plane-wave reflection analysis described in Section C.4; the more complex spherical-wave reflection coefficient calculation is unnecessary. In addition, all rays arriving from various paths can be combined at the receiver incoherently, so the entire analysis becomes much simpler. It is no longer necessary to calculate accurate propagation times and distances because the relative phase of the rays arriving at the receiver vary randomly with time and thus no longer need be taken into account.

5.2.7 Infrasound and Low-frequency Noise

Because infrasound and low-frequency noise (IFLN) is not attenuated significantly by atmospheric absorption nor by ground reflections, and mid- and high-frequency noise *is* significantly attenuated, then at distances of two or more kilometres from a wind turbine, the noise spectrum tends to be dominated by low-frequency energy and infrasound, in contrast to the spectrum close to the turbines, which is dominated by mid-frequency sound. When there is a strong downwind component from the source to the receiver and/or a temperature inversion, there will be a minimum distance beyond which more than one sound ray will reach the receiver (see Section 5.2.4). For the case of a wind turbine, this distance can be several kilometres, a distance at which the noise is dominated by low-frequencies and infrasound. In addition, as discussed in Section 5.2.4, more and more rays will affect the noise level as the distance from the source increases, resulting in the noise levels decreasing at a rate of 3 dB for each doubling of distance from the source once the minimum distance of the receiver from the source for more than one ray arriving is exceeded. In summary, the reasons why it seems that infrasound and low-frequency noise generated by wind farms decay at a slower rate than mid- and high-frequency sound are:

- low-frequency noise is insignificantly attenuated on reflection from the ground
- low-frequency noise is insignificantly attenuated due to air absorption
- multiple reflections (see Section 5.2.4) begin to occur around 1.5–2 km from a wind turbine, where the sound energy is dominated by low-frequencies and infrasound.

5.2.8 Propagation Modelling Procedure

For most propagation models, the process consists of the following steps.

1. Determine the octave band sound power levels L_{wi} of each wind turbine in the wind farm plus any other sources, such as transformers in substations, associated with the wind farm. Generally, these sound power levels are the worst-case levels for the particular turbine type and generally occur in the 10–14 m/s range of hub-height wind speed. Some early regulations and guidelines based on ETSU (1996), refer to normalisation of wind speeds to a standard 10-m height, for which the hub-height wind speed range of 10–14 m/s corresponds to a range of 6–10 m/s standardised wind speed at a height of 10 m above the ground. However, more recent regulations and standards are written in terms of sound power level and sound pressure level as a function of hub-height wind speed. The wind speed, U_{ref}, at any height, h_{ref}, may be determined approximately from wind speed measurements with an anemometer at height, h, as described in Section 5.2.4.
2. For a given environment and a given receiver location, calculate the total excess attenuation A_{Ei} in each octave band for sound propagating to the receiver location from all

turbines and any other sources such as transformers, with sound sources numbered from $i = 1$ to N.

3. Compute the resulting sound pressure levels, L_{pi}, at each of the selected locations for each of the individual turbines, i, and other significant sound sources using

$$L_{pi} = L_{wi} - 20 \log_{10} r - 11 - A_{Ei} \quad \text{(dB)} \tag{5.68}$$

where L_{wi} is the sound power level of the ith source.

4. Compute the predicted sound pressure levels produced by all of the individual sources at the selected locations in the environment by converting the sound pressure level due to each source to pressure squared, adding the squared pressure contributions and converting the total to sound pressure level (see Section 2.2.8).

Since the early 1970s many propagation models of varying complexity have been developed and validated to varying degrees. The implementation of the various models in commercial software programs has not been consistent, so that the results obtained for the same model from various commercial programs are not identical for the more complex models, which leads to confusion and doubt about the results obtained. Perhaps for this reason, most practitioners in the past have used the simpler but less accurate and less reliable models.

In an attempt to improve model accuracy and consistency, the European Commission of the European Union, in the context of the European Environmental Noise Directive 2002/49/EC (END), prepared the Common NOise aSSessment methOdS (CNOSSOS-EU) for road, railway, aircraft and industrial noise in order to improve the reliability and the comparability of results obtained by different organisations and different commercial software packages. CNOSSOS investigated three commonly used and complex noise propagation models in detail (Harmonoise, NMPB-2008 and ISO9613-2), as described in European Commission (2010b), Kephalopoulos et al. (2012) and European Commission (2010a), respectively. In 2015, CNOSSOS published software modules, algorithms and documentation for implementation of all three models, noting that the model based on the NMPB-2008 algorithm had been selected as the preferred one.

In the sections to follow, the three CNOSSOS models, two other well-known and extensively used propagation models, and some much simpler models will be discussed in order of increasing complexity. The simple models to be discussed include the 1991 Danish simplified noise model, the Danish low-frequency noise model, and the 2001 Swedish noise model (updated in 2012, see SEPA (2012)). Standard models include the CONCAWE model, the ISO9613-1 model and the NMPB-2008 model and complex models include the NORD2000 and Harmonoise models.

All models have limitations, as it is impossible to accurately model the ground or atmospheric conditions. These limitations result in uncertainty limits being placed on the predictions. Although predictions for a large number of receiver locations are generally better on average than indicated by the uncertainty limits, the predictions at any single location are well described by the uncertainty limits. This means that all wind farm noise predictions should be accompanied by uncertainty estimates and the upper level of the uncertainty estimate should not be permitted to exceed the allowable level. In fact, Verheijen et al. (2011) showed that the standard models, which are used in many current studies, can substantially underestimate noise levels for locations at large distances during temperature inversions. Uncertainty is discussed in detail in Section 5.12.

5.3 Simplest Noise Propagation Models

In 1991, the Danish Environmental Protection Agency (Danish Environmental Protection Agency 1991) published a statutory order prescribing how to predict A-weighted noise levels generated at residences near land-based wind turbines. Although this was written in Danish, an excellent summary in English is given by Søndergaard and Plovsing (2005). The A-weighted sound pressure level, L_{pA}, at a residence located a horizontal distance of d metres from the base of a turbine, which has a hub height of h_h above the receiver and which is radiating an A-weighted sound power level of L_{wA}, is calculated using,

$$L_{pA} = L_{wA} - 10 \log_{10} (d^2 + h_h^2) - 8 - \Delta L_{\text{atm}} \quad \text{(dB)} \tag{5.69}$$

where the constant, 8 (instead of the usual 11), implies that a hard ground is assumed, with the reflected ray adding 3 dB to the direct ray at the receiver.

The excess attenuation due to atmospheric absorption, $\Delta L_{\text{atm}} = A_{\text{atm}}$ in decibels per metre, can be calculated for each octave band centre frequency according to Eq. (5.9) in the next section. However, the Danish statutory order allows the atmospheric absorption for the overall A-weighted sound pressure level to be approximated by 0.005 dB/m (Søndergaard and Plovsing 2005), and if calculated in octave bands, the values are different to those obtained using Eq. (5.9). However, Eq. (5.9) represents the method outlined in the most recent ISO and ANSI standards and is recommended if the Danish method is adopted.

In 1998, EU Project JOR3-CT95-0065 (Kragh 1998) produced a model for calculation of noise levels at distances of up to 600 m for a land-based turbine and up to 350 m for an offshore turbine.

In 2001, a Swedish prediction scheme was proposed for propagation over both land and water (SEPA 2001). This model was updated in 2012 (SEPA 2012). For land based turbines and propagation distances less than 1000 m the Swedish scheme is the same as the Danish one of Eq. (5.69), except that the calculation of atmospheric absorption is different. Also, the Swedish method calculates atmospheric absorption down to 63 Hz, in contrast to the Danish method, which only goes down to 250 Hz. It is recommended that, if the Swedish method is used, Eq. (5.9) is used to calculate atmospheric absorption. For land-based turbines and propagation distances greater than 1000 m, the Swedish method requires that 2 dB be subtracted from the sound pressure level calculated using Eq. (5.69). For propagation over water, for distances greater than 200 m, the sound pressure level at a receiver location is calculated using

$$L_{pA} = L_{wA} - 10 \log_{10} (d^2 + h^2) - 8 - \Delta L_{\text{atm}} + 10 \log_{10} \left(\frac{\sqrt{d^2 + h^2}}{200} \right) \quad \text{(dB)} \tag{5.70}$$

The last term in Eq. (5.70) causes the sound pressure level, L_{pA} to decay by 3 dB per doubling of distance (cylindrical spreading) rather than 6 dB (spherical spreading) after the distance from the turbine exceeds 200 m. The reason for this is that when there is a positive atmospheric sonic gradient (such as in downwind conditions) rays emanating from a source will be refracted towards the ground. At a particular distance from the turbine noise source, the noise will begin to decay at a rate of 3 dB rather than 6 dB per doubling of distance (see Section 5.2.4). This effect is more apparent for sound propagation over hard surfaces such as water and concrete, and also for low-frequency noise and infrasound for which atmospheric absorption is negligible, as well as the ground absorption being negligible. The suggestion in the Swedish model that cylindrical spreading begins at 200 m from a turbine over water may only apply to turbines with hub heights between 20 and 40 m (Søndergaard and Plovsing 2005). According to Søndergaard and Plovsing (2005), for higher hub heights (80–100 m), the effect of multiple reflections only becomes apparent at much greater distances (2000–3000 m). This is discussed in Section 5.2.4.

5.4 Danish Low-frequency Propagation Model

In January 2012, the Danish Government introduced an updated procedure for the calculation of A-weighted low-frequency noise inside Danish houses in the frequency range from 10–160 Hz. Calculations of A-weighted noise levels in each 1/3-octave band, i, between 10 and 160 Hz inclusive are carried out using Eq. (5.71).

$$L_{pA,i} = L_{WA,i} - 10\log_{10}(d^2 + h^2) - 11 - A_{\text{atm}} + \Delta L_g - \Delta L_\sigma \quad \text{(dBA)} \tag{5.71}$$

where d (m) is the horizontal distance from the turbine to the residence, h (m) is the height difference between the turbine nacelle and a standing resident and the quantities, ΔL_g (ground effect correction), A_{atm} (atmospheric absorption correction $= \alpha_a d$) and ΔL_σ (sound transmission loss (or attenuation) of house correction) are listed in Table 5.6 for each 1/3-octave band.

The total low-frequency A-weighted noise level, $L_{pA,LF}$ is then calculated by logarithmic addition of the thirteen 1/3-octave band levels, $L_{pA,i}$, which have previously been determined using Eq. (5.71):

$$L_{pA,LF} = 10\log_{10}\sum_{i=1}^{13} 10^{L_{pA,i}/10} \quad \text{(dBA)} \tag{5.72}$$

Table 5.6 Constants used in the Danish low-frequency noise regulation (2012)

Frequency (Hz)	10.0	12.5	16	20	25	31.5	40
ΔL_g (land) (dB)	6.0	6.0	5.8	5.6	5.4	5.2	5.0
ΔL_g (sea) (dB)	6.0	6.0	6.0	6.0	6.0	5.9	5.9
ΔL_σ (dB)	4.9	5.9	4.6	6.6	8.4	10.8	11.4
α_a (dB/m)	0.00	0.00	0.00	0.00	0.02	0.03	0.05

Frequency (Hz)	50	63	80	100	125	160
ΔL_g (land) (dB)	4.7	4.3	3.7	3.0	1.8	0.0
ΔL_g (sea) (dB)	5.8	5.7	5.5	5.2	4.7	4.0
ΔL_σ (dB)	13.0	16.6	19.7	21.2	20.2	21.2
α_a (dB/m)	0.07	0.11	0.17	0.26	0.38	0.55

Møller et al. (2012) explained that the sound attenuation values listed in Table 5.6 and used in this regulation are incorrect because they are based on average noise levels measured inside typical rooms rather than the maximum noise levels, which would be measured near the room corners. Thus for wind farm noise assessment purposes, Møller et al. (2012) recommend that 5 dBA be added to the total A-weighted noise levels calculated by adding all the 1/3-octave band levels using Eq. (5.72).

5.5 CONCAWE (1981)

CONCAWE is a noise propagation model developed for environmental noise estimations for noise radiated by petroleum and petrochemical complexes to surrounding communities (Manning and Bijl 1981). Since that time, the original version or its modified form (Marsh 1982) have been used in almost all commercially available software for calculating the level of noise radiated into surrounding communities by any sound source, including wind turbines.

The equation used in the CONCAWE model is a derivative of Eq. (5.68) and may be written for the ith source producing a sound pressure level, L_{pik}, at the kth community location as,

$$L_{pik} = L_{wi} + DI_{ik} - A_{Eik} \quad \text{(dB)} \tag{5.73}$$

where L_{pik} is the octave band sound pressure level at community location k due to the ith source, L_{wi} is the octave band sound power level radiated by the ith source. DI_{ik}, is the octave band directivity index of source i in the direction of community location, k. It is usually assumed to be 0 dB for wind turbines, as calculations are often done in the downwind direction for which the sound power level measurement includes the directivity index in the downwind direction (see Section 4.6).

Calculations are done in octave bands from 63 Hz to 8 kHz and the overall A-weighted sound pressure level is calculated by applying the A-weighting correction (see Table 2.2) to each octave band level and then summing the levels logarithmically (see Section 2.2.8). C-weighted sound levels can be calculated in a similar way.

A_{Eik} is the excess attenuation experienced by a sound pressure disturbance travelling from source i to community location k, and is given by

$$A_{Eik} = (K_1 + K_2 + K_3 + K_4 + K_5 + K_6 + K_7)_{ik} \tag{5.74}$$

Each of these attenuation factors are discussed in the following paragraphs.

5.5.1 Spherical Spreading, K_1

The attenuation as a result of the sound waves spreading out as they travel away from the sound source is discussed in Section 5.2.1, and $K_1 = A_{\text{div}}$ (Eq. (5.5)).

5.5.2 Atmospheric Absorption, K_2

This is discussed in Section 5.2.2, and $K_2 = A_{\text{atm}}$ (Eq. (5.8)).

5.5.3 Ground Effects, K_3

For an acoustically hard surface such as asphalt, concrete or water, $K_3 = -3$ dB. For all other surfaces, a set of empirical curves is used (see Figure 5.12). However, these curves were developed for noise sources close to the ground and will over-predict the actual ground effect for high sources such as wind turbines. The error will become larger as distance from the turbine increases. A conservative approach would be to use the hard ground value of −3 dB for K_3 for all ground surfaces.

5.5.4 Meteorological Effects, K_4

Accounting for meteorological effects is perhaps the most difficult of all the excess attenuations. The CONCAWE procedure accounts for turbulence effects as well as wind and temperature-gradient effects. In this procedure, meteorological effects have been graded into six categories based upon a combined vertical wind and temperature gradient. In Table 5.7, incoming solar radiation is defined for use in Table 5.8 (full cloud cover is 8 octas, half cloud cover is 4 octas, and so on). In Table 5.8, the temperature gradient is coded in terms of Pasquill stability categories A–G. Category A represents a strong lapse condition (large temperature decrease with height). Categories E, F and G, on the other hand, represent a weak, moderate and strong temperature inversion respectively, with the strong inversion being that which may be observed early on a clear morning. Thus category G represents very stable atmospheric conditions while category A represents very unstable conditions and category D represents neutral atmospheric conditions. The wind speed in this table is a non-vector quantity and is included for the effect it has on the temperature gradient. Wind-gradient effects are included in Table 5.9.

The Pasquill stability category is combined with the magnitude of the wind vector using Table 5.9 to determine one of the six meteorological categories (CAT) for which

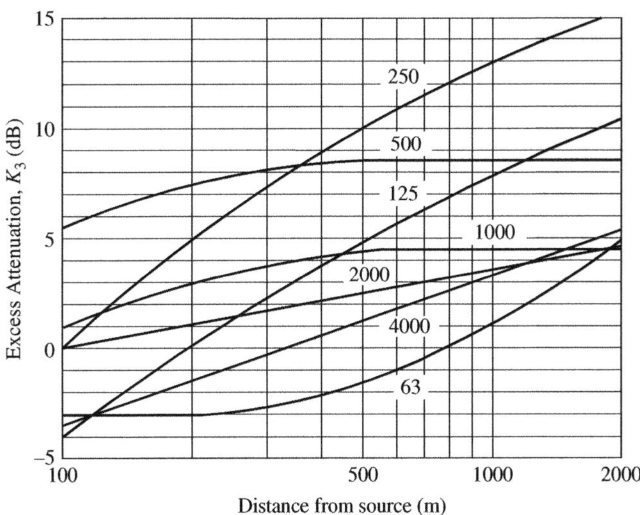

Figure 5.12 Excess attenuation due to the ground. The octave band centre frequency (Hz) corresponding to each curve is indicated on the figure.

Table 5.7 Daytime incoming solar radiation

Latitude of sun	Cloud cover (octas)	Incoming solar radiation
<25°	0–7	slight
25°–45°	<4	moderate
>45°	5–7	moderate
>45°	<4	strong

Table 5.8 CONCAWE determination of Pasquill stability category from meteorological information

Wind speed[a)] m/s	Daytime incoming solar radiation: Strong	Moderate	Slight	Over-cast	1 h before sunset or after sunrise	Nighttime cloud cover (octas): 0–3	4–7	8
<2	A	A–B	B	C	D	F or G[b)]	F	D
2.0–2.5	A–B	B	C	C	D	F	E	D
3.0–4.5	B	B–C	C	C	D	E	D	D
5.0–6.0	C	C–D	D	D	D	D	D	D
>6.0	C	D	D	D	D	D	D	D

After Marsh (1982).
a) Wind speed is measured to the nearest 0.5 m/s and is measured at a height of 10 m above the ground;
b) Category G is restricted to nighttime with less than 1 octa of cloud and a wind speed less than 0.5 m/s.

Table 5.9 CONCAWE determination of meteorological category

Meteorological category	A, B	Pasquill stability category C, D, E wind speed, v (m/s)	F, G
1	$v < -3.0$	–	–
2	$-3.0 < v < -0.5$	$v < -3.0$	–
3	$-0.5 < v < +0.5$	$-3.0 < v < -0.5$	$v < -3.0$
4[a)]	$+0.5 < v < +3.0$	$-0.5 < v < +0.5$	$-3.0 < v < -0.5$
5	$v > +3.0$	$+0.5 < v < +3.0$	$-0.5 < v < +0.5$
6	–	$v > 3.0$	$+0.5 < v < +3.0$

Negative values represent the receiver upwind of the source and positive values are for the receiver downwind of the source.
a) Category with assumed zero meteorological influence. After Marsh (1982).

attenuations, $K_4 = A_{met}$, were experimentally determined from surveys of three European process plants (Manning and Bijl 1981). Attenuations are shown in Figure 5.13 for the octave bands from 63 Hz to 4 kHz. The wind speed used in Tables 5.8 and 5.9 is measured at ground level and is the proportion of the wind vector pointing from the source to the receiver.

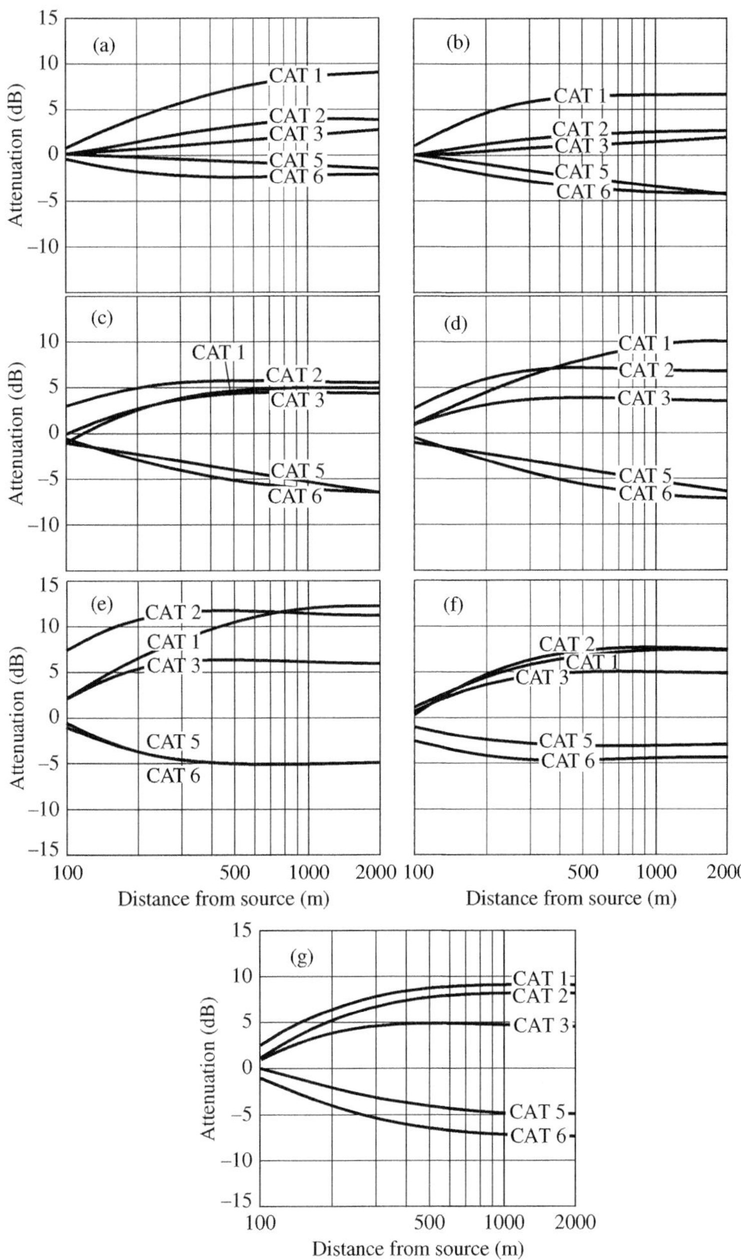

Figure 5.13 CONCAWE meteorological curves for various octave bands. Categories 1 and 2 correspond to upwind propagation whereas categories 5 and 6 correspond to downwind propagation. (a) 63 Hz, (b) 125 Hz, (c) 250 Hz, (d) 500 Hz, (e) 1000 Hz, (f) 2000 Hz, (g) 4000 Hz.

5.5.5 Source-height Effects, K_5

If Figure 5.12 is used to calculate the ground effect, then for source heights greater than 2 m an additional correction K_5 must be added to K_3. The correction, K_5, is given by:

$$K_5 = \begin{cases} (K_3 + K_4 + 3)(\gamma - 1) \text{ (dB)} & \text{if } (K_3 + K_4) > -3 \text{ (dB)} \\ 0 \text{ (dB)} & \text{if } (K_3 + K_4) \leq -3 \text{ (dB)} \end{cases} \tag{5.75}$$

where

$$\gamma = 1.08 - 0.478(90 - \theta) + 0.068(90 - \theta)^2 - 0.0029(90 - \theta)^3 \tag{5.76}$$

where θ is in degrees and $\gamma_{max} = 1$

Note that K_5 is always negative, which means that it acts to reduce the excess attenuation. The angle, θ, is in degrees and is defined in Figure 5.2. If propagation is to a receiver located on a hillside, or across a valley floor, the value of K_5 should be made more negative (greater in magnitude) by up to 3 dB to account for multiple reflections from the hillside.

5.5.6 Barrier Attenuation, K_6

Barriers are any obstacles that represent interruptions of the line of site from the turbine sound source to the community location. In the CONCAWE model, barriers are modelled as thin screens and the corresponding attenuation is calculated using a procedure according to Maekawa (1968, 1977), with no allowance made for bending of sound over the barrier as a result of atmospheric wind and temperature gradients. For wind turbines, barrier attenuation effects are often a result of terrain such as hills and this type of barrier requires a different treatment to the traditional treatment of thin screen barriers, as modelling of terrain shielding as a thin barrier of height equal to the highest terrain point that interrupts the line of sight between the source and receiver significantly overestimates the attenuation. The CONCAWE model is restricted to the use of the barrier calculation procedure due to Maekawa (Maekawa 1968, 1977), which is strictly only suitable for relatively thin walls. In this case, the attenuation can be calculated as described next.

Referring to Figure 5.14, a Fresnel number, N, is calculated as

$$N = \pm(2/\lambda)(A + B - d) \tag{5.77}$$

where

$$\begin{aligned} d &= (X^2 + Y^2 + (h_R - h_S)^2)^{1/2} \\ A &= (X_S^2 + Y_S^2 + (h_b - h_S)^2)^{1/2} \\ B &= (X_R^2 + Y_R^2 + (h_b - h_R)^2)^{1/2} \end{aligned} \tag{5.78}$$

and

$$Y_R = YX_R/X \quad \text{and} \quad Y_S = X_S Y_R / X_R \tag{5.79}$$

For a barrier that is very long compared to its height, there will be four sound ray paths that need to be taken into account as shown in Figure 5.15. The attenuation, A_{bi} for each ray path is then calculated as

$$\{A_{bi} = \Delta_{bi} + 20 \log_{10} [(A_i + B_i)/d_i]\}_{(i=1,4)} \tag{5.80}$$

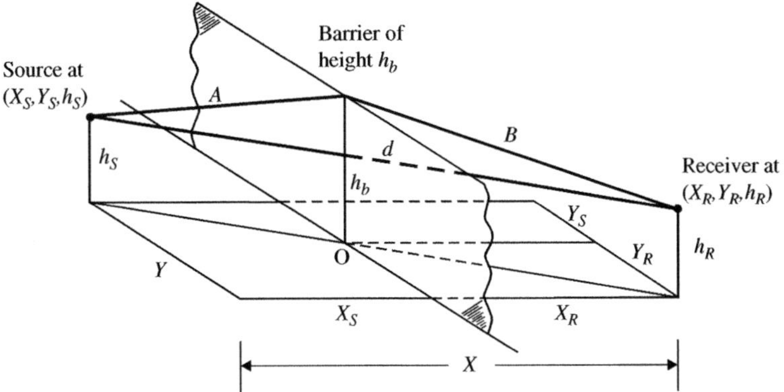

Figure 5.14 Geometry for sound propagation over a single edge barrier.

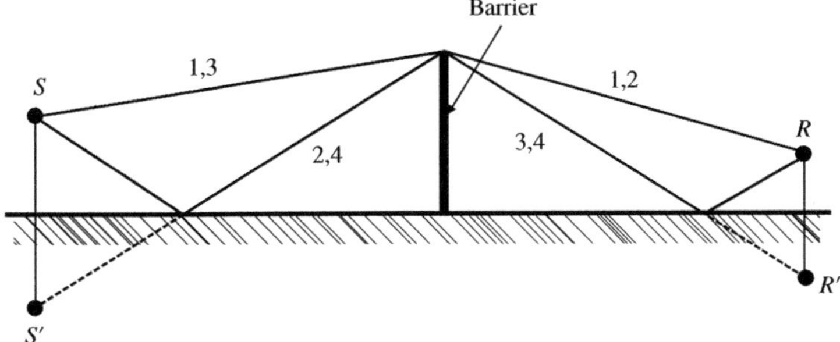

Figure 5.15 Diffraction paths (labelled 1 to 4) over a long barrier.

where Δ_{bi} is calculated as a function of Fresnel number, N_i for path i using Figure 5.16. The horizontal scale in the figure is logarithmic for values of Fresnel number, N, greater than one, but it has been adjusted for values less than one to provide the straight-line representation shown. For the four paths over the top of the barrier, the overall noise reduction due to the barrier is calculated using

$$A_b = 10 \log_{10} [1 + 10^{-(A_{Rw}/10)}] - 10 \log_{10} \sum_{i=1}^{n_A} 10^{-(A_{bi}+A_{Ri})/10} \tag{5.81}$$

where the reflection loss, A_R, due to the ground is added to each path that involves a ground reflection. The subscript, i, refers to the ith path around the barrier and the subscript, w, refers to the ground-reflected path in the absence of the barrier.

For ground that is not uniform between the source and receiver, the reflection loss at the point of the ground corresponding to specular reflection is used. For plane-wave reflection, A_R (equal to A_{Ri} or A_{Rw}) is given by:

$$A_R = -20 \log_{10} |R_p| \tag{5.82}$$

where R_p is the plane-wave reflection coefficient defined in Section C.4.

If the barrier is short compared to its height, the path around each end of the barrier would also have to be taken into account, as described in Section 5.6.2.

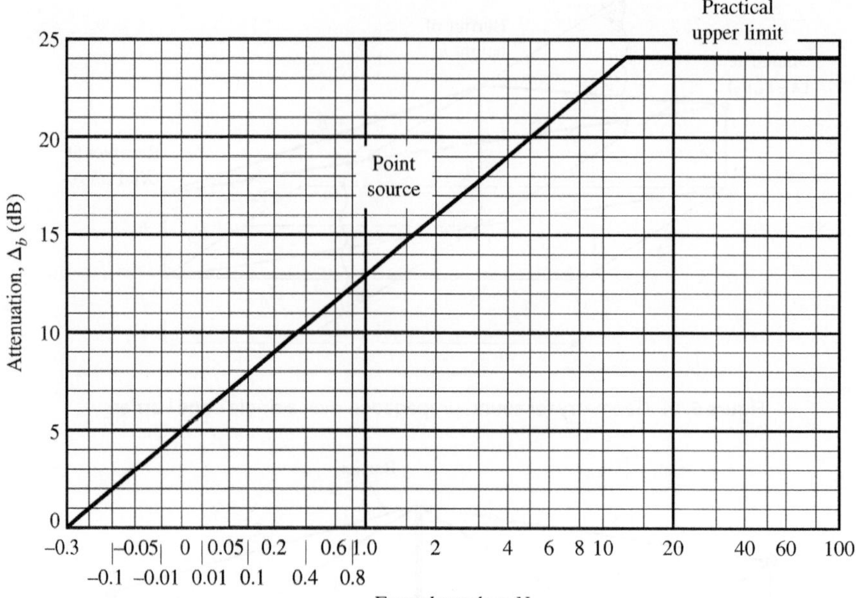

Figure 5.16 Sound attenuation of a semi-infinite screen in free space. If there is a direct line of sight between the source and receiver, N is set negative.

Although the above barrier attenuation calculations are applicable only to thin walls, they may be applied to buildings provided a thickness correction is used (see Bies and Hansen (2009, Ch. 8)) or the obstacle is treated as a double diffraction barrier, as described in Section 5.6.2. However, this treatment is not strictly part of the original CONCAWE model, but may be included in some commercial software implementations. If the shielding is by terrain rather than a manmade barrier, the equivalent construction to find the Fresnel number is as illustrated in Figure 5.17. However, it should be pointed out that the calculation of barrier attenuation for terrain in this way may be an overestimate.

The Fresnel number for the situation in Figure 5.17, where $N > 0$, is given by

$$N = \pm(2/\lambda)(A + B + C - d) \tag{5.83}$$

and for the case, $N \leq 0$, Eq. (5.77) applies.

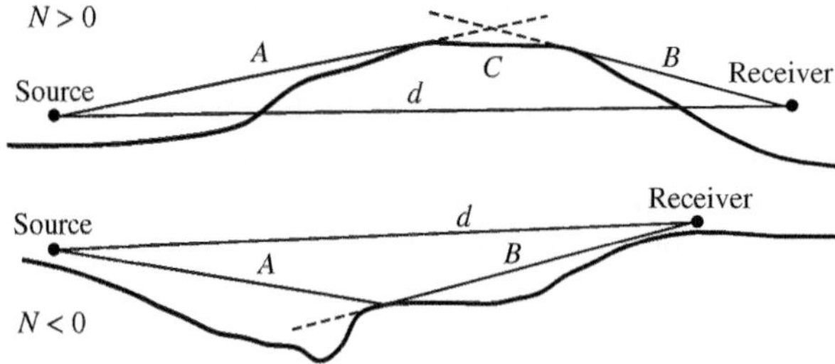

Figure 5.17 Parameters used for calculation of the Fresnel number for propagation over a hill or across a valley. In the lower figure the Fresnel number is negative and the barrier effect does not include the contribution of reflected paths.

5.5.7 In-plant Screening, K_7

Manning and Bijl (1981) found that in-plant screening was only significant for large petrochemical plants and for these a value for K_7 was difficult to estimate due to the limited amount of data available. However, in-plant screening was found to be negligible for the small plants, such as transformer substations, associated with wind farms, so K_7 should be assumed to be zero for most wind farm noise predictions, as wind farms are usually located in the countryside away from large industrial plants.

5.5.8 Vegetation Screening, K_v

Marsh (1982) suggested that the following equation could be used to estimate the excess attenuation of sound (in an octave band of centre frequency, f) caused by its travel a distance r through vegetation, such as a forest. Marsh (1982) did point out that this equation was likely to underestimate the attenuation in European forests.

$$K_v = 0.01rf^{1/3} \tag{5.84}$$

For sound propagating a distance r through long grass or shrubs, the attenuation may be calculated using

$$K_v = (0.18 \log_{10} f - 0.31)r \tag{5.85}$$

5.5.9 Limitations of the CONCAWE Model

The limitations of the CONCAWE model are:

- The model has been validated only for distances from source to receiver between 100 m and 2000 m.
- The model is only valid for wind speeds less than 7 m/s.
- The model is only valid for octave band analysis for octave band centre frequencies ranging from 63 Hz to 4000 Hz. Of course it is also valid for calculating overall A-weighted sound pressure levels over the frequency range covered by the 63 Hz to 4000 Hz octave bands.
- The attenuation due to the ground is based on a very simplified model derived from experimental data taken at two air fields. Its accuracy is questionable for propagation over other types of ground surface.
- The model used to calculate the excess attenuation due to vegetation is not reliable.
- The calculation of the attenuation due to shielding by terrain is oversimplified and may overestimate the terrain-shielding effect.
- Most software packages that implement the CONCAWE model only implement its standard form, in which case the following aspects discussed in this section are excluded:
  - vegetation effects (or if they included, they are limited to forests)
  - when considering barrier attenuation, ground-reflected rays (only sound rays travelling over the top of the barrier with no ground reflection are included)
  - the reduced attenuation of barriers in the presence of a downward-refracting atmosphere (downwind and/or temperature inversion conditions)
  - terrain shielding.

- A detailed statistical analysis has been carried out on comparisons between predictions and measurements with the 95% confidence limits shown in Table 5.10 (the average of many observed minus predicted values are also shown in brackets, indicating the extent of bias in the model). These limits are for use of the model over the range from 100–2000 m. For smaller distances the confidence limits would be smaller than shown in the table but for distances over 1000 m, they are larger (Marsh 1982). It is worth noting that the confidence limits are smaller for meteorological categories 5 and 6, which represent downwind propagation. Marsh (1982) points out that these numbers may be even smaller if the analysis were limited to a range of ±45° from the directly downwind direction.
- The CONCAWE model assumes that the turbine sound power levels as a function of wind speed at rotor height are accurate. In most cases, turbine sound power levels are derived from measurements made on flat ground with a smooth horizontal air flow into the turbine rotor. These conditions deviate considerably from those generally found in practice. In many cases, turbines are mounted on hilltops in hilly terrain, so the inflowing air can deviate considerably from horizontal. In addition, local turbulence due to the uneven terrain and uneven heating of the air near the ground results in an air flow into the turbines that is often far from smooth. Inflow turbulence and non-horizontal inflow can both add to the rated turbine sound power levels. Marsh (1982) also showed that the bias in the CONCAWE model was small, as evidenced by comparing the mean of a large number of observed minus predicted measurements and calculations, as shown in Table 5.10. The numbers in brackets represent the average difference between predictions and measurements for many measurement locations. The numbers outside of brackets are the 95% confidence limits for

Table 5.10 95% confidence limits of the CONCAWE model, which is representative of the expected reliability of a single calculation

Meteorological category	dB(A)	Octave-band centre frequency (Hz)		
		63	125	250
2	6.8 (0.5)	5.4 (0.1)	5.4 (0.1)	9.1 (2.0)
3	6.9 (0.6)	5.0 (0.0)	6.2 (0.5)	9.4 (1.6)
4	5.7 (0.5)	4.8 (0.3)	6.5 (0.8)	8.7 (−1.2)
5	4.7 (0.0)	3.9 (−0.1)	5.4 (0.0)	8.4 (−2.3)
6	4.5 (0.5)	5.2 (−0.8)	6.1 (−0.3)	6.7 (−1.7)

Meteorological category	Octave-band centre frequency (Hz)			
	500	1000	2000	4000
2	9.4 (2.2)	7.8 (2.2)	9.8 (−0.2)	12.4 (0.4)
3	10.1 (0.4)	8.5 (0.8)	8.5 (0.8)	9.4 (0.4)
4	9.8 (−0.2)	6.6 (0.1)	5.6 (1.4)	6.7 (0.2)
5	8.1 (0.4)	5.2 (−0.6)	5.6 (0.9)	6.7 (−0.9)
6	9.3 (1.2)	4.9 (−0.2)	5.5 (0.1)	8.2 (−0.9)

After Marsh (1982).

the accuracy of the measurement at a single location, so that we are 95% certain that the difference will be less than this number. For downwind propagation only, the 95% confidence limits for the CONCAWE model are approximately ±4.5 dBA, (expanded uncertainty with 95% confidence limits – see Section 6.2.16).

5.6 ISO9613-2 (1996) Noise Propagation Model

This standard recommends that calculations be done in octave bands with centre frequencies between 63 Hz and 8 kHz, and that for each octave band, the same basic equation as used by the CONCAWE model (Eq. (5.73)) is used. This equation is used to calculate the sound pressure level, L_{pik} at location, k, due to the octave band sound power level, L_{wi} radiated by sound source, i, with an octave band directivity index in the direction of the receiver location of DI_{ik}. It is given by

$$L_{pik} = L_{wi} + DI_{ik} - A_{Eik} \quad \text{(dB)} \tag{5.86}$$

The overall A-weighted sound pressure level is calculated by applying the A-weighting correction (see Table 2.2) to each octave band level and then summing the levels logarithmically (see Section 2.2.8). The overall C-weighted sound pressure level can be calculated by following a similar procedure.

Normally for wind turbines, the sound power level is measured in the downwind direction, so the directivity index, DI_{ik}, is inherently included in the sound power measurement for downwind noise level calculations.

In the standard, the excess attenuation A_{Eik} is given by

$$A_{Eik} = [A_{\text{div}} + A_{\text{atm}} + A_{\text{gr}} + A_{\text{bar}} + A_{\text{misc}}]_{ik} \tag{5.87}$$

The attenuation component, A_{div} is identical to the CONCAWE attenuation component, K_1, and is calculated assuming a point source using Eq. (5.5). The attenuation component, A_{atm} is identical to the CONCAWE attenuation component, K_2, and is calculated using Eq. (5.8)) as described in Section 5.2.2. Both of these excess attenuation effects are applied to the final total calculated sound pressure level at the receiver for each sound source.

The attenuation components, A_{gr} and A_{bar} are equivalent to the CONCAWE attenuation components, K_3 and K_6, respectively, but are calculated differently. The ISO model does not explicitly include an excess attenuation due to meteorological effects, because its calculation of A_{gr} is for downwind sound propagation only and this would be the worst-case situation for most cases. The attenuation components, A_{gr} and A_{bar}, are discussed in Sections 5.6.1 and 5.6.2, respectively.

The attenuation component, A_{misc} is made up of the following subcomponents:

- the attenuation due to propagation of sound through an industrial site on its way from the source to the receiver (identical to the quantity, K_7 in the CONCAWE model); this is not generally relevant for the propagation of wind farm noise
- the attenuation due to propagation of the sound through a housing estate on its way from the source to the receiver; again this is usually not of importance for wind farm noise calculations, as criteria are based on satisfying noise levels at the houses nearest to the wind farm

- the attenuation due to propagation of sound through foliage, (A_f); this is the same as the quantity, K_v that is estimated in Section 5.5.8, but the calculation of A_f relies on a different procedure, which is outlined in Section 5.6.3.

5.6.1 Ground Effects, A_{gr}

For the ISO9613 method, the space between the source and receiver is divided into three zones: source, middle and receiver zones. The source zone extends a distance of $30h_S$ from the source towards the receiver and the receiver zone extends $30h_R$ from the receiver towards the source. The middle zone includes the remainder of the path between the source and receiver and will not exist if the source–receiver separation (projected on to the ground plane) is less than $d = 30h_S + 30h_R$, where h_S and h_R are defined in Figure 5.2. The acoustic properties of each zone are quantified using the parameter, G. This parameter has a value of 0.0 for hard ground, 1.0 for soft (or porous) ground, and for a mixture of hard and soft ground it is equal to the fraction of ground that is soft. It is assumed that for downwind propagation, most of the ground effect is produced by the ground in the vicinity of the source and receiver so the middle part does not contribute much to the overall value of A_{gr}. In practice, it has been found that for wind farm noise predictions, it is best to set $G = 0$ for propagation over paved ground or over bodies of water and $G = 0.5$ for propagation over grass-covered ground (IOA 2013).

The total excess attenuation due to the ground is the sum of the excess attenuations for each of the three zones. That is,

$$A_{gr} = A_S + A_{mid} + A_{Rec} \tag{5.88}$$

Values for each of the three quantities on the right-hand side of Eq. (5.88) may be calculated using Table 5.11. If the source–receiver separation distance is much larger than their heights above the ground, then $d \approx r$. See Figure 5.2 for a definition of the quantities d and r.

Table 5.11 Octave-band ground attenuation contributions, A_S, A_{Rec} and A_{mid}

	Octave-band centre frequency (Hz)			
	63	125	250	500
A_S (dB)	-1.5	$-1.5 + G_S a_S$	$-1.5 + G_S b_S$	$-1.5 + G_S c_S$
A_{Rec} (dB)	-1.5	$-1.5 + G_R a_R$	$-1.5 + G_R b_R$	$-1.5 + G_R c_R$
A_{mid} (dB)	$-3q$	$-3q(1 - G_m)$	$-3q(1 - G_m)$	$-3q(1 - G_m)$
	Octave-band centre frequency (Hz)			
	1000	**2000**	**4000**	**8000**
A_S (dB)	$-1.5 + G_S d_S$	$-1.5(1 - G_S)$	$-1.5(1 - G_S)$	$-1.5(1 - G_S)$
A_{Rec} (dB)	$-1.5 + G_R d_R$	$-1.5(1 - G_R)$	$-1.5(1 - G_R)$	$-1.5(1 - G_R)$
A_{mid} (dB)	$-3q(1 - G_m)$	$-3q(1 - G_m)$	$-3q(1 - G_m)$	$-3q(1 - G_m)$

After ISO9613-2 (1996).

In Table 5.11, G_S, G_R and G_m are the values of G corresponding to the source zone, the receiver zone and the middle zone, respectively. The quantity, A_{mid} is zero for source–receiver separations of less than $r = 30h_S + 30h_R$, and for greater separation distances it is calculated using the lines labelled A_{mid} in Table 5.11 with

$$q = 1 - \frac{30(h_S + h_R)}{d} \tag{5.89}$$

The coefficients, a_S, b_S, c_S and d_S and the coefficients a_R, b_R, c_R and d_R of Table 5.11 may be calculated using the following equations. In each equation, $h_{S,R}$ is replaced with h_S for calculations of A_S and by h_R for calculations of A_{Rec}.

$$a_S, a_R = 1.5 + 3.0\text{e}^{-0.12(h_{S,R}-5)^2}\left(1 - \text{e}^{-d/50}\right) + 5.7\text{e}^{-0.09h_{S,R}^2}\left(1 - \text{e}^{-2.8\times10^{-6}\times d^2}\right) \tag{5.90}$$

$$b_S, b_R = 1.5 + 8.6\text{e}^{-0.09h_{S,R}^2}\left(1 - \text{e}^{-d/50}\right) \tag{5.91}$$

$$c_S, c_R = 1.5 + 14.0\text{e}^{-0.46h_{S,R}^2}\left(1 - \text{e}^{-d/50}\right) \tag{5.92}$$

$$d_S, d_R = 1.5 + 5.0\text{e}^{-0.9h_{S,R}^2}\left(1 - \text{e}^{-d/50}\right) \tag{5.93}$$

5.6.2 Barrier Attenuation, A_{bar}

It should be noted that the procedures outlined in this section considerably overestimate shielding effects of wind farm noise due to the terrain profile; that is, if hills block the line of sight between turbine and receiver (IOA 2013). For an obstacle to be classified as a barrier for which the calculations in this section are valid, it must satisfy the following conditions:

- Obstacle surface density must be greater than 10 kg/m^2.
- No large cracks or gaps that would allow sound to travel through them may exist.
- The obstacle length normal to the line between the sound source and receiver should be greater than a wavelength at the frequency corresponding to the lower frequency limit of the octave band centre frequency of interest.

Obstacles that fulfil the above conditions are replaced for the purposes of the calculation with an equivalent flat rectangular panel with height above the ground equal to the average height of the obstacle.

Because the barrier interrupts the ground-reflected wave, for downwind propagation (the only condition addressed by ISO9613-2), the ground-effect term is replaced in the barrier calculation by an expression that includes a new ground-reflection term. Thus, for diffraction over the top edge of the barrier, the octave band excess attenuation is given by

$$A_{\text{bar}} = D_{zi} - A_{\text{gr}} > 0 \quad \text{(dB)} \quad (i = 1) \tag{5.94}$$

and for diffraction around the two vertical ends of the barrier

$$A_{\text{bar}} = D_{zi} > 0 \quad \text{(dB)} \quad (i = 2, 3) \tag{5.95}$$

where A_{gr} is the ground attenuation in the *absence* of the barrier and, in both Eqs. (5.94) and (5.95), $A_{\text{bar}} > 0$. The term A_{gr} is still included in Eq. (5.87) because it is cancelled by the A_{gr} term in Eq. (5.94) and an allowance for ground reflection on either side of the barrier is included in the barrier excess attenuation, D_{zi}.

Normally, only the propagation path over the top of the barrier is considered. In cases where it is expected that propagation around one or more ends of the barrier should be considered then the excess attenuations due to each propagation path should be combined to obtain an overall noise reduction, as shown in Eq. (5.81). The excess attenuation, D_{zi}, for the propagation paths over and around the ends of the barrier is given by

$$D_{zi} = 10 \log_{10} [3 + (C_2/\lambda)C_3 K_{\text{met}} \Delta z_i] \quad \text{(dB)} \tag{5.96}$$

where λ is the wavelength of sound at the octave band centre frequency, $C_2 = 20$ for diffraction over the barrier top ($i = 1$), and $C_2 = 40$ for diffraction around the ends of the barrier ($i = 2, 3$). Δz_i is the difference between the line-of-sight distance between source and receiver and the path length over the top or around the side of the barrier, depending on the current calculation being undertaken. The path lengths for diffraction over the top of the barrier may be calculated using Figure 5.14 for single diffraction (a thin obstacle) or Figure 5.18 for double diffraction (a wide obstacle).

From Figure 5.14 for single diffraction, the path difference, Δz_i is given by

$$\Delta z_i = A + B - d \tag{5.97}$$

where

$$d = \left(X^2 + Y^2 + (h_R - h_S)^2\right)^{1/2}$$
$$A = \left(X_S^2 + Y_S^2 + (h_b - h_S)^2\right)^{1/2} \tag{5.98}$$
$$B = \left(X_R^2 + Y_R^2 + (h_b - h_R)^2\right)^{1/2}$$

and

$$Y_R = YX_R/X \quad \text{and} \quad Y_S = X_S Y_R/X_R \tag{5.99}$$

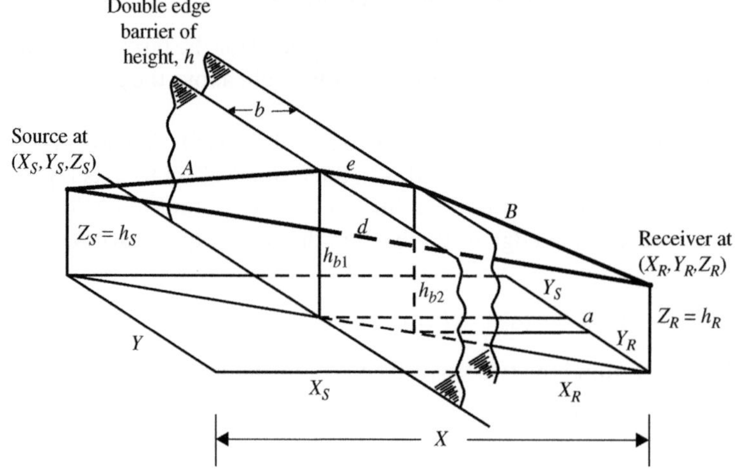

Figure 5.18 Geometry for sound propagation over a double edge (or thick) barrier.

From Figure 5.18 for double-edge diffraction, the path difference, Δz_i, for the two edges of heights, h_{b1} and h_{b2}, is given by

$$\Delta z_i = A + B + e - d \tag{5.100}$$

or, alternatively,

$$\Delta z_i = \left\{ \left[\left[(X_S^2 + (h_{b1} - Z_S)^2)^{1/2} + ((X_R^2 + (h_{b2} - Z_R)^2)^{1/2} + b \right]^2 + Y^2 \right]^{1/2} - d \right\} \tag{5.101}$$

where for barriers of the same height, $e = (a^2 + b^2)^{1/2}$ and for barriers of different heights, h_{b1} and h_{b2}, $e = [a^2 + b^2 + (h_{b1} - h_{b2})^2]^{1/2}$.

K_{met} of Eq. (5.96) is a meteorological correction factor for downwind propagation and is given by

$$K_{\mathrm{met}} = \exp\left[-\frac{1}{2000}\sqrt{\frac{ABd}{2(A + B + e - d)}} \right] \tag{5.102}$$

where all dimensions are in metres and $\exp(x) = \mathrm{e}^x$ (see Section A.4).

5.6.3 Vegetation Screening, A_f

ISO9613-2 (1996) gives the attenuation values in Table 5.12 for sound propagation through dense foliage. For distances less than 20 m, the values given are absolute decibels. For distances between 20 and 200 m, the values given are decibels per metre and for distances greater than 200 m, the value for 200 m is used.

The distance of travel through the foliage is not equal to the extent of the foliage between the source and receiver. It depends on the height of the source and receiver and the radius of curvature of the propagating wave as a result of wind and temperature gradients. ISO 9613-2 (1996) recommends that a radius of 5 km be used for downwind propagation. The centre (always below the ground plane) of the circular arc, representing the sound ray path from source to receiver, is easily found, using a scaled drawing, as the intersection of two lines of length equal to 5 km, with one line intersecting the source location and the other intersecting the receiver location. The distance $r_f = r_1 + r_2$, where r_1 and r_2 are the distances travelled through foliage as defined in Figure 5.19.

Table 5.12 Octave band attenuation, A_f due to dense foliage (after ISO9613-2 (1996)).

	Octave-band centre frequency (Hz)							
	63	**125**	**250**	**500**	**1000**	**2000**	**4000**	**8000**
A_f (dB) for 10 m $\leq r_f \leq$ 20 m	0	0	1	1	1	1	2	3
A_f (dB/m) for 20 m $\leq r_f \leq$ 200 m	0	0.015	0.025	0.025	0.02	0.02	0.015	0.015

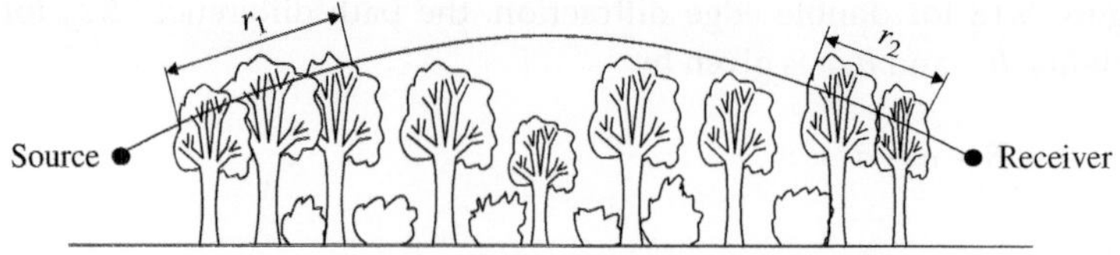

Figure 5.19 Path lengths for sound propagation through foliage.

5.6.4 Effect of Reflections other than Ground Reflections

If the receiver is close to the wall of a building, the expected sound pressure level will be increased as a result of reflection from the wall. ISO9613-2 provides a means for calculating this effect and the procedure is summarised below.

As can be seen from inspection of Figure 5.2, if a reflecting plane is present, sound arrives at the receiver by a direct path and also by a reflected path. If the reflecting plane is the ground, the effect of the reflected wave is taken into account in the ISO model by the excess attenuation due to ground term, A_{gr}. However, if there is an additional reflecting surface such as a tank or wall of a house, then it can be taken into account using the procedure outlined below provided the reflecting surface satisfies the following conditions:

- the magnitude of the surface absorption coefficient, α_r, is less than 0.8.
- the surface is sufficiently large in extent such that the following equation is satisfied.

$$\frac{1}{\lambda} > \left[\frac{2}{(L_{\min}\cos\theta)^2}\right]\left[\frac{r_S r_R}{r_S + r_R}\right] \tag{5.103}$$

where $L_{\min}$ is the minimum dimension of the surface (usually the smallest of the height or width).

The sound pressure level at the receiver location is determined by calculating the sound pressure levels due to the direct and reflected rays separately (using all the excess attenuation parameters for each of the direct and reflected rays) and then adding the two results logarithmically (see Section 2.2.8). The sound power used for the reflected ray, L_{wr} is derived from the sound power level of the source, L_w, which is the sound power level used for the direct ray. Thus,

$$L_{wr} = L_w + 10\log_{10}(1-\alpha_r) + DI_r \quad \text{(dB)} \tag{5.104}$$

where DI_r is the source directivity index in the direction of the reflecting surface and α_r is the absorption coefficient of the surface. If measured absorption coefficient data are unavailable, then ISO recommends using the values in Table 5.13.

5.6.5 Recommended Adjustments to the ISO9613-2 Model for Wind Farm Noise

Perkins et al. (2013) suggested adjustments to the ISO9613-2 propagation model to make it more accurate for wind farm noise predictions. These are listed below.

1. Equation 9 of the standard should be used to calculate ground effects. However, if no sound power spectral data for the turbines are available, the ground effect A_{gr} = −3 dB and the air absorption is that for 250 Hz.
2. A soft ground factor of $G = 1$ should not be used.

Table 5.13 Estimates of sound absorption coefficient

Object	α_r
Flat, hard walls with no openings	0.0
Walls of building with windows and small additions or bay	0.2
Cylinders with hard surfaces (tanks, silos)[a]	$\left[1 - \frac{D\sin(180-\theta)}{2r_S}\right]$ where D = cylinder diameter (m) r_S is the distance from the source to the cylinder centre and the cylinder centre is at location O in Figure 5.2

a) Applies only if the cylinder is much closer to the source than it is to the receiver. Derived from ISO9613-2 (1996).

3. With the exception of propagation over large bodies of water or in urban areas (where $G = 0$), it is recommended to use $G = 0.5$.
4. Uncertainty should be included in turbine sound power levels (±2 dB should be used in the absence of any data).
5. A receiver height of 4.0 m should be used to minimise the potential of over-sensitivity to the ground effect in the vicinity of the receiver.
6. Atmospheric conditions of 10°C and 70% relative humidity should be used to represent a reasonably low level of air absorption.
7. Topographic screening effects of the terrain (ISO 9613-2, Eq. 12) should be limited to a reduction of no more than 2 dB, and then only if there is no direct line of sight between the highest point on the turbine rotor and the receiver location.
8. A further correction of +3 dB (or +1.5 dB if using $G = 0$) should be added to the calculated overall A-weighted noise level for propagation across a concave ground profile, as a result of additional reflected rays reaching the receiver. A ground profile for which the mean height, h_m, of the direct line from source to receiver above the ground is defined as a valley if the following inequality is satisfied

$$h_m \geq 1.5 \,|\, (h_S - h_R)/2 \tag{5.105}$$

In addition, a number of adjustments were suggested during investigations of the most appropriate EU noise model (European Commission 2010a). The most appropriate ones are listed below.

- The sentence under Note 2 on page 2: 'Two different situations are considered in this part of ISO 9613, namely short-term downwind and long-term overall averages'. should be deleted.
- The values proposed for G should be replaced by Table 5.14. However, these G-values have been superseded by those recommended in Table C.2 for the NMPB-2008 propagation model. Overall G values are to be weighted by the relative contributions of the different ground types between source and receiver.
- Section 7.3.2, which is an alternative method for calculating A-weighted noise levels at the receiver, is to be entirely disregarded.

Table 5.14 Values for the parameter, *G* (not to be confused with *G*-weighting)

Surface description	*G*
Very soft	1.0
Uncompacted, loose ground	0.8
Normal uncompacted ground	0.6
Compacted field, lawns, gravel	0.4
Compacted dense ground (unpaved road)	0.2
Hard and very hard (asphalt, concrete, water)	0.0

- Under Section 7.4:
  - Equation (12) in the standard (see Eq. (5.94) above) should be replaced with

$$A_{gr} + A_{bar} = \max(A_{gr}, D_z) \tag{5.106}$$

  - A_{gr} is defined as the ground attenuation in the absence of a barrier.
  - The value of D_z calculated using Equation (14) (D_{zi} in Eq. (5.96) above) cannot be less than 0.
  - The quantity, C_2 in Equation (14) (Eq. (5.96) above) is always equal to 20.

5.6.6 Limitations of the ISO9613-2 Model

The ISO9613-2 model when applied to the prediction of wind farm noise in surrounding communities has the following uncertainties and issues associated with it.

- It has only been validated for distances between source and receiver of less than 1 km.
- The model is only valid for octave band analysis for octave band centre frequencies ranging from 63–4000 Hz. Of course, it is also valid for calculating overall A-weighted sound pressure levels over the frequency range covered by the 63–4000 Hz octave bands.
- Downwind propagation is assumed by the ISO model, but only wind speeds between 1 and 5 m/s (measured between 3 and 11 m above the ground) are valid.
- Significant deviations from the ISO model may be expected for wind speeds above the 5 m/s limit (Kalapinski and Pellerin 2009).
- The ISO model has only been validated for source heights of less than 30 m above the ground.
- No allowance is made for the attenuation effects of scattering due to atmospheric turbulence.
- The ISO standard states that the expected accuracy of sound level predictions are ±3 dB for source heights less than 30 m and for distances between the source and receiver that are less than 1000 m. It is expected that the errors will increase as the source–receiver distance increases. Also as the source height increases above 30 m the calculation of the excess attenuation due to the ground becomes less accurate.
- The ISO model assumes that sound propagation is from point sources, which makes the calculated sound levels within about 200 m of the closest turbine subject to greater errors than those estimated above.

- As for all models, the ISO model assumes that the turbine sound power levels as a function of wind speed at rotor height are accurate. In most cases, turbine sound power levels are derived from measurements made on flat ground with a smooth horizontal air flow into the turbine rotor. These conditions deviate considerably from those generally found in practice. In many cases, turbines are mounted on hilltops in hilly terrain, so the inflowing air can deviate considerably from horizontal. In addition, local turbulence due to the uneven terrain and uneven heating of the air near the ground results in an air flow into the turbines that is often far from smooth. Inflow turbulence and non-horizontal inflow can both add to the rated turbine sound power levels.

5.7 NMPB-2008 Noise Propagation Model

The NMPB-2008 model (sometimes called the NMPB-Routes-2008 model) is a revised French model that was originally intended for the prediction of the propagation of road traffic noise. However, it has recently been selected by the European Union as the preferred model for propagation from all noise sources, including industrial sources such as wind turbines (see Dutilleux et al. (2010) and Kephalopoulos et al. (2012)). The calculation procedure is implemented for the centre frequencies of the octave bands from 63 Hz to 8 kHz. The propagation effect is considered to be the same for all frequencies in any one octave band. The recommended calculation procedure to determine the noise level at a receiver due to all relevant sound sources is as follows:

1. Determine equivalent point sources for all noise sources.
2. Determine the sound power level, L_{wi} for the ith source in the direction of the kth receiver. This will then include the directional properties of the source.
3. Determine all propagation paths between the ith source and the receiver, including reflected and diffracted paths.
4. For each propagation path calculate:
   - the attenuation for neutral (or homogeneous) (no wind or temperature gradients) atmospheric conditions
   - the attenuation for downward-refracting (favourable) atmospheric conditions
   - the percentage of time that favourable conditions occur
   - the long-term sound level arriving at the receiver via each path.
5. Determine the total contribution arising from combining the contributions from each contributing path. This is done by assuming incoherent addition of contributions, as discussed in Section 2.2.8.

The basic equation for calculating the octave band sound pressure level L_{pik} at location k due to the octave band sound power level L_{wi} radiated by sound source i with an octave band directivity index in the direction of the receiver location of DI_{ik}, is similar to Eq. (5.73) for the CONCAWE model and is given by

$$L_{pik} = L_{wi} - A_{\text{div},ik} + DI_{ik} - A_{Eik} \quad \text{(dB)} \tag{5.107}$$

where normally for wind turbines, the sound power level is measured in the downwind direction, so the directivity index, DI_{ik}, is inherently included in the sound power measurement for downwind noise level calculations.

The total octave band sound pressure level due to all sound sources is calculated by incoherent addition (see Eq. (2.27)) of the sound pressure level generated by each source. This total is denoted L_H for neutral or homogeneous atmospheric conditions for which no vertical sonic gradient exists, and L_F for downward-refracting conditions for which the speed of sound increases with increasing height above the ground.

The long-term average octave band sound level L_{LT}, is obtained by summing the sound levels due to the two atmospheric conditions, with the sound levels weighted by the fraction of time that each condition is deemed to exist.

$$L_{LT} = 10\log_{10}\left[w10^{(L_F/10)} + (1-w)\,10^{(L_H/10)}\right] \quad \text{(dB)} \tag{5.108}$$

where $w \leq 1.0$ is the fraction of time that downward-refraction conditions exist and the remaining time is allocated to the homogeneous atmospheric condition.

The overall A-weighted sound pressure level is calculated by applying the A-weighting correction (see Table 2.2) to each octave band level and then summing the levels logarithmically (see Section 2.2.8). The overall C-weighted sound pressure level can be calculated by following a similar procedure.

The excess attenuation, A_{Eik} is defined by

$$A_{Eik} = \left[A_{\text{atm}} + A_{\text{gr+bar}} + A_r\right]_{ik} \quad \text{(dB)} \tag{5.109}$$

where $A_{\text{gr+bar}}$ is the only term affected by wind and temperature gradients and atmospheric turbulence, and the term, A_{misc}, used in the ISO9613-2 model is excluded from this model.

The attenuation component $A_{\text{div},ik}$ in Eq. (5.107) is calculated assuming a point source using Eq. (5.5) and the component A_{atm} is calculated using Eq. (5.8), as described in Section 5.2.2. Both of these excess attenuation effects are applied to the final total calculated sound pressure level at the receiver for each sound source.

The attenuation component, A_r, only applies to rays that are reflected from vertical (or near vertical; less than 15° from vertical) surfaces such as buildings. The attenuation is calculated as described in Section 5.7.1 and is applied to the ray prior to its sound pressure level at the receiver being incoherently summed with the sound pressure levels from non-reflected rays. The A_r term does not include ground reflections as these are included in the $A_{\text{gr+bar}}$ term, which is discussed in the next subsection. This model does not take into account the effect of multiple ground reflections when the receiver is at a large distance from the source in comparison with the source or receiver height.

5.7.1 Ground, Barrier and Terrain Excess Attenuation, $A_{\text{gr+bar}}$

In comparing Eqs. (5.109) with (5.87), it is noted that in the NMPB-2008 model, the excess attenuation due to the ground is combined with that due to barriers, whereas these two attenuations are treated separately in the ISO9613-2 model. Two atmospheric conditions, resulting in two calculation procedures, are considered in the calculation of the $A_{\text{gr+bar}}$ term: a neutral atmosphere characterised by no refraction and a downward-refracting atmosphere (referred to in the NMPB-2008 documentation as 'favourable'). For each atmospheric condition, three situations are considered in the NMPB-2008 model:

1. Ground effect with no diffraction, so that $A_{\text{gr+bar}} = A_{\text{gr},H}$ for a homogeneous atmosphere (no sonic gradient) and $A_{\text{gr+bar}} = A_{\text{gr},F}$ for a downward-refracting atmosphere.

2. Diffraction over a barrier with no ground effect so that $A_{\text{gr+bar}} = A_{\text{bar},H}$ for a homogeneous atmosphere and $A_{\text{gr+bar}} = A_{\text{bar},F}$ for a downward-refracting atmosphere.
3. Diffraction over or around a barrier with the ground effect included, so that $A_{\text{gr+bar}} = A_{\text{gr+bar},H}$ for a homogeneous atmosphere and $A_{\text{gr+bar}} = A_{\text{gr+bar},F}$ for a downward-refracting atmosphere.

The calculation of the quantities in items 1–3 is discussed in the following subsections.

Mean Ground Plane

Before continuing, it is necessary to define a mean ground plane and the heights of the source and receiver above this mean plane, as illustrated in Figure 5.20. The equivalent height of a ground point is the height above the mean plane. If a height is negative, it is set equal to zero. Real heights above the real ground are denoted with the symbol h and heights above the mean ground plane are denoted with the symbol z.

The location of the mean plane shown in Figure 5.20 is calculated as follows. The cross-section of the topography between the source S and receiver R is defined by a set of discrete points (x_k, z_k); $k = 1, 2, \ldots, n$, where x is the coordinate in the horizontal direction and z is the coordinate in the vertical direction, *referenced to an arbitrary origin*. For convenience, the origin is usually chosen as the point on the ground directly beneath the source so that this point corresponds to $k = 1$. The distance in the x-direction between data points should be 1 m or less. Each discrete point is joined to the next by a straight line to form a poly-line, the segments of which are defined by

$$z_k = a_k x_k + b_k \tag{5.110}$$

where

$$a_k = \frac{z_{k+1} - z_k}{x_{k+1} - x_k} \quad \text{and} \quad b_k = \frac{z_k x_{k+1} - z_{k+1} x_k}{x_{k+1} - x_k} \tag{5.111}$$

The equation of the mean line is then,

$$z = ax + b; \quad x \in [x_1, \ldots x_n] \tag{5.112}$$

where

$$a = \frac{3[2A - B(x_n + x_1)]}{(x_n - x_1)^3} \tag{5.113}$$

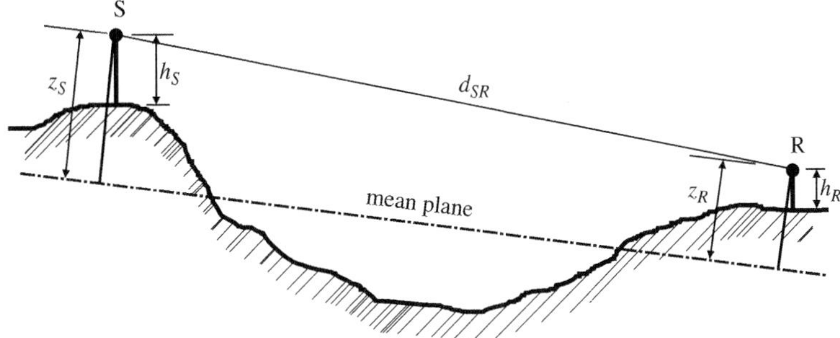

Figure 5.20 Source and receiver heights above the ground and the mean ground plane.

and

$$b = \frac{2B(x_n^3 - x_1^3)}{(x_n - x_1)^4} - \frac{3A(x_n + x_1)}{(x_n - x_1)^3} \tag{5.114}$$

where

$$A = \frac{2}{3}\sum_{k=1}^{n-1} a_k(x_{k+1}^3 - x_k^3) + \sum_{k=1}^{n-1} b_k(x_{k+1}^2 - x_k^2) \tag{5.115}$$

and

$$B = \sum_{k=1}^{n-1} a_k(x_{k+1}^2 - x_k^2) + 2\sum_{k=1}^{n-1} b_k(x_{k+1} - x_k) \tag{5.116}$$

Ground Effect with no Diffraction

The ground effect is a function of the flow resistivity of the ground surface, where the extent of different surfaces between the source and receiver must also be taken into account by simple averaging. The 'ground effect with no diffraction' condition can be divided into two categories, which are discussed in the following two sections. The first is for a homogeneous atmosphere and the second is for a downward-diffracting atmosphere. Expressions for the excess attenuation for each of these categories are provided in those sections.

The purpose of this section is to introduce the factor, G, which is used to characterise the flow resistivity of the ground surface in a similar way as done in the ISO9613-2 model. Values of G to be used in the NMPB-2008 model for various ground surfaces are listed in Table C.2.

To account for varying ground surfaces between the source and receiver, a quantity, G_{path} is defined and used in place of G in the following analysis.

$$G_{\text{path}} = \frac{\sum_{i=1}^{\ell} G_i d_i}{d_p} \tag{5.117}$$

where ℓ is the number of different ground surfaces between the source and receiver, G_i is the value of G for the ith ground surface, d_i is the length of the ith ground surface and

$$d_p = \sum_{i=1}^{\ell} d_i \tag{5.118}$$

Some adjustment is necessary for relatively small source receiver distances, so a new quantity, G'_{path} is used to replace G_{path} and this new quantity is defined as

$$G'_{\text{path}} = \begin{cases} G_{\text{path}}\dfrac{d_p}{30(z_S + z_R)} + G_S\left(1 - \dfrac{d_p}{30(z_S + z_R)}\right) & d_p \le 30(z_S + z_R) \\ G_{\text{path}} & d_p > 30(z_S + z_R) \end{cases} \tag{5.119}$$

where G_S is the value of G in the vicinity of the source.

Ground Effect for Homogeneous Atmosphere and no Diffraction

The excess attenuation, $A_{gr,H}$, due to the ground effect for homogeneous atmospheric conditions is calculated as follows for $G'_{path} \neq 0$.

$$A_{gr,H} = \max \begin{cases} -10 \log_{10} \left[\dfrac{4k^2}{d_p^2} \left(z_S^2 - z_S \sqrt{\dfrac{2C_f}{k}} + \dfrac{C_f}{k} \right) \left(z_R^2 - z_R \sqrt{\dfrac{2C_f}{k}} + \dfrac{C_f}{k} \right) \right] \\ A_{gr,H,\min} \end{cases} \tag{5.120}$$

where the wave number, $k = 2\pi f_m/c$, f_m is the octave band centre frequency, c is the speed of sound in air of 343 m/s and

$$A_{gr,H,\min} = -3(1 - G_m) \tag{5.121}$$

The quantity, C_f is defined as

$$C_f = d_p \frac{1 + 3wd_p \exp(-\sqrt{wd_p})}{1 + wd_p} \tag{5.122}$$

where

$$w = 0.0185 \frac{f^{2.5} G_w^{2.6}}{f^{1.5} G_w^{2.6} + 1.3 \times 10^3 f^{0.75} G_w^{1.3} + 1.16 \times 10^6} \tag{5.123}$$

and where for a homogeneous atmosphere and no diffraction,

$$G_w = G_m = G'_{path} \tag{5.124}$$

If $G'_{path} = 0$, then $A_{gr,H} = -3$ dB.

Ground Effect for a Downward-refracting Atmosphere and no Diffraction

For a downward-refracting atmosphere and for $G'_{path} \neq 0$, Eq. (5.120) can be used but with $A_{gr,H}$ replaced with $A_{gr,F}$ on the left-hand side of the equation and with $A_{gr,H,\min}$ replaced with $A_{gr,F,\min}$ on the left-hand side of the equation. Also, the quantities, z_S and z_R are replaced by $z_S + \delta z_S + \delta z_T$ and $z_R + \delta z_R + \delta z_T$, respectively, where δz_S and δz_R account for the bending of the sound rays due to refraction and δz_T accounts for the effect of atmospheric turbulence. These quantities are defined as,

$$\delta z_S = a_0 \left(\frac{z_S}{z_S + z_R} \right)^2 \frac{d_p^2}{2} \tag{5.125}$$

$$\delta z_R = a_0 \left(\frac{z_R}{z_S + z_R} \right)^2 \frac{d_p^2}{2} \tag{5.126}$$

$$\delta z_T = 6 \times 10^{-3} \left(\frac{d_p}{z_S + z_R} \right) \tag{5.127}$$

where $a_0 = 2 \times 10^{-4}$ m^{-1} is the inverse of the radius of curvature of the sound ray. This corresponds to an assumed mean value of the sound-speed gradient of $B_m = 2c \times 10^{-4}$ m^{-1} $= 0.07$ s^{-1}. This value of a_0 is an assumed quantity for the purposes of this analysis and is considered to represent a typical downward-refracting atmosphere but is not representative of what occurs in the presence of a low jet.

For the downward-refracting case, the lower bound, $A_{\mathrm{gr},F,\min}$ for Eq. (5.120) is defined as

$$A_{\mathrm{gr},F,\min} = \begin{cases} -3(1-G_m); & d_p \leq 30(z_S + z_R) \\ -3(1-G_m)\left\{1 + 2\left[1 - \dfrac{30(z_S+z_R)}{d_p}\right]\right\}; & d_p > 30(z_S + z_R) \end{cases} \tag{5.128}$$

If $G'_{\mathrm{path}} = 0$, then $A_{\mathrm{gr},F} = A_{\mathrm{gr},F,\min}$.

Equations (5.121) and (5.123) and are also used for the downward-refracting case in the absence of diffraction, but the quantities, G_w and G_m are re-defined as

$$\begin{cases} G_w = G_{\mathrm{path}} \\ G_m = G'_{\mathrm{path}} \end{cases} \tag{5.129}$$

Diffraction with no Ground Effect

For pure diffraction with no ground effects, the excess attenuation is given by,

$$A_{\mathrm{bar}} = \begin{cases} 10C_h \log_{10}\left(3 + \dfrac{40}{\lambda}C''\delta\right); & \dfrac{40}{\lambda}C''\delta \geq -2 \\ 0; & \dfrac{40}{\lambda}C''\delta < -2 \end{cases} \tag{5.130}$$

where

$$C_h = \min\begin{cases} fh_0/250 \\ 1 \end{cases} \tag{5.131}$$

and $h_0 = \max(z_{PR}, z_{PS})$ (see Figure 5.23), λ is the wavelength at the octave band centre frequency, f, δ is the path difference between direct and diffracted sound rays (see Figures 5.21 and 5.22), C'' is a coefficient to take into account multiple diffractions and $C'' = 1$ for a single diffraction, as shown in Figures 5.21a,b and 5.22a–c. For multiple diffractions, as shown in Figures 5.21c–e and 5.22d–f,

$$C'' = \frac{1 + (5\lambda/e)^2}{1/3 + (5\lambda/e)^2} \tag{5.132}$$

If $A_{\mathrm{bar}} < 0$, then A_{bar} is set equal to 0. The path difference δ for various relationships between source, receiver and diffraction edges is illustrated in Figure 5.21 for a homogeneous atmosphere (no vertical sonic gradient and straight sound rays) and in Figure 5.22 for a downward-refracting atmosphere (positive vertical sonic gradient and curved sound rays). For the curved sound ray case, all path lengths are along the curved paths and are thus greater than the straight-line distance between the two points joined by the curved ray. For the multiple diffraction cases, involving more than two diffraction edges, a convex hull is drawn and any diffraction edges below the hull outline are ignored as shown in Figures 5.21e,f.

For the homogoneous atmosphere case, the pure diffraction excess attenuation, $A_{\mathrm{bar},H} = A_{\mathrm{bar}}$, with δ in Eq. (5.130) corresponding to the path length differences in

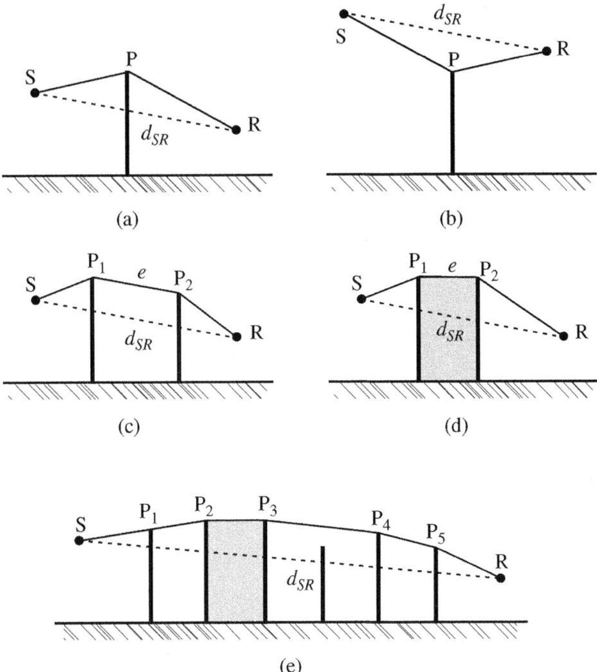

Figure 5.21 Various geometries for determining the path length differences between direct and diffracted rays for a homogeneous atmosphere. In part (e), $e = P_1P_2 + P_2P_3 + P_3P_4 + P_4P_5$.

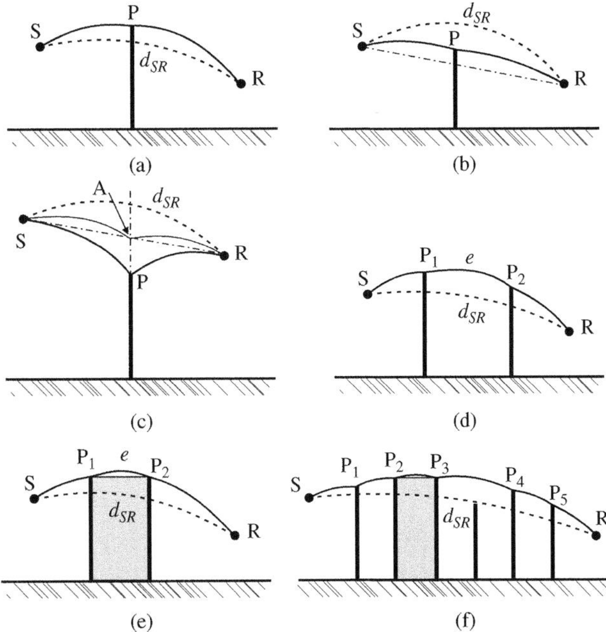

Figure 5.22 Various geometries for determining the path length differences between direct and diffracted rays for a downward-refracting atmosphere. In part (f), e is the sum of the curved path lengths, $P_1P_2 + P_2P_3 + P_3P_4 + P_4P_5$.

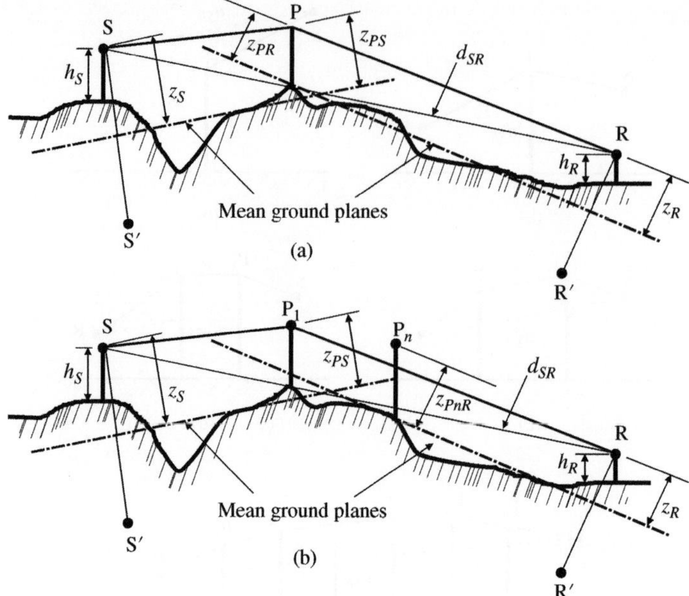

Figure 5.23 Geometry for diffraction over single and multiple diffracting edges. For more than one diffracting edge, the ground effect between two adjacent diffracting edges is ignored, as shown in part (b), where only diffraction edge P_1 and diffraction edge P_n, are shown.

Figure 5.21. For the downward-diffracting atmosphere case, the pure diffraction excess attenuation, $A_{\text{bar},F} = A_{\text{bar}}$, with δ in Eq. (5.130) corresponding to the path length differences in Figure 5.22.

The path differences, δ, for use in Eq. (5.130) are defined for a homogeneous atmosphere represented by the various parts of Figure 5.21 as,

$$\delta = \begin{cases} \text{SP} + \text{PR} - d_{SR} & \text{Figure 5.21a} \\ -(\text{SP} + \text{PR} - d_{SR}) & \text{Figure 5.21b} \\ \text{SP}_1 + e + \text{P}_2\text{R} - d_{SR} & \text{Figure 5.21c} \\ \text{SP}_1 + e + \text{P}_2\text{R} - d_{SR} & \text{Figure 5.21d} \\ \text{SP}_1 + \text{P}_1\text{P}_2 + \text{P}_2\text{P}_3 + \text{P}_3\text{P}_4 + \text{P}_4\text{P}_5 + \text{P}_5\text{R} - d_{SR} & \text{Figure 5.21e} \end{cases} \tag{5.133}$$

and for a downward-refracting atmosphere, represented by the various parts of Figure 5.22, δ is defined as

$$\delta = \begin{cases} \text{SP} + \text{PR} - d_{SR} & \text{Figure 5.22a} \\ \text{SP} + \text{PR} - d_{SR} & \text{Figure 5.22b} \\ 2SA + 2AR - \text{SP} - PR - d_{SR}) & \text{Figure 5.22c} \\ \text{SP}_1 + e + \text{P}_2\text{R} - d_{SR} & \text{Figure 5.22d} \\ \text{SP}_1 + e + \text{P}_2\text{R} - d_{SR} & \text{Figure 5.22e} \\ \text{SP}_1 + \text{P}_1\text{P}_2 + \text{P}_2\text{P}_3 + \text{P}_3\text{P}_4 + \text{P}_4\text{P}_5 + \text{P}_5\text{R} - d_{SR} & \text{Figure 5.22f} \end{cases} \tag{5.134}$$

The path lengths in Eq. (5.134) are all curved path lengths, as shown in Figure 5.22. In part (c), point A is intersection of the straight line between the source and receiver and the vertical extension of the diffracting obstacle.

If the path length difference, δ, is less than $-\lambda/20$ for either of the two cases discussed above, then the diffraction effect is ignored and the excess attenuation due to diffraction is 0 dB.

The definition of the path length difference, δ, for the general case of n diffraction edges on the convex hull between the source and receiver is,

$$\delta = \mathrm{SP}_1 + \sum_{i=1}^{i=n-1} (\mathrm{P}_i\mathrm{P}_{i+1}) + \mathrm{P}_n\mathrm{R} - d_{SR} \tag{5.135}$$

where the paths are straight lines for the homogeneous atmosphere case and curved lines for the downward-refracting case. The radius of curvature, R, of the curved rays is given by,

$$R_c = \max(1000, 8d_{SR}) \tag{5.136}$$

where d_{SR} is the straight line distance between the source and receiver.

The length, ℓ_c, of a curved ray path between any point P_i and P_{i+1} is given by

$$\ell_c = 2R_c \arcsin\left(\frac{\mathrm{P}_i\mathrm{P}_{i+1}}{2R_c}\right) \tag{5.137}$$

where $\mathrm{P}_i\mathrm{P}_{i+1}$ is the straight line distance between points P_i and P_{i+1}.

Diffraction with Ground Effect

When the ground effect is included with the diffraction effect the following quantities are calculated:

- Excess attenuation due to the ground between the source and the closest diffraction edge, $A_{\mathrm{gr,SP}_1}$. A new mean ground plane must be calculated for the ground between the source and the nearest diffraction edge in order to calculate the quantity, z_S, as shown in Figure 5.23.
- Excess attenuation due to the ground between the receiver and the closest diffraction edge, $A_{\mathrm{gr,P}_n\mathrm{R}}$. A new mean ground plane must be calculated for the ground between the receiver and the nearest diffraction edge, in order to calculate the quantity, z_R.
- Diffraction between the source, S, and the receiver, R, by considering the path difference, δ, between the direct path from source to receiver and the path over all obstacles between the source and receiver.
- Diffraction between the image source, S′, and the receiver, R, by considering the path difference, δ, between the direct path from the image source to the receiver and the path over all obstacles between the image source and receiver. A new mean ground plane must be calculated for the ground between the source and the nearest diffraction edge, in order to calculate the quantity, z_S, and locate the image source (as a reflection in the mean ground plane), as shown in Figure 5.23.
- Diffraction between the source, S, and the image receiver, R′, by considering the path difference, δ, between the direct path from the source to the image receiver and the path over all obstacles between the source and image receiver. A new mean ground

plane must be calculated for the ground between the receiver and the nearest diffraction edge, in order to calculate the quantity, z_R and locate the image receiver (as a reflection in the mean ground plane), as shown in Figure 5.23.

To calculate the excess attenuation due to diffraction, it is necessary to replace the mean ground plane between the source and receiver with two mean ground planes: one between the source and the nearest diffraction edge and one between the receiver and the nearest diffraction edge, as shown in Figure 5.23, resulting in a re-location of the image source, S′ and image receiver, R′.

The calculation of the excess attenuation, $A_{\mathrm{gr+bar}}$, due to diffraction with ground reflections only takes into account the ground between the source and its nearest diffraction edge, P_1, and the ground between the receiver and its nearest diffraction edge, P_n (see Figure 5.23). $A_{\mathrm{gr+bar}}$ is calculated as follows.

$$A_{\mathrm{gr+bar}} = A_{\mathrm{bar(SR)}} + \Delta_{\mathrm{gr(SP_1)}} + \Delta_{\mathrm{gr(P}_n\mathrm{R)}} \tag{5.138}$$

where P_1 is the top of the diffraction edge nearest the source and P_n is the top of the diffraction edge nearest the receiver and

$$\Delta_{\mathrm{gr(SP_1)}} = -20\log_{10}\left[1 + \left(10^{-A_{\mathrm{gr(SP_1)}}/20} - 1\right) \times 10^{-\left(A_{\mathrm{bar(S'R)}} - A_{\mathrm{bar(SR)}}\right)/20}\right] \tag{5.139}$$

and

$$\Delta_{\mathrm{gr(P}_n\mathrm{R)}} = -20\log_{10}\left[1 + (10^{-A_{\mathrm{gr(P}_n\mathrm{R)}}/20} - 1) \times 10^{-\left(A_{\mathrm{bar(SR')}} - A_{\mathrm{bar(SR)}}\right)/20}\right] \tag{5.140}$$

where

- $A_{\mathrm{gr(SP_1)}}$ is the excess attenuation due to the ground between the source and diffraction edge, P_1, calculated as described in the ground effect section, but with the following conditions:
  - o z_R is replaced with z_{PS}– see Figure 5.23;
  - o the height, z_S is calculated relative to the mean ground plane representing the ground between the source and nearest diffraction edge, as shown in Figure 5.23;
  - o the value of G_{path} of Eq. (5.117) is calculated only for the ground between the source and the nearest diffraction edge;
  - o in homogeneous atmospheric conditions, G_w of Eq. (5.123) is replaced by G'_{path} of Eq. (5.119);
  - o in downward-refracting atmospheric conditions, G_w of Eq. (5.123) is replaced by G_{path};
  - o in both homogeneous and downward-refracting atmospheric conditions, G_m of Eqs. (5.121) and (5.128) is replaced by G'_{path}.
- $A_{\mathrm{gr(P}_n\mathrm{R)}}$ is the excess attenuation due to the ground between the receiver and diffraction edge, P_n, calculated as described in the ground effect section, but with the following conditions:
  - o z_S is replaced with $z_{\mathrm{P}_n\mathrm{R}}$– see Figure 5.23;
  - o the height, z_R is calculated relative to the mean ground plane representing the ground between the receiver and nearest diffraction edge as shown in Figure 5.23;
  - o the value of G_{path} is calculated only for the ground between the receiver and the nearest diffraction edge;

  - as the source has been replaced with the diffraction edge nearest to the receiver, the G'_{path} correction is not used;
  - in both homogeneous and downward-refracting atmospheric conditions, G_w of Eq. (5.123) is replaced by G_{path};
  - in both homogeneous and downward-refracting atmospheric conditions, G_m of Eqs. (5.121) and (5.128) is replaced by G_{path}.
- $A_{\text{bar(SR)}}$ is the attenuation due to pure diffraction between S and R, calculated as described in the previous section, Eq. (5.130).
- $A_{\text{bar(S'R)}}$ is the attenuation due to pure diffraction between the image source, S′ and R, calculated as described in the previous section, where the image source location is shown in Figure 5.23.
- $A_{\text{bar(SR')}}$ is the attenuation due to pure diffraction between S and the image receiver, R′, calculated as described in the previous section, where the image receiver location is shown in Figure 5.23.

The ray from the image source to the image receiver is ignored in this model, probably as its contribution is considered negligible in almost all situations.

Vertical Edge Diffraction with Ground Effect

In some cases it is of interest to be able to calculate the sound pressure level arriving at a receiver after diffraction around a vertical edge such as the side of a building that lies between the source and receiver. This sound pressure level can then be added to the level calculated due to diffraction over the top of the building. In this case, the excess attenuation is still calculated using Eq. (5.109). However, the term, A_{div} is calculated from the direct distance, d_{SR}, while the terms, A_{atm} and A_{gr} are calculated using the total length of the propagation path around the side of the obstacle. The term, $A_{\text{gr+bar}}$, is calculated using

$$A_{\text{gr+bar}} = A_{\text{gr}} + A_{\text{bar(SR)}} \tag{5.141}$$

where $A_{\text{bar(SR)}}$ is calculated by considering the path difference between the direct path from source to receiver and the path around the diffraction edge and using Eq. (5.130).

5.7.2 Reflections from Vertical Surfaces

An obstacle is considered to be vertical if its slope is less than 15°. The obstacle is ignored if any dimension is less than 0.5 m. The sound pressure arriving at the receiver due to the reflected ray is added incoherently to the ray arriving via the non-reflected path. The additional attenuation due to reflection must be added to the excess attenuations due to spherical spreading, scattering and atmospheric absorption. The resulting sound pressure level at the receiver is then added incoherently to the sound pressure level due to other rays that have not been reflected from vertical surfaces. If both a ground-reflected and direct ray are reflected from a vertical surface, then each of these is treated separately and the contribution of each at the receiver is added incoherently at the receiver. The additional attenuation due to reflection is given by

$$A_r = -10 \log_{10} (1 - \alpha_r) \tag{5.142}$$

where $0 < \alpha_r < 1$ is the absorption coefficient of the surface.

5.7.3 Limitations of the NMPB-2008 Model

NMPB-2008 has the following limitations:

1. The model is valid for octave band analysis for octave band centre frequencies ranging from 63–8000 Hz. Of course it is also valid for calculating overall A-weighted sound pressure levels over the frequency range covered by the 63–8000 Hz octave bands.
2. It is valid only for point sources. Where necessary, other sources are modelled as collections of point sources with the same total sound power output. However, even large sources appear as point sources at distances greater than about 2 source dimensions from them, so this should not result in significant errors in almost all situations.
3. The terrain between source and receiver is approximated by a number of straight lines and this may not model actual situations accurately. However, it is superior to the approaches used in the CONCAWE and ISO models.
4. The geometry used for the terrain effects is based on two- rather than three-dimensional modelling, which may be inaccurate in some cases.
5. Any barriers are generally assumed to be infinite in length normal to the line between the source and receiver. However, if the effect of finite width on the barrier attenuation needs to be taken into account, then refraction around the vertical edges of the barrier may be included.
6. Only two atmospheric sonic profiles are considered: one with zero velocity gradient and one with a downward-refracting gradient that produces a specified radius of curvature of the sound rays emitted from the source.
7. The model does not account for multiple ground reflections for the downward-refracting atmosphere case, so its accuracy may not be sufficient for large distances between the source and receiver.
8. Only reflecting surfaces angled less than 15° from the vertical are considered.

5.8 Nord2000 Noise Propagation Model

The Nord2000 model was developed more recently than the CONCAWE and ISO9613-2 models and is purported to be more accurate than the earlier models. Unlike the models described in previous sections above, which are applicable to octave bands with centre frequencies between 63 Hz and 8 kHz, the Nord2000 model is applicable to 1/3-octave bands with centre frequencies between 25 Hz and 10 kHz. The Nord2000 model also uses more complex calculations of the excess attenuation due to meteorological effects, ground effects and terrain-shape effects, as will be seen in the following discussion. All calculations for the Nord2000 model are done at 1/3-octave band centre frequencies and the complex sound pressure (amplitude and phase) is calculated for a single tone at each 1/3-octave band centre frequency. The model then allows for corrections, which include the effect of averaging over a 1/3-octave band of frequencies.

Although the Nord2000 model is based on the same fundamental equation as the CONCAWE, ISO9613-2 and NMPB-2008 models, which is repeated below for convenience, the developers (Plovsing 2006a,b, 2014) added some confusion by changing the sign of the excess attenuation term, A_{Eik}, and then changing the sign of the terms that make up the excess attenuation to eventually achieve the same result as if there were no sign changes. To avoid confusion for readers of this book, the following description uses the same sign for the excess attenuation as was used in the CONCAWE, ISO9613-2 and NMPB-2008 models. Thus the basic equation for calculating the 1/3-octave band sound pressure level, L_{pik} at location, k, due to the 1/3-octave band sound power level, L_{wi} radiated by sound source, i with a 1/3-octave band directivity index in the direction of the receiver location of DI_{ik}, is the same as for the CONCAWE model (Eq. (5.73)) and is given by

$$L_{pik} = L_{wi} - A_{\text{div},ik} + DI_{ik} - A_{Eik} \quad \text{(dB)} \tag{5.143}$$

where the minus sign replaces the plus sign in the Nord2000 model documentation. Normally for wind turbines, the sound power level is measured in the downwind direction, so the directivity index, DI_{ik}, is inherently included in the sound power measurement for downwind noise level calculations.

The total sound pressure level due to all sound sources is calculated by incoherent addition (see Eq. (2.27)) of the sound pressure level generated by each source. The overall A-weighted sound pressure level is calculated by applying the 1/3-octave band centre frequency A-weighting correction (see Table 2.2) to each 1/3-octave band level and then summing the levels logarithmically (see Section 2.2.8). The overall C-weighted sound pressure level can be calculated by following a similar procedure.

The excess attenuation, A_{Eik} is defined by

$$A_{Eik} = \left[A_{\text{atm}} + A_{\text{gr+bar}} + A_{\text{sc}} + A_r\right]_{ik} \quad \text{(dB)} \tag{5.144}$$

where the signs of all the above excess attenuation terms will be opposite to what they are in the Nord2000 documentation, where they are defined as 'propagation effects' rather than excess attenuation. In Eq. (5.144), the term, A_{misc} in the ISO model has been replaced with the term, A_{sc}, which also calculates the excess attenuation due to scattering effects but by using an entirely different approach.

The attenuation component, $A_{\text{div},ik}$ is calculated using Eq. (5.5), and assuming a point source for the turbine noise source. The distance used in the calculation is the distance between the source and receiver. The additional spreading due to ground-reflected and diffracted paths being longer than the direct path is taken into account in the excess attenuation term, $A_{\text{gr+bar}}$.

The excess attenuation component, A_{atm} is calculated using Eq. (5.8), as described in Section 5.2.2. Both of these excess attenuation effects are applied to the final total calculated sound pressure level in each 1/3-octave band at the receiver for each sound source. Although the calculations of excess attenuation due to atmospheric absorption are only for single frequencies, the CONCAWE and ISO models rely on the octave band centre frequency absorption being representative of the entire octave frequency band, resulting in increasing errors as the calculated absorption value increases. The Nord2000 model takes this error into account in the same way as the Harmonoise model (see Section 5.9) and applies a correction that is valid for 1/3-octave band analysis, which is calculated using Eq. (5.193):

$$A_{\text{atm}} = A_0(1.0053255 - 0.00122622A_0)^{1.6} \quad \text{(dB)} \tag{5.145}$$

where A_0 is the excess attenuation due to atmospheric absorption at the 1/3-octave band centre frequency, calculated using the same procedure as for the CONCAWE and ISO9613 methods (see Eq. (5.8)).

Calculation of the excess attenuation terms, A_{gr+bar}, A_{sc} and A_r, of Eq. (5.144) will be discussed in subsections to follow. When calculating the excess attenuation term, A_{gr+bar}, it is necessary to combine rays arriving at the receiver via different paths, which include the direct path, the ground-reflected path and any diffracted paths. The rays are combined coherently, which means that there must be a phase relationship between them. To enable a phase relationship to exist, the calculations are done for pure tones only and these are the centre frequencies of the 1/3-octave bands from 25 Hz to 10 kHz. The excess attenuation results obtained for the pure tones are then applied to the corresponding 1/3-octave band. The effect of spreading the frequency bandwidth of the tone to a 1/3-octave is taken into account by reducing the coherence between the different rays when they are combined. This is referred to as 'partial coherence', and means for calculating the partial coherence and combining the different rays together to obtain the 1/3-octave band sound pressure level are described in Section 5.8.1.

The excess attenuation due to scattering is discussed in Section 5.8.4, and its value may be different for direct and ground-reflected rays as they will be travelling along different paths between the source and receiver. The effect of scattering on the coherence between the direct ray and the other rays (ground-reflected and diffracted) is taken into account in the coherent analysis in Section 5.8.1. However, the amplitude, $p_{i,\,sc}$, of ray, i, arriving at the receiver via the ith path and travelling through a scattering zone is given by,

$$p_{i,\,sc} = p_i 10^{-A_{sc}/10} \quad \text{(Pa)} \tag{5.146}$$

where p_i is the sound pressure arriving at the receiver in the absence of scattering and the calculation of A_{sc} is discussed in Section 5.8.4.

The excess attenuation, A_r, to account for the loss on reflection from a vertical surface is only applied to rays that are reflected from vertical surfaces. Rays reflected from vertical surfaces are combined incoherently with rays that are not reflected from vertical surfaces. In addition to A_r, the reflected ray also has to have applied to the attenuation components in Eq. (5.144) that are due to atmospheric absorption, scattering and spherical spreading. Calculation of A_r is discussed in Section 5.8.5.

5.8.1 Combination of Sound Rays from the Same Source Arriving at the Receiver via Different Paths

When calculating the excess attenuation term, A_{gr+bar}, it is necessary to combine coherently or partially coherently the direct, ground-reflected, and diffracted sound rays arriving at the receiver. In the Nord2000 model, the propagation times of each path are calculated precisely and the rays representing tones at each 1/3-octave band centre frequency are combined coherently or partially coherently by taking into account their relative phase, to obtain a total sound pressure for each 1/3-octave band.

In order to add coherently two or more acoustic rays at the same frequency, it is necessary to know both the sound pressure amplitudes of each ray as well as their relative phase. Phases must be referenced, so for convenience, we will reference all phases to one of the rays which we will denote as having a time-varying pressure, p_1 and frequency, ω, given by

$$p_1 = |p_1| e^{j\omega t} \quad \text{(Pa)} \tag{5.147}$$

where $|p_1|$ is the amplitude of the ray and its relative phase is set equal to 0 because it is used as the reference for the relative phases of other rays.

We can write the following expression for a second tonal ray, p_2, with a phase β_2 relative to ray 1.

$$p_2 = |p_2|e^{j\omega t}e^{j\beta_2} \tag{5.148}$$

The mean square sound pressure, $\langle p_t^2 \rangle$, resulting from the coherent combination of rays 1 and 2 is then,

$$\langle p_t^2 \rangle = \langle p_1^2 \rangle + \langle p_2^2 \rangle + 2\langle p_1 p_2 \rangle \cos(\beta_2) \quad (\text{Pa}^2) \tag{5.149}$$

where $\langle p_1^2 \rangle$ and $\langle p_2^2 \rangle$ are the mean-square sound pressures of rays 1 and 2 respectively, $\langle p_1 p_2 \rangle$ is the mean-square value of the instantaneous product of p_1 and p_2, and the phase (β_t) between the combined ray t and ray 1 is

$$\beta_t = \tan^{-1}\left(\frac{p_2 \sin(\beta_2)}{p_1 + p_2 \cos(\beta_2)}\right) \tag{5.150}$$

The sound pressure level, L_{pt}, obtained from the coherent combination of rays 1 and 2 is then given by,

$$L_{pt} = 10 \log_{10}\left(\frac{\langle p_t^2 \rangle}{p_{\text{ref}}^2}\right) \quad (\text{dB}) \tag{5.151}$$

where $p_{\text{ref}} = 20\mu\text{Pa}$.

A third ray, represented by p_3, may be added using Eq. (5.149), by substituting:

- p_t, obtained by combining rays 1 and 2, for p_1
- p_3 for p_2
- $\beta_3 - \beta_t$ for β_2.

The new result is denoted p_{t3} to distinguish it from p_t, which was obtained by coherently adding rays 1 and 2.

In practice, noise arriving at a receiver from one path is not entirely coherent with noise arriving from other paths, especially when it is not a pure tone. This is taken into account in the Nord2000 model by using partial coherence. The Nord2000 model takes into account incoherence effects caused by:

- atmospheric turbulence
- frequency band averaging
- fluctuating atmospheric refraction (due to fluctuating wind and temperature gradients)
- surface roughness
- scattering by vegetation and barriers.

The Nord2000 model only considers reduction in coherence between the primary sound ray (one with the largest sound pressure level at the receiver and denoted, p_1) and all other rays, but not between two or more non-primary rays. This is considered to be sufficiently accurate, although not technically rigorous. The ray usually considered as the primary ray is the direct ray from the sound source to the receiver location, if it exists. If the line of sight is blocked by a screen or barrier, then the direct ray over the top of the barrier with no ground reflections is considered to be the primary ray.

The Nord2000 model takes into account the incoherence between the primary ray, $i = 1$ and the ith ray ($i \geq 2$), due to the above listed causes, using an overall coefficient of coherence, F_i, for each path, $i, i \geq 2$, as follows

$$\frac{|p_t|^2}{|p_1|^2} = \left|\frac{p_t}{p_1}\right|^2 = \left[\left|1 + \sum_{i=2}^{N} F_i \frac{p_i}{p_1}\right|^2 + \sum_{i=2}^{N}(1 - F_i^2)\left|\frac{p_i}{p_1}\right|^2\right] \tag{5.152}$$

Note that

$$A_{\text{gr+bar}} = -10 \log_{10} \left|\frac{p_t}{p_1}\right|^2 = -10 \log_{10} \left(\frac{\langle p_t^2 \rangle}{\langle p_1^2 \rangle}\right) \tag{5.153}$$

and

$$\frac{p_i}{p_1} = \left|\frac{p_i}{p_1}\right| \mathrm{e}^{\,\mathrm{j}\beta_i} \tag{5.154}$$

where β_i is the relative phase between ray i and ray 1. To evaluate $A_{\text{gr+bar}}$, it is necessary to calculate the complex quantity, p_i/p_1 (real and imaginary parts), for each ray path i, and this is discussed in Section E.9 for cases with and without diffraction.

The overall coefficient of coherence for the ith ray, F_i, is made up of the product of the components, F_f due to frequency averaging in the 1/3-octave band, $F_{\Delta\tau}$ due to fluctuating refraction, F_t due to turbulence, F_r due to terrain roughness, and F_s due to scattering through vegetation, where each component refers only to the ith path. That is,

$$F_i = F_f F_{\Delta\tau} F_t F_r F_s \tag{5.155}$$

The component, F_f, of the *coefficient of coherence* as a result of *frequency averaging* over a 1/3-octave band, of centre frequency, f, is given by

$$F_f = \begin{cases} 1; & X = 0 \\ \dfrac{\sin X}{X}; & 0 < X < \pi \\ 0; & X \geq \pi \end{cases} \tag{5.156}$$

where

$$X = 0.23\pi f \Delta\tau \tag{5.157}$$

and where $\Delta\tau > 0$ is the travel time difference between the primary ray and the secondary ray that is under consideration.

Fluctuating refraction is the short time variation in refraction of a sound ray travelling from source to receiver as a result of variations in the atmospheric wind and temperature gradients. The component, $F_{\Delta\tau}$, of the *coefficient of coherence* as a result of *fluctuating refraction* over a 1/3-octave band is given by

$$F_{\Delta\tau} = \begin{cases} 1; & X = 0 \\ \dfrac{\sin X}{X}; & 0 < X < \pi \\ 0; & X \geq \pi \end{cases} \tag{5.158}$$

where

$$X = 2\pi f |\Delta\tau_+ - \Delta\tau| \tag{5.159}$$

where τ_+ is the upper end of the expected range of the travel time due to the range of expected variation in the wind and temperature gradients and $\Delta\tau_+ = |\tau_{1+} - \tau_{2+}|$ is the upper end of the variation in the difference in travel times of rays 1 and 2. The value of τ_+ is determined by calculating the upper expected value of the speed of sound as a function of height and then using this value instead of the average value to determine the travel times of the two rays. The Nord2000 model uses Eq. (5.26) for calculating the speed of sound as a function of height above the ground.

When the weather coefficients of Eq. (5.26), A_m and B_m are fluctuating ($s_{Am} > 0$ and $s_{Bm} > 0$) upper values of A_m and B_m are denoted A_{m+} and B_{m+} and are determined as follows (where a Gaussian distribution is assumed):

$$\begin{aligned} B_{m+} &= B_m + 1.7 s_{Bm} \\ A_{m+} &= A_m + 1.7 s_{Am} \end{aligned} \tag{5.160}$$

where s_{Am} and s_{Bm} are the standard deviations of the fluctuations in A_m and B_m, respectively.

The *coherence coefficient* due to *turbulence* is given by

$$F_t = \begin{cases} e^X; & X \geq -1 \\ (2+X)e^{-1}; & -2 < X < -1 \\ 0; & X \leq -2 \end{cases} \tag{5.161}$$

where

$$X = -0.1365 \gamma_T \left(\frac{f}{\bar{c}}\right)^2 q^{5/3} d \tag{5.162}$$

where the constant 0.1365 is as provided by Salomons and Janssen (2011) and Plovsing (2006a) (but differs considerably from the possibly incorrect value of 5.3888 provided by Plovsing (2014)), and

$$\gamma_T = \frac{C_T^2}{\overline{T} + 273.15} + \frac{22 C_v^2}{3\bar{c}^2} \tag{5.163}$$

where $\overline{T}$ is the average temperature (in degrees C) over the sound propagation path, $\bar{c}$ is the average speed of sound corresponding to the average temperature, d is the horizontal distance between source and receiver, C_T^2 is the turbulence strength due to temperature effects (sometimes called the temperature turbulence structure parameter), and C_v^2 is the turbulence strength due to wind effects (sometimes called the velocity turbulence structure parameter). The quantity q is half the mean separation of the direct and reflected ray paths and is defined as

$$q = \frac{h_S h_R}{h_S + h_R} \tag{5.164}$$

where h_S and h_R are the source and receiver heights above the ground. The turbulence strength parameters can be measured using a SODAR system or an ultrasonic anemometer, and will usually be found to be altitude dependent. For the purposes of the calculations described here, we only need to know an approximate estimate of the turbulence strength below about 200 m. Typical values of C_T^2 and C_v^2, measured approximately 3 m above the ground, were provided by Daigle (1982). He measured values of C_v^2 ranging from 2.0 in the afternoon to 0.5 in the evening and night, with 1.0 being the most common value (although he did measure a value as low as 0.1 in the early evening). He also measured values of C_T^2 ranging from 8.0 to 10.0 in the afternoon to 0.2 at night. As values of C_T^2 and C_v^2 are difficult to estimate, Salomons and Janssen (2011) give a value of γ_T for moderate turbulence as 5×10^{-6}, although Bullen (2012) suggests that a value of 10^{-5} gives better agreement with measurements and other models.

The *coherence coefficient* component, F_r, due to *surface roughness* is defined as

$$F_r = \begin{cases} e^{0.5g(X)}; & -2 \le g(X) \le 0 \\ (2 + 0.5g(X))e^{-1}; & -4 < g(X) < -2 \\ 0; & g(X) \le -4 \end{cases} \tag{5.165}$$

where

$$g(X) = \begin{cases} 0; & X \le 0.026686 \\ 0.55988(0.115448 - X) - 0.049696; & 0.026686 < X < 0.115448 \\ -0.066 + 1.066X - 8.543X^2 + 4.71X^3 - 0.83X^4; & X \ge 0.115448 \end{cases} \tag{5.166}$$

and where

$$X = k\sigma_r \sin(90 - \theta) \tag{5.167}$$

where k is the wavenumber, σ_r is the RMS value of the surface height variations (in metres), and θ is in degrees and is defined in Figure 5.2.

The *coherence coefficient* due to *scattering by forests* is given by

$$F_s = 1 - k_f\left(\frac{r_s n d_t}{1.75}\right)^2 \tag{5.168}$$

where r_s is the distance that the ray travels through a forest, n is the number of trees per square metre, d_t is the mean tree trunk diameter, and k_f is defined in Table 5.15 as a function of the product of wavenumber, k, and mean tree trunk diameter, d_t. A similar

Table 5.15 Estimates of the coefficient, k_f

$kd_t/2$	k_f	$kd_t/2$	k_f	$kd_t/2$	k_f
0	0.00	1.5	0.2	10	0.95
0.7	0.00	3	0.7	20	1.00
1	0.05	5	0.82		

expression for propagation through a housing estate can be found in Plovsing (2006a). If $r_s n d_t > 1.75$, then $r_s n d_t$ is set equal to 1.75.

5.8.2 Ground, Barrier and Terrain Excess Attenuation, A_{gr+bar}

The Nord2000 model treats barriers and screens differently to the ISO9613-2 model, taking the relative phase of the different rays into account when combining all of the direct and ground-reflected rays, as well as diffracted rays, arriving at the receiver from the same source. This is done by combining the complex sound pressures for each path, taking into account their full or partial coherence, as discussed in Section 5.8.1. The excess attenuation component, A_{gr+bar} in Eqs. (5.143) and (5.144) is then calculated for each 1/3-octave band, as explained in Section E.9.

For wind farms, the downwind propagation case with an atmospheric temperature inversion is of most interest, as this results in the loudest noise levels at the receivers. For this reason, the analysis in Appendix E excludes the case of any combination of upwind propagation or normal atmospheric vertical temperature profile that results in sound rays that are upward curving. Only downward-curving or straight sound paths are considered. Straight sound paths occur if there are no significant vertical atmospheric wind or temperature gradients present, and this corresponds to the homogeneous (or neutral) condition used in the NMPB-2008 model, as discussed in Section 5.7.

All of the above calculations are done in 1/3-octave bands (between and including 25 Hz and 10 kHz) and the 1/3-octave band centre frequency is used whenever a calculation involves a frequency or wavelength term. Sound pressure levels from rays arriving at the receiver due to reflections from obstacles are added incoherently, as are sound pressure levels due to other sound sources. This is implemented as explained in Section 2.2.8 for addition of decibels.

In the Nord2000 model, the calculation of the distance along each ray path is based on the assumption that the curvature introduced into the ray path as a result of vertical atmospheric wind and temperature gradients is circular. Unfortunately the log-lin sound-speed profile of Eq. (5.26) does not produce a circular path, so before ground and barrier effects can be calculated, it is necessary to find an *equivalent linear vertical sound-speed profile*, as discussed in Section D.2, which leads approximately to the same propagation effects as the more complex profile.

The model used in Nord2000 for calculating the effect of terrain is a two-dimensional model and relies on a cross-section of the terrain between the source and receiver. Any manmade barriers placed in the path between source and receiver are generally treated as infinitely wide, although there is a capability in the model to consider finite-width screens (Plovsing 2006a). However, in wind farm noise calculations, manmade barriers rarely have to be taken into account and any screening effects are usually the result of hills between the wind farm and the receiver location.

To calculate the combined effects of terrain and atmospheric refraction, A_{gr+bar}, the following steps need to be followed:

1. The first step is to divide the terrain cross-section between the sound source and the receiver into a number of straight-line segments, as described in Section E.3.

2. The second step involves using the segmented approximation to the actual terrain shape to find the two most efficient diffraction edges, as described in Section E.7. The most efficient diffraction edge is the one that results in the largest noise reduction and is referred to as the *first diffraction edge.* The second most efficient diffraction edge is referred to as the *second diffraction edge.* These edges will only exist in hilly terrain, not across valleys or over flat terrain, unless manmade barriers have been constructed between the sound source and receiver locations. If a manmade barrier has been constructed, then the diffraction efficiency of that also must be considered when finding the two most efficient diffracting edges. If the second diffracting edge is between the first diffracting edge and the receiver, then for calculating the attenuation due to diffraction over the second edge, the source location is moved to the first diffracting edge and this location is referred to as the *secondary source.* If the second diffracting edge is between the first diffracting edge and the source, then for the purpose of calculating the attenuation due to diffraction over the second edge, the receiver location is moved to the first diffracting edge and this location is referred to as the *secondary receiver.*
3. The third step is to calculate the propagation path lengths and propagation times for each ray, as described in Section D.3.
4. The fourth step is to calculate the sound pressure (amplitude and phase) arriving at the receiver due to each diffracted and reflected sound ray. This includes rays that are reflected off the ground on either side of the diffraction edge or rays reflected from the ground in the absence of a diffracted edge. The treatment of reflected and diffracted rays are explained separately below:
   - The reflected sound pressure contribution from each ground segment between the source (or secondary source) and receiver (or secondary receiver) is calculated using the Fresnel zone analysis of Section F.3 or Section F.4, depending on the type of ground segment (see Section E.2). The point of specular reflection (where angle of incidence = angle of reflection) is found for each ground segment (or its imaginary extension). The term, *imaginary extension* refers to extending the ground segment past its actual physical end point when necessary, so that for a ray reflected from this extension, the angle of incidence is equal to the angle of reflection. The contribution of each ground segment to the total reflected sound is defined by a Fresnel coefficient, calculated as described in Section F.3 or Section F.4. The concept of a Fresnel zone (see Appendix F) is used to quantify the extent of ground surface that is involved in reflection of a sound ray travelling from the source to the receiver, and is determined by construction of a Fresnel ellipsoid or ellipse, as discussed in Appendix F. The finite size of the wavelength of sound means that the ground on either side of the specular reflection point will affect the amplitude and phase of the reflected ray. The concept of a Fresnel weighting is used to estimate the contribution of each ground segment to the total amplitude of the reflected sound ray. The specular reflection point for each ground segment is found by extending each ground segment so that the image source for that ground segment can be defined. Then the Fresnel ellipse, with the receiver and image source as the two foci, is constructed, as explained in Appendix F, so that it intersects the extension of the ground segment in two places (see Figure F.3). The proportion of the actual ground segment that exists between the two intersection points is used to calculate the Fresnel weighting for that ground segment. In the case of a downward-refracting

atmosphere, the curvature of the sound rays must be taken into account when finding this reflection point (see Section D.3.2). This step requires calculations of the spherical-wave reflection coefficient for the ground surface (see Section C.5).

- Where there are no terrain or manmade barriers interrupting the line-of-sight between the source and receiver, there are two scenarios that need to be considered:
  a. When there is flat ground, in which case there is a single specular reflection point; the excess attenuation due to the ground effect is calculated as shown in Section E.4.
  b. When there is a valley between the source and receiver, in which case the ground between the source and receiver is divided into segments, each of which contribute to the reflected sound ray an amount determined by its Fresnel weighting. In this case, the amplitude attenuation of the ground-reflected ray is calculated by summing over all ground segments between the source and receiver, the product of the Fresnel weight and the attenuation (in decibels) due to reflection for each ground segment, as discussed in Section E.6 and Appendix F. Note that each ground segment contributes only one reflected ray.
- The reduction in sound pressure due to diffraction of a sound ray around a diffracting edge is a function of wavelength and difference in path lengths between the diffracted-ray path and the direct path from source to receiver. The path length difference for straight-line propagation with no ground reflections was discussed in detail in Section 5.6. However, for the Nord2000 model, the curvature of the sound rays must be considered, and the rays reflected from the ground are treated separately from the non-ground-reflected rays. Ground-reflected rays have their amplitude and phase affected on reflection and this effect is calculated as described in Section C.5. This effect is added to the effect due to the length of the path travelled by the ray. The propagation times and path lengths (see Section D.3) for each ray, based on the corresponding path length and speed of sound, need to be calculated for the Nord2000 model to allow proper combination of the rays from the same source at the receiver location. These lengths and times are based on the reflection point being at the location of specular reflection (angle of incidence of the sound ray on the ground equal to the angle of reflection). The location of the reflection point P is calculated as described in Section D.3.2. Any additional phase delays due to ground reflections also have to be added to the phase delays due to propagation-time differences when coherently combining different rays from the same source arriving at the receiver via different propagation paths.

5. The fifth step is to combine all the rays arriving at the receiver (direct, ground-reflected and diffracted), by taking into account relative phases as well as propagation effects and coherence-reducing effects. This is done using complex pressures (made up of a real and imaginary part) and combining them as described in Section E.9.

The contribution due to a ray reflected from a vertical surface may also include a ground-reflected ray that was also reflected from the same vertical surface, and these are combined incoherently, which is different to the coherent combination of rays not reflected from a vertical surface. The contribution from rays reflected from different vertical surfaces, which include the spherical-spreading, scattering and atmospheric-absorption excess attenuation effects of Eqs. (5.143) and (5.144), are

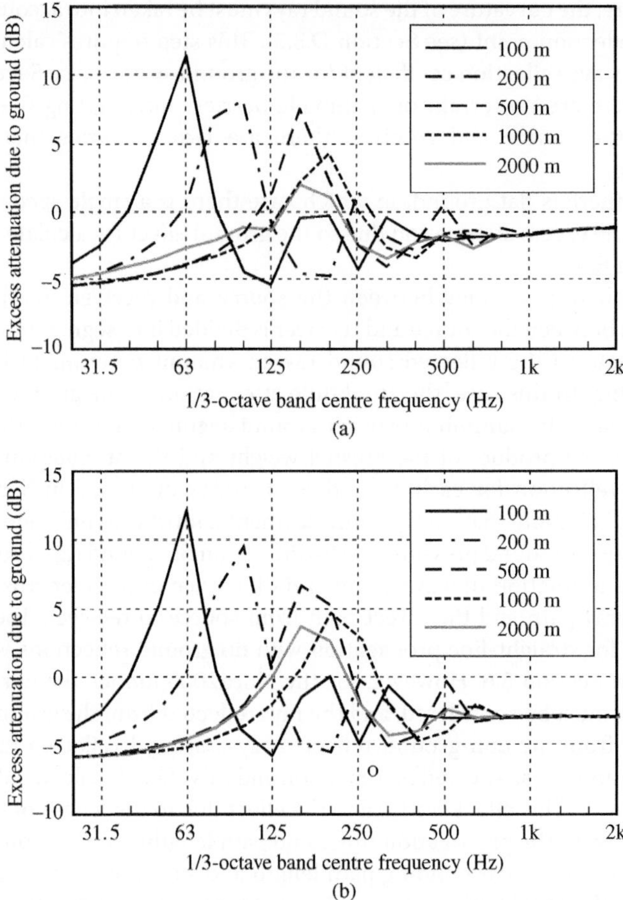

Figure 5.24 Excess attenuation in 1/3-octave bands due to the ground, for propagation over flat ground, from a source at height 100 m to a receiver at height 2 m downwind (wind speed 8 m/s at 10 m) for various horizontal propagation distances, *d*: (a) grass covered ground; (b) hard ground (water) and for grass-covered ground. Data taken from Søndergaard et al. (2007).

combined incoherently (see Section 2.2.8) together and also incoherently with the total contribution from rays that are not reflected from vertical surfaces.

An example of the excess attenuation due to the ground in the presence of a downward-refracting atmosphere, calculated using the Nord2000 model for a receiver at various distances from a source over grass-covered ground, is shown in Figure 5.24.

5.8.3 Multiple Ground Reflections

For propagation in the presence of a downward-refracting atmosphere, at a sufficient distance between the source and receiver, more than one ground-reflected ray will arrive at the receiver. This phenomenon is discussed in detail in Section 5.2.4. Here, the quantification of this effect on sound levels calculated using the Nord2000 model is outlined (Plovsing 2006b, 2014). The intention is to calculate the additional noise level at the receiver due to sound rays that have been reflected from the ground more than once. This contribution is added incoherently to the sound pressure levels (that is, addition of mean-square sound pressures; see Section 2.2.8).

The Nord2000 method assumes that the vertical atmospheric sound-speed profile is logarithmic, as shown in Eq. (5.26), with $A_m = 0$. The first step is to define N, which is the approximate number of rays reaching the receiver, including the direct ray and rays that have only experienced a single ground reflection. The contribution from additional rays is ignored for $N < 4$, as these rays have one or no reflections and have already been included. The number of rays, N, that are possible between the source and receiver for a logarithmic velocity profile defined by Eq. (5.26) is

$$N = \frac{4d}{h_{\max}}\sqrt{\frac{B_m}{2\pi c_0}} \tag{5.169}$$

where $h_{\max}$ is the maximum value of h_S and h_R (and the barrier height if a barrier is present). N is not necessarily an integer, and is divided into an integer, N_i, and non-integer, ΔN, part such that,

$$N = N_i + \Delta N \tag{5.170}$$

Reflected rays are grouped into orders, with the maximum order, n_{mx} given by

$$n_{\text{mx}} = \text{Int}\left(\frac{N-1}{4}\right) + 1 \tag{5.171}$$

where Int(x) is the rounded down integer value of x.

The contribution, $\Delta L_{\text{mult},\ell}$, to the sound pressure level at the receiver due to multiple reflections for sound that contains a significant low-frequency content (less than 200 Hz) is then calculated using a coherent summation of the ray contributions as follows:

$$\Delta L_{\text{mult},\ell} = 20\log_{10}\left\{\Delta N[R_f(n_{\text{mx}})]^{n_{\text{mx}}}A_{\text{ray}}(n_{\text{mx}}) + \sum_{M=5}^{N_i}[R_f(n)]^n A_{\text{ray}}(n)\right\} \tag{5.172}$$

The high-frequency expression uses an incoherent summation and is only used when the noise is dominated by mid- and high-frequencies, and is

$$\Delta L_{\text{mult},h} = 10\log_{10}\left\{\{\Delta N[R_f(n_{\text{mx}})]^{n_{\text{mx}}}A_{\text{ray}}(n_{\text{mx}})\}^2 + \sum_{M=5}^{N_i}\{[R_f(n)]^n A_{\text{ray}}(n)\}^2\right\} \tag{5.173}$$

where n, $A_{\text{ray}}(n)$, and $R_f(n)$ are defined as

$$n = \text{Int}\left(\frac{M-1}{4}\right) + 1 \tag{5.174}$$

where M is the index in the sums of Eqs. (5.172) and (5.173),

$$A_{\text{ray}}(n) = 10^{-0.1\sqrt{\frac{|\psi_G(n) - \psi_G'(n)|}{0.00001\sin\psi_G(n)}}} \tag{5.175}$$

and

$$R_f(n) = \sum_{i=1}^{N_{ts}} |R_p| \frac{x_{i+1} - x_i}{x_{RGv} - x_{SGv}} \tag{5.176}$$

where x_{RGv} is the horizontal location of the ground point immediately below the receiver, x_{SGv} is the horizontal coordinate location of the ground point immediately below the source, x_i is the horizontal coordinate location of the end of the ground segment nearest the source, x_{i+1} is the horizontal coordinate location of the end of the ground segment nearest the receiver, N_{ts} is the number of terrain segments between the source and receiver, $|R_p|$ is the modulus of the plane-wave reflection coefficient, calculated using an incident and reflected angle of $\psi_G(n)$ (see Figure D.2) and

$$\psi_G(n) = \arccos\left\{\frac{c_0}{c[h_{\max}(n)]}\right\} \tag{5.177}$$

$$\psi'_G(n) = \arccos\left\{\frac{c_0}{c[1.00001 h_{\max}(n)]}\right\} \tag{5.178}$$

$$h_{\max}(n) = \frac{d}{n}\sqrt{\frac{B_m}{2\pi c_0}} \tag{5.179}$$

where d is the horizontal distance between the source and receiver, and $c[h_{\max}(n)]$ is the speed of sound at height $h_{\max}(n)$. The above equations also apply when $n = n_{mx}$.

The excess attenuation, A_{mult}, due to multiple reflections (to be added to the total excess attenuation, A_E) is then,

$$A_{\text{mult}} = -\Delta L_{\text{mult},\ell} \tag{5.180}$$

with a similar expression for mid- and high-frequency noise.

The preceding analysis applies for the case of a logarithmic atmospheric sound-speed profile, usually as a result of an atmospheric wind gradient. The temperature-gradient effect is to produce a linear profile, and this contribution is taken into account by calculating a new value of B_m, which accounts for both the linear and logarithmic sound-speed profiles. This is done by calculating the radius of curvature of the sound ray resulting from the combined effect of the linear and logarithmic sound-speed profiles, and then using this to calculate a new equivalent B_m. The new expression for B_m is

$$B_m = \frac{\pi c_0 d^2}{32 R_c^2} \tag{5.181}$$

where

$$R_c = \left[\frac{1}{R_A} + \frac{1}{R_B}\right]^{-1} \tag{5.182}$$

and R_A and R_B are given by Eqs. (5.36) and (5.56), respectively.

If the terrain is not flat, the heights of the source, receiver and barriers (if present) are determined relative to the terrain baseline. For a valley where the average deviation of the terrain from the terrain baseline is negative (see Section E.2), the numerical value of the deviation is added to $H_{\max}$.

5.8.4 Excess Attenuation, A_{sc}, due to a Ray Travelling Through a Scattering Zone

A scattering zone is usually a forest, and sound rays travelling through a forest or band of trees suffer a scattering effect. Although the effect of scattering has been partially accounted for in calculating the effect of the scattering zone on the coherence of a sound ray passing through it, its effect on the amplitude of the scattered ray arriving at the receiver must also be accounted for and this is the purpose of this section. The important parameter that has to be determined first is the length of the path, $r_s = (r_1 + r_2)$, travelled through the scattering zone (see Figure 5.19). If a barrier or terrain screen exists in the scattering zone or between two scattering zones, the calculation of the path travelled through the scattering zone on the source side of the barrier is based on the receiver being replaced by the top of the barrier and the calculation of the path travelled through the scattering zone on the receiver side of the barrier is based on the source being replaced by the top of the barrier. In the presence of atmospheric refraction, the sound rays will be characterised by a radius of curvature.

The excess attenuation due to scattering, A_{sc}, is subtracted from the final sound pressure level at the receiver, calculated by considering all the other excess attenuation effects. The length of path travelled through a scattering zone is based on the direct path from source to receiver or the path over the top of a barrier when the barrier blocks the line of sight between the source and receiver. Paths involving reflection from the ground are ignored, although the scattering excess attenuation should be applied to paths involving reflection from a vertical surface prior to the contribution of that ray being logarithmically added to the sound level at the receiver resulting from the paths *not* reflected from vertical surfaces. The effect of scattering on the coherence of the direct and ground-reflected rays has already been taken into account in the calculation of A_{gr+bar}.

The propagation effect, ΔL_s, which has a minimum allowed value of −15 dB, for a scattering zone is calculated using (Plovsing 2006a),

$$\Delta L_s = k_f \left(\frac{r_s n d_t}{1.75} \right)^2 k_p [\Delta L(h', \alpha, r') + 20 \log_{10} (8r')] \tag{5.183}$$

where k_f is given as a function of $kd_t/2$ in Table 5.15, k is the wave number, k_p is a constant and equal to 1.25 for trees, d_t is the mean trunk diameter, n is the number of trees per square metre, $\Delta L(h', \alpha, r')$ can be calculated by interpolation (see below) between values listed in Table 5.16 and

$$r' = n d_t r_s \tag{5.184}$$

The excess attenuation due to scattering is then given by

$$A_{sc} = -\Delta L_s \tag{5.185}$$

Interpolation is used to find values of $\Delta L(h', \alpha, r')$ for parameters between those listed on the table and implies the following steps (where the subscripts, i, j and k refer to values listed in the table). All steps must be followed, except in cases where one or more of the parameters is a match for an entry in the table.

1. Using $\Delta L(h'_i, \alpha_j, r'_k)$ and quadratic interpolation, interpolate for h' using h'_i values in the table larger and smaller than h', to obtain values of $\Delta L(h', \alpha_j, r'_k)$. This interpolation is done for all values of α_j combined with a sufficient number of r'_k values to enable step 2 below to be undertaken.

Table 5.16 Estimates of the coefficient, $\Delta L(h', \alpha, r')$

	$h' = 0.01$			$h' = 0.1$			$h' = 1$		
r'	$\alpha = 0$	$\alpha = 0.2$	$\alpha = 0.4$	$\alpha = 0$	$\alpha = 0.2$	$\alpha = 0.4$	$\alpha = 0$	$\alpha = 0.2$	$\alpha = 0.4$
0.0625	6.0	6.0	6.0	6.0	6.0	6.0	6.0	6.0	6.0
0.125	0.0	0.0	0.0	0.0	0.0	0.0	0.0	0.0	0.0
0.25	−7.5	−7.5	−7.5	−6.0	−7.0	−7.5	−6.0	−7.0	−7.5
0.5	−14.0	−14.25	−14.5	−12.5	−13.5	−14.5	−12.5	−13.0	−14.0
0.75	−18.0	−18.8	−19.5	−17.3	−18.0	−19.0	−16.0	−16.8	−17.7
1.0	−21.5	−22.5	−23.5	−20.5	−21.6	−22.8	−19.3	−20.5	−21.3
1.5	−26.3	−27.5	−29.5	−25.5	−27.2	−29.0	−24.0	−25.5	−26.3
2.0	−31.0	−32.5	−34.5	−30.0	−32.0	−33.3	−27.5	−29.5	−30.8
3.0	−40.0	−42.5	−45.5	−37.5	−40.5	−42.9	−34.2	−36.0	−37.8
4.0	−49.5	−52.5	−56.3	−45.5	−49.5	−52.5	−40.4	−42.8	−45.5
6.0	−67.0	−72.5	−78.0	−62.0	−67.0	−72.0	−52.5	−56.2	−60.0
10.0	−102.5	−113.0	−122.5	−94.7	−103.7	−112.5	−78.8	−84.0	−89.7

r' per Eq. (5.184); $h'=nd_th$, h, average height of scattering obstacle above the ground; α, absorption coefficient of the scattering obstacles (normally in the range 0.1–0.4).

2. Using $\Delta L(h', \alpha_j, r'_k)$ values determined in step 1 above and quadratic interpolation, interpolate for α using values of α_j in the table larger and smaller than α, to obtain values of $\Delta L(h', \alpha, r'_k)$. This interpolation is done for a sufficient number of r'_k values to enable step 3 below to be undertaken.
3. Using $\Delta L(h', \alpha, r'_k)$ values determined in step 2 above and cubic interpolation, interpolate for r' using two larger and two smaller values of r'_k to obtain $\Delta L(h, \alpha, r')$. At the boundaries of r', only quadratic interpolation can be used.

A similar approach to that described in this section can be used to estimate the attenuation due to scattering by housing, as outlined by Plovsing (2006a).

5.8.5 Excess Attenuation, A_r, due to Reflection from a Facade or Building

The attenuation component, A_r, only applies to rays that are reflected from vertical (or near-vertical) surfaces such as buildings. The sound pressure level at the receiver is calculated for the reflected ray, taking into account the spherical spreading, scattering and atmospheric absorption excess attenuation effects. A_r is then subtracted from the result prior to it being incoherently summed with the sound pressure levels from non-reflected rays, as shown in Eq. (2.27). The excess attenuation, A_r, is a result of absorption by the reflecting surface and its calculation is discussed in the following paragraphs.

The excess attenuation, A_r, is given by

$$A_r = -\Delta L_r + \Delta L_d \tag{5.186}$$

where ΔL_d is the attenuation due to the additional distance travelled by the reflected ray (over the distance travelled by the direct ray) and is given by

$$\Delta L_d = 20 \log_{10} \frac{\sqrt{d_1^2 + (h_O - h_S)^2} + \sqrt{d_2^2 + (h_O - h_R)^2}}{\sqrt{d^2 + (h_R - h_S)^2}} \tag{5.187}$$

where d is the distance between the source and receiver in the horizontal plane, h_S is the source height coordinate, h_R is the receiver height coordinate and h_O is the height coordinate of the reflection point. ΔL_r is a measure of the efficiency of reflection and is calculated as follows:

$$\Delta L_r = 20 \log_{10} \overline{\Re}_S + 20 \log_{10} \left(\frac{S_{\text{ref}}}{S_F} \right) \tag{5.188}$$

where $\overline{\Re}_S$ is the incoherent reflection coefficient (see Section C.6), S_{ref} is the area of reflecting surface that lies within the Fresnel zone and S_F is the area of the Fresnel zone calculated using Eq. (F.12).

The Fresnel zone is associated with reflection in the plane of the reflecting surface and the two foci of the Fresnel ellipsoid are the receiver and the image of the source in the vertical reflection surface. This is best understood by referring to Figure G.6. Although this figure is for reflection from a horizontal ground plane, it can be easily rotated so that it represents reflection from a vertically oriented plane.

The point of reflection is obtained by first finding the image source in the plane of the reflecting surface and then noting where the straight line from the image source to the receiver intersects the plane containing the reflecting surface. This intersection, which may lay outside the reflecting surface, is the point of reflection. Even if the point of reflection lies outside the actual reflecting surface, part of the Fresnel zone associated with the reflection may lie on the reflecting surface and thus result in a contribution to the sound pressure at the receiver. However, in the Nord2000 model, a reflecting surface is ignored if the point of reflection is more than 2 m away from the actual surface.

The calculation of S_{ref} and S_F requires that the point of reflection be located. The horizontal location of the reflection point, O, in the plane of the reflecting surface is obtained as shown in Figure 5.25a. However, due to the curvature of the ray as a result of the atmospheric sonic gradient, the location of the vertical coordinate, h_O, of the reflection is a bit more complicated, as shown in Figure 5.25b. However, it can be calculated as described in Section 5.2.4, by considering a ray travelling from the source to the image receiver, where the image is at the same height as the receiver and as far behind the reflecting surface as the receiver is in front.

The size of the reflecting surface that lies within the Fresnel zone requires calculation of the coordinates of the four corners of the equivalent rectangular Fresnel zone relative to the coordinates of the reflection point, and this is discussed in Appendix F. The Fresnel zone is defined by distances a_1, a_2 and $b/2$, which represent the distances of the Fresnel zone edges from the reflection point, O. Two edges are shown superimposed in Figure F.1 and represented by the line, P1–O–P2, with one line spaced $b/2$ out of the page and one spaced $b/2$ into the page. The other two edges complete the rectangle defined by the two lines P1–O–P2. Distances a_1, a_2 and b are calculated using Eqs. (F.1) and (F.11). The reflection plane considered here is shown as the ground plane in Figure F.1. To represent a vertical reflecting plane, the figure should be rotated anti-clockwise by 90°, so that the ground plane in the figure becomes the vertical reflecting plane.

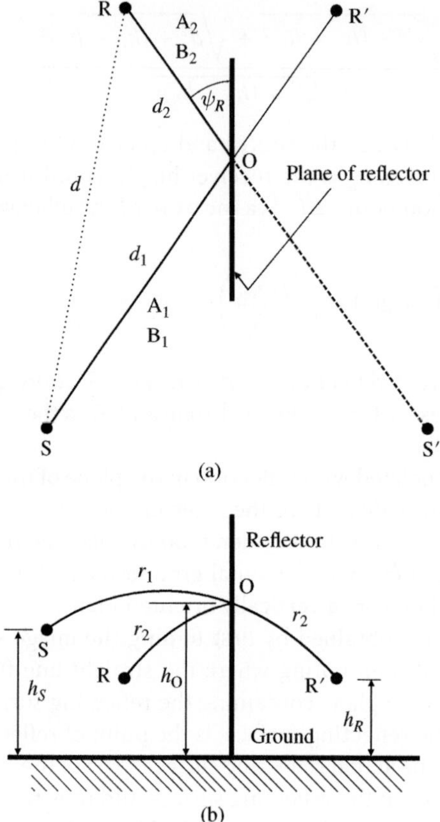

Figure 5.25 Ray path for reflection from a vertical surface: (a) plan view; (b) elevation view.

Once the location and size, S_F, of the rectangular Fresnel zone are defined with respect to the reflection point, the size S_{ref} of reflecting surface that lies within the Fresnel zone can be determined, as the coordinates of the reflecting surface are also known with respect to the reflection point. The process for doing this is explained in Section F.2. Then the values of S_F and S_{ref} can be substituted into Eq. (5.188) to find ΔL_r.

As the incident and reflected rays are travelling in different directions, the velocity component in each direction is likely to be different. For this reason, an equivalent log-lin velocity profile has to be calculated based on the profiles for the incident and reflected rays. This requires calculation of parameters, A_m and B_m, based on the parameters A_1 and B_1 for the incident ray and A_2 and B_2 for the reflected ray. A_1 and A_2 are calculated by substituting A_1 and A_2, respectively, for A_m in Eq. (5.22) and B_1 and B_2 are calculated by substituting B_1 and B_2, respectively, for B_m in Eq. (5.30). The new values of A_m and B_m for the reflected ray are,

$$A_m = \frac{d_1 A_1 + d_2 A_2}{d_1 + d_2} \tag{5.189}$$

and

$$B_m = \frac{d_1 B_1 + d_2 B_2}{d_1 + d_2} \tag{5.190}$$

where d_1 and d_2 are defined in Figure 5.25a.

The consideration of reflection from non-vertical surfaces is quite complex and is not included as part of the Nord2000 model.

5.8.6 Limitations of the Nord2000 model

Nord2000 has the following limitations:

1. The model is only valid for 1/3-octave band analysis for 1/3-octave band centre frequencies ranging from 25–10 000 Hz, and can thus provide A-weighted sound pressure levels over this frequency range by applying the A-weighting correction (see Table 2.2) to each 1/3-octave band level and then summing the levels logarithmically (see Section 2.2.8). The overall C-weighted sound pressure level can be calculated by following a similar procedure. Of course the Nord2000 model is also valid for calculating overall A-weighted sound pressure levels over the frequency range covered by the 31.5–8000 Hz octave bands.
2. Nord2000 is valid only for point sources. However, even large sources appear as point sources at distances greater than about two source dimensions from them, so this should not result in significant errors in almost all situations.
3. The terrain between source and receiver is approximated by a number of straight line segments, and this may not model actual situations accurately. However, it is superior to the approaches used for the CONCAWE and ISO models described in the preceding sections.
4. The geometry used for the terrain effects is based on two- rather than three-dimensional modelling, which may be inaccurate in some cases. This means that the calculation of the excess attenuation due to terrain effects is based on the cross-section of terrain between the source and receiver so that the terrain outside of the direct line from source to receiver has no effect on the result obtained.
5. Any barriers are generally assumed to be infinite in length normal to the line between the source and receiver. However, if the effect of finite width on the barrier attenuation needs to be taken into account, then refraction around the vertical edges of the barrier may be included in much the same way as done in the ISO model.
6. Only the two most significant diffraction edges are included for rays propagating from a source to a receiver past barriers or hilly terrain. This approximation probably results in only very small errors.
7. Only the coherence between the primary ray arriving at the receiver and all other rays are taken into account. Coherence effects between rays other than the primary ray are not taken into account. This approximation probably results in only very small errors.
8. Lin-log vertical atmospheric sonic profiles are assumed to approximate actual sonic profiles. It is also assumed that there are no sonic gradients in the horizontal direction and that the atmosphere has the same sonic gradient profile at all locations

between the source and receiver. This is why the model should be used to provide a worst-case noise prediction or a range of predicted noise levels based on expected average vertical atmospheric sonic profiles. It is not possible to use this model or any model to accurately predict receiver noise levels for a complex wind and temperature profile that varies significantly between the source and receiver locations.

9. The model is valid only up to 1000 m for sources close to the ground, but for high sound sources such as wind turbines, the model is claimed by its author to be valid for source–receiver distances of several kilometres.
10. There is an error in one of the equations used for the calculation of the incoherent reflection coefficient (Plovsing 2006a), which results in a significant overestimation of the incoherent reflection coefficient. This is used for high-frequency calculations where the ground-reflected ray is considered to be incoherent with the direct ray. The error results in a slight overestimation of the sound pressure level at high frequencies (above 1–2 kHz).
11. Only reflecting surfaces that are vertically oriented are considered.

The Nord2000 model has been validated for wind farm noise predictions at distances up to 1.4 km (Søndergaard and Plovsing 2009) and for low-frequency noise at distances up to 3.4 km (Søndergaard and Sorensen 2013).

5.9 Harmonoise (2002) Noise Propagation Engineering Model

There are two Harmonoise models: the Engineering Model and the Reference Model. The Engineering Model is intended for general use as a propagation model for calculating environmental noise levels. The Reference Model is intended to be used as a validation tool for the Engineering Model for relatively simple situations. The Reference Model (see Defrance et al. (2007)) is based on a combination of the boundary element method (BEM), the parabolic equation (PE) method and a ray-tracing method. Computations using these methods are time consuming and the required input data for accurate results are detailed and extensive, which is why the Reference Model is only used for validation purposes. Thus, only the Harmonoise Engineering Model is discussed in the following paragraphs.

The Harmonoise model borrows many of its procedures from the Nord2000 model. The main difference lies in its treatment of refraction through the atmosphere as a result of wind and temperature gradients. Refraction of sound in the atmosphere is the bending of sound rays as a result of the sound speed varying with height above the ground. The Harmonoise model uses a coordinate transformation to simulate atmospheric refraction with a corresponding curved ground profile, so that the sound rays can be straight rather than curved.

Harmonoise is a more recent development than the Nord2000 model, but it is not generally considered to be more accurate, although some aspects are simpler to implement. Nevertheless, the calculations of excess attenuation due to meteorological effects, ground effects, and terrain-shape effects are still quite complex to implement, as will be seen in the following discussion. As with the Nord2000 model, the Harmonoise model calculates the sound pressure level at a receiver location for 1/3-octave bands between

25 Hz and 10 kHz, and the 1/3-octave band centre frequency is used whenever a calculation involves a frequency or wavelength term. The overall A-weighted sound pressure level is calculated by applying the A-weighting correction at the 1/3-octave band centre frequency (see Table 2.2) to each 1/3-octave band level and then summing the levels logarithmically (see Section 2.2.8). The overall C-weighted sound pressure level can be calculated by following a similar procedure.

The basic equation for calculating the sound pressure level, L_{pik} at location, k, due to the sound power, L_{wi} radiated by sound source, i with a directivity index in the direction of the receiver location of DI_{ik}, is given by

$$L_{pik} = L_{wi} + DI_{ik} - A_{Eik} \quad \text{(dB)} \tag{5.191}$$

Normally for wind turbines, the sound power level is measured in the downwind direction, so the directivity index, DI_{ik}, is inherently included in the sound power measurement for downwind noise level calculations.

The excess attenuation, A_{Eik} is defined by

$$A_{Eik} = [A_{\text{div}} + A_{\text{atm}} + A_{\text{gr+bar}} + A_{\text{sc}} + A_r]_{ik} \quad \text{(dB)} \tag{5.192}$$

The attenuation component A_{div} is calculated, assuming a point source, using Eq. (5.5); the component A_{atm} is calculated using Eq. (5.8), as described in Section 5.2.2. Both of these excess attenuation effects are applied to the final total calculated sound pressure level at the receiver for each sound source. Although the calculations of excess attenuation due to atmospheric absorption are only for single frequencies, the CONCAWE and ISO models rely on the octave band centre frequency absorption being representative of the entire octave frequency band, resulting in increasing errors as the calculated absorption value increases. The Harmonoise model takes this error into account in the same way as the Nord2000 model and applies a correction that is valid for 1/3-octave band analysis and calculated using Eq. (5.193).

$$A_{\text{atm}} = A_0(1.0053255 - 0.00122622A_0)^{1.6} \quad \text{(dB)} \tag{5.193}$$

In comparing Eq. (5.192) with Eq. (5.87), it is noted that in the NMPB-2008, Harmonoise and Nord2000 models, the excess attenuation due to the ground is combined with that due to barriers, whereas these two attenuations are treated separately in the CONCAWE and ISO9613-2 models. Also, the Harmonoise and Nord2000 models include calculation of the effects of terrain geometry and atmospheric wind and temperature gradients, which are included in the $A_{\text{gr+bar}}$ term, whereas the ISO9613-2 model is only applicable to a ‘worst case’ downward-refracting atmospheric wind and/or temperature vertical gradient.

The term, A_{misc} in the ISO model has been replaced with the term, A_{sc} in the Harmonoise model, which is the excess attenuation due to scattering by forests and atmospheric turbulence.

The A_r term is only applied to rays that are reflected from the vertical surface of an obstacle and accounts for the obstacle absorption properties and dimensions. As for the Nord2000 and ISO models, the Harmonoise model treats rays reflected from objects separately and calculates separately their corresponding sound pressure levels at the receiver location. These are added incoherently according to Eq. (2.27). In addition to A_r, the reflected ray also has to have the attenuation components for atmospheric absorption, scattering and spherical spreading applied. Calculation of A_r is discussed in Section 5.8.5.

As with the Nord2000 and ISO models, there is no explicit factor in Harmonoise to account for meteorological effects. The Harmonoise model includes the meteorological effects in the $A_{\text{gr+bar}}$ term by using the sonic gradient (resulting from the meteorological conditions) to calculate the radius of curvature of the direct and ground-reflected rays arriving at the receiver (see Section 5.2.4). However, the Harmonoise model does not use curved sound rays in its calculations. Instead it uses a coordinate transformation to curve the ground, with a radius calculated as described in Section 5.2.4. It can then use straight sound rays. Unfortunately this restricts the distance over which the Harmonoise model is valid to distances that satisfy the condition that $d_{SR} < 0.2R$, where R is the radius of curvature of the sound ray before the coordinate transformation is done and d_{SR} is the straight-line distance between the source and receiver prior to the coordinate transformation. The coordinate transformation used by Harmonoise will be discussed in more detail below when we discuss terrain effects and the $A_{\text{gr+bar}}$ term in Section 5.9.3.

Although the coordinate transformation and curved ground do not have much effect on the attenuation due to spherical spreading (A_{div} term), they do affect the relative phase difference between the direct and reflected sound rays, and this influences the amount of constructive or destructive interference that may occur between rays arriving at the receiver from the same sound source. The effects that act to reduce the coherence (see Eqs. (5.195)–(5.201)) between the various rays arriving at the receiver are also included in the Harmonoise model in the calculation of the $A_{\text{gr+bar}}$ term.

The Harmonoise model treats barriers and screens in a similar way to the Nord2000 model, except that the equations are simplified by making a small approximation, as will be discussed in detail in the following subsection. In addition, the Harmonoise model is able to use straight ray paths and a modified ground profile, whereas the Nord2000 model uses curved ray paths and the actual ground profile. As with the Nord2000 model, the relative phase of the different rays are taken into account when combining all of the direct and ground-reflected rays arriving at the receiver from the same source.

The Harmonoise model calculates each of the excess attenuation effects separately and then calculates the overall sound pressure level due to a particular source at the receiver using Eq. (5.191). When scattering due to vegetation affects different ray paths differently, the scattering effect is included for each ray path in the calculation of the $A_{\text{gr+bar}}$ term.

Applying the Harmonoise model to the calculation of the excess attenuation terms, $A_{\text{gr+bar}}$, A_{sc} and A_r, relies on calculating the difference between the sound pressure level of a direct ray arriving from the source (with no excess attenuation effects except for spherical spreading) and the sound pressure level of a ray with the effect of one of the above terms included in addition to the A_{div} term.

In a similar way to the Nord2000 model, the Harmonoise model allows for full or partial coherent addition of the various rays arriving at the receiver from a particular source. These rays include the direct ray, the ground-reflected ray, and any rays diffracted by barriers. However, the Harmonoise model uses a smaller range of coherence effects than does the Nord2000 model.

The total sound pressure level due to all sound sources is calculated by incoherent addition (see Eq. (2.27)) of the sound pressure level generated by each source.

5.9.1 Combination of Sound Rays from the Same Source Arriving at the Receiver via Different Paths (for Calculating A_{gr+bar})

The reduction in coherence between different rays arriving at a receiver location from the same source is caused mainly by atmospheric turbulence. The Harmonoise model takes into account incoherence effects caused by:

- atmospheric turbulence
- frequency-band averaging
- uncertainty in distance between source and receiver
- uncertainty in source and receiver heights.

The coherence coefficients are taken into account when the complex sound pressures (consisting of a real and imaginary part) due to rays arriving at the receiver from the same source along different paths are combined. The Harmonoise model uses the coherence coefficients due to the above effects to combine the individual complex sound pressures into a single sound pressure amplitude that represents the combined effect of all the contributing rays. The Harmonoise model only considers the reduction in coherence between the primary sound ray (the one with the largest sound pressure level at the receiver) and all other rays, but not between two or more non-primary rays. This is considered to be sufficiently accurate, although not technically rigorous. The ray usually considered the primary ray is the direct ray from the sound source to the receiver location, if it exists. If the line of sight is blocked by a screen or barrier, then the direct ray over the top of the barrier with no ground reflections is considered to be the primary ray.

The Harmonoise model takes into account the incoherence between the primary ray, $i = 1$, and the ith ray ($i \geq 2$), due to the above listed causes using an overall coefficient of coherence, F_i, for each path, $i, i \geq 2$, as follows:

$$\frac{|p_t|^2}{|p_1|^2} = \left[\left| 1 + \sum_{i=2}^{N} F_i \frac{p_i}{p_1} \right|^2 + \sum_{i=2}^{N} (1 - F_i^2) \left| \frac{p_i}{p_1} \right|^2 \right] \tag{5.194}$$

where p_1 is the sound pressure due to the primary ray and p_t is the total sound pressure at the receiver due to all rays. A ratio of pressures is used as that allows the excess attenuation to be calculated without requiring a knowledge of the absolute value of the sound pressure (which depends on the source sound power).

The incoherence between all rays other than the primary ray is not included in the Harmonoise model and would be unlikely to have a significant effect on the results.

The overall coefficient of coherence for the ith ray, F_i, is made up of the product of the components, F_a, due to the combined effects of frequency averaging in the 1/3-octave band, uncertainty in the distance between source and receiver, uncertainty in the source and receiver heights, and F_t due to turbulence, where each component refers only to the ith path. That is,

$$F_i = F_a F_t \tag{5.195}$$

The component, F_a, of the *coefficient of coherence* is given by (Salomons and Janssen 2011),

$$F_a = \mathrm{e}^{-0.5\sigma_\phi^2} \tag{5.196}$$

where σ_ϕ is the standard deviation of the fluctuation of the phase difference, ϕ,

$$\phi = k\Delta d = \frac{2\pi f}{c_0}\frac{2h_S h_R}{d} \tag{5.197}$$

and where

$$\sigma_\phi^2 = \phi^2\left[\left(\frac{\sigma_f}{f}\right)^2 + \left(\frac{\sigma_d}{d}\right)^2 + \left(\frac{\sigma_{h_S}}{h_S}\right)^2 + \left(\frac{\sigma_{h_R}}{h_R}\right)^2\right] \tag{5.198}$$

and k is the wave number, d is the horizontal separation distance between source and receiver, σ_d is the standard deviation of the uncertainty in that distance, h_S is the source height above the local ground, with σ_{h_S} the standard deviation of its uncertainty, and h_R is the receiver height above the local ground, with σ_{h_R} the standard deviation of its uncertainty. Including uncertainties in the source–receiver distance and source–receiver heights results in a smoother variation of the ground effect with frequency, which is a more realistic result (van Maercke and Defrance 2007).

The first term on the right-hand side of Eq. (5.198) accounts for the effect of 1/3-octave frequency band integration and is given by

$$\frac{\sigma_f}{f} = \frac{1}{3}\frac{\Delta f}{f} = \frac{1}{3}(2^{B/2} - 2^{-B/2}) \tag{5.199}$$

where $B = 1/3$ for 1/3-octave band averaging and $B = 1$ for octave band averaging.

In the absence of better information, the following standard deviation values are used.

$$\begin{aligned}\frac{\sigma_d}{d} &= 0\\ \frac{\sigma_{h_S}}{h_S} &= \frac{h_S}{10}\\ \frac{\sigma_{h_R}}{h_R} &= \frac{h_R}{10}\end{aligned} \tag{5.200}$$

The standard deviations for source and receiver heights are only non-zero for actual source and receiver locations and are zero when the source or receiver is moved to the top of a diffracting edge (see analysis in Section 5.9.3).

The *coherence coefficient* due to *turbulence* is given by

$$F_t = \mathrm{e}^X \tag{5.201}$$

where X is defined by Eq. (5.162).

5.9.2 Coordinate Transformation for the Ground Profile

In the Harmonoise model, the effect of atmospheric vertical wind and temperature gradients is taken into account by using these gradients to calculate the radius of curvature of the ground in a transformed coordinate system so that it has the same effect as the

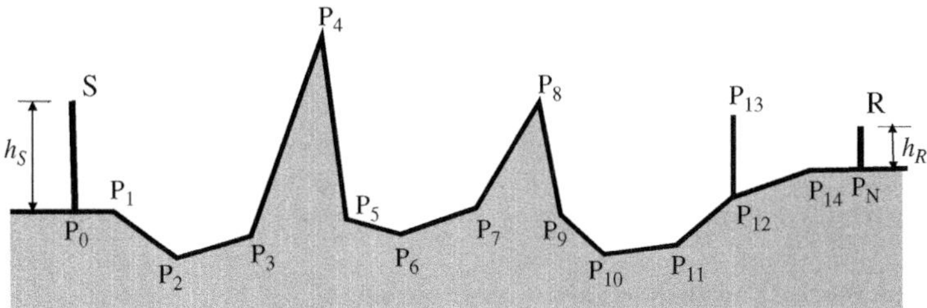

Figure 5.26 Typical ground profile divided into segments using the method described in Section E.3.

radius of curvature of the sound rays on the sound pressure arriving at the receiver from the source. To begin, the atmospheric sonic gradient is linearised to produce a sound ray in the shape of a circular arc and the radius of curvature of the sound ray is calculated as described in Section 5.2.4. The curvature of the ground is thus circular in the transformed coordinate system. The calculation of the excess attenuation due to ground, barrier and terrain effects is then calculated using straight ray paths with terrain points redefined in the transformed coordinate system.

Prior to transforming the ground profile into the new coordinate system, it is necessary to divide it into segments as explained in Section E.3. A typical segmented ground profile is shown in Figure 5.26 and the coordinate transformation begins with this, as explained by Salomons and Janssen (2011). Each point in the profile where the slope changes is defined by the horizontal coordinate x and the vertical coordinate h. The maximum allowed length of each ground segment (between two slope changes) is $d/3$ for source–receiver distances d less than 150 m, 50 m for source–receiver distances between 150 m and 1 km, and $d/20$ for distances greater than 1 km. The coordinate transformation is applied to the (x, h) coordinate of each point (P_i, $i = 0, \ldots N$) on the ground profile that defines the beginning or end of a ground segment. The coordinate location of the ground point, P_0, immediately below the source is denoted (x_0, h_0) and the coordinate location, P_N, of the ground immediately below the receiver is denoted (x_N, h_N), for the case of N ground segments. The coordinates (x, h) for the beginning and end points of each ground segment are replaced with the coordinates (x', h') via a coordinate transformation such that

$$x' = \frac{C_0^2 x''}{x''^2 + (C_0 + h'')^2} \tag{5.202}$$

$$h' = \frac{C_0(x''^2 + h''^2 + h'' C_0)}{x''^2 + (C_0 + h'')^2} \tag{5.203}$$

$$x'' = x - \frac{x_0 + x_N}{2} \tag{5.204}$$

$$h'' = h - \frac{h_0 + h_S + h_N + h_R}{2} \tag{5.205}$$

$$C_0 = 2\left(R_c + \frac{h_S + h_R}{2}\right) \tag{5.206}$$

where h_S and h_R are the actual original source and receiver heights and R_c is the radius of curvature of the sound ray in the original coordinate system as a result of the linearised atmospheric sound-speed profile (see Section 5.2.4). The source and receiver heights above the new transformed coordinates are still h_S and h_R, respectively.

The new coordinates, (x', h') are then used as the (x, h) coordinates in the analysis to follow. The use of the coordinate transformation is valid provided $|R_c| > 5d_{SR}$, where d_{SR} is the source–receiver separation distance. This is a fundamental limitation of Harmonoise, preventing it from being applied to large source–receiver separation distances.

5.9.3 Ground, Barrier and Terrain Excess Attenuation, A_{gr+bar}

The excess attenuation, $A_{\text{gr+bar}}$ is made up of two components: that due to diffraction over one or more barriers (if they exist) with no ground reflections, and that due to paths that include a ground reflection, even if the ray also travels over the top of the barrier.

Thus,

$$A_{\text{gr+bar}} = -20 \log_{10} \left| \frac{p_t}{p_F} \right| = -(\Delta L_D + \Delta L_g) \quad \text{(dB)} \tag{5.207}$$

where p_t is the total sound pressure at the receiver including the direct ray, diffracted rays, ground-reflected rays and the effects of atmospheric refraction, p_F is the free-field sound pressure at the receiver for a homogeneous atmosphere in the absence of any diffraction or ground effects, ΔL_D (dB) is the increase in sound level (usually negative, signifying a decrease) at the receiver due to diffraction over obstacles between the source and receiver, and ΔL_g (dB) is the increase in sound level at the receiver due to ground reflections. The increase or decrease is relative to the sound level that would exist in a homogeneous atmosphere in the absence of ground or diffraction obstacles (free-field sound pressure). The calculation of ΔL_D and ΔL_g is outlined in Appendix G for a single source.

5.9.4 Excess Attenuation due to Scattering

The Harmonoise model considers scattering loss due to rays travelling through forests and turbulence. Each of these two effects is calculated separately and the results (in dB) are added arithmetically to obtain the total loss due to scattering that must be subtracted from the overall sound pressure level calculated at the receiver after taking into account all other effects. The total excess attenuation, A_{sc}, due to scattering is then given by,

$$A_{\text{sc}} = A_{\text{sc},f} + A_{\text{sc},t} \quad \text{(dB)} \tag{5.208}$$

where $A_{\text{sc},f}$ is the loss (in dB) due to scattering by forests and $A_{\text{sc},t}$ is the loss (usually negative) due to scattering by turbulence.

Excess Attenuation, $A_{sc,f}$, due to Scattering Through Trees

To calculate $A_{\text{sc},f}$, the important parameter that has to be determined first is the length of the path, $r_s = (r_1 + r_2)$, travelled through the scattering zone (see Figure 5.19). If a barrier or terrain screen exists in the scattering zone or between two scattering zones, the calculation of the path travelled through the scattering zone on the source side of the barrier is based on the receiver being replaced by the top of the barrier; the calculation of the path travelled through the scattering zone on the receiver side of the barrier is based on the source being replaced by the top of the barrier. In the presence of atmospheric refraction, the sound rays will be characterised by a radius of curvature.

The excess attenuation, $A_{sc,f}$, due to propagation through trees has a maximum allowed value of 15 dB, and is calculated using (Nota et al. 2005):

$$A_{sc,f} = 1.25k_f\left(\frac{r_s}{43.75}\right)^2\left[A_e(r_s, h_s') + 20\log_{10}\left(\frac{8r_s}{25}\right)\right] \tag{5.209}$$

where k_f is given as a function of $kd_t/2$ in Table 5.17, k is the wave number, h_s' is the average height of the tree tops above the ground, d_t is the mean trunk diameter, and $A_e(r_s, h_s')$ can be calculated by interpolation (see below) between the values listed in Table 5.18.

A similar approach to that described in this section can be used to estimate the attenuation due to scattering by housing, as outlined by Plovsing (2006a).

Excess Attenuation $A_{sc,t}$, due to Dcattering from Atmospheric Turbulence

To calculate $A_{sc,\,t}$, the following expression is used.

$$A_{sc,t} = -25 - 10\log_{10}\gamma_T - 3\log_{10}\frac{f}{1000} - 10\log_{10}\frac{d}{100} \tag{5.210}$$

where f is the 1/3-octave band centre frequency, d is the horizontal distance between source and receiver and γ_T is defined in Eq. (5.163). Scattering by turbulence adds to the sound level at the receiver whereas scattering by trees reduces it.

5.9.5 Excess Attenuation, A_r, due to Reflection from a Facade or Building

Sound reflected from an obstacle or building facade on its way to the receiver is treated in the same way as a non-reflected ray, except that an attenuation (in decibels) due to

Table 5.17 Values for the parameter, k_f

$kd_t/2$	≤ 0.7	1.0	1.5	3.0	5.0	10.0	≥ 20
k_f	0.00	0.05	0.20	0.70	0.82	0.95	1.00

Table 5.18 Values for the parameter, $A_e(r_s, h_s')$

r_s[m]	h_s' =2.5 m	$h_s' \geq 25$ m
≤ 3.13	0.0	0.0
6.25	6.5	6.5
12.5	13.0	12.8
18.75	17.7	16.4
25.0	21.1	19.9
37.5	26.4	24.8
50.0	31.0	28.5
75.0	39.0	35.1
100.0	47.5	41.6
150.0	64.5	54.4
≥ 250	99.2	81.4

the efficiency of the reflection is subtracted from the sound pressure level arriving at the receiver. The procedure is the same as described above for the Nord2000 model (see Section 5.8.5).

5.9.6 Limitations of the Harmonoise Model

Harmonoise has the following limitations:

1. The model is only valid for 1/3-octave band analysis for 1/3-octave band centre frequencies ranging from 25–10 000 Hz, and can thus provide A-weighted sound pressure levels over this frequency range by applying the A-weighting correction (see Table 2.2) to each 1/3-octave band level and then summing the levels logarithmically (see Section 2.2.8). The overall C-weighted sound pressure level can be calculated by following a similar procedure. Of course the Harmonoise model is also valid for calculating overall A-weighted sound pressure levels over the frequency range covered by the 31.5–8 000 Hz octave bands.
2. Harmonoise is strictly limited to the prediction of sound propagation from point sources and is not valid for line sources. However, this limitation for application to line sources can be overcome by dividing line sources into a number of sources of small length that can be treated as point sources.
3. The terrain between source and receiver is approximated by a number of straight-line segments and this may not model actual situations accurately. However, it is superior to the approaches used for the CONCAWE and ISO models.
4. Any barriers are generally assumed to be infinite in length normal to the line between the source and receiver. However, if the effect of finite width on the barrier attenuation needs to be taken into account, refraction around the vertical edges of the barrier may be included in much the same way as done in the ISO model.
5. Only the coherence between the primary ray arriving at the receiver and all other rays are taken into account. Coherence effects between rays other than the primary ray are not taken into account. This approximation probably results in only very small errors.
6. Lin-log vertical atmospheric sonic profiles are assumed to approximate actual sonic profiles. It is also assumed that there are no sonic gradients in the horizontal direction and that the atmosphere has the same sonic gradient profile at all locations between the source and receiver. This is why the model should be used to provide a worst-case noise prediction or a range of predicted noise levels based on expected average vertical atmospheric sonic profiles. It is not possible to use this model or any model to accurately predict receiver noise levels for a complex wind and temperature profile that varies significantly between the source and receiver locations.
7. The model is valid only up to 1000 m for sources close to the ground. For high sound sources, such as wind turbines, the model should be valid for source–receiver distances of several kilometres.
8. Only vertically oriented reflecting surfaces are considered.
9. The use of a curved ground profile rather than curved sound rays to account for atmospheric refraction means that the Harmonoise model is not accurate when distances from the source exceed the point at which multiple ground reflections would occur.

5.10 Required Input Data for the Various Propagation Models

One of the most difficult aspects of applying the various noise propagation models in practice is obtaining the required input data. The required data become more detailed as the propagation model becomes more complex. In addition, if noise level predictions are required for other than worst-case downwind propagation, then the specification of atmospheric wind and temperature gradients as well as turbulence is needed for the more complex models. The advantage of the more complex Nord2000 and Harmonoise models is that noise level predictions can be made for any atmospheric condition, which allows a range of expected noise levels to be produced that reflect the range of atmospheric conditions that are likely in practice. No models allow for different atmospheric conditions to exist at different locations along the path from the source to the receiver. In hilly terrain, it is quite possible that non-uniform atmospheric conditions will exist, and this will result in a degree of uncertainty in the final predicted noise levels. Each of the models discussed in this chapter will now be considered in terms of the input data needed.

5.10.1 CONCAWE

The input data needed for implementation of a CONCAWE propagation model are listed below:

- height of source, h_S, and receiver, h_R, above a horizontal reference line passing through the base of the source or receiver and the vertical extension of the line joining the ground to the other (see Figure 5.27)
- horizontal distance, *d*, between source and receiver
- cloud cover (octas);
- wind speed (in direction source to receiver) at 10 m above the ground
- average atmospheric temperature and relative humidity between the source and receiver
- whether the ground is hard (concrete, asphalt, water, packed earth) or other
- location, length and height of any barriers between the source and receiver and the difference in path lengths between the straight-line source–receiver distance and the source–receiver path via the top of the barrier; terrain obstacles such as hills should be included in addition to any manmade barriers.

5.10.2 ISO9613-2

The input data needed for implementation of an ISO9613-2 propagation model are listed below:

- height of source, h_S, and receiver, h_R, above a horizontal reference line passing through the base of the source or receiver and the vertical extension of the line joining the ground to the other (see Figure 5.27)
- horizontal distance, *d*, between source and receiver
- average atmospheric temperature and relative humidity between the source and receiver

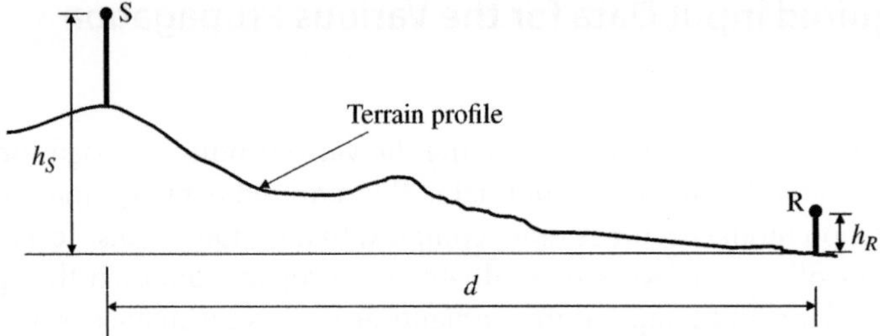

Figure 5.27 Definition of source and receiver heights for the CONCAWE model.

- whether the ground is hard (concrete, asphalt, water, packed earth) or soft; for this purpose, three assessments are needed:
  - one for ground extending from the source in the direction of the receiver for a distance of 30 times the source height
  - one for ground extending from the receiver in the direction of the source for a distance of 30 times the receiver height
  - one for the ground in between, which is not included in the two ground sections near the source and receiver described above.
- location, length and height of any barriers between the source and receiver and the difference in path lengths between the straight-line source–receiver distance and the source–receiver path via the top of the barrier for both single and double diffraction situations (in the event of multiple barriers existing, only the two most significant are considered); terrain obstacles such as hills should be included in addition to any manmade barriers
- length of the ray path that travels through foliage based on a radius of curvature of the sound ray of 5 km (see Eq. (5.61))
- location and absorption coefficient of any vertically reflecting objects that could affect the sound pressure level at the receiver.

5.10.3 NMPB-2008

The input data needed for implementation of an NMPB-2008 propagation model are listed below:

- average atmospheric temperature and relative humidity between the source and receiver
- category (A–G, see Table C.2) of ground sections between the source and receiver and extent of each ground-section type so a weighted average value of G can be calculated for use in the model
- terrain profile from source to receiver represented as a series of joined straight lines with defined end points; equivalent barriers to represent any hills between the source and receiver
- height, h_S, of source and receiver, h_R, above the local ground immediately below
- horizontal distance, d, between source and receiver
- fraction of the time that downward-refracting atmospheric conditions exist
- location, length and height of any manmade barriers between the source and receiver

- location and absorption coefficient of any vertically reflecting objects that could affect the sound pressure level at the receiver.

5.10.4 Nord2000

The input data needed for implementation of a Nord2000 propagation model are listed below:

- segmented terrain profile from source to receiver (see Figure E.2)
- category (A–H, see Table C.2) of each ground segment between the source and receiver
- height, h_S, of source and height, h_R, of receiver above the local ground immediately below the source and receiver, respectively (see Figure E.2)
- height, h'_S, of source and height, h'_R, of receiver above each ground segment or its extension (see Figure E.13)
- horizontal distance, *d*, between source and receiver (see Figure E.2)
- wind speed component from source to receiver at a height of 10 m
- cloud cover (octas)
- temperature at a height of 1 m above the ground
- average atmospheric temperature and relative humidity between the source and receiver
- location, length and height of any manmade barriers between the source and receiver
- forest scattering zone sizes and locations, including the average tree-trunk diameter and the tree density
- location and absorption coefficient of any vertically reflecting objects that could affect the sound pressure level at the receiver.

5.10.5 Harmonoise

The input data needed for implementation of a Harmonoise propagation model are listed below:

- segmented terrain profile from source to receiver (see Figure E.2)
- category (A–H, see Table C.2) of each ground segment between the source and receiver
- height, h_S, of source and height, h_R, of receiver above the local ground immediately below the source and receiver, respectively (see Figure E.2)
- standard deviation of the uncertainty in source and receiver heights and separation distances
- height, h'_S, of source and height, h'_R, of receiver above each ground segment or its extension (see Figure G.4)
- horizontal distance, *d*, between source and receiver (see Figure E.2)
- wind speed component from source to receiver at a height of 10 m
- temperature at a height of 1 m above the ground
- average atmospheric temperature and relative humidity between the source and receiver
- location, length and height of any manmade barriers between the source and receiver.
- forest scattering zone sizes and locations, including the average tree-trunk diameter and the average height of the tree tops above the ground.

- location and absorption coefficient of any vertically reflecting objects that could affect the sound pressure level at the receiver.

5.11 Offshore Wind Farm Propagation Models

The calculation of sound pressure levels at an onshore receiver due to an offshore wind farm follows the same procedures as above for land-based wind farms. The only difference is that the ground surface is modelled as very hard (flow resistivity of 20 000 000 Nsm^{-4}) and there are no barrier effects due to the terrain, unless the receiver site is located inland in hilly terrain. An additional difference is that there are often LLJs (Baas et al. 2009) that result in more severe wind gradients, and hence smaller radii of curvature of diffracted rays (see Eq. (5.183)) than occur over land, thus resulting in a shorter propagation distance before multiple ground reflections become important (see Section 5.2.4). The main effect of the phenomenon of multiple ground reflections is that it reduces the attenuation due to spherical spreading from 6 dB per doubling of distance to 3 dB per doubling of distance (Salomons 2001) for low frequencies, where there is little attenuation due to water and ground absorption.

For downwind conditions, the effect of multiple reflections combining incoherently at the receiver at low frequencies must be taken into account. This can be done as described in Section 5.2.4. However a simplified approach (IOA 2014) assumes a standard distance of 700 m for the point at which the attenuation of sound due to spherical spreading changes from 6 dB per doubling of distance from the turbine to 3 dB, resulting in the following expression for the calculation of the sound pressure level at a distance, *d*, from the turbine, provided that at least 700 m of propagation is over water.

$$L_p = L_w - 20\log_{10} d - 11 + 3 - A_{\text{atm}} + 10\log_{10}\left(\frac{\sqrt{h^2 + d^2}}{700}\right) \quad \text{(dB)} \qquad (5.211)$$

where d is the horizontal distance between the turbine and receiver, h is the height difference between the turbine source and receiver and the last term on the right-hand side is only used for distances greater than 700 m.

5.12 Propagation Model Prediction Uncertainty

The accuracy of noise propagation modelling is often the subject of discussions in court, especially when predicted noise levels are close to allowable noise levels. Generally, most noise models allow noise predictions to be made for the worst meteorological conditions that are expected to occur on a regular basis. However, the best accuracy that can realistically be expected for sound pressure level predictions at distances of 1 km or more from a wind farm is ±3 dB, although ±4 dB is probably more realistic when uncertainties in the turbine sound power levels are taken into account. Some practitioners claim an accuracy of better than ±2 dB, but there is insufficient data available to confirm that. The difficulty in obtaining accurate predictions is mainly associated with the variability of the atmospheric wind and temperature profiles over time and geographic location.

The most extensive study of propagation model prediction uncertainty was done by Marsh (1982) for the CONCAWE model for source–receiver separations spanning distances of 200–2000 m. His results are summarised in Table 5.10 for the various meteorological categories discussed in Section 5.5. For downwind propagation only, the 95% confidence limits for the CONCAWE model are approximately ±4.5 dBA, (expanded uncertainty with 95% confidence limits – see Section 6.2.16).

The ISO9613-2 model (see Section 5.6) includes an estimate of the accuracy of the predicted overall A-weighted noise level. For a mean source and receiver height less than 5 m ($[h_S + h_R]/2 < 5$m), the estimated accuracy is ±3 dB. For a mean source and receiver height between 5 m and 30 m, the estimated accuracy is ±1 dB for source–receiver distances of less than 100 m, and ±3 dB for source–receiver distances between 100 m and 1000 m. It is expected that for higher sources, such as wind turbines, the upper distance limits mentioned above may be extended further.

If the valley effect recommended by the IOA (2013) is included, then there may be cases at the margins where a small change in receiver height results in the propagation path being defined as a valley, with its associated 3-dB increases in noise levels at the receiver. This adds further uncertainty to the estimated noise levels.

An example of the inaccuracy of the ISO9613-2 prediction method is provided by Stigwood et al. (2015) for the Cotton Farm Wind Farm. In their Figure 1, Stigwood et al. (2015) show that predictions are consistently 2–4 dBA below the average of the measured data over a range of wind speeds in the downwind direction. The difference between the level exceeded 10% of the time and the predictions is even greater (by about a factor of 2).

Perhaps the greatest uncertainty lies in the input data used in the models, especially the more complex Nord2000 and Harmonoise models. However, assuming that the input data are accurate, the Nord2000 model aims for 'good' accuracy for source–receiver separations of up to 1000 m, and acceptable up to 3000 m. 'Good' and 'acceptable' have not been defined, but it seems that 'good' implies an accuracy of ±2 dB for overall A-weighted sound levels (Kragh et al. 2001). Generally, 'acceptable' would mean an accuracy of ±3 dB in overall A-weighted sound levels, although the uncertainty for individual octave band levels is likely to be greater than this.

The uncertainty in the Harmonoise model has not been addressed in detail in the literature, but one may assume that it is similar to the Nord2000 model.

All propagation models are more accurate for higher sound sources such as wind turbines, as the distance from the source at which multiple ground reflections begin to occur increases with an increasing difference between the source and receiver height. This would imply that, for wind turbine noise, the uncertainties mentioned above would extend for greater distances from the turbine noise source, thus also making the models useful for noise level predictions at greater distances from the source. In addition, when multiple wind turbines contribute to the total sound level at a particular location, the expected difference between measurement and prediction should decrease as the number of turbines contributing to the noise level increases. The approximate standard uncertainty, u_{tot}, can be calculated using Equation (5.212), where the standard uncertainty can be approximately substituted for 'accuracy/1.73' in the ISO9613-2 standard:

$$u_{\mathrm{tot}} = \frac{\sqrt{\sum_{i=1}^{N} (u_i \times 10^{L_i/10})^2}}{\sum_{i=1}^{N} 10^{L_i/10}} \tag{5.212}$$

where u_i is the standard uncertainty expected in the sound pressure level, L_i, at the receiver due to turbine, i, N is the number of turbines, and u_{tot} is the expected standard uncertainty in the predicted sound pressure level, L_{tot}, at the receiver due to all turbines.

For wind farm noise, a significant part of the uncertainty is associated with the measured sound power level of the turbines making up the wind farm. It is important that the standard uncertainty, u, of any sound power measurements be reported, and that the measurements be carried out in accordance with the most recent version of the standard, IEC 61400-11 Ed.3.0 (2012), as described in Section 1.5.2. Sound power levels are usually provided by turbine manufacturers.

If standard uncertainty data are available, it is recommended (IOA 2013) that an amount of $1.645u$ should be added to the reported sound power levels and if no standard uncertainty data are available, an amount of 2 dB should be added, which would increase the predicted sound pressure level at the receiver by the same amount. In some cases noise predictions are made in terms of L_{A90}, whereas sound power levels of turbines are usually reported as L_{Aeq} values. In this case, it is usual to subtract 2 dB from the sound power levels prior to calculating the L_{A90} levels at residential properties (IOA 2013).

Sound power levels are generally determined for incident air flow arriving horizontally at the turbine after propagating a considerable distance over flat ground, and in the absence of other turbines, so that inflow turbulence (and resulting noise generation) is at a minimum. As discussed in detail in Section 4.6, the sound power level is obtained from a single sound pressure measurement on the ground, directly downwind from the turbine at a horizontal distance from the base of the tower equal to the height of the centre of the rotor above the ground plus half of the rotor diameter, as illustrated in Figure 4.3. This position is referred to as the 'reference position'. The veracity of this single measurement in determining the wind turbine sound power is based on the assumption that wind turbine noise is not at all directional in nature in the vertical plane and if it is directional in the horizontal plane, then the downwind measurement will represent the worst case. This may be true for low-frequency noise, but the use of the single measurement location is likely to result in an underestimate of the sound power levels in the mid- to high- frequency range due to directivity of the generated sound in the vertical plane. The IEC standard referenced above allows for manufacturers to measure the sound pressure level at three other positions spaced horizontally at 90° intervals from the reference position and at the same distance from the wind turbine. These measurements could be used to estimate environmental noise levels for wind conditions other than directly downwind, but instances of this use are rare.

The sound pressure level measurement, L_{pr}, made at the reference position is used to calculate the reference sound power level, L_{wr} using the following equation (IEC 61400-11 Ed.3.0 2012).

$$L_{wr} = L_{pr} + 10 \log_{10} (4\pi r^2) - 6 \tag{5.213}$$

where r is the distance in metres from the turbine hub to the reference position and the constant of 6 in the equation accounts for the reference measurement position being on the ground so that direct and ground-reflected rays add coherently. The microphone is mounted in the centre of a metal or wooden disk, 1 m in diameter, which simulates a hard ground in the mid- to high-frequency range. Although the specified disk is too small to be effective at low frequencies, most ground surfaces are almost 100% reflective at low frequencies so it is not usually a problem.

The IEC standard requires the time-averaged sound pressure level to be measured over a range of wind speeds. The measured data are corrected for background noise (see Eq. (2.29)) and then plotted on a graph of energy averaged, background corrected, A-weighted sound pressure level (L_{Aeq}) versus wind speed (at least covering the range 6–10 m/s) and a regression line of best fit is drawn through the data. The regression line is then used to determine the L_{Aeq} value as a function of wind speed. The L_{Aeq} value is then substituted for L_{pr} in Eq. (5.213) to obtain the A-weighted sound power level.

The use of the distance from the turbine hub to the measurement position to calculate the turbine sound power level assumes that the noise source is centred at the hub. As aerodynamic noise radiation is centred on the blade centres of pressure, and low-frequency noise is associated with blade/tower interaction, the assumption of the hub being the source centre for all frequencies is only a rough approximation, and results in the measured sound power level having a significant degree of uncertainty associated with it. The standard seems to acknowledge the uncertainty in the measurement by referring to the result as 'the apparent sound power'.

The sound power level, L_{w1} to be used in calculating the sound pressure level at residences around the wind farm for directions other than directly downwind is related to the reference sound power level, L_{wr}, by

$$L_{w1} = L_{wr} + L_{p1} - L_{pr} + 20 \log_{10} (r_1/r) \tag{5.214}$$

where r is the distance in metres from the turbine hub to the reference position, L_{p1} is the sound pressure level measured on the ground at distance r_1 from the turbine hub in the direction of interest and r_1 is, as close as possible, equal to r.

The uncertainty associated with the turbine sound power data can be combined with the uncertainty due to the propagation model when estimating noise levels at the receiver. For a standard uncertainty of u_{turb} in the turbine sound power data and a standard uncertainty of u_{prop} in the sound propagation model, the overall standard uncertainty, u_i, in the sound pressure level at a receiver location, due to one turbine, i, is

$$u_i = \sqrt{u_{\text{turb},i}^2 + u_{\text{prop}}^2} \tag{5.215}$$

Typical values for the standard uncertainty of turbine sound power measurements are ±2 dBA (Cox et al. 2012; Harrison 2011; Keith et al. 2016) for each turbine. The uncertainty is usually lowest in the 1/3-octave bands between 250 and 2000 Hz, and gradually increases with increasing frequency for frequencies above 2000 Hz and gradually increases with reducing frequency for frequencies below 250 Hz. When multiple turbines are involved, the overall turbine sound power standard uncertainty can be calculated using

$$u_{\text{turb},t} = \frac{\sqrt{\sum\limits_{i=1}^{N} (u_{\text{turb},i} \times 10^{L_i/10})^2}}{\sum\limits_{i=1}^{N} 10^{L_i/10}} \tag{5.216}$$

where $u_{\text{turb},i}$ is the standard uncertainty (in decibels) expected in the sound pressure level, L_i (dB), at the receiver due to turbine, i, as a result of only the uncertainty in the sound power level, N is the number of turbines, and $u_{\text{turb},t}$ is the expected standard uncertainty in the predicted sound pressure level, L_{turb}, at the receiver due to all turbines, where only the uncertainty in the sound power levels of the turbines is taken into account. This result is then combined with the uncertainty in the propagation modelling using Eq. (5.215) (in which $u_{\text{turb},i}$ is replaced with $u_{\text{turb},t}$). The overall expanded uncertainty, corresponding to 95% confidence limits of the prediction at any given receiver location, is greater than the standard uncertainty by a factor of 2 (assuming that the uncertainty is normally distributed). If the uncertainty is described by a rectangular distribution, the factor of 2 is replaced with 1.65. Thus for most A-weighted environmental noise predictions, it would be wise to suggest that the variation between prediction and measurement for any particular location in a downward-refracting atmosphere is of the order of ±4 dBA, depending on the number of turbines involved.

5.13 Outside versus Inside Noise at Residences

The construction of a residence has a significant effect on the sound transmission from the outside to the inside. All residences are more effective at reducing mid- and high-frequency sound than ILFN. This means that inside a residence, low-frequency sound will be more noticeable and potentially more annoying than expected, as it is not as well masked by mid- and high-frequency noise as it is outside of a residence. Data obtained for the noise reduction as a result of transmission into a house vary considerably depending on the particular study. The main reasons lie in the methods used to define both the exterior and interior levels, the source of sound used for the tests, the housing construction tested and the closed/open state of any windows.

For determining the housing structure noise reduction, the standard method adopted in some countries is to use a loudspeaker source located 5 m from the facade of a house that is closest to the potentially annoying noise. The loudspeaker is placed on the ground and directed toward the facade of the house at a horizontal angle of incidence of approximately 45° to the normal projected from the centre of the facade, as discussed in Section 6.2.9. This arrangement would overestimate the noise reduction achieved for wind farm noise, as low-frequency sound from the wind farm some distance away would be evenly distributed over the roof, front (nearest the wind farm) and side walls of the house, whereas the loudspeaker would only produce significant energy on the wall adjacent to its location.

The outdoor sound pressure level is determined by measurements made very close to the facade at a height of approximately 1.5 m above the floor level of the receiving room. Free-field sound pressure levels are obtained by subtracting 6 dB from these measured levels.

Determining indoor sound pressure levels for comparison with the outdoor levels is subject to considerable variation because, especially at low frequencies, noise levels vary markedly throughout a room. Some studies use interior noise levels averaged over a number of interior locations, whereas others use data measured in the noisier parts of the room such as the room corners, where the floor or ceiling joins two walls. Methods adopted in various countries for determining the interior noise level are listed in Table 5.19. In cases where there is more than one measurement, the data from the various N positions are energy averaged as follows.

Table 5.19 Interior noise level measurement requirements

Country	Frequency range (Hz)	Octave or 1/3-octave bands	Number of measurement positions	Measurement positions
Denmark	5–160	1/3-octave	3	Floor corner. Two at height 1–1.5 m. All no closer than 0.5 m to wall.
Sweden	31.5–200	1/3-octave	3	Corner with highest level. Two at representative ear positions, separated by more than 1.5 m and at height 0.6, 1.2 or 1.6 m. All no closer than 0.5 m to any room surface
Germany/Austria	10–80	1/3-octave	1	Where noise is most annoying.
Netherlands	20–100	1/3-octave	1	Chosen by complainant or in room corner, 0.2 to 0.5 m from walls
Japan	10–80	1/3-octave	1	Chosen by complainant.

$$L_{p,av} = 10 \log_{10} \frac{1}{N}(10^{L_{p1}/10} + 10^{L_{p2}/10} + ... + 10^{L_{pN}/10}) \quad (5.217)$$

For the purpose of assessing annoyance caused by wind farm noise, Pedersen et al. (2007) suggested that the noise level that is exceeded in 10% of the room should be used and that this is approximated by taking the average of measurements in four room corners, at a distance of less than 0.1 m from the corners. All room surfaces must be represented by the choice of corners; that is, four ceiling corners or four floor corners are not acceptable choices.

Measurements of sound transmission have been made by the authors for some residences in the vicinity of the Waterloo wind farm in South Australia, and the results corresponding to an indoor microphone mounted in a room corner for two residences are shown in Figure 5.28, where the exterior noise source was a wind farm. Residence 1 was unoccupied at the time. Only data corresponding to nighttime periods, between midnight and 5am, when the wind farm was particularly noisy were used, and all data were recorded in real time for later analysis and listening to make sure no extraneous noise sources were included. Data corresponding to wind speeds exterior to the houses that exceeded 2 m/s were excluded. Exterior and interior noise data were averaged over 10 min to obtain one data point. Up to 50 data points were used to establish the range of data points shown as the shaded areas in Figure 5.28.

The negative noise reduction for both houses in the 3–4 Hz range is most likely due to the resonance associated with the mass of the walls and roof acting with the stiffness of the air volume inside the house. Resonance will occur when the sum of the impedance of the wall mass and the enclosed volume is equal to zero. The acoustic impedance,

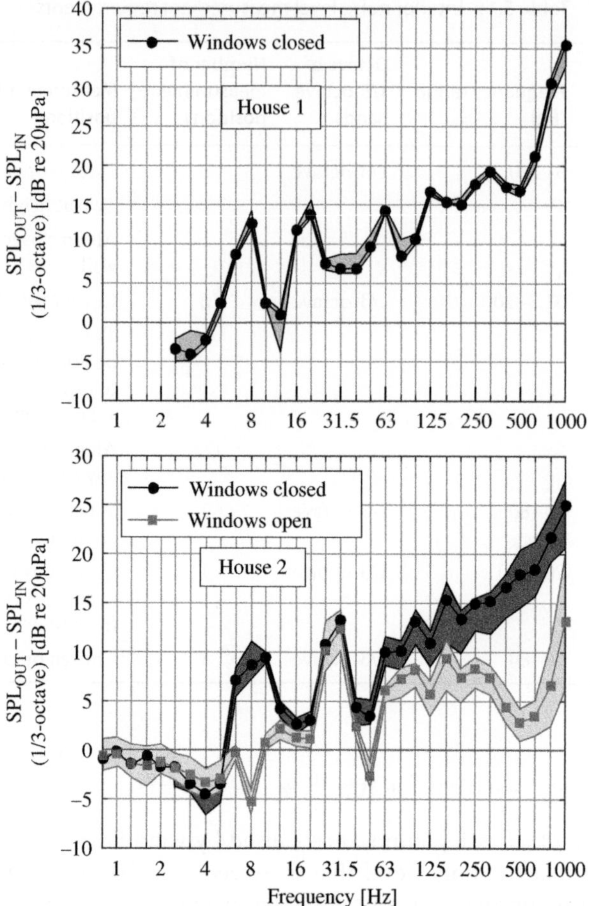

Figure 5.28 Noise reduction measurements for two residences in the vicinity of the Waterloo wind farm. SPL = sound pressure level.

Z_{wr}, of the wall and roof mass at frequency, f (Hz), is given approximately by

$$Z_{wr} = \frac{\mathrm{j}2\pi f m_{\mathrm{av}}}{S_{wr}} \tag{5.218}$$

where m_{av} is the average mass per unit area of the walls, roof and ceiling, and S_{wr} is the total surface area of the walls and roof. As the roof is usually sloped, a good approximation is to increase the plan area of the roof by 10–20%, depending on the slope angle. The mass per unit area of the roof, m_r, should include the ceiling and the average mass of the ceiling/roof and walls is an area-weighted average of the two, calculated using

$$m_{\mathrm{av}} = \frac{m_r S_r + m_w S_w}{S_{wr}} \tag{5.219}$$

where the subscript, r, refers to the roof and the subscript, w, refers to the walls.

The impedance of the air volume corresponding to the interior of the house is given by

$$Z_v = -\mathrm{j}\frac{\rho c^2}{2V\pi f} \tag{5.220}$$

where c is the speed of sound in air (=343 m/s), ρ is the density of air (= 1.2kg/m^3) and V is the air volume in the house (calculated by reducing the actual volume by 10–15% to account for interior furnishings).

The resonance frequency, f_0, is the frequency (Hz) corresponding to the sum of the above two impedances being equal to zero and is given by

$$f_0 = \frac{c}{2\pi}\sqrt{\frac{\rho S_w}{m_{\mathrm{av}} V}} \tag{5.221}$$

For the cases considered here, house 1 was larger than house 2, which explains why the resonance frequency was lower for house 1.

The dip in the noise reduction curve at 10 Hz for house 2 with the windows open is associated with the Helmholtz resonance effect of the open window interacting with the volume of the house. This resonance frequency is calculated using (Bies and Hansen 2009)

$$f_0 = \frac{c}{2\pi}\sqrt{\frac{S_o}{\ell_e V}} \tag{5.222}$$

where S_o is the open area of the window, V is the volume of air in the room containing the window (calculated by reducing the actual volume by 10–15% to account for interior furnishings) and ℓ_e is the thickness, w, of the wall containing the window plus two end corrections, given by Bies and Hansen (2009) as,

$$\ell_e = w + 0.34\sqrt{S_o} + 0.46(1 - 1.33\sqrt{S_o/S_1})\sqrt{S_o} \tag{5.223}$$

where S_1 is the area of wall containing the open window.

The dip at 63 Hz in the noise reduction curve for house 2 with an open window is probably associated with a higher-order Helmholtz resonance. The dip at 12.5 Hz for house 1 and 20 Hz for house 2 with windows closed could be associated with a structural resonance of the house. The noise reduction below 2.5 Hz is approximately zero as may be expected. However, there are less data for this frequency range (in fact, no data for house 1), as the outside noise measurements are contaminated by wind noise over the microphone for wind speeds greater than 0.5 m/s, even though both primary and secondary wind screens were used (see Section 6.2.3).

Møller and Pedersen (2011) presented data for measurements of noise reduction from outside to inside in two rooms in each of five houses in Denmark with windows closed. One set of data has been discarded for our purposes here as it was for a store room. Indoor sound pressure levels were obtained as the power average of measurements in four arbitrary room corners, and outdoor levels were generated by a loudspeaker mounted as described above. The range of noise reduction values obtained as a function of 1/3-octave band centre frequency is shown in Figure 5.29 as the shaded grey area.

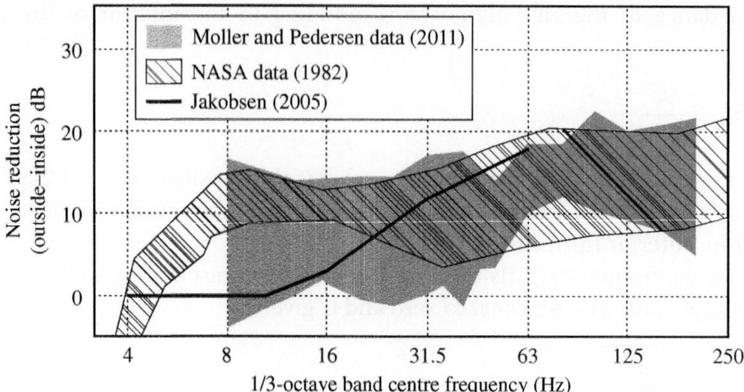

Figure 5.29 Noise reduction measurements for five residences (total of 9 rooms) in Denmark (data from Møller and Pedersen (2011)) and from a number of separate surveys reported by Stephens et al. (1982).

Measurements reported by Stephens et al. (1982) for houses with windows closed are shown in Figure 5.26 as a hatched area. The residence constructions ranged from concrete, brick walls and wooden walls, and with and without insulation. Details of the noise source locations are not available. However, the data generally lie within the range of data reported by Møller and Pedersen (2011).

The noise reductions reported for closed windows would decrease somewhat for open windows, depending on the open area of any windows and the noise reduction of the wall. For a noise reduction, N_R dB with the windows closed, and ratio, S_R (less than 1), of open window area to total wall area (excluding the open window area), the noise reduction, N_{RO} with the window open would be,

$$N_{RO} = -10 \log_{10} \left[\frac{S_R + 10^{-N_R/10)}}{1 + S_R} \right] \tag{5.224}$$

5.14 Vibration Propagation

The propagation of vibration from wind turbines via the ground to a residence is difficult to quantify. Essentially, the support tower for each turbine is anchored rigidly to a large concrete foundation slab, which is set in a deep hole in the ground. Any vibration levels in the tower will be transmitted into the ground and will then propagate to surrounding areas. The problem can be divided into three main areas:

- vibration generation
- vibration transmission
- vibration detection.

Each of these will be discussed in the following subsections.

5.14.1 Vibration Generation

There are a number of possible sources that can result in vibration being experienced at residences in the vicinity of a wind farm, and these are discussed in the following subsections.

Gearbox Vibration Transmitted Through the Tower to the Ground

Most existing wind farms consist of turbines with the turbine rotor connected to the generator via a planetary gearbox, with the tooth mesh frequency of the planetary gear in the range 20–40 Hz. The pitch of the turbine blades is controlled to ensure that the rotational speed of the turbine rotor and blades remains as close to constant as possible. As the gearbox is usually attached to the top of the support tower with little or no vibration isolation, gearbox vibrations are readily transmitted into the tower and thus into the ground. The vibration of the tower also results in a significant level of noise radiation, although more research is needed to determine exactly how significant. As an aside, the gearbox vibration is also transmitted into the turbine rotor blades, resulting in noise radiation from the blades.

Turbines are now being manufactured that are variable speed, with the turbine rotor speed being dependent on the wind strength and again being adjusted by controlling the blade pitch. The advantage of variable speed turbines is that they can be controlled to produce more power than fixed-speed turbines over an extended period. A possible advantage from the noise generation point of view is that any gear-mesh noise will be less noticeable, as it will not be at a fixed tonal frequency but will cover a small range about a centre frequency.

Turbines are also being manufactured that are 'direct drive' (for example the Enercon E82 turbine), which means that the turbine rotor is connected directly to the generator with no gearbox in between, so these turbines obviously radiate no gearbox noise, and do not transmit any gearbox vibrations into the tower and hence into the ground. Similar to variable speed turbines, it is necessary to use a power converter to convert the generator power output frequency to the transmission line frequency (50 Hz or 60 Hz, depending on which country).

Infrasound Generated by the Blades Passing the Tower

Infrasound generated by the blades passing the tower can be transmitting through the air and then couple into the ground at the receiver. The work undertaken by Styles et al. (2005b) demonstrated conclusively that vibration levels as a result of this mechanism would be much smaller than the levels resulting from the transmission of tower vibrations through the ground.

5.14.2 Vibration Propagation

Vibration is transmitted along the ground via surface waves (Rayleigh waves), resulting in both vertical and horizontal velocities at the receiver. Detailed seismic measurements were made in Scotland near wind farms containing Vestas V47 turbines with a rated power of 0.66 MW (Styles et al. 2005b) and the conclusion reached was that both the vertical and horizontal vibration velocities generally decayed at a rate proportional to the square root of distance (although there was a significant level of variation from this rule), and the horizontal and vertical velocities were of the same order of magnitude. For example, if a velocity vibration amplitude of $v_{100} = 20$ mm/s were measured at 100 m from a turbine, that turbine would contribute a velocity, v_{1000} at a distance of 1000 m given by

$$v_{1000} = v_{100}\sqrt{100/1000} = 20/3.16 = 6.3 \text{ mm/s} \tag{5.225}$$

Styles et al. (2005b) also found that significant vibration levels existed in the ground even when the turbines were not operating, suggesting that aero-elastic excitation of the blades and/or tower could be significant contributors to vibration levels experienced by residents in the near vicinity of wind farms. The vibration levels measured by Styles et al. (2005b) are much lower than might be expected for wind farms with modern turbines rated at 3–5 MW, compared to the 0.66-MW turbines of their study. Styles et al. (2005b) also showed that the vibration amplitude measured at a location as a result of the operation of multiple turbines scales as the square root of the number of turbines; that is, vibrational energy adds rather than vibration amplitude, a result that would be expected from the combination of randomly phased ground waves.

5.14.3 Vibration Detection

Our own measurements show that at distances of 3.3–3.5 km from the nearest turbine in a 37-unit 3-MW per turbine wind farm, vibration levels measured on windows are sufficiently high in the 50 Hz 1/3-octave band to be felt if a hand is placed on the window. However, vibration levels measured on the floor and bed are at least 20 dB below what would be detectable by the most sensitive individual and there is no significant difference between vibration levels measured with the wind farm operating close to full power and with the wind farm shut down for the same wind conditions at the residence. It is highly likely that the window vibrations are a result of excitation by sound waves rather than by ground-borne vibration.

The authors have spoken to a number of people who say that they can feel vibrations through their pillow when the wind farm is operating and that this makes it difficult to sleep. A possible explanation for this is that people may sense 'vibration' through their pillow, when one ear is against it, as the pillow acts as a sound attenuator for mid- and high-frequency sound, resulting in a lowering of the infrasound hearing threshold for the ear against the pillow. As the pillow has no effect on infrasound transmission, the ear against the pillow will respond more to the incoming infrasound than the other ear and such low-frequency infrasound can easily be perceived as vibration. This may explain the results reported by Bakker et al. (2012), who suggest that some wind farms may produce vibrations that people may feel when lying down.

According to Stephens et al. (1982), the mean vibration threshold for people subjected to vibration, whether in the sitting, standing or lying position, is as the 'whole body vibration curve' shown in Figure 5.30. This curve is very close to the 1/3-octave band values listed in the ANSI/ASA S2.71 (R2012) standard for between 1 and 80 Hz, which are stated to represent the approximate threshold of perception for 'most-sensitive humans', which is 'half the mean threshold of perception' (in m/s^2). It is highly likely that most-sensitive humans would have a threshold below half the mean, especially as there seems to be no reason to assume that the distribution of sensitivities does not even approximately follow a normal distribution. The tactile perception curve represents the mean (or average) vibration level that people can only just feel through their fingers on walls, windows and doors.

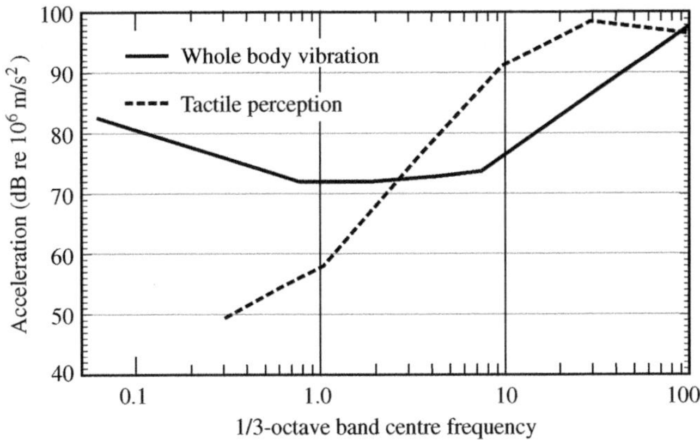

Figure 5.30 Vibration perception threshold in the most sensitive direction.

The more recent standard, ISO2631-1 (1997), suggests that the overall weighted RMS whole-body vibration acceleration, a_w, that 50% of normal, alert and fit people will detect is 0.015 m/s^2, with the range between the first and fourth quartiles extending from 0.01 to 0.02 m/s^2. It would not be unexpected for some sensitive people to be able to detect a weighted RMS vibration acceleration of 0.002 m/s^2, but even this is above any measurements the authors made on bed frames and floors in residences.

For a given 1/3-octave frequency band of centre frequency, f (Hz), a vibration velocity, v can be calculated from a vibration acceleration a using $v = (2\pi f)a$. The overall weighted acceleration a_w can be calculated according to ISO2631-1 (1997), for crest factors less than 9, using measured data in 1/3-octave bands as follows.

$$a_w = \left[\sum_i (W_i a_i)^2 \right]^{1/2} \tag{5.226}$$

where W_i is the weighting to be applied to 1/3-octave band, i, and a_i is the acceleration (in m/s^2) measured in 1/3-octave band i (see Table 5.20). For vertical body acceleration (whether lying down or standing up or seated), $W_i = W_k$, and for head acceleration, $W_i = W_j$ for the particular 1/3-octave band, i.

Wind farms have also been shown to be problematic in causing disturbance at low frequencies (below about 7 Hz) to seismic arrays used to detect earthquakes and nuclear explosions at large distances from the array, as well as gravitational waves (Flori et al. 2009; Styles et al. 2005a,b). Styles et al. (2011) discuss an seismically-quiet tower system that reduces the extent of vibration transmission into the ground over the frequency range, 2–6 Hz and that can be installed during construction or retrofitted to existing towers.

5.15 Summary

This chapter was primarily concerned with the estimation of noise levels produced by wind farms near residences as well as inside residences. The calculation of outside noise

Table 5.20 Weighting values to be applied to measured 1/3-octave band vibration levels to calculate the overall vibration level using Eq. (5.226)

1/3-octave band centre frequency (Hz)	0.1	0.125	0.16	0.2	0.25	0.315	0.4	0.5	0.63	0.8	1
$W_k \times 1000$	31	49	79	121	182	263	352	418	459	477	482
$W_j \times 1000$	31	48	79	120	181	262	351	417	458	478	484
1/3-octave band centre frequency (Hz)	**1.25**	**1.6**	**2.0**	**2.5**	**3.15**	**4**	**5**	**6.3**	**8**	**10**	**12.5**
$W_k \times 1000$	484	494	531	631	804	967	1039	1054	1036	988	902
$W_j \times 1000$	485	483	482	489	524	628	793	946	1017	1030	1026
1/3-octave band centre frequency (Hz)	**16**	**20**	**25**	**31.5**	**40**	**50**	**63**	**80**	**100**	**125**	**160**
$W_k \times 1000$	768	636	513	405	314	246	186	132	89	54	29
$W_j \times 1000$	1018	1012	1007	1001	991	972	931	843	708	539	364

Values derived from ISO2631-1 (1997).

levels is complex and there are various models of varying complexity that may be used to undertake the calculations. As wind turbine noise sources are a considerable height above the ground, it is often only necessary to know the characteristics of the ground in the vicinity of the receiver and in many cases it is difficult to justify the application of the more complex propagation models such as Nord2000 and Harmonoise. In fact the NMPB-2008 model, which is the one recommended by the European Union for transportation and industrial noise sources is perhaps the most suitable, although many practitioners claim that either the CONCAWE model or the ISO9613 model modified for wind farm noise is adequate, provided that the ground is set to 'hard' ($G = 0$). The problem with the CONCAWE method is that the meteorological and ground-attenuation curves are only available up to a distance of 2 km, and it is often necessary to estimate wind farm noise levels at greater distances. The problem with the ISO9613-2 method is that it is only valid for receivers closer than 1 km to the source. However, as wind turbine noise sources are so high, the validity of the model can probably be extended to 3 km or more.

When estimating interior noise levels for the purpose of assessing sleep disturbance according to the WHO guidelines, it is important to take into account the dominance of wind farm noise by low frequencies at distances exceeding 1–1.5 km from the nearest turbine in a wind farm. This means that the A-weighted overall noise level threshold for interior noise should be lower than it would be for traffic noise and, in addition, the allowed exterior noise should be much lower than for traffic noise due to the poorer transmission loss of the house structure for low frequencies.

References

ANSI/ASA S1.26 2014 Methods for calculation of the absorption of sound by the atmosphere.

ANSI/ASA S2.71 (R2012) 1983 Guide to the evaluation of human exposure to vibration in buildings.

Baas P, Bosveld F, Baltink H and Holtslag A 2009 A climatology of nocturnal low-level jets at Cabauw. *Journal of Applied Meteorology and Climatology*, **48**(8), 1627–1642.

Bakker R, Pedersen E, van den Berg G, Stewart R, Lok W and Bouma J 2012 Impact of wind turbine sound on annoyance, self-reported sleep disturbance and psychological distress. *The Science of the Total Environment*, **425**, 42–51.

Bass J 2011 Investigation of the "Den Brook" amplitude modulation methodology for wind turbine noise. *Acoustics Bulletin*, **36**(6), 18–24.

Bies D and Hansen C 2009 *Engineering Noise Control: Theory and Practice*, 4th edn. Spon Press.

Bowdler R 2009 Wind shear and its effect on noise assessment. *Third International Meeting on Wind Turbine Noise*, Aalborg, Denmark.

Bullen R 2012 The Harmonoise noise prediction algorithm: validation and use under Australian conditions. *Acoustics 2012*, Australian Acoustical Society, Freemantle, Australia.

Cox R, Unwin D and Sherman T 2012 Wind turbine noise impact assessment: where ETSU is silent. Technical report, National Wind Watch.

Daigle G 1982 Diffraction of sound by a noise barrier in the presence of atmospheric turbulence. *Journal of the Acoustical Society of America*, **71**(4), 847–854.

Danish Environmental Protection Agency 1991 Statutory order 304. noise from wind turbines (in Danish). Technical report, Danish Ministry of Environment.

Davenport A 1960 Rationale for determining design wind velocities. *ASCE Journal of the Structural Division*, **86**(5), 39–68.

De Jong R and Stusnik E 1976 Scale model studies of the effects of wind on acoustic barrier performance. *Noise Control Engineering Journal*, **6**(3), 101–109.

de Roo F, Salomons E, Heimann D and Hullah P 2007 IMAGINE-reference and engineering models for aircraft noise sources, vol. 1. report no. Technical report IMA4DR-070323-EEC-10, Harmonoise.

Defrance J, Salomons E, Noordhoek I, Heimann D, Plovsing B, Watts G, Jonasson H, Zhang X, Premat E and Schmich I 2007 Outdoor sound propagation reference model developed in the European Harmonoise project. *Acta Acustica United with Acustica*, **93**(2), 213–227.

Dutilleux G, Defrance J, Ecotière D, Gauvreau B, Bérengier M, Besnard F and Duc EL 2010 NMBP-routes-2008: the revision of the French method for road traffic noise prediction. *Acta Acustica United with Acustica*, **96**(3), 452–462.

ETSU 1996 ETSU-R-97 Assessment and rating of noise from wind farms. Technical report, UK Department of Trade and Industry.

European Commission 2010a Common noise assessment methods in Europe (EU Directive 2002/49/EC, Annex II: Assessment Methods for Noise Indicators, Annex II-v9d-ISO9613). Technical report, European Parliament.

European Commission 2010b Common noise assessment methods in Europe (EU Directive 2002/49/EC, Annex II: Assessment Methods for Noise Indicators, Annex II-v9d-JRC2010). Technical report, European Parliament.

Flori I, Giordano L, Hild S, Losurdo G, Marchetti E, Mayer G and Paoletti F 2009 A study of the seismic disturbance produced by the wind park near the gravitational wave detector GEO-600. *Third International Meeting on Wind Turbine Noise*, Aalborg, Denmark.

Harrison J 2011 Wind turbine noise. *Bulletin of Science, Technology & Society*, **31**(4), 256–261.

IEC 61400-11 Ed.3.0 2012 Wind turbines – Part 11: Acoustic noise measurement techniques. International Electrotechnical Commission.

IOA 2013 A good practice guide to the application of ETSU-R-97 for the assessment and rating of wind turbine noise. Technical report, Institute of Acoustics.

IOA 2014 A good practice guide to the application of ETSU-R-97 for the assessment and rating of wind turbine noise. Supplementary guidance note 6: noise propagation over water for on-shore wind turbines. Technical report, Institute of Acoustics.

Irwin J 1979 A theoretical variation of the wind profile power-law exponent as a function of surface roughness and stability. *Atmospheric Environment (1967)*, **13**(1), 191–194.

ISO9613-2 1996 Acoustics: Attenuation of sound during propagation outdoors. International Standards Organisation.

ISO2631-1 1997 Mechanical vibration and shock – evaluation of human exposure to whole-body vibration – part 1: General requirements. International Standards Organisation.

Jakobsen J 2012 Danish regulation of low frequency noise from wind turbines. *Journal of Low Frequency Noise, Vibration and Active Control*, **31**(4), 239–246.

Kalapinski E and Pellerin T 2009 Wind turbine acoustic modeling with the ISO 9613-2 Standard: methodologies to address constraints. *Third International Meeting on Wind Turbine Noise*, Aalborg, Denmark.

Keith S, Feder K, Voicescue S, Soukhoftsev V, Denning A, Tsang J, Broner N, Richarz W and van den Berg F 2016 Wind turbine sound power measurements. *Journal of the Acoustical Society of America*, **139**(3), 1431–1435.

Kephalopoulos S, Paviotti M and Anfosso-Ledee F 2012 Common noise assessment methods in Europe (EU Commission Directive 2015/1996). Technical report, European Commission Joint Research Centre Institute for Health and Consumer Protection.

Kragh J 1998 Noise immission from wind turbines. Final report of project JOR3-CT95-0065. Technical report, Delta Acoustics.

Kragh J, Plovsing B, Storeheier S, Taraldsen Gand Jonasson H 2001 Nordic environmental noise prediction methods, Nord2000 summary report. general nordic sound propagation model and applications in source-related prediction methods. Technical report, Delta Acoustics.

Maekawa Z 1968 Noise reduction by screens. *Applied Acoustics*, **1**(3), 157–173.

Maekawa Z 1977 Shielding highway noise. *Noise Control Engineering*, **9**(1), 38–44.

Makarewicz R 2011 Is a wind turbine a point source? *Journal of the Acoustical Society of America*, **129**(2), 579–581.

Manning C and Bijl L 1981 *The propagation of noise from petroleum and petrochemical complexes to neighbouring communities*. CONCAWE.

Marsh K 1982 The CONCAWE model for calculating the propagation of noise from open-air industrial plants. *Applied Acoustics*, **15**, 411–428.

Møller H and Pedersen CS 2011 Low-frequency noise from large wind turbines. *Journal of the Acoustical Society of America*, **129**(6), 3727–3744.

Møller H, Pedersen S, Staunstrup J and Pedersen C 2012 Assessment of low-frequency noise from wind turbines in maastricht. Technical report, Danish Department of Development and Planning.

Nota R and van Leeuwen H 2004 Harmonoise: noise predictions and the new European harmonised prediction model. *Proceedings of the Institute of Acoustics*, **26**(7), 9–16.

Nota R, Barelds R and van Maercke D 2005 Harmonoise WP 3 engineering method for road traffic and railway noise after validation and fine-tuning. Technical Report HAR32TR-040922-DGMR20, Harmonoise.

Nunalee C and Basu S 2014 Mesoscale modeling of coastal low-level jets: implications for offshore wind resource estimation. *Wind Energy*, **17**(8), 1199–1216.

Pedersen S, Møller H and Persson-Waye K 2007 Indoor measurements of noise at low frequencies – problems and solutions. *Journal of Low Frequency Noise, Vibration and Active Control*, **26**(4), 249–270.

Perkins R, Cand M, Davis R, Hayes M and Jordan C 2013 The production of a good practice guide to assess wind turbine noise in the United Kingdom using ETSU-R-97. *5th International Meeting on Wind Turbine Noise*, Denver, Colorado.

Piercy J, Embleton T and Sutherland L 1977 Review of noise propagation in the atmosphere. *Journal of the Acoustical Society of America*, **61**(6), 1403–1418.

Plovsing B 2006a Comprehensive outdoor sound propagation model. Part 1: Propagation in an atmosphere without significant refraction. Technical Report AV 1849/00, Delta.

Plovsing B 2006b Comprehensive outdoor sound propagation model. Part 2: Propagation in an atmosphere with refraction. Technical Report AV1851/00, Delta.

Plovsing B 2014 Proposal for Nordtest method: Nord2000 –prediction of outdoor sound propagation. Technical report, Delta.

Renewable UK 2013 Wind turbine amplitude modulation: research to improve understanding as to its cause and effect. Technical report, Renewable UK.

Rudnik I 1957 Propagation of sound in the open air. In: Harris CM, *Handbook of Noise Control*. McGraw Hill.

Salomons E 2001 *Computational Atmospheric Acoustics*. Springer Science & Business Media.

Salomons E and Janssen S 2011 Practical ranges of loudness levels of various types of environmental noise, including traffic noise, aircraft noise, and industrial noise. *International Journal of Environmental Research and Public Health*, **8**(6), 1847–1864.

SEPA 2001 Ljud fraån vindkraftverk (noise from wind turbines). Technical Report Report 6241 (in Swedish), Swedish Environmental Protection Agency, Stockholm.

SEPA 2012 Ljud fraån vindkraftverk (noise from wind turbines). Technical Report Report 6241, version 3 (in Swedish), Swedish Environmental Protection Agency, Stockholm.

Søndergaard B and Plovsing B 2005 Noise from offshore wind turbines. Technical report, Danish Ministry of Environment, EPA.

Søndergaard B and Plovsing B 2009 Validation of the Nord2000 propagation model for use on wind turbine noise. Technical report, Delta.

Søndergaard B, Hoffmeyer D and Plovsing B 2007 Low frequency noise from large wind turbines. *Second International Meeting on Wind Turbine Noise*, Lyon, France.

Søndergaard L and Sorensen T 2013 Validation of WindPRO implementation of Nord2000 for low frequency wind turbine noise. *5th International Meeting on Wind Turbine Noise*, Denver, Colorado.

Stephens D, Shepherd K, Hubbard H and Grosveld F 1982 Guide to the evaluation of human exposure to noise from large wind turbines. Technical report, NASA.

Stigwood M, Large S and Stigwood D 2015 Cotton Farm wind farm - long term community noise monitoring project - 2 years on. *6th International Meeting on Wind Turbine Noise*, Glasgow, UK.

Styles P, England R, Stimpson I, Toon S, Bowers D and Hayes M 2005a A detailed study of the propagation and modelling of the effects of low frequency seismic vibration and infrasound from wind turbines. *First International Meeting on Wind Turbine Noise: Perspectives for Control*, Berlin, Germany.

Styles P, Stimpson I, Toon S, England R and Wright M 2005b Microseismic and infrasound monitoring of low frequency noise and vibrations from windfarms. Technical report, Keele University.

Styles P, Westwood R, Toon S, Buckingham M, Marmo B and Carruthers B 2011 Monitoring and mitigation of low frequency noise from wind turbines to protect comprehensive test ban seismic monitoring stations. *Fourth International Meeting on Wind Turbine Noise*, Rome, Italy.

Sutton O 1953 *Micrometeorology*. McGraw Hill.

van Maercke D and Defrance J 2007 Development of an analytical model for outdoor sound propagation within the Harmonoise project. *Acta Acustica United with Acustica*, **93**(2), 201–212.

Verheijen E, Jabben J, Schreurs Eand Smith K 2011 Impact of wind turbine noise in the Netherlands. *Noise and Health*, **13**, 459–463.

Wieringa J 1980 Representativeness of wind observations at airports. *Bulletin of the American Meteorological Society*, **61**(9), 962–971.

Wieringa J 1992 Updating the Davenport roughness classification. *Journal of Wind Engineering and Industrial Aerodynamics*, **41**(1), 357–368.

6

Measurement

6.1 Introduction

The measurement of wind farm noise at affected residential locations presents some unique challenges due to the significant contribution of low-frequency noise, the requirement to measure in windy conditions and the influence of atmospheric conditions on both turbine sound power output and propagation characteristics from the turbines to the residence. To minimise the influence of wind-induced noise, which is also heavily weighted to low frequencies, specialised wind screens are required to reduce the atmospheric turbulence incident on the microphone.

Atmospheric conditions that affect the sound level arriving at a receiver include wind speed and direction, ambient temperature and relative humidity, amount of cloud cover and atmospheric turbulence. Since there is a large variation in atmospheric conditions over time, measurements often last for several days, weeks or even months. This requires implementation of an unattended monitoring system where all instruments are synchronised to allow comparison between the various data sets. Care should be taken to ensure that the allocated timestamp refers to the same measurement. Some instruments may designate the timestamp to the start of the measurement while others may designate it to the end.

The occurrence of rain should be recorded and all associated data should be discarded because this is a source of extraneous noise. In the case of heavy rainfall, it may be necessary to discard data recorded after the rain stopped if the microphone and/or wind screen became excessively wet. Microphones can recover from water contamination but will not give correct results when they are wet. Wind screens that are not water repellent may suffer changes in their insertion-loss properties when wet. Long-term monitoring also demands system reliability, as data loss can be costly and can also complicate the analysis process. Some systems can be monitored via the internet to check functionality, but this is not feasible in some rural areas where internet access is limited.

The large amount of instrumentation required for wind farm noise characterisation can be time consuming to set-up, depending on the level of detail that is required in the analysis. One important feature that should therefore be considered when selecting instrumentation is the efficiency with which it can be assembled. One method of improving efficiency is to use a system that provides wireless transfer of noise, meteorological and wind farm operational data to a central server that can be accessed remotely. Such systems have already been developed (Bradley 2015; McDonald et al. 2011; Thompsen and Nielsen 2013) and are commercially available.

Wind Farm Noise: Measurement, Assessment and Control, First Edition.
Colin H. Hansen, Con J. Doolan, Kristy L. Hansen.

It is important to ensure that instrumentation is robust, as there is potential for animals to inflict damage, particularly in the rural environments where wind farms are often located. Therefore, sensitive devices should be placed in suitably robust cases that can be fastened, have adequate ventilation (if necessary) and are waterproof. Measurements can be contaminated by water ingress into the cables and therefore all cable and connector joints should be made as waterproof as possible by taping them with duct tape, or preferably with self-amalgamating silicone tape.

Some aspects of wind turbine noise generation can be best observed and understood by taking measurements in wind tunnels of noise from models of wind turbines complete with wind turbine blades and turbine blade sections. These tests are referred to as *fixed-blade testing*. Wind tunnel testing can remove some of the complicating factors related to atmospheric conditions. Also, scaling requirements result in higher rotor speeds and hence the frequency range of interest is not so low as to require specialised low-frequency microphones. On the other hand specialised wind screens are still required to minimise the wind-induced noise on the microphones. As aerodynamic and acoustic measurements are important when evaluating rotor blade sections, there is a discussion in Section 6.6.2 on wind tunnel testing and how to take acoustic measurements in wind tunnels. This includes single-microphone and acoustic-array measurements plus a summary of some recent wind tunnel measurements relevant to wind turbines. A small section is devoted to recent acoustic-array testing of full-scale turbines.

6.2 Measurement of Environmental Noise Near Wind Farms

Wind turbine noise is measured for various reasons, including wind farm noise source characterisation, establishment of noise levels that existed prior to the wind farm development, compliance measurements and complaints investigation. For each of these scenarios it is important to consider such parameters as microphone positioning, extraneous noise sources and environmental conditions. It is also imperative to measure the background noise levels that occur when the wind farm is not operating to establish the difference between various aspects of the noise in operational and non-operational conditions. Large differences are likely to lead to increased levels of perception and annoyance.

To ensure that wind turbine noise arriving at sensitive locations is adequately characterised, measurements should be continuous and span several weeks. The sound level and statistical levels can be measured in time blocks spanning 10–15 min (depending on requirements of the local regulatory authority) and stored for later analysis. Preferably, 1/3-octave data and continuous time data should also be recorded to enable wind turbine noise to be distinguished from extraneous noise sources and to allow investigation of the character of the wind farm noise.

Apart from the low-frequency dominance of wind farm noise indoors at most residential receiver locations, another characteristic that needs special consideration is the existence of special audible characteristics, such as the presence of tonality, amplitude modulation and/or amplitude variation for specific atmospheric conditions. If such special audible characteristics do occur, it is essential that they are recorded during the measurement period and that sufficient data are collected to determine the statistical probability of their occurrence.

To identify the presence of special audible characteristics, it is necessary to record the time series data using a microphone and a suitable data-acquisition system. The memory requirements can be large, so the sampling frequency may need to be limited in some circumstances. At propagation distances greater than 1–1.5 km from the nearest turbine in a wind farm, noise detected by residents in their homes is mainly low-frequency (usually less than 160 Hz) as a result of propagation effects and house transmission effects, so low sampling frequencies are acceptable. Where memory limitations exist, some sound level meters provide the option of recording time data for a given period in response to exceedance of a user-defined threshold. This approach can prove problematic in areas where extraneous noise occurs frequently.

Where circumstances permit, it is useful to be able to control the noise-recording instrument remotely. This enables recording to be paused for several hours or days when the weather is unsuitable and continued again when weather conditions become acceptable. Also, selected data can be downloaded for analysis at regular intervals for processing and to check that there are no obvious faults in the measurement system.

6.2.1 Instrumentation

Microphones

The most commonly used device for measuring wind farm noise is a condenser microphone, which is a high-precision transducer that converts sound pressure fluctuations to measurable electrical signals (Bies and Hansen 2009). A condenser microphone consists of a small thin diaphragm that is separated from a polarised backing plate by a very narrow gap. In the presence of an oscillating pressure the diaphragm is displaced, thus changing the gap width between the diaphragm and the backplate. This produces an oscillating voltage, which is proportional to the original pressure oscillation. Condenser microphones are used because of their very uniform frequency response and long-term stability. There are many different types of condenser microphone available and therefore when choosing the optimum configuration for a particular application it is important to consider parameters such as frequency response, polarisation type, dynamic range, sensitivity, directional characteristics and temperature range. These parameters will be discussed below in the context of wind farm noise measurement.

The frequency response of a microphone is the usable frequency range. Smaller diameter microphones are generally most suitable for measuring higher frequencies, and larger ones for measuring lower frequencies. To measure the entire spectrum of wind turbine noise, specialised microphones that have a flat frequency response down to at least 0.5 Hz are required. For most modern wind turbines, this lower frequency limit is sufficient to measure the fundamental blade-pass frequency. The upper limit of the frequency response for wind turbine noise measurement is covered adequately by most commercially available, precision condenser microphones, which are capable of measuring up to 20 kHz.

Microphones can be either externally polarised by application of a bias voltage in the power supply or prepolarised internally through use of an electret, which is a dielectric material that has a permanent electric charge. The advantage of a prepolarised type is that it is less sensitive to dust and moisture, which is advantageous for outdoor measurements. Also, the constant current supply that powers these microphones is relatively inexpensive compared to an externally polarised power supply, and can be supplied by most modern data acquisition systems, thereby eliminating one component in the measurement chain. With electret microphones, standard coaxial cables with BNC connectors can be used, instead of LEMO 7-pin connectors and cables, which are more expensive. This is an important consideration since it can sometimes be necessary to extend cables over distances exceeding 100 m. An advantage of externally polarised microphones is that they can operate reliably at elevated temperatures, although this characteristic is not necessary in wind farm noise measurement. Despite the fact that the prepolarised microphone type appears to be best suited to wind farm noise measurement, there are some microphone models that are only available as an externally polarised unit, such as microphones that can measure very low sound pressure levels.

The dynamic range of a microphone is the range between the upper and lower limit of the sound pressure level that the microphone can measure. The lower limit of the dynamic range for a microphone, often referred to as the *noise floor*, is dictated by the electrical noise floor of the electronics used to condition and amplify the microphone signal. A microphone with higher sensitivity will have a lower noise floor in decibel terms, as it will produce a higher level electrical signal for the same sound pressure level, while the electrical noise expressed in volts will be the same.

The sensitivity of a microphone is primarily governed by the size of the microphone and the tension of its diaphragm. In general, a large diameter microphone with a loose diaphragm will have a high sensitivity and a small diameter microphone with a stiff diaphragm will have a low sensitivity. In addition, a higher polarisation voltage results in a higher microphone output for the same sound pressure level input, thus resulting in a lower noise floor in terms of decibels. This is why low-noise microphones are externally polarised with 200 V, to a level that cannot be achieved with a prepolarised electret.

A microphone will always be attached to a pre-amplifier in practice and therefore it is important to determine the inherent noise of the microphone and pre-amplifier combination to find the lower limit of the dynamic range. The upper limit of the dynamic range is dictated by the amount by which the microphone diaphragm can move before the signal becomes excessively distorted (usually 3% distortion). In simple terms, the distortion of a signal is a measure of the percentage of the sound pressure input to the microphone at a particular frequency that appears in error as a microphone output at higher frequencies. For the measurement of wind farm noise, it is only necessary to consider the lower limit of the dynamic range otherwise known as the 'noise floor'. This is because the ambient noise level in some locations near wind farms is extremely low, and to appreciate the contrast between operational and non-operational conditions, it is necessary to measure these levels accurately.

At high frequencies, where the measured sound has a wavelength that is comparable to the diameter of the microphone diaphragm, diffraction effects can occur, whereby the microphone response becomes sensitive to the direction of incidence. This effect is relevant for frequencies with wavelengths shorter than 10 times the diameter of the microphone, which corresponds to a frequency of approximately 3 kHz for a half-inch microphone. To compensate for these effects, the frequency response of a microphone can be adjusted to take into account the expected angle of incidence in a particular sound field. There are three types of sound field that are encountered in acoustic measurements: free field, pressure field and diffuse field. Microphones are optimised to have a flat frequency response in one of these fields and therefore a suitable microphone type should be selected for a given application. For wind farm noise measurement where the signal is dominated by low frequencies, any of these microphone types are suitable because they all have a flat frequency response below 1 kHz.

For environmental measurements of wind farm noise, the maximum temperature specifications of the microphone will rarely be encountered. Nonetheless, it is useful to take note of the temperature-range specifications and to be aware that a microphone can be adversely affected under extreme operating or storage temperatures and will also need to be calibrated more often in these conditions.

Pre-amplifiers

The output signal from a microphone is a tiny voltage that cannot drive low-impedance devices such as cables. Therefore a high-impedance pre-amplifier needs to be attached directly to the microphone or placed in very close proximity to it. Rather than amplifying the microphone signal, the pre-amplifier converts the high-impedance output from the microphone into a low-impedance voltage signal that can drive cables. Hence another name for a pre-amplifier is an impedance converter. When choosing an appropriate pre-amplifier for a particular microphone, it is important to verify that its electrical noise floor is sufficiently low and that it has a flat frequency response in the frequency range of interest. Some compromise may be unavoidable and in the context of wind farm noise measurement the influence of the pre-amplifier on the lower limit of the frequency response and dynamic range is particularly important. The pre-amplifier has its own low-frequency roll-off and also the noise floor of pre-amplifiers is greater at low frequencies.

Cables

Noise signals measured by the microphone need to be recorded and stored for later analysis. The microphone is therefore connected via a cable either to a recording sound level meter or to a data-acquisition device. The type of cable chosen is dependent on whether the microphone is prepolarised or externally polarised. In the former case, standard coaxial cables with BNC or TNC connectors can be used, whereas in the latter case LEMO 7-pin connectors and cables must be used. The advantage of using standard coaxial cables is that they are less expensive and more readily available.

Other points to consider when selecting an appropriate cable are the distance over which it is extended and the potential for contamination of the signal by electrical noise. Electrical noise will show up in the spectrum as discrete peaks at 50 or 60 Hz and associated harmonics. Thus it is recommended that a sample spectrum is recorded and analysed before the instrumentation is left unattended for long-term monitoring. In cases where electrical noise is an issue, it is possible to use double-shielded coaxial cable but this type of cable is much more expensive and less readily available than standard, single-shielded coaxial cable. Therefore, before deciding to use such a cable, other potential sources of electrical noise should be identified. These can include but are not limited to:

- coiled or crossed cables that act as an antenna
- a cheap power-supply cable for a laptop computer
- earth leakage from mains power supplies to instrumentation.

The latter problem can be tackled by grounding the data-acquisition system or sound level meter, or alternatively, using battery power exclusively. Grounding can be achieved by hammering a copper rod as far as possible into the earth and connecting this rod to the recording device via an earth cable. The effectiveness of the grounding can sometimes be improved by pouring some water into the hole in which the rod is located. The earth cable should only be connected to a single place in the measurement chain to avoid unwanted 'ground loops'.

Recording Devices

The most common and convenient instrument for recording sound pressure is a sound level meter. Currently available sound level meters are digital and have a dynamic range up to 100 dB. They consist of an analogue-to-digital converter that is connected to a digital signal processor board, which converts the time-varying signal to a user-specified output. The desired output is specified via a graphical user interface. For the measurement of wind turbine noise, most regulations specify that sound level meters have attained at least Class 2 certification in accordance with IEC 61672-1 (2013). In the future, there is likely to be a requirement to use only Class 1 sound level meters (Teague and Leonard 2011). While the two classes of sound level meter have the same design goals, the main differences lie in the tolerance limits and the range of operational temperatures (IEC 61672-1 2013). The tolerance limits are important for wind farm noise measurement because there can be a small margin between compliance and non-compliance in some situations, so a Class 1 sound level meter is recommended. The required specifications for a sound level meter will ultimately depend on the wind farm noise measurement methodology outlined by the local regulatory authority. However, to fully characterise wind turbine noise some useful capabilities are listed below:

1. The ability to calibrate the microphone.
2. A-, C- and G-weighted internal filter networks (see Section 2.2.11).
3. The option to measure unweighted noise levels with minimal roll-off at low frequencies. It should be noted that while the Z-weighting (also referred to as 'linear') has a flat frequency response down to 10 Hz, the roll-off below this frequency is unspecified by IEC 61672-1 (2013), and is thus chosen by the designer of the sound level meter. It is preferable to use no weighting rather than Z-weighting but if this is not possible, then the manufacturer's specifications should be sought to determine the exact frequency roll-off below 10 Hz.
4. The ability to measure continuous averages (that is, L_{eq}, L_{Aeq}) over a user-defined period with negligible time delay between each measurement. Since wind farm noise measurements often span several weeks, the instrument should provide the ability to measure these averages for a long time without significant drift (in practice the amount of data that can be measured is constrained by the storage capacity of the instrument and/or compatible memory storage device such as a USB flash drive or memory card).
5. A means of storing data measured over several weeks. For a sound level meter with the option of memory storage via USB flash drive, it is important to ensure that adequate power is available for chosen USB flash drive. This may limit the number of devices available for USB storage and it is generally impossible to use portable hard drives due to their relatively slow data-writing speed, therefore restricting storage capacity. It is recommended that USB compatibility is confirmed before the sound level meter is used in the field.

6. The capability of measuring statistical quantities such as L_{A90}, which is commonly required for background noise measurements.
7. 1/3-octave and octave filters that can be implemented on the raw signal or after applying A-, C- and G-weighting.
8. A choice of meter response times, often referred to as 'fast', 'slow' and 'impulse'. The 'fast' response time is 100 ms or 125 ms and is designed to approximate the response of the human ear. The 'slow' response time is 1 s and is used in methods to determine the existence of amplitude modulation, as discussed by Fukushima et al. (2013). The 'impulse' response has a standard time constant of 35 ms and can be useful in determining the degree of impulsiveness of the noise. Impulsive noise is characterised by relatively short duration pulses and is more annoying than continuous noise. Blade–tower interaction is an example of impulsive noise in the context of wind turbines, as discussed in Section 3.3.7.
9. FFT or narrowband analysis option to enable identification of tonal and/or AM characteristics.
10. The opportunity for firmware updates to ensure that the device remains consistent with the most recent standards. Moreover, the manufacturer may develop additional capabilities over time and it is useful if these can be conveniently installed. Firmware updates should be provided in a format that can be easily installed by the user without having to return the instrument to the manufacturer.
11. A display that can show the overall A-, C- and G-weighted and unweighted sound pressure levels as well as the 1/3-octave and octave band levels (not necessarily simultaneously). This ability is important as it is always necessary to check that the instrumentation is functioning properly before leaving it unattended for several days or weeks. The user should have a reasonable idea of what reading is appropriate for the overall level based on their experience with listening to noise at different levels. Any unusual peaks should be investigated, particularly those occurring at 50 Hz (or 60 Hz) and 100 Hz (or 120 Hz). Peaks at 50 Hz are often caused by electrical noise and peaks at 100 Hz can be caused by a refrigerator or a transformer. On the other hand, sometimes wind farm noise is produced in the 50 Hz and 100 Hz 1/3-octave bands and in the vicinity of the Waterloo wind farm in South Australia, a peak in the 50 Hz 1/3-octave band was strongly correlated with wind farm operation (Hansen et al. 2014). In such cases, it may be necessary to use the FFT or narrowband function of the sound level meter to determine the narrowness of the peak (electrical noise is characterised by a discrete peak occurring in a very narrow band). It is also useful to observe if there are harmonics of the 50 Hz (or 60 Hz) peak as this is also an indication of electrical noise.

The main disadvantage of using a sound level meter is that data-storage space is generally limited and therefore it is not possible to continuously measure raw time data over several weeks with multiple inputs. A more suitable instrument for this purpose is a data acquisition system. Data acquisition systems digitise an analogue input signal for storage on a memory card or computer. These devices are commonly available in modules containing 4–6 input channels and these modules can be combined if additional channels are required. Most systems also require software to drive the hardware but it is often possible to purchase a limited licence if an alternative analysis software is available to the user. There is also the option for the user to develop specialised code to communicate with the data-acquisition system; however, this can be time-consuming and requires a good understanding of the principles involved. There are some important considerations to take into account regarding the format of the file to be saved. The most efficient format in which to save the data is binary, with minimal information other than the raw data. Binary files consist of binary digits (bits) grouped in eights to form bytes. When interpreting a binary file, it is necessary to know the byte ordering (big endian or little endian), the encoding scheme and the overall structure of the file. Data can be saved into a single file, which can later be split into segments of a desired length, or it can be saved in separate files of user-specified length. The latter format makes analysis more efficient, particularly when the user desires averages over a fixed time interval. It is very important to ensure that the system does not introduce a delay between each measurement as this will result in lost data and loss of synchronisation with other instrumentation such as weather stations.

It is important that the self-noise of the sound level meter or data-acquisition system is lower than that of the microphone/pre-amplifier combination to ensure that it does not result in a significant increase in the noise floor. A useful feature is the ability to select an input range based on the expected values to be measured. If this option is not available, it is also possible to amplify the signal from the microphone and pre-amplifier before it enters the recording system. In this case, an amplifier should be selected that does not limit the frequency response of the system nor introduce excessive self-noise. Provided that the sound pressure levels of the measured data are more than 3 dB above the instrumentation noise floor, the noise floor can be subtracted logarithmically from the measured data using Eq. (2.29). If the measured data are within 3 dB of the noise floor, they must be discarded.

The sample rate of the data acquisition system should be chosen carefully depending on the specific application. Ideally, the sample rate should be as fast as possible to avoid limitations during post-processing. One such limitation is that A- and C-weighting filters require 10 times oversampling to be accurate, as discussed in Section 6.2.6. Therefore, the chosen sampling rate should be as fast as possible within the constraints of memory limitations. As measurement complexity increases through increasing the number of measurement channels and measurement periods, the chosen sample rate will almost always represent a compromise.

Calibration

Calibration is an important procedure used to verify that the instrumentation is working properly. It should be performed before and after a set of measurements using the same configuration as required for the measurements (that is, the same cable lengths, and so on). It should also be carried out regularly during long-term measurements to confirm that all instrumentation is working properly and that there is no significant drift. Any drift should be noted and considered when determining the margin of error associated with the results, as discussed in Section 6.2.16. Random fluctuations of $\pm$ 0.5 dB are considered normal and acceptable. Ideally, the instrumentation should be tested over the entire useful frequency range, but this is not feasible in practice as it is a time-consuming procedure requiring specialised equipment. Therefore, calibration is generally performed at a single frequency using a noise-generating device that is placed over the microphone. The noise-generating device is usually called a *pistonphone* or a *calibrator*. Pistonphones are more accurate (0.1 dB or 0.2 dB, depending on the manufacturer, and require corrections for atmospheric pressure at the measurement location. Depending on the model, pistonphones generate a calibration level of 114, 124 or 134 dB at 250 Hz. Calibrators are more convenient to use as they do not require atmospheric pressure corrections. However, their accuracy is not as good as pistonphones and varies between 0.2 and 0.5 dB, depending on the manufacturer.

Calibrators produce a calibration tone of 94 dB at 1000 Hz, which is very convenient, as it allows the instrument to be calibrated with the A-weighting filter turned on or off since the A-weighting is zero at 1000 Hz. Care should be taken to ensure that there is a good seal between the microphone housing and the calibrator or pistonphone cavity and it may be necessary to use a special insert for a microphone with a small diameter. The calibrator should meet Class 1 specifications, as outlined by IEC 60942 (2003) and this information is generally provided by the manufacturer.

During calibration, the sensitivity of the sound level meter or data-acquisition device is adjusted so that the output reading is equal to the sound pressure level generated by the calibration device. If the sensitivity requires a much greater adjustment than previously used, the first thing to do is to check that all cables have been connected correctly and that there are no loose contacts. As mentioned above, it is also important to ensure that the microphone is inserted properly into the calibrator. Some calibrators time out to save batteries and therefore it is necessary to double check that the calibrator is still generating a tone and that the tone is at the desired level. If calibration is successful on one channel of the recording device but not another, it may be necessary to interchange microphones and cables to identify whether either one is faulty. This procedure will also allow the user to ascertain whether the recording device has a faulty channel. Recording devices and microphones that do not achieve successful calibration should be returned to the manufacturer for checking.

Calibration can also be carried out by introducing an electrical signal of known amplitude and frequency to the position in the circuit corresponding to the microphone input. This enables calibration of recording devices independently of the microphone and this can be useful for determining the corresponding frequency response and noise floor. This procedure may also prove useful for troubleshooting when the recording system is displaying unexpected outputs for a given input signal.

It is recommended that the microphone calibrator and sound recording device be returned to the manufacturer for checking at least once every year. Microphones should also be calibrated over their entire useful frequency range every year. This can be done through use of a specialised calibrator that exposes the microphone to a range of frequencies. For the measurement of wind farm noise, the low-frequency range is most important and thus the use of a low-frequency calibrator is recommended. Calibration below 50 Hz requires the use of a specific apparatus that exposes the entire microphone to the pressure variations of the calibration signal, rather than the microphone diaphragm alone. The reason for this is that in order to maintain a satisfactory response at very low frequencies, a microphone must have an equalisation system that exposes the rear of the diaphragm to atmospheric pressure. This system must act in a way that is fast enough to compensate for changes in altitude and barometric conditions but slow enough not to affect the response of the microphone at the lowest frequencies for which it is intended to be used. Equalisation can be achieved through incorporating a small hole in the side of the microphone to expose the rear of the diaphragm to outside air. As the frequency response requirements shift to lower frequencies, the diameter of this hole must decrease. The smaller the hole, the greater is the potential for it to become blocked by dirt, so low-frequency microphones should be handled with care. Also, it is recommended that low-frequency microphones be regularly calibrated at one very low frequency at least, to verify that the equalisation system is functioning properly.

Power Requirements of the System

For long-term measurements, it is important to ensure that the instrumentation is provided with a constant source of power. Depending on the length of time that the instrumentation is left unattended and its associated power requirements, it may be possible to use battery power exclusively. This is advantageous when measurements are taken in an open field, away from any residences. For continuous measurements spanning several weeks, it is important that sufficient power is available for the measurement duration and that the battery can provide good performance when operating at partial states of charge. Other desirable characteristics include low cost, long life, easy maintenance and long proven history of use. These specifications are most suitably met by a deep-cycle flooded lead-acid battery. When choosing a specific battery, it is important to consider the capacity, which indicates how much energy can be stored by the battery and the cycle life, which is the number of discharge/charge cycles that the battery can provide before its capacity falls to a specified percentage of rated capacity. Of course, the voltage provided by the battery must match the system requirements (usually 12 V, but two 12 V batteries can be connected together to produce 24 V).

The power requirements of the system can be determined by measuring the current drawn by each instrument. The most convenient method of determining the current drawn is to use a clamp-on ammeter (current clamp) as this device is clamped around an insulated current-carrying wire rather than having to be connected through physical contact with the wire. Batteries provide an amp-hour rating, which can be used to determine the amount of operational time that is available.

Solar panels can be used to extend the measurement duration and increase the lifespan of the battery by reducing the number of discharges that occur. Deep-cycle flooded lead-acid batteries are also optimal for use with solar panels and therefore minimal adaptation of the off-grid power system is required. Alternatively, lithium-ion batteries can be used. Solar panels increase the cost and complexity of the system and may not be required for measurements of short duration. To calculate the power rating, P_s (watts), required of the solar panels, the total current draw of the instrumentation in amps, A_i, is multiplied by the battery voltage, V_b. This value is then multiplied by the number of hours per day that the system is in use, t_i (24 h for continuous measurements) and then divided by the number of hours of sunlight estimated for the measurement location at the time of year that the measurements are carried out, t_s. Finally, the result is multiplied by the charge controller efficiency (%) divided by 100, $E/100$. The calculation is as follows:

$$P_s = \frac{A_i \times t_i \times V_b \times E}{t_s \times 100} \quad \text{(W)} \tag{6.1}$$

The number of hours of sunlight, t_s, should be estimated through consultation of climate statistics for the area, including daylight hours and average cloud cover, and a conservative value should be chosen to allow for deviations from expected trends. Once the value of P_s has been calculated, the next largest solar panels should be selected to allow for any uncertainties in cloud cover estimates, uncertainties in the value of E and any losses in the cables or battery.

The charge controller is an essential component in the system, as it prevents the battery from overcharging and ensures that the solar panels do not drain current from the system when there is no sunlight available. Charge controllers incorporate varying degrees of sophistication and modern devices tend to use either pulse width modulation or maximum power point tracking (MPPT). There are various advantages and disadvantages associated with these types of controller but MPPT systems are more recent and offer greater charging efficiency with the drawback of being more expensive. To select the appropriate-sized charge controller, the input and output current from the solar panels and load current, respectively, need to be determined. The minimum input current is calculated by multiplying the solar panel short-circuit current by the number of panels in parallel and then multiplying by a safety factor of 1.25. The minimum output (or load) current corresponds to the current drawn by the instrumentation. It is possible to connect the instrumentation to the charge controller or the battery, depending on user preference. The former arrangement is chosen to protect the battery from too much discharge since the charge controller can be programmed to disconnect the load when the battery charge drops below a certain level. This is desirable if prolonging the battery's lifespan is a priority. On the other hand, if the load is connected directly to the battery, it is possible to measure for a longer duration. In either case, it is useful to incorporate an alarm system that notifies the user when the battery charge has reduced below a specified amount.

Where power is available from the grid, it is still recommended that the deep-cycle flooded lead-acid battery is used as a backup power source, since power interruptions are a regular occurrence in rural areas. A commercial uninterruptible power supply usually generates too much noise, either as a result of continuous beeping when the power goes out or noise generated by the fan used to cool the system. A less expensive and more effective solution is to connect all instrumentation to the battery and then to connect the battery to a charger that has no cooling fan to produce noise. The charger is then plugged into the mains and should provide adequate charge to ensure that the battery maintains its level of charge when power from the grid is available. This set-up effectively performs the same function as a commercial UPS.

The connectors that are used between the battery terminals and the instrumentation should be considered carefully. The cheapest and most readily available solution is to use a cable with clamps at one end to connect to the battery terminals and a cigarette lighter socket at the other end. To connect multiple instruments to the same battery, an adapter is available that has one plug and multiple sockets. Several of these adapters can be joined together to provide further inputs if necessary.

There are some disadvantages associated with the above solution. Since the continual provision of power to the measurement system is of vital importance, failures can be very expensive as a large amount of data can be lost. The use of clamps on the battery terminals brings the possibility of the cables being dislodged by people or animals. The connection between a socket and plug can be easily compromised through movement of the cables and many available sockets do not lock into place, which exacerbates this problem. The types of connection systems described above are also cheap and thus subject to manufacturing faults and hence failures. Therefore, the recommended solution is to use screw-on terminal connectors rather than clamps and to use Anderson connectors rather than sockets. Regardless of the connector type chosen, the power cord of the instrument must be modified to ensure compatibility and the voltage input to the instrument must be correct, which may require the use of a voltage converter.

6.2.2 Effects of Wind

Wind-induced noise is the noise measured by a microphone as a result of the combination of pseudo-noise and acoustic noise caused by wind. Pseudo-noise is the false indication of the sound pressure level resulting from pressure fluctuations incident on the microphone that are unrelated to acoustic sources. These pressure fluctuations can originate from both atmospheric turbulence and the disturbance caused by the air flow over the microphone. This latter effect is manifested as pressure fluctuations generated by the boundary layer on the wind screen as well as the turbulent pressure fluctuations on the lee side of the microphone due to separation/vortex shedding. Pressure fluctuations are also generated by vortex shedding in the wake of associated accessories (such as the wind screen). Despite the physical differences between turbulent fluctuations and acoustic waves, their effects on the microphone are indistinguishable. There are three main sources of pressure fluctuations due to atmospheric turbulence and these are referred to as the *stagnation pressure interaction, turbulence–turbulence interaction* and *shear–turbulence interaction* (Raspet et al. 2008). The stagnation pressure interaction describes the local interaction between velocity fluctuations in the atmosphere with the surface of the microphone/wind screen and this component is the dominant source of pseudo-noise at the microphone (Jackson et al. 2014). The associated power spectral density (see Section 2.4.1) is proportional to $k^{-5/3}$, where k is the wave number (Raspet et al. 2008).

The other two sources of pressure fluctuations due to atmospheric turbulence are intrinsic to a turbulent flow and would be measured even if the wind screen design were ideal (Raspet et al. 2008). They arise as a result of interaction between turbulence structures (turbulence–turbulence interaction) as well as interaction between the turbulence structures and velocity gradients caused by shearing in the flow (shear–turbulence interaction) (Jackson et al. 2014). The turbulence–turbulence interaction follows a $k^{-7/3}$ power relationship in the inertial range (where the inertia or mass effects of the air are much larger than the viscous effects) and dominates in the high-frequency region, whereas the shear–turbulence follows a $k^{-11/3}$ power law in the inertial range (George et al. 1984). For measurements in windy conditions with the microphone mounted at a height of the order of 1 m, the inertial range is from a few hertz up to 10^4 Hz (Raspet et al. 2005).

For outdoor measurements, pressure fluctuations at the microphone due to vortex shedding from microphones and accessories are less important than those caused by atmospheric turbulence fluctuations (Morgan and Raspet 1991). On the other hand, this contribution can become significant in low-turbulence environments such as wind tunnels (Strasberg 1988).

The turbulence spectra of wind are dominated by low frequencies and therefore the levels of wind-induced noise are higher at low frequencies. On average, approximately 90% of energy in wind-induced noise is concentrated below 15 Hz and 95% below 30 Hz (Hessler 1996). This presents challenges in the measurement of wind farm noise, which is also heavily dominated by lower frequencies. Reduction of wind-induced noise is of particular importance in this context to ensure that the signal-to-noise ratio is optimised under windy conditions. The standard mitigation strategy for reducing the pseudo-noise component of wind-induced noise is to use a wind screen, as will be discussed in the next section.

In contrast to pseudo-noise, acoustic noise is generated when objects such as trees are put into motion by the wind or when air movement past stationary objects gives rise to vortex shedding and associated noise phenomena such as whistling. The most effective way to minimise wind-induced noise caused by interaction of the wind with objects is to place the microphone in a large, open field, away from such sources. To reduce vortex shedding around the microphone mount, the commonly-used camera tripod can be replaced with a star-dropper.[1]

To estimate the contribution of wind-induced noise, L_{wind}, to the measured sound pressure level, Kamiakito et al. (2015) developed a regression function that is dependent on the mean wind speed, U_0, turbulence intensity, Iu, and 1/3-octave band centre frequency, f. This function is,

$$L_{\text{wind}} = A(f) + B(f)\ \log_{10}(1 + Iu)^2 + C(f)\ \log_{10}{U_0}^2\ (\text{dB}) \tag{6.2}$$

The coefficients A, B and C are determined from measurements of wind noise, so they are difficult to determine if there are other significant sources of noise contributing to the microphone signal at frequencies below 100 Hz. Kamiakito et al. (2015) determined these coefficients at four different measurement sites and concluded that they appear to be dependent on the surface roughness of the surrounding area as well as on frequency. Therefore, when determining the coefficients in Eq. (6.2), the type of terrain at the measurement location should be representative, as it is not possible to do the associated regression analysis accurately in the vicinity of an operating wind farm due to the influence of the wind-farm-generated noise.

6.2.3 Wind Screens for Microphones

Wind screens are acoustically transparent shields that are used in noise measurement to protect the microphone from wind-induced noise. The basic noise reduction mechanism at low frequencies is that the variation in phase between the pressure fluctuations at the surface of the wind screen results in cancellation of certain components in the area-averaged pressure measured by the microphone (Raspet et al. 2005). A spherically-shaped wind screen is believed to be the most effective at low frequencies, as pressure fluctuations at the front and rear of the sphere are opposite in sign for a given velocity fluctuation and hence cancel one another (Raspet et al. 2005).

For wind tunnel measurements, where the wind direction is constant and the turbulence intensity is low, the best protection against wind-induced noise will be given by a nose cone (Wuttke 2005) or aerodynamic cap, as discussed in Section 6.6.2. However, these conditions are seldom encountered outdoors and, since the wind direction is variable, particularly for long-term measurements, a spherical design is optimal. As the size of the wind screen increases, the averaging area also increases, improving the effectiveness of the wind screen (Raspet et al. 2008). For measurements of wind turbine noise, where instrumentation needs to be transported to various measurement sites, the upper limit in wind screen size is based on issues of practicality.

---

1 A thin steel post with a star-shaped cross-section. It is used for low cost fencing in rural areas.

The material used for wind screens is selected to reduce velocity fluctuations at the microphone diaphragm whilst ensuring that there is minimal noise attenuation. The most common type of wind screen is a flexible open-cell foam type that fits snugly over the microphone. The popularity of this design is based on convenience, durability, low cost, well-documented acoustic characteristics and effectiveness for A-weighted measurements. Various other materials have been used in the past, including fine silk mesh or nylon mesh stretched over a frame, metal mesh, sintered metal, fur and other porous materials (Wuttke 2005). The advantage of fur is that it is a good absorber of turbulent energy and does not create any additional noise (Wuttke 2005). It is important that wind screen materials are securely attached to the mounting arrangement and/or microphone because loose materials and an exposed frame can create undesirable noise. Obviously, a wind screen that has blown off the microphone is useless and all associated measurements would need to be discarded.

For outdoor measurements in wind speeds less than 5 m/s, it is common to use a standard wind screen consisting of a 90-mm diameter, solid sphere of open-cell foam that fits snugly over the microphone, as recommended in ANSI/ASA S12.9-3 (1993). One of the reasons that a standard 90-mm wind screen is deemed acceptable for wind speeds up to 5 m/s is that the commonly used A-weighting filter applies a heavy penalty to low-frequency noise, thus minimising the effect of wind-induced noise, which is dominant at lower frequencies. On the other hand, it is debatable that application of the A-weighting filter is appropriate for measurements of wind farm noise at a residence since the wind turbine noise spectrum indoors is heavily biased towards lower frequencies. It has been shown that at low frequencies, the signal-to-noise ratio reduces with increasing wind speed and that the lowest valid frequency of measurement is directly related to the wind speed (ANSI/ASA S12.9-7 draft 2014). At 1/3-octave band centre frequencies less than or equal to 31.5 Hz, valid measurements can only be made under very light or zero wind conditions when using standard 90-mm wind screens (ANSI/ASA S12.9-7 draft 2014). Therefore, a secondary wind screen concept has been developed and a description is included in the IEC 61400-11 (2012) standard, which describes a methodology for measuring the noise emissions from a wind turbine. This standard specifies that the outdoor microphone should be mounted in the centre of a 1-m diameter ground board and covered with a standard 90-mm wind screen. The secondary wind screen should have a diameter of at least 450 mm and be located symmetrically over the smaller 90-mm primary wind screen. While the use of the secondary wind screen and ground-mounting arrangement is a requirement for sound power measurements (IEC 61400-11 2012), there is no such requirement for noise measurements at a receiver location in most standards and guidelines.

A number of different infrasonic and low-frequency wind screen designs have been developed with the aim of reducing the pseudo-noise at the microphone to minimise the threshold for the lowest measurable noise level at low frequencies. Some of these designs consist of two or more wind screen layers, where the wind screen material is attached to a metal frame and an air gap exists between each layer. These wind screens will be hitherto referred to as *frame-type* constructions. These often specify use of a 90-mm open-cell foam wind screen to fit over the microphone to enhance the effectiveness of the secondary wind screen layers. Other designs consist of a large, solid, open-cell foam wind screen, which can be conveniently manufactured in different shapes. The effectiveness of such a wind screen can be enhanced at low frequencies by hollowing out the foam to form a chamber around the microphone (Wuttke 2005).

As mentioned in Section 6.2.2, the dominant source of turbulent pressure fluctuations experienced by an outdoor microphone is the intrinsic turbulence in a flow rather than the fluctuating wake behind the wind screen. Therefore streamlining the wind screen is of secondary importance to the ability of wind screens to diffuse the vortical structures in the turbulent flow. Larger wind screens are more successful at reducing wind-induced noise because the incoherent wind noise is averaged over a larger surface (Strasberg 1979). Also, there is more time for viscous dissipation to occur before the turbulence reaches the microphone. For a multi-layered wind screen design, viscous dissipation is provided by the volume of air between wind screen layers (Morgan and Raspet 1991). Turbulence effects can also be minimised by locating the microphone in close proximity to the ground, where the wind speed is lower. In this case, the vortex shedding about the wind screen will also be reduced.

Several investigations have been carried out to determine the performance of wind screens with various shapes, sizes, materials and internal construction and some of these laboratory studies and field measurements are summarised below. The first part of this review concerns performance of wind screens in general and the second part focusses on designs specifically targeted to reduce low-frequency noise and infrasound.

A comprehensive analysis of the effectiveness of frame-type spherical wind screens of various sizes, materials and number of layers was carried out by Bleazey (1961). This study included a theoretical analysis, indoor measurements in an anechoic chamber with a wind generator and outdoor measurements. The results indicated that the best performance was achieved by the largest wind screen and that fine mesh silk provided optimal attenuation characteristics. The highest overall attenuation was just over 30 dB, which was achieved with a 12-inch ($\approx$305 mm) diameter wind screen. Other materials were equally successful in attenuating wind-induced noise but introduced attenuation of noise at higher frequencies. It was found that wind screen effectiveness was only slightly improved by increasing the number of layers and that more than three layers gave no further performance enhancement. Outdoor measurements taken using the Brüel and Kjær Type UA 0082, which consists of a 240 mm diameter spherical wire frame covered with two layers of nylon mesh, also showed attenuations in wind noise of up to 25 dB over a frequency range of 20 Hz–20 kHz when the microphone diaphragm was oriented parallel to the wind (Skøde 1966). The attenuation was shown to decrease with increasing wind speed and spectral analysis revealed that the main contributing factor to this trend was the large increase in the low-frequency wind-induced noise for higher wind speeds (Skøde 1966). Doubling of the wind screen diameter to 480 mm was shown to improve the wind noise attenuation by 2 dB but it was concluded that this small improvement in performance did not justify the impracticality of such a large design.

Wind tunnel results for nine different solid, open-foam wind screens of various shapes and sizes also revealed that the largest design was the most successful in reducing wind-induced noise (Hessler 2008). This design was spherical, with a diameter of 175 mm. A comparison of the wind noise measured in a wind tunnel for a range of spherical and cylindrical wind screens with diameter less than 100 mm was made by Strasberg (1979). A linear relationship was found between the normalised and dimensionless 1/3-octave band sound pressure level, $20 \log p_{1/3}/\rho U^2$, generated by wind interaction with the various types of wind screens and the dimensionless frequency,

fD/U, indicating a direct relationship between the sound pressure level and vortex shedding frequency. In these relationships, $p_{1/3}$ is the 1/3 octave band RMS sound pressure, ρ is the density of air, U is the speed of the wind blowing over the wind screen, f is the centre frequency of the 1/3-octave band and D is the wind screen diameter. Some departure from linearity for values of $fD/U > 5$ was attributed to porosity effects. The conclusions in this paper are relevant for conditions of low turbulence intensity where pseudo-noise at the microphone is dominated by pressure fluctuations due to vortex shedding. This is generally not the case for measurements made outdoors with a significant amount of wind.

Noise reductions at lower frequencies were demonstrated for outdoor measurements in a large field by Imaizumi and Takahashi (2010) using a multi-layer wind screen. These researchers placed a hot-wire on the inside of the wind screen and noticed a significant reduction in wind speed and associated fluctuations compared to what was measured with no wind screen. The reduction in turbulence was more pronounced when the number of wind screen layers was increased from two to three. According to the data presented in this study, noise reductions of at least 5 dB were observed over the frequency range from 1 to 40 Hz when the wind speed was greater than 4 m/s. The reduction in wind-induced noise for a double-layer design was also quantified by Tachibana et al. (2013) and was found to be greater than 15 dB at infrasonic frequencies and wind speeds up to 7 m/s, compared to a microphone with a 70-mm wind screen. One unique aspect of this design was that the outer, secondary wind screen was dodecahedral in shape, consisting of 12 pentagons. Improvements in the measurement of impulse noise from artillery firings was also achieved using a multi-layer wind screen, where the overall C-weighted level was reduced by up to 40 dB in comparison to results obtained using a standard wind screen (Nakajima et al. 2001).

An alternative wind screen design for low-frequency and infrasonic measurement is an underground box with an open lid, which has been replaced by open-cell foam (Betke et al. 1996). This design minimises exposure to wind-induced noise as well as eliminating wake-induced turbulence, and has been successful in reducing wind-induced noise (Hansen et al. 2014).

In addition to using specialised wind screens for wind-induced noise reduction, it is also possible to use an array of microphones and then to reduce the wind-induced noise contamination through signal processing techniques. This method takes advantage of the fact that low-frequency and infrasonic acoustic waves propagating from wind turbines are well-correlated over several metres, whereas wind noise becomes uncorrelated within a few metres. The simplest example is a two-microphone array which would be aligned with the nearest wind turbine, whilst ensuring that there are no reflecting surfaces nearby. The cross power spectral density between the microphone signals is calculated and the result represents the average of the two microphone signals, where the uncorrelated wind-induced noise has been averaged out. A discussion of the cross power spectral density is provided in Section 2.4.11. Figure 6.1 shows a plot of the cross power spectral density between signals measured with two microphones that were positioned on the ground 10 m apart and aligned with the nearest wind turbine, which was 3.5 km away. The wind screens were designed according to IEC 61400-11 (2012), the wind speed at 1.5 m was 2 m/s and the power output of the wind farm was 70%. The plot of the cross power spectral density shows the blade-pass peaks more distinctly than the power spectral density plots of the separate microphone signals. The correlation between the

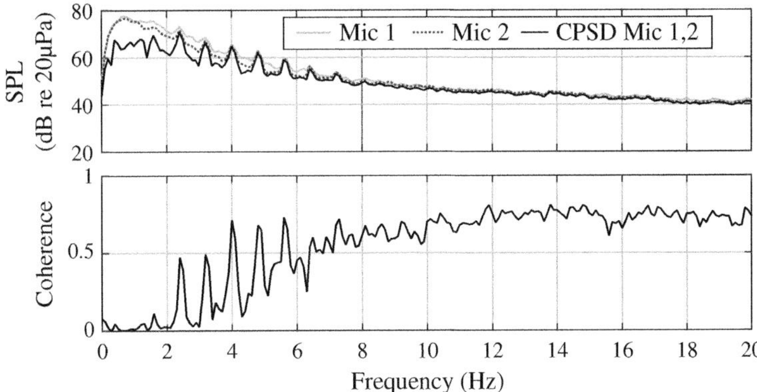

Figure 6.1 Sound measurement with two-microphone array: (top) cross power spectral density; (bottom) coherence (see Section 2.4.12). SPL, sound pressure level.

microphone signals is poor at frequencies below 2 Hz, where uncorrelated wind noise is dominant, but is good at blade-pass peaks above 1.6 Hz. This technique is more suited to low and infrasonic frequencies, where correlation between the microphone signals is expected to be good due to the large associated wavelengths of the noise.

Another technique for reducing wind-induced noise at the microphone is to use a wind fence enclosure, which is particularly useful for wind noise reduction at infrasonic frequencies. Wind fence enclosures reduce wind-induced noise by separating the microphone from direct stagnation pressure interactions, which are instead transferred to the surface of the enclosure (Abbott and Raspet 2015). Generally, the enclosure is uniformly shaped so that the wind direction does not influence the noise reduction; the dimensions are large to increase reductions at low frequencies. The height is sufficiently large to ensure that turbulence entering over the walls of the enclosure produces a negligible effect at the microphone. Alternatively, the enclosure could have a flat roof made of the same material as the walls. This method of reducing wind-induced noise was investigated by Abbott and Raspet (2015) for detecting infrasound in the presence of relatively high winds. It was found that the most important parameters in achieving significant noise reduction were the size and porosity of the wind fence. Best results were obtained when the wind fence enclosure was supplemented by a secondary wind screen. Noise reductions were found to be frequency dependent, with the best results for the study occurring at 3 Hz, where the noise reduction was between 25 and 27 dB. In general, the typical noise reduction in the frequency range of 0.1–20 Hz was 5 dB. Wind fence enclosures are not often used for wind turbine measurements because they are relatively large structures, which makes them difficult to transport and erect. They are therefore not usually a realistic option for compliance measurements, which must be made at a number of different locations.

Various other techniques have been developed for wind-induced noise reduction in the infrasonic frequency range, as several important applications require signals to be measured in this range. These applications generally involve measuring signals from large events over large distances, for example volcanoes, hurricanes and explosions. One such method is the spatial filtering technique, whereby an array of non-porous pipes is distributed over a large area and is focussed onto a central manifold at which the incident pressures are summed (Walker and Hedlin 2010). This technique also takes advantage of the fact that infrasound wavefronts propagating from wind turbines remain well-correlated over several metres, while wind noise becomes uncorrelated over a few metres (Annan et al. 2015). The most standard design of spatial filter is the rosette filter, which consists of several clusters, or rosettes, of low-impedance inlets and was first proposed by Alcoverro (1998). These rosette filters can provide wind noise reduction of the order of 15–20 dB (Walker and Hedlin 2010). Use of these filters for wind farm noise measurement is limited by their sensitivity to grazing angle of the incident acoustic wave, resonance, complexity and cost. A comprehensive evaluation of rosette filters was carried out by Hedlin et al. (1981) and other more recent technologies for reducing wind-induced noise to improve signal-to-noise ratios for infrasonic measurements were surveyed by Walker and Hedlin (2010).

Insertion Loss of Wind Screens

When using custom-made wind screens for microphones, it is important to measure the insertion loss, which is the amount by which sound is attenuated as it passes through the wind screen to the microphone. In general, the insertion loss is not likely to be an issue for measurement of noise at low frequencies, provided a porous wind screen is used. Insertion loss can be measured in an anechoic chamber by using a speaker and positioning the microphone directly in front and as far away as possible to minimise the effects of wind-induced noise caused by movement of the speaker diaphragm. To achieve maximum sound pressure levels at low frequencies, it is useful to generate noise in each 1/3-octave band separately; to ensure that equal energy is produced in each 1-Hz frequency bin, white noise is used. It is also possible to use pink noise (equal energy per 1/3-octave band) and generate noise across the entire frequency spectrum of interest as specified in IEC 61400-11 (2012).

The noise produced by the speaker is measured using a microphone with a primary wind screen (that is, a standard 90 mm one) and a microphone protected by the custom-made wind screen as well as the primary wind screen. The microphone should remain in the same position for both measurements and optimally the position of the microphone and speaker should be interchanged as well, to verify the results.

Insertion loss can also be measured outdoors provided extraneous noise sources are negligible and the wind is light. A method for measuring insertion loss outdoors is discussed in Annex E of IEC 61400-11 (2012). According to this method, a speaker is mounted at a height of 4 m and is used to generate pink noise (see Section 2.3). The test microphone is placed on a measurement board on the ground (zero height) at horizontal distances from the speaker of 4.8, 6 and 7.2 m. An additional 'control' microphone is placed alongside the test microphone and protected using a secondary wind screen. The level difference between the wind screen 'on' case and the wind screen 'off' case is found and then normalised by the difference between the measurements taken with the control microphone for each of the two test cases. This ensures that variations in ambient noise level do not effect the results. Some other specifications are listed below.

- All measurements are made in 1/3-octave bands.
- The background noise in each 1/3-octave band must be at least 3 dB below the noise measured with the loudspeaker on and all valid measurements must be corrected to account for the contribution of background noise. This is done by logarithmically subtracting the background noise from the total measured noise (see Section 2.2.9).

- Measurements are 1–2 min long.
- If background noise has influenced measurements at 1/3-octave frequencies below 100 Hz then the insertion loss can be assumed equal to the 125-Hz 1/3-octave band insertion loss.

Insertion loss measurements were undertaken for a large number of commercially available wind screens by Kaufmann and Kock (2015), and it was found that the insertion loss and standard deviation varied between designs, particularly at high frequencies. The implication of this result is that measured turbine sound power levels could vary by as much as 1 dB if this were not taken into account. Other factors such as water contamination, poor centring of the primary wind screen and using an octopus strap to secure the wind screen did *not* have a noticeable affect on the sound power measurement. Errors could also result in noise level measurements at sensitive receiver locations, but as the insertion loss is usually negligible at low frequencies, this will only cause errors in the overall A-weighted noise levels by a similar amount to that obtained for turbine sound power level measurements (±1 dB).

6.2.4 Microphone Height

The most common height chosen for the outdoor measurement microphone is 1.5 m, as this corresponds approximately to the ear height of an average receiver. The main disadvantage of measuring at this height is that wind-induced noise can affect the results, particularly at low frequencies. As was discussed in Section 6.2.3, wind-induced noise caused by wind interaction with the microphone can be reduced by measuring closer to the ground, and this explains the use of alternative mounting arrangements such as the 1-m diameter ground board described in IEC 61400-11 (2012) and the underground box developed by Betke et al. (1996).

On the other hand, results obtained from any height that does not represent the human receiver should be interpreted with caution, particularly at low frequencies when large propagation distances are involved. The reason for this is that the difference in noise level at various heights above the ground is a function of the frequency, ground-surface properties, distance from the source and the height of the source. Some examples are given by Salomons (2001, Ch. 4). Therefore, it is not sufficient to assume that the noise level at ground level and at 1.5 m height is the same for low frequencies with corresponding large wavelengths. Also, the noise-level difference between these heights at high frequencies may not be exactly 3 dB over large propagation distances. Further research in this area is necessary before measurement at ground level or underground becomes standardised. Another issue to consider in the case of the underground box is that there will be resonant acoustic modes in the space inside the box, the resonance frequencies of which are governed by the box dimensions. It is therefore important to calculate the resonance frequencies of these modes using Eq. (2.35) and to ensure that the associated range of frequencies are not part of the analysis. Also, it can be time-consuming to dig a hole for the box, particularly during summer months in hot countries.

6.2.5 Ambient or Background Noise Assessment

Ambient or background noise levels are determined in the planning phase of a proposed wind farm development so that the relative contribution of wind turbine noise can be determined once the wind farm is operational, as discussed in Section 6.2.14. Background noise measurements should be undertaken at relevant receivers, which are residences that are most likely to be impacted by the proposed development. Landowners who have entered into an agreement with the wind farm developer are classified as non-relevant receivers. Special consideration should be reserved for receivers that are sheltered by terrain when they are downwind from the wind farm because the local wind-induced noise can be relatively low at such locations, whilst the noise generated by the wind farm can be simultaneously high.

Prior to the commencement of monitoring, the wind direction(s) that will result in maximum noise impact at the relevant receiver and possible seasonal variations in wind direction should be considered, to ensure that worst-case conditions are included in the background measurements. Seasonal variations in background noise also need to be taken into account to ensure that the measurements are representative. These variations can be caused by changes in the local environment such as varying flow in rivers and streams, the presence or absence of leaves on trees and season-specific bird, insect and animal activity. Noise generated by human activity is also season-dependent, particularly in regions that are used predominantly for farming. During some periods of the year, such as harvesting, farming activity may create more noise than usual and therefore such periods should not be selected for background measurements.

The possibility of extraneous noise sources affecting the data should be considered according to AS 4959 (2010), and measurements contaminated by extraneous noise should be removed from the data set according to NZS 6808 (2010). However, this becomes a subjective exercise, as the assessor must make a distinction between which sources produce the characteristic background noise for the area and which sources are extraneous (or intrusive). The ETSU (1996) guidelines consider that only data affected by rainfall can be excluded from the analysis as other data are representative of the background noise environment. However, these guidelines also specify days and times corresponding to quiet periods for collection of background data. Nighttime and daytime data should also be separated in the analysis, as it is expected that the background noise level is generally lower during the night.

Measurement

Monitoring of background noise involves measuring sequential 10-min averages of the noise level exceeded for 90% of the time, $L_{A90,10\mathrm{min}}$. The wind speed at a specified reference height is measured simultaneously. The measurement period of 10 min is believed to give the best correlation between background noise and wind speed, according to the ETSU guidelines (ETSU 1996). Use of the L_{A90} metric minimises the influence of transient noise events such as wind gusts, dogs barking, aircraft flyovers and so on, which are not constant sources in the background noise environment. Nevertheless, the ambient noise varies significantly over time and therefore most standards and guidelines recommend collecting at least 2000 data points (AS 4959 2010) so that a regression analysis can be carried out. This involves plotting the measured sound pressure level against the wind speed at a specified reference point and finding the line of best fit through the data. The reference wind speed is generally the wind speed at hub height (AS 4959 2010; NZS 6808 2010; SAEPA 2009), although the ETSU guidelines specify a measurement height of 10 m (ETSU 1996). It is also possible to measure the wind speed and direction at a different height and to extrapolate the data to hub height by using wind shear data for the same measurement periods. The Australian standard, AS 4959 (2010), specifies that this alternative measurement height must not be less than half of the hub height. The hub-height wind speed is measured at a location on the wind farm site where maximum wind speeds are expected to occur, to ensure that predicted noise levels are conservative. If the wind speed at a particular wind turbine were significantly higher than that at the measurement location, it could lead to difficulties in demonstrating compliance because the wind turbine may emit higher noise levels than expected.

The selected measurement location should also be unaffected by wind farm operation. The reason for the above requirements is that the same location is used for background noise measurements, noise predictions and compliance checking and therefore it must satisfy the requirements for each of these procedures. Although not a requirement of the standards (AS 4959 2010; NZS 6808 2010), it is also recommended that the wind speed at the residence at heights of 1.5 m and 10 m above the ground should be measured. It is important to quantify the local wind speed in the vicinity of the microphone to ensure that measurements that are taken during periods of excessive wind-induced noise can be identified and possibly removed.

Typically, the background noise level is measured at a height of 1.2–1.5 m above ground level, with a single microphone protected by a 90-mm wind screen. The microphone is placed away from reflecting surfaces (other than the ground) to avoid interference between the reflected and direct sound rays, which can lead to an increase in the measured noise levels. The microphone should be located close to the residence to ensure that the measured noise levels are representative, but it is more important to ensure that the line of sight between the microphone and wind farm is not obstructed by any structures, foliage or other obstacles. The microphone should also be placed as far as possible from any extraneous sources of noise that could contaminate the measurements, such as pumps, air-conditioning units, foliage, electrical transformers or loose objects that rattle in the wind. Standards and guidelines (AS 4959 2010; ETSU 1996; NZS 6808 2010) typically specify 20 m as the maximum distance that the microphone can be placed from the receiver, although this is unnecessarily restrictive in some cases. Locating the microphone at greater distances from a residence will generally result in less error in the measurements than selecting a location within the specified range where the measurements will be influenced by obstructions and/or extraneous noise, as discussed above.

The wind speed at microphone height is measured simultaneously with the sound pressure level in 10-min periods using a weather station that is placed in the vicinity of the microphone with adequate spacing to minimise additional weather station noise (mainly rotating cup noise) at the microphone. For A-weighted measurements using a microphone protected with a 90-mm wind screen, data measured during local wind speeds greater than 5 m/s should be discarded. Higher wind speeds may be permissible using specialised wind screens, depending on the manufacturer's specifications. Information about the local wind speed also provides the opportunity for subtracting the background level associated with a particular local wind speed from wind turbine noise measurements obtained at a later date, as discussed in Section 6.2.14. The weather station should also possess the capability of measuring rainfall so that this information can be used to identify data to be removed from the analysis.

The collected data should be representative of a large range of wind speeds and directions corresponding to the operating conditions of the wind farm. Obtaining such an extensive data set requires continuous monitoring over several weeks and therefore it is important that the instrumentation provides sufficient memory to store data for the entire measurement period. Data-storage requirements can be minimised by measuring the $L_{A90,10\text{min}}$ exclusively, as this quantity is most representative of the actual ambient conditions according to several standards and guidelines (AS 4959 2010; NZS 6808 2010; SAEPA 2009). On the other hand, measuring raw time data assists in the removal of data contaminated by extraneous noise. Some instrumentation provides the option of specifying a threshold level above which a trigger is activated to commence measurement for a user-defined period of time.

The noise floor of the instrumentation is a critical factor for background noise measurements, because the ambient noise levels in rural areas can be extremely low, particularly at nighttime. Low ambient noise levels are also associated with periods of minimal wind and therefore, if the noise floor of the instrumentation introduces a lower limit to the data, the regression line will not be accurate, as shown in Figure 6.2. In the left-hand figure, the instrumentation noise floor of 17 dBA is shown to be sufficient for these measurements. The right-hand figure shows the effect of introducing an artificial noise floor of 25 dBA for the purposes of demonstration. In this case, the instrumentation noise floor changes the position of the line of best fit in the regression analysis, which leads to higher values of $L_{A90,10\text{min}}$ as a function of wind speed at hub height. This would ultimately result in higher allowable limits of wind farm noise according to standards, such as NZS 6808 (2010), that specify a 'background + 5dB' limit at relatively high wind speeds where the base limit is exceeded. It can be seen that the coefficient of determination (R^2 value) also reduces as a result of the artificial noise floor, which indicates a poorer fit to the data. The significance of the R^2 value is discussed in more detail in Section 6.2.5.

The measurement set-up should be documented in detail, which includes taking photographs of the instrumentation from multiple directions and measuring the distance between two points of reference, and the microphone and weather station, to allow the same position to be used for compliance measurements. Photographs of the local surroundings will also allow monitoring of any changes that may occur between the background and compliance measurement periods, which could affect the measured noise levels.

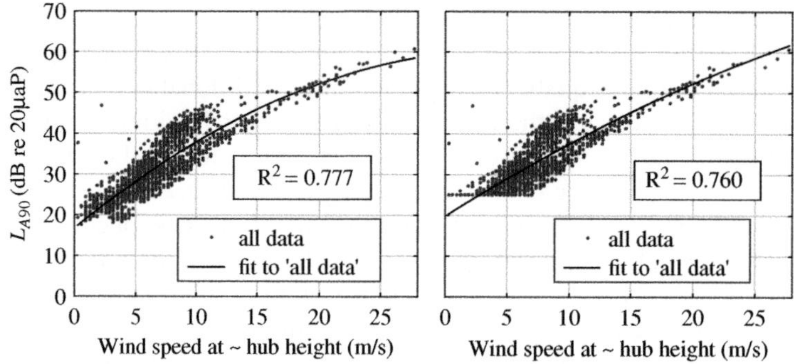

Figure 6.2 Regression analysis of $L_{A90,10\text{min}}$ against wind speed at hub height, derived using Eq. (5.17) from the measurements at 10-m height, showing acceptable data (left) and an issue with instrumentation noise floor (right). The linear regression fit is to all data in both figures.

Analysis

The level of background noise in quiet rural areas generally correlates well with the local wind speed, but the local wind speed may not correlate well with the hub-height wind speed, especially at times of high wind shear (wind speed increasing strongly with altitude), which often occurs in rural areas at night. However, in the analysis of background noise levels, it is common practice to ignore the lack of correlation that occurs at times between the local wind speed and the hub-height wind speed and to take a large number of measurements so that this effect is reduced to an acceptable level. The resulting $L_{A90,\,10\text{min}}$ data are then plotted against the wind speed at hub height. The $L_{A90,\,10\text{min}}$ metric is used because it minimises the influence of transient noise sources.

The specific function used to find the best fit to the data depends on the guidelines of the local regulatory authority. The SAEPA (2009) guidelines specify the use of a polynomial function from linear up to third order, which is consistent with AS 4959 (2010). On the other hand, a specific function is not referred to in NZS 6808 (2010), and the ETSU (1996) method recommends the use of a logarithmic function, although examples with polynomial fits are shown. The accuracy of the curve fit can be quantified by calculating the coefficient of determination R^2, which indicates the degree to which the regression plot, f_i, fits the data, y_i. The general equation used to determine the R^2 value is given by:

$$R^2 = 1 - \frac{\sum_{i=1}^{n} (y_i - f_i)^2}{\sum_{i=1}^{n} (y_i - \overline{y})^2} \tag{6.3}$$

where, $\overline{y}$ is the mean value of data set $y_1, y_2, \ldots, y_n$. An R^2 value of 1 represents a perfect fit to the data whereas an R^2 value of 0 indicates that the line does not fit the data at all, and this usually occurs when the data are completely random. More information about regression analysis is provided in Draper and Smith (2014).

An acceptable magnitude for the R^2 value is not specified in any standards and guidelines, which is an unfortunate omission as it allows an element of subjectivity in the analysis. On the other hand, the noise level data are expected to increase with increasing wind speed and there should not be more than one distribution according to AS 4959 (2010). Where a regression relationship is not evident in the plot of background $L_{A90,\,10\text{min}}$ against wind speed at hub height, the New Zealand standard (NZS 6808 2010) suggests that separate scatter plots can be generated for different times of day and/or wind direction. The degree of correlation can also be improved by changing the wind speed range used in the analysis and/or removing outliers. In cases where the correlation remains poor, regardless of data separation, it may be necessary to gain more insight into the wind conditions. The New Zealand standard (NZS 6808 2010) suggests the use of wind-flow modelling, site observations, local knowledge and local wind monitoring to investigate instances of poor correlation. This information enables differences in wind conditions at hub height and at the measurement location to be identified.

Discussion and Implementation

The following analysis presents results from unattended noise measurements that were made at a residence located in the vicinity of a proposed wind farm development. Results are shown for a period of 12 days from 19 September to 1 October 2014, during which 1692 valid data points were collected. Noise was monitored using several outdoor microphones with various wind screen configurations to protect against wind-induced noise and there was good agreement between the data measured by the

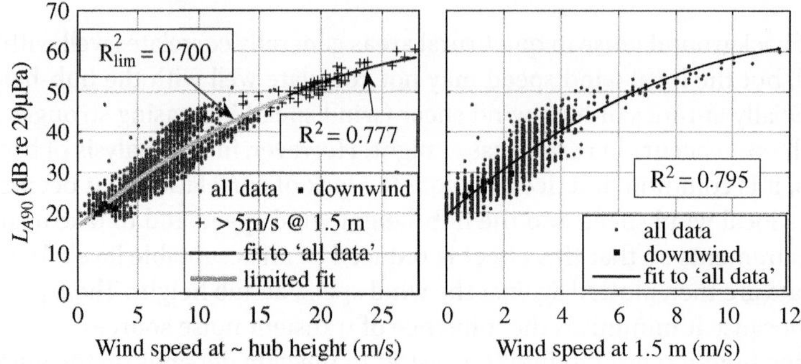

Figure 6.3 Regression analysis of $L_{A90,\,10\text{min}}$ against (left) hub-height wind speed derived from measurements at 10 m; (right) measured wind speed at 1.5 m (right). There are currently no wind turbines within 100 km of this location. The regression fit is to all data in both figures.

various microphones. The results presented in Figure 6.3 were recorded using a single G.R.A.S. 40AZ microphone with a SVANTEK SV 12L pre-amplifier connected to a SVANTEK 958 sound level meter. This microphone was mounted on the ground and protected from wind-induced noise by a hemispherical wind screen designed to be consistent with the IEC 61-400 standard (IEC 61400-11 2012).

Weather data were collected in close proximity to the outdoor microphones using Davis weather stations mounted at 1.5 and 10 m. Actual hub-height wind speed data were not available for these measurements and therefore the wind speed at 10 m has been extrapolated up to the proposed hub height of 95 m using Eq. (5.17). Note that the wind speed at hub height calculated using this equation may not be a very good estimate, especially during periods of high wind shear. A number of scatter plots are shown to illustrate the various possibilities for data separation in background regression analyses. The R^2 value is shown on each plot where applicable, and all regression fits are quadratic for consistency. In each figure, the regression fit has been applied over all wind speeds in the range over which data exists, since the 'cut-in' and 'rated' wind speed specifications of the proposed wind turbines are not readily available.

A comparison between the $L_{A90,\,10\text{min}}$ plotted against the derived (using Eq. (5.17)) and measured wind speeds at hub height and 1.5 m, respectively is shown in Figure 6.3. Data that were measured during downwind conditions (as defined by the proposed wind farm site) are indicated by solid square symbols and it is evident that during the measurement period, the residence was 'downwind' from the nearest proposed turbine in only four instances. On many occasions, strong winds greater than 5 m/s occurred at the measurement site, but due to the ground-mounted, secondary wind screen configuration, the corresponding data points fit well to the regression curve and even improve the R^2 value, as shown in the left-hand plot of Figure 6.3. Therefore, data points measured during wind speeds greater than 5 m/s are included in the remaining analysis of data from this residence. Comparison between the left- and right-hand plots reveals that a slightly better regression fit, indicated by a higher R^2 value, is obtained by plotting the $L_{A90,\,10\text{min}}$ against the wind speed at a height of 1.5 m than by plotting the $L_{A90,\,10\text{min}}$ against the derived wind speed at hub height. The reason for this is that the wind speed measured at a height of 1.5 m is more representative of the local wind conditions at the microphone, which leads to improved correlation between the wind-induced noise and the measured noise level. If actual hub-height data were available, it would be expected that the R^2 value in the associated plot would be lower still, due to wind-shear variations over the measurement period.

Alternative methods for determining the relationship between the $L_{A90,10\text{min}}$ and the wind speed at hub height are shown in Figure 6.4. The left-hand plot shows solid black circles that correspond to arithmetic averaging in 1-m/s wind-speed bins (centred on integer wind speeds minus 0.5 m/s), so that the first bin spans 0–1 m/s. The standard deviation for each data point is indicated by the error bars. The right-hand figure includes large black data points that lie at the mid-point of linear regression curves (multiple short grey lines) that are determined for each 1-m/s wind-speed bin. The individual data points used to derive what is shown on the figures have been omitted for clarity. Further details of these methods are discussed by Smith and Chiles (2012). The largest difference between the arithmetic average and regression curve is 3–4 dB and this occurs at very low wind speeds and very high wind speeds. Similar characteristics are observed for the 1-m/s bin regression curves. These extreme ranges of wind velocity are less critical than the mid-range of wind speeds where wind farm noise is generally highest. Other discrepancies between the arithmetic averaging, 1-m/s regression bins and standard regression analysis occur in the wind speed range between 12 and 16 m/s where there are fewer data points. In this region, the slope of the 1-m/s regression curves varies significantly from one bin to the next, even becoming negative in some instances. This does not reflect the expected correlation between wind speed and noise level. The standard regression curve shows the most consistent relationship between the derived wind speed at hub height and noise level and therefore, in this analysis, the standard regression method appears to be more robust in the face of sparse data.

In some instances, it is problematic to fit a regression curve across a wide wind speed range, since three specific regions can be observed in the regression plot.

- At low wind speeds the noise level is independent of the wind speed.
- At medium wind speeds, the noise level increases as a function of wind speed.
- At higher wind speeds, the rate of increase in noise level with wind speed is reduced.

Therefore, since the medium wind speed range is often the most critical, it has been suggested that the regression fit should include wind speeds from cut-in to 95% of rated

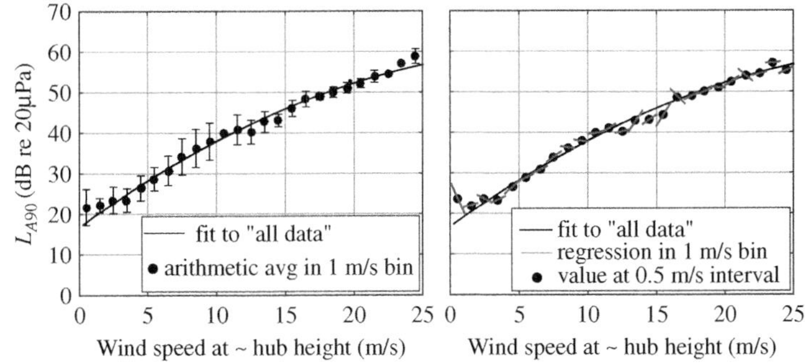

Figure 6.4 Arithmetic averaging (left) and linear regression analysis in bins of width 1 m/s (right) for $L_{A90,10\text{min}}$ against wind speed at hub height, derived from the measurements at 10 m height, are shown as possible alternatives to a standard regression fit. The analyses have been carried out on all data. Data were measured at the same location as for Figure 6.3.

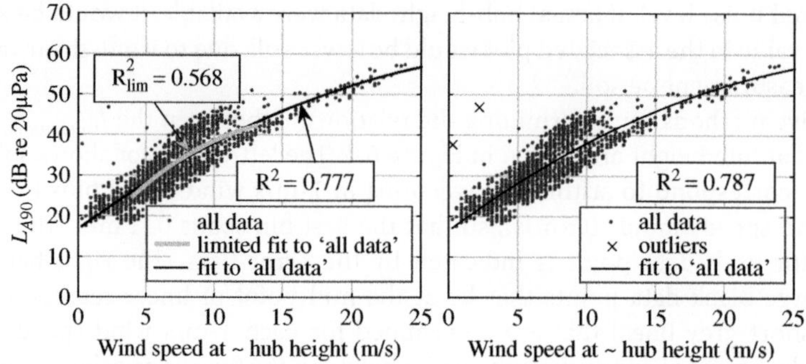

Figure 6.5 Effect of limiting the velocity range (left) and removing outliers (right) on the regression analysis for $L_{A90,10min}$ against wind speed at hub height, derived from the measurements at 10 m height. There is no discernable difference between the regression fits obtained by including and excluding outliers from the analysis. Data were measured at the same location as for Figure 6.3.

power (Smith and Chiles 2012). For the data presented here, the R^2 value significantly decreased for the regression fit over this narrow wind speed range and therefore this method does not appear to offer any advantages in this case. These results are presented in the left-hand plot of Figure 6.5, where it can be seen that the regression fit plotted over this limited velocity range is almost the same as the original that was plotted over the entire operating wind speed range. In the right-hand plot of this figure, the effect of eliminating outliers from the regression analysis is investigated. It can be seen that the R^2 value has increased as expected, however, there is negligible effect on the regression fit itself.

6.2.6 A- and C-weighted Levels

Measurement

An important point to consider when measuring unweighted sound pressure levels with a sound level meter is that the noise levels can be dominated by low frequencies to such an extent that the A-weighted levels, which are dominated by higher frequencies, cannot be determined accurately by post-processing the unweighed data. Therefore it is recommended to always record the A-weighted values as well as unweighted values during the measurement program. Since the C-weighting applies a more constant weighting across the frequency range, a similar problem will not be encountered when the C-weighted sound pressure level is measured.

Analysis

The A- and C-weighting filters are the most common filters required for acoustic applications relating to wind farm noise, so they will be discussed in more detail here. The G-weighting filter will be discussed in more detail in Section 6.2.7. Filter response curves and tabulated correction values are provided in Section 2.2.11. According to IEC 61672-1 (2013), the A- and C-weighting filters as a function of frequency, A(f) and C(f) are calculated in decibels according to Eqs. (6.4) and (6.5).

$$A(f) = 20\log_{10}\left[\frac{f_4^2 f^4}{(f^2+f_1^2)(f^2+f_2^2)^{1/2}(f^2+f_3^2)^{1/2}(f^2+f_4^2)}\right] - A_{1000} \tag{6.4}$$

$$C(f) = 20\log_{10}\left[\frac{f_4^2 f^2}{(f^2+f_1^2)(f^2+f_4^2)}\right] - C_{1000} \tag{6.5}$$

where approximate values for frequencies f_1 to f_4 in Eqs. (6.4) and (6.5) are:

- $f_1 = 20.60$ Hz
- $f_2 = 107.7$ Hz
- $f_3 = 737.9$ Hz
- $f_4 = 12\ 194$ Hz.

A_{1000} and C_{1000} are normalisation constants, in decibels, representing the electrical gain needed to provide frequency weightings of zero decibels at 1 kHz. The values of the normalisation constants, A_{1000} and C_{1000}, rounded to the nearest 0.001 dB, are − 2.000 dB and − 0.062 dB, respectively.

While it is possible to apply the A- and C-weighting to 1/3-octave and octave levels in the frequency domain, it is more accurate to apply these filters in the time domain. This is a requirement for Class 1 and 2 sound level meters according to IEC 61672-1 (2013). The design criteria for both the frequency-domain response (amplitude and phase) and time-domain response of A- and C-weighting filters used in measurements are given in ANSI/ASA S1.42 (2006). Therefore, application of the A- and C-weighting in the context of wind farm noise measurement is only relevant where unweighted time series data are to be analysed.

Since most filter theory was developed in the analogue era, it is often necessary to create a digital filter that best approximates the analogue filter characteristics; there is no exact transformation (Brandt 2011). An important consideration is that the digital approximation to the analogue filter becomes poorer as the Nyquist frequency (see Section 2.3.1) is approached and therefore it is recommended to use 10-times oversampling of the highest octave band centre frequency in the frequency range of interest (Brandt 2011). Implementation of a digital filter requires knowledge of the locations of poles and zeroes in the Laplace or s-plane and this information is presented in Table 1 of ANSI/ASA S1.42 (2006). The A- and C-weighting filters can be expressed in the s-plane as follows (Rimmel et al. 2015):

$$A(s) = G_A \frac{\omega_4^2 s^4}{(s+\omega_1)^2(s+\omega_2)(s+\omega_3)(s+\omega_4)^2} \tag{6.6}$$

$$C(s) = G_C \frac{\omega_4^2 s^2}{(s+\omega_1)^2(s+\omega_4)^2} \tag{6.7}$$

where $\omega_n = 2\pi f_n$, $n = 1,2,3,4$ and $G_A = 10^{(-A1000/20))}$ and $G_C = 10^{(-C1000/20))}$. Values for these frequencies and constants are given in the text following Eq. (6.5).

Equations (6.6) and (6.7) then need to be converted from analogue (s-domain) to digital (z-domain) as discussed by Rimmel et al. (2015). The recommended method for converting to the digital domain is to use a bilinear transform (Brandt 2011; Rimmel et al. 2015) and this can be easily implemented using MATLAB®. Codes for calculating the filter coefficients, *a* and *b*, of a digital infinite impulse response (IIR) filter (see Section 2.3.1) for A- and C-weighting filters are provided in the 'Octave' package provided by Christophe Couvreur on MATLAB® Central. For convenience, these codes are given below, and the values of f_1, f_2, f_3, f_4, A_{1000} and C_{1000} are consistent with those described above. F_s is the sampling frequency, which should be chosen to be ten times the highest octave band centre frequency, as described above. The frequency response can be calculated from the filter coefficients *a* and *b* and then plotted according to the last few lines of the A-weighting filter code.

```
% A-weighting filter
% Sampling Rate
Fs = 44100;
% Pole frequencies
f1 = 20.598997; f2 = 107.65265; f3 = 737.86223;
f4 = 12194.217;
% Normalisation constant, chosen so weighted level is 0 dB
at 1000 Hz
A1000 = 1.9997;
% Finding numerator and denominator transfer function coef-
ficients
NUM = [ (2*pi*f4)^2*(10^(A1000/20)) 0 0 0 0 ];
DEN = conv([1 4*pi*f4 (2*pi*f4)^2],
[1 4*pi*f1 (2*pi*f1)^2]);
DEN = conv(conv(DEN,[1 2*pi*f3]),[1 2*pi*f2]);
% Bilinear transformation of analogue design to get the dig-
ital filter.
[b,a] = bilinear(NUM,DEN,Fs);
% Plot frequency response with log frequency scale
f = 1:20000;
figure
freqz(b,a,f,Fs)
ax = findall(gcf, 'Type', 'axes');
set(ax, 'XScale', 'log');

% C-weighting filter
% Sampling Rate
Fs = 44100;
% Pole frequencies
f1 = 20.598997; f4 = 12194.217;
% Normalisation constant, chosen so weighted level is 0 dB
at 1000 Hz
C1000 = 0.0619;
% Finding numerator and denominator transfer function coef-
ficients
NUM = [ (2*pi*f4)^2*(10^(C1000/20)) 0 0 ];
DEN = conv([1 +4*pi*f4 (2*pi*f4)^2],
[1 +4*pi*f1 (2*pi*f1)^2]);
% Bilinear transformation of analogue design to get the dig-
ital filter
[b,a] = bilinear(NUM,DEN,Fs);
```

6.2.7 Infrasound and Low-frequency Noise

Measurement

The measurement of infrasound and low-frequency noise requires the use of specialised microphones, pre-amplifiers and sound recording devices that have a flat frequency response down to at least 0.5 Hz, as discussed in Section 6.2.1. Outdoor measurements also require the use of specialised wind screen designs to minimise the effects of wind-induced noise, which is most problematic at low frequencies. This is discussed in more detail in Sections 6.2.2 and 6.2.3.

An alternative device for measuring noise at infrasonic and low frequencies is an *infrasound sensor*, which is also referred to as a *microbarometer* and has been described by Woolworth et al. (2013). Despite its name, this sensor can have a flat frequency response to within 1% in the frequency range from 0.02 to 150 Hz. The sensor system described by Woolworth et al. (2013) incorporates an analogue sensor, 24-bit digitiser, GPS system for accurate timing and a computer with a LINUX operating system that can be accessed wirelessly. The sensor is rugged, with low power consumption, allowing it to be used over extended periods in all weather conditions. At present, infrasound sensors are located worldwide and can be used to detect a range of phenomena related to environmental factors and human activity. Through use of beamforming, infrasound sensors placed along the US coastline were able to detect Hurricane Katia (Woolworth et al. 2013). Hence, it is entirely plausible that an array of these sensors can be used to localise infrasonic and low-frequency noise sources for both a wind turbine and an entire wind farm. An array shape for such a purpose was proposed by Woolworth et al. (2013).

The infrasound monitoring system described by Annan et al. (2015) consists of infrasound sensors, a wind-filtering apparatus, data-logging equipment and signal processing capabilities. It was demonstrated that this system performed well in isolating wind turbine noise from other sources of infrasound such as marine storms, aeroplanes, urban noise and wind noise. Through use of four infrasound sensors arranged in a square topology with 50-m spacing, the direction of arrival of infrasound signals could be monitored and it was found that this correlated well with the direction of the wind farm.

A range of infrasound sensors (such as Infiltec) are commercially available and in cases where the infrasonic and low-frequency range is of prime importance (such as at large distances from a wind farm), they can prove to be more cost effective than a microphone. Other advantages of these sensors are their low power consumption and ruggedness, as discussed above. The main disadvantage is that they can only be used to measure noise at low and infrasonic frequencies.

Analysis

To assess the annoyance and direct perception of infrasound, the G-weighting has been developed, as discussed in Section 2.2.11. The G-weighting can be applied by a sound level meter at the time of measurement or it can be applied during post-processing in either the frequency or time domain. As mentioned in Section 6.2.6, it is more accurate to apply the filter in the time domain and a digital filter can be designed for this purpose. ISO 7196 (1995) specifies the poles and zeros of the G-weighting filter and defines a gain of 0 dB at 10 Hz. This information can be used to devise a MATLAB® code that calculates the filter coefficients for the G-weighting and an example code is provided below. It should be noted that the original data file may need to be downsampled using the **decimate** function in MATLAB® before the G-weighting filter can be used.

```
% G-weighting filter
% Sampling Rate
Fs = 1000;
% Poles and zeroes according to ISO 7196
z = [0+0*1i; 0+0*1i; 0+0*1i; 0+0*1i];
p = [2*pi*(-0.707 + 0.707*1i); 2*pi*(-0.707 - 0.707*1i);...
     2*pi*(-19.27 + 5.16*1i); 2*pi*(-19.27 - 5.16*1i);...
     2*pi*(-14.11 + 14.11*1i); 2*pi*(-14.11 - 14.11*1i);...
     2*pi*(-5.16 + 19.27*1i); 2*pi*(-5.16 - 19.27*1i)];
% Normalisation constant, chosen so weighted level is 0 dB
at 10 Hz
k = 9.825e8;
% Zero-pole to continuous transfer function conversion
[bc,ac] = zp2tf(z,p,k);
% Plot in analogue domain to check frequency response
f = 1:200;
h = bodeplot(tf(bc,ac),f); grid on
setoptions(h,'FreqUnits','Hz','PhaseVisible','off');
% Bilinear transformation of analogue design to get the dig-
ital filter
[b,a] = bilinear(bc,ac,Fs);
% Plot frequency response with log frequency scale
figure
freqz(b,a,f,Fs)
ax = findall(gcf, 'Type', 'axes');
set(ax, 'XScale', 'log');
```

To visualise the infrasonic and low-frequency components of wind turbine noise, it is useful to generate a narrowband plot, as discussed in Section 6.2.13 and shown in Figure 6.26. On the other hand, Bray and James (2011) argue that the sound pressure levels in the narrowband plot do not reflect the true sensation perceived by the human ear. The reason for this is that the frequency resolution required to distinguish the blade-pass frequency and harmonics is very high and, conversely, the time resolution is very low (see Section 2.4.8). So to identify high crest-factor short-term events at low and infrasonic frequencies, it is recommended that analyses be done using frequency bandwidths equal to the widths of the critical bands that characterise the ear.

Critical bands are employed in psychoacoustics to describe hearing sensations (Fastl and Zwicker 2007) and are referred to in DIN 45631/A1 (2010) for the assessment of time-varying loudness. The critical bandwidth of frequencies below about 500 Hz is 100 Hz, so the lowest possible band spans the range from 0 to 100 Hz (Bark 0.5). Above 500 Hz, the bandwidth is about 20% of the centre frequency, f_c. The critical bands are defined in a similar way to 1/3-octave bands, but they are interpreted slightly differently. Rather than being considered as fixed, the critical bands can be thought of as varying along the Bark 0.5 to Bark 24 limits (Fastl and Zwicker 2007), but defined entirely by an arbitrarily specified centre frequency and a corresponding bandwidth, which is a

Table 6.1 Critical frequency bands corresponding to integer and half-integer Bark values.

Bark	f_c	f_ℓ	f_u	Δf	Bark	f_c	f_ℓ	f_u	Δf
0.5	50	0	100	100	12.5	1850	1720	2000	280
1.0	100	50	150	100	13.0	2000	1850	2150	300
1.5	150	100	200	100	13.5	2150	2000	2320	320
2.0	200	150	250	100	14.0	2320	2150	2500	350
2.5	250	200	300	100	14.5	2500	2320	2700	380
3.0	300	250	350	100	15.0	2700	2500	2900	400
3.5	350	300	400	100	15.5	2900	2700	3150	450
4.0	400	350	450	100	16.0	3150	2900	3400	500
4.5	450	400	510	110	16.5	3400	3150	3700	550
5.0	510	450	570	120	17.0	3700	3400	4000	600
5.5	570	510	630	120	17.5	4000	3700	4400	700
6.0	630	570	700	130	18.0	4400	4000	4800	800
6.5	700	630	770	140	18.5	4800	4400	5300	900
7.0	770	700	840	140	19.0	5300	4800	5800	1000
7.5	840	770	920	150	19.5	5800	5300	6400	1100
8.0	920	840	1000	160	20.0	6400	5800	7000	1200
8.5	1000	920	1080	160	20.5	7000	6400	7700	1300
9.0	1080	1000	1170	170	21.0	7700	7000	8500	1500
9.5	1170	1080	1270	190	21.5	8500	7700	9500	1800
10.0	1270	1170	1370	200	22.0	9500	8500	10 500	2000
10.5	1370	1270	1480	210	22.5	10 500	9500	12 000	2500
11.0	1480	1370	1600	230	23.0	12 000	10 500	13 500	3000
11.5	1600	1480	1720	240	23.5	13 500	12 000	15 500	3500
12.0	1720	1600	1850	250	24.0	15 500	13 500	17 500	4000

function of frequency, and given by Eq. (6.8) and shown in Figure 6.23 (IEC 61400-11 2012). Thus, any frequency band can be defined as a fractional Bark, with its Bark value dependent on the chosen centre frequency, f_c. The bands corresponding to integer and half integer Bark values between Bark 0.5 and Bark 24 (the upper limit) are listed in Table 6.1.

Unlike 1/3-octave bands, which have specified centre frequencies and bandwidths, critical bands are defined by choosing the centre frequency, f_c, to be the frequency of interest, and once this has been specified, the critical bandwidth is defined using Eq. (6.8) and Figure 6.23 (IEC 61400-11 2012)).

$$\text{Critical bandwidth} = 25 + 75\left(1 + 1.4\left[\frac{f_c}{1000}\right]^2\right)^{0.69} \tag{6.8}$$

As discussed in Section 2.3.3, the rise time of a fourth-order 'Bark 0.5' Butterworth band-pass filter centred on 50 Hz is approximately 8.8 ms. The use of critical bands rather than 1/3-octave bands for analysis thus results in more accurate assessment of the potential for annoyance of high-crest-factor short-term events that are associated with wind farm noise.

The short-duration, high-peak infrasonic noise levels associated with wind farm noise can be best understood by representing the wind farm noise in this frequency range as a series of single-frequency cosine waves with frequencies corresponding to the blade-pass frequency and its harmonics. The relative amplitudes of the cosine waves can be defined by a sinc ($\sin(f)/f$) function (where f is the cosine wave frequency) to approximate the characteristics of typical infrasonic wind turbine noise spectra (Kelley et al. 1985). Although the measurements reported by Kelley et al. (1985) were for a downwind turbine, our own measurements have confirmed the presence of a similar type of impulsive noise for upwind turbines, albeit at a lower level. The purpose of the following analysis is to show that the amplitudes of the total sound pressure levels obtained by combining the fundamental and higher harmonic contributions are dependent on the relative phase between the various harmonics. It will be shown that the sound pressure levels of the individual harmonics are not necessarily indicative of the maximum sound pressure levels that may exist when the harmonics are combined with a particular phase relationship.

For the purposes of the analysis and based on our measurements 3.3 km from a wind farm, it is convenient as well as a good approximation to represent the amplitudes of the blade-pass frequency (in this case, 1 Hz) and its harmonics as a sinc function in the frequency domain, as shown in the right-hand part of Figure 6.6. The amplitude of the time-domain signal obtained by combining the various harmonics in a phase-correlated way is indicated by the solid circles on the left-hand part of Figure 6.6. The amplitude obtained by combining the harmonics in a random way is indicated by the open circles. Both the phase-correlated and random-phase combined signals shown in the left-hand part of Figure 6.6 produce the same spectrum in the frequency domain, shown in the right-hand part of the figure. However, in the time domain, the peaks are much higher for the phase-correlated signal than for the random signal, as indicated by the vertical difference between the open circle and solid circle symbols in Figures 6.6–6.9.

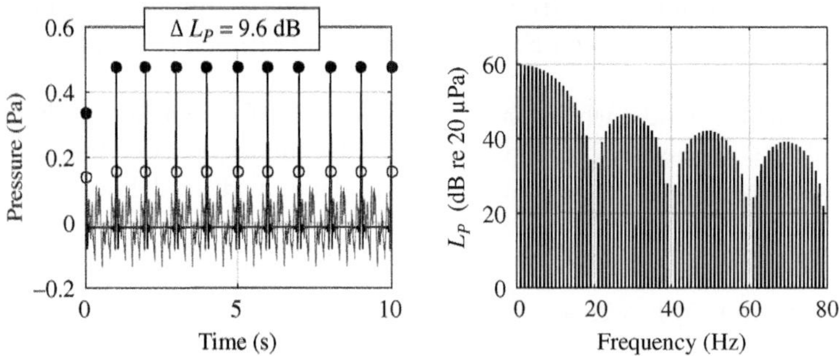

Figure 6.6 ($\sin(f)/f$) function in the frequency domain (right) used to approximate typical amplitudes of the blade-pass frequency (1 Hz) and harmonics up to 80 Hz. The time-domain equivalent (left) is obtained by adding the various harmonics in a phase-correlated way (black curve with peaks represented by solid black circles) or in a random way (grey curve with peaks represented by open circles). ΔL_p is the maximum difference between any two vertically aligned solid and open circles.

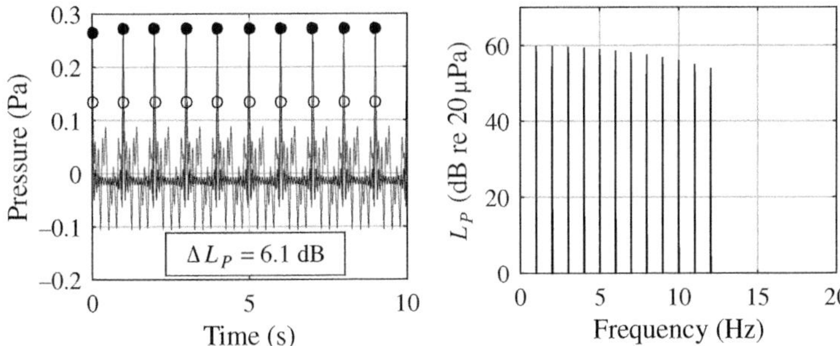

Figure 6.7 (sin(f)/f) function (right figure) used to approximate typical amplitudes of the blade-pass frequency (1 Hz) and the first twelve harmonics. The time-domain equivalent (left) is obtained by adding the various harmonics in a phase-correlated way (black curve with peaks represented by solid black circles) or in a random way (grey curves with peaks represented by open circles). ΔL_p is the maximum difference between any two vertically aligned solid and open circles.

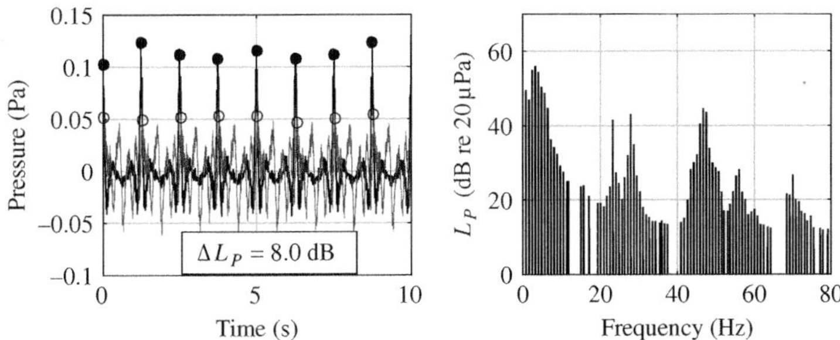

Figure 6.8 Measured levels used to approximate typical amplitudes of the blade-pass frequency 0.8 Hz and harmonics in the frequency domain (right) for frequencies up to 80 Hz. The time-domain equivalent (left) is obtained by adding the various harmonics in a phase-correlated way (black curve with peaks represented by solid black circles) or in a random way (grey curve with peaks represented by open circles). ΔL_p is the maximum difference between any two vertically aligned solid and open circles.

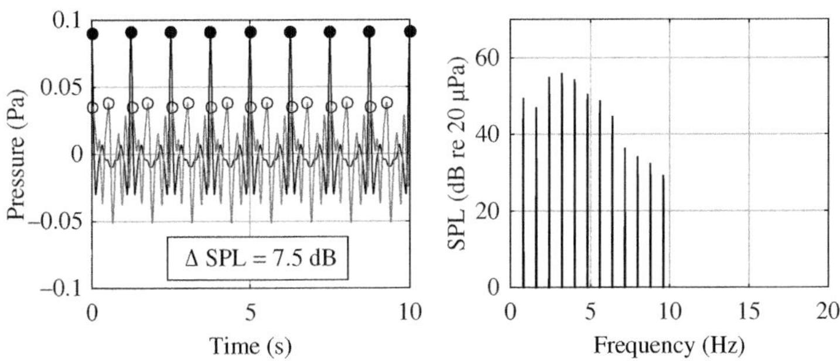

Figure 6.9 Measured levels used to approximate typical amplitudes of the blade-pass frequency and harmonics in the time-domain for the first twelve harmonics. The time-domain equivalent (left) is obtained by adding the various harmonics in a phase-correlated way (black curve with peaks represented by solid black circles) or in a random way (grey curve with peaks represented by open circles). ΔL_p is the maximum difference between any two vertically aligned solid and open circles.

The reason for this is that when the cosine waves are phase-correlated, they do not add randomly but rather linearly, with individual maxima coinciding.

Comparison between time-domain signals that include the first 80 harmonics and those that only include the first 12 harmonics, shows that the more harmonics that are included, the greater will be the difference between the maximum values in the time domain for the phase-correlated signal and the random signal. The maximum value of this difference is indicated on each figure by ΔL_p, which is the maximum difference in sound pressure level between any two vertically aligned solid and open circles in the figures.

6.2.8 Indoor Measurements

To adequately characterise the low-frequency wind farm noise experienced by a resident inside their dwelling, it is not sufficient to measure outdoor noise exclusively, for a number of reasons:

1. Measurement contamination from wind-induced noise occurs even during light winds at low frequencies, and this can be minimised by measuring indoors.
2. Outdoor measurements do not consider structure-borne noise as a result of the structure being excited by the external sound field and then re-radiating the energy indoors as sound.
3. Outdoor-to-indoor noise reduction values that are considered representative for A-weighted noise (Hurtley 2009) are not representative for noise dominated by low-frequencies.
4. Standing waves occur when the wavelength of sound is comparable to the room dimensions and this can result in higher indoor noise levels at some frequencies in specific locations (Hubbard and Shepherd 1991).
5. The housing structure selectively attenuates noise at mid and high frequencies, resulting in an inside spectrum that is more heavily weighted towards lower frequencies. This is more annoying than a well-balanced spectrum (Blazier Jr 1997).

Due to the presence of standing waves, the indoor noise level can vary by as much as 20–30 dB as a function of room location at low frequencies (Pedersen et al. 2007). Therefore, when measuring indoors it is important to select multiple locations and to ensure that the associated noise level is representative of the highest level experienced by the resident. This can be achieved by choosing measurement locations that have been indicated by the resident as problematic or, more conservatively, noise can be measured in room corners. The room corner is an anti-node for all room resonant acoustic modes and is therefore the location with the highest associated sound pressure level. However, at frequencies in the infrasonic range, well below the first room acoustic resonance, the sound field in the room will be relatively uniform and the variability will gradually increase as the resonance frequency of the first room acoustic mode is approached (usually between 20 and 30 Hz).

The Danish guideline, (DEPA 1997) for measuring indoor low-frequency noise and infrasound in rooms, specifies the use of at least three measurement locations in a room, the associated levels then being energy averaged (Jakobsen 2001). These locations include a corner, 0.5–1 m from the adjoining walls and 1–1.5 m above the floor, and two other locations that are at least 0.5 m from walls and large pieces of furniture and 1–1.5 m above the floor. Preferably these latter two locations are indicated by the occupant as regions in which the noise level is highest. If the occupant is not able to specify such a location, then the measuring points should be chosen based on experience, keeping in mind that midpoints in the room are to be avoided, since the noise level is lowest at such positions. In small rooms with area less than 20 m^2, the noise can be measured at two corner positions with the same specifications as mentioned above.

The Swedish procedure for measuring low-frequency noise and infrasound is explained by Pedersen et al. (2007). This method involves measuring in three room positions and calculating the energy average. However, these positions vary from the Danish method. One position is the room corner that has the highest associated C-weighted level and the other two positions are representative of normal room usage. The corner position at which the measurement should be made is identified by vertically scanning the vertical parts of the room-edge perimeter, between the heights of 0.5 m and 1.5 m, at a distance of 0.5 m from the walls. The other two positions are not closer than 0.5 m to the walls and the associated height must be either 0.6, 1.2 or 1.6 m. Positions around 1/4-, 2/4- and 3/4-fractions along the length and width of the room are avoided. The distance between all three positions should be greater than 1.5 m, although this is considered less important than selecting locations that are representative of actual room usage.

The performance of the Danish and Swedish methods was evaluated through measurements in three rooms in a study by Pedersen et al. (2007). The target sound pressure level was defined as the level exceeded in 10% of the room and this was determined using a scanning technique to accurately characterise the sound field in each room. The energy average calculated using the Danish and Swedish methods was compared to this value. It was found that the Danish method could give values substantially below the target and that the Swedish method gave results close to the target. On the other hand, it was pointed out that the component of noise that contributes most to the C-weighted level may not be the most annoying component of noise and therefore this may not be the most accurate way of choosing the most problematic corner in the room. It was therefore proposed that the most reliable results could be attained by measuring in four three-dimensional room corners and then calculating the energy average of these measurements. The reason for measuring in multiple corners when all corners are expected to have the highest sound pressure level is that the room may not be exactly rectangular, objects and furniture may distort the sound field and the microphone is not positioned exactly in the room corner (this becomes important for frequencies where the wavelength is shorter than ten times the distance from the wall).

During the measurement period, windows and doors should be opened or closed, depending on normal room usage. It is important to note the open area of doors and windows, as this may assist in interpretation of the peaks in the frequency spectra. Helmholtz resonators are formed by room volumes in combination with openings such as doors and windows and this phenomenon can lead to higher indoor levels than outdoor levels at low frequencies (Hansen et al. 2015a; Hubbard and Shepherd 1991). The dimensions of the measurement room should be recorded to allow prediction of the room resonances for comparison with the measured data. The room resonances, f_n can be calculated according to Eq. (2.35).

6.2.9 Outdoor-to-indoor Noise Reduction

Compliance limits for wind farm noise are usually based on allowable outdoor sound pressure levels. Since these limits are designed to protect the amenity of the people residing indoors, the outdoor-to-indoor noise reduction is an important consideration. The World Health Organisation (Berglund et al. 1999; Hurtley 2009) outlines acceptable levels for outdoor noise, but these recommendations are based on outdoor-to-indoor noise reductions for traffic noise. Traffic noise is dominated by mid-frequency energy, whereas wind farm noise is dominated by low-frequency energy for which noise reductions are expected to be much lower. At lower frequencies, the sound field inside a room also becomes more complicated due to structural resonances, Helmholtz resonances, room modes and coupling between the air volume inside the residence and the stiffness of the walls, roof and ceiling. In some cases this can lead to negative outdoor-to-indoor noise reductions at specific frequencies (Hansen et al. 2015a).

Measurement

To determine the outdoor-to-indoor reduction of wind farm noise at a residence, it is necessary to position one microphone outdoors and at least three microphones indoors. The outdoor microphone must be protected by a suitable secondary wind screen and placed at a height of 1.5 m (ANSI/ASA S12.9-7 draft 2014). Where possible, the indoor microphones should be placed in locations that have been identified as problematic, and at least one microphone should be placed in a three-dimensional room corner. If the resident is not able to identify problematic locations within the room then all indoor microphones should be placed in room corners to measure worst-case conditions.

The outdoor and indoor microphones used in the analysis should preferably be of the same type and connected to the same sound level meter or data-acquisition system to ensure that the measurement parameters are as consistent as possible. Since the ambient noise level inside a residence can be very low, particularly in rural areas, a low noise floor microphone should be used indoors. On the other hand, to achieve a noise floor that is low enough, a compromise must be made with regards to the frequency response. For example, a microphone with a noise floor of 6.5 dBA has a flat frequency response only down to 6 Hz. To overcome this issue, a low-frequency microphone (flat response down to at least 0.5 Hz) can also be placed in the room and the results from this microphone can be used for frequencies below 6 Hz, in cases where noise levels are well above its noise floor. It is also possible to calculate a correction factor in the frequency domain corresponding to the difference between the signals measured using the low-noise and low-frequency microphones at frequencies below 6 Hz. This should be done using data measured during periods when the noise was well above the noise floor of both microphones. Since frequencies of 6 Hz and below have wavelengths that are much larger than the dimensions of the room, the placement of each microphone relative to the other is not important in an average-sized room.

To characterise the outdoor noise level at a residence, the microphone can either be located at the facade or greater than 10 m from any structure (ANSI/ASA S12.9-7 draft 2014). For measurements at the facade, the microphone is located flush on a solid surface that has an area of at least 1 m^2 and 6 dB is subtracted from the overall level or the level measured in each 1/3-octave band (ANSI/ASA S12.9-7 draft 2014) to account for coherent addition of the incident and reflected waves. One drawback to this measurement location is that there are often trees, shrubs and loose objects that rattle in the wind in close proximity to the residence. Therefore, measuring greater than 10 m away reduces the influence of wind-induced noise. On the other hand, the house still acts as a reflector at such distances but in this case the incident and reflected waves add incoherently and the measured level will have to be corrected by subtracting the reflected wave contribution from the total sound pressure level. It is expected that when the wavelength of the incident sound waves is longer than the critical dimension of the facade by a factor of three, reflection will not occur and therefore no correction is needed. However, when the wavelength of the incident sound waves is shorter than the critical dimension of the facade by a factor of three, 3 dB should be subtracted from the overall level. Assuming that an average size house has a facade length of 10 m and wall height of 3 m, the critical dimension is the square root of the length squared plus the height squared. Therefore, in this case, a correction would not be required at frequencies below 10 Hz, a 3 dB correction would be required at frequencies above 100 Hz and at frequencies between, the correction would range from 0 to 3 dB. Further investigation is required to determine more exact correction values in this low-frequency range from 10 –100 Hz. The method for determining the indoor noise level at a residence is discussed in Section 6.2.8 and this method is also relevant in ascertaining the outdoor-to-indoor noise reduction.

To facilitate interpretation of the sound field within a room, it is important to record details about the residence construction and critical dimensions. This includes information about the thickness and material of the walls, roof, ceiling and windows. The presence of ceiling insulation should be noted and associated details recorded. Critical dimensions of the residence include the width, length and height of the measurement room, the approximate volume of the entire residence (taking into account the volume occupied by furniture), the surface area of the roof and outside walls and the window area. See Section 5.13 for a discussion on how these data are used. The external window area associated with the measurement room is of particular importance, especially if the windows are partially open during the measurements. In this case, the total open and closed area should be calculated. This calculation should also be undertaken for any external doors of the measurement room that were open or partially open.

It is most useful to determine the outdoor-to-indoor noise reduction for wind farm noise under worst-case conditions, where the difference between wind farm noise and ambient noise is maximised. This can be achieved through selective analysis of specific measurements in a continuous data set that includes sequential measurements recorded over several weeks. Measurements can be selected based on various criteria, including time of day, wind speed at the outdoor microphone, wind direction at hub height and power output of the nearest wind turbine (or of the entire wind farm if this information is not available). Hansen et al. (2015a) achieved good results using the following criteria:

- nighttime (12am to 5am). to maximise the contrast between wind farm and ambient noise
- maximum allowed wind speed of 2 m/s at a height of 1.5 m to minimise wind-induced noise at the outdoor microphones

- minimum power output from the closest wind farm cluster of 40%
- downwind conditions, defined as ±45° either side of the closest wind turbine.

An alternative method for measuring the outdoor-to-indoor noise reduction at a residence is to use a loudspeaker, as described in Søndergaard et al. (2007). The disadvantage of this method is that the loudspeaker does not correctly simulate a wind turbine due to differences in the location of the incident rays at the residence. In addition, sound rays from a wind turbine are incident on the entire house, including the roof and at low frequencies, the back of the house. However, when using a loudspeaker placed near a facade, these other surfaces receive less energy, as a result of directivity and larger relative distances from the source (Møller and Pedersen 2011). Typically, a loudspeaker will not be able to generate sufficient noise at low frequencies to exceed the corresponding ambient noise. Nevertheless, this method requires less measurement time and enables greater control of the source properties. A brief summary of this method is provided below:

1. The loudspeaker is placed on the ground outside of the building at a distance of at least 5 m and an incidence angle of approximately 45° to the centre of the facade.
2. The loudspeaker emits broadband noise with an upper frequency limit of 250 Hz. Equalisation is used to compensate for the loudspeaker frequency response characteristics.
3. The outdoor microphone is mounted directly to the facade of the receiving room, at a height of 1.5 m.
4. The indoor microphones are located in three-dimensional room corners, as far as possible from facade windows and doors. Four corners are chosen and these include more than one wall and both the ceiling and floor in the room.
5. The sound pressure levels from the four indoor positions are averaged on an energy basis.
6. Background noise is measured, and corrections for background noise are applied using logarithmic subtraction as discussed in Section 2.2.9.
7. The outdoor-to-indoor noise reduction is calculated as the difference between the background-noise-adjusted outdoor level minus 6 dB and the background-noise-adjusted average indoor level.

Analysis

The outdoor-to-indoor noise reduction is usually assessed using 1/3-octave bands. This is a straightforward procedure and therefore the most challenging aspect of analysis is in the interpretation of the peaks and troughs in the results, particularly at low frequencies. The troughs are of primary importance, as they represent 1/3-octave frequencies where the outdoor-to-indoor noise reduction is poor. As mentioned earlier, various phenomena can affect low-frequency sound waves within a room and hence this section provides guidance for interpretation of results.

In an average-sized room in a house, room acoustic resonances occur between 30 and 200 Hz. The frequencies associated with these resonances can be calculated through knowledge of the width, length and height of the room using Eq. (2.35). Structural resonances generally occur between 12 and 30 Hz according to data collected for a number of different house structures (Hubbard 1982). Detailed calculation procedures for determining structural resonances are given in (Bies and Hansen 2009, Ch. 8). Other resonances are associated with the mass of the walls and roof acting with the stiffness of the air volume inside the house and also as a result of open doors and windows (Helmholtz resonances). The calculation of frequencies corresponding to these resonances is discussed in Section 5.13.

Discussion and Implementation

The outdoor-to-indoor 1/3-octave noise reduction measured near the Hallett wind farm in South Australia during operational conditions is shown in Figure 6.10 and these data have been filtered using the criteria defined in Section 6.2.9. The outdoor levels were measured using a ground-mounted microphone protected by a hemispherical secondary wind screen and the indoor levels correspond to a microphone located in a room corner. More details about the measurement set-up are provided in Hansen et al. (2015a). For the 1/3-octave bands below 2.5 Hz, results are only shown for measurements where the wind speed was less than 0.5 m/s, since the remaining data were affected by wind-induced noise on the outdoor microphone, even though the microphone was protected by both a primary and a secondary wind screen. It can be seen in Figure 6.10 that troughs occur in the 4, 16, 40 and 50 Hz 1/3-octave bands. Reduced attenuation occurs when the windows are open and a new trough appears at 8 Hz, while the trough at 50 Hz becomes deeper by up to 7 dB.

Using Eq. (2.35), it can be found that room resonances occur at 41 and 49 Hz, leading to increased noise levels in the 40- and 50-Hz 1/3-octave bands. Increased noise levels in the 16-Hz 1/3-octave band can be attributed to structural resonances. The negative noise reduction in the 4-Hz 1/3-octave band is most likely due to resonance associated

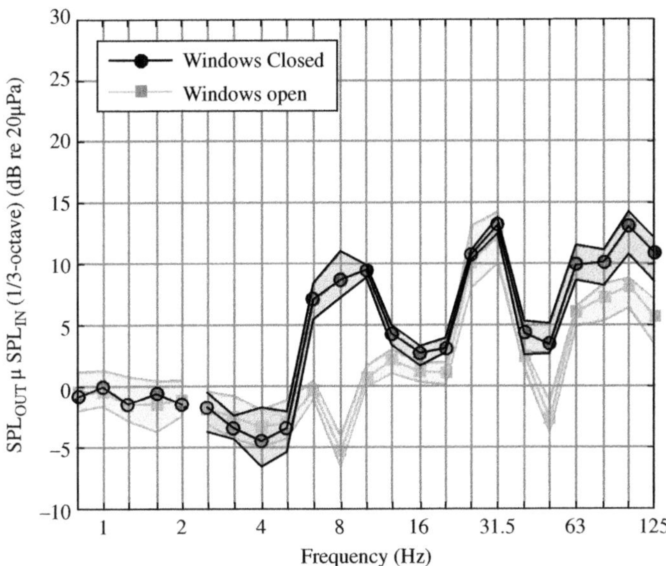

Figure 6.10 Outdoor-to-indoor 1/3-octave band differences between a ground-mounted outdoor microphone and indoor microphone(s), showing the range from measured maximum to minimum as a shaded area and the mean as the data points joined by a solid line. SPL, sound pressure level.

with the mass of the walls and roof acting with the stiffness of the air volume inside the house. For the windows open measurements, increased indoor levels at 50 Hz are caused by the attenuation loss related to the open area of the window. The change from a peak to a trough in the outdoor-to-indoor noise reduction value at 8 Hz is related to a Helmholtz resonator effect, which is facilitated by the open window (Leventhall 2003). A more detailed discussion of these results is given in Hansen et al. (2015a).

Measurement results corresponding to the outdoor-to-indoor noise reductions for a living room and small-sized room in five different Danish houses are presented by Søndergaard and Madsen (2008). In this study, a similar methodology to the one described by Søndergaard et al. (2007) was employed. The results indicate that negative noise reductions occur at some frequencies and in general, there is a large spread in the noise reduction data at low frequencies for different rooms/residences. Above 50 Hz, the outdoor-to-indoor noise reduction is greater than 10 dB for the majority of rooms/residences, but below 50 Hz, there are several 1/3-octave frequency bands for which the outdoor-to-indoor noise reduction is less than 10 dB for the majority of rooms/residences.

Møller and Pedersen (2011) identified some shortcomings of the loudspeaker method outlined above and recommended some improvements. As mentioned at the beginning of this section, the facade is not fully reflecting at low frequencies and therefore subtracting 6 dB for measurements made on the facade surface is not applicable at these frequencies. Hence Møller and Pedersen (2011) proposed that the free-field level of the loudspeaker should be determined at a position without reflective surfaces other than the ground. In this case, it should be ensured that the loudspeaker output is identical to that used in the outdoor-to-indoor noise reduction measurements.

Another issue is that the difference between the indoor noise level with the speaker on and off could be very small, particularly at low frequencies. In such a case, it is not possible to determine the relative contribution of loudspeaker noise accurately. Hence Møller and Pedersen (2011) suggested using the maximum-length-sequence technique (see Section 2.4.17) to generate the signal to drive the loudspeaker, resulting in measurements of outdoor-to-indoor noise reduction that are not significantly affected by background noise. Another suggested approach is to generate noise in each 1/3-octave band separately to maximise speaker output in the associated frequency range (Møller and Pedersen 2011).

When adjusting the indoor noise level to account for background noise, it is recommended that the procedure is dependent on the magnitude of the difference between the combined speaker and background noise and background noise only. The recommended procedure is as follows:

- Differences greater than 10 dB: no adjustment
- Differences from 3 to 10 dB: logarithmic subtraction of background noise only from combined speaker and background noise (see Eq. (2.29))
- Differences less than 3 dB: data discarded.

6.2.10 Amplitude Modulation and Variation

The annoyance associated with exposure to wind farm noise is not only related to the overall noise level and low-frequency content but also to the time variability of the noise. This time variability can be either random or periodic, depending on the modulating factors at the source. According to listening tests conducted by Lee et al. (2011), amplitude modulation (AM; periodic variation) of wind farm noise significantly contributes to noise annoyance. Random variation in the amplitude of wind farm noise is also likely to contribute to annoyance, but listening tests using wind farm noise containing random amplitude variation have not been published so far, to the best of the authors' knowledge.

Research by Stigwood et al. (2013), Large and Stigwood (2014) and Stigwood et al. (2015) suggests that the highly erratic nature of wind farm noise, with its rapidly changing beats and rhythms, is likely to result in increased perception and annoyance. The significant time variability of wind farm noise is due to several factors, including meteorological conditions, blade loading variations, noise directivity variations and interactions between noise from two or more turbines.

Random variations in noise emissions can be attributed to transient bursts of coherent turbulent energy in the turbine inflow (Kelley et al. 2000), often referred to as inflow turbulence, and to regions of constructive/destructive interference that are defined by the relative phase of the incoming sound waves from different turbines, which varies with wind direction and blade rotation rate, leading to a time-dependent amplitude of sound at a given location. A demonstration of this phenomenon is presented in Section 6.2.7. Measurements carried out by Larsson and Öhlund (2014), also suggest that this effect is relevant, as they found little correlation between the degree of AM measured at an emission and immission point (near a residence) over a measurement period of approximately 30 h. Additionally, Cooper et al. (2015) noted that a tone showing clear AM at the blade-pass frequency near a wind turbine demonstrated similar levels of amplitude variation at a residence located 1.5 km away but that the amplitude variation was more random and 'could not be described as blade-pass frequency modulated'. Amplitude variation could also be caused by tonal emissions from wind turbines at slightly different frequencies, resulting in the beating phenomenon discussed in Section 2.2.6.

In addition to the random amplitude variations discussed above, periodic variations in the loudness of wind turbine noise (in other words, AM) also occur, with the frequency of variation generally equal to the rate at which the blades pass the tower. For large wind turbines in the 0.5–3 MW range, the blade-pass frequency is typically in the range 0.5–1 Hz. Within a few hundred metres of a wind turbine, the dominant source of AM is the 'swishing' noise (Oerlemans and Schepers 2009), but since it occurs in the frequency range of 400 Hz–1 kHz (Renewable UK 2013), it is highly attenuated during propagation due to atmospheric and ground absorption. In addition, any remaining noise at these frequencies tends to be masked by background noise at a typical receiver. Therefore, AM at low frequencies is of greatest interest at receivers located at distances greater than 1–1.5 km from a wind farm. The corresponding noise is often described by listeners as 'thumping' or 'rumbling'.

Low-frequency AM has been attributed to wind shear (Renewable UK 2013), where the inflow velocity is greater at the top of the blade trajectory, hence increasing the local angle of attack and susceptibility to stall. The angle of attack of the blade is defined by the difference between the pitch angle of the blade and the angle formed by the vector sum of the incoming wind velocity and blade rotational velocity. If the incoming wind velocity increases rapidly from the bottom of the rotor disc to the top, the angle of attack may be pushed past its stall angle. This type of AM is generally referred to as EAM or OAM and is discussed in more detail in Section 3.3.6.

Measurements on a full-scale wind turbine have shown that transient stall is likely to occur over part of the rotor revolution, particularly during operation in a wake of another turbine (Madsen et al. 2014). For operation during strong wind shear, it has been found that modulation of the surface spectra can be as high as 14 dB at frequencies below 200 Hz (Madsen et al. 2014). Stall noise is characterised by an increase in noise level (Paterson et al. 1975) and reduction in frequency of the noise (Moreau et al. 2009). Even for the case where the blade does not undergo stall, the increased angle of attack gives rise to a thicker or possibly separated turbulent boundary layer, which in turn increases the noise level and reduces the frequency of the noise (Renewable UK 2013). In contrast to attached flow trailing-edge noise, which shows most significant AM in the crosswind direction at large distances (Oerlemans and Schepers 2009), AM associated with stall noise is highest in the upstream and downstream directions (Renewable UK 2013).

Another proposed AM mechanism is interference between multiple direct and reflected ray paths arriving at a fixed location from the three rotating blades (Bradley 2015), and this phenomenon can occur at all frequencies. Both of the above-mentioned mechanisms refer to AM of broadband noise.

There are two main mechanisms responsible for AM of tonal noise: periodic variation of gearbox noise; and blade–tower interaction. Since the dominant noise-generating mechanisms on modern wind turbines are aerodynamic in origin, gearbox noise is not generally problematic unless there is a fault in the gearbox (Moorhouse et al. 2007). On the other hand, it appears that the unsteady thickness component of blade–tower interaction noise can give rise to the generation of amplitude-modulated tonal components that exceed the generally accepted threshold of hearing for normal ears subjected to steady tonal noise, as defined in ISO389-7 (2005) (Hansen et al. 2014). Amplitude modulation of tonal noise caused by blade–tower interaction can occur in the infrasonic frequency range as well as the low-frequency range and this was observed by DiNapoli (2011). When it occurs in the low-frequency range, the noise is often perceived as a 'rumbling' noise.

Due to the high dependence of AM and amplitude variation on meteorological conditions, the duration of these phenomena can range from several seconds to several hours. It has been postulated that persistent AM or amplitude variation is caused by steady wind direction and strength as well as prolonged high wind shear (Stigwood et al. 2013).

Measurement

The most important consideration in the measurement of AM and amplitude variation is the noise floor of the instrumentation, which was discussed in Section 6.2.1. If the noise floor is not sufficiently low then the troughs in an amplitude-modulated signal can be lost in the noise floor. As a consequence, the level of AM appears lower due to the reduced difference between the peak and trough noise levels.

To ensure maximum flexibility in the analysis of wind farm AM and amplitude variation, it is recommended to measure continuous time data using a data-acquisition system. If this option is not available, a sound level meter can be used to measure the required parameters for algorithms such as those outlined by Fukushima et al. (2013) and Cooper et al. (2015), which are discussed in more detail in the next subsection. The sound level meter should conform to Class 1 standards (IEC 61672-1 2013) and should be capable of measuring 1/3-octave spectra at least over the range of 10–800 Hz. There should also be the option to measure and store statistical levels, such as L_{A90}, as well as $L_{A,125ms}$ or $L_{A,100ms}$, which correspond to the 'fast' time weighting and $L_{A,1s}$, which corresponds to the 'slow' time weighting.

The sound level meter should be capable of tracking the amplitude modulating signal properly and this may not be achieved through simple adherence to IEC 61672-1 (2013), which specifies a minimum fall time of 25 dB/s. Sound level meters having a faster fall time will be able to track the time-varying signal more faithfully, leading to possible differences in results measured using different sound level meters compliant with IEC 61672-1 (2013) (Huson 2015). The capabilities mentioned above should be available for unweighted levels as well as A-weighted levels.

The type of microphone selected for the detection of AM and amplitude variation can be an important factor, depending on the metric selected for analysis. Some metrics (Fukushima et al. 2013; RES Developments 2014) specify use of the A-weighted level and in these cases, the low-frequency response of a standard microphone will be sufficient. On the other hand, where the unweighted levels are measured, it is important that the microphone has a flat frequency response down to 0.5 Hz to capture AM over the low-frequency and infrasonic ranges.

The measurement location should be selected carefully when assessing the degree of AM or amplitude variation associated with a wind farm. The reason for this is that some locations regularly experience greater AM than others as a result of the influence of wind direction, meteorology (particularly wind shear) and synchronicity between two or more wind turbines (Stigwood et al. 2013). Also, the microphone should not be placed between two or more reflective surfaces, as this may result in an increase in the modulation depth by as much as ±6 dB(A) peak to trough (ETSU 1996).

There is currently no standard that defines an acceptable measurement period or specific meteorological conditions that are required for AM and amplitude variation detection and quantification. However, it is known that AM and amplitude variation are highly dependent on meteorological conditions and therefore it is recommended that measurements should span several weeks, and include periods of time during which stable, worst-case conditions occur. These periods most often occur at night (Van den Berg 2005) and since this is also the period where extraneous noise is lowest, detection of AM will be maximised. To approximate the atmospheric stability, the wind speed and direction should be measured for at least two different heights. More accurate results can be obtained through measuring wind speed and direction at multiple heights and measuring additional variables as well. This will be discussed in more detail in Section 6.4.

Analysis

When assessing the potential for annoyance in a time-varying signal, it is important to implement an objective measure of the signal variability. This enables quantification of the time-varying effects, facilitating comparisons with subjective assessments of annoyance. A large number of metrics have been proposed to quantify the degree of variability of a signal and the majority target periodically varying components. There is currently no universally accepted standard metric for the analysis of the variation in wind turbine noise, which is possibly due to the fact that different metrics may be suitable for different situations, depending on the nature of the wind farm noise at a given location. Nevertheless, the various metrics that have been proposed share a number of common objectives:

- determining the modulation depth, the modulated frequencies and the frequency of modulation

- rating the noise in terms of its intrusiveness
- minimising the influence of extraneous noise on detection of AM and amplitude variation
- identifying the prevalence of AM.

The metrics that will be considered in this section are presented in Table 6.2. Each metric focuses on different aspects of time-varying effects and a not-to-be-exceeded critical value is often defined, which defines when penalties may need to be applied to overall levels to account for additional annoyance. The critical value is generally based on the results of listening tests. Several of the metrics were developed with the intention of identifying 'swishing' noise, which is amplitude-modulated, broadband, mid- to high-frequency noise, and therefore their application to low-frequency amplitude-modulated tonal noise is questionable (Hansen et al. 2015b). Several variations of the metrics presented in Table 6.2 exist and the list is not exhaustive, although it is believed that it is representative of the available methods. More details are provided for each method in the list that follows the table.

D_{AM} Method The first step in applying this method is to calculate the difference between the fast and slow weighted values that are shown in Figure 6.11a. The time interval between samples for these quantities must be the same, so that they can be subtracted from one another. Since the time interval between samples is not defined in the context of the D_{AM} method, it is suggested here that a value between 10 and 25 ms, as adopted by the Nordtest method (Nordtest 2002), should be used to sample ΔL_A. The 5th percentile and 95th percentile values of ΔL_A are then determined as shown in Figure 6.11b. Since these values correspond to levels that are exceeded 5% and 95% of the time, respectively, 95% and 5% of levels, respectively, are lower than these values, as reflected by the intersection point with the cumulative distribution curve in Figure 6.11. The degree of amplitude variation, D_{AM}, is determined by subtracting the $\Delta L_{A,95}$ value from the $\Delta L_{A,5}$ value. The analysis period is 3 min and the critical value shown in the table has been determined on the basis of 'fluctuation sensation'. This appears to be the threshold at which perception of AM/amplitude variation occurs and therefore annoyance may occur above this threshold for some people. According to previous studies (Bowdler and Leventhall 2011), $D_{AM} > 2$ is the value at which 'fluctuation sensation' initiates. This value was confirmed through listening tests, where subjects were exposed to measured and synthesised wind farm noise (Yokoyama et al. 2013). The D_{AM} method detects any type of amplitude variation, including AM.

Den Brook Condition (Condition 20) This method considers cases only where $L_{Aeq,1\text{min}} >$ 28 dB(A). Periods of two seconds length are analysed to identify occurrences of an increase in the $L_{Aeq,125\text{ms}}$ of more than 3 dB(A) and a subsequent fall of more than 3 dB(A) within that period as shown in Figure 6.12. These events signify amplitude variation and should occur at least five times in six or more 1-min periods over an interval of 1 h. This method detects any type of amplitude variation, including AM. The definition of modulation depth using this method is consistent with the one introduced in Section 2.2.7.

Hilbert Transform Method For this method, the time-domain signal is band-pass filtered using 1/3-octave filters. The Hilbert transform is calculated to determine the envelope signal in the time domain for each 1/3-octave band, as shown in Figure 6.13, and the fast

Table 6.2 Metrics used for detection of amplitude modulation, amplitude variation and impulsiveness.

	Metric	Citation	Procedure	Critical value defining unacceptable AM	Noise character
1	D_{AM} method	Fukushima et al. (2013)	$D_{AM} = \Delta L_{A,5} - \Delta L_{A,95}$ $\Delta L_A = L_{A,125\text{ms}}(t) - L_{A,1\text{s}}(t)$	D_{AM} >2 dB	AV
2	Den Brook Cond. 20	Analysed by Bass (2011)	Peak-to-trough difference $L_{Aeq,1\text{min}} > 28\text{dB(A)}$	Multiple cases of peak-to-trough >3 dB	AV
3	Hilbert transform method	McCabe (2011)	$m = 2A_1/A_0$	Not available	AM
4	Lee et al. method	Lee et al. (2011)	$\Delta L = 20\log_{10}(\frac{p_o+p_f}{p_o-p_f})$	Not available	AM
5	RES/RUK methods		$A = 2\sqrt{2\Delta f E_c}$	$A \geq 2.5$ dB	AM
	–Denbrook Cond. 21	RES Developments (2014)			
	–RUK	Renewable UK (2013)			
6	Lundmark method	Lundmark (2011)	$\max[AMS(f)]$	$AMS \geq 0.4$ dB	AM
7	Hybrid method 1	Cooper and Evans (2013)	BPF detection Peak-to-trough difference	Peak-to-trough >5–6 dB	AM
8	Hybrid method 2	AMWG (2015)	BPF detection Peak mean–trough mean	Not available	AM
9	Nordtest method	Nordtest (2002)	$P = 3\log_{10}(\text{onset rate[dB/s]})+$ $2\log_{10}[\text{level diff. [dB]}]$	$P > 5$	Imp
10	Lenchine method	Lenchine (2016)	Cross correlation of L_p with cosine function of frequency equal to blade-pass frequency	Not available	AM

AM, amplitude modulation; AV, amplitude variation; Imp, impulsiveness.

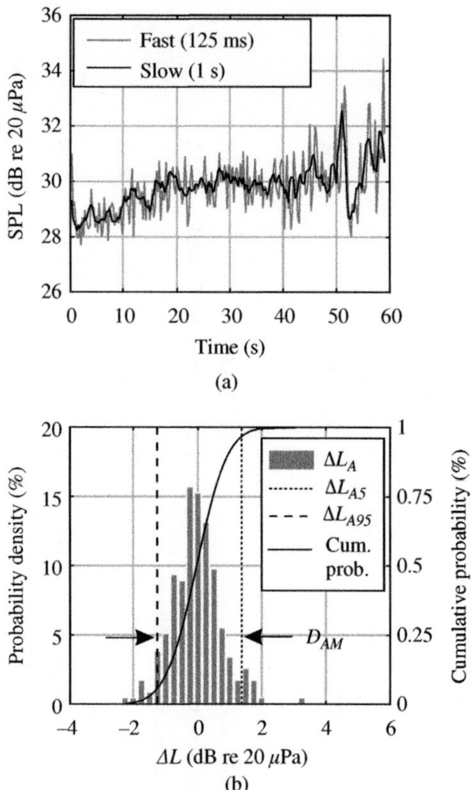

Figure 6.11 D_{AM} method of amplitude modulation and amplitude variation detection.

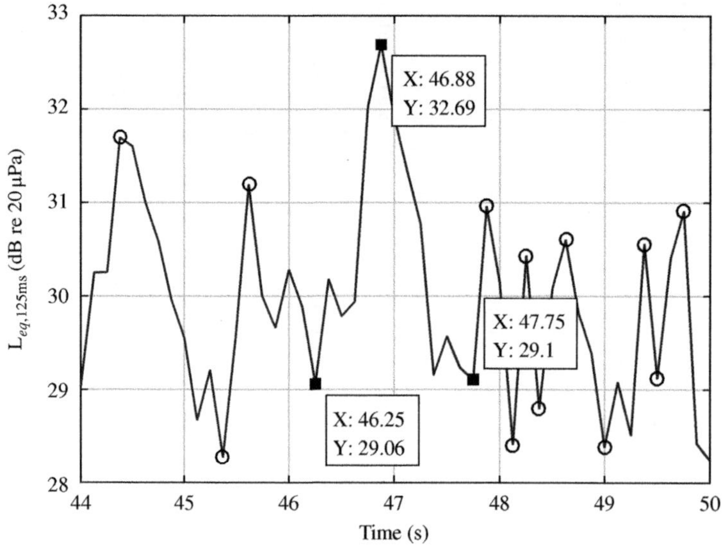

Figure 6.12 An example of a 3-dB variation in peak-to-trough level occurring in a 2-s period that would be identified as amplitude variation according to the Den Brook condition.

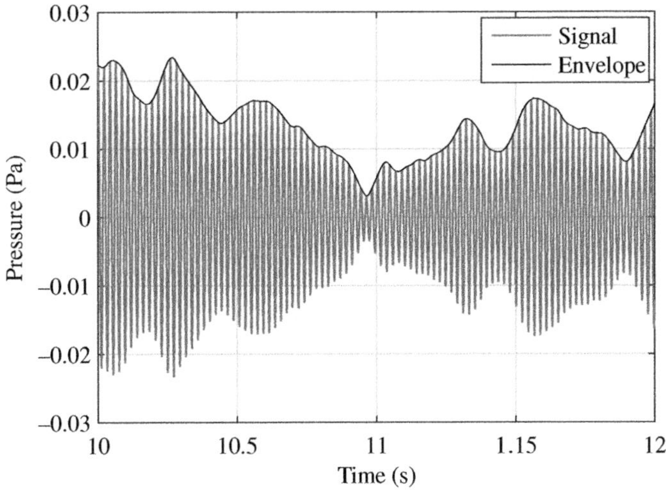

Figure 6.13 Signal envelope obtained by calculating the Hilbert transform.

Fourier transform (FFT) of the signal envelope is computed, as shown in Figure 6.14. The modulation factor for each 1/3-octave band is defined as $m = 2A_1/A_0$, where A_1 is the amplitude of the signal at the modulating frequency and A_0 is the signal amplitude at frequency, $f = 0$ Hz. This is non-zero because only the positive part of the signal envelope is used for the FFT analysis, as shown in Figure 6.13. The modulation factor differs from the modulation depth because it is calculated in the frequency domain rather than the time domain. The Hilbert transform can be calculated in MATLAB® using the **hilbert** command; the absolute value is then found using the **abs** command and subsequently the FFT is calculated, resulting in the spectrum shown in Figure 6.14. Using

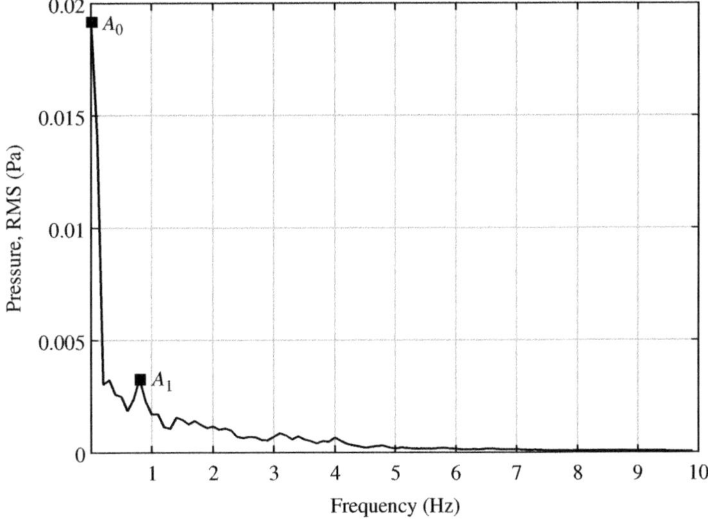

Figure 6.14 FFT of signal envelope showing a peak at the blade-pass frequency.

the Hilbert transform (see Section 2.4.10) method on data measured 450 m from a wind farm, McCabe (2011) found that there was a good correlation between the modulation factor and wind shear, whereas the correlation between modulation factor and wind speed at heights of 1.5 m and 80 m was poor.

Lee et al. Method This method is based on the construction of a spectrogram, which is a contour plot of the sound pressure level as a function of frequency and time, as shown in Figure 6.15a. The spectrogram is obtained by first dividing the time series data into segments and then applying an FFT to each segment to obtain a frequency spectrum. The segment lengths are chosen to be sufficiently short so that there are sufficient frequency spectra to resolve the AM. This usually means that ten spectra should exist in the time taken between each blade pass, so if the blade-pass frequency is 1 Hz (1 s between each blade pass), each spectrum should represent a time interval of 0.1 s. For a frequency resolution of 5 Hz, the uncertainty principle (see Section 2.4.8) requires a minimum of 0.2-s sample time to acquire the data for one frequency spectrum. If frequency spectra have to be obtained at intervals of 0.1 s, then each data segment must overlap the preceding segment by 50%. This means that all data will be used twice.

The next step is to take the data along each horizontal line in the sonogram (illustrated in Figure 6.15a. An FFT is then applied to each line. As an example, an FFT of the 45-Hz line (representing frequencies between 42.5 and 47.5 Hz) is shown in Figure 6.15b. It can be seen in Figure 6.15b that the variation in the sonogram along this horizontal line occurs at frequencies of 0.8 Hz (the blade-pass frequency) and the first few harmonics. The modulation factor for this method is defined as $\Delta L = 20\log_{10}\frac{p_o+p_f}{p_o-p_f}$. Here, p_o and p_f are the RMS sound pressures at 0 Hz and the modulation frequency respectively, and are calculated from L_{po} and L_{pf} (shown in Figure 6.15b) as follows:

$$p_o = p_{\text{ref}} \times 10^{L_{po}/20} \quad \text{and} \quad p_f = p_{\text{ref}} \times 10^{L_{pf}/20} \tag{6.9}$$

where $p_{\text{ref}} = 20\mu Pa$.

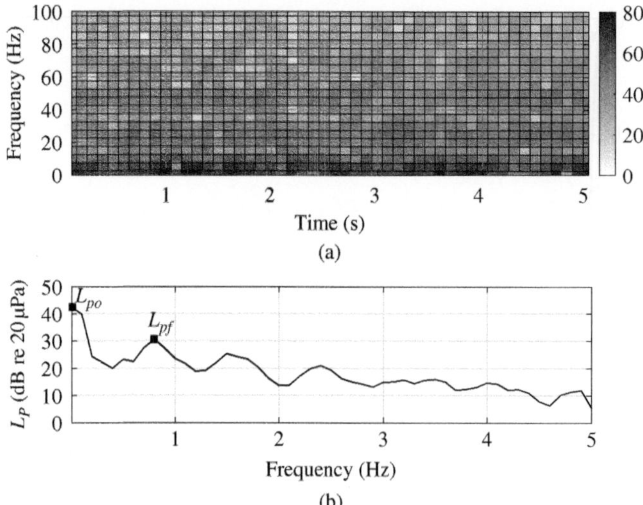

Figure 6.15 (a) Sonogram and (b) FFT of the 45-Hz line in the sonogram, representing the variation in L_p in the frequency range, 42.5–47.5 Hz.

RES (Revised Condition 21)/RUK Methods For these methods, the time series data are divided into 10-s non-overlapping blocks, converted to 60 data points of $L_{Aeq,100ms}$ and then detrended by subtracting the mean value. This mean value is determined by fitting a fifth-order polynomial to the 60 data points of $L_{Aeq,100ms}$ in each block. The fifth-order polynomial is used rather than just taking the mean of all the data because the mean could be varying upwards or downwards over time and this must be taken into account.

The power spectral density for the $L_{Aeq,100ms}$ data is calculated using a rectangular window and a frequency resolution, Δf, of 5/128 Hz, giving *raw spectra*. To obtain such fine resolution with 10 s of $L_{Aeq,100ms}$ data, it is necessary to use zero-padding; the associated procedure is discussed in Section 2.4.7. Spectra are then integrated in constant-width frequency bands of width 0.9 to 1.1 of the blade-pass frequency in hertz, giving *integrated spectra*, E_c. The raw spectrum and integrated spectrum are shown in Figure 6.16 for a signal containing AM in the 50-Hz 1/3-octave band.

The level of AM, A, is determined using the equation $A = 2\sqrt{2\Delta f E_c}$. It can be seen that for detecting AM in this particular signal (measured at a distance of approximately 3.3 km from a wind farm), it was more useful to analyse unweighted data rather than A-weighted data. Both the RES and RUK methods require that there is consistency between the blade-pass frequency identified from the power spectral density plot and the blade-pass frequency measured at the turbine when AM is identified. There are some differences between the RES and RUK methods and these are described below.

- Valid periods for analysis using the RES method are defined as at least four periods of 1 h during a period of 14 days, where rainfall and extraneous noise are not present. For 1-min periods for which the condition, $L_{Aeq,1min}$ >28 dB(A) is not met, analysis is not carried out and therefore it is assumed that the noise level must be sufficiently low for AM not to be annoying. For the RUK method, all 10-min measurement periods that are not contaminated by rain or extraneous noise are assessed for AM.
- Using the RES method, the Den Brook condition described above is implemented when the AM, defined as $A = 2\sqrt{2\Delta f E_c}$, is greater than 2.5 dB in at least six separate 10-s periods. The RUK method specifies that an overall 10-min AM value is determined by finding the arithmetic mean of the 12 highest AM values from the 60 10-s periods. Overall AM values for each 10-min period are then plotted against the average wind speed for that period and a line of best fit is drawn to determine the

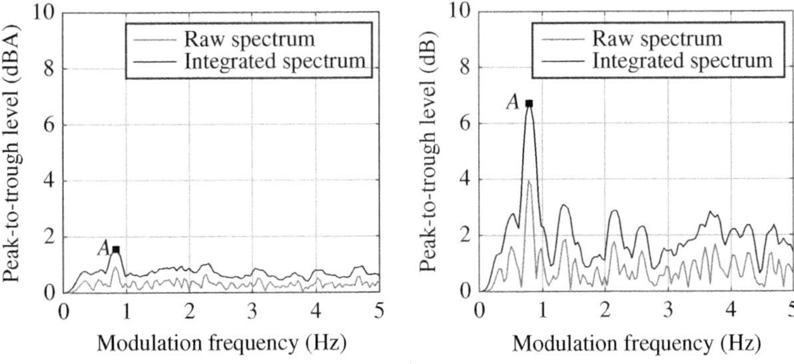

Figure 6.16 Raw spectra and integrated spectra obtained using the RES method of AM detection, showing the level of AM, *A*, for A-weighted spectra (left) and unweighted spectra (right).

applicable penalty, which is added to the measured noise level (prior to comparison with the allowed level) at each integer wind speed when AM is greater than 3 dBA. Therefore, the applied AM penalty is a function of wind speed. If there is no apparent trend with wind speed then the arithmetic mean of the values is used. RUK specifies that a penalty is to be applied when the AM is greater than 3 dBA. The penalty increases linearly from a value of 3 dBA when the AM is equal to 3 dBA, up to a value of 5 dBA when the AM is greater than or equal to 10 dBA.

Lundmark Method With this method, values of $L_{Aeq,125\text{ms}}$ are determined and then an FFT ($F\{L_{A,125\text{ms}}\}$) is carried out with 16 s of data, resulting in a frequency resolution of 0.0625 Hz. The AM spectrum, *AMS*, is calculated according to $AMS = \sqrt{2} * |F\{L_{A,125\text{ms}}\}|/N$, where N is the number of samples of $L_{Aeq,125\text{ms}}$ and is equal to 128 for a sample period of 16 s. The peak in the resulting spectrum, max[*AMS*(*f*)], provides an indication of the degree of AM. The critical value of $AMS = 0.4$ dB was proposed by Larsson and Öhlund (2014) to identify periods where AM was present. Larsson and Öhlund (2014) extended the method by specifying some additional criteria. Noise in 1/3-octave bands from 10–630 Hz was considered exclusively and the modulation frequency was constrained to the range of 0.6–1 Hz. Data points corresponding to low power output of the wind farm or shut-down conditions were discarded. Data were only analysed when the predicted noise level at the receiver was greater than 30 dBA. It was found that the method was not triggered by passing cars, birds and aircraft but detection of wind farm AM was poor during conditions of strong masking (Larsson and Öhlund 2014).

Hybrid Method 1 According to this method, 2-min data samples of overall $L_{Aeq,100\text{ms}}$ (a total of 1200 samples) are analysed. To identify the blade-pass frequency for each 2-min period, the original A-weighted time-domain signal is further filtered using 1/3-octave band-pass filters and the A-weighted power spectrum is calculated for each 1/3-octave band using time windows of length 24 s and an overlap of 50%, resulting in a frequency resolution of 0.042 Hz (see Section 2.4.8). The blade-pass frequency can then be identified from peaks in the spectra that fall within the range of operating speeds.

Considering a frequency range of 250–1000 Hz, the total number of spectra for a 2-min period is 63 (seven 1/3-octave bands with 9 windows for each 2-min period). The blade-pass frequency is identified as that corresponding to the maximum number of peaks in a possible blade-pass frequency bin (frequency bins are defined as possible blade-pass frequencies). Maxima and minima in both the overall A-weighted time-domain signal ($L_{Aeq,100\text{ms}}$) and each A-weighted, 1/3-octave band filtered time-domain signal (sampled at 100 ms) are identified, ensuring that their spacing corresponds to the modulation period associated with the blade-pass frequency. For each 2-min period, the peak-to-trough differences are linearly averaged, resulting in the modulation depth for that period as defined in Section 2.2.7. The critical value for the overall signal is a peak-to-trough level greater than or equal to 5 dB and the critical value for the individual 1/3-octave bands is a peak-to-trough level greater than or equal to 6 dB. The variation in peak-to-trough level must occur on a regular basis at a rate equal to the blade-pass frequency. These critical values are based on guidance from NZS 6808 (2010), which specifies a corresponding penalty of 5 dB if the critical value is exceeded.

Hybrid Method 2 This method involves analysis of 10-min periods, divided into non-overlapping segments which are 10 s long and consist of $L_{Aeq,100ms}$ data. Initially, the blade-pass frequency is identified by either using SCADA (System Control And Data Acquisition) data or by determining the applicable range of blade-pass frequencies at a given location. According to the latter method, the maximum value on the modulation spectrum that lies within the expected range is the blade-pass frequency (AMWG 2015). Once the blade-pass frequency has been identified, filters centred on the blade-pass frequency and first two harmonics are applied separately to the time-domain signal. The upper and lower boundary frequencies for the filters are determined using the relevant equations in BS EN 61260-1 (2014). This results in three separate time-domain signals, as shown in Figure 6.17a–c, which are added together linearly to create a cleaner signal for analysis in the time domain. This combined signal is shown in Figure 6.17d; the original $L_{Aeq,100ms}$ data are shown for comparison. Since the data are analysed in 10-s blocks and the bandwidth of the filters is narrow, the band-pass filter is applied to 30 s of data. This includes 10 s of data either side of the 10-s block of interest plus the 10-s block of interest, to allow for filter rise and fall time. The filters must be applied twice, whereby the sequence of samples that make up the output of the first IIR filter are reversed for input into the second IIR filter, producing an

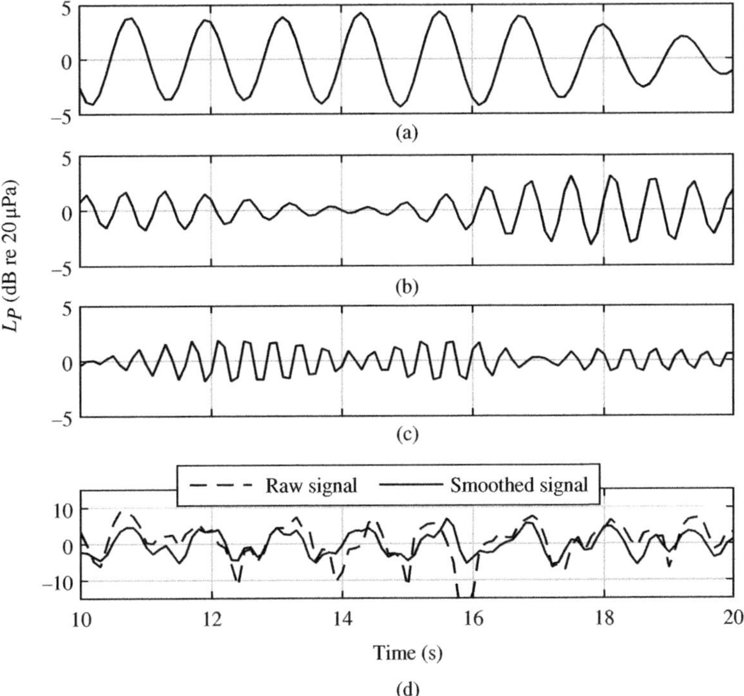

Figure 6.17 Signals obtained by band-pass filtering: (a) output from the narrowband filter with centre frequency equal to the blade-pass frequency; (b) output from the narrowband filter with centre frequency equal to the first harmonic of the blade-pass frequency; (c) output from the narrowband filter with centre frequency equal to the second harmonic of the blade-pass frequency; (d) sum of curves (a)–(c) compared to the unfiltered signal. Sample rate is 10 samples/s.

output with zero-phase distortion. In each 10-s block of interest, peaks and troughs are identified as the maximum and minimum levels occurring in a time window of length equal to the blade-pass period. The difference between the mean value of all peaks and the mean value of all troughs in each 10-s block corresponds to the modulation depth in that 10-s period. This definition of modulation depth is slightly different to the peak-to-trough difference defined in Section 2.2.7. The modulation depth for each 10-min block of data is proposed as the 90th percentile of the distribution of AM values calculated in the 10-s blocks. To reduce the identification of false positives, some conditions have been proposed, which include ensuring the following:

- The maximum value minus the minimum value of the time-series data filtered around the blade-pass frequency, $L_{pk-pk,0}$, should be greater than 1.5 dB.
- The maximum value minus the minimum value of the time-series data filtered around the harmonic, i, in question, $L_{pk-pk,i}$, should be greater than 1 dB.
- $L_{pk-pk,0}$ should be greater than $L_{pk-pk,i}$ for all harmonics i considered.

Nordtest Method The analysis period is 30 min and a moving average of $L_{Aeq,125\text{ms}}$, sampled in the range of 10–25 ms, is computed. The onset rate (how quickly the noise rises in level) is equal to the slope in decibels of the line of best fit between the onset starting point and end point, as shown in Figure 6.18. The starting point, L_s, is defined as the point where the gradient first exceeds 10 dB/s and the ending point, L_e, is where the gradient decreases to less than 10 dB/s. The level difference (how much the noise changes, in dB) is $LD = L_e - L_s$. Onsets less than 50 ms in length are disregarded. K_I is added to the overall measured sound pressure level if $P > 5$, where $K_I = 1.8(P - 5)$.

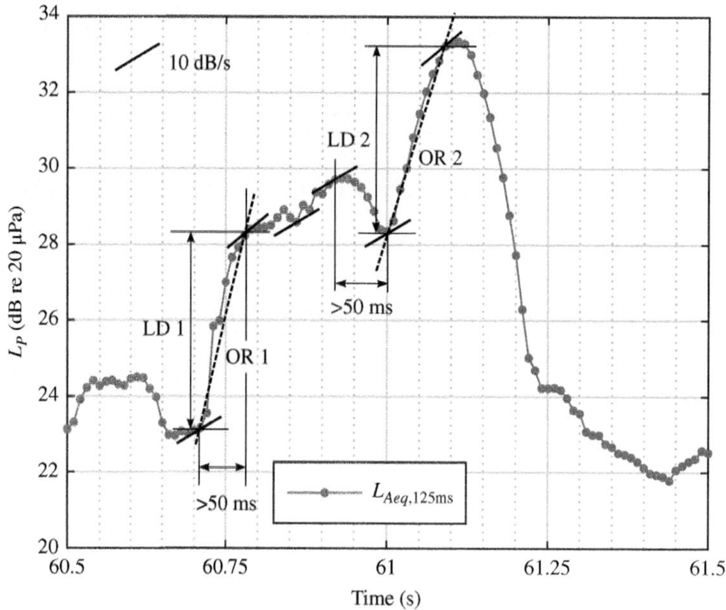

Figure 6.18 Definition of onset rate and level difference used in the Nordtest method. OR, onset rate; SPL, sound pressure level.

Lenchine Method The Lenchine method (Lenchine 2016) is only intended to provide an accurate estimate of the modulation amplitude, A. It is based on the cyclostationary method of Cheong and Joseph (2014) and does not offer any suggestions for acceptable levels of A. Lenchine begins by writing an expression for the overall sound pressure level variation as a function of time,

$$L_p = L_{p0} + A\cos(\omega t + \alpha) + n(t) \tag{6.10}$$

where the values of sound pressure level as a function of time are usually 100-ms or 125-ms averages, L_{p0} is the mean sound pressure level, α is a phase angle, $n(t)$ is time-varying pseudo-noise and A is the modulation amplitude in decibels, with $A = R_{mm}/2$, and R_{mm} defined by Eq. (2.26). Lenchine then performs a cross-correlation between the sound pressure level data and a cosine function of frequency equal to the modulation frequency. He shows that the resulting cross-correlation function plotted against the time is a cosine function with an amplitude equal to $R_{mm}/4$. The modulation frequency is assumed to be equal to the blade-passing frequency of the turbines and can be found using FFT analysis of the time-domain sound pressure signal from the turbines. This method is able to accurately determine the modulation amplitude in decibels, even in the presence of considerable levels of other noise.

Discussion and Implementation

The success of AM and amplitude variation detection algorithms can be improved by removing data contaminated by extraneous noise and by selecting periods of analysis when AM and/or amplitude variation are expected to occur. This can be achieved as follows:

- Analyse data only between 12am and 5am when contamination from extraneous noise is expected to be lowest and stable conditions are most likely to occur.
- Select periods for analysis based on the occurrence of stable meteorological conditions.
- Filter data to include hub height speeds at and above those at which the noise emissions are expected to be maximum. The data can also be filtered to include only cases where the wind speed at the residence at a height of 1.5 m is simultaneously less than 1 m/s.
- Analyse periods that have been identified by residents as annoying.
- Remove data with obvious contamination by extraneous noise.
- Use On/Off testing.

There is no current consensus on which metric is most suitable for detecting and quantifying AM and the choice of metric may depend on the signal being analysed. When selecting periods for analysis, it is important to clarify whether the objective is to identify worst-case conditions or whether it is considered to be more important to quantify how often a given metric exceeds the critical value. It is also important to clarify whether the identification of random amplitude variation is to be considered as well as periodic AM.

A decision must be made as to whether the chosen detection algorithm will give a reasonable result when automated or whether it is necessary to listen to the audio files. When listening to audio files, it is important that the original signal is recreated exactly and hence the frequency response of the listening apparatus must be known and appropriate corrections applied to ensure that low-frequency roll-off is not introduced into the signal. It should be noted that most audio equipment is designed to operate in the audible range (20 Hz–20 kHz) and such equipment would not be capable of reproducing sound outside of this frequency range. It is also important to ensure that the ambient noise in the listening room is sufficiently low so that the replayed signal is not significantly affected by background noise.

There are two methods that are commonly used for assessing audio files:

- use of a computer and headphones
- use of a speaker and listening room.

In both cases, human test subjects are used and are referred to as *listeners*. If using a computer and headphones, low-frequency roll-off can be introduced by both the sound card of the computer and the headphones. Prior to engaging a human listener, it is necessary to verify that the signal has been properly replicated. The best way to do this is to use a head form with a microphone embedded in each ear. Alternatively, a microphone could be inserted through a well-sealed hole in the earphone cup worn by the test subject, so that the signal experienced by the ear is measured directly.

The recommended methodology for listening to the audio files is to replay them in an anechoic chamber with an appropriate speaker system. In this case a microphone would need to be positioned at the location of the listener to ensure that the noise had been reproduced with correct amplification at all frequencies. This would also be challenging, since the frequency response of standard speakers rolls off at low frequencies, which means that a graphic equaliser capable of amplifying the very low frequencies is needed. Alternatively, audio files could be modified using MATLAB® to compensate for the frequency response of the reproduction system.

For algorithms that rely on identification of the blade-pass frequency, such as the RES method (Davis and Smith 2014; Fiumicelli 2011; item 5 in Table 6.2) and the hybrid method Cooper and Evans 2013; item 7 in Table 6.2), results can be improved through use of tachometer data. This entails comparison between the calculated blade-pass frequency and the blade-pass frequency measured by a tachometer located on the wind turbine. Agreement between the results provides verification that the AM is caused by the wind farm. Lack of agreement implies that the AM is caused by extraneous noise sources and is therefore not relevant in the assessment of wind farm AM. On the other hand, caution is advised when the wind turbines have variable rotational speeds (Stigwood et al. 2015), as data averaging may result in discrepancies between acoustic measurements and tachometer data, particularly if these data recordings are not properly synchronised. Also, the acoustic signal measured at a residence often consists of noise that has propagated from multiple turbines that may not be rotating at the same rate.

Some of the methods of AM detection described in Table 6.2 were applied (Hansen et al. 2015b) to signals containing low-frequency amplitude-modulated tonal noise that was measured at four different residences located between 1.3 and 8.7 km from the Waterloo wind farm in Australia. The results were compared to analysis with signals containing only ambient noise, which were measured at the same locations. It was found that for this specific wind farm noise source, the Hilbert transform method (item 3 in Table 6.2) and the Lee et al. method (item 4 in Table 6.2) most successfully captured the periodic AM. The method proposed in item 5 in Table 6.2 was also successful in detecting AM when used with unweighted data rather than with A-weighted data.

The Den Brook method (item 2 in Table 6.2), which is designed to detect any type of amplitude variation, did not identify AM for signals in which the AM was clearly present in the narrowband plot. The method also indicated that amplitude variation was present when the wind farm was shut-down. Therefore, in this analysis, the Den Brook method was unreliable for detection of wind-turbine-induced AM. According to results obtained using the D_{AM} method (item 1 in Table 6.2), the cases with the most severe AM had the largest associated modulation depths, which were greater than the 'fluctuation sensation' value of 2 dB. On the other hand, the modulation depth associated with data measured during non-operational conditions was also greater than 2 dB, which indicates that this method can return false positives.

A comparison between various methods of AM and amplitude variation detection was also carried out by Large and Stigwood (2014). The three measurement locations investigated in their study were at residences located between 350 and 470 m from the nearest turbine. These locations are much closer to the wind farm than those investigated in Hansen et al. (2015b) and thus the most prominent AM occurs at relatively higher frequencies. Large and Stigwood (2014) suggested that the most disturbing components of the noise at these distances were a pulsating hum at 280 Hz and a whine at 1 kHz, where the former was the dominant source of annoyance. The results, obtained through application of the D_{AM}, Den Brook and Nordtest metrics (items 1, 2 and 9 respectively, in Table 6.2), indicated that all three metrics successfully identified AM in the cases investigated. The Nordtest method was also implemented by DiNapoli (2011) for measurements up to 2 km from a wind farm and the predicted prominence of AM was consistently above the value of $P = 5$.

In a subsequent publication, Stigwood et al. (2015) analysed the Den Brook, D_{AM}, RUK and RES metrics and provided a block diagram comparison of the associated methodology. One difference that was accentuated is that the Den Brook and D_{AM} methods require subjective confirmation that AM is present and that it is generated by the wind farm. This is because these methods do not distinguish between AM and random amplitude variation, which is a significant weakness. The RUK and RES methods require minimal pre-analysis and can therefore be easily automated. Despite this convenience, Stigwood et al. (2015) conclude that the quickest and most effective way of identifying AM or amplitude variation is through manual assessor checks. The reason for this is that the RUK and RES methods were found to perform poorly when the signal was corrupted by extraneous noise or when there were contributions from multiple turbines. Stigwood et al. (2015) also mentioned that the modulation depth identified using the RUK, RES and D_{AM} methods bore little resemblance to the actual peak-to-trough value that could be observed in the time-domain trace. The Den Brook criterion was found to perform well on signals with identified wind farm AM, giving a peak-to-trough value that is believed to be more representative of the actual modulation depth.

Further work has recently been contributed (Large 2015), in which the advantages and disadvantages of various methods of AM detection have been discussed. In this report, it is mentioned that both the RUK and RES methods contain several flaws, which include imprecise condition wording, inability to filter extraneous noise and the detection of false negatives. The RES method was also found to detect false positives. Here, a 'false positive' refers to the detection of AM when there is none and a 'false negative' refers to the inability to identify AM when it exists. The imprecise condition wording referred to ambiguity of terms such as 'corrupted data' and the vagueness of the statement 'checking for consistency with SCADA data' (Stigwood et al. 2015). On the whole, the RUK method was deemed unfit for purpose as it failed to enable protection against adverse impacts in a case where the degree of AM was considered unacceptable.

The RES method was considered to be a useful tool in the identification of AM as long as it was not used as a standalone method. The D_{AM} method was found to successfully identify periods of AM and to generally provide a good indication of the degree of modulation. However, some limitations were noted in the analysis of signals contaminated by extraneous noise and signals containing erratic AM.

Large (2015) appear to favour the Den Brook method, as this method was reported to identify AM without being adversely affected by extraneous noise. On the other hand, it was mentioned that successful implementation of the method requires correct interpretation and implementation. This implies that an element of subjectivity is required in the analysis. The Den Brook method has been criticised by the wind farm industry (Large 2015) and in an analysis of noise recorded in an area where there were no wind turbines, Bass (2011) concluded that the method is subject to a large number of false positives, indicating that the Den Brook method is in fact affected by extraneous noise.

Broad application of the methods of AM and amplitude variation detection would require standardisation of parameters involved in the analysis, which differ depending on the chosen algorithm. However, in general the following parameters should be considered:

- data sampling rate and the type of data used in the analysis, such as overall $L_{Aeq,125\text{ms}}$ or $L_{eq,125\text{ms}}$ or bandpass filtered $L_{Aeq,125\text{ms}}$ or filtered $L_{eq,125\text{ms}}$ data
- filtering: 1/3-octave bands, or bands centred on the frequency being modulated
- analysis interval used to detect instances of AM or amplitude variation
- block size (number of data samples used in the time series signal to perform an FFT analysis – see Section 2.4; this is relevant for methods involving FFT analysis); the choice of this parameter may depend on how constant the blade-pass frequency is
- frequency resolution (relevant for methods involving FFT analysis)
- total analysis period: the period needed to acquire sufficient data to determine whether AM is present or not
- critical value at and above which AM would be at an unacceptable level and would thus attract a penalty; this would need to be established via listening tests
- percentage of time that the critical value can be exceeded for daytime and nighttime periods (if at all).

It is important that the data sample rate is fast enough to capture the peaks and troughs in the time-varying signal. As mentioned by Huson (2015), the 'fast' time weighting does not necessarily reflect the response of the human ear, so it may not be the best measure for assessment of AM and amplitude variation. According to Bray and James (2011), the time response of the ear is approximately 10 ms, so this may be a more appropriate averaging time for obtaining L_{Aeq} or L_{eq} values (Huson 2015).

When considering the option of filtering, it should be noted that the overall signal does not fluctuate to the same extent as the individual frequency bands and so the true variation in level will only be apparent if smaller bandwidths are analysed and no A-weighting filter is used. Also, filtering of the signal into smaller bandwidths is likely to lead to enhanced detection of AM and amplitude variation and the bandwidth recommended is 1/3-octave. A decision must then be made as to whether to quantify the AM or amplitude variation based on the 1/3-octave frequency band that is most severely modulated or whether to take an average of the results across all of the frequency bands, as suggested by McCabe (2011). From the perspective of perception, the most conservative approach is to use the results corresponding to the frequency band with the highest modulation. As the most significantly modulated frequency band may change with time (Stigwood et al. 2013), the detection algorithm should be capable of focussing on the relevant frequency band for each time period analysed.

For situations where amplitude variation is essentially AM, which is directly related to the blade-pass frequency, it is important to realise that modulations are never exactly sinusoidal and therefore the FFT of the $L_{Aeq,125\text{ms}}$ (items 5 and 6 in Table 6.2) or the instantaneous sound pressure level (items 3 and 4 in Table 6.2), results in a series of peaks at harmonic frequencies (Renewable UK 2013). Therefore, it is possible to combine these peaks by logarithmically (or energy) summing (see Section 2.2.8) a user-defined number of harmonics, thus producing an increase in the calculated modulation depth, which may provide a more accurate estimate of the degree of AM. Further information is provided by Renewable UK (2013).

As modulation of low-frequency noise is usually more annoying than modulation of mid- and high-frequency noise, it is useful to focus AM measurements in the frequency range below about 400 Hz.

The Institute of Acoustics Noise Working Group has released a discussion document (AMWG 2015) on methods for rating AM in wind turbine noise. In this document, a large number of methods have been reviewed and placed into three main categories: time-series methods, frequency-domain methods and hybrid methods. These methods have been summarised in Table 6.3. Hybrid methods incorporate analysis in both frequency and time domains. Methods for the assessment of impulsivity, such as the Nordtest method (Nordtest 2002), and psychoacoustic descriptors, which are discussed in Section 6.2.11, were not considered relevant in the assessment of AM of wind farm noise by this group. The three categories are next discussed in detail.

Time-series Methods The time series methods that were studied by AMWG (2015) were the Denbrook method and the D_{AM} method. The Den Brook method is convenient to implement and can even be used by individuals with no acoustics training, provided

Table 6.3 Advantages and disadvantages of time-series, frequency domain and hybrid methods (mainly from AMWG (2015)).

	Classification (Examples)	Advantages	Disadvantages
1.	Time-series (Den Brook, D_{AM})	Easy to implement Minimal post-processing	Batch processing unreliable Data check reqd. (audio/visual) Influenced by external noise
2	Frequency-domain (McCabe, Lee, RES RUK, Lundmark)	AM identified using BPF More reliable batch processing	Ignores detailed character of signal Calculated AM depth not meaningful Cannot detect random AV
3	Hybrid (Cooper, AMWG)	Identifies variations at BPF Preserves signal shape More directly related to signal More reliable batch processing	Complex post-processing Cannot detect random AV

they have access to the $L_{Aeq,125ms}$ data. Another advantage of this method is that the peak-to-trough value represents the true magnitude of the amplitude variation of the $L_{Aeq,125ms}$ data. The D_{AM} method is slightly more complex to implement but does not require signal-processing skills, provided the data are measured using a sound level meter that can output the fast and slow sampled values of $L_{Aeq,125ms}$ and $L_{Aeq,1s}$, respectively. The modulation depth obtained using this method does not reflect the true magnitude of amplitude variation and while the fluctuation sensation has been found to become important when D_{AM} >2 dB, the definition of fluctuation sensation is rather vague. The report by AMWG (2015) has therefore included plots showing the relationship between the D_{AM} metric (representing metric 1 in Table 6.3) and design modulation depths (defined following item 3 below) as shown in Figure 6.19. In theory, batch processing can be applied to the time-series methods listed in Table 6.3. However, it has been noted that these methods do not perform well on data that are contaminated by extraneous noise and therefore data containing AM or random variation should be identified manually prior to application of either the Den Brook or D_{AM} method. This can be achieved by selecting periods identified by residents as problematic, checking time data for recognisable patterns, identifying wind turbine contributions from 1/3-octave and narrowband plots and listening to the audio file (Stigwood et al. 2015).

Frequency-domain Methods The frequency domain methods can be batch-processed more reliably that time series methods to detect AM, as extraneous noise seldom varies periodically at the blade-pass frequency. However, due to the averaging process

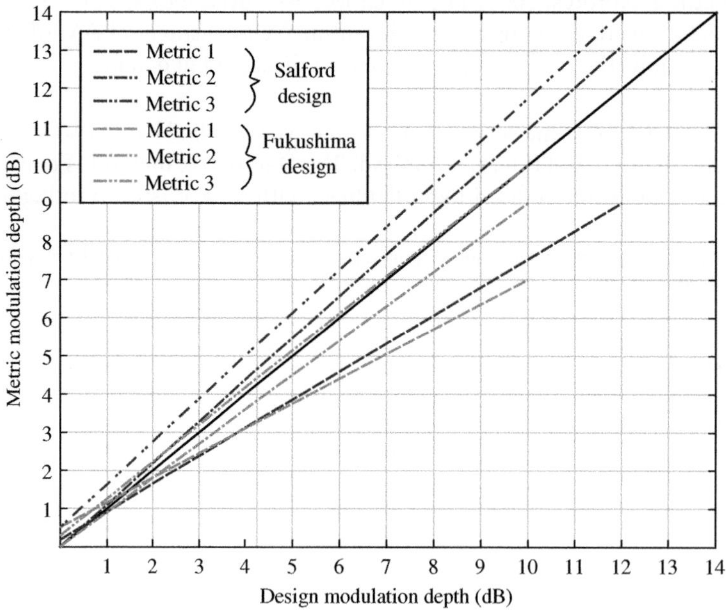

Figure 6.19 Comparison between the modulation depth obtained using three different metrics and applying the Salford and Fukushima methods described below. Numbering of AM metrics is consistent with Table 6.3, where Metric 1 is represented by the D_{AM} method, Metric 2 is represented by the RES/RUK methods and Metric 3 is represented by the AMWG method.

associated with the frequency-domain methods, the resulting spectra contain only frequencies at which periodic amplitude variation (AM) occurs and it is thus not possible to detect random amplitude variations using these methods. Also, the majority of the frequency-domain methods restrict the modulation frequency to the range below 10 Hz (Lee, RES, RUK, Lundmark) due to the processes involved and the RES and RUK methods consider modulation at the blade-pass frequency exclusively. An additional concern is that the amplitude of the peak identified in the modulation spectrum is not related to the peak-to-trough level in the $L_{Aeq,125\text{ms}}$ time-domain signal and therefore listening tests would be required for each method to determine the correlation between annoyance and the modulation factor, which is defined for the different methods in items 3–6 in Table 6.2. Alternatively, if a relationship is established between the modulation depth obtained using each method and a standard measure of modulation depth, then results from listening tests using the latter definition of modulation depth could be used as a good approximation. This procedure has been carried out for a method similar to the RES and RUK methods in the report by (AMWG 2015), as shown in Figure 6.19.

Hybrid Methods The hybrid methods have the advantage of reliable identification of modulation at the blade-pass frequency in the frequency domain, whilst maintaining a realistic measure of the modulation depth through analysis in the time domain. These methods can be used to batch process data, but the coding can be challenging as it requires implementation of all the requirements outlined in items 7 and 8 of Table 6.2 and described in detail in the text following the table. Since both hybrid methods focus on modulation at the blade-pass frequency and harmonics (in the case of the AMWG (2015) method), these methods do not enable detection of random amplitude variation.

Modulation Depth

In Figure 6.19 it can be seen that the relationship between the design modulation depth and the modulation depth calculated using different metrics depends on the definition of modulation depth. In the Salford study, the design modulation depth is the difference between the mean peak and the mean trough in the A-weighted RMS time series for any consecutive group of 12 pulses that occur during each 20-s block, as defined in Renewable UK (2013) and von Hunerbein and Piper (2015). In the study by Fukushima et al. (2013), referred to as the 'Tachibana method' in AMWG (2015), the modulation index (design modulation depth), m_{idx}, is found using Eq. (6.11).

$$m_{\text{idx}} = \frac{10^{\Delta L/20} - 1}{10^{\Delta L/20} + 1} \tag{6.11}$$

where $\Delta L = 10\log_{10}(A_{\max}^2/A_{\min}^2)$ and $A_{\max}$ and $A_{\min}$ are defined in Figure 6.20. Differences in the results obtained using the Salford and Fukushima definitions of design modulation depth can also be explained by the fact that the stimuli used by Fukushima et al. (2013) were more broadband and regular.

Conveniently, the relationship between the design modulation depth and the modulation depth calculated using different metrics is linear. Using data obtained from the AMWG (2015) report, the modulation depths calculated using the three types of metric described in Table 6.3 can be compared, as shown in Figure 6.21. This figure shows that slightly different results are obtained depending on whether the Salford or Fukushima

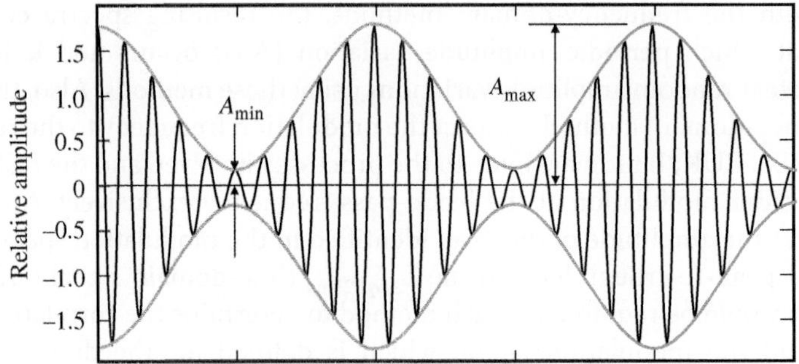

Figure 6.20 Definition of A_{max} and A_{min} used to define modulation depth according to Fukushima et al. (2013)

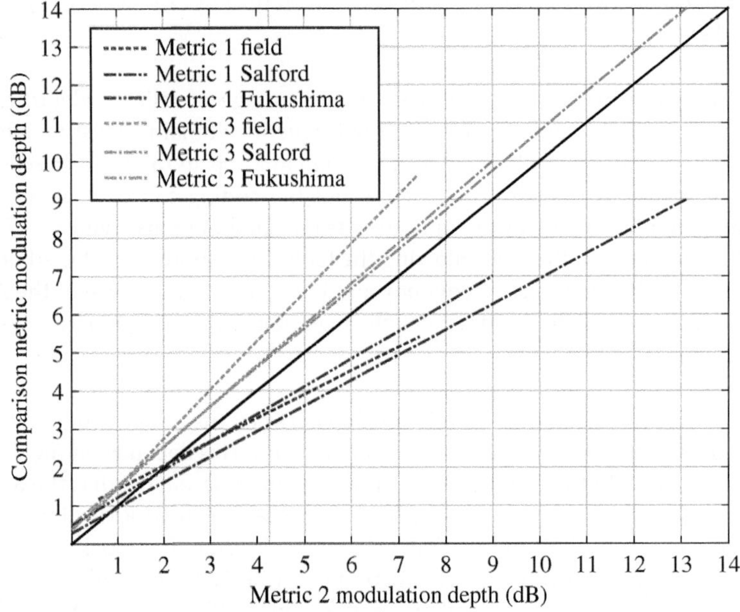

Figure 6.21 Comparison of the relative values of modulation depth obtained from the three different metrics using data from the Salford and Fukushima design methods as well as field data with extraneous noise filtered out of the analysis. Numbering of AM metrics is consistent with Table 6.3, where Metric 1 is represented by the D_{AM} method, Metric 2 is represented by the RES/RUK methods and Metric 3 is represented by the AMWG method. Each plotted curve represents data of the same type (i.e. the first curve shows a comparison between field data analysed using Metric 1 and 2).

definitions of design modulation depth are used or whether the data are measured in the field and direct comparison is made between the AM metrics. The general trends are the same though, and it can be seen that the modulation depth is consistently highest using Metric 3 (represented by the AMWG method) and consistently lowest using Metric 1 (represented by the D_{AM} method). Figure 6.21 provides a convenient reference for translating modulation depth results from one metric to another.

Comparison of Modulation Assessment Methods

The AMWG (2015) report also includes a comprehensive assessment of these three methods (Metrics 1, 2 and 3 of Table 6.3), using data measured near wind farms. Specific parameters that were common to all three modulation assessment methods were:

- sample using $L_{Aeq,100\text{ms}}$
- filter into 1/3-octave bands (Metrics 2 and 3).
- signal band limited to 100–400 Hz (Metrics 2 and 3)
- sample length (minor) = 10 s; this is the period of analysis to detect possible occurrences of AM/amplitude variation
- sample length (major) = 10 min, to maintain consistency with other monitoring procedures.

Statistics are obtained for the 'major' interval based on the number of occurrences of AM/amplitude variation in the non-overlapping 'minor' intervals. The modulation depth for each 10-min block of data is proposed as the 90th percentile of the distribution of AM values calculated in the 10-s blocks.

Free evaluation software for each of these three methods is available online on the Institute of Acoustics website (IOA 2015).

6.2.11 Psychoacoustic Descriptors

The character of wind farm noise is an important parameter that may not be measurable through metrics designed for identifying AM or amplitude variation. The concept was addressed by Large and Stigwood (2014), but a quantitative alternative for assessing noise character was not proposed. Persson-Waye and Öhrström (2002) obtained a psychoacoustic profile for different wind turbine noise signals but found that none of the psychoacoustic parameters such as sharpness, loudness, roughness, fluctuation strength or modulation could explain the differences in annoyance response for signals of equal equivalent sound pressure level. On the other hand, it is known that the character of wind farm noise varies between wind farms and is influenced by meteorological conditions, so psychoacoustic models may be relevant in other cases. Some of these models are therefore described briefly in this section. More details are provided in Fastl and Zwicker (2007).

The psychoacoustic model for fluctuation strength proposed by Fastl and Zwicker (2007) describes the perception of a periodically fluctuating sound (AM) and is relevant for modulation frequencies up to 20 Hz. The expression for calculating the fluctuation strength differs depending on whether the modulated noise is broadband or tonal, but the units are vacil in both cases. A unit of 1 vacil corresponds to subjective perception of a 60-dB, 1-kHz tone that is 100% amplitude modulated (that is, $A_{\min} = 0$ in Figure 6.20) at a frequency of 4 Hz. For broadband noise, the fluctuation strength, F, is calculated using Eq. (6.12).

$$F = \frac{5.8(1.25\mu - 0.25)(0.05L - 1)}{(f_m/5)^2 + (4/f_m) + 1.5} \tag{6.12}$$

Where f_m is the modulation frequency in Hz, L is the broadband sound pressure level in dB and μ is the modulation factor. The modulation factor can be determined from the depth of modulation (or modulation amplitude in dB), R_{mm} (see Section 2.2.7),

$$R_{mm} = 20\log_{10}\left(\frac{1+\mu}{1-\mu}\right) \quad \text{(dB)} \tag{6.13}$$

Here, the fluctuation strength is a non-linear function that is equal to zero until the modulation depth, R_{mm}, reaches approximately 3 dB (Lenchine 2009). The fluctuation strength increases with overall level and modulation depth and a maximum is reached for modulation frequencies of 4 Hz (Fastl and Zwicker 2007). For instance, the fluctuation strength corresponding to modulation at 1 Hz is approximately 50% of that at 4 Hz (Renewable UK 2013). The modulation becomes discernible above a fluctuation strength of 0.2 vacil according to Lenchine (2009). Measurements by DiNapoli (2011) showed that the fluctuation strength of broadband noise varied with distance from a wind farm consisting of 5×600 kW, stall-regulated wind turbines. A peak in the fluctuation strength occurred at approximately 630 m downwind from the nearest wind turbine and the fluctuation strength exceeded 0.2 vacil using both unweighted and overall A-weighted sound pressure level in Eq. (6.12). At other distances up to 2 km, the value of 0.2 vacil was only exceeded using the unweighted sound pressure level in Eq. (6.12).

The fluctuation strength, F, for amplitude-modulated tonal noise can be calculated by integrating the temporal masking depth, ΔL, in units of dB Bark, along the critical-band rate, z, as defined in Eq. (6.14) (Fastl and Zwicker 2007).

$$F = \frac{0.008\int_0^{24\,\text{Bark}}(\Delta L)\,dz}{(f_m/4)+(4/f_m)} \tag{6.14}$$

where f_m is the modulation frequency (usually the blade-pass frequency), ΔL is the masking depth (see next paragraph) and the critical band rate, z is equivalent to the Bark number, so the integration is over the Bark numbers from 0 to 24.

Limited guidance is provided by Fastl and Zwicker (2007) in calculating the masking depth, ΔL, so practical application of Eq. (6.14) to a measured wind farm noise signal is questionable. As mentioned by Persson-Waye and Öhrström (2002), since calculations of psychoacoustic parameters other than loudness are not standardised, calculations made using different commercial software packages may differ.

Loudness is a subjective scale that depends on the noise level and frequency for a stationary sound that is not varying in amplitude. Loudness is measured in sones, a unit based on sensory level. This scale was established from so-called direct measurement methods (Stevens 1956), asking listeners in an experiment to give proportional numbers to the loudness of sounds presented to them at different intensities. For non-stationary sounds, a loudness model has been developed by Fastl and Zwicker (2007) and this provides the loudness of a noise signal versus time. The calculation of loudness has been standardised in ANSI/ASA S3.4 (2007), ISO1996-1 (1975) and DIN 45631/A1 (2010). The first of these standards is based on the model of Moore and Glasberg and the latter two on the model by Zwicker, as described in Scheuren (2014).

At present, ISO1996-1 (1975) is under revision and will be released as two separate standards, ISO 532-1 and ISO 532-2, both defining alternative methods for calculation of loudness according to the Zwicker model and the Moore/Glasberg model (Scheuren 2014). For convenience, the ANSI/ASA S3.4 (2007) standard provides a computer program that can be used in the loudness calculation. A loudness toolbox has been developed for MATLAB® and was provided online by Genesis Acoustics in 2009. This toolbox also provides users with access to methods for calculating loudness of time-varying and impulsive sounds that were developed by Fastl and Zwicker (2007), Glasberg and Moore (2002), Boullet (2005) and Boullet et al. (2006). In Fastl and Zwicker (2007), loudness is denoted by the symbol N and N' is the loudness in each critical band. For time varying sound, the level exceeded or reached over a specified percentage of time can be specified. For example, N_5 is the loudness value exceeded or reached for 5% of the measurement time.

Rhythm has been identified as a meaningful characteristic component of wind farm noise (Large and Stigwood 2014). Fastl and Zwicker (2007) developed a model for rhythm, which is based on the temporal variation of loudness. Maxima are taken into account when they meet specific criteria; that is, when they are more than 120 ms apart, lie above 0.43 of the highest maxima, N_M, within a relevant time, and produce an increase in loudness with respect to the preceding maximum of at least 0.12 times the loudness corresponding to N_M. Further analysis would involve comparison between rhythmic patterns associated with wind turbine noise and patterns that invoke annoyance, which requires development of appropriate listening tests.

Other psychoacoustic descriptors that were considered by Persson-Waye and Öhrström (2002), such as sharpness and roughness, are not considered relevant in the analysis of low-frequency amplitude-modulated noise. According to Fastl and Zwicker (2007), the sharpness parameter is most influenced by the spectral content and the centre frequency of narrowband sounds, and the degree of sharpness reduces rapidly as the centre frequency decreases. The sensation of roughness starts to increase at modulation frequencies above 15 Hz, attaining a maximum at 70 Hz modulation frequency. Wind turbine noise is amplitude modulated at the blade-pass frequency and the first two harmonics. The blade-pass frequency is 0.5–1 Hz for large modern wind turbines and hence roughness is not a relevant psychoacoustic descriptor for wind turbine noise.

A measure of sound quality that takes into account aspects of fluctuation strength, loudness, sharpness and roughness associated with a noise signal is termed *psychoacoustic annoyance*. The psychoacoustic annoyance, PA, is defined as,

$$PA = N_5(1 + \sqrt{w_S{}^2 + w_{FR}{}^2}) \tag{6.15}$$

where

$$w_S = (S - 1.75)\ 0.25 \log_{10}(N_5 + 10) \quad \text{for} \quad S > 1.75 \tag{6.16}$$

and

$$w_{FR} = \frac{2.18}{(N_5)^{0.4}}(0.4F + 0.6R) \tag{6.17}$$

It is suggested that the parameters of sharpness, S, and roughness, R, can be set to zero for low-frequency-dominated wind farm noise as discussed earlier in this section. A comparison between the psychoacoustic annoyance of various car sounds is given in (Fastl and Zwicker 2007, Table 16.2) and this provides a basis for comparison.

6.2.12 Tonality

Sound with tonal components can create an adverse community response at lower levels than sound without such characteristics. Typically, these tonal components are not associated with wind turbines that have been well-designed and are maintained on a regular basis. Their presence can suggest a maintenance issue, often with the gearbox. Several standards are available that outline specific methodology for tonality assessment, and the most widely used of these are: IEC 61400-11 (2012), Annex C of ISO1996-2 (2007) and DIN 45681 (2005). It should be noted that ISO1996-2 (2007) is an updated version of the Joint Nordic method of assessment (DEPA 1984), which has also been adapted for implementation with computer software (Søndergaard and Sorensen 2013).

Since the method of tonality assessment described by ETSU (1996) was also based on the Joint Nordic method, it is reasonable to assume that it has now been superseded. All of the aforementioned methods of tonality assessment employ the psychoacoustic concept of critical frequency bands, which are frequency bands defined in such a way that tones outside a given critical band do not contribute significantly to tonal audibility inside that critical band. The noise level of potential tones within a critical band is compared to the noise level of masking noise in that same critical band to determine the audibility of the tone. In cases where a tone is clearly audible, a penalty is applied, often to the overall noise level. The penalty usually takes the form of an arithmetic addition of either a specified number of decibels to the measured noise or a number of decibels that is determined using a sliding scale based on tonal audibility (AS 4959 2010).

The IEC 61400-11 (2012) method of tonality assessment is applicable to measurements taken at a horizontal distance from the wind turbine equal to the hub height plus the rotor diameter. It has also been referenced in guidelines, such as those published by the SAEPA (2009), as a recommended method for determining the presence of tonality during compliance measurements at a residence, although it should be noted that it was not developed for this purpose. In the SAEPA (2009) guidelines, a penalty of 5 dBA is applied where the tonal audibility, ΔL_a, exceeds 0 dB, using the assessment outlined in IEC 61400-11 (2012).

Measurement

Currently, the most widely used method for tonal assessment of wind turbine noise is described in IEC 61400-11 (2012), which provides means to assess tones at frequencies as low as 20 Hz, a capability not matched by the other standards. Additionally, this standard specifically refers to the analysis of tonal noise from wind turbines, whereas other standards cover tonal noise from other sources as well. For these reasons, the IEC 61400-11 standard will be discussed in detail and then important differences between the three commonly used standards of tonality assessment will be highlighted. The main limitation of IEC 61400-11 is that assessments of tonality are based on measurements close to a wind turbine and no guidance is provided for the assessment of tonality either inside or outside of residences where complaints may originate.

According to IEC 61400-11, characterising tonality requires measurements of noise spectra as well as wind speed. It is recommended that measured noise spectra are allocated to a wind-speed bin. Each bin spans a wind speed of 0.5 m/s and the wind speed corresponding to the centre of each bin is an integer or half integer value in terms of metres per second. So the lowest wind-speed bin spans 0.25 to 0.75 m/s and has a centre speed of 0.5 m/s. All noise spectra recorded with a wind speed between 0.25 and 0.75 m/s will be allocated to this bin. Masking noise is the level of noise in the spectrum that does not include tones and its quantification is described in detail later in this section.

The requirements for measurement instrumentation according to IEC 61400-11 (2012) are the same as for turbine sound power measurements and are described in Section 4.6.1.

Analysis

The IEC 61400-11 (2012) methodology involves analysis of at least ten A-weighted spectra that are 10 s in length. The spectra, j, are calculated over 10-s time intervals using a Hanning window (see Section 2.4.4), with an overlap of at least 50% and a frequency resolution, Δf, between 1 and 2 Hz. All spectra are sorted into wind-speed bins, k, of 0.5 m/s width and centred around integer and half-integer wind speeds. For a wind-speed bin to be considered for tonality analysis, the same tone must appear in at least six out of the ten or more spectra under consideration for that bin. Tones in different spectra are considered to be the same tone (or 'tones of the same origin', as defined in IEC 61400-11 (2012)), if their frequencies lie within a range of ±25% of the critical bandwidth, corresponding to the frequency, f_c of the tone, where

$$\text{Critical bandwidth} = 25 + 75\left(1 + 1.4\left[\frac{f_c}{1000}\right]^2\right)^{0.69} \tag{6.18}$$

Next, the tonality for tones of the same origin is determined for each wind-speed bin, which is the difference between the tonal sound pressure level and a logarithmic (or energy) average of the remaining parts of the spectrum classified as masking noise. Finally, the tonal audibility is found by applying a frequency-dependent correction to the calculated tonality, which compensates for the response of the human ear to tones of different frequency. Where the average tonal audibility, ΔL_a, is greater than or equal to -3 dB in at least 20% of spectra and at least six spectra satisfy these conditions in a given wind-speed bin, the tone is reported, along with its audibility.

Procedures for calculating the tonal noise level $L_{t,j,k}$ for spectrum j in wind-speed bin, k and the corresponding masking level $L_{pn,j,k}$ are provided below under the appropriate headings. A flow chart is provided in Figure 6.22 to illustrate the procedure.

1. **Identifying possible tones from local maxima**
   - Local maxima are identified in each spectrum
   - The width of the critical band centred on each local maximum at frequency, f_c Hz, is calculated according to Eq. (6.18). For possible tones with frequencies, f_c, between 20 and 70 Hz, the critical band is defined from 20 to 120 Hz. There is no definition included in the IEC 61400-12-1 (2005) standard for infrasonic frequencies (<20 Hz). The width of the critical band as a function of frequency is illustrated in Figure 6.23.
   - The average energy in the critical band is calculated excluding contributions from the frequency line of the local maximum and the two adjacent lines. The average energy, $L_{p,av,j,k}$ for spectrum, j, in wind-speed bin, k, with a total of N lines included in the average is calculated as follows.

$$L_{p,av,j,k} = 10\log_{10}\frac{1}{N}\sum_{\ell=1}^{N}\left(10^{L_{p,j,k,\ell}/10}\right) \tag{6.19}$$

   - The local maxima are possible tones if they are 6 dB above the average energy in their respective critical band.

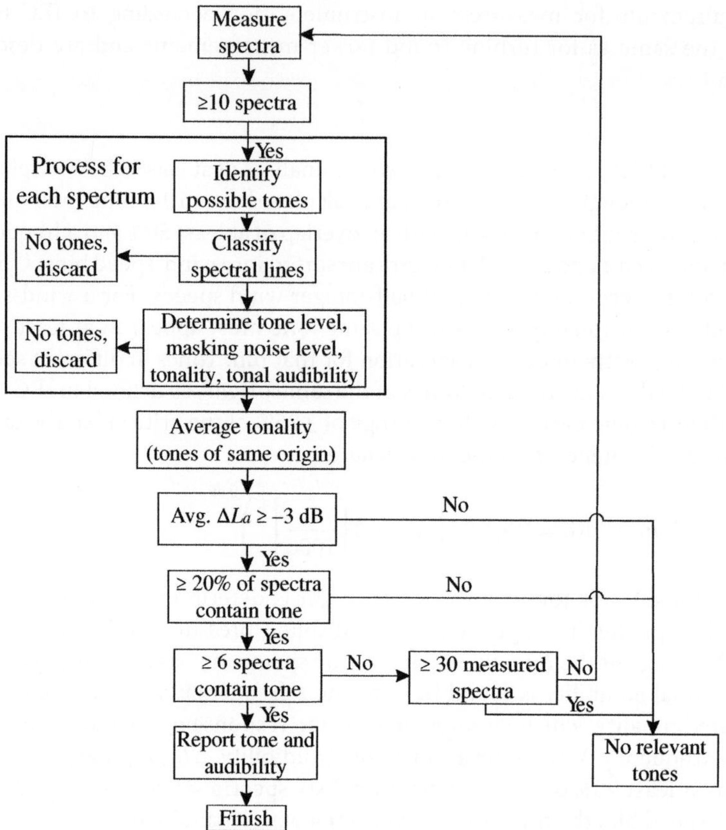

Figure 6.22 Flow chart illustrating tonality assessment process according to IEC 61400-11 (adapted from IEC 61400-11 (2012)).

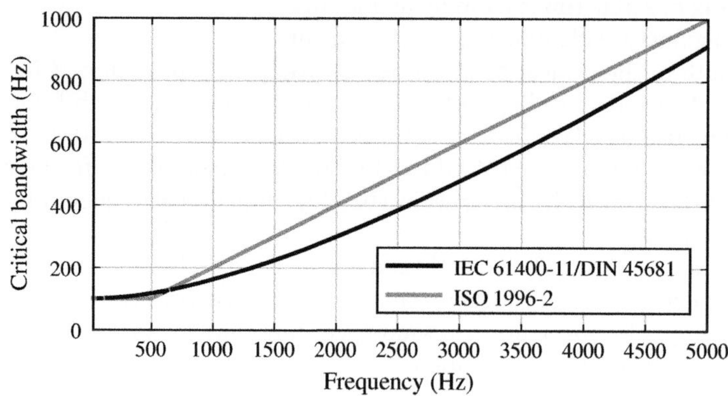

Figure 6.23 Critical bandwidths proposed in the IEC 61400-11 (2012), ISO1996-2 (2007) and DIN 45681 (2005) standards.

2. **Classifying spectral lines**
   For all spectra identified as having a tone that is common (of the 'same origin') in at least six spectra, the following procedure is followed to classify each of the frequency lines in the spectrum:
   - Within each critical band, all spectral (or frequency) lines are classified as a 'tone', 'masking' or 'neither'.
   - The $L_{70\%}$ is calculated, which is the energy average of the spectral lines which contain the lowest 70% of levels. For example, the energy-averaged sound pressure level, $L_{70\%,j,k}$, of N spectral lines in the critical band corresponding to spectrum j in wind-speed bin k is calculated as follows.

$$L_{70\%,j,k} = 10\log_{10}\frac{1}{N}\sum_{\ell=1}^{N}(10^{L_{70\%,j,k,\ell}/10}) \tag{6.20}$$

   - A spectral line is classified as 'masking' if its level is below $L_{70\%,j,k}$ + 6 dB. The corresponding masking sound pressure level is referred to as $L_{pn,j,k}$. The energy average (see Eq. (6.20)) of lines classified as masking is $L_{pn,\text{avg},j,k}$.
   - A spectral line is classified as 'tone' if its sound pressure level, L_{pt}, exceeds $L_{pn,\text{avg},j,k}$ + 6 dB. The spectral line in the critical band with the highest value of L_{pt} is identified as the tonal frequency of sound pressure level, $L_{pt,\max}$. Adjacent spectral lines are only classified as a tone if their levels are greater than $L_{pt,\max}-10$ dB. Note that this method includes all tones lying in the same critical band as the tone being considered.
   - All spectral lines not classified as 'masking' or 'tone' are classified as 'neither' and are subsequently excluded from the analysis.

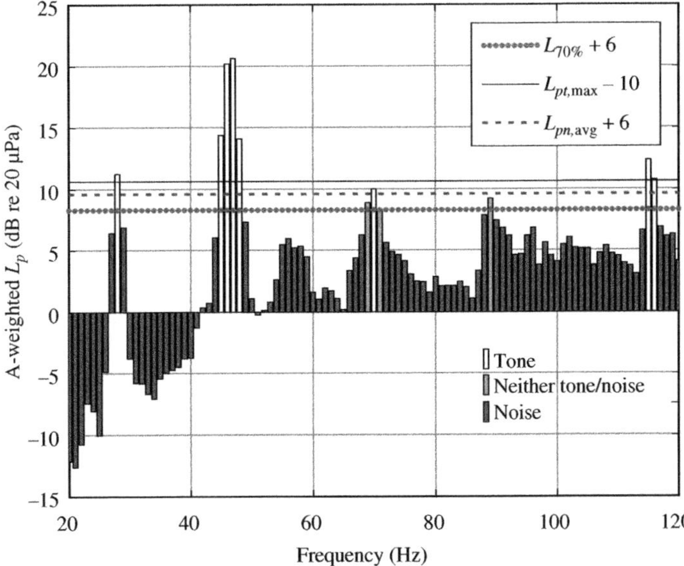

Figure 6.24 Criteria levels used for classification of spectral lines (frequency resolution: 1 Hz). SPL, sound pressure level.

- The criteria levels used in the classification of spectral lines are illustrated in Figure 6.24.

3. **Finding the tone level**
   - The tone sound pressure level, $L_{pt,j,k}$, is determined by summing the energy of all spectral lines that have been classified as tones in a given critical band. For example, if N spectral lines have been labelled 'tone', then the tone sound pressure level, $L_{pt,j,k}$, is given by,

$$L_{pt,j,k} = 10 \log_{10} \sum_{\ell=1}^{N} (10^{L_{pt,j,k,\ell}/10}) \tag{6.21}$$

   - A correction is necessary when there are two or more adjacent spectral lines labelled as 'tone', and this involves dividing these components in the energy sum on the right-hand side of Eq. (6.21) by 1.5 to account for use of the Hanning window. Note that only adjacent lines are adjusted in this way; other lines that are not adjacent to any others are not corrected.
   - Where tones in different spectra, j, occur within 25% of the critical band centred at f_c, they are considered tones of the same origin and are thus treated and reported as one tone.

4. **Finding the masking level**
   - The masking noise level, $L_{pn,j,k}$ is defined according to Eq. (6.22):

$$L_{pn,j,k} = L_{pn,\text{avg},j,k} + 10 \log_{10} \left[\frac{\text{Critical bandwidth}}{\text{Effective noise bandwidth}} \right] \tag{6.22}$$

   where the effective noise bandwidth $= 1.5\Delta f$, which includes a correction for use of the Hanning window and where Δf is the frequency resolution of the spectrum being considered.

5. **Determining tonality**
   - The tonality, $\Delta L_{tn,j,k}$, is defined as the difference between the tone level and the masking level in the corresponding critical band, $\Delta L_{tn,j,k} = L_{pt,j,k} - L_{pn,j,k}$.

6. **Calculating audibility**
   - Each value of $\Delta L_{tn,j,k}$ is corrected to compensate for the varying response of the human ear to tones of different frequency. This frequency-dependent audibility criterion, L_a, is given by Eq. (6.23).

$$L_a = -2 - \log_{10} \left[1 + \left(\frac{f_c}{502} \right)^{2.5} \right] \tag{6.23}$$

   - The tonal audibility, $\Delta L_{a,j,k}$, is defined as $\Delta L_{a,j,k} = \Delta L_{tn,j,k} - L_a$
   - The $\Delta L_{a,j,k}$ are energy averaged, using a similar procedure to that illustrated in Eq. (6.20), to give $\Delta L_{a,k}$ for each tone of the same origin. Only spectra with identified tones are included in the average.
   - Tones are reported when $\Delta L_{a,k} \geq -3$ dB and when an adequate number of spectra indicate the presence of the tone (as defined in Figure 6.24).
   - A tone is considered audible when $\Delta L_{a,k} > 0$ dB.

DIN45681 The DIN 45681 (2005) tonality assessment method is very similar to the IEC 61400-11 (2012) standard, although it is designed for analysis of tonal noise problems in general rather than those specific only to wind turbines. One limitation of this standard in the context of wind farm noise is that it can only be used for frequencies greater than or equal to 90 Hz; for analysis of lower frequencies, the DIN 45680 (2013) standard is recommended instead. This low-frequency standard is discussed later in this section. Some of the main differences between IEC 61400-11 (2012) and DIN 45681 (2005) are

- the method used for classification of masking noise
- the specifications of bandwidth of the tone and edge steepness
- inclusion of spectra without tones in the energy average of tonal audibility (as negligible values).

The other parts of DIN 45681 (2005) that are different to IEC 61400-11 (2012) are listed below, where the nomenclature adopted here is consistent with IEC 61400-11 (2012) to avoid confusion.

- It is valid only for frequencies greater than or equal to 90 Hz.
- 12 spectra are required and the measurement time for each is 3 s.
- Wind speed and the allowed deviation in direction from due downwind are not specified.
- Frequency resolution, Δf, of the spectra, is 1.9–4 Hz.
- Instead of calculating $L_{70\%}$+6 dB to classify spectral lines as masking, an iterative process is employed. The energy average of all spectral lines is calculated, excluding the spectral line corresponding to the local maximum. Then spectral lines that are 6 dB above the calculated value are removed and the energy average is calculated again. This procedure is repeated until the energy average does not change or until there are fewer than five spectral lines to the left or right of the local maximum.
- The distinctness of the tone must be evaluated to verify that the tone is audible to individuals with normal hearing. This parameter depends on the bandwidth of the tone and its edge steepness. The bandwidth of the tone, Δf_R, is found by multiplying the number of tonal spectral lines remaining after the previous step by the frequency resolution, Δf. The bandwidth must satisfy Eq. (6.24), where f_c is the frequency of the tone.

  $$\Delta f_R \leq 26(1 + 0.001\, f_c) \tag{6.24}$$

  The edge steepness must be at least 24 dB/octave for the peak to be identified as a tone. This is evaluated for frequencies below and above the tone using Eqs. (6.25) and (6.26):

  $$\Delta L_u = \frac{f_c}{\sqrt{2}} \frac{L_{tn,\max} - L_u}{f_c - f_u} \geq 24 \text{ dB} \tag{6.25}$$

  $$\Delta L_o = \frac{f_c}{\sqrt{2}} \frac{L_{tn,\max} - L_o}{f_o - f_c} \geq 24 \text{ dB} \tag{6.26}$$

  where $L_{tn,\max}$ is the maximum narrowband level of the tone, L_u and L_o are the levels of the first spectral lines above and below the tone, respectively, and f_u and f_o are the corresponding frequencies of these adjacent spectral lines.
- If exactly two tones occur in the same critical band at frequencies f_{c1} and f_{c2}, which are both less than 1000 Hz, they may need to be considered as separate tones. This is true if Eq. (6.27) is satisfied.

  $$|f_{c1} - f_{c2}| > 21 \times 10^A \tag{6.27}$$

where

$$A = 1.2\left|\log_{10}\left(\frac{\max(f_{c1}, f_{c2})}{212}\right)\right|^{1.8} \tag{6.28}$$

- The $\Delta L_{a,j,k}$ are energy averaged (as illustrated in Eq. (6.21)) to give $\Delta L_{a,k}$ for each tone of the same origin but spectra without identified tones are given a value of $\Delta L_{a,j,k} = -10$ dB and included in the average. This process would result in a lower overall tonal audibility, $\Delta L_{a,k}$, than calculated using IEC 61400-11 (2012).
- The uncertainty of $\Delta L_{a,j,k}$ cannot exceed 1.4 dB. An uncertainty analysis procedure is provided in Annex G of DIN 45681 (2005). This uncertainty is generally achieved, even with strongly fluctuating signals such as wind turbine noise, when 12 spectra are analysed. Therefore, it is only required when fewer than 12 spectra are analysed.
- A variable tonal adjustment, K_T, is specified as an addition to measured values of L_{Aeq}. The adjusted L_{Aeq} level is then compared to the allowed level as a compliance check. The value of K_T is dependent upon the value of $\Delta L_{a,j,k}$, where the maximum adjustment is 6 dB.

The ISO1996-2 (2007) Annex C method of tonality assessment is applicable to general environmental noise. In principle, it is similar to the IEC 61400-11 (2012) and DIN 45681 (2005) standards, the main differences being the classification of masking noise and the method used for detecting tones, which involves the identification of *noise pauses*. These differences are expanded upon below and some other minor differences are also highlighted.

- Narrowband analysis must be carried out on at least one 1-min data sample, which may be divided into smaller segments for tones that vary in frequency between the original segments by more than 10% of the critical bandwidth, centred on the tone.
- Critical bandwidths are 100 Hz for tones between 50 and 500 Hz. Above 500 Hz, the critical bandwidth is equal to $0.2f_c$, where f_c is the frequency of the tone, and f_c lies at the centre of the critical band. If two or more tones lie in the critical band defined for one of the tones, the critical band shall be positioned with the most significant tones near the centre of the band in such a way that the difference between the sound pressure level of the tone and that of the masking noise is maximised. For these purposes a significant tone must have a peak sound pressure level within 10 dB of the level corresponding to the highest-level tone.
- The effective analysis bandwidth is required to be less than 5% of the critical bandwidth. This defines the maximum frequency resolution, Δf, which is equal to the effective analysis bandwidth divided by 1.5 for a Hanning window.
- Wind speed and direction are not specified.
- Identification of tones can be achieved through visual inspection or via a computer-implemented calculation process. Visually, local maxima in the frequency spectrum under consideration are considered as tones if the 3-dB bandwidth is less than 10% of the critical band, as illustrated in Figure 6.25.
- For computer implementation, local maxima with a probability of being a tone are defined as noise pauses. The start of a noise pause is found on the positive slope of a local maximum and is defined as the spectral line s that satisfies Eqs. (6.29) and (6.30). The end of a noise pause is found on the negative slope of a local maximum

and is defined as the spectral line e that satisfies Eqs. (6.31) and (6.32). A noise pause is illustrated in Figure 6.25.

$$L_s - L_{s-1} \geq \Delta \tag{6.29}$$

$$L_{s-1} - L_{s-2} < \Delta \tag{6.30}$$

$$L_e - L_{e+1} \geq \Delta \tag{6.31}$$

$$L_{e+1} - L_{e+2} < \Delta \tag{6.32}$$

The tone seek criterion, Δ, is usually set to 1 dB, but it may be adjusted up to 3–4 dB for short averaging times.

- Tones are found within noise pauses and are classified as such when any spectral line is 6 dB or more above the levels of the spectral lines $s - 1$ and $e + 1$.
- The above criterion is overruled when tones vary in frequency by more than 10% of the critical band within the averaging period, which is 1 min to begin with. In this situation, a shorter averaging time is recommended.
- All spectral lines adjacent to the local maxima and within 6 dB of it are classified as 'tone', for the purposes of determining the tonal sound pressure level. The tonal sound pressure level is obtained by logarithmically summing the sound pressure levels for all the spectral lines classified as tones, in the same way as was done for the IEC 61400-11 (2012) method, taking into account the Hanning window effect for adjacent spectral lines classified as tones; see Eq. (6.21) and the discussion that follows.
- If the total tone level in a critical band is below the hearing threshold, this tone is not included in the tonal audibility assessment.
- All spectral lines that are not classified as noise pauses are considered to contribute to masking noise. To find the average value of these spectral lines, denoted the 'average noise level', $L_{pn,\text{avg}}$, a linear regression fit is found, as shown in Figure 6.25. The range of the regression analysis is usually ± 0.75 of a critical bandwidth around the centre frequency, f_c, but can be extended to ± 1–2 of a critical bandwidth where appropriate. The average noise level is then calculated based on the integral of the noise level under the regression curve in the critical band, using Eq. (6.33), for the case of N spectral lines classified as contributing to masking noise:

$$L_{pn,\text{avg},j,k} = 10 \log_{10} \frac{1}{N} \sum_{\ell=1}^{N} (10^{L_{p,\ell}/10}) \tag{6.33}$$

The masking noise sound pressure level, L_{pn} is then found from the average noise level using Eq. (6.22).

- To account for the effect of the tone on annoyance, the measured L_{Aeq} for a time interval is adjusted upwards by K_T dB. Values of K_T are specified in ISO1996-2 (2007) and depend on the value of the tonal audibility, ΔL_a, and the centre frequency of the critical band containing the tone (usually the tonal frequency). Where multiple tones exist, the adjustment is calculated for each tone and the resulting highest value of K_T

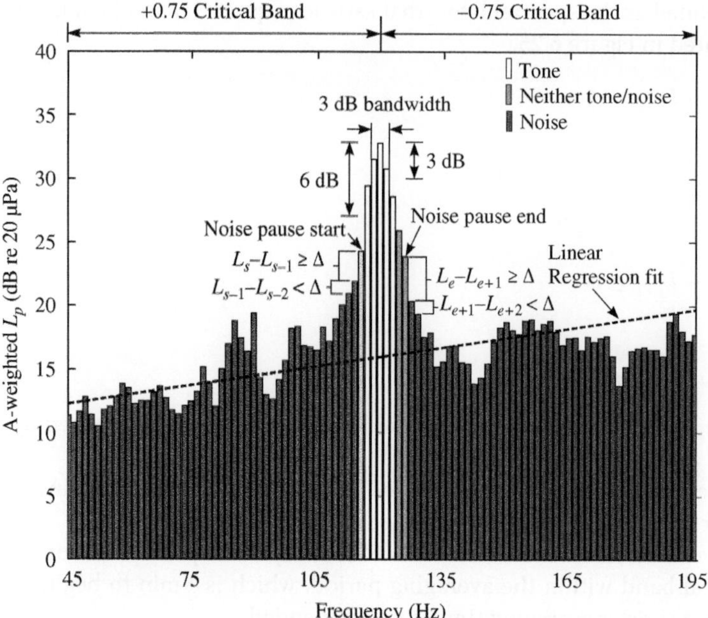

Figure 6.25 Definition of tones, noise pause and noise, where Δ is the tone seek criterion, usually set to 1 dB. In this figure, the local maximum is shown in white and the 6 dB point is shown to illustrate which spectral lines would be considered as tones. Spectral lines corresponding to the boundaries of the noise pause, *s* and *e*, are shaded in grey. SPL, sound pressure level.

is used to adjust the L_{Aeq} measurement. Figure C.1 in ISO1996-2 can be used to derive the following relationships to calculate K_T directly.

$$K_T = \begin{cases} \Delta L_{a,j,k} - 2 & f_c \leq 250 \text{ Hz} \\ \Delta L_{a,j,k} - 1.7 & f_c = 500 \text{ Hz} \\ \Delta L_{a,j,k} - 1.2 & f_c = 1000 \text{ Hz} \\ \Delta L_{a,j,k} - 2.44 \log_{10} f_c + 4.5 & f_c > 1000 \text{ Hz} \end{cases} \tag{6.34}$$

Values of K_T are constrained to lie in the range between 0 and 6 dB.

A simplified method of tonality assessment is also provided in Annex D of ISO1996-2 (2007). This method involves comparison between the 1/3-octave band where the possible tone exists and the adjacent 1/3-octave bands. Level differences that are indicative of a possible tone are as follows:

- 15 dB in the low-frequency 1/3-octave bands (25–125 Hz).
- 8 dB in middle-frequency bands (160–400 Hz).
- 5 dB in high-frequency bands (500–10 000 Hz).

This procedure does not always identify tones that are found to be audible using other methods (Evans and Cooper 2015; Gange 2013) and therefore it should be used with caution. One reason that this method can prove unsuccessful in the analysis of wind farm tonal noise is that the tone is often amplitude modulated at the blade-pass frequency, resulting in multiple side bands adjacent to the main tonal peak. As a consequence, the noise levels in 1/3-octave bands adjacent to the one containing the tone may contain higher noise levels than would be present if the wind farm were not operating, resulting in the tone being less prominent than it should be. A similar simplified method method for tonality assessment is specified in DIN 45680 (2013), which covers the frequency range from 10 to 80 Hz. The difference between the assessment of DIN 45680 (2013) and the assessment of ISO1996-2 (2007) is that in the former case:

- the relationship between the time that the tone exists in a measurement period and the duration of the measurement period is taken into account;
- a tone is deemed to be present when the difference between adjacent 1/3-octave bands is 5 dB;
- guidance is provided for the case where a tone lies within the transition band of two 1/3-octave bands.

Discussion and Implementation

Søndergaard and Pedersen (2013) performed listening tests using different high quality headphones with a frequency response down to at least 16 Hz. A double-blind testing scheme was implemented and the assessors comprised a listening expert panel (23 assessors) and acoustic experts (4 assessors), all with normal hearing for their age. The stimuli used in the listening tests included 24 samples, 20 s in length, that were taken from recordings of wind turbines. Results from the listening tests indicated good agreement between the perceptual assessment of tonality carried out by listeners and the tonal audibility calculated using both the IEC 61400-11 (2012) and ISO1996-2 (2007) methods. This implies that both methods of tonality assessment work well in the assessment of tonality associated with short-term measurements of wind farm noise. On the other hand, in the analysis of longer measurements, the IEC 61400-11 (2012) and ISO1996-2 (2007) methods followed the same trends but yielded different magnitudes. Differences in the results were attributed to non-stationary aspects of the tones that were accentuated when the analysis period was extended to 1 min. Comparison between the IEC 61400-11 (2012) and ISO1996-2 (2007) methods was also carried out by Evans and Cooper (2015), and it was found that the ISO1996-2 (2007) method resulted in a higher number of detections of tonality for measurements at a residence. Furthermore, Kobayashi et al. (2015) observed that the tonal audibility calculated using the ISO1996-2 (2007) method was higher than that calculated using IEC 61400-11 (2012).

Measurements of tonality at seven different wind turbines were carried out by Evans and Cooper (2015) and it was found that for each turbine, an audible tone occurred that may not have been detected through assessment carried out in strict accordance with IEC 61400-11 (2012). The reason for this is that the tones frequently occurred outside of the wind speed and direction ranges specified by the standards. For example, results measured at one wind turbine indicated that a tone occurred more frequently during upwind/crosswind, low wind speed conditions. At another wind turbine, a tone occurred more frequently during crosswind, high wind speed conditions. Therefore it was suggested that a wider range of wind directions and wind speeds should be specified in future versions of IEC 61400-11 (2012). Analysis of tonality at a residence located 1.5 km from a wind farm was carried out by Cooper et al. (2015) and similar results were obtained. The identified tone appeared to occur most often during crosswind conditions at hub-height wind speeds between 6 and 7 m/s. Cooper et al. (2015) also observed that the tone occurred more frequently at night and therefore this should be taken into account when extending the use of IEC 61400-11 (2012) to measurements at a residence. Although the South Australian guidelines (SAEPA 2009) specify use of IEC 61400-11 (2012) for tonal assessment near residences, they do not elaborate on how this standard should be adapted for measurements at a residence. Guidance for measurement at a residence has been provided by the DEPA (2011). According to this document, tonality assessment is undertaken at the receiver with the highest noise level using the Joint Nordic method of assessment (which is similar to the ISO1996-2 (2007) method). The wind speed and direction are measured near the wind turbine at a height of 10 m, and tonality assessment can be undertaken when this wind speed is in the range 6–8 m/s and the residence is downwind (within ±45°) from the nearest wind turbine. Data are assessed when the tone is most prominent and, when there is a clearly audible tone, a penalty of 5 dB is applied. The definition of 'clearly audible' is not expanded upon. According to the results of Cooper et al. (2015) and Evans and Cooper (2015), a tone could still go undetected using this method due to the limited range of wind speeds and directions analysed.

A tonality assessment was undertaken by Kobayashi et al. (2015) on data measured at a large number of locations in Japan using IEC 61400-11 (2012). At all 29 emission points (close to a turbine) that were investigated, tonal audibility of greater than − 3 dB was detected. At the immission points (close to residences) that were located between 100 m and 1 km from the nearest wind turbine, a tonal audibility of greater than − 3 dB was detected at 129 measurement points from a total of 164 positions. At both the emission and immission points, the tonal audibility ranged from greater than − 3 dB up to 14 dB and occurred in the frequency range between 50 and 1000 Hz. As part of the study, listening tests were designed whereby tones were added to synthesised wind turbine noise (a spectrum shape of − 4 dB/octave) and then listeners were asked to evaluate the difference between the baseline signal and signals containing tones of varying audibility at a number of different frequencies. An unexpected result was that for evaluation responses described as 'considerably different' and 'definitely different', the tonal audibility was approximately 6 dB for tonal frequencies of 50, 100, 200 and 800 Hz, whereas it was only 2 dB for a tonal frequency of 400 Hz. This is unexpected since the IEC 61400-11 (2012) method includes a correction to compensate for the varying response of the human ear to tones of different frequency.

An alternative method of tonality assessment was developed by Liu et al. (2012) to improve the consistency and accuracy of tone identification. The success of this approach is based on the reduction of background noise from the wind turbine noise signal as well as optimal segmentation of data blocks for analysis. The method uses order tracking to reduce the background noise while preserving the tonal components of interest. Order tracking (see Section 2.4.9) is an established technique for analysing signals relating to the rotational speed and harmonics of rotating machines using a tachometer signal. Rather than using uniform 10-s segments as specified in IEC 61400-11 (2012), the signal is divided into relatively stationary segments based on the rotational speed of the wind turbine. This technique takes advantage of the close relationship between the rotor speed and the tonal noise associated with a wind turbine. The drawback of the method is that a tachometer output from the wind turbine is required and this may not be available to some investigators.

6.2.13 Additional Turbine Noise Analysis Techniques

To ascertain more detailed information about the spectral content of acoustic signals measured in the vicinity of a wind farm, a narrowband analysis can be carried out. Results measured at a residence 3.3 km from the nearest wind turbine in the Waterloo wind farm are shown in Figure 6.26. This plot was obtained using a frequency resolution of 0.1 Hz, a Hanning window and 50% overlap. The frequency resolution was chosen to ensure that the blade-pass frequency could be distinguished. The narrowband analysis reveals many important details that can be overlooked through focussing solely on 1/3-octave band levels. The most crucial advantage of viewing the narrowband spectra is that AM of a spectral peak can be identified. AM is characterised by sidebands adjacent to the spectral peaks that have spacing equal to the blade-pass frequency. In Figure 6.26, the most prominent amplitude modulated spectral peaks occur at 23.1, 28 and 46.6 Hz and the corresponding blade-pass frequency is 0.8 Hz.

A sonogram is a useful way of showing the variation in sound pressure level as a function of frequency and time simultaneously. An example of a sonogram is shown in Figure 6.27, where data in the upper and lower plots were obtained during shut-down and operational conditions, respectively. The x- and y-axes of a sonogram show the time and frequency, respectively, and the contours indicate the corresponding sound pressure level in decibels. The sonogram is obtained by calculating the FFT over segment lengths that are chosen depending on the desired frequency resolution, as discussed in Section 2.4.8, so that the product of the frequency resolution in hertz and the time in seconds used to acquire one segment of data, is one. The amount of overlap used between segments determines the time resolution. Comparison between the upper and lower sonograms in Figure 6.27 indicates that there are three obvious tones between 10 and 100 Hz that are related to wind farm operation. Also, the level of infrasound above 2 Hz is clearly higher during operational conditions and the noise is amplitude modulated at the blade-pass frequency, as indicated by the periodic change in sound pressure

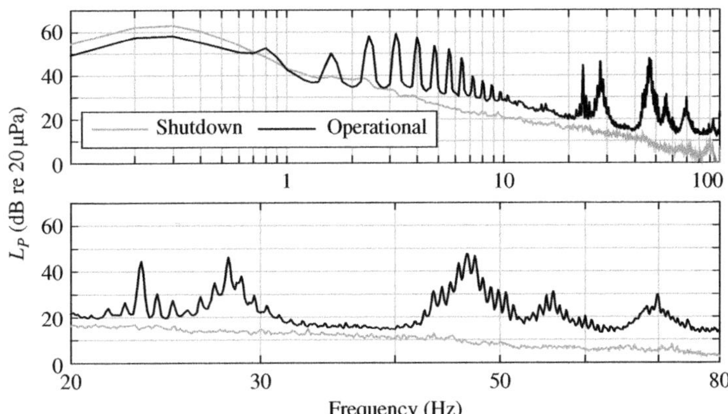

Figure 6.26 Narrowband plots corresponding to noise measured 3.3 km from the nearest wind turbine in the Waterloo wind farm during shut-down and operational conditions between 12 midnight and 5am. Frequency resolution is 0.1 Hz. Blade-pass frequency of 0.8 Hz is clearly evident in the plot representing operational conditions. The lower plot is zoomed in with respect to frequency to show tonal peaks with side-bands spaced at the blade-pass frequency, which indicates the presence of AM. SPL, sound pressure level.

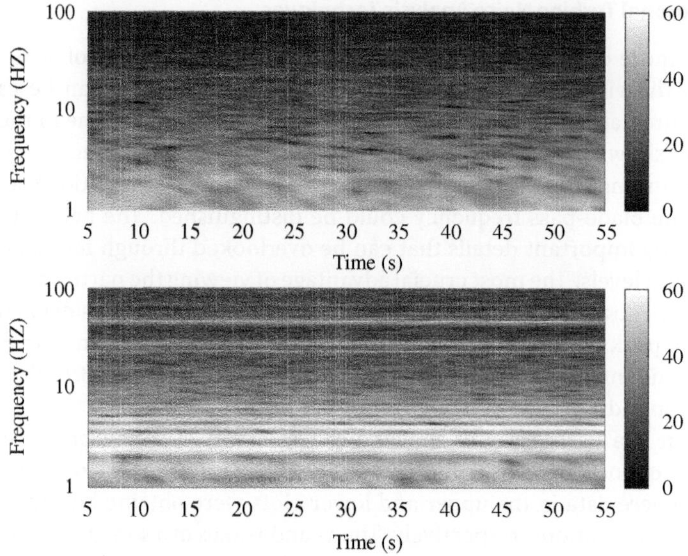

Figure 6.27 Sonogram plot using the same data as for Figure 6.26 where shut-down conditions are shown in the upper plot and operational conditions shown in the lower plot. The frequency resolution is $\Delta f = 0.8$ Hz and the time resolution is $\Delta t = 0.625$ s.

level as a function of time. The AM of the tones at 23.3, 28 and 46.6 Hz is indicated by an increase in sound pressure level at the sideband frequency, although this is difficult to see in Figure 6.27. Therefore, alternative techniques such as the Hilbert transform (see Section 2.4.10) are recommended for identifying AM at frequencies above approximately 10 Hz.

A method for extracting a periodic infrasound pulse from measurements taken near a wind turbine whose rotation is not precisely periodic was proposed by Vanderkooy and Mann (2014). This technique is based on order tracking (see Section 2.4.9), which is an established technique for analysing signals relating to the rotational speed and harmonics of rotating machines using a tachometer signal. The method of Vanderkooy and Mann (2014) involves the use of an optical telescope fitted with a photodetector that produces a pulse (equivalent to a tachometer signal) each time a blade passes through the telescopic field of view. The optical telescope is directed towards the point of rotation at which maximum contrast is achieved between the blade and the sky and acoustic data are measured simultaneously. The tachometer signal is then used as a time base to resample the acoustic data (using interpolation between actual samples) so that samples corresponding to all records are aligned in time. Extraction of the first 15 harmonic lines in the resulting averaged spectrum by removing the noise between spectral lines, and then taking an inverse FFT allows identification of an infrasound pulse that the authors postulate is generated each time a blade passes the support tower. Vanderkooy and Mann (2014) observed that in some cases, the infrasonic pulse would not be detected without use of the method. It is anticipated that in the future a variation of this method can be used to isolate the infrasonic pulses generated by individual turbines within a wind farm (Vanderkooy and Mann 2014).

6.2.14 Compliance Testing

The aim of compliance testing is to verify that a new wind farm development adheres to noise limits. This is a requirement of some regulatory authorities and it is also a form of reassurance for members of the local community that wind farm noise emissions do not exceed predicted values. In some cases, such as when a conservative prediction model has been used, the relevant regulatory authority may not require compliance testing (AS 4959 2010). In these instances, measurements of the wind turbine generator sound power level may be required to verify that the levels do not exceed those used in the predictions of sound pressure level at sensitive receiver locations. Additionally, compliance measurements may still be required in such areas if there are complaints related to noise from the wind farm operation.

The relevant receiver locations chosen for compliance measurements should be residences or sensitive receivers that were identified by the propagation model as the most likely to be subjected to the highest wind farm noise levels. It is preferable that compliance measurements are taken at the same locations as background measurements to allow accurate comparison to be made between operational and non-operational conditions. However, in some cases it is not possible to shut-down the wind farm during weather conditions that are similar to those occurring when operational noise measurements were obtained. In these cases, background data measured prior to wind farm construction from the nearest location are used instead.

Measurement

The measurement methodology used in compliance measurements is identical to that specified for background measurements as described in Section 6.2.5. Any additional information that can be gained from the wind farm operator is beneficial to the analysis during operational conditions. These data include but are not limited to: wind speed and direction at hub height, power output of the nearest wind turbine and of the wind farm as a whole, and rotor rotational speed.

As with the measurement of wind turbine noise in general, a significant challenge associated with wind farm noise compliance measurements is that wind turbines are only operational during windy conditions, meaning that measurements can often be contaminated by wind-induced noise at the microphone location. This is true especially during the daytime, when most compliance measurements tend to be made as a matter of convenience. When there is a significant amount of wind at the measurement location, it is difficult to separate wind farm noise levels from background noise levels for the purposes of checking compliance with the noise limits. Background noise levels are also highly susceptible to change across seasons and years, particularly in a rural environment, due to variations in noise from sources such as foliage, streams, surf, livestock, insects, agricultural machinery and others.

Analysis

Compliance procedures outlined in standards and guidelines take issues of background noise variability into account to varying degrees, and some of the adopted methodology is discussed below. The Australian standard (AS 4959 2010) specifies two possible methodologies for compliance testing: using the background noise-monitoring procedure and on/off testing.

Using the Background Noise Monitoring Procedure While it is emphasised that it would be desirable to measure the $L_{Aeq,10min}$ values, it is argued that due to the occurrence of extraneous noise, it is more practical to measure the total $L_{A90,10min}$ level with the wind farm operational and then use this value for the wind-farm-only $L_{Aeq,10min}$ level. In a somewhat ambiguous manner, AS 4959 (2010) suggests that the $L_{A90,10min}$ level (wind farm plus background) is equivalent to the $L_{Aeq,10min}$ level (wind farm only) due to the well-known difference of 1.5–2.5 dBA between the $L_{Aeq,10min}$ and $L_{A90,10min}$ levels for wind farm noise. This implies that the background level increases the total level by at least 1.5 dB, which corresponds to a difference between wind farm only and background only levels of only 4 dB. For example, if the wind farm were compliant according to the South Australian EPA guidelines (SAEPA 2009), the measured noise level could be 40 dBA during compliance measurements. In this case, the equivalence between $L_{Aeq,10min}$ (turbine only) and $L_{A90,10min}$ (turbine plus background) implies that the background level is 36 dBA, which is quite unrealistic for rural communities (where many wind farms are located), particularly at night. In these cases, the $L_{A90,10min}$ and $L_{Aeq,10min}$ equivalence would underestimate the $L_{Aeq,10min}$ by 1.5–2.5 dBA and possibly more.

The $L_{A90,10min}$ data collected from compliance testing according to AS 4959 (2010), thus represent the $L_{Aeq,10min}$ levels corresponding to wind farm noise only. These levels are plotted against the wind speed (generally at hub height) to enable regression analysis. Data corresponding to the wind direction from the wind farm towards the relevant receiver (downwind direction) would normally be of primary interest to the local regulatory authority (AS 4959 2010). The wind speed data should include a range of wind speeds between 'cut in' and 'rated' wind speeds (that is, from the wind speed at which the turbines start to produce useful power output up to the wind speed corresponding to the rated power of the turbine model). The line of best fit is usually of the same order polynomial or lower as was used in the background-only regression analysis. At this point any relevant penalties for special audible characteristics are added to the regression curve obtained during wind farm operating conditions. Comparison can then be made with the allowable $L_{Aeq,10min}$ limit and also between the wind farm noise and background noise regression curves to determine compliance.

The regression analysis approach is also followed in the New Zealand standard (NZS 6808 2010), although there are some differences in the methodology. In this standard, the $L_{A90,10min}$ is considered as a 'consistent objective measure' of wind farm noise since wind farm noise has 'regular temporal characteristics'. This does not agree with the results of a significant amount of research that has been carried out on wind farm AM, as summarised by Renewable UK (2013). Based on the assumption that the $L_{A90,10min}$ accurately characterises wind farm noise, NZS 6808 (2010) specifies logarithmic subtraction of the background $L_{A90,10min}$ noise levels from the operational $L_{A90,10min}$ levels at each integer hub-height wind speed. This procedure is carried out on the two regression curves, not the individual data points, for both the background and operational measurements. The $L_{A90,10min}$ levels given by the resulting regression curve are then compared to allowable L_{Aeq} levels to establish compliance. There are a number of problems with this approach as follows.

1. It is not valid to directly use $L_{A90,10min}$ data as a comparison with $L_{Aeq,10min}$ criteria as these quantities are not equivalent.
2. Subtracting averaged data is not equivalent to subtracting the simultaneously existing background noise from individual measurements and then calculating an average

of the results. However, it is not possible to do this because the background noise existing at the time of the measurements is unknown; subtracting averaged data may be the best compromise possible.

3. Background noise levels at a residence are usually correlated with the wind speed at the residence, but the wind speed at the residence is not necessarily correlated with the wind speed at hub height so it should be of no surprise that there is usually a large scatter in the plots of background noise versus wind speed at hub height.
4. Logarithmic subtraction is an invalid operation for percentile data (that is, $L_{A90,10\text{min}}$) and readers are referred to Nelson (1973) in which it is discussed how to combine noise from different time-varying sources to arrive at wind-farm-noise-only data from a combination of wind farm and background noise. To implement this method, the $L_{A10,10\text{min}}$ needs to be measured concurrently with the $L_{A90,10\text{min}}$ and the noise level distribution must be Gaussian. To verify that the distribution is in fact Gaussian, the L_{Aeq} should be averaged over shorter intervals, where the averaging time is selected to give enough data points to show the nature of the distribution for each 10-min measurement period.

The ETSU guidelines (ETSU 1996) present a similar methodology to the Australian and New Zealand standards (AS 4959 2010; NZS 6808 2010) for determining compliance, and the $L_{A90,10\text{min}}$ is the level of interest. However, one difference is that the wind speed is either measured at 10-m height or standardised to 10-m height from measurements at an alternative height. This would lead to differences in turbine electrical power output for a given 10-m wind speed due to wind-shear variations. It is expected that this would reduce the value of the coefficient of determination (R^2 value) and hence the reliability of the analysis. These limitations are more relevant for modern wind turbines, which are much taller than the wind turbines that existed when the standard was written. The difference between the $L_{Aeq,10\text{min}}$ and $L_{A90,10\text{min}}$ levels is accounted for when setting compliance limits by arithmetically subtracting the value of 2 dB from the allowable $L_{Aeq,10\text{min}}$ level to obtain the allowable $L_{A90,10\text{min}}$ level. This is a different approach to that adopted in the Australian and New Zealand standards discussed above. In cases where non-compliance is identified, the levels are corrected by logarithmically subtracting the background noise from the combination of wind farm noise and background noise. Alternatively, the measurements are repeated at times of lower background noise.

The wind speed range considered in the analysis is limited to 12 m/s at 10-m height in ETSU (1996), but this appears to cover the range of wind speeds at least to the 'rated' wind speed. According to ETSU (1996), if a wind farm demonstrates compliance at wind speeds less than 12 m/s measured at 10-m height, it is unlikely to cause any greater loss in amenity at higher wind speeds. Some additional details provided in ETSU (1996), include the specification of an expected measurement duration of at least 1 week and the recommendation that at least 20–30 measurements are taken within ±2 m/s of the 'critical wind speed' (which is not explicitly defined in the main document, but may be assumed to be the wind speed at which the predicted receiver noise levels are highest) and that at least ten measurements lie either side of the critical wind speed. The good practice guide to the application of ETSU-R-97 (IOA 2014) recommends that monitoring should cover a duration of at least 1 month.

On/Off Testing The second methodology specified by AS 4959 (2010) for compliance testing is to obtain $L_{Aeq,10min}$ data when the wind farm is both shut-down and operational ('on/off' testing) during worst-case wind conditions at the receiver of interest. A critical wind speed is defined, which is the wind speed with the smallest predicted margin of compliance, which usually corresponds to the highest predicted noise level at the receiver under consideration. Measurements are required to cover a wind speed range of at least 6 m/s, centred on this wind speed. A minimum of ten measurement intervals above and below the critical wind speed are specified for the operational measurements and a similar number of measurements with similar wind speeds are required for shut-down conditions.

The regression analysis is the same as for the other methodology, except that the plotted variable is $L_{Aeq,\ 10min}$, corresponding to wind farm noise only. Nevertheless, the $L_{A90,\ 10min}$ is still measured and where differences greater than 3 dBA occur between the $L_{Aeq,10min}$ and the $L_{A90,10min}$ it is assumed that the measurement was dominated by extraneous noise and the $L_{A90,10min}$ + 1.5 dB measurement is used. In this second methodology, the derived background $L_{Aeq,10min}$ levels are logarithmically subtracted from the derived operational levels at each integer wind speed. In this context, 'derived' refers to the value that lies on the relevant regression curve.

The New Zealand standard (NZS 6808 2010) also prescribes methodology for 'on/off' testing, but it is considerably different from the Australian standard. More specifically, the measurement periods are 2 min rather than 10 min, the $L_{A90,2min}$ is measured exclusively rather than the $L_{Aeq,10min}$ and wind speed measurements are not required. Background measurements are taken immediately after operational measurements, as wind conditions are assumed to be similar during 'on' and 'off' conditions. This assumption may not be valid during variable wind conditions. To ensure that a suitable range of operating conditions are covered, the power output data are used in place of the wind speed data. Wind turbines that generate combined levels that are 10 dBA lower than the highest contributing turbine can continue to operate during 'shut-down' conditions. The good practice guide to application of ETSU-R-97 (IOA 2014) reserves 'on/off' testing for situations of non-compliance.

Extraneous Noise In both of the methodologies discussed above, it is standard procedure that measurements contaminated by extraneous noise are excluded from the analysis. The most obvious sources of contamination are rain, as discussed in Section 6.1, and wind-induced noise. Associated data points can be identified and excluded in an automated manner. Other extraneous sources may be identified through listening to the audio files associated with outliers in the analysis. However, these outliers may not have a notable effect on the regression fit and hence it is recommended to examine their effects before investing time in listening to audio files. An example showing the influence of outliers on the regression fit is provided later in this section.

Discussion

Demonstration of compliance is a necessary condition for operation of a wind farm, but in spite of this a number of wind farms around the world are subject to complaints from residents living in the near vicinity. Aside from psychological factors, the most logical reason for these complaints is that compliance methodologies that are currently used contain inadequacies. These inadequacies originate from four main issues:

1. In the case of long-term compliance measurements, the specification of data points to be included in the analysis is not specific enough and this leaves scope for variation and possible bias in interpretation.
2. The 'average' noise level as defined by the regression curve does not necessarily represent the level of wind turbine noise and therefore the data points that lie above this curve may represent 10-min periods where the levels of wind farm noise exceeded the relevant limits, sometimes by quite large amounts.
3. The wind speed at hub height is not necessarily correlated with the levels of background noise at the residence. This is only applicable to compliance methodology such as ETSU (1996) and NZS 6808 (2010), which specify subtracting background levels from compliance measurements, where the relevant background levels are determined based on hub-height wind speed.
4. Wind turbine noise is time-varying and is therefore not adequately characterised using the $L_{A90,10\text{min}}$ parameter.

These four issues are discussed in detail below:

Inadequate Specification of Data Points The various compliance methodologies prescribed in the standards differ in the areas in which they are specific and non-specific with regards to inclusion of data points. For example, ETSU (1996) requires that data are collected during specific wind speeds but the wind direction need only be 'representative' for the region. The Australian and New Zealand standards (AS 4959 2010; NZS 6808 2010) require that a range of data points are collected between 'cut-in' and 'rated' wind speeds such that there is a uniform distribution of data points. The required wind directions are not explicitly stated in either of these standards but the Australian standard (AS 4959 2010) states that data collected during downwind conditions would be of 'primary interest'. This statement can be interpreted in a number of different ways during data analysis. For example, downwind points could be considered exclusively or could make up a percentage of the analysed data and these variations would affect the final result.

While ETSU (1996) acknowledges that measuring during the nighttime minimises the influence of extraneous noise, it still only states that it 'may be necessary' to perform measurements at night, which means that this is not a requirement. The Australian and New Zealand standards (AS 4959 2010; NZS 6808 2010) do not mention a suitable measurement period, which implies that all times of day are valid. When it comes to removal of extraneous noise, ETSU (1996) specifies that data collected during periods of rainfall should be removed but that measurements affected by human and animal activity are representative of the local noise environment and therefore should be included in the analysis.

The New Zealand standard (NZS 6808 2010) specifies removal of all data influenced by extraneous noise where practical, whereas the Australian standard (AS 4959 2010) requires only that data points influenced by rain and wind-induced noise are removed from the analysis. It is stated that 'the possibility of insects, fauna or other extraneous noise sources affecting data should also be considered' but there is no guidance on what should be done with the associated data points. The possibility of seasonal variations in the level of background noise is widely acknowledged (AS 4959 2010; ETSU 1996; NZS 6808 2010), but there is no requirement that the background and compliance measurements be taken during the same season. Seasonal variations are only considered if they have a significant effect on the analysis and therefore subtle differences can pass unnoticed.

The reason that it is important for compliance methodologies to be specific with regards to data filtering is that removal of large numbers of data points will affect the position of the regression curve. Removal of data points affected by extraneous noise can be a subjective process and must be consistent over the background and compliance data sets, which is difficult to implement. Rather than removing data points randomly, it is more sensible to analyse quiet periods only, such as nighttime, where data contamination by extraneous noise is expected to be minimal.

Further justification for considering nighttime data only is that annoyance and sleep-disturbance effects are likely to be most critical during this time. The quietest nighttime period generally occurs between 12am and 5am, although this period can be extended, depending on insect, bird and animal activity. The morning bird chorus should be avoided as this will have a significant impact on the measurements. Insect noise may be seasonal and thus avoidable, but it is also possible to introduce a low-pass filter at 1000 Hz for measurements affected by insect noise. In most cases, wind farm noise will be insignificant at frequencies above 1000 Hz for distances greater than 1 km, due to atmospheric and ground absorption, although this should be confirmed for the specific case using propagation analysis.

The wind direction is another important parameter that affects the analysis, because wind farm noise is expected to be dominant under downwind conditions. Therefore, the compliance methodology should specify that downwind data are to be used exclusively. Analysing data from other wind directions effectively averages out the worst case wind farm levels, which is not useful. The regression method should maximise the contribution of wind farm noise and minimise the contribution of extraneous noise and then advise on an acceptable number of data points that can exceed the regression curve, as discussed in the next subsection.

To minimise the effect of seasonal and yearly variations, background noise measurements should be made just prior to the wind farm operating for the first time. Compliance measurements should then be made as soon as the wind farm becomes operational.

Another methodology that reduces the impact of variations in background noise is the on/off testing discussed above. The disadvantage of this technique is that it results in large power variations, which need to be managed within the electricity system for large grid-connected wind farms. Also, shut-downs result in reduced power output, which can be costly for the wind farm operator. Another drawback is that this type of testing often involves attended measurements (AS 4959 2010) and therefore shut-downs are scheduled for daytime hours, which is not useful because the maximum difference between wind farm noise and background noise is at night.

Representativeness of the Regression Curve The second assumption is that the 'average' noise level indicated by the regression fit is indicative of the level of wind turbine noise. This implies either that up to 50% of 10-min data points that lie above the regression curve can be attributed to extraneous noise sources and/or that it is acceptable for the wind farm to exceed compliance limits up to 50% of the time. This method appears to be a compromise that is used to account for the existence of extraneous noise, but it could result in a large number of exceedances of compliance limits. Therefore, as mentioned in the last subsection, nighttime data collected under downwind conditions should be plotted exclusively. Then the regression analysis procedure would need to be modified to account for the reduced presence of data points affected by extraneous noise. This could involve translating the regression curve in the positive-dB direction to a position where a smaller proportion of the data exceeded the curve.

Correlation of Hub-height Wind Speed and Noise Levels The third assumption – that the wind speed at hub height is correlated with background noise levels at residences – is incorrect because there is no consistent relationship between the wind speed and direction at hub height and the wind speed and direction at the residence where the measurement microphone is located. This is due to the variation in atmospheric conditions over time, which results in differences in the wind shear and directional shear (Van den Berg 2005). This concept is discussed in more detail in Section 6.4.2. The variation in wind speed and direction is not only a function of height, but also of location, due to influences of the surrounding topography. For example, a residence that is located in the vicinity of a wind farm situated on a ridge-top can experience lower wind speeds in the downwind direction than in the upwind direction due to being located in the lee of a hill for the former case.

The lack of correlation between the wind speed and direction at hub height and at the residence means that it is generally not valid to use the hub-height wind speed to determine the background noise level for the purposes of logarithmic subtraction from the wind farm noise level (including background) as specified in standards and guidelines (NZS 6808 2010; SAEPA 2009). It is more correct to use the wind speed and direction at the residence to determine the associated levels of background noise, as these levels will be influenced by wind-induced noise, which is governed by the local wind speed. In summary, although wind farm noise at a residence is generally well correlated with the combination of wind speed and direction at hub height, background noise levels are not. Thus logarithmic subtraction of background noise levels from operational noise levels using regression curves based on hub-height wind speed is generally an invalid process that cannot be justified.

Measurements should be made using a suitable weather station positioned at a height of 1.5 m, as close as possible to the microphone but not so close that the weather station produces significant levels of noise. In cases where there are tall trees in the near vicinity of the microphone, it may be better to use the wind speed at a height of 10 m rather than 1.5 m, as this may be better correlated with background noise generated by the local wind.

It has been suggested that regression analyses should use the local wind speed at the residence exclusively, as this allows compliance testing on behalf of the community when hub-height data are not provided by the operator (Cooper and Evans 2013). However, the lack of correlation between the wind speed and direction at hub height and at the receiver would also lead to poor correlation of turbine noise with wind speed at the residence. Therefore, the best compromise is to measure the wind speed and direction at both hub height and at microphone height near the measurement microphone and to plot the level of wind turbine noise (determined by subtracting the background level at the corresponding integer wind speed measured at the residence) against the wind speed at hub height. Thus wind speed at the receiver is used to determine background noise levels (from the corresponding regression curve) and wind speed at hub height at the turbine is used to determine wind farm noise levels (from the regression curve of wind speed at hub height). In this instance, the local regulatory authority should make it compulsory for the wind farm to provide operational data, in relevant time intervals, such as the wind speed at hub height and the power output of each wind turbine, in a spreadsheet format that can be readily accessed by anyone.

Changes in Noise Levels over Time The fourth assumption – that wind turbine noise does not vary rapidly with time – is inconsistent with extensive literature on the subject. This assumption is reflected in the specification of the $L_{A90,\,10\text{min}}$ as an acceptable parameter to characterise wind turbine noise. Wind farm noise can be significantly underestimated by $L_{A90,\,10\text{min}}$ levels because peaks that are associated with unsteady effects such as AM are not present for 90% of the time and hence their disturbance potential is never evaluated. Also, thresholds of audibility are not dependent on the 10- or 15-min average of the RMS value of the noise signal alone. However, these are the overall dBA levels that are measured and reported according to wind farm guidelines and regulations, and are often erroneously compared with laboratory-determined hearing threshold levels. This type of comparison ignores differences in character between the measured noise and the noise used in the laboratory to determine threshold levels. The main differences in character that are important include the presence of multiple harmonics of the blade-pass frequency and the crest factor of the noise. The measured average noise levels for wind farm noise have been shown to contain peaks that are up to 15 dB above the reported average level (Zajamšek et al. 2016). Even for 'compliant' wind farms, such peaks are well above the levels required to disturb sleep (Hurtley 2009). It is also worth noting that traffic noise, on which the WHO document on nighttime noise levels is based, is not characterised by such high crest factors and thus has less potential for disturbing sleep. In addition, it is well known that noise dominated by low-frequency energy (such as wind farm noise) is more disturbing than noise with a broad frequency spectrum, such as traffic noise, on which the WHO findings are based.

Suggested Improvements in Approaches to Compliance Testing

There are several suggestions that could be considered for improvements to the current methodology of compliance testing and some of these have been described in the literature. The suggestions described below are not an exhaustive list.

Taking Measurements at a Location other than the Residence This is suggested for cases where it has been established that background noise is adversely impacting the results.

Concurrent Measurement of Background Noise in Four Cardinal Directions These directions need not be aligned with the compass axes, but measurements should be taken at distances far enough from the wind farm that its impact is negligible but close enough that the background levels are representative (Hessler 2011). Locations should be representative of the environment in which compliance measurements are taken. The locations are chosen to be as far apart as possible from one another (that is, 90°) to minimise the occurrence of extraneous noise events that affect more than one location at the same time.

Filtering Data by Wind Direction See Smith and Chiles (2012).

Use of Bin Analysis This involves collating noise level data into wind-speed bins that are 1 m/s wide (Smith and Chiles 2012). Thus, all noise data corresponding to a range of wind speeds are allocated to a wind speed corresponding to the nominal bin speed. For example, all sound pressure level data collected for wind speeds between 5.5 and 6.5 m/s would be associated with the 6 m/s wind-speed bin. The data can then be treated in one of two ways. The first would be to average all sound pressure level data in each wind-speed bin to obtain a single data point for the regression line. The average could be arithmetic, as is implied by the current regression analysis procedure, or it could be logarithmic (see Section 2.2.8), which would be more appropriate as it represents an energy average, which is how the L_{Aeq} metric is calculated. The second way would be to fit separate regression lines to the data in each bin and then find the sound pressure level corresponding to the nominal wind speed for the bin (6 m/s in the above example). This effectively represents an arithmetic averaging of data.

Approximation from Measurements at Reduced Wind Speeds Only At low wind speeds, the background sound pressure level is independent of wind speed and sound level meters may be close to their noise floor. At higher wind speeds, the rate of increase in sound pressure level with wind speed begins to reduce. Therefore, the quality of the curve fit applied in the regression analysis may be improved by considering a narrower wind speed range (Smith and Chiles (2012)). This concept has already been introduced to the IEC 61400-11 (2012) standard, in which analysis occurs over the wind speed range from 0.8 to 1.3 times the wind speed at 85% of maximum power.

Others Other suggestions include:

- removal of outliers (Smith and Chiles 2012)
- collection of 1/3-octave data and raw time data for further analysis
- calculating the proportion of time that the hearing threshold is exceeded in each 1/3-octave band
- determining the proportion of time that the allowable A-weighted level is exceeded by wind farm noise.

It is unlikely that measurements in a location other than the residence or not including the full range of operational wind speeds would provide confidence to residents concerned about possible non-compliance. Suitable methods should be used to both minimise the influence of background noise on the results and to distinguish between components of the noise that can be attributed to wind farm operation and background noise. Data filtering is an effective tool because it provides a means of separating data to highlight trends and to select conditions under which wind turbine noise is likely to be most important.

Implementation of Alternative Methods of Compliance Testing

In the following investigation, data collected at an unoccupied residence located 3.1 km from the nearest wind turbine in the Stage 5 Hallett wind farm are analysed below using some of the methods that were discussed in the previous subsection. The measurement period at this residence spanned eight weeks and during this time, noise was monitored using several microphones located in an open field, approximately 50 m from the residence, with various wind screen configurations to protect against wind-induced noise. The results presented below were recorded using a single GRAS 40AZ microphone with a SVANTEK SV 12L pre-amplifier connected to a SVANTEK 958 sound level meter. This microphone was mounted on the ground and protected from wind-induced noise by a hemispherical wind screen designed to be consistent with IEC 61400-11 (2012). Weather data were collected in the near vicinity of the outdoor microphones using Davis weather stations mounted at heights of 1.5 m and 10 m. Davis weather stations were also positioned on top of the nearest hill at heights above the ground of 1.5 m and 10 m and these measurement heights were close to hub height for the nearest wind turbine, located approximately 4 km from the hill. This set-up was used because hub-height wind speed and direction data were not available from the wind farm operator. While it would be ideal to measure the weather conditions at the wind turbine location, the data are still considered representative.

Figure 6.28 shows results for a period of 3 days and 4 nights, illustrating the large variations in noise level, wind speed and wind direction that occur over time. Comparison between the $L_{Aeq,10\mathrm{min}}$ and the $L_{A90,\,10\mathrm{min}}$ indicates that there is a significant amount of extraneous noise during the daytime, particularly within daylight hours. This is evident when there is a large difference in the $L_{Aeq,\,10\mathrm{min}}$ and the $L_{A90,\,10\mathrm{min}}$ levels and is expected since the residence is located in a farming area, where farming machinery is in use. During the night, there are some peaks in the $L_{Aeq,10\mathrm{min}}$ data but generally there is a consistent difference between the $L_{Aeq,\,10\mathrm{min}}$ and the $L_{A90,\,10\mathrm{min}}$ levels. Another notable feature in this plot is that the wind speed and direction at hub height and at the residence are often uncorrelated, which further confirms that it is not suitable to subtract background noise as a function of the wind speed at hub height, as discussed in the previous section. It is also interesting to note that a resident living on the same farm found

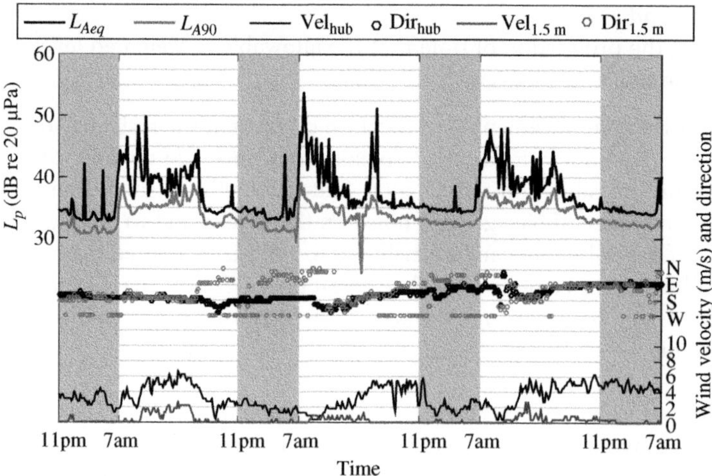

Figure 6.28 Variation over 3 days and 4 nights of $L_{Aeq,10\mathrm{min}}$, $L_{A90,10\mathrm{min}}$ and wind speed/direction (hub height and 1.5 m) for a residence located 3.1 km from the nearest turbine of the Hallett stage 5 wind farm. SPL, sound pressure level.

the wind farm noise annoying in the early hours of the morning during this period, but the wind farm would have been compliant according to local regulations (SAEPA 2009), which specify an allowable $L_{Aeq,\ 10\text{min}}$ limit of 40 dBA.

Noise data collected between 29 May and 16 July 2014 at all times of the day and night are plotted against the wind speed at hub height in Figures 6.29 to 6.34. There are 6893 data points in total, with 2681 data points corresponding to downwind conditions (light grey points). In each figure, these same data are analysed using various methods and where the entire data set is plotted, 'all data' is displayed in the legend. A quadratic regression fit has been used in all cases, where applicable, and the corresponding R^2 value is displayed on the plot.

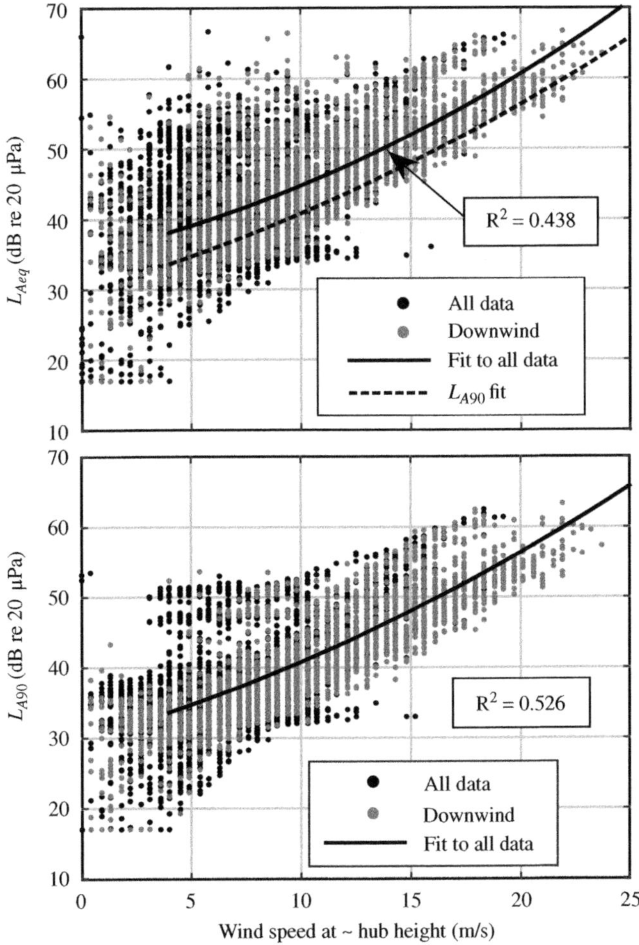

Figure 6.29 Regression analysis of (top) $L_{Aeq,\ 10\text{min}}$ and (bottom) $L_{A90,\ 10\text{min}}$ against wind speed measured at the approximate hub height. Data were measured at a residence located 3.1 km from the nearest turbine of the Hallett stage 5 wind farm during operational conditions. The regression fit is to all data in both figures, but excludes data where the wind speed at 1.5 m is greater than 5 m/s. The L_{A90} regression fit is also shown as a dashed line in the upper figure for comparison with the $L_{Aeq,\ 10\text{min}}$ line.

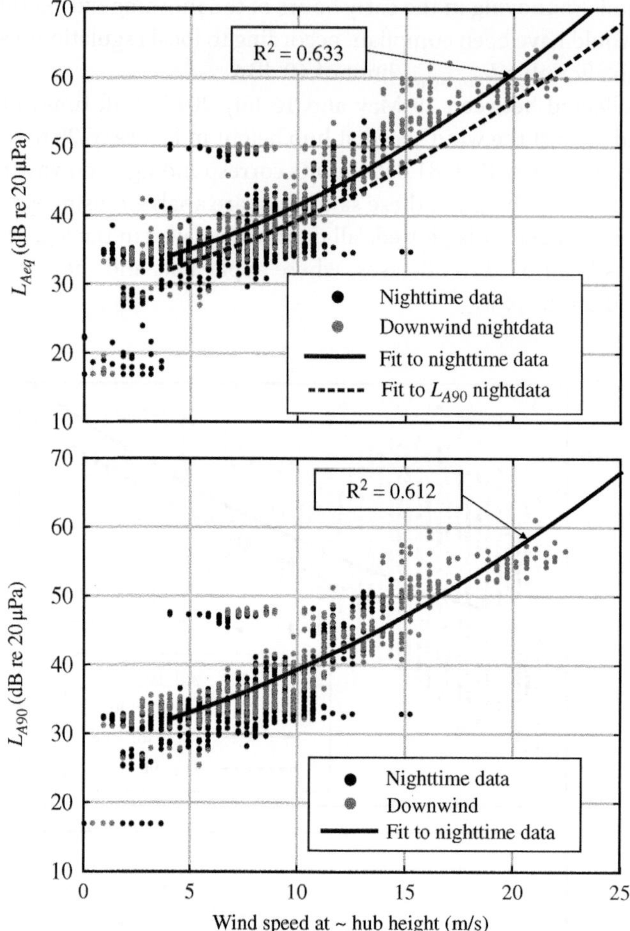

Figure 6.30 Regression analysis of (top) $L_{Aeq,10\mathrm{min}}$ and (bottom) $L_{A90,10\mathrm{min}}$ against wind speed measured at the approximate hub height during nighttime (in this case 12am–5am). Data were measured at a residence located 3.1 km from the nearest turbine of the Hallett stage 5 wind farm during operational conditions. The regression fit is to all data in both figures, but excludes data where the wind speed at 1.5 m is greater than 5 m/s. The L_{A90} regression fit is also shown as a dashed line in the upper figure for comparison with the $L_{Aeq,10\mathrm{min}}$ line.

The results of a standard regression analysis are presented in Figure 6.29, where it is evident that the regression fit between the $L_{A90,10\mathrm{min}}$ data and the wind speed at hub height is better than the corresponding fit for the $L_{Aeq,10\mathrm{min}}$ data, as expected. On the other hand, the R^2 value in both cases is relatively low and this is attributed to extraneous noise sources that occur during the day, as shown in Figure 6.28. The noise floor of the instrumentation of 17 dBA can be seen in the plots, although this does not affect the regression fit, which is applied to data measured above the wind turbine cut-on speed of 4 m/s. At this residence, the data do not reflect any consistent increase in noise level under downwind conditions.

Comparing the regression curves, it can be seen that there is a consistent 4-dB difference between the $L_{Aeq,10\mathrm{min}}$ and the $L_{A90,10\mathrm{min}}$ data. This is higher than the

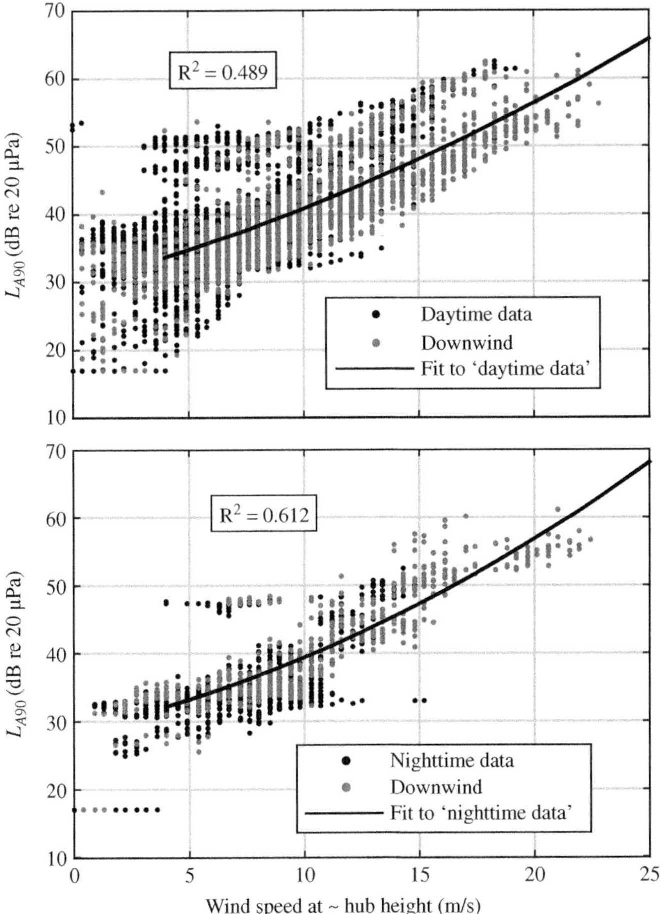

Figure 6.31 Regression analysis of $L_{A90,\,10\text{min}}$ for daytime and nighttime (12am–5am) against wind speed at the approximate hub height. Data were measured at a residence located 3.1 km from the nearest turbine of the Hallett stage 5 wind farm during operational conditions. The regression fit is to daytime data and nighttime data in the upper and lower figures, respectively.

expected difference of 1.5 dB between the $L_{A90,\,10\text{min}}$ and the $L_{Aeq,\,10\text{min}}$ levels as proposed by AS 4959 (2010). On the other hand, the true significance of this observation can be investigated by plotting the data measured between 12am and 5am, which has been minimally affected by extraneous noise sources. This plot is shown in Figure 6.30 and it can be seen that in the critical wind speed range between 'cut-in' and 'rated' power (4–14 m/s) (Suzlon, Ltd 2012), the difference between the $L_{A90,\,10\text{min}}$ and the $L_{Aeq,\,10\text{min}}$ levels ranges from 2–3 dB. This is still higher than the expected difference of 1.5 dB.

The regression fit obtained by plotting nighttime (in this case 12am–5am) $L_{A90,\,10\text{min}}$ data against the wind speed at hub height is significantly better than the daytime regression fit. This is not surprising when considering the large variations in noise level that occur during the day in this region, as shown in Figure 6.28. In spite of this, the regression fit itself is very similar in both cases. Nevertheless, it cannot be assumed that this would always be the case and, based on the increase in the R^2 value for nighttime

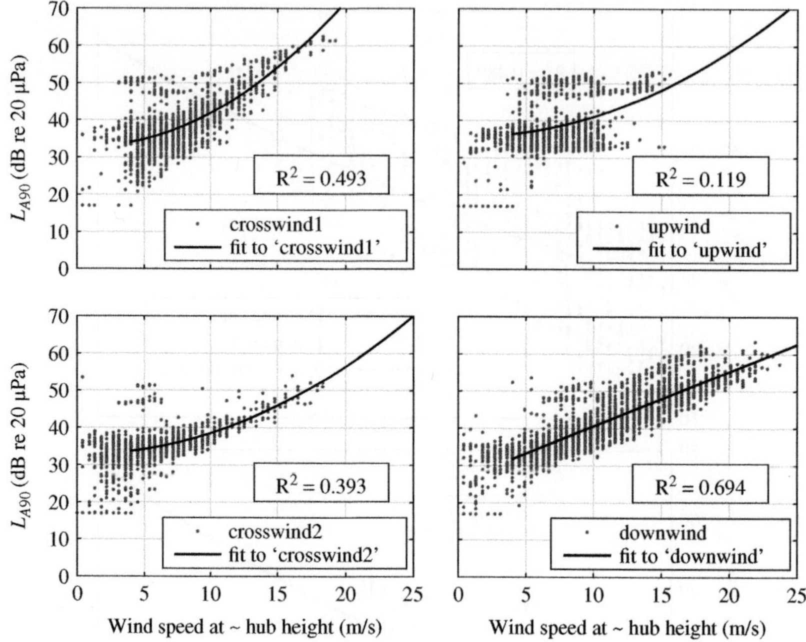

Figure 6.32 Regression analysis for data divided into four directions, 90° apart. Data were measured at a residence located 3.1 km from the nearest turbine of the Hallett stage 5 wind farm during operational conditions.

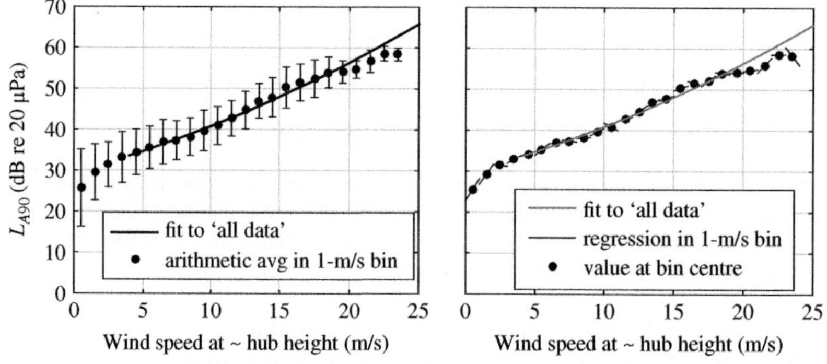

Figure 6.33 Arithmetic averaging and linear regression analysis in bins of width 1 m/s for $L_{A90,10min}$ against wind speed measured at the approximate hub height are shown as possible alternatives to a standard regression fit. Data were measured at a residence located 3.1 km from the nearest turbine of the Hallett stage 5 wind farm during operational conditions. The analyses have been carried out on all data.

conditions, it is suggested that nighttime data could be measured exclusively. This also corresponds to the time when people are most annoyed by wind farm noise as they are either trying to relax or sleep. The definition of nighttime would be specific to a given region and it is perhaps better to define this loosely to allow attainment of the best regression fit. At this residence, the R^2 value decreased when the nighttime hours were extended beyond the 12am–5am period.

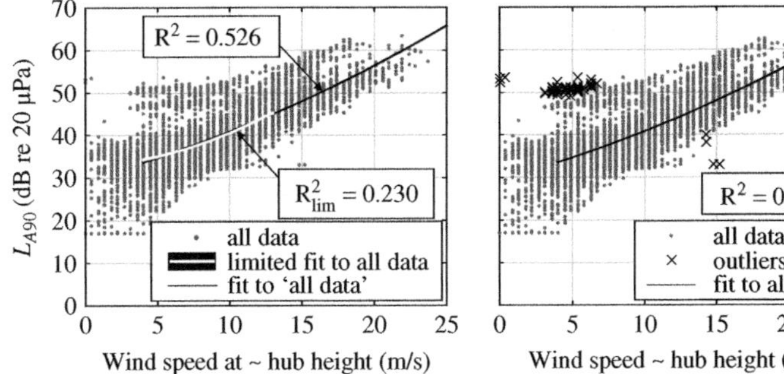

Figure 6.34 Effect of limiting the hub-height wind speed range (left figure) and removing outliers on the regression analysis (right figure) for $L_{A90,10\text{min}}$ against wind speed measured at the approximate hub height. There is no discernible difference between the regression fits obtained by including and excluding outliers from the analysis. Data were measured at a residence located 3.1 km from the nearest turbine of the Hallett stage 5 wind farm during operational conditions.

To determine the effect of wind direction on the regression fit, $L_{A90,10\text{min}}$ data were divided into four equally-spaced directions and plotted against the wind speed at hub height as shown in Figure 6.32. Here, the regression fit is significantly better for downwind conditions, as would be expected. This means that the apparent lack of direction dependence in Figure 6.29 does not reflect the true situation in this case.

Arithmetic averaging and linear regression analysis carried out in 1-m/s wind-speed bins are shown in the left- and right-hand plots of Figure 6.33 as alternative methods to the standard regression analysis carried out on the entire data set. Bin analysis involves collating groups of data that correspond to a range of wind speeds, with the usual range spanning 1 m/s; for example, one bin may include all wind speeds from 2 to 3 m/s and have a nominal speed of 2.5 m/s. For the data presented here, there are no clear advantages of these alternative methods using bin analysis, since the data points align well with the standard regression curves. This occurs over the majority of the operational range, which includes hub-height wind speeds from 4–25 m/s. Some discrepancies are noted above 20 m/s, although wind farm noise is not likely to be an issue at such high wind speeds due to a combination of lower predicted noise emissions and higher wind-induced noise.

The effect of limiting the wind speed range used to determine the regression curve is shown in the left-hand plot of Figure 6.34. For this data set, the regression curves calculated over the operational and limited wind speed ranges, respectively, are almost indistinguishable, although the R^2 value is significantly lower for the regression fit over the narrow wind speed range. Therefore, limiting the wind speed range used in the regression analysis does not appear to offer any advantages in this case. In the right-hand plot of Figure 6.34, selected data points that have been identified as possible outliers are indicated by black crosses. Removing these data points from the analysis has a negligible effect on the position of the regression curve, although there is a corresponding increase in the R^2 value, as expected.

Another possible method for determining compliance is to analyse the frequency spectra measured in the vicinity of a wind farm and this is most conveniently achieved using 1/3-octave band analysis. Figure 6.35 shows a 1/3-octave plot of the noise measured outdoors during shut-down and operational conditions near a residence located

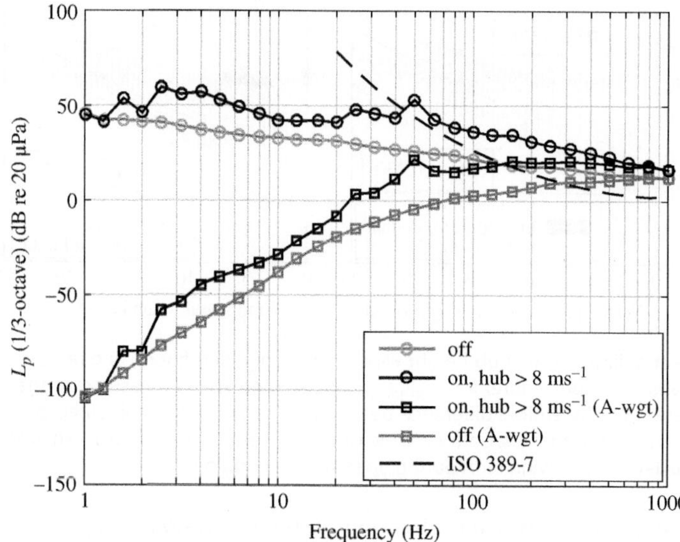

Figure 6.35 Comparison between 1/3-octave unweighted (circles) and A-weighted (squares) spectra for operational and shut-down conditions when measured approximately 3.3 km from the Waterloo wind farm.

3.3 km from the Waterloo wind farm. Details of the measurement set-up are discussed in Hansen et al. (2014). In Figure 6.35, the A-weighting has been applied in the time domain for frequencies above 10 Hz. For frequencies of 10 Hz and below, applying the A-weighted filter in the time domain results in the signal dropping below the instrumentation noise floor. Therefore, the plotted data at 10 Hz and below correspond to the A-weighted level calculated manually by arithmetically adding to each unweighted 1/3-octave band level the A-weighting corresponding to that 1/3-octave band centre frequency.

The unweighted 1/3-octave sound pressure levels can be compared to the generally accepted threshold of hearing for normal ears subjected to steady tonal noise (ISO389-7 2005). It is evident that the largest difference between shut-down and operational conditions occurs in the 50 Hz 1/3-octave band and that this noise would be audible to a person with normal hearing according to ISO389-7 (2005). The noise in this 1/3-octave band is clearly associated with wind farm operation, as it is absent when the wind farm is shut-down and also the magnitude of the peak is highest when the hub-height wind speed is greater than 8 m/s. According to specifications (Vestas 2006), the Vestas v90, 3-MW wind turbines that are installed at Waterloo would be expected to produce maximum noise emissions at these wind speeds. Further analysis of the time data indicates that the noise in this 1/3-octave band is amplitude modulated and appears to be tonal as well. According to the relevant criteria (SAEPA 2009), this wind farm would be compliant despite a large number of complaints that have been recorded by the local residents. There are two possible reasons:

1. The large reduction applied by the A-weighting filter at low frequencies (see Figure 2.6), including the 50-Hz 1/3-octave band does not reflect the extent of the annoyance associated with the noise.

2. The overall average noise levels associated with wind farm operation are relatively low and do not reflect the occurrence of special audible characteristics that would make the noise more annoying.

An alternative method of compliance testing would be to measure 1/3-octave levels and time data continuously over several weeks and to analyse the data for downwind, nighttime (12am–5am) conditions only. Background 1/3-octave data would also be required for comparison. The unweighted spectra could be compared to the hearing threshold in ISO389-7 (2005), as shown in Figure 6.35, and an acceptable limit above this threshold could be established. Penalties could be applied to 1/3-octave bands containing excessive tonality and AM. There could also be a specification on the proportion of time that the limit applicable to a given 1/3-octave band could be exceeded, if at all.

A possible method to calculate the proportion of time that the allowed A-weighted level is exceeded during a compliance monitoring period has been developed by Ashtiani (2013) and was mentioned in Section 1.6.1. The method was expanded to include 1/3-octave band analysis by Ashtiani (2015) for comparison with hearing thresholds. However, extension to use with hearing thresholds is difficult due to the background noise masking the turbine noise for situations where the two levels are similar or when the background noise contribution exceeds the turbine contribution.

The method of Ashtiani (2013) determines the cumulative percentages of time that the noise generated at a residence by a wind farm exceeds a specific criterion. This method will be illustrated here using data collected by the authors near a local wind farm. The method involves collecting a minimum number (which is substantial, requiring continuous monitoring for at least three weeks) of data points for operational and non-operational conditions for each integer wind-speed bin and undertaking the analysis described below for each. The wind-speed bins are centred on integer wind speeds and are 1 m/s wide. Noise and weather data are collected simultaneously and averaged over 1-min intervals. The measurement period is restricted to nighttime hours between 10pm and 5am and data are excluded in the case of precipitation within an hour, temperatures below − 10°C and maximum or minimum wind speeds differing from the average by more than 2 m/s (Ashtiani 2015).

To minimise contamination from extraneous noise events, data are also excluded if $L_{Aeq,\,1\text{min}}$ >80 dBA or if $L_{Aeq,\,1\text{min}} - L_{A90,\,1\text{min}}$ >6 dBA (Ashtiani 2015). For each integer wind-speed bin, the probability density is plotted against the overall A-weighted sound pressure level for both operational (on) and non-operational (off) conditions. The sound pressure level data are plotted at 1-dBA intervals so that, for example, the value plotted for 40 dBA includes all values between 39.5 and 40.5 dBA. Data can also be plotted for 1/3-octave band sound pressure levels if the frequency content of the turbine noise is of interest, and the following procedure is then repeated for each 1/3-octave band.

Figure 6.36 shows the overall A-weighted sound pressure level data measured when the wind speed at 10 m was in the integer wind-speed bin centred at 4 m/s. The figure shows the probability of the sound pressure level falling into a particular integer decibel bin. For example, the probability corresponding to 40 dBA is the probability of the sound pressure level being between 39.5 and 40.5 dBA. Since the noise level of interest is the turbine-only level, it is convenient to consult a look-up table (which can be included in a computer code) that shows the resulting values after logarithmic subtraction of the turbine-off level from the turbine-on level. An example is shown in Table 6.4, and the

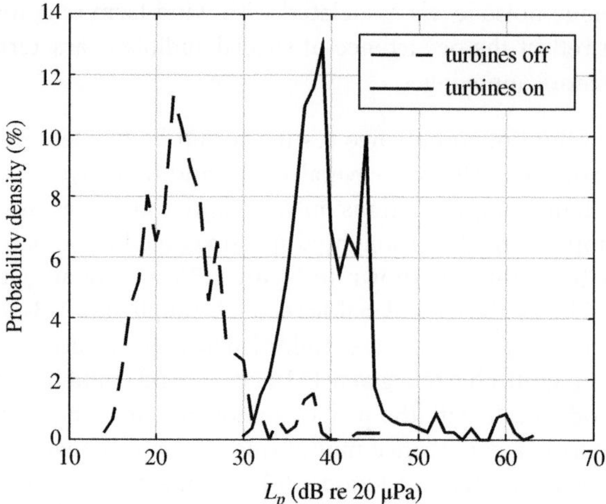

Figure 6.36 Probability density function for overall A-weighted sound pressure levels for turbine-on and off conditions. SPL, sound pressure level.

range of values is chosen based on the range of noise levels measured during on and off conditions. In Table 6.4, the range has been reduced for presentation purposes. The next step in the analysis is to import the data represented in Figure 6.36 into two tables corresponding to each of the two curves in the figure (see Tables 6.5 and 6.6). Then the data in Table 6.4 are used to determine the turbine-only sound pressure level for each combination of turbine-off and turbine-on levels. Then, based on the probability of the existence of particular turbine-off and turbine-on sound pressure levels, the probability of a particular turbine-on level co-existing with a particular turbine-off level can be obtained by multiplying the probabilities of the two levels existing alone and dividing the result by 100, thus producing the results in Table 6.7. Since a number of combinations can result in a given turbine-only level, the probabilities corresponding to all possible combinations must be added to give the final result for each integer dBA value.

To illustrate this method, we will continue analysis on the data that were presented in Figure 6.36. In this case, we are interested in the percentage of time that a specified A-weighted level is exceeded by the turbine-only noise. To demonstrate the analysis process, we will focus on the instances where the turbine-only level is equal to 34 dBA. As can be seen in Table 6.4, this can be achieved with a number of combinations of turbine-on and turbine-off data. These combinations are highlighted in bold font in the table. Note that this is only a section of the entire table, which consists of columns spanning the range of turbine-on levels and rows spanning the range of turbine-off levels. The pairs of array values, representing turbine-off levels (Table 6.6) and turbine-on levels (Table 6.5), can be multiplied together and the result divided by 100 to determine the combined probability, giving rise to Table 6.7. This produces the same result as converting the two probabilities to fractions (by dividing each by 100), multiplying the two fractions together to produce a combined probability and then multiplying the result by 100 to produce a combined probability.

The final probability that turbine-only levels will be equal to 34 dBA is obtained by adding the values in those cells of Table 6.7 that have been highlighted in bold and this

Table 6.4 Logarithmic subtraction of L_{Aeq} turbine-off levels from L_{Aeq} turbine-on levels to obtain turbine-only sound pressure levels in 1 dBA bins.

Turbines OFF A-weighted SPL (dB re 20 μPa)	Turbines on A-weighted SPL (dB re 20 μPa)							
	34	**35**	**36**	**37**	**38**	**39**	**40**	**41**
14	**34**	35	36	37	38	39	40	41
15	**34**	35	36	37	38	39	40	41
16	**34**	35	36	37	38	39	40	41
17	**34**	35	36	37	38	39	40	41
18	**34**	35	36	37	38	39	40	41
19	**34**	35	36	37	38	39	40	41
20	**34**	35	36	37	38	39	40	41
21	**34**	35	36	37	38	39	40	41
22	**34**	35	36	37	38	39	40	41
23	**34**	35	36	37	38	39	40	41
24	**34**	35	36	37	38	39	40	41
25	33	35	36	37	38	39	40	41
26	33	**34**	36	37	38	39	40	41
27	33	**34**	35	37	38	39	40	41
28	33	**34**	35	36	38	39	40	41
29	32	**34**	35	36	37	39	40	41
30	32	33	35	36	37	38	40	41
31	31	33	**34**	36	37	38	39	41
32	30	32	33	35	37	38	39	40
33	27	31	33	35	36	38	39	40
34	0	28	32	**34**	36	37	39	40
35	0	0	29	33	35	37	38	40
36	0	0	0	30	**34**	36	38	39
37	0	0	0	0	31	35	37	39
38	0	0	0	0	0	32	36	38
39	0	0	0	0	0	0	33	37
40	0	0	0	0	0	0	0	**34**

Table 6.5 Turbine on probabilities for overall A-weighted sound pressure levels between 34 and 41 dB.

Sound pressure level	34	35	36	37	38	39	40	41
% probability of occurrence	3.6	5.3	7.9	11.0	11.6	13.0	6.9	5.4

Table 6.6 Turbine off probabilities for overall A-weighted sound pressure levels between 14 and 40 dB.

Sound pressure level	14	15	16	17	18	19	20	21	22	23	24	25	26	27
% probability of occurrence	0.2	0.7	2.2	4.3	5.2	8.0	6.5	7.6	11.3	10.2	8.9	8.0	4.6	6.5
Sound pressure level	28	29	30	31	32	33	34	35	36	37	38	39	40	
% probability of occurrence	3.3	2.8	2.6	0.7	0.9	0.0	0.7	0.2	0.4	1.3	1.5	0.2	0.0	

results in a final answer of 3.9%. Using data that have not been shown in Tables 6.4–6.7 due to space considerations, the probability that turbine-only levels will exceed 34 dBA can also be determined. This is done by considering all combinations of turbine-on and turbine-off levels of Table 6.6 that produce a turbine-only level of greater than 34 dBA. The result is 92.0%.

To implement the method described above in regulations, it would be necessary to decide how to combine the results from the various integer wind-speed bins. One simple way would be to estimate the probability of the wind speed falling into each wind-speed bin and then multiply this fractional probability by the probability that the allowed noise level would be exceeded for wind speeds lying in that particular bin. This would be done for each wind-speed bin and the combined probabilities so obtained would be added together to arrive at the overall probability of the wind farm noise level exceeding the allowed level. This could then be converted to a percentage probability by multiplying by 100 and then compared to a regulation that, for example, allowed the specified allowed level to be exceeded for say 5% of the time at night.

Summary

Compliance testing is an important stage in the commissioning of a new wind farm and provides an opportunity to verify noise (sound pressure) level predictions at noise-sensitive locations. Current methods of compliance testing do not appear to protect the amenity of the surrounding community and the main reason for this is that annoying characteristics of wind farm noise are not reflected in overall $L_{A90,10\text{min}}$ levels that are generally used to test compliance. Given the inherent difficulties associated with conducting conclusive wind farm compliance measurements, it seems appropriate to adopt a conservative approach for sound pressure level predictions at sensitive receivers, so as to reduce the likelihood of any non-compliance occurring. It is also suggested that predictions could be expressed as unweighted 1/3-octave levels that can be compared to the hearing threshold in ISO389-7 (2005).

Current methods of compliance methodology can be improved through the use of data filtering. It is therefore recommended that worst-case downwind conditions during nighttime hours are analysed exclusively to ensure that the contribution from the wind farm is maximised and the contribution from extraneous noise sources is minimised. The number of allowable exceedances should also be considered and this should be incorporated into the regression analysis. The current methodology that allows 50%

Table 6.7 Probability of turbine-only overall A-weighted sound pressure levels for various combinations of turbine ON/OFF levels.

Turbines OFF A-weighted SPL (dB re 20 μPa)	Turbines on A-weighted SPL (dB re 20 μPa)							
	34	35	36	37	38	39	40	41
14	**0.01**	0.01	0.02	0.02	0.03	0.03	0.02	0.01
15	**0.02**	0.03	0.05	0.07	0.08	0.08	0.05	0.04
16	**0.08**	0.12	0.17	0.24	0.25	0.28	0.15	0.12
17	**0.16**	0.23	0.34	0.48	0.50	0.56	0.30	0.24
18	**0.19**	0.28	0.41	0.57	0.61	0.68	0.36	0.28
19	**0.29**	0.43	0.64	0.88	0.93	1.04	0.56	0.44
20	**0.23**	0.35	0.52	0.72	0.76	0.85	0.45	0.35
21	**0.27**	0.40	0.60	0.84	0.88	0.99	0.53	0.41
22	**0.40**	0.60	0.89	1.24	1.31	1.47	0.78	0.61
23	**0.37**	0.54	0.81	1.12	1.19	1.32	0.71	0.56
24	**0.32**	0.47	0.70	0.98	1.03	1.16	0.62	0.48
25	0.29	0.43	0.64	0.88	0.93	1.04	0.56	0.44
26	0.16	**0.24**	0.36	0.50	0.53	0.59	0.32	0.25
27	0.23	**0.35**	0.52	0.72	0.76	0.85	0.45	0.35
28	0.12	**0.17**	0.26	0.36	0.38	0.42	0.23	0.18
29	0.10	**0.15**	0.22	0.31	0.33	0.37	0.20	0.15
30	0.09	0.14	0.21	0.29	0.30	0.34	0.18	0.14
31	0.02	0.03	**0.05**	0.07	0.08	0.08	0.05	0.04
32	0.03	0.05	**0.07**	0.10	0.10	0.11	0.06	0.05
33	0	0	0	0	0	0	0	0
34	0.02	0.03	0.05	**0.07**	0.08	0.08	0.05	0.04
35	0.01	0.01	0.02	0.02	0.03	0.03	0.02	0.01
36	0.02	0.02	0.03	0.05	**0.05**	0.06	0.03	0.02
37	0.05	0.07	0.10	0.14	0.15	0.17	0.09	0.07
38	0.05	0.08	0.12	0.17	0.18	0.20	0.11	0.08
39	0.01	0.01	0.02	0.02	0.03	0.03	0.02	0.01
40	0	0	0	0	0	0	0	**0**

of data points to exceed the regression curve does not seem reasonable if wind farm noise is the dominant source. It is also recommended that a regression analysis be carried out for the local receiver noise level plotted against wind speed at hub height as well as the noise level plotted against the local wind speed at microphone height at the receiver for both background and compliance measurements. The reason for this is that while it is generally believed that most of the background noise is caused by the wind, the wind speed at hub height is not always proportional to the wind speed at the residence. Particularly at nighttime, during conditions of high wind shear, the wind speed at hub height can be relatively high while the wind speed at the residence can be negligible. In such cases, the masking noise provided by the vegetation is insignificant, but this would not be reflected in the plot of noise level against hub-height wind speed. To ensure the success of this method, operational measurements would only be included when the wind farm was generating a reasonable power output, such as 40% of full capacity. Compliance would be determined based on the most conservative result from the two regression analyses.

Another recommendation is that background measurements should be carried out just prior to wind farm operation so that the compliance measurements can be taken at a similar time when the wind farm is operational for the first time. This procedure minimises seasonal and yearly variations in background noise. Alternatively, background noise data could be acquired after wind farm operations begin by turning the turbines off for a few nights. However, this option is not popular with wind farm operators due to the significant loss of revenue when the turbines are not operating.

An alternative method of compliance testing would be to use a statistical approach that would give a result in terms of the percentage of time that the wind farm only noise exceeded a specified level. This could be done in terms of overall A-weighted noise level (see Section 6.2.14 and Ashtiani (2013)) or in terms of 1/3-octave bands (Ashtiani 2015). Then regulators could specify the allowed percentage of time that a particular level may be exceeded during the sensitive nighttime hours for example. This method also requires background noise level data with the turbines not operating, which can be obtained prior to the wind farm beginning operations or after the beginning of operations by turning the turbines off.

When a wind farm is shown to be non-compliant, it is necessary to develop a mitigation strategy (IOA 2014). This may involve limiting the power output or switching off specific wind turbines for wind speeds and directions for which compliance is not demonstrated. The specific wind turbines are those that dominate the noise immissions at the receiver where compliance is not met. With this mitigation strategy in place, compliance measurements need to be repeated to establish whether adequate measures of noise reduction have been achieved (IOA 2014).

6.2.15 Beamforming for Source Localisation on Full-scale Wind Turbines

Beamforming is a technique that is used to identify the relative importance of different parts of a sound source to the far-field sound pressure level. In the context of full-scale wind turbines, this technique can be used to reveal the distribution of sound sources in the rotor plane, their relative amplitude and the dominant frequencies. The principle behind beamforming is that the summed output from a large number of microphones placed in an array is maximised in a specified direction and minimised in other directions. The direction of maximum response can be varied by adding an adjustable time delay to the signal from each microphone, thus 'steering' the array to determine the relative intensity of sound arriving from different directions.

The number and arrangement of microphones in a beamforming array influences the dynamic range, depth of focus, frequency range and the ability to reject sound sources that are located away from the focus point. To avoid spatial aliasing (maxima in the array response in directions other than the intended direction) while minimising the number of required microphones, microphones are spaced at irregular intervals and this can be achieved by using spiral designs. More detailed information about array designs is provided in Section 6.6.2.

An advantage of beamforming is that it can image distant sources as well as moving sources, which is ideal for the measurement of wind turbine noise. On the other hand, the spatial resolution of beamformers is poor and they do not perform well at low frequencies. For a beamforming array of largest dimension D, located at a distance L from the source, the resolution Res, or smallest distance between two separate sources that can be resolved, is given by,

$$Res = 1.22\frac{L}{D}\lambda \tag{6.35}$$

To ensure that the entire source is mapped, the array should be placed far enough away from the source that its extremities do not subtend an angle to the source extremities greater than 30°. Ideally, the array should be placed at a distance slightly greater than or equal to one array diameter from the source.

The dynamic range of a beamforming array is greatest for broadband noise sources and lowest for low-frequency tonal noise sources. This makes beamforming a useful technique for measuring mid-frequency, broadband trailing-edge noise, which was the focus of the majority of the studies summarised in Section 6.6.3. On the other hand, for measuring low-frequency blade–tower-interaction noise, inflow turbulence noise and stall noise (OAM), the accuracy with which the source can be located will diminish significantly. Therefore, in these cases, beamforming can be used to verify that low-frequency noise originates from a given wind turbine or the wind farm in general, rather than pinpointing the actual source location in the rotor plane.

Details of the theory underpinning beamforming are given in Section 6.6.2. More information on work done by others in beamforming on full-scale wind turbines is given in Section 6.6.3.

6.2.16 Measurement Uncertainty

To provide an indication of measurement quality, it is imperative to carry out an uncertainty analysis. This procedure is often not mandatory in relevant guidelines and standards and therefore it is not yet routinely considered. The implications of this omission are that it is not possible to determine if sound pressure level predictions are reliable or to ascertain whether there is a possibility that compliance limits could be exceeded at sensitive receiver locations. The latter is relevant in cases where borderline compliance is achieved, particularly where complaints are involved. Determination of measurement uncertainty also facilitates comparison of results measured at different wind farms. There are several sources of uncertainty in wind farm measurements and these are outlined in the following subsections, along with a method for estimating the overall measurement uncertainty. Good practice, such as traceable calibration, detailed record keeping, appropriate choice of instrumentation and positioning, and careful calculation can reduce measurement uncertainties.

Sources of Measurement Uncertainty

When considering measurement uncertainty in the context of wind farm noise, it is important to distinguish between:

- variations in measurements that are caused by changes in source and propagation characteristics due to meteorological conditions

- uncertainties that can be attributed to the choice of instrumentation and measurement locations at a given receiver.

The position of the microphone relative to the wind farm may be an important consideration if certain weather patterns are prominent in an area, leading to increased probability that particular wind speeds and directions will occur. In these cases, certain measurement positions may have higher associated sound pressure levels and/or greater AM due to repeatable interaction patterns between the noise from multiple wind turbines. It is important to measure at such positions, where possible, particularly if they occur in an area where the resident spends considerable time.

As mentioned in Section 6.2.4, the height at which the microphone is placed above ground level influences the measured sound pressure level, and the magnitude of this effect varies with frequency, ground surface properties, distance from the source and the elevation of the source relative to the measurement location. When choosing the measurement height, a compromise must be made between minimising uncertainties related to wind-induced noise and minimising uncertainties related to microphone height. Most standards specify a measurement height of 1.5 m above ground level as this approximates the ear height of an average person standing, although many standards specify 1.2 m above ground level, which approximates the ear height of an average person seated.

As discussed in Section 6.2.2, wind-induced noise can cause significant errors for outdoor measurement of low-frequency noise and this is the reason for the use of specialised wind screens as described in Section 6.2.3. To properly quantify the errors associated with wind interaction with the microphone or pseudo-noise, it is recommended that the measurement system is deployed in a rural area as far as possible from any wind farm. In this way, the response of the microphone/wind-screen combination can be determined in the presence of winds at various speeds to ascertain the lowest levels of wind farm noise that can be measured at different frequencies in the presence of wind at varying strengths. Results from measurements of wind-induced noise in the Mojave desert are presented in ANSI/ASA S12.9-7 draft (2014) and can be used as a reference.

Another potential error is the insertion loss (or reduction in the sound pressure level reaching the microphone) caused by the presence of the wind screen. Insertion loss is not likely to be an issue for the measurement of noise at low frequencies provided a porous wind screen is used. Nevertheless, the insertion loss should always be checked for a custom-made wind screen and a description of the procedure is given in Section 6.2.3.

One possible source of error that is relevant in some applications is the directivity of the microphone, when the orientation of the microphone relative to the source becomes important. However, this is not usually a concern for measurement of wind farm noise in the community due to the low-frequency dominance of wind farm noise and the relatively large separation distances between the source and receiver. Free-field corrections for different angles of incidence of noise at the microphone are usually published by the microphone manufacturer and standard microphones do not require a directivity correction for frequencies up to 1 kHz.

A typical microphone for wind turbine noise measurement is a GRAS 40 AZ and this particular microphone requires negligible correction for frequencies up to 1 kHz and less than 1 dB correction for frequencies up to 3 kHz. Therefore, microphones can point in any direction for most measurements of wind farm noise. Possible exceptions to this rule are sound power measurements and situations where the receiver is within a few hundred metres of the nearest wind turbine, where significant levels of noise occur at frequencies above 1 kHz.

Proximity to reflective surfaces (other than the ground) can also lead to measurement uncertainties. Therefore, microphones should be placed as far away from reflective surfaces as possible. In cases where reflective surfaces cannot be avoided, it is important to consider the frequency range of interest at the measurement location. Sound energy is only reflected when the wavelength is small compared to the dimension of the surface. Low-frequency noise, which is of greatest interest at distances greater than 1–1.5 km from a wind farm, will diffract around all but the largest surfaces due to the large associated wavelengths. Also, if the reflective surface consists of lightweight material, the reflections of low-frequency noise may be weak (Craven and Kerry 2007). So for measurement of low-frequency noise, the measurement errors associated with small- or medium-sized surfaces may be negligible.

The calibrator used to check that the microphone is performing adequately also has an associated error and this will affect the accuracy of the microphone. This error is usually ±0.2 dB for Class 1 calibrators and for other calibrators it is not greater than ±0.5 dB.

Another point to consider is that the tolerance ratings on Type 1 sound level meters at low frequencies are high when using A-, C- and Z-weightings (IEC 61672-1 2013). This could result in measurement errors as well as variations in the noise levels measured with different instrumentation, particularly where the measured signal is dominated by low-frequency noise. As can be seen in Table 6.8, the allowable tolerances at very low frequencies are much greater than those at frequencies higher than 100 Hz, for all

Table 6.8 Tolerance limits including maximum expanded uncertainties of measurement for sound level meters with A-, C- or Z-weighted filters for nominal frequencies from 10 to 160 Hz.

Nominal frequency	Tolerance (dB)	
	Class 1	Class 2
10	+3.5; −∞	+5.5; −∞
12.5	+3 ; −∞	+5.5; −∞
16	+2.5; −4.5	+5.5; −∞
20	±2.5	±3.5
25	+2.5; −2	±3.5
31.5	±2	±3.5
40	±1.5	±2.5
50	±1.5	±2.5
63	±1.5	±2.5
80	±1.5	±2.5
100	±1.5	±2
125	±1.5	±2
160	±1.5	±2

Adapted from IEC 61672-1 (2013).

filter weightings. This table is relevant when considering 1/3-octave frequencies separately, such as when compliance with the DEFRA criterion is investigated, as discussed in Section 1.4.4. It is also relevant when logarithmically summing the noise level in each 1/3-octave band from 10 to 160 Hz to establish whether or not compliance with the Danish regulations has been achieved, as discussed in Section 5.4. The magnitude of the error in the latter case will depend on which frequencies dominate the noise signal. To determine the tolerances associated with overall noise measurements, where the sound energy is integrated over a larger bandwidth, it is more relevant to consult Table 6.9.

It should also be recalled that the roll-off for the Z-weighting is unspecified by IEC 61672-1 (2013) below 10 Hz and therefore the Z-weighting is not necessarily representative of the unweighted noise level, as discussed in Section 6.2.1. Therefore to properly quantify the unweighted level it is necessary to use no weighting or, if this is not possible, a correction for the frequency roll-off below 10 Hz needs to be applied.

As discussed in Section 2.3.3, the application of 1/3-octave or octave filters to an acoustic signal that is varying rapidly can result in considerable measurement errors. This is most relevant to measurement of low-frequency noise, where the sampling time required to obtain the true value becomes larger due to the required increase in filter rise and settling times. Without knowledge of the actual signal, it is impossible to quantify the contribution of this error to the measurement uncertainty. The best approach is to minimise the error by increasing the bandwidth, as suggested in Section 2.3.3. To quantify the random error associated with applying 1/3-octave or octave filters to a time-domain signal of finite length, Eq. (2.44) is utilised. To minimise this error, the sample time should be increased.

Uncertainties that occur during data processing should also be taken into account. For example, when carrying out an FFT on noise with tonal components, there is an error associated with applying a window to the data. This error is a maximum when the measured tone occurs half-way between two frequency bins. For a Hanning window, which

Table 6.9 Summary of some of the relevant sources of error for noise measured at frequencies <1000 Hz.

Source of error	Maximum expanded uncertainty (dB) (see next section)
Calibrator	±0.2
Directional response >250 Hz	±0.3
Frequency weightings A, C, Z, unweighted (<200 Hz)	±0.5
Frequency weightings A, C, Z, unweighted (≥200 Hz)	±0.4
Level linearity error	±0.3
1–10 dB change in level between measurement and calibration	±0.3
Electrical output	±0.1
Static pressure influence	±0.3
Air temperature influence	±0.3
Humidity influence	±0.3

Source: IEC 61672-1 (2013) and IEC 61400-11 (2012).

is recommended for general-purpose frequency analysis, the maximum associated error is 1.4 dB. Errors associated with other types of window are shown in Table 2.5.

Seasonal variations in background noise levels caused by changes in farming activity, vegetation, surf size, river flow and so on, can contribute to measurement error when comparing background noise levels with operational noise levels. To minimise this error, background noise measurements should be made just prior to the wind farm operating for the first time and compliance measurements should be made directly afterwards, as discussed in Section 6.2.14. Alternatively, measurements of background and operational noise levels should at least be made in the same season. However, since background levels vary from year to year, this is not an ideal solution. Accounting for the measurement uncertainty due to seasonal variations is practically impossible and this reiterates the fact that these measurements should be taken as close together in time as possible, although sometimes this may be difficult due to the length of time required to construct a large wind farm.

Quantifying Measurement Uncertainty

Measurement uncertainty is determined through identifying all relevant contributions, calculating the associated uncertainty and then combining all results according to set statistical procedures to give an overall value. In some cases, there is insufficient information to determine a particular uncertainty and consequently it is necessary to make an educated guess.

There are three different ways of expressing uncertainty:

- a range of expected values (which is the same as a tolerance limit)
- standard uncertainty
- expanded uncertainty.

These latter two uncertainty types are defined below. When comparing uncertainties or when calculating an overall measurement uncertainty from a list of individual uncertainties that make up the entire measurement chain, it is important that they are in the same units and are of the same type. The result usually of interest is the expanded uncertainty, because it is associated with a confidence interval, which is usually 95% (Craven and Kerry 2007), giving a 95% confidence that the measured value will be in the specified range. This is more meaningful to most practitioners than a standard deviation, which is associated with standard uncertainty. Thus it is important to be able to convert a range of expected values and a standard uncertainty to an expanded uncertainty. The standard uncertainty, u, is a deviation from the mean with a size of plus or minus one standard deviation (Bell 1999).

In addition to the three types of uncertainty measure, there are two different types of uncertainty, which are referred to as Type A and Type B uncertainties.

Type A Uncertainties Type A uncertainties are evaluated using statistics, usually based on repeated readings and usually expressed as a standard uncertainty, u. For wind turbine noise measurements, these uncertainties can be calculated for both the measured sound pressure level and the measured wind speed. Usually a different uncertainty is calculated for the sound pressure level corresponding to each wind-speed bin. The standard uncertainty (or standard deviation of the mean) can be expressed as,

$$u = \frac{\sigma}{\sqrt{N}} \tag{6.36}$$

where N is the number of measurements and the standard deviation, σ, of the series of N sound pressure level measurements, $L_{p,j}\,(j = 1, N)$, in a particular wind-speed bin is given by

$$\sigma = \sqrt{\frac{\sum_{j=1}^{N} (L_{p,j} - \overline{L}_p)^2}{N - 1}} \quad \text{(dB)} \tag{6.37}$$

where

$$\overline{L}_p = 10 \log_{10}\left(\frac{1}{N}\sum_{j=1}^{N} 10^{(L_{p,j}/10)}\right) \tag{6.38}$$

A similar approach is used to calculate the standard deviation σ_u in wind speed, U. Again this is usually done for each wind-speed bin. In this case, the standard deviation is given by,

$$\sigma_u = \sqrt{\frac{\sum_{j=1}^{N} (U_j - \overline{U})^2}{N - 1}} \quad \text{(m/s)} \tag{6.39}$$

where

$$\overline{U} = \frac{1}{N}\sum_{j=1}^{N} U_j \tag{6.40}$$

Standard Type A uncertainties u can be converted to expanded Type A uncertainties u_e, by multiplying by a constant, q, which is equal to 2 for a 95% confidence limit (assuming that the standard uncertainty is normally distributed). Other percentage confidence limits or different distribution types for the standard uncertainty would require a different multiplying factor (such as 1.65 for a rectangular distribution). Thus,

$$u_e = qu \tag{6.41}$$

Type B Uncertainties Type B uncertainties are estimates of the uncertainty in quantities such as the sound pressure level, which are taken from other information such as calibration certificates, manufacturer's' specifications or published material. In some cases, Type B estimates can be provided by the practitioner based on experience.

For Type B uncertainties, the quantity is often given as a range of possible values that completely encompasses all expected values and which thus approximates a rectangular distribution. In this case, the standard uncertainty is calculated as (JCGM 2008):

$$u = \frac{a}{\sqrt{3}} \tag{6.42}$$

where a is the range between the upper and lower limits of uncertainty divided by 2.

Usually, data from calibration certificates are normally distributed, in which case the divisor in Eq. (6.42) is 2 rather than $\sqrt{3}$.

The expanded uncertainty is then calculated from the standard uncertainty using Eq. (6.41) for each case and the result of this calculation is what is provided in Table 6.9 for various causes of uncertainty.

Overall Expanded Uncertainty Finally, what is needed is an overall expanded uncertainty, $u_{e,t}$, (that is the 95% confidence limit), which is calculated from the individual Type A and Type B expanded uncertainties, $u_{e,i}$, as follows:

$$u_{e,t} = \sqrt{\sum_{i=1}^{n} (U_{e,i})^2} \tag{6.43}$$

where n is the number of individual Type A and Type B uncertainties to be combined and $u_{e,i}$ are the individual uncertainties in decibels.

It should be noted that the above analysis has been simplified and does not take into account relative influences of each error on the measured value of interest. Also, the confidence associated with the error estimates is not taken into account in the above analysis. Therefore, if these aspects are predicted to affect the accuracy of the uncertainty estimate, then further guidance should be sought from the JCGM (2008).

6.3 Vibration

For a typical residence located in the vicinity of a wind farm, any vibration measured indoors at the residence is almost always the result of acoustic-induced vibration rather than vibration that has propagated from the wind turbine via the ground (Kelley et al. 1985). Acoustic-induced structural vibration of a house is a result of low-frequency sound incident on the walls and roof and is particularly relevant in the range of 12–30 Hz, according to data measured for several different housing structures (Hubbard 1982). In the case of the MOD-1 wind turbine described by Kelley et al. (1985), the acoustic-induced structural vibration was strong enough to cause loose objects to rattle and loose dust to fall from high ceilings. This vibration was described by residents as an intermittent phenomenon that accompanied the low-frequency thumping associated with this turbine. Despite these observations, the vibration measured on the floor was below the perception threshold. Thus, Kelley et al. (1982) hypothesised that the acoustic field coupled with the human body resonances created a sensation of whole-body vibration. It should be noted that the vibration levels associated with modern upwind rotor wind turbines are expected to be lower than those associated with the downwind rotor turbines discussed above and therefore the likelihood of ground vibration exceeding the perception threshold is low because the acoustic energy in the 12–30 Hz range generated by the turbines is less. Although modern turbines do generate ground-borne vibrations, these are below the threshold of human perception. However, ground vibrations generated by turbines propagate for considerable distances and can interfere with sensitive instruments used for remote monitoring of seismic activity and nuclear tests (Styles et al. 2011). This is corroborated by a small body of literature on the subject and the authors' own observations from measurements. Although the authors have not measured levels of vibration on floors and beds that should be detectable, vibrations measured on the glass of windows facing a wind farm are easily detectable and, in some cases, residents have complained that they can feel vibration sensations, caused by wind farm operation, through their pillows when they are lying down. Therefore, to be thorough, it is prudent to include vibration measurement as part of the assessment procedure for wind farms.

6.3.1 Instrumentation

The transducer most commonly used to measure vibration is an accelerometer and this device generally consists of a small piezoelectric crystal loaded with a small weight that has been designed to operate in a range well below its natural resonance frequency. The quantity measured by an accelerometer is acceleration (generally expressed in m/s^2). As vibrations at a residence located in the vicinity of a wind farm are expected to be relatively low, it is important that the accelerometer has a high sensitivity and a low noise floor. It is also important that the frequency response of the accelerometer is flat down to at least 0.5 Hz, to ensure that potential vibrations at the blade-pass frequency and harmonics can be measured.

The output signal from an accelerometer is very small and therefore some accelerometers are designed with a built-in charge amplifier. The output from such an accelerometer should be consistent with the IEPE standard[2] to ensure that it is compatible with other instruments. Accelerometers that do not have a built-in charge amplifier require an external one. The effect of the charge amplifier on the signal is minimal, as due to its very high input impedance, it does not load the accelerometer output. Charge amplifiers also allow measurement of acceleration down to 0.2 Hz, can be used with cable lengths up to 500 m and are relatively insensitive to cable movement.

Accelerometers generally measure the vibration that is perpendicular to the surface on which they are mounted. However, there are also accelerometers that can measure vibration occurring in three orthogonal directions simultaneously and these devices are referred to as tri-axial accelerometers. Tri-axial accelerometers can be relevant for measurements near wind farms since acoustic-induced vibration does not necessarily occur in the vertical direction. According to Kelley et al. (1985), the horizontal floor motion was found to be more strongly coupled with the room acoustic pressure than the vertical floor motion for measurements taken near the MOD-1 wind turbine.

6.3.2 Measurement

At the beginning and end of measurements, accelerometers should be calibrated. This involves attaching the accelerometer to an appropriate calibrator, consisting of a shaker that produces a known value of acceleration at a given reference frequency. As mentioned in Section 6.2.1, the sensitivity of the attached vibration meter or data-acquisition device is then adjusted accordingly. If the calibration is unsuccessful, a similar procedure to that outlined in Section 6.2.1 should be used to identify the cause.

Accelerometers need to be attached firmly to the surface of interest and for measurements inside people's homes, it is also required that no permanent damage is done to the surface. Therefore, the most suitable mounting solutions in this context are double-sided

---

2 IEPE accelerometers are a piezoelectric type with integrated electronics. The IEPE interface is specified in the IEEE 1451.4 (2004) standard.

tape or beeswax. Where beeswax is hard and unmalleable, a small amount can be softened using a soldering iron. The double-sided tape or beeswax should be evenly distributed to ensure that the accelerometer is correctly mounted in the desired direction.

Vibration measurement is not only relevant for floors but also for beds and chairs and any other locations where the resident spends considerable time. Results from the measurements can then be compared to perception thresholds (see Section 5.14).

6.3.3 Analysis

Vibration measurements can be analysed in terms of overall values and/or 1/3-octave band and narrowband spectra. The A-weighting is not applied to the measured acceleration since this weighting relates to hearing rather than perception of vibration. Rather, there are weightings that are specific to the measurement of vibration, as shown in Table 5.20. These weightings are used to calculate the overall weighted acceleration according to Eq. (5.226). The calculated value can then be compared to the detection thresholds expressed in the ISO2631-1 (1997) standard, as discussed in Section 5.14.3. The 1/3-octave acceleration is determined by applying 1/3-octave filters to the data, as discussed in Section 2.3.2. The data can then be converted to a vibration level, L_a, using Eq. (6.44).

$$L_a = 20 \log_{10} \left(\frac{a}{a_{\text{ref}}} \right) \tag{6.44}$$

where $a_{\text{ref}} = 10^{-6}\text{m/s}^2$. Measured 1/3-octave vibration levels can then be compared to Figure 5.30 (Stephens et al. 1982), which indicates the threshold of perception for 'most sensitive humans' as discussed in Section 5.14.3.

Narrowband analysis of vibration reveals more detailed information about the vibration spectra, which can be used to determine whether or not the vibration is wind farm related. Characteristics such as tonal peaks at the blade-pass frequency and harmonics and tonal peaks with sidebands spaced at the blade-pass frequency, generally provide clear indications that the vibration is related to wind farm operation.

6.4 Wind, Wind Shear and Turbulence

The measurement of wind, temperature and atmospheric turbulence is of importance in wind farm noise investigations, as these parameters affect the source characteristics, propagation characteristics and measurement results. The wind is commonly measured as a two-component vector in the horizontal plane, because the vertical component is generally negligible near the surface of the earth. However, more advanced measurement systems provide the option of measuring the vertical component of velocity as well. The wind speed is typically averaged over a period of 10 min; it has been observed that this period is adequate to obtain relatively stable spectra of turbulence fluctuations in the surface layer of the atmosphere (Oboukov 1962).

6.4.1 Instrumentation

For measuring wind speed and direction during wind farm assessment, a cup anemometer together with a wind vane are generally used because they provide a sufficiently accurate reading and they are relatively inexpensive compared to other anemometers. For higher accuracy but at greater cost, an ultrasonic anemometer can be used to measure both the wind speed and direction. Due to the relatively high sample rate of this device, the turbulence intensity can also be determined from the wind speed data, allowing a more detailed characterisation of the meteorological conditions. Ultrasonic anemometers are commonly installed on the wind turbine nacelle because, having no moving parts, they require less maintenance than cup anemometers.

Cup anemometers and ultrasonic anemometers must be mounted at the height at which wind speed and direction data are required and this is a disadvantage for measurements at high altitudes such as turbine hub height, where data are often required. The cost of meteorological towers that are high enough to reach the hub height of modern wind turbines is substantial and maintenance of anemometers installed at such heights is costly and time consuming. Therefore, a more convenient method of acquiring hub-height wind speed and direction data is to use a sonic detection and ranging (SODAR) instrument or a light detection and ranging (LIDAR) instrument. These systems are usually located on a trailer and can be efficiently deployed with relatively low set-up costs. On the other hand, a large proportion of data measured by these systems may be invalid for various reasons, including reflections from obstacles such as trees and wind turbines, low signal-to-noise ratio, precipitation and sources of electrical noise (SODAR only). This can be problematic when the wind speed and direction at a specific time is required. Also, specific expertise is required to interpret the data, so it is common practice for the instrument supplier to offer data analysis, which of course comes at a price.

All of the above instruments are supplied with appropriate technology for measuring the temperature at the position of the anemometer. Therefore, if a large number of anemometers are mounted on a meteorological tower at various heights, it is possible to determine the atmospheric temperature as a function of height, which can be used to obtain the temperature gradient. Another method of determining the atmospheric temperature profile, without the need of significant infrastructure, is to use a balloon system with an on-board sensor.

Standard Cup Anemometer

The most common instrument used to measure the wind speed and direction for wind farm noise measurements is a cup anemometer with a wind vane, as this is the lowest cost system and provides reasonable results. The cup anemometer rotates in response to horizontal wind and the rate of rotation is proportional to the wind speed. A transducer converts the rotational movement into an electrical signal, which is transferred to a data logger. The standard configuration consists of three cups, this arrangement providing the most constant torque and best response to wind gusts. The wind vane consists of a pointer and a fin, where the latter is orientated into the direction of least wind resistance. This results in the fin pointing into the direction that the wind is travelling and the pointer indicating the direction from which the wind is coming, which is the parameter of interest.

The disadvantage of cup anemometers is that there is friction associated with the bearing and the wind must overcome it before the cups begin to rotate. Therefore, cup anemometers do not respond in very light wind conditions. Also, cup anemometers have a finite response time to changes in wind speed and therefore do not perform well in gusty conditions. Cup anemometers accelerate faster in response to an increase in wind speed than they decelerate in response to a decrease in wind speed, resulting in a higher mean wind speed than the true value (Srivastava 2008). On the other hand, despite previous concerns, it is now accepted that this is not a major source of error, except under highly turbulent conditions (Zhang 2015). During freezing conditions, cup anemometers can cease to operate due to icing, so if these conditions are anticipated, it is recommended that a heated version is used.

Propeller Anemometer

One of the main differences between a cup anemometer and a propeller anemometer is the orientation of the axis of rotation, which is vertical and horizontal, respectively. As with the cup anemometer, the wind speed is proportional to the rate of rotation of the propeller. A propeller also has a fin, but in this case it is mounted on the same axis as the propeller, ensuring that the propeller is always pointing into the wind, whilst simultaneously providing wind-direction information.

The advantage of a propeller anemometer is that it has a very low starting speed due to the high amount of aerodynamic torque that can be generated. A significant disadvantage is that it may not always successfully track the wind, particularly when the wind direction is highly variable. This can result in directional overshoot, causing the propeller to be off-axis relative to the incoming wind. In such conditions, the wind speed readings can appear lower than the true value. A further disadvantage that has been observed for propeller anemometers is that they can stall in highly turbulent conditions (Zhang 2015).

Ultrasonic Anemometer

The most significant advantages to using an ultrasonic anemometer are that it has excellent resolution and accuracy, a fast response time to changes in the wind conditions and the ability to measure turbulence intensity due to the relatively high sample rates available. Many ultrasonic anemometers also provide the option of heaters that enable the instrument to operate in freezing conditions. The main disadvantage of using an ultrasonic anemometer is the relatively high cost of the device compared with a cup anemometer.

Ultrasonic anemometers emit inaudible sound waves at ultrasonic frequencies greater than 20 kHz. In the presence of a wind, the speed of sound changes slightly, and this characteristic is used to determine the wind speed. Ultrasonic anemometers consist of two pairs of transmitters/receivers that are placed opposite one another, at a specified separation distance. Ultrasonic pulses are emitted alternately from a transmitter in each pair at specified time intervals and the wind speed can be calculated based on the propagation time and distance between the transmitter and receiver. The instantaneous wind speed and direction can then be determined through vector addition and trigonometry. The calculation is independent of temperature. Ultrasonic anemometers are capable of resolving turbulence at frequencies up to 20–30 Hz. Local temperature can also be provided by the instrument (see Eq. (2.11)), because it can calculate the speed of sound from the time taken for a pulse to traverse the distance from the emitter to the receiver.

SODAR

A SODAR system is located at ground level and can monitor the wind speed and direction up to a height of 200–300 m above the ground, with a vertical resolution down to

5 m and a minimum height of 10–40 m. For the purposes of wind farm measurements, it is desirable that measurements span at least the maximum and minimum heights reached by the blade-tip. The SODAR system uses a powerful loudspeaker to generate audible acoustic pulses (in the range of 3.5–7.5 kHz), which are transmitted into the atmosphere via an antenna. The sound waves are scattered by temperature fluctuations in the air and a proportion of the acoustic energy is scattered back to the device to be detected by a microphone (Stull 1988). Since turbulent fluctuations travel at the same velocity as the wind, the Doppler effect shifts the frequency of the sound pulse during the scattering process (Antoniou et al. 2003). The amount of frequency shift is proportional to the wind velocity in the direction that the acoustic pulse is emitted. Therefore, to ensure that all three components of velocity can be resolved, three sound beams are released each cycle. These are equi-spaced and tilted at a small angle from the vertical. Other components of the SODAR system include a sound-focusing enclosure, signal processing unit, power supply and a central computer.

Commercially available SODAR systems are often purchased with data processing included in the costs since this procedure requires specific expertise. The supplier may also assemble the system to ensure that optimal performance is achieved for a given application. Siting of the SODAR unit is important to minimise the proportion of invalid data. The unit should be positioned in a relatively flat area (the unit can be levelled manually during installation) at least 40 m from surrounding trees. Rustling of leaves and branches creates background noise, which reduces the signal-to-noise ratio of the SODAR unit. Other natural and manmade sources of background noise should also be avoided and it must be ensured that the operating frequency (3.5–7.5 kHz) is set in a range where the impact from background noise is minimised. The SODAR unit should be sited as far as possible from tall objects to minimise reflections, which are referred to as *fixed echoes*. For example, if the SODAR is sited near a meteorological mast for validation purposes, it should be located at a distance at least 1.5 times the height of the mast (Fulcrum3D Pty Ltd 2014). The SODAR unit should be positioned as far as possible from nearby residents because the high-frequency acoustic pulses may cause a disturbance. The minimum recommended distance between the SODAR and closest residence is 100 m (Fulcrum3D Pty Ltd 2014).

LIDAR

The operating principles of LIDAR and SODAR systems used for measurement of wind speed and direction in the atmospheric boundary layer are the same, except that a LIDAR emits light waves instead of acoustic waves. These light waves are scattered by air molecules, cloud droplets and aerosols in the boundary layer and the returned light is collected in a telescope and focussed onto a photomultiplier detector (Stull 1988). Since light can be more precisely focused and spreads at a reduced rate in the atmosphere compared to sound, LIDAR systems have a higher accuracy and signal-to-noise ratio than SODAR systems (Courtney et al. 2008). On the other hand, LIDAR systems are more expensive than SODAR systems. However, some LIDAR systems can also produce an atmospheric temperature profile, which is a useful input for propagation models.

Balloon Systems

Balloon systems with on-board sensors can measure temperature, humidity, pressure, wind speed and wind direction as a function of height and GPS position. These systems are referred to as upper-air sounding systems and consist of a radiosonde that measures and transmits meteorological data and GPS position data to a ground station. Theoretically, balloon systems can either be tethered or free, since the radiosonde is relatively inexpensive and therefore disposable. Releasing the balloon system without a tether is the only method for determining the wind speed and direction with this system. On the other hand, to release weather balloons into the atmosphere, special permission is required from local aviation authorities, which often presents logistical problems. Also, the wind speed and direction data acquired by the balloon system is not as accurate as that acquired using the other methods described in Section 6.4.2. Therefore, for wind turbine applications, balloon systems are most useful when they are tethered and the most useful data that they can provide is temperature data, which is not available through use of other techniques (except where a very high meteorological tower is available) and is an input required for propagation models. One disadvantage of using a tethered balloon system is that the position of the balloon will change depending on wind strength and direction. On the other hand, the on-board GPS provides an accurate record of the measurement position for each temperature reading.

6.4.2 Measurement

This section outlines techniques to measure meteorological quantities using the instrumentation described in Section 6.4.1. The most important aspects to consider are the positioning of the anemometer and the mounting arrangement. In general, anemometers should be placed in a clear, open area where turbulence and shadowing from nearby structures and foliage is minimised. Mounting requirements increase in complexity, cost and availability as the required height of the supporting structure increases. For measurements at 1.5 m, anemometers can be placed on a tripod or star-dropper; for measurements at 10 m, a tower is required, but it is possible to purchase one that can be assembled by 1 or 2 people. For measurements at hub height and above, the infrastructure required for physically mounting the anemometer at the required height is considerably more complex, requires greater manpower to construct and therefore is more permanent. Remote sensing devices such as SODAR and LIDAR provide a solution to these issues since they can be mounted on the ground and transported relatively easily. These measurement solutions also provide data at a range of heights, avoiding the need for multiple anemometers mounted at multiple heights to obtain wind-velocity profiles.

Meteorological Mast

Meteorological masts can be installed at various heights above ground level, providing a means of mounting a cup or ultrasonic anemometer with relatively low disturbance to the flow compared to a nacelle anemometer. Care should be taken to ensure that the mast is located in an area where the wake effects from wind turbines and other objects are minimised. Also, the anemometer should be mounted such that it is clear of the top of a vertical mast or, where this is impractical, it should be positioned on a horizontal boom with sufficient length to ensure that disturbance from the supporting structure is minimised.

Historically, 10-m meteorological masts have been used to measure the wind speed and direction and the data have been extrapolated to hub height using Eq. (5.17). However, since this equation applies to a neutral atmosphere, extrapolation is only valid under specific meteorological conditions, as discussed by Van den Berg (2005). Therefore, it is important that the meteorological mast is tall enough to allow measurement at hub height, where hub-height data are required. Data from anemometers mounted on masts that are lower than hub height can be used to estimate the wind shear in the vicinity of the wind turbines and also at a residence. If two anemometers are mounted at different heights, the wind shear coefficient, ξ, can be determined using Eq. (5.16) and the same equation can be used to determine the hub-height wind speed. However, this method is also prone to errors since the wind shear may vary with height.

The disadvantage of meteorological masts is that they are often permanent structures, requiring several people with appropriate training for installation. A temporary 10-m mast was used by the authors for wind farm measurements and required approximately 30 min and two people to erect. This telescopic mast was originally designed to mount a television antenna and was intended to be secured using concrete and guy wires. A base plate was therefore designed that was secured to the ground using heavy-duty tent pegs. A pin attachment from this base plate to the mast enabled the mast to be pivoted from a position parallel to the ground to the final, vertical position. The guy wires were then secured with heavy-duty tent pegs and the whole arrangement was able to sustain impact from the very strong winds that were encountered during the measurement program. Meteorological masts are commercially available, but they are costly and their large size makes them costly to transport as well.

Nacelle Anemometer

Most large wind turbines have a cup or ultrasonic anemometer mounted to the nacelle to determine wind speed and direction for control purposes. The output data from this anemometer is often used in wind farm noise assessments. However, it is important that the data are corrected for the presence of the passing blades. Data measured during shut-down conditions should be treated with caution because false readings can occur due to the presence of a stationary blade in front of the anemometer (Broneske 2014).

Wind Speed Derived from the Turbine Power Curve

Derivation of wind speed from the turbine electrical output power curve, as described in Section 4.6.1 and specified in IEC 61400-11 (2012), is a more accurate method for determining the wind speed than using a nacelle-mounted anemometer, as the latter is affected by the presence of the turbine on which it is mounted as well as by upstream turbines. However, where there are multiple wind speeds that correspond to a specific value of output power, this method cannot be used. IEC 61400-11 (2012) specifies that the nacelle anemometer can be used in these cases provided it is corrected by multiplying by the average ratio of the power curve wind speed to the nacelle wind speed, as discussed in Section 4.6.1.

Ground-mounted Anemometers

Remote sensing systems such as SODAR and LIDAR can measure wind speed and direction at a wide range of heights without the requirement of a tower. Their mobility also allows them to be placed on a wind farm site in a position where the influence from obstacles and wind turbines is minimal.

6.4.3 Analysis

In the analysis of meteorological measurements, it is useful to compare data measured using different instruments and mounting arrangements to eliminate possible outliers from the analysis process. For example, in Figure 6.37 data are shown for measurements taken using an ultrasonic anemometer, mounted on a wind turbine hub and a SODAR located on a ridge-top near the wind turbines, as shown in Figure 6.38. The data measured using the two techniques generally agree well, but some data points obtained using the SODAR indicate unrealistically low wind speeds. Therefore, in further analysis of the SODAR data, it would be recommended that these data points be discarded. The positioning of the SODAR should also be reassessed in this instance, to verify that the optimum location has been chosen.

Another factor to consider is that the wind speed at higher altitudes cannot be determined accurately using the logarithmic profile, particularly in areas with complex terrain and during conditions of high stability (Van den Berg 2007). A logarithmic curve fit is shown in Figure 6.39, which includes wind speed data from weather stations (cup anemometers) mounted at 1.5 and 10 m, hub-height wind speed data from the nacelle anemometer (ultrasonic anemometer) and wind profile data from 210–310 m (SODAR). These data were collected over a period of approximately one week and the average curves are plotted in Figure 6.39 for daytime and nighttime (12am–5pm). Note that all heights are expressed relative to the height of the residence where the 1.5-m and 10-m weather stations were located. The nacelle anemometer data were taken from the southernmost wind turbine in Figure 6.38 because this was the nearest wind turbine to the residence ($\approx$2.7 km away) where concurrent noise measurements were taken. The SODAR was positioned 10 km north of the nacelle anemometer and $\approx$12 km from the residence. The quantity B_m in Eq. (4.30) has been adjusted so that the resulting logarithmic curves for $U(h)$ (one for day and one for night), calculated using Eqs. (5.17) and (5.30), are the best fit to all of the data. The values of B_m corresponding to the curves

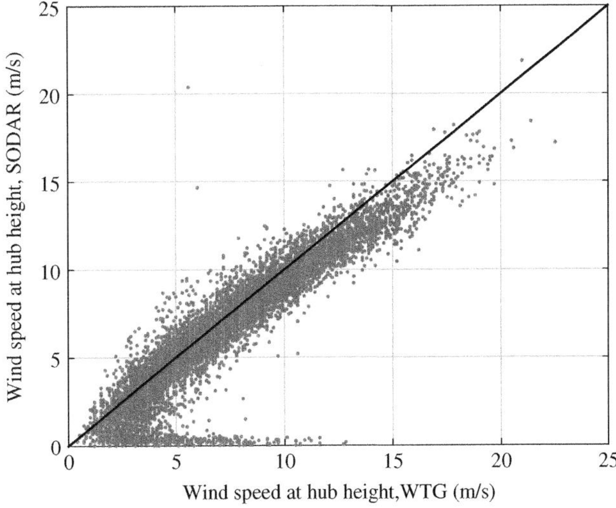

Figure 6.37 Comparison between hub-height wind speed measurements using data from an ultrasonic anemometer located on the wind turbine generator (WTG) and SODAR data.

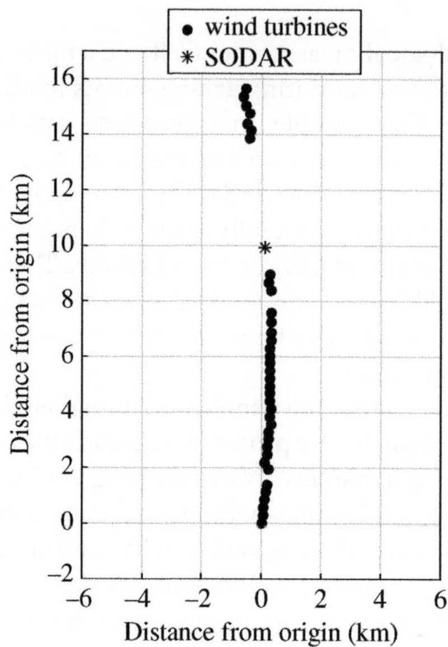

Figure 6.38 Waterloo wind farm layout showing the position of the SODAR, which is located on the ridge-top.

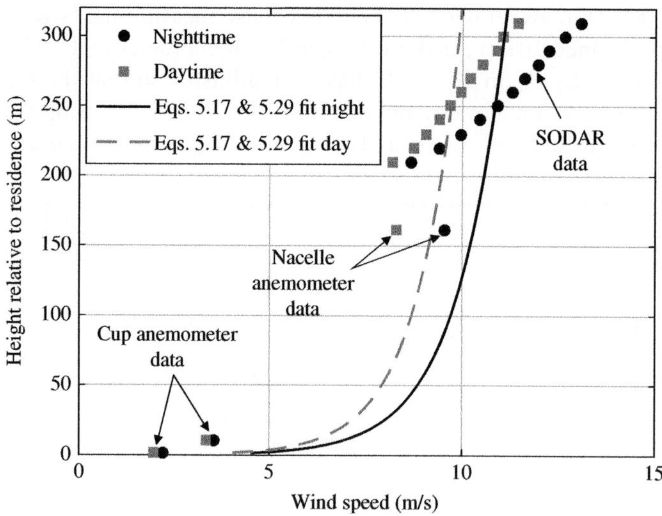

Figure 6.39 Velocity profiles for night (12am–5pm) and day, determined using data from cup anemometers (1.5 m and 10 m), nacelle ultrasonic anemometer (161 m) and SODAR (210–310 m). Note that all heights are expressed relative to the elevation of the residence at which the 1.5 m and 10 m measurements were made. The logarithmic curves were determined by adjusting B_m in Eq. (5.30) to obtain the best least squares curve fit when substituted into Eq. (5.17). $B_m = 1.11$ for daytime and 1.24 for nighttime.

shown in Figure 6.39 are 1.11 for daytime and 1.24 for nighttime. However, it can be seen that the logarithmic curve fits do not align well with the data, particularly at night, when the atmospheric stability would be expected to be higher (Van den Berg 2005). It is also evident that data measured using different instruments can vary and therefore, when determining velocity profiles, it is important to have as many data points as possible. The results could improve if the SODAR were located closer to the anemometers, although the wind speed in the range between 50 and 150 m above the continuous ridge between the SODAR and the southernmost wind turbine is expected to be similar, especially at night.

To further emphasise the potential inaccuracies involved with using the logarithmic profile to derive the hub-height wind speed, Figure 6.40 shows a comparison between the measured wind speed at hub height and higher altitudes (the same data as Figure 6.39) compared to the velocity profile calculated using Eq. (5.17) with the wind speed at 10 m as the input. It is evident that the wind speed is significantly underestimated using the logarithmic extrapolation, particularly for the nighttime data. A wind speed profile with a better fit to the measured data for atmospheric conditions that are not neutral can be achieved using Eq. (5.16) and Figure 5.3. The profiles calculated using this latter method are shown on Figure 6.40 for both the daytime and nighttime data. The profiles are based on the average measurement at 10 m height and a value of $\zeta = 0.325$ for the daytime and $\zeta = 0.355$ for the nighttime, indicating quite stable conditions for both (see Figure 5.3).

Meteorological data can be analysed over a long period before a wind farm is constructed to ascertain trends that are specific to the region of interest. This

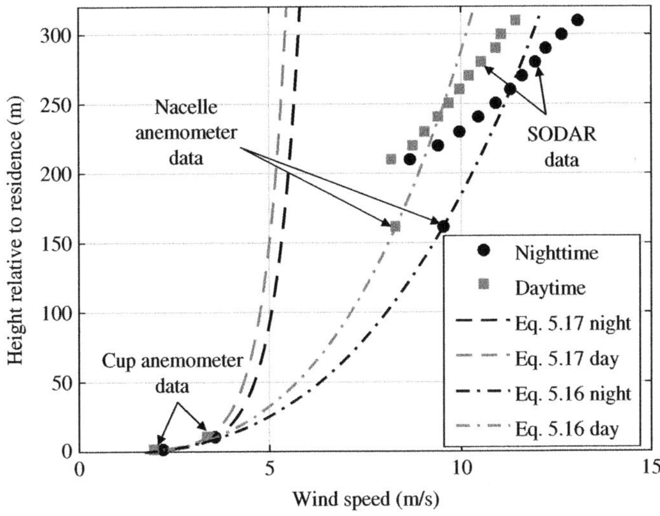

Figure 6.40 Comparison of measured wind speed data with wind speed profiles for nighttime (12 am–5 pm) and daytime, calculated using two different methods with the wind speed at 10 m as the input: Eq. (5.17); and Eq. (5.16), with ξ adjusted to obtain the best fit to the measured data. Note that all heights are expressed relative to the elevation of the residence at which the 1.5 m and 10 m measurements were made.

ensures that realistic inputs are used for propagation modelling. Long-term meteorological data are also required for compliance measurements. Wind speed profile data are useful in determining whether there is a potential for AM due to high wind shear in the wind turbine rotor plane. Figure 6.41 shows a plot of such a wind profile and it is shown that the wind shear in the rotor plane is greater during the nighttime (here 5pm–7am) than the daytime, on average, as indicated by the steeper velocity profile (Zajamšek et al. 2016). Note that in this figure, the height is referenced to the SODAR elevation, which is approximately the same as the base height of the nearest wind turbine. The increased wind shear during the night could be a source of stall noise and may also have some significant effects on the blade-pass frequency and harmonic peaks, according to Hubbard and Shepherd (1991). Comparison indicates that better agreement is achieved between the logarithmic profile and the daytime data than between the nighttime data and the logarithmic profile.

Figure 6.42 shows the variation in atmospheric stability over four seasons in rural South Australia according to data measured using the SODAR system that was located in the position shown in Figure 6.38 (Zajamšek et al. 2016). According to Figure 6.42, very stable conditions are more prevalent in the winter, and this is partially attributed to the reduced 'sun-up' period (Zajamšek et al. 2016), during which the earth is heated by the sun, warming the surrounding air and thus promoting instability. The data in

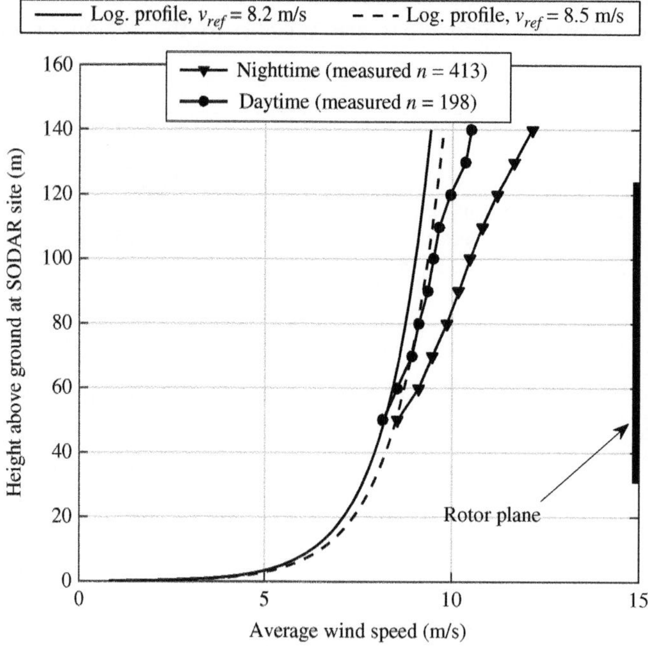

Figure 6.41 Typical measured wind velocity profiles at nighttime and daytime across the rotor plane, and calculated logarithmic profile according to Eq. (5.17), with $z_0 = 0.05$m. The parameter, n, is the number of wind speed profiles used to calculate the average wind speed profile (adapted from Zajamšek et al. (2016)). All heights are relative to the SODAR height.

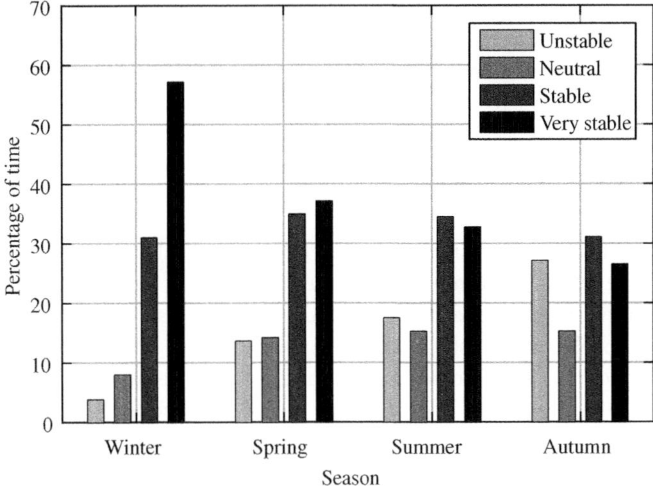

Figure 6.42 Variation of atmospheric stability over four seasons in 2013.The parameter *n* indicates the total number of samples included in the analysis for each season (Zajamšek et al. 2016).

Figure 6.42 indicates that very stable conditions could occur in this region up to ≈30% of the nighttime during spring, autumn and summer and up to ≈60% of the nighttime during winter. This analysis considered a constant definition of nighttime over the seasons of 5pm–7am. On the other hand, van den Berg (2013) showed that as a percentage of actual nighttime hours, a very stable atmosphere occurred more often in summer in the western part of the Netherlands, suggesting that the seasonal variation in atmospheric stability is dependent on the climate zone as well as the definition of nighttime. The effects of atmospheric stability on the source and propagation characteristics are discussed in Section 3.3.6 and Section 5.2.4, respectively.

6.5 Reporting on Noise, Vibration and Meteorological Conditions

The items described below are strongly recommended for inclusion in reports on measurements of wind farm noise and many of these points are minimum requirements of local regulatory authorities.

General instrumentation

- Serial numbers and copies of calibration certificates for the measurement equipment to demonstrate that all instruments were calibrated at the time of monitoring.
- Instrument name, type and name of manufacturer.
- Relevant instrumentation tolerances.
- Description of outdoor instrument positions in terms of height above the ground, distance from the wind farm, distance from the residence and proximity to reflective surfaces such as buildings and sources of extraneous noise such as trees, bushes and streams. Inclusion of justification for the choice of measurement locations.

- Details of indoor instrument positions including specification of mounting height, location within the room and justification for these positions.
- Photos of each instrument showing position relative to the residence and other instruments; position relative to the wind farm, where possible, and position within the room, where applicable.

General data

- Time and date that the measurements were taken, interval used for averaging (that is, 10 min) and the total duration of the measurements.
- Confirmation that data from different instruments (that is, microphones, accelerometers and meteorological masts) is synchronised in time.
- The number of data points included in the analysis and details of data point removal, including the number of data points removed and the associated reason.
- A description of calculation methods.
- Specification of the sample rate used in measurements, where applicable.

Acoustic-related instrumentation

- Electrical noise floor of the sound level meter and the degree of compliance with IEC 61672-1 (2013).
- Microphone/pre-amplifier specifications such as noise floor and frequency response.
- Details of primary wind screen used.
- Details of secondary wind screen used (if any) and the associated insertion loss.
- Orientation of the microphone(s) relative to the ground and to the source.

Acoustic data

- Relevant standards/guidelines and applicable allowable limits.
- Results of calibration.
- Measured overall noise levels (unweighted, A-weighted and C-weighted) and associated measurement interval.
- Measured 1/3-octave noise levels (unweighted, A-weighted and C-weighted) and associated measurement interval.
- Measured L_n (unweighted, A-weighted and C-weighted), where n denotes the percentage of time that a given level is exceeded (usually 90, but data corresponding to other values of n may also be reported).
- Background noise levels, including a description of the time of year that the measurements were taken, number of valid data points, possible sources of data contamination, location where the measurements were taken.
- A statement of whether the noise signal has been assessed for AM/variation and if so, include details on the following:
  - time and date that measurements were assessed for AM/variation
  - amplitude modulation/variation metric used in the analysis
  - data sampling rate
  - filtering; 1/3-octave bands, low-pass or band-pass
  - analysis interval to detect instances of AM or variation
  - total analysis period

  - critical value of the AM metric used and whether or not is has been exceeded
  - adjustment, if relevant.
- A statement of whether a tonality assessment has been undertaken and, if so, provision of the following information:
  - time and date that measurements were assessed for tonality
  - standard used in the analysis
  - whether the results were obtained by visual inspection or by automatic calculation
  - range of wind speeds and directions analysed
  - frequency limits of the relevant critical band(s)
  - frequency resolution of the analysis
  - selected range for visual averaging or linear regression (relevant for analysis using ISO1996-2 (2007))
  - at least one typical spectrum showing a plot of the masking noise and tones in the critical band range
  - frequency and levels (L_{pt}) of identified tone(s)
  - level of masking noise (L_{pn}) in the critical band(s)
  - tonal audibility (ΔL_a)
  - adjustment (K_T in dB), if relevant (see Section 6.2.12).

Vibration-related instrumentation

- Electrical noise floor of the vibration meter and the degree of compliance with IEC 61672-1 (2013).
- Noise floor and frequency response of accelerometer.
- Whether the accelerometer contained a built-in charge amplifier or an external one was used.
- Axis of measured vibration, including justification.
- Description of the surface on which the accelerometer was mounted.
- Details of the method used for attaching the accelerometer to the relevant surface.

Vibration data

- Relevant standards/guidelines and applicable allowable limits.
- Results of calibration.
- Measured overall weighted acceleration and associated measurement interval.
- Measured 1/3-octave vibration levels and associated measurement interval.
- Background vibration levels, including a description of the time of year that the measurements were taken, number of valid data points, possible sources of data contamination, location where the measurements were taken.

Meteorological and operational data

- Time and date that the measurements were taken and the total duration of the measurements.
- Method(s) used to determine wind speed and direction.
- Wind speeds and directions recorded during the measurement period at hub height at the wind farm.

- Wind speeds and directions recorded during the measurement at heights of 1.5 m and 10 m at the residence.
- Wind shear calculation details, including measurement heights and method used.
- Method(s) used to extrapolate data to other conditions.
- Number of data points collected during downwind conditions.
- Air temperature, atmospheric pressure, humidity, rainfall and locations of sensors to measure these variables.
- Cloud cover.
- Power output of the wind farm and operational data for individual wind turbines.
- Rotor rotational speed (if available).

Details of the surrounding environment

- Map/schematic showing wind farm layout and measurement positions.
- Photos of the wind farm and local surroundings.
- Description of the wind farm, including number of wind turbines, power rating and their positioning (that is, on a ridge-top, and so on), layout (linear, grid, and so on) and approximate distance between adjacent wind turbines.
- Description of local topography and ground surface characteristics.
- Estimated roughness length of the ground surface (see Section 5.2.4).
- Details of nearby reflective surfaces, including buildings or other structures, cliffs, water surfaces and trees.
- Details of nearby man-made noise sources not related to the wind farm, such as highways, airports, industrial activities, farming operations and other wind farms.
- Details of nearby natural sound sources, such as rivers, animals, surf or foliage.

Details of residence (for indoor measurements)

- Photos of the residence and local surroundings.
- Age of residence.
- Specifications of construction materials and their associated thicknesses.
- Dimensions of the measurement room, including width, length and height of the walls and the width and height of the windows.
- Description of any openings (such as doors and windows) and the associated open area during measurements.
- Estimate of the percentage of the room occupied by furniture.
- Estimate of the total room volume, wall surface area and roof surface area of the entire residence.

Uncertainty

- Estimate of the measurement uncertainty and associated confidence level for noise, vibration and meteorological measurements (as discussed in Section 6.2.16).

6.6 Wind Tunnel Testing

Wind tunnels are usually associated with aerodynamic measurements, but they have been increasingly used to measure noise generated by airfoils, bluff bodies and other components placed in fluid flow. Wind tunnels are useful for the design and understanding of wind turbine noise in that they can be used to take controlled and detailed measurements of both the flow field and the radiated noise from isolated airfoils and scaled full-rotors (provided the facility is large enough). Such testing allows comprehensive understanding of the noise-generating physics as well as testing of novel noise-control ideas, new airfoil shapes and new rotor configurations.

This section will give an overview of wind tunnels, and their use in aeroacoustic testing. Instrumentation and measurement techniques are summarised along with a review of current noise measurements in wind tunnels.

6.6.1 Wind Tunnel Techniques

Open-return Wind Tunnels

There are a wide variety of wind tunnel designs and operational methods (Bradshaw and Pankhurst 1964; Cattafesta et al. 2010), but most of the time these can be classified into two basic types: closed-return or open-return. Figure 6.43 illustrates the basic operating principles of the open return type. Here, a fan draws air through the tunnel, thus reducing the effects of the fan exhaust and other unsteadiness on the test flow in the working section. The air is drawn initially through the inlet with a bell mouth or other shape that eases the flow into the tunnel without unsteadiness, flow separation or other non-uniformity. The flow then passes through flow conditioning, which consists of a honeycomb section and a series of screens.

Honeycomb can be thought of a matrix of small cells or 'tubes' (with a length-to-diameter ratio of approximately 6–8), the purpose of which is to remove swirl and lateral velocity variations (Mehta and Bradshaw 1979). There is a proviso that the mean flow incident angle upon the honeycomb must not be greater that 10°, otherwise the cells will stall and create flow unsteadiness. The shape of the honeycomb appears to be relatively unimportant (Bradshaw and Pankhurst 1964), with hexagonal, circular and triangular shapes having been used, but with practical considerations of honeycomb availability, strength and materials playing a greater role in their choice than the cross-sectional shape. The number of cells should be approximately 150 per settling chamber 'diameter', so approximately $150 \times 150 = 22\,500$ is the recommended range (Mehta and Bradshaw 1979). These values are chosen so that the cell diameter is smaller than the smallest lateral wavelength of velocity variation, thus helping to produce an even flow.

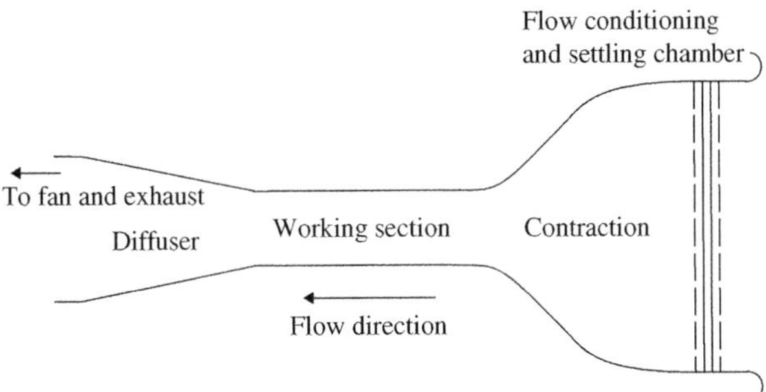

Figure 6.43 Schematic of open-return wind tunnel.

Screens are also required to condition the flow. They serve two purposes: to improve flow uniformity by imposing a static pressure drop and to reduce turbulence levels (Bradshaw and Pankhurst 1964; Cattafesta et al. 2010; Mehta and Bradshaw 1979; Mehta 1985). The fine screens break up turbulence in the flow into smaller eddies that subsequently decay in the wake of the screen. There are many guidelines for the choice of screen properties (such as open area ratio) and spacing (Mehta and Bradshaw 1979, for example), but Cattafesta et al. (2010) say that the wire size should be set so that the Reynolds number (based on wire diameter) is $Re_D = DU_\infty/\nu < 60$ (where D is the wire diameter, U_∞ is the free stream velocity and ν is the kinematic viscosity of the air) and that the screens should be spaced to allow the turbulence to decay sufficiently before encountering the next screen. Bradshaw and Pankhurst (1964) suggest that 500 wire diameters be used for screen spacing in most cases. A settling chamber is needed after the screens to allow the small-scale turbulence to decay before reaching the contraction.

A contraction is an essential part of a wind tunnel design (Doolan 2007). It is mainly required to accelerate the flow to the test conditions. Practically, it is a good idea to have most of the flow in the wind tunnel be at a low speed. This reduces the magnitude of the pressure losses through screens and bends, as they are proportional to velocity squared, which reduces the power requirement of the fan. Also very important is the effect that a contraction has on the turbulence intensity of the test flow. The acceleration of flow through the contraction reduces the fluctuating quantities with respect to the mean. In fact, the contraction will reduce streamwise turbulence intensity by a factor of $1/CR$ (where CR is the contraction ratio or ratio of inlet to outlet areas of the contraction) and $1/\sqrt{CR}$ for the cross-stream turbulence intensities (Mehta and Bradshaw 1979). Contraction design can be performed by eye, theory, by matched polynomials or through numerical means. The objective of the contraction shape is to provide a uniform, low-turbulence-intensity flow with boundary layers that are as small as possible. Usually contraction ratios vary between 4 and 20, with 8 considered to be a practical value that trades size with flow uniformity. A recent comparison of different wall shapes (Doolan 2007) shows that the fifth-order polynomial of Bell and Mehta (1988) provides a good design solution.

The flow passes from the contraction to the working section, where the tests are performed. The exact requirements of the working section depend upon the application, and this will be treated in more detail below for aeroacoustic measurements. For closed-section wind tunnels, useful information can be found in Cattafesta et al. (2010), Mehta and Bradshaw (1979) and Bradshaw and Pankhurst (1964). After the working section, the flow enters a diffuser, the purpose of which is to reduce the velocity of the flow before it enters the fan. This again reduces the power requirement of the fan.

Open-return wind tunnels are used widely, especially in university research. The advantages are that they are relatively inexpensive to manufacture and are well-suited to small test facilities. They are, however, susceptible to drafts and other disturbances in the air of the laboratory (or intake-air, which in some cases is the air outside the building) where the flow is harvested. They may need screens or flow straighteners in the laboratory where they are housed. They also need to be placed well away from walls to minimise flow disturbances affecting the working section flow quality. Further, the open-return system may also be configured as a blower-type, where the fan is placed

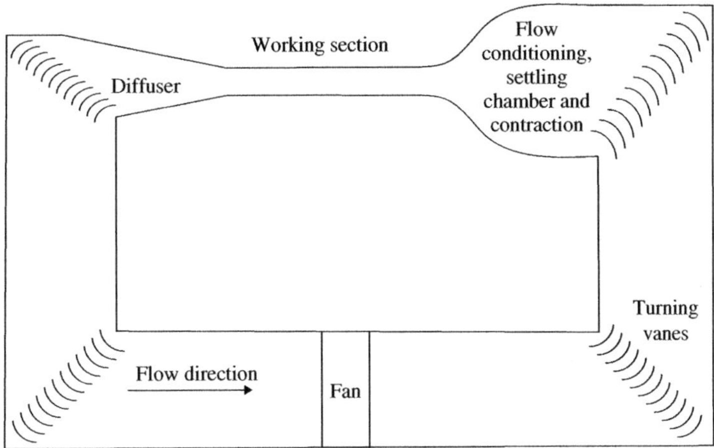

Figure 6.44 Schematic of closed-return type wind tunnel.

upstream of the flow conditioning systems. This is sometimes used for anechoic wind tunnels (see below).

Closed-return Wind Tunnels

Figure 6.44 shows a schematic of the closed-return type of wind tunnel. Many of the same design principles apply as discussed for the open-return wind tunnel, but the closed-return type provides the benefit of reduced power requirements, as the kinetic energy of the flow is not destroyed at the exit of the facility. There are power losses associated with the corners, but these can be minimised with proper design (Bradshaw and Pankhurst 1964; Mehta and Bradshaw 1979). Another difficulty that must be overcome, especially with larger facilities, is excessive heating. With the same air circulating through the facility, cooling is required to overcome a build up of heat, mainly due to viscous heating and energy addition from the fans. This can be expensive, but may be necessary to ensure control of temperature for repeatability and Reynolds and Mach number matching.[3] It may also be required to ensure the safety of operators. Closed-return tunnels are able to utilise less seeding for laser-diagnostics than open-return types, but it is more difficult to remove the seeding once the testing is complete (Cattafesta et al. 2010).

Open-jet Anechoic Wind Tunnels

The most common type of wind tunnel used for aeroacoustic testing is the open-jet variety. Figure 6.45 illustrates an open-jet anechoic wind tunnel of the suction type. Examples of this type of wind tunnel can be found at the Universities of Florida (Mathew et al. 2005) and Notre Dame (Mueller 2002, Ch. 5). Alternatively, blower type anechoic wind tunnels are also used, such as those used at the Universities of Adelaide (Moreau et al. 2011) and Southampton (Chong et al. 2009). Regardless of the type of tunnel used,

3 To perform scaled wind tunnel tests with the correct flow physics, the engineer must match the Reynolds number (flow velocity × model dimension/viscosity, where the model dimension is a characteristic dimension such as wing chord, *b*) and the Mach number (flow velocity/speed of sound). As the air temperature affects both the viscosity of the air and its speed of sound, care must be taken to control and monitor it during wind tunnel testing.

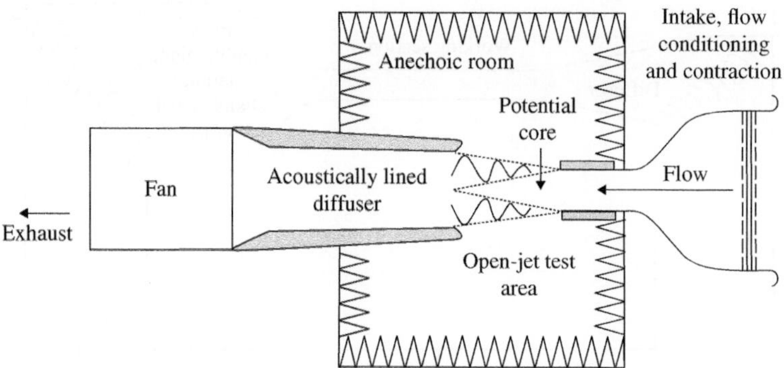

Figure 6.45 Schematic of an open-jet anechoic wind tunnel.

testing is done in an open-jet, usually in an anechoic or semi-anechoic room. Open-jet facilities are ideal for acoustic testing, as they allow the acoustic waves to radiate away from the test model and be recorded on microphones placed in the surrounding anechoic room. As the microphones are not in the flow, they are not contaminated by air flowing over them. Provided the surrounding anechoic chamber is large enough, the microphones can be placed far enough away from the test model, so they can considered to be in the acoustic far field.

The air enters the test area via the flow-conditioning systems and the contraction, as is normally done and described earlier in this chapter. The air then passes out of the contraction exit nozzle and forms a free jet. The test model (for example, a wing section) is placed within the potential core of the free jet; that is, the central zone that remains relatively unaffected by the growing turbulent shear layers formed on the outside of the free jet. This is illustrated in Figure 6.45 as the approximately triangular region immediately downstream of the contraction.

While testing with the model in the potential core of the free jet is useful for acoustic testing, it has some drawbacks. First, the flow conditions within the potential core vary with downstream position. This usually increases the free-stream turbulence levels, so these must be monitored if important for the particular testing being done. Second, the free jet will affect the aerodynamics of the test object. If an airfoil is tested at an angle of attack, the free jet itself will be deflected, chaining the effective angle of attack. While this effect can be quantified experimentally, a useful correction factor that allows an experimenter to obtain the effective angle of attack for a test airfoil in a free jet was determined by Brooks et al. (1989).

Third, the mean velocity profiles of the turbulent shear layers either side of the potential core will refract the acoustic waves produced within it, causing uncertainty in level and position. This can be taken into account using a variety of methods, and this topic is discussed in Section 6.6.2.

The quality of the acoustic measurements depends largely on the anechoic environment that the tests are conducted in. The anechoic room needs to designed so that it provides a near reflection-free environment for as wide a frequency range as possible. Typically, lower cut-off frequencies for the anechoic room can vary between 150 Hz (Mueller 2002, Ch. 5) and 250 Hz (Moreau et al. 2011). The construction of the room is ideally made from reinforced concrete blocks, isolated from ground vibrations and lined with rock wool, fibreglass or acoustic foam wedges (Chong et al. 2009; Mueller 2002). Smaller-budget facilities may construct double-walled plywood rooms, lined with acoustic foam wedges (Moreau et al. 2011). Openings need to be provided for the contraction and the diffuser. Safety must also be considered in the design of anechoic rooms, with door interlock systems installed to make sure the tunnel cannot run with personnel inside and in some cases a fire safety system to suppress fires if they should occur.

The size of the anechoic room is also important. Sound measurements should be made in the acoustic far-field, which is normally taken to mean that the distance between the model under test and the microphone is larger than the wavelength of the lowest frequency sound that must be measured. The distance between the centre of the source and the microphone must also be larger than twice the size of the source to avoid aerodynamic pressure fluctuations, (Mueller 2002, Ch. 1), which are not acoustic pressure fluctuations (because they do not travel at the speed of sound) but are still measured by a microphone. Measuring within a wavelength will introduce the effects of near-field (evanescent) waves. This may or may not be a problem for the test program, but needs to be taken into account when analysing the data. The size of the anechoic room will also affect the number of microphones that can be used, as well as their angle to the test flow. This will limit the resolution of a directivity test, which is useful for interpreting the source type and for ensuring that design changes are actually reducing noise, not redirecting the energy to another angle.

The diffuser has the important role of collecting the turbulent free jet and passing it to the fan or exhaust system. It should also incorporate an acoustic silencer to isolate the test section from fan noise. Diffuser silencer designs can be complex, due to the requirement for low pressure loss and high transmission loss. Example designs have been published by Mueller (2002) and Mathew et al. (2005). For the blower-type tunnels, the silencer must be placed downstream of the blower and upstream of the contraction and flow-conditioning devices. An example design is published by Chong et al. (2009), but this will have high pressure loss and in many cases a duct lined with acoustic foam will suffice, provided it is long enough (Mathew et al. 2005).

Finally, vibration isolation must be used as far as practicable. This includes effectively isolating the fan and drive motor from the ground and tunnel as well as providing flexible joints between ductwork, such as the connection between the diffuser and fan.

6.6.2 Noise Measurements in Wind Tunnels

Microphone Measurements

Microphones are the most common form of transducer used to measure acoustic pressure in an aeroacoustic test. Typically, condenser microphones are used to measure sound, and good descriptions of their operational principles and types are common (Bies and Hansen 2009; Mueller 2002). In open-jet anechoic wind tunnel testing, free-field microphones are typically used outside of the flow and sufficiently far away from the source to be considered in the free field. A free-field microphone has been specially designed to remove the effects of the microphone on the acoustic field. It does this through a carefully designed microphone cap. These microphones have a so-called flat frequency response, typically from 20 Hz to 20 kHz when the acoustic wave is normally incident to the diaphragm. This frequency range is normally sufficient to study both scaled and full-size wind turbine blades at low Mach number flow conditions. Specialist microphones and/or frequency-dependent calibration are normally required for measurements outside of this frequency range.

Although microphones are placed well outside the flow, it is advisable to place wind screens over the face of the microphones during testing, especially in university facilities that may use small anechoic chambers. Small flow velocities may develop in the chamber due to the shear forces created by the free jet. This can induce wind noise on the microphones, especially at low frequencies, and a wind screen is a sensible precaution to improve signal-to-noise ratios of the measurements.

In some circumstances, microphones are used within the flow itself. This is normally done in hard-walled (or closed-jet) wind tunnels or in situations where it is desirable to place the transducer as close as possible to the noise source. An excellent summary of inflow, strut-mounted microphone measurement techniques can be found in (Mueller 2002, Ch. 1). Inflow sound measurements can be successfully performed if the microphone is placed in a special aerodynamic cap. This reduces the noise from the flow and protects the diaphragm so that it is responsive to pressure fluctuations about the local static pressure. The design of the cap is extremely important to minimise self-noise as well as turbulence-interaction noise. Many microphone manufacturers supply aerodynamic caps for inflow microphone measurements.

Refraction through Shear Layers

When testing in a free-jet anechoic wind tunnel, the effect of the free shear layer must be taken into account. Figure 6.46 illustrates how the ray path is affected by the free shear layers that form on either side of the free jet. The wave created by the source within the uniform flow of the jet is initially convected by the flow field. When the acoustic wave encounters the edge of the jet, it is refracted by the velocity gradient within the shear layer. Once the wave has passed through the shear layer, it propagates through the quiescent air about the jet and passes over the microphones, which record the acoustic pressure time history at discrete locations in the free field.

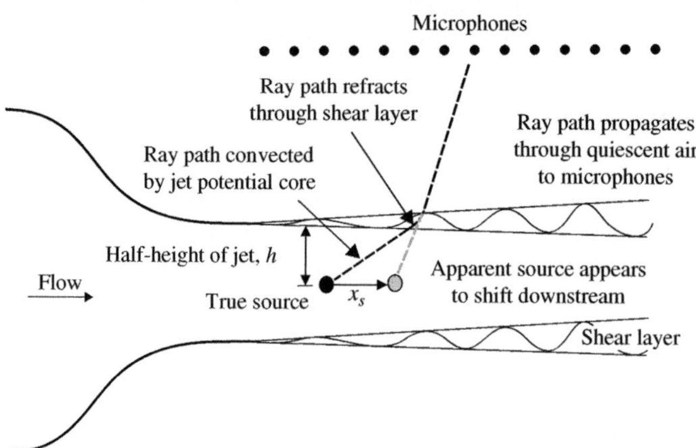

Figure 6.46 Illustration of acoustic refraction through shear layers in an open-jet anechoic wind tunnel.

The interaction of acoustic waves through the shear layer changes the direction of propagation as well the sound pressure level. It makes the source appear to be located slightly downstream of the true source location. Therefore, understanding shear-layer refraction is important when determining the directivity of the aeroacoustic source and relating the results to applications in practice, such as to a full-scale wind turbine.

The most popular method to correct for the effect of the shear layer is to model it as an infinitely thin sheet of vorticity (Amiet 1978; Mueller 2002). These formulations are well documented and provide corrections for the angle of propagation as well as changes in the level. A good summary of methods is found in Padois et al. (2013), who provide a comprehensive comparison of various shear-layer correction techniques with a numerical method. They show, via theoretical arguments based on the convected wave equation, validated numerically and experimentally, that the apparent source location can be approximately located at a distance, x_s, downstream of the true source location, where

$$x_s = Mh \tag{6.45}$$

Here, M is the Mach number of the free-jet potential core and h is the half-height of the jet (the distance from the jet centreline to the nozzle lip). This approximation was shown to be valid for Mach numbers $M < 0.4$ only. Further, amplitude corrections using the methods outlined by Amiet (1978) and Amiet and Schlinker (1980) were used and found to be necessary only above a Mach number $M > 0.2$.

Hence, for most low Mach number aeroacoustic testing of wind turbine blade sections, it would appear that the relatively straightforward method of Padois et al. (2013) is sufficient to correct for the source location.

Source Location using Microphone Arrays

The measurement of noise in anechoic wind tunnels is sometimes difficult, particularly in the case of broadband airfoil noise that may be only a few decibels above the background noise level of the tunnel itself, even when great care has been taken to reduce extraneous noise sources. It is also very important to be able to discriminate noise created by different parts of the airfoil or test object, for example to determine what part of the spectrum is dominated by leading-edge noise or trailing-edge noise. One way to improve the signal-to-noise ratio and to localise sound sources is to use a microphone array. Arrays are also used in hard-wall wind tunnels that are not normally used for acoustic testing. Thus the signal-to-noise improvements of these devices extend the capabilities of existing facilities.

A microphone array is simply a collection of microphones located accurately in space that record sound simultaneously during a test in a wind tunnel or other noise-generating situation. By recording the sound simultaneously and being able to accurately account for the phase differences between the microphones, noise-source location and strength can be acquired even in the presence of flow noise and reverberation.

The signal-processing technique that is used to analyse array data is called beamforming.[4] This post-experimental numerical processing procedure 'focusses' the array at various points of interest in the test area or, more typically, at uniformly spaced points

4 Michel (2006) produced an interesting history of beamforming.

on a pre-determined grid. This produces a 'sound-map' that describes the distribution of source strength over the test area of interest and enables the discrimination of noise-source location and strength.

There are many useful design guides to assist engineers and scientists develop their own microphone arrays (Mueller 2002; Prime and Doolan 2013, for example). In order to keep the number of microphones reasonable and avoid the spatial aliasing problems of uniformly spaced arrays, irregularly spaced arrays are used. In fact, to reduce spatial aliasing and sidelobe contamination (regions on the source map that indicate false noise sources), irregular arrays should have unique spacings between each microphone to ensure different phase delays between each microphone-pair.

Popular array designs, as recently reviewed by Prime and Doolan (2013), include the Archemedian spiral, Dougherty log-spiral (Dougherty 1998), Arcondoulis spiral (Arcondoulis et al. 2011), multi-spiral (Mueller 2002, Ch. 3), Brüel and Kjær style (Christensen and Hald 2006) and the Underbrink array design (Mueller 2002, Ch. 3). An analysis by Prime and Doolan (2013) showed that the Underbrink-style array appears to offer the best all-around performance, with the best array resolution and low maximum sidelobe (MSL) levels.

An example of an Underbrink-style array design is shown in Figure 6.47, which shows a 31-microphone array with each microphone spaced according to a pattern developed by Underbrink (Mueller 2002, Ch. 3). The pattern design parameters are described in the figure caption. The Underbrink design places microphones in the centre of equal-area segments that spiral away from the array centre. The procedure for calculating the microphone locations is to select the maximum and minimum (microphone) radii, r_{max} and r_0, the number of spiral arms, N_a, the number of microphones per spiral N_m, and the spiral angle, v. The area of the array is then separated into $N_m - 1$ equal-area annuli, which are further subdivided into equal-area segments, with microphones placed at the centre of these segments. Finally, an inner circle of microphones is added at r_0 to improve the high-frequency MSL and, if desired, an extra microphone can be placed at

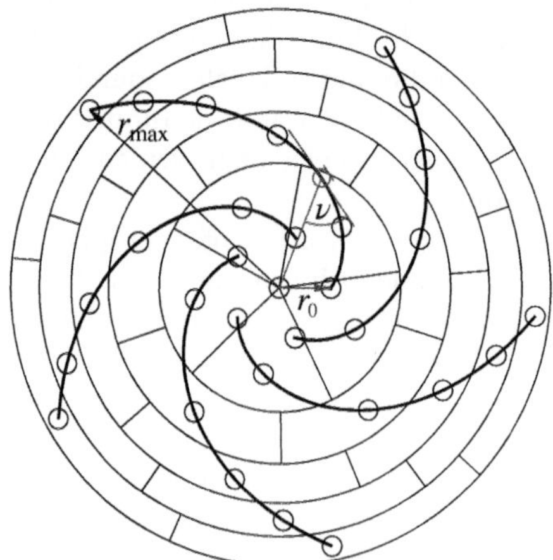

Figure 6.47 Underbrink array design with $N = 31$, $N_a = 5$, $N_m = 6$, $v = 5\pi/16$ rad, and $r_0/r_{max} = 0.2$. Each open circle represents a microphone location. Source Prime and Doolan (2013).

the origin. The radii of the microphones are:

$$r_{m,1} = r_0, \quad m = 1, \dots, N_a \tag{6.46}$$

$$r_{m,n} = \sqrt{\frac{2n-3}{2N_m - 3}} r_{\max}, \quad m = 1, \dots, N_a, \quad n = 2, \dots N_m \tag{6.47}$$

With the radii of the microphones known, the angles are calculated by placing each microphone along a log spiral, and rotating the spiral around the origin so that there are N_a spiral arms. Thus the angles of the microphones are:

$$\theta_{m,n} = \frac{\log_e\left(\frac{r_{m,n}}{r_0}\right)}{\cot(\nu)} + 2\pi\frac{m-1}{N_a}, \quad m = 1, \dots, N_a, \quad n = 1, \dots, N_m \tag{6.48}$$

Theory The theory underpinning beamforming arrays has been comprehensively described elsewhere (Dudgeon and Johnson 1993; Mueller 2002) and will only be briefly summarised below. Beamforming must assume the nature of the acoustic source it wishes to localise. The experimentalist can assume plane waves (sometimes known as *far-field beamforming*) or spherical waves (so-called *near-field beamforming*; not to be confused with acoustic holography techniques that measure near-field or evanescent waves). It is common in aeroacoustic wind tunnel testing to assume near-field or spherical acoustic waves because the array is normally placed at a position such that the waves have significant curvature as they pass through the array plane.

An array of N microphones will record pressure on each array microphone during a test so that $p_m(t)$ is the sound pressure signal recorded on the mth sensor. The Fourier transform of each signal is $P_m(f)$ and a matrix of cross-spectra $\hat{G}$ can be formed as follows,

$$\hat{G} = \begin{bmatrix} G_{11} & G_{12} & \dots & G_{1N} \\ \vdots & G_{22} & \dots & \vdots \\ \vdots & & \ddots & \vdots \\ G_{N1} & G_{N2} & \dots & G_{NN} \end{bmatrix} \tag{6.49}$$

where each element is defined as,

$$G_{m\ n} = \frac{S_A}{Q} \sum_{i=1}^{Q} P_m^*(f) \cdot P_n(f) \tag{6.50}$$

$P^*(f)$ is the complex conjugate of $P(f)$ and $\hat{G}$ is known as the *cross-spectral matrix*, which is used in so-called conventional beamforming; each element of $\hat{G}$ is averaged using Q spectra with scaling factor S_A, as defined in Chapter 2. Following Brooks and Humphreys (2006), steering vectors can be determined for each microphone, defined as follows for microphone m in the array

$$e_m = a_m \frac{r_m}{r_c} \exp\ \{j\omega\tau_m\}; \quad \omega = 2\pi f \tag{6.51}$$

The steering vector is the means by which the phase differences between microphones can be taken into account when analysing a particular point in space (the array is 'steered' to that point). The phase is a function of the propagation time, τ_m, from the steering location (location where the array is focussed) to the microphone, which must take into account any delays from shear-layer correction, if testing is done in an open-jet anechoic facility.

$$\omega\tau_m = kr_m + \omega\Delta t_{m,\text{shear}} \tag{6.52}$$

where $k = \omega/c$ is the wavenumber. The additional time correction $\Delta t_{m,\text{shear}}$, is the additional time that the acoustic wave takes to arrive at the microphone due to convection through the open-jet shear layer, or through the uniform flow field in a closed-section tunnel. There are a few different ways to calculate the time delay due to refraction through a shear layer and these were summarised earlier in this chapter: Padois et al. (2013) describes various methods and a simple correction technique. Amplitude correction is taken into account using the term a_m (Amiet 1978; Padois et al. 2013) and the array response is normalised to the centre position using the ratio r_m/r_c, where r_m is the distance from microphone m to the steering location and r_c is the distance from the centre of the array to the steering location.

The array output power spectrum is a function of steering location

$$\hat{Y}(\hat{e}) = \frac{\hat{e}^{\text{H}}\hat{G}\hat{e}}{N^2} \tag{6.53}$$

where the superscript H represents the Hermitian transpose (transpose of complex conjugate) and $\hat{e}$ is a vector of steering locations for each microphone

$$\hat{e} = \begin{bmatrix} e_1 \\ e_2 \\ \dots \\ e_N \end{bmatrix} \tag{6.54}$$

As noted by Brooks and Humphreys (2006), $\hat{G}$ often has the background level (obtained by removing the model from the wind tunnel and performing the same test) subtracted from it to improve results. Microphone self-noise can be removed from the array output to improve signal-to-noise ratio using a 'diagonal removal' technique, which means that the diagonal terms of $\hat{G}$ are replaced with zeros. This is justified by the fact that self-noise is not present in the off-diagonal terms, due to the cross-correlation procedure. Further, an analysis (Mueller 2002, Ch. 2), also shows that the diagonal terms do not improve resolution, due to lack of phase information on these microphones. Hence when diagonal terms are removed, the beamformer output becomes

$$\hat{Y}(\hat{e})_{\text{diag}=0} = \frac{\hat{e}^{\text{H}}\hat{G}_{\text{diag}=0}\hat{e}}{N^2 - N} \tag{6.55}$$

where $\hat{G}_{\text{diag}=0}$ is $\hat{G}$ with diagonals set to zero (note the denominator takes into account the effect of diagonal removal on the amplitude of the output of the array). Weighting functions (Brooks and Humphreys 2006; Mueller 2002; Padois et al. 2013) may also be used to shade various microphones and improve array resolution for varying frequencies. Sound maps are obtained at frequencies (or frequency bands) of choice by evaluating the array output $\hat{Y}(\hat{e})$ over a scanning grid and plotting the levels over that grid as a contour map.

Improving Resolution The approximate resolution of an array is set by the Rayleigh criterion,

$$\frac{\Delta \ell}{\lambda} = \frac{Wz}{D} \tag{6.56}$$

where $\Delta \ell$ is the approximate separation distance between two sources emitting sound of wavelength $\lambda = c/f$, z is the distance between the array and the source, D is the array diameter and $W = 1.22$ is the Rayleigh limit (which is the distance between the peak and the first zero of an ideal diffraction pattern, divided by wavelength). Conventional beamforming algorithms will only be able to resolve sources separated by a distance set by Eq. (6.56), which depends on array dimensions and wavelength. Resolution is also quantified using the beamwidth of the array, defined as the width of the pattern created on the beamformer output at a particular frequency, measured from its peak to a point 3 dB down in amplitude. Thus the low-frequency limit of the array is dependent on the source separation resolution that is required according to Eq. (6.56).

The high-frequency limit of an array is set by the minimum microphone spacing. Spatial aliasing limits the smallest resolved wavelength to double the minimum microphone spacing. Attempting to resolve noise sources at higher frequencies (smaller wavelengths) creates unwanted peaks called side lobes on the beamforming output. Array designs are thus a compromise between aperture size (the diameter, D) and the number of microphones, which trades frequency range against size and cost. The actual response of any proposed array design should be determined using Eq. (6.55) with different source locations and microphone patterns. Such a method will give a much better indication of resolution and spatial-aliasing effects of a given array design.

In an effort to improve array resolution and reduce sidelobe contamination, deconvolution techniques (see Section 2.4.15) have gained popularity over the last decade and have now become a standard way to improve beamforming images. Deconvolution will improve a 'dirty map' (contaminated results) $b(x, y)$ measured over a grid defined by points at locations (x, y) with respect to a predefined origin (such as the array centre). The goal is to find a source distribution $q(x', y')$ over the source region (defined by points at (x', y')) by solving the following equation

$$b(x, y) = \sum_{(x', y')} q(x', y') p(x, y|x', y') \tag{6.57}$$

which is often written in matrix form

$$\hat{A}\hat{q} = \hat{b} \tag{6.58}$$

where $\hat{A}$ is a matrix whose elements are the point-spread function, $p(x, y|x', y')$. The point-spread function is the response of the array when it is focussed at a steering location (x, y), to a point source with unit strength placed at an arbitrary location (x', y') on the source observation grid. It is determined analytically by placing ideal monopole sources at multiple locations in the focussing plane and computing the beamformer output, corresponding to the beamformer being focussed at various steering locations, using Eq. (6.55); the point-spread function therefore contains all resolution and spatial-aliasing information of the array. The point-spread function thus distorts the true acoustic image and deconvolution is the process by which this distortion is removed. Mathematically, the vectors $\hat{q}$ and $\hat{b}$ contain the desired source strengths and measured (or 'dirty') map respectively. Deconvolution methods solve for $\hat{q}$ using Eq. (6.58).

Ehrenfried and Koop (2007) provide a recent review of common deconvolution techniques used in aeroacoustics. Some deconvolution methods stem from earlier applications in astronomy, such as the CLEAN algorithm (Högbom 1974; Schwarz 1978), applied by Dougherty and Stoker (1998) to aeroacoustics. The most popular and well-known deconvolution method is DAMAS (deconvolution approach for the mapping of acoustic sources) (Brooks and Humphreys 2006). Here, a numerical method has been devised that maintains positivity of the source terms, even though the $\hat{A}$ matrix is singular. While CLEAN and DAMAS have been shown to provide good results, they are computationally expensive because they use the exact point-spread function from each source (scanning grid) point to the array. This high computational cost can sometimes make the use of DAMAS undesirable.

Significant reductions in computational expense can be gained if the point-spread function can be simplified. A common assumption is to make the point-spread function shift-invariant. This means that the point-spread function is only a function of the distance from the array, and the point-spread function is defined using a single representative unit source placed in the region of interest,[5] at (x_s', y_s'). That is,

$$p_{\text{shift invariant}} = p(x, y|x_s', y_s') \tag{6.59}$$

The reduction in accuracy of this assumption is often justified by the reduction of computational cost needed to obtain a clean sound map. Common shift-invariant deconvolution methods include DAMAS2 and DAMAS3 (Dougherty 2005), which have been used widely.

Another, very useful technique is a method called CLEAN-SC (Sijtsma 2007). The creators of this technique make the useful observation that sources may not be represented as a point monopole source, as is assumed when calculating the point-spread function. Real aeroacoustic sources may exist over a region of space (the trailing edge or tip of a wind turbine rotor, for instance) and may not have uniform directivity. To this end, the CLEAN algorithm is modified so that instead of performing a deconvolution based on peak levels and synthetic point-spread functions, the spatial coherence (hence the acronym CLEAN-SC) of the sound map is analysed (Oerlemans and Sijtsma 2002). As the peak source will be spatially coherent with its side lobes, a beam pattern based on the measured spatial coherence is subtracted from the sound map. Thus the sound map has been 'cleaned' of the response of the actual source. This avoids the use of synthetic point-spread functions (based on ideal monopoles) and vastly improves the source map. To visualise the sources, a 'clean-beam' of sound replaces the spatially coherent information that was removed. Great care is taken by Sijtsma (2007) to ensure the power of this beam correctly represents the true source strength. CLEAN-SC is also not very computationally intensive, in that it takes only about twice the computational cost of conventional beamforming, in contrast to CLEAN and DAMAS.

Improving the resolution of aeroacoustic beamforming arrays is an active area of research. New deconvolution schemes that provide enhanced resolution include the mapping of acoustic correlated sources method (Yardibi et al. 2010b), the

5 The region of interest is where the engineer believes the source is likely to be located. It is often a position aligned with the centre of the array.

linear programming technique (Dougherty et al. 2013) and functional beamforming (Dougherty 2014).

Practical Issues There are many important practical issues when designing and using an acoustic array for aeroacoustic applications. Many of these issues, including data-acquisition requirements, are covered by Underbrink (Mueller 2002, Ch. 3) and the reader is referred to this comprehensive and still very relevant and useful work. From the experience of the authors, particular care must be taken to rigidly mount the array and to have a low positional error for each microphone location. Management of large numbers of microphones, cables (including their weight) and synchronising data-acquisition channels is very important. Unwanted reflections from the array or its support structure must also be minimised.

There are a few calibration methods available, including building a small anechoic chamber to place over the array, to building small portable calibration sources to place over each microphone. Phase matching of microphones is not as important as it may first seem. With many (maybe hundreds) of microphones, phase errors along the entire measurement chain must be accounted for, rather than just the phases between individual microphone outputs; these errors may differ due to cable lengths, signal conditioning and the data-acquisition system. End-to-end level and phase calibration of the system is desirable to understand phase differences and to correct for them in post-processing.

Yardibi et al. (2010a) have provided an extensive uncertainty analysis of the Dougherty's array calibration technique (Mueller 2002, Ch. 2) as well as conventional beamforming (via an analysis of an existing array at the University of Florida). They show that calibration of an array and data-acquisition chain is essential (both level and relative phase) in order to minimise error. If the Dougherty's calibration method is used, then errors less than ± 1 dB can be expected with 95% confidence.

Precision microphones are usually the sensors of choice for experimenters, but many low-cost electret condenser and MEMS[6] microphones, for example the Panasonic WM-61A electret condenser microphone, are making their way into aeroacoustic arrays. These low-cost sensors allow larger-channel-count systems to be deployed in wind tunnels. Such large microphone systems are now able to be used by university researchers as well, given the continued lower cost of high-speed multi-channel data-acquisition systems. MEMS sensors encourage researchers to buy many more sensors than they need, and perform their own quality assurance checking, only choosing sensors that have the necessary dynamic range and/or frequency response for the array.

Microphone mounting is important, and will differ depending on the application. Beamforming on hard-walled wind tunnel facilities presents the designer with two choices. The first is to flush mount the sensors with the wall. This will expose the microphone to turbulent boundary-layer pressure fluctuations that must be removed during signal processing. The second is to recess mount the microphones behind a tensioned kevlar wall, that physically removes the sensors from the evanescent waves generated by the boundary layer. Both techniques are discussed in some detail by Underbrink (Mueller 2002, Chs 2 and 3).

6 Micro-electro-mechanical systems

6.6.3 Review of some Recent Measurements

Aeroacoustic beamforming has proven to be a powerful technique for wind tunnel testing of stationary airfoil sections and many other components. It is also becoming widely used for full-scale wind turbines, as well as model rotors in wind tunnels. To obtain a complete understanding of wind turbine noise, testing should be done under both controlled conditions and in the field. Using a lab-scale model rotor in a wind tunnel gives high-quality measurements that enable researchers to link flow, operational and design conditions to noise output, location and type. This allows complete understanding and validation of numerical models. Field testing on full-scale turbines is also required, in order to confirm the performance of new rotor designs, understand the influence of repairs or other modifications to rotors on noise and to investigate and diagnose noise problems.

Fixed-blade Aeroacoustic Tests

Fixed blade testing is done for practical reasons, mainly to do with size and spatial resolution of the flow field (that is, it is easier to obtain data from a larger-chord fixed airfoil than a smaller-chord, rotating-blade system). Nevertheless, there are important differences between fixed and rotating blades, so controlled rotating-blade testing is also necessary in order to isolate the specific physical causes of noise production.

There have been many fixed-blade or airfoil tests previously reported upon in the literature. A recent comprehensive review of airfoil trailing-edge noise was performed by Doolan and Moreau (2015). It was found that despite the long history of airfoil trailing-edge noise research, there is still uncertainty regarding the accuracy of current experimental data sets and how they compare against each other. The consensus is that trailing-edge noise is controlled by the turbulent eddies in the outer boundary layer and the wall shear stress. Hence the effects of surface roughness, tripping, free-stream turbulence and of course airfoil shape will influence noise generation. The review also indicates that acoustic scattering from the leading edge and other parts of a test facility may cause some uncertainty in resolving the peak radiating noise level and frequency.

Doolan and Moreau (2015) compare trailing-edge noise data sets and show that there are important and unexplained differences, thus suggesting that a new comprehensive experimental program is needed to re-evaluate trailing-edge noise and the parameters that affect its production. Experiments are required across a wider range of Reynolds numbers, with improvements in instrumentation that enable better resolution of the low-frequency peak in the noise spectrum. The role of outer-boundary-layer eddies in noise production needs further investigation, as does the influence of acoustic scattering from the leading edge and other parts of the facility. Comprehensive cross-facility comparisons need to be expanded (although some work has already been performed by Migliore and Oerlemans (2004), Devenport et al. (2013) and Bahr et al. (2008)) as do the number of computational studies. The role of computational work is essential, as it can eliminate the effects of extraneous noise sources and facility scattering.

Scaled Rotor Tests in Wind Tunnels

The number of rotating blade wind turbine acoustic tests performed in a wind tunnel are relatively few. Oerlemans et al. (2001) performed an early study as part of the European DATA[7] program. Experiments were performed in the DNW-LLF wind tunnel on

7 Design and Testing of Acoustically Optimized Airfoils for Wind Turbines.

a model scale wind turbine. The purpose of the tests was to understand and verify the noise-reduction capabilities of optimised airfoil shapes on a rotating blade. This was needed because the low-noise airfoil shapes were developed using two-dimensional testing and computation. Tests were performed using a two-bladed 4.48 m diameter rotor with the rotor blade having optimised airfoil shapes, as well as serrated blades. Acoustic measurements were performed using a 136-microphone array with conventional beamforming used to process the array signals. Further, a special de-rotational method, known as ROSI[8] was used to visualise the noise sources at the tip in the frame of the blade itself. Results showed that optimised blades were able to reduce noise on average by about 4 dB. The use of serrated blades obtained a further 2–3 dB reduction.

Cho et al. (2010) performed aeroacoustic beamforming measurements on a 12% scale NREL wind turbine model placed in a wind tunnel. A 144-microphone array was placed on the floor of the wind tunnel working section, upstream of the turbine. The wind tunnel was a closed-return, hard-walled facility with a 4 × 3 m test section. The tunnel was operated so as to produce a 1–13 m/s test flow and the turbine operated at 600 RPM. The model turbine had a rotor diameter of approximately 1.2 m. The results of these tests showed that the source location moved outboard (towards the blade tip) as the frequency increased and the noise level increased significantly below 2 kHz when the rotor blade was stalled.

Ryi et al. (2014) developed a method of scaling small-scale turbine noise data from wind tunnels so they could predict noise generated by full-scale wind turbines. The scaling was verified using wind turbine rotor noise measurements in an open-jet anechoic wind tunnel with a 1.8 × 1.8 m jet outlet. A 1.4 m diameter turbine was placed in the potential core and operated at 491 to 1473 RPM, with free-stream velocities up to 35 m/s. Noise data were recorded using a small array of six microphones. The results confirmed that small-scale turbine noise data can be used to predict noise generated by full-scale ones. Lee et al. (2013) used the same wind tunnel and rotor test setup to test the noise generated by optimised rotor blades. They showed that 2–3 dB noise reduction was possible using an optimised shape, developed using their numerical procedure.

Bale and Johnson (2013) developed a low-cost MEMS microphone array for testing small wind turbines, either in wind tunnels or in the field. In this array, 30 low-cost (Knowles) microphones were arranged in an Underbrink spiral, with an aperture size (twice r_{max} in Figure 6.47) of 940 mm. Acoustic tests were performed using an open-jet wind tunnel with a 610 mm square outlet. A three-bladed, 1.3 m diameter turbine was placed 1.6 m downstream of the outlet. The array successfully located the trailing-edge noise source on the turbine.

An interesting study was presented in Pearson and Graham (2013), where the noise generated by a scaled vertical axis wind turbine was measured in a hard-walled closed-return wind tunnel at the University of Cambridge. The working section of the tunnel was 1.679 × 1.220 m and the diameter of the rotor and the blade span were approximately 0.5m. Noise measurements were made using a phased acoustic array mounted in the floor of the tunnel, consisting of two nested arrays, one high-frequency and one low-frequency, each with 48 microphones. Testing was performed at wind speeds up to 21 m/s. Noise from the turbine was successfully measured and analysed,

8 ROtating Source Identifier

showing that many features of the noise spectrum can be related to unsteady loading on the turbine blades.

Beamforming Full-scale Wind Turbines

Oerlemans et al. (2007) published seminal and highly-cited work on full-scale wind turbine beamforming. They used an array of 148 Panasonic WM-61 electret condenser microphones, to localise acoustic sources on a GAMESA G58 turbine. The array was mounted 58 m upwind of the turbine on a 15 × 18 m mounting board, with the microphone diaphragms mounted flush with the surface. The array was configured in an elliptical shape so that it maintained similar resolution in the vertical and horizontal directions of the rotor plane. As with the wind tunnel study of Oerlemans et al. (2001), both conventional and ROSI de-rotational beamforming were applied to the array data. No deconvolution (adjustment for distortion caused by the point-spread function of the array) was used during post-processing. This paper convincingly proved that trailing-edge noise is the dominant aerodynamic noise source – for frequencies 500 Hz and above, as the frequency range of the measurements was set by the requirement to filter out problematic low-frequency noise – on a turbine. It also showed the roles that convective amplification and trailing-edge noise directivity have on the measured source pattern on the rotor plane.

Buck et al. (2013) of the US National Renewable Energy Laboratory (NREL) have developed a large, all-weather beamforming array for full-scale turbine testing over extended periods of time. This is a 63-microphone array that uses relatively low-cost electret condenser microphones (Panasonic WM-64BC or PUI Audio POM-2242P-C33-R) mounted in specially designed individual weatherproof enclosures that contain the power supply and signal-conditioning electronics. Each microphone is mounted on a wooden board (in a similar manner to that recommended in IEC 61400-11 (2012)) with provision for wind screens. The data-acquisition system is also deployed in the field in an aluminium enclosure with heating and dehumidifying functions to protect the electronics. Such protection allows the array to be deployed in the harsh conditions of NREL's test site, which includes snow, rain, high wind and wind-borne debris, and temperatures between about −25°C and 40°C. The microphone pattern of this array was determined using an optimisation procedure as documented by Oerlemans et al. (2007). Each microphone board was mounted on a dirt/mortar base and was precisely located using a laser scanning system. The size of the array (approximately 30 × 30 m) allows a lower-frequency resolution of 750 Hz (corresponding to a beamwidth of 3.53 m). Recent results from the array show clear resolution of trailing-edge noise and the existence of scattering of rotor noise from the support tower.

More recently, Ramachandran et al. (2014) performed acoustic beamforming on a full-scale GE 1.5-MW wind turbine. A commercial OptiNav 24 microphone compact array was used and placed at 50 m and 85 m from the tower and results were presented for frequencies between 155 and 8944 Hz. The main aim of the study was to compare a range of advanced deconvolution methods for full-scale wind turbine testing. The methods tested and compared were conventional beamforming, DAMAS (Brooks and Humphreys 2006), LP (Dougherty et al. 2013), TIDY (Dougherty and Podboy 2009) and CLEAN-SC (Sijtsma 2007). Their findings were that DAMAS and LP provided the best resolution of distributed trailing-edge noise sources and that compact arrays can be used for full-scale acoustic beamforming if advanced deconvolution is used to improve the results. Interestingly, they showed (via beamforming and deconvolution) that during yawing motion noise is radiated from the tower, which appears to act as a resonator.

For beamforming at low frequencies, a large-aperture acoustic array is currently being developed by DTU in Denmark (Bradley et al. 2015). The array consists of 48 custom-built radio microphones, which have a low noise floor and a flat frequency response down to approximately 20 Hz. An Underbrink multi-arm spiral array design has been chosen (as illustrated in Figure 6.47) and near-field array projection is used to achieve a more symmetrical beam pattern. To ensure that the measurement range extends to frequencies down to 100 Hz, at least two arrays will be used in the project.

6.7 Conclusions

This chapter has presented a wide variety of measurement techniques for characterising the sound and vibration generated by wind turbines. The most challenging aspects are associated with the measurement and analysis of turbine noise arriving at sensitive receiver locations, and treatment of this topic has occupied the bulk of the chapter. A detailed discussion on the quantification of AM and tonality has also been included along with techniques to measure and estimate the level of wind shear.

The operation and use of wind tunnels are also important in developing an understanding of how wind turbines generate noise and this aspect of measurement has been summarised here as well. The use of acoustic arrays for characterising aerodynamic noise is becoming standard due to the falling costs of high-speed data-acquisition systems, computers and, to some extent, microphones themselves. So part of this chapter has been concerned with a discussion of the optimisation of these arrays and the analysis of the data they generate in order to identify sources of noise on a rotating airfoil such as a wind turbine blade.

References

Abbott J and Raspet R 2015 Calculated wind noise for an infrasonic wind noise enclosure. *Journal of the Acoustical Society of America*, **138**(1), 332–343.

Alcoverro B 1998 Proposition d'un systeme de filtrage acoustique pour une station infrason IMS. Technical report No. 241, CEA-DASE (in French).

Amiet R 1978 Refraction of sound by a shear layer. *Journal of Sound and Vibration*, **58**(4), 467–482.

Amiet R and Schlinker R 1980 Refraction and scattering of sound by a shear layer. *6th AIAA Aeroacoustics Conference*, Hartford, USA.

AMWG 2015 Discussion document: Methods for rating amplitude modulation in wind turbine noise. Technical report, IOA Noise Working Group (Wind Turbine Noise): Amplitude Modulation Working Group.

Annan A, Manwell J, McLaughlin D and Pepyne D 2015 A system for measuring wind turbine infrasound emissions. *Sixth International Meeting on Wind Turbine Noise*, Glasgow, UK.

ANSI/ASA S12.9-3 (R2003) 1993 Quantities and procedures for description and measurement of environmental sound – Part 3: Short-term measurements with an observer present. Technical report, ANSI.

ANSI/ASA S12.9-7 draft 2014 Measurement of low frequency noise and infrasound outdoors in the presence of wind and indoors in occupied spaces. Technical report, ANSI.

ANSI/ASA S1.42 2006 Design response of weighting networks for acoustical measurements. Technical report, ANSI.

ANSI/ASA S3.4 2007 Procedure for the computation of loudness of steady sounds. Technical report, ANSI.

Antoniou I, Jørgensen HE, Ormel F, Bradley S, von Hünerbein S, Emeis S and Warmbier G 2003 On the theory of SODAR measurement techniques. Technical report, Risoe National Laboratory.

Arcondoulis E, Doolan CJ, Brooks LA and Zander A 2011 A modification to logarithmic spiral beamforming arrays for aeroacoustic applications. *Proceedings of the 17th AIAA/CEAS Aeroacoustics Conference*, Portland, Oregon.

AS 4959 2010 Acoustics – Measurement, prediction and assessment of noise from wind turbine generators. Standards Australia.

Ashtiani P 2013 Generating a better picture of noise immissions in post construction monitoring using statistical analysis. *5th International Meeting on Wind Turbine Noise*, Denver, Colorado.

Ashtiani P 2015 Spectral discrete probability density function of measured wind turbine noise in the far field. *6th International Meeting on Wind Turbine Noise*, Glasgow, UK.

Bahr C, Yardibi T, Liu F and Cattafesta L 2008 An analysis of different measurement techniques for airfoil trailing edge noise number. AIAA 2008-2957 in *14th AIAA/CEAS Aeroacoustics Conference*, Vancouver, Canada.

Bale A and Johnson DA 2013 The application of a MEMS microphone phased array to aeroacoustics of small wind turbines. *Wind Engineering*, **37**(6), 637–658.

Bass J 2011 Investigation of the 'Den Brook' amplitude modulation methodology for wind turbine noise. *Acoustics Bulletin*, **36**(6), 18–24.

Bell J and Mehta R 1988 Contraction design for small low speed wind tunnels. Technical report CR-177488, NASA.

Bell S 1999 A beginner's guide to uncertainty of measurement. Measurement Good Practice Guide No. 11 (Issue 2), National Physical Laboratory, UK.

Berglund B, Lindvall T and Schwela D 1999 Guidelines for community noise. WHO.

Betke K, Schultz-von Glahn M, Goos O and Remmers H 1996 Messung der infraschallabstrahlung von windkraftanlagen. *Proceedings of DEWEK '96, 3rd German Wind Power Conference*, Wilhelmshaven, Germany (in German).

Bies D and Hansen C 2009 *Engineering Noise Control: Theory and Practice*, 4th edn. Spon Press.

Blazier Jr, W 1997 RC Mark II: A refined procedure for rating the noise of heating, ventilating, and air-conditioning (HVAC) systems in buildings. *Noise Control Engineering Journal*, **45**(6), 243–150.

Bleazey J 1961 Experimental determination of the effectiveness of microphone wind screens. *Journal of the Audio Engineering Society*, **9**, 48–54.

Boullet I 2005 La Sonie des Sons Impulsionnels: Perception, Mesures et Modales. PhD thesis, LMA-CNRS, Marseille (in French).

Boullet I, Gagneaux F, Rabau G, Meunier S and Boussard P 2006 Un estimateur de sonie d'impulsion: élaboration et validation. *French Congress of Acoustics CFA'06*, Tours, France (in French).

Bowdler R and Leventhall G 2011 *Wind Turbine Noise*. Multi-Science.

Bradley S 2015 Time-dependent interference: the mechanism causing amplitude modulation noise? *6th International Meeting on Wind Turbine Noise*, Glasgow, UK.

Bradley S, Mikkelsen T and Legg M 2015 The DTU wind energy WTN test facility. *6th International Meeting on Wind Turbine Noise*, Glasgow, UK.

Bradshaw P and Pankhurst RC 1964 The design of low-speed wind tunnels. *Progress in Aerospace Sciences*, **5**, 1–69.

Brandt A 2011 Noise and Vibration Analysis: Signal Analysis and Experimental Procedures. John Wiley & Sons.

Bray W and James R 2011 Dynamic measurements of wind turbine acoustic signals, employing sound quality engineering methods considering the time and frequency sensitivities of human perception. *Proceedings of Noise-Con*, Portland, Oregon, USA, pp. 25–7.

Broneske S 2014 Wind turbine noise measurements – how are results influenced by different methods of deriving wind speed? *Internoise 2014*, Melbourne, Australia.

Brooks T and Humphreys W 2006 A deconvolution approach for the mapping of acoustic sources (DAMAS) determined from phased microphone arrays. *Journal of Sound and Vibration*, **294**(4–5), 856–879.

Brooks T, Pope D and Marcolini M 1989 Airfoil self-noise and prediction. Reference publication 1218, NASA.

BS EN 61260-1 2014 Electroacoustics. octave-band and fractional-octave-band filters. specifications. British Standards Institute.

Buck S, Roadman J, Moriarty P and Palo S 2013 Acoustic array development for wind turbine noise characterization. Technical report NREL/TP-5000-60457, National Renewable Energy Laboratory, USA.

Cattafesta L, Bahr C and Mathew J 2010 Fundamentals of wind-tunnel design. In: Blockley R and Shyy W, *Encyclopedia of Aerospace Engineering*. John Wiley and Sons.

Cheong C and Joseph P 2014 Cyclostationary spectral analysis for the measurement and prediction of wind turbine swishing noise. *Journal of Sound and Vibration*, **333**, 3153–3176.

Cho T, Kim C and Lee D 2010 Acoustic measurement for 12% scaled model of NREL Phase VI wind turbine by using beamforming. *Current Applied Physics*, **10**(S), S320–S325.

Chong T, Joseph P and Davies P 2009 Design and performance of an open jet wind tunnel for aero-acoustic measurement. *Applied Acoustics*, **70**(4), 605–614.

Christensen JJ and Hald J 2006 Beam forming array of transducers. US Patent 7,098,865.

Cooper S, 2013 Hiding wind farm noise in ambient measurements-noise floor, wind direction and frequency limitations. *5th International Meeting on Wind Turbine Noise*, Denver, Colorado.

Cooper J and Evans T 2013 Automated detection and analysis of amplitude modulation at a residence and wind turbine, *Acoustics 2013*, Victor Harbor, Australia.

Cooper J, Evans T and Petersen D 2015 Method for assessing tonality at residences near wind farms. *International Journal of Aeroacoustics*, **14**(5/6), 903–908.

Courtney M, Wagner R and Lindelöw P 2008 Testing and comparison of LIDARS for profile and turbulence measurements in wind energy. *14th International Symposium for the Advancement of Boundary Layer Remote Sensing*, Copenhagen, Denmark.

Craven N and Kerry G 2007 A good practice guide on the sources and magnitude of uncertainty arising in the practical measurement of environmental noise. Technical report, University of Salford.

DEPA 1984 Measurement of environmental noise from industry; the Joint Nordic method for the evaluation of tones in broadband noise. Technical report guideline No 6, Danish Environmental Protection Agency, Danish Ministry of Environment.

DEPA 1997 Information no. 9/1997. orientering om lavfrekvent støj, infralyd og vibrationer i eksternt miljø (in Danish). Technical report, Danish Environmental Protection Agency, Danish Ministry of Environment.

DEPA 2011 Statutory order 1284. Noise from wind turbines (in Danish). Technical report, Danish Environmental Protection Agency, Danish Ministry of Environment.

Davis R and Smith M 2014 Assessment of proposed Den Brook Condition 21 scheme for the implementation of Condition 20 (amplitude modulation of wind turbine noise). Technical report 9097-R02, University of Southampton, ISVR consulting.

Devenport WJ, Burdisso RA, Borgoltz A, Ravetta PA, Barone MF, Brown KA and Morton MA 2013 The kevlar-walled anechoic wind tunnel. *Journal of Sound and Vibration*, **332**(17), 3971–3991.

DIN 45631/A1 2010 Calculation of loudness level and loudness from the sound spectrum – Zwicker method – Amendment 1: Calculation of the loudness of time-variant sound. German Institute for Standardization.

DIN 45680 2013 Measurement and assessment of low-frequency noise immissions. German Institute for Standardization.

DIN 45681 2005 Acoustics – determination of tonal components of noise and determination of a tone adjustment for the assessment of noise immissions. German Institute for Standardization.

DiNapoli C 2011 Long distance amplitude modulation of wind turbine noise. *Fourth International Meeting on Wind Turbine Noise*, Rome, Italy.

Doolan CJ 2007 Numerical evaluation of contemporary low-speed wind tunnel contraction designs. *Journal of Fluids Engineering*, **129**(9), 1241–1244.

Doolan CJ and Moreau DJ 2015 Review of airfoil trailing edge noise and the implications for wind turbines. *International Journal of Aeroacoustics*, **14**(5–6), 811–832.

Dougherty RP 1998 Spiral-shaped array for broadband imaging. US Patent 5,838,284.

Dougherty RP 2005 Extensions of DAMAS and benefits and limitations of deconvolution in beamforming. AIAA 2005-2961 in *11th AIAA/CEAS Aeroacoustics Conference*, Monterey, CA.

Dougherty RP 2014 Functional beamforming for aeroacoustic source distributions. AIAA 2014-3066 in *20th AIAA/CEAS Aeroacoustics Conference*, Atlanta, GA.

Dougherty RP and Podboy GG 2009 Improved phased array imaging of a model jet. AIAA 2009 3186 in. *15th AIAA/CEAS Aeroacoustics Conference*, Miami, FL.

Dougherty RP and Stoker RW 1998 Sidelobe suppression for phased array aeroacoustic measurements. AIAA 1998 2242 in *4th AIAA/CEAS Aeroacoustics Conference*, Toulouse, France.

Dougherty RP, Ramachandran RC and Raman G 2013 Deconvolution of sources in aeroacoustic images from phased microphone arrays using linear programming. *International Journal of Aeroacoustics*, **12**(7), 699–718.

Draper N and Smith H 2014 Applied Regression Analysis, 3rd edn. John Wiley & Sons.

Dudgeon DE and Johnson DH 1993 Array Signal Processing: Concepts and Techniques. PRT Prentice Hall.

Ehrenfried K and Koop L 2007 Comparison of iterative deconvolution algorithms for the mapping of acoustic sources. *AIAA Journal*, **45**(7), 1584–1595.

ETSU 1996 ETSU-R-97 Assessment and rating of noise from wind farms. Technical report, Energy Technology Support Unit, UK Department of Trade and Industry.

Evans T and Cooper J 2015 Tonal noise from wind turbines. *6th International Meeting on Wind Turbine Noise*, Glasgow, UK.

Fastl H and Zwicker E 2007 *Psychoacoustics: Facts and Models.* Springer.

Fiumicelli D 2011 Den Brook wind farm Den Brook Revised Condition 20. Technical report T1890, Temple.

Fukushima A, Yamamoto K, Uchida H, Sueoka S, Kobayashi T and Tachibana H 2013 Study on the amplitude modulation of wind turbine noise: Part 1 – Physical investigation. *Internoise 2013*, Innsbruck, Austria.

Fulcrum3D Pty Ltd 2014 Fulcrum3D FS1 SODAR operations manual: SODAR introduction and operations. Technical report, Fulcrum3D Pty Ltd.

Gange M 2013 Low-frequency and tonal characteristics of transformer noise. *Acoustics 2013*, Gold Coast, Australia.

George W, Thom A and Yang A 1984 Pressure spectra in turbulent free shear flows. *Journal of Fluid Mechanics*, **148**, 155–191.

Glasberg B and Moore B 2002 A model of loudness applicable to time-varying sounds. *Journal of the Audio Engineering Society*, **50**(5), 331–342.

Hansen K, C. H and Zajamšek B 2015a Outdoor to indoor reduction of wind farm noise for rural residences. *Building and Environment*, **94**, 764–772.

Hansen K, Zajamšek B and Hansen C 2014 Comparison of the noise levels measured in the vicinity of a wind farm for shutdown and operational conditions. *Internoise 2014*, Melbourne, Australia.

Hansen K, Zajamšek B and Hansen C 2015b Quantifying the character of wind farm noise. *ICSV 22*, Florence, Italy.

Hedlin M, Alcoverro B and D'Spain G 1981 Evaluation of rosette infrasonic noise-reducing spatial filters. *Journal of the Acoustical Society of America*, **114**(4), 1807–1820.

Hessler D 2011 Accounting for background noise when measuring operational sound levels from wind turbine projects. *Fourth International Meeting on Wind Turbine Noise*, Rome, Italy.

Hessler G 1996 A real-time blast noise detection and wind noise rejection system. *Noise Control Engineering Journal*, **44**(6), 306–314.

Hessler G 2008 The noise perception index (NPI) for assessing noise impact from major industrial facilities and power plants in the US. *Noise Control Engineering Journal*, **56**(5), 374–385.

Högbom J 1974 Aperture synthesis with a non-regular distribution of interferometer baselines. *Astronomy and Astrophysics Supplement Series*, **15**, 417.

Hubbard H 1982 Noise induced house vibrations and human perception. *Noise Control Engineering Journal*, **19**(2), 49–55.

Hubbard H and Shepherd K 1991 Aeroacoustics of large wind turbines. *Journal of the Acoustical Society of America*, **89**(6), 2495–2508.

Hurtley C 2009 Night noise guidelines for Europe. WHO Regional Office Europe.

Huson W 2015 Constraints imposed by and limitations of IEC 61672 for the measurement of wind farm sound emissions. *6th International Meeting on Wind Turbine Noise,* Glasgow, UK.

IEC 60942 2003 Electroacoustics – sound calibrators, edn 1.0. International Electrotechnical Commission.

IEC 61400-11 2012 Wind turbines – part 11: Acoustic noise measurement techniques, edn 3.0. International Electrotechnical Commission.

IEC 61400-12-1 2005 Wind turbines – part 12-1: Power performance measurements of electricity producing wind turbines. International Electrotechnical Commission.

IEC 61672-1 2013 Electroacoustics – sound level meters – part 1: Specifications. International Electrotechnical Commission.

IEEE 1451.4 (2004) Standard for a smart transducer interface for sensors and actuators. IEEE Standards Association.

Imaizumi H and Takahashi Y 2010 Improvement of wind noise reduction performance of microphone-enclosed type windscreen. *Internoise 2010,* Lisbon, Portugal.

IOA 2014 A good practice guide to the application of ETSU-R-97 for the assessment and rating of wind turbine noise. Supplementary guidance note 5: Post completion measurements. Technical report, Institute of Acoustics.

IOA 2015 Amplitude modulation application. Institute of Acoustics. http://www.ioa.org.uk/news/amplitude-modulation-application-released.

ISO 7196 1995 Acoustics, frequency weighting characteristics for infrasound measurement. International Standards Organization.

ISO1996-1 1975 Acoustics: Method for calculating loudness level (Reviewed 2012). International Standards Organization.

ISO1996-2 2007 Acoustics: Description, measurement and assessment of environmental noise – Part 1: Determination of environmental noise levels. International Standards Organization.

ISO2631-1 1997 Mechanical vibration and shock – evaluation of human exposure to whole-body vibration – Part 1: General requirements. International Standards Organization.

ISO389-7 2005 Acoustics: Reference zero for the calibration of audiometric equipment – Part 7: Reference threshold of hearing under free-field and diffuse-field listening conditions. International Standards Organization.

Jackson I, Kendrick P, Cox T, Fazenda B and Li F 2014 Perception and automatic detection of wind-induced microphone noise. *Journal of the Acoustical Society of America,* **136**(3), 1176–1186.

Jakobsen J 2001 Danish guidelines on environmental low frequency noise, infrasound and vibration. *Journal of Low Frequency Noise, Vibration and Active Control,* **20**(3), 141–148.

JCGM 2008 Evaluation of measurement data: guide to the expression of uncertainty in measurement. Joint Committee for Guides in Metrology.

Kamiakito N, Shimura M, Nomura T, Hiroshi H, Osafune T and Iwabuki H 2015 Wind noise estimation functions for low frequency sound measurement in natural wind at different topography types. *6th International Meeting on Wind Turbine Noise,* Glasgow, UK.

Kaufmann A and Kock U 2015 Investigation into the influence of windscreens during sound emission measurements in accordance with IEC 61400-11 ed. 3.0. *Sixth International Meeting on Wind Turbine Noise,* Glasgow, UK.

Kelley N, Hemphill R and McKenna H 1982 A methodology for assessment of wind turbine noise generation. *Journal of Solar Energy Engineering*, **104**, 112–120.

Kelley N, McKenna H, Hemphill R, Etter C, Garrelts R and Linn N 1985 Acoustic noise associated with the MOD-1 wind turbine: its source, impact, and control. Technical report, US Department of Energy.

Kelley N, Osgood R, Bialasiewicz J and Jakubowski A 2000 Using wavelet analysis to assess turbulence/rotor interactions. *Wind Energy*, **3**, 121–134.

Kobayashi T, Yokoyama S, Fukushima A, Ohshima T, Sakamoto S and Tachibana H 2015 Assessment of tonal components contained in wind turbine noise in immission areas. *6th International Meeting on Wind Turbine Noise*, Glasgow, UK.

Large S 2015 Wind turbine amplitude modulation & planning control study. Work package 5 – towards a draft AM condition. Technical report, Independent Noise Working Group, MAS Environmental.

Large S and Stigwood M 2014 The noise characteristics of 'compliant' wind farms that adversely affect its neighbours. *Internoise 2014*, Melbourne, Australia.

Larsson C and Öhlund O 2014 Amplitude modulation of sound from wind turbines under various meteorological conditions. *Journal of the Acoustical Society of America*, **135**(1), 67–73.

Lee S, Kim K, Choi W and Lee S 2011 Annoyance caused by amplitude modulation of wind turbine noise. *Noise Control Engineering Journal*, **59**(1), 38–46.

Lee S, Lee S, Ryi J and Choi JS 2013 Design optimization of wind turbine blades for reduction of airfoil self-noise. *Journal of Mechanical Science and Technology*, **27**(2), 413–420.

Lenchine V 2009 Amplitude modulation in wind turbine noise. *Acoustics 2009*, Adelaide, Australia.

Lenchine V 2016 Assessment of amplitude modulation in environmental noise measurements. *Applied Acoustics*, **104**, 152–157.

Leventhall G 2003 A review of published research on low frequency noise and its effects. Technical report, DEFRA.

Liu X, Bo L and Veidt M 2012 Tonality evaluation of wind turbine noise by filter-segmentation. *Measurement*, **45**(4), 711–718.

Lundmark G 2011 Measurement of swish noise: a new method. *Fourth International Meeting on Wind Turbine Noise*, Rome, Italy.

Madsen H, Bertagnolio F, Fischer A and Bak C 2014 Correlation of amplitude modulation to inflow characteristics. *Internoise 2014*, Melbourne, Australia.

Mathew J, Bahr C, Sheplak M, Carroll B and Cattafesta LN 2005 Characterization of an anechoic wind tunnel facility. *ASME 2005 International Mechanical Engineering Congress and Exposition*, Chicago, IL.

McCabe J 2011 Detection and quantification of amplitude modulation in wind turbine noise. *Fourth International Meeting on Wind Turbine Noise*, Rome, Italy.

McDonald P, Geraghty D and Humphreys I 2011 A long term noise measurement system for wind farms. *Fourth International Meeting on Wind Turbine Noise*, Rome, Italy.

Mehta R and Bradshaw P 1979 Design rules for small low speed wind tunnels. *Aeronautical Journal*, **73**, 443–449.

Mehta RD 1985 Turbulent boundary layer perturbed by a screen. *AIAA Journal*, **23**(9), 1335–1342.

Michel U 2006 History of acoustic beamforming. *Proceedings of the Berlin Beamforming Conference*, Berlin, Germany.

Migliore P and Oerlemans S 2004 Wind tunnel aeroacoustic tests of six airfoils for use on small wind turbines. *Journal of Solar Energy Engineering*, **126**, 974.

Møller H and Pedersen C 2011 Low-frequency noise from large wind turbines. *Journal of the Acoustical Society of America*, **129**, 3727.

Moorhouse A, Hayes M, Von Hünerbein S, Piper B and Adams M 2007 Research into aerodynamic modulation of wind turbine noise: final report. Technical report, Department for Business, Enterprise and Regulatory Reform.

Moreau D, Brooks L and Doolan C 2011 Broadband trailing edge noise from a sharp-edged strut. *Journal of the Acoustical Society of America*, **129**(5), 2820–2829.

Moreau S, Roger M and Christophe J 2009 Flow features and self-noise of airfoils near stall or in stall. *30th AIAA Aeroacoustics Conference*, Miami, Florida.

Morgan S and Raspet R 1991 Low frequency wind noise contributions in measurement microphones. *Journal of the Acoustical Society of America*, **92**(2), 1180–1183.

Mueller T 2002 Aeroacoustic Measurements. Springer Verlag.

Nakajima Y, Fukushima K, Yokota A, Tsukioka H, Makino K, Ochiai H and Yamada I 2001 Unattended monitoring and source identification of low frequency impulse sound. *Internoise 2001*, The Hague, Netherlands.

Nelson P 1973 The combination of noise from separate time varying sources. *Applied Acoustics*, **6**, 1–21.

Nordtest 2002 NT-ACOU-112: prominence of impulsive sounds and for adjustment of L_{aeq}. Nordic Innovation Centre, Oslo, Norway.

NZS 6808 2010 Acoustics – wind farm noise. Standards New Zealand.

Oboukov A 1962 Some specific features of atmospheric turbulence. *Journal of Fluid Mechanics*, **13**, 77–81.

Oerlemans S and Schepers J 2009 Prediction of wind turbine noise and validation against experiment. *International Journal of Aeroacoustics*, **8**(6), 555–584.

Oerlemans S and Sijtsma P 2002 Determination of absolute levels from phased array measurements using spatial source coherence. Technical Publication NLR-TP-2002-226, National Aerospace Laboratory NLR.

Oerlemans S, Schepers J, Guidati G and Wagner S 2001 Experimental demonstration of wind turbine noise reduction through optimized airfoil shape and trailing-edge serrations. Technical report NLR-TP-2001-324, National Aerospace Laboratory NLR.

Oerlemans S, Sijtsma P and Mendez Lopez B 2007 Location and quantification of noise sources on a wind turbine. *Journal of Sound and Vibration*, **299**(4–5), 869–883.

Padois T, Prax C and Valeau V 2013 Numerical validation of shear flow corrections for beamforming acoustic source localisation in open wind-tunnels. *Applied Acoustics*, **74**(4), 591–601.

Paterson RW, Amiet RK and Munch CL 1975 Isolated airfoil-tip vortex interaction noise. *Journal of Aircraft*, **12**(1), 34–40.

Pearson C and Graham W 2013 Investigation of the noise sources on a vertical axis wind turbine using an acoustic array. AIAA 2013-2179 in *19th AIAA/CEAS Aeroacoustics Conference*.

Pedersen S, Møller H and Waye K 2007 Indoor measurements of noise at low frequencies-problems and solutions. *Journal of Low Frequency Noise, Vibration and Active Control*, **26**(4), 249–270.

Persson-Waye K and Öhrström E 2002 Psycho-acoustic characters of relevance for annoyance of wind turbine noise. *Journal of Sound and Vibration*, **250**(1), 65–73.

Prime Z and Doolan CJ 2013 A comparison of popular beamforming arrays. *Acoustics 2013*, Victor Harbor, Australia.

Ramachandran RC, Raman G and Dougherty RP 2014 Wind turbine noise measurement using a compact microphone array with advanced deconvolution algorithms. *Journal of Sound and Vibration*, **333**(14), 3058–3080.

Raspet R, Webster J and Dillon K 2005 Framework for wind noise studies. *Journal of the Acoustical Society of America*, **119**(2), 834–843.

Raspet R, Yu J and Webster J 2008 Low frequency wind noise contributions in measurement microphones. *Journal of the Acoustical Society of America*, **123**(3), 1260–1269.

Renewable UK 2013 Wind turbine amplitude modulation: research to improve understanding as to its cause and effect. Technical report, Renewable UK.

RES Developments 2014 Written scheme relating to Condition 21 Den Brook wind farm. Implementation of Condition 20 for the identification of greater than expected amplitude modulation. Technical report, RES Developments Ltd.

Rimmel A, Mansfield N and Paddan G 2015 Design of digital filters for frequency weightings (A and C) required for risk assessments of workers exposed to noise. *Industrial Health*, **53**, 21–27.

Ryi J, Choi JS, Lee S and Lee S 2014 A full-scale prediction method for wind turbine rotor noise by using wind tunnel test data. *Renewable Energy*, **65**, 257–264.

SAEPA 2009 Wind farms: environmental noise guidelines. South Australian Environmental Protection Authority.

Salomons E 2001 *Computational Atmospheric Acoustics*. Springer Science & Business Media.

Scheuren J 2014 ISO 532 – living and working with alternative loudness standards. *Internoise 2014*, Melbourne, Australia.

Schwarz U 1978 Mathematical-statistical description of the iterative beam removing technique (method clean). *Astronomy and Astrophysics*, **65**(2), 345–356.

Sijtsma P 2007 CLEAN based on spatial source coherence. *International Journal of Aeroacoustics*, **6**(4), 357–374.

Skøde F 1966 Windscreening of outdoor microphones. Technical report, Brüel and Kjær.

Smith M and Chiles S 2012 Analysis techniques for wind farm sound level measurements. *Acoustics Australia*, **40**(1), 51–56.

Søndergaard B and Madsen K 2008 Low frequency noise from large wind turbines: summary and conclusions on measurements and methods. Technical report, Delta.

Søndergaard B, Hoffmeyer D and Plovsing B 2007 Low frequency noise from large wind turbines. *Second International Meeting on Wind Turbine Noise*, Lyon, France.

Søndergaard L and Pedersen T 2013 Tonality in wind turbine noise. IEC 61400-11 ver. 2.1 and 3.0 and the Danish/Joint Nordic method compared with listening tests. *5th International Meeting on Wind Turbine Noise*, Denver, Colorado.

Søndergaard L and Sorensen T 2013 Validation of WindPRO implementation of Nord2000 for low frequency wind turbine noise. *5th International Meeting on Wind Turbine Noise*, Denver, Colorado.

Srivastava G 2008 *Surface Meteorological Instruments and Measurement Practices*. Atlantic Publishers.

Stephens D, Shepherd K, Hubbard H and Grosveld F 1982 Guide to the evaluation of human exposure to noise from large wind turbines. Technical report, NASA.

Stevens S 1956 Calculation of the loudness of complex noise. *Journal of the Acoustical Society of America*, **28**(5), 807–832.

Stigwood M, Large S and Stigwood D 2013 Audible amplitude modulation: results of field measurements and investigations compared to psychoacoustical assessment and theoretical research. *5th International Meeting on Wind Turbine Noise*, Denver, Colorado.

Stigwood M, Large S and Stigwood D 2015 Cotton farm wind farm – long term community noise monitoring project – 2 years on. *6th International Meeting on Wind Turbine Noise*, Glasgow, UK.

Strasberg M 1979 Nonacoustic noise interference in measurements of infrasonic ambient noise. *Journal of the Acoustical Society of America*, **66**, 1487–1493.

Strasberg M 1988 Dimensional analysis of windscreen noise. *Journal of the Acoustical Society of America*, **83**, 544–548.

Stull R 1988 An Introduction to Boundary Layer Meteorology. Kluwer Academic Publishers.

Styles P, Westwood R, Toon S, Buckingham M, Marmo B and Carruthers B 2011 Monitoring and mitigation of low frequency noise from wind turbines to protect comprehensive test ban seismic monitoring stations. *Fourth International Meeting on Wind Turbine Noise*, Rome, Italy.

Suzlon, Ltd 2012 S88 – 2.1 MW specifications. Technical report, Suzlon, Ltd.

Tachibana H, Yano H, Sakamoto S and Sueoka S 2013 Nationwide field measurements of wind turbine noise in Japan. *Internoise 2013*, Melbourne, Australia.

Teague P and Leonard A 2011 Prediction, validation, assessment and compliance of wind farm noise in Australia. *Fourth International Meeting on Wind Turbine Noise*, Rome, Italy.

Thompsen C and Nielsen S 2013 Wind turbine noise measurements in practice. *5th International Meeting on Wind Turbine Noise*, Denver, Colorado.

van den Berg F 2013 Wind turbine noise: an overview of acoustical performance and effects on residents. *Acoustics 2013*, Victor Harbor, Australia.

Van den Berg G 2005 Wind gradient statistics up to 200 m altitude over flat ground. *First International Meeting on Wind Turbine Noise: Perspectives for Control*, Berlin, Germany.

Van den Berg G 2007 Wind profiles over complex terrain. *Second International Meeting on Wind Turbine Noise*, Lyon, France.

Vanderkooy J and Mann R 2014 Measuring wind turbine coherent infrasound. Technical report, University of Waterloo, Dept of Computer Science.

Vestas 2006 General specification v90 – 3.0 MW VCRS. Technical report 950010.R5, Vestas.

von Hunerbein, S. and Piper, B. 2015 Affective response to amplitude modulated wind turbine noise. *Sixth International Meeting on Wind Turbine Noise*, Glasgow, UK.

Walker K and Hedlin M 2010 A review of wind-noise reduction technologies. In: Le Pichon A, Blanc E, and Hauchecorne A. Infrasound Monitoring for Atmospheric Studies. Springer.

Woolworth D, Waxler R and Webster J 2013 Proposed method for characterizing wind turbine noise and their dependence on meteorological effects for validation of existing studies. *5th International Meeting on Wind Turbine Noise*, Denver, Colorado.

Wuttke J 2005 Microphones and wind. *Journal of the Audio Engineering Society*, **40**(10), 809–817.

Yardibi T, Bahr C, Zawodny N, Liu F, Cattafesta III, L and Li J 2010a Uncertainty analysis of the standard delay-and-sum beamformer and array calibration. *Journal of Sound and Vibration*, **329**(13), 2654–2682.

Yardibi T, Li J, Stoica P, Zawodny NS and Cattafesta III, LN 2010b A covariance fitting approach for correlated acoustic source mapping. *Journal of the Acoustical Society of America*, **127**(5), 2920–2931.

Yokoyama S, Sakamoto S and Tachibana H 2013 Perception of low frequency components contained in wind turbine noise. *5th International Meeting on Wind Turbine Noise*, Denver, Colorado.

Zajamšek B, Doolan C, Hansen K and Hansen C 2016 Characterisation of wind farm infrasound and low-frequency noise. *Journal of Sound and Vibration*, **370**, 176–190.

Zhang M 2015 *Wind Resource Assessment and Micro-siting*. John Wiley & Sons.

7

Effects of Wind Farm Noise and Vibration on People

7.1 Introduction

In recent years, there have been a number of public enquiries on the adverse effects of wind farm noise on some people (see Section 1.7) and much has been written on the topic in the general literature. Many articles have been peer reviewed and many have not. Some writers claim that wind farm noise could not be of a sufficient level to affect any person. Others suggest that wind farm noise is annoying to some people and could lead to sleep disturbance and stress in affected individuals. Within this latter group, most believe that the noise that causes disturbance is low-frequency and time-varying in nature. Some within this group believe that the possibility of infrasound contributing to annoyance should not be ruled out, even though the levels of infrasound are below the audibility threshold for most people. Yet another group of writers believe that infrasound produced by wind farms can cause adverse physiological effects in some people, although this group of writers is in the minority. It is worth noting that wind farms are not the only sources of environmental low-frequency noise and infrasound, with the resulting adverse health effects and community complaints. Adverse effects similar to those suffered by residents near wind farms have been reported by people living near compressor or pumping stations and other industrial operations that generate low-frequency noise and infrasound.

The World Health Organisation (WHO) defines noise as unwanted sound and states that excessive noise seriously harms human health as well as interfering with daily activities. Health is defined by WHO as 'a state of complete physical, mental and social well-being and not merely the absence of disease or infirmity'. There are two aspects of the response of people to unwanted sound. The first is defined generally as 'annoyance' and the second is a direct physiological response resulting in an adverse health effect. Annoyance represents more than a slight irritation; it can lead to a significant degradation in the quality of life and can lead to some adverse health effects. This makes it difficult to distinguish direct physiological effects from indirect effects resulting from annoyance and recurring sleep disturbance.

Several studies on the adverse health effects caused by noise generated by wind farms have been conducted by the Australian Medical research body, NHMRC and Health Canada. The results of these studies were discussed in detail in Section 1.7 and all of these have concluded that there is no evidence to support the view that wind farms cause adverse health effects, in spite of a significant body of anecdotal evidence and case study

Wind Farm Noise: Measurement, Assessment and Control, First Edition.
Colin H. Hansen, Con J. Doolan, Kristy L. Hansen.

results suggesting that wind farm noise is a problem for between 10 and 20% of the population. The reviews mentioned above cited many problems with studies reported in the literature, including low response rates to questionnaires leading to selection bias, as in the study undertaken by Krogh et al. (2011). On the other hand, the symptoms reported in many studies are still relevant for a significant proportion of the population and results concerning the majority of people are not necessarily relevant. It is also possible that the percentages of people found with adverse health effects in some surveys are lower than they might actually be, as some people may not have developed symptoms at the time the survey was done, they may not have linked their symptoms to wind farm noise or they may have signed a 'Good Neighbour Agreement' that prevented them from filling out a survey in a way that was negative towards wind turbines (see Finlaysons (2014); HWL Ebsworth Lawyers (2013)).

Symptoms reported by people exposed to wind farm noise vary considerably between individuals. For some, symptoms begin immediately on exposure to the noise. For others, the symptoms gradually develop and become worse as the exposure duration increases, and for others, no symptoms occur at all, even after a long exposure. Often when severely distressed, people leave their homes for a holiday or permanently, and their symptoms usually disappear after a few days, although in some cases it may take a few weeks, which tends to suggest that the wind farm was the cause, especially as symptoms did not exist prior to the wind farm commencing operations. On the other hand, there are those who believe that the possibility of the involvement of psychological factors should not be discounted.

Three recent studies conducted in rural areas in Australia are worth mentioning. Morris (2012) wrote a report on a self-reporting survey that was sent to all 230 residences within 10 km of the Water loo wind farm (37 turbines of 3 MW rated power). The response rate was 40% and of these, 40% reported nighttime disturbance and 29% reported sleep disturbance. If it is assumed that the non-respondents were unaffected, then the above percentages become 16% and 12% of the population, respectively. However, these percentages could be greater because an unknown number of residents (including turbine hosts) have signed 'friendly neighbour' agreements, which prevent them from responding to such surveys.

In another Australian study, Schneider (2012) sent questionnaires to 19 residences within 5 km of the Cullerin Range wind farm and achieved a response rate of 73%, of which 78.5% reported sleep disturbance. Schneider (2013) repeated the study for residents living within 10 km of the same wind farm and sent surveys to 35 residences, receiving a response rate of 68.5%. Of the 40 people responding:

- 20% were within 0–2 km of the wind farm, with 100% reporting sleep disturbance due to the wind farm
- 30% were within 2–5 km, with 90% reporting sleep disturbance
- 50% were within 5–8 km, with 92% reporting sleep disturbance.

There are many well-documented cases of adverse health effects resulting from both short- and long-term exposure to wind turbine noise. These symptoms include nausea, dizziness, pressure (or fullness) in the ears and recurring sleep disturbance; in most cases medical professionals are unable to otherwise explain them. There are also a significant number of people who are so badly affected that they have had to leave their homes or have had to undergo medical treatment to address their symptoms. Sleep disturbance experienced by people living in the vicinity of wind farms manifests as prevention of the onset of sleep (insomnia) or the prevention of a return to sleep after being awakened. Ten to twenty percent of the population are easily disturbed from their sleep and a higher percentage can be expected to be found in rural areas (Hanning and Nissenbaum, 2011). Of course, a smaller percentage of the population (generally 5–10%, according to Thorne (2013)), report serious health effects, although a considerably greater percentage suffer from annoyance and recurring sleep disturbance. In some cases, some members of the same family suffer terrible effects while others suffer no ill effects at all. The actual percentages mentioned above depend on how far the residents are from the nearest turbine. In cases where they are within 500 m of the nearest turbine, up to 30% of people may be highly annoyed and suffer some form of adverse health effect in addition to sleep deprivation. As distance from the nearest turbine increases, the percentages of affected people decreases.

Thorne (2012, 2013) found that adverse health effects occur in some people when wind farm noise levels exceed $L_{Aeq,\,10\mathrm{min}}$ levels of either:

- 32 dBA outside the residence
- 22 dBA indoors, if wind farm noise is at a level that is perceptible inside the home.

He also concluded that noise that exhibited excessive levels of 'fluctuation' could contribute to adverse effects. He defined fluctuation in terms of L_{Aeq} and L_{Zeq} levels in 1/3-octave bands averaged over 0.1–0.125 ms. A difference of 2 dB or less between the peak and trough of the measured 1/3-octave band sound level (plotted as a function of time) was considered to be acceptable, a difference of 4 dB was considered to be unreasonable, and a difference of 6 dB or more was considered to be excessive. The L_{Zeq} metric is used to determine if AM exists for low-frequency noise or infrasound, where the L_{Aeq} metric is more likely to identify AM in the mid-frequency range. AM and its measurement is discussed in Section 6.2.10.

In a recent study, McBride et al. (2014) found that some people living near wind farms had a lower quality of life than others due to increased levels of stress and poorer sleep. It also became apparent that the 'objective manifestation of health effects associated with noise-related annoyance may emerge after some years' of exposure.

The Cape Bridgewater wind farm has been the subject of studies by Thorne (2012) and Cooper (2015), who both found poor sleep quality and adverse health effects in residents. Cooper surveyed six occupants of three houses located between 650 and 1600 m from the nearest turbine. He found that these residents suffered from sleep disturbance, headache, ear pressure, tinnitus and an elevated pulse rate. The onset of these latter four 'sensations' was found to be well correlated with *changes* in the turbine output power, caused by the following operating scenarios:

- turbines starting up
- increase in power of 20% or more
- decrease in power of 20% or more
- turbines operating at maximum power and wind increased above 12 m/s.

Cooper also found that the infrasound environment with the wind farm operational was characterised by a mix of tonal noise and random noise, and was very different to the natural environment, which is characterised by random noise only. He found that the total sound pressure level (denoted WTS) of the first 11 harmonics of the blade-passing frequency of the wind turbines was correlated with unacceptable sensations, such as those listed above. Cooper used narrowband analysis with a 0.0625 Hz frequency resolution and recorded the peak levels of the first 11 harmonics. He then added them together logarithmically (see Section 2.2.8) to obtain what he referred to as the WTS (dB). He found that if the WTS (dB) exceeded 51 dB, an unacceptable sensation level would exist. He pointed out that all of the residents involved in his study have been exposed to wind farm noise for six years and had a heightened sensitivity to it. Cooper found that the use of dBA or dBC weightings for indoor measurements did not successfully separate wind farm noise from wind noise for exposed houses and that his WTS (dB) measure was the only reliable one for predicting the onset of unacceptable sensations. However Leventhall (2015b), in his submission to the Australian Senate inquiry, pointed out that the large changes in wind farm output power would produce a noticeable change in audible noise produced by the wind farm and this change would draw people's attention to it and attract the recorded responses. He also pointed out that people in the Cooper study recorded increases in audible noise at the same time as they recorded increases in sensations, leading Leventhall to conclude that the adverse health effects experienced by residences were a result of audible noise, not inaudible infrasound, and that 'residents are reacting in the typical manner of highly stressed persons, who are affected by a noise which they do not wish to hear'.

Apart from the Cooper study, there are only limited data available that include simultaneous measurement of wind farm noise and annoyance or wind farm noise and adverse health effects. Some studies have employed simultaneous outdoor and indoor monitoring of noise levels in conjunction with questionnaires (Bockstael et al., 2012; Magari et al., 2014; SAEPA, 2013). An online reporting system was implemented by Bockstael et al. (2012) as a means by which residents could report their annoyance to wind farm noise in real time. Doolan and Moreau (2013) conceived a novel system that measured time data in response to a resident registering their annoyance via a computer. The advantage of this system is that time data are recorded at the time of the noise event, rather than at set intervals of 1 min of recorded data in each 15-min time block, as used by Bockstael et al. (2012). The disadvantage of all of the studies discussed above is that they require self-reporting on the part of the residents and are thus prone to information bias. Also, people do not always register their annoyance as they may be too tired or busy, and they may be drowsy or be unwilling to get out of bed, or they may not be near a computer at the time of the noise event, particularly if they are farmers. Not surprisingly, the corresponding results have been inconsistent, with some studies concluding that a dose–response relationship was not supported (Magari et al., 2014) while another study observed that annoyance to noise exposure depended on wind speed, wind direction and power output of the wind turbines (Bockstael et al., 2012).

In another study involving a questionnaire (Nissenbaum et al., 2012), clear dose–response relationships were demonstrated between exposure to wind farm noise and sleep quality, daytime sleepiness, and mental health. Distance from the wind farm was found to significantly affect the results. In yet another study, Bakker et al. (2012) investigated self-reported sleep disturbance and psychological distress via a questionnaire. It was found that exposure to wind turbine noise was related to sleep disturbance and psychological distress among those who reported that they could hear the sound, but that these effects were caused by annoyance. No direct effects of wind turbine noise on sleep disturbance or psychological stress could be established.

There have been a number of literature reviews on adverse health effects of wind farm noise in addition to those undertaken by government bodies, and the results from some of these will be summarised here.

The first was authored by Doolan (2013), who reviewed literature concerning wind turbine noise perception, annoyance and low-frequency emission. He concluded the following:

- There is a similarity between findings of studies on low-frequency noise annoyance and wind farm noise annoyance.
- There may be a link between personal and social moderating factors and annoyance.
- Noise levels that satisfy acceptable A-weighted criteria (see Section 2.2.11) can still cause annoyance, suggesting that the A-weighted noise level is an inappropriate criterion for specifying allowable wind farm noise levels.
- Spectral balance and temporal qualities of the noise should be included in any criteria of acceptability.
- It is possible that a person's susceptibility to wind farm noise may change as the length of exposure increases.
- A method is required that will acquire wind farm noise signatures simultaneously with self-reported annoyance levels and sleep disturbance over extended periods of time, so that relationships between wind farm characteristics and annoyance can be established. This will assist with the design and planning of future wind farms.

Almost simultaneously with the Doolan review, Farboud et al. (2013) reviewed papers published on the topic in the previous ten years. They concluded the following:

> There is ample evidence of symptoms arising in individuals exposed to wind turbine noise. Some researchers maintain that the effects of wind turbine syndrome are clearly just examples of the well known stress effects of exposure to noise, as displayed by a small proportion of the population. However, there is an increasing body of evidence suggesting that infrasound and low-frequency noise have physiological effects on the ear. Until these effects are fully understood, it is impossible to state conclusively that exposure to wind turbine noise does not cause any of the symptoms described. The effects of infrasound and low-frequency noise require further investigation.

In 2014, three literature reviews were published: by Jeffery et al. (2014), by Schmidt and Klokker (2014) and by Arra et al. (2014). The Jeffery study concluded the following:

> If placed too close to residents, industrial wind turbines (IWTs) can negatively affect the physical, mental and social well-being of people. There is sufficient evidence to support the conclusion that noise from audible IWTs is a potential cause of health effects. Inaudible low-frequency noise and infrasound from IWTs cannot be ruled out as plausible causes of health effects. Amplitude modulation of IWTs, audible LFN, and tonal, impulse and nighttime noise can contribute to annoyance and other effects on health. In addition, there is emerging evidence that suggests inaudible LFN or infrasound from IWTs may result in negative health effects.

The Schmidt study concluded the following:

> Wind turbines emit noise, including low-frequency noise, which decreases incrementally with increases in distance from the wind turbines. Likewise, evidence of a dose-response relationship between wind turbine noise linked to noise annoyance, sleep disturbance and possibly even psychological distress was present in the literature. Currently, there is no further existing statistically-significant evidence indicating any association between wind turbine noise exposure and tinnitus, hearing loss, vertigo or headache. At present it seems reasonable to conclude that noise from wind turbines increases the risk of annoyance and disturbed sleep in exposed subjects in a dose-response relationship. There seems to be a tolerable limit of around L_{Aeq} of 35 dB.

The Arra study reviewed 18 previous studies on the possible adverse health effects of wind farms on people, undertaken between 2004 and 2012 in the USA, UK, Sweden, New Zealand, the Netherlands and Canada, and concluded that there is…

> …reasonable evidence (Levels Four and Five) supporting the existence of an association between wind turbines and distress in humans. The existence of a dose-response relationship between distance from wind turbines and distress as well as the consistency of association across studies found in the scientific literature argues for the credibility of this association. Future research in this area is warranted.

There was one review study, published by Onakpoya et al. (2015), which concluded that…

> …evidence from cross-sectional studies suggests that exposure to wind turbine noise may be associated with increased frequency of annoyance and sleep problems. Evidence also suggests that living in proximity to WTGs (wind turbine generators) could be associated with changes in the quality of life. Individual attitudes could influence the type of response to noise from WTGs.

The remainder of this chapter is concerned with shedding some light on why some people are adversely affected by wind farm noise and others are not, as well as the sort of symptoms that may be experienced by affected people and how these symptoms manifest as a function of exposure time.To assist with the understanding of how low-frequency noise and infrasound may affect some people, a description of our hearing mechanism is also included.

7.2 Annoyance and Adverse Health Effects

Well before wind farms were developed, researchers were working on evaluating the effects of low-frequency noise and infrasound on people. It is not feasible to include here all the work that was done before the 1990s, so only a selection of the findings will be discussed. Karpova et al. (1970) exposed male volunteers to infrasound at 5 and 10 Hz at levels of 100 and 135 dB for 15 min and recorded feelings of apathy, depression, fatigue, pressure in the ears, drowsiness, loss of concentration and vibration of internal organs. Tempest (1973) investigated low-frequency noise in a car, a diesel locomotive, traffic noise indoors, an oil furnace and a ventilation system and found that the number of complaints far exceeded what would be expected from the A-weighted (see Section 2.2.11) noise level. Broner (1978) reviewed earlier literature and concluded that the effects of noise in the 20–100 Hz range were more apparent than effects due to the same sound pressure levels in the frequency range below 20 Hz. Møller (1984) reported on a study in which 16 subjects were exposed for 3 h to inaudible and audible infrasound and found no effects for infrasound at levels below the hearing threshold (see Section 7.3.4), but he did find annoyance and a feeling of pressure on the ear at levels above the hearing threshold.

Although these early studies produced a considerable amount of data on the effects of short term exposure to infrasound at both high and low levels, no studies have investigated the effects of long-term exposure of weeks and months to low-level infrasound, such as produced by wind farms at nearby residences.

Kelley (1987) reported on the results of a study that involved seven test subjects in a laboratory setting and from the results developed a proposed metric for assessing community annoyance from wind turbine noise. He investigated the correlation of annoyance with noise weighted by various measures, including the G-weighting, C-weighting, A-weighting and LSL weighting. These weightings are defined in Section 2.2.11. Kelley (1987) obtained the best correlation with annoyance by using the C-weighting and the LSL weighting. However, as noise levels were measured outdoors and annoyance was measured indoors, Kelley (1987) proposed adjustments to the weighting curves to account for the noise reduction due to a typical residence and also to account for the for the type of noise being emitted. Values of the C-weighting and LSL weighting, corrected for the noise reduction due to transmission into a residence, are listed in Table 7.1.

Kelley (1987) proposed the following procedure to determine the likelihood of wind turbine noise causing annoyance. However, this was based on studies in a laboratory with seven test subjects and short-term exposures and is likely to under-predict annoyance for long-term exposure of residents near wind farms.

1. Obtain a series of representative, unweighted, averaged 1/3-octave band pressure spectra over a range of 2–160 Hz for a range of operating conditions and sufficiently close to the wind farm that a sufficient signal-to-noise ratio for this frequency range can be reasonably obtained. Use recording periods of at least 2 min but not more than 10 min.
2. Establish whether or not the turbine exhibits impulsive noise.
3. Using Table 7.1, determine the equivalent weighted indoor 1/3-octave band noise levels, both C-weighted and LSL-weighted.
4. Calculate the equivalent LSL or C levels at a reference distance of 1 km by assuming spherical divergence (−6 dB per doubling of distance).
5. Add 15 dB to the results of step 4.
6. Calculate the overall weighted C-weighted and LSL-weighted level by adding logarithmically (see Section 2.2.8) all the weighted 1/3-octave band levels calculated in steps 4 and 5.
7. Compare the result of step 6 to the values in Table 7.2 to assess the annoyance potential.

Table 7.1 Corrected weightings at 1/3-octave band centre frequencies to include the effect of transmission from outside to inside a residence (after Kelley (1987))

	Impulsive noise		Non-impulsive noise	
	LSL	C	LSL	C
Frequency (Hz)	(dB)	(dB)	(dB)	(dB)
2.0	−61	−45	−61	−45
2.5	−56	−40	−56	−40
3.15	−50	−34	−50	−34
4.0	−41	−25	−41	−25
5.0	−30	−14	−32	−16
6.3	−25	−11	−28	−12
8.0	−24	−8	−24	−8
10.0	−20	−5	−22	−7
12.5	−16	−2	−20	−6
16.0	−12	−0	−22	−10
20.0	−14	−4	−23	−13
25.0	−12	−4	−19	−11
31.5	−8	−3	−15	−10
40	−3	−1	−11	−9
50	+6	+5	−5	−4
63	−3	+2	−12	−5
80	−12	−1	−21	−8
100	−18	−0	−25	−7
125	−20	+4	−32	−8
160	−30	+0	−35	−5

Berglund et al. (1996) reviewed a considerable body of previous work on the effects of low-frequency industrial noise on people. They concluded that the effects of low-frequency noise on humans is difficult to establish for many reasons, including unethical exposure of test subjects. However, they did conclude that based on the considerable previous work that had been done, the 'balance of probability would appear to favour the conclusion that low-frequency noise has a variety of adverse effects on humans, both physiological and psychological'.

There are certain aspects of wind farm noise that are similar to other sources of low-frequency noise, such as ventilation noise, compressor noise, noise from heavy vehicles and so on. For example, the audible "rumbling" noise that is generated during specific atmospheric conditions at the Waterloo wind farm in South Australia is very similar to a noise sample that was used in a laboratory study by Persson-Waye (2004), where subjects were exposed to ventilation noise that included a tone at 50 Hz, which was amplitude modulated at 2 Hz. Compared to a reference night with no exposure to noise, it was found that subjects took nearly twice as long to fall asleep when exposed to the low-frequency ventilation noise. A significant attenuation in salivary

Table 7.2 Interior annoyance level criteria as a function of noise type

Stimuli class	Perception threshold		Annoyance threshold		Unacceptable annoyance	
	LSL	C	LSL	C	LSL	C
Non-impulsive, periodic random	58	68	65	75	68	77
Periodic, impulsive	53	63	57	67	60	68
Random, periodic	59	67	68	76	70	78

After Kelley (1987).

cortisol was measured for subjects exposed to the low-frequency noise and lower cortisol levels were found to be related to tiredness and negative mood (Persson-Waye et al., 2003). A study by Verzini et al. (1999) found that exposure to low-frequency noise (20–160 Hz) from installations, air conditioning units, industrial processes and traffic noise from tunnels was found to be significantly related to sleep disturbance, concentration difficulties, irritability, anxiety and tiredness. A cross-sectional study involving a comparison between children aged 7–10 who were exposed to 24-h truck traffic and children from the same age group who lived in quiet areas revealed that the former group experienced disturbance of the circadian rhythm of cortisol (Ising and Ising, 2002). Exposure to low-frequency noise while undertaking specific tasks has been found to result in reduced cognitive ability, indicating the potential to negatively interfere with work performance (Persson-Waye et al., 1997). Moderate levels of low-frequency noise have also been found to adversely affect performance in tasks requiring attention (Bengtsson et al., 2004). Persson-Waye (2004) reviewed available epidemiological and experimental studies and concluded that sleep disturbance due to low-frequency noise warrants further investigation.

James (2012) reviewed a large body of work that had involved field investigations of the effects of low-frequency noise and infrasound on people, mainly from sound sources such as building ventilation systems, industrial gas turbines and other large rotating machinery such as compressor and pumping installations. Regarding sick building syndrome, James (2012) pointed out that researchers in the 1970s found that the lack of productivity, tiredness, headaches and difficulty in concentration experienced by staff was linked to ventilation-system-generated low-frequency noise and infrasound, which was sometimes at a level below the hearing threshold, making it inaudible to staff who were affected. In referring to the past work on the effects of low-frequency noise and infrasound on people caused by ventilation systems, compressor stations, pump stations and industrial gas turbines, James (2012) concludes…

> …that had past experience and information, which was available prior to the widespread implementation of the modern upwind industrial-scale wind turbine, been incorporated into the government and industry guidelines and regulations used for siting wind turbine utilities, many of the complaints and adverse health effects currently reported would have been avoided.

There are many factors that can influence the level of annoyance experienced by residents near wind farms, especially as wind farms operate through the night as well as the day. Some of the factors are listed below.

- The extent to which those affected believe the noise is harmful to their health.
- The predictability of the noise: noise that is highly variable is much more annoying.
- Length of exposure time: although people can become used to some types of noise, it seems that annoyance with wind farm noise increases as the exposure time increases.
- Insufficient community consultation prior to installation of a wind farm: this makes residents feel 'out of control' and results in them being more annoyed.
- The type of neighbourhood: people in rural areas are generally not used to high levels of background noise and generally they will find wind farm noise more intrusive than people in suburban environments, where traffic noise is already present.
- General sensitivity to noise: people vary considerably in their sensitivity to noise, with some not noticing a noise that others find greatly disturbing.
- Feelings about the importance and necessity of the noise.
- The extent of financial compensation.
- Attitudes to green energy and the visual effect of the turbines.
- The type of terrain between the wind farm and residences.
- Distance from the wind farm.

A notable feature of wind turbine noise is that residents living near a wind farm report annoyance, even when the measured noise levels are relatively low. One reason for this is that wind turbines are often located in rural areas where background noise levels are very low, particularly at night, so that even low levels of wind farm noise are intrusive. The contrast between ambient noise and noise due to wind farm operation is also exacerbated during the evening and night due to stable atmospheric conditions (van den Berg, 2004). During these conditions, the wind turbines continue to operate, while the wind speed at the residence can be negligible, giving rise to low levels of background noise. Stable atmospheric conditions are also characterised by high wind shear, which is suggested to be a major factor responsible for the AM of wind turbine noise (Renewable UK, 2013; van den Berg, 2005). Residents living in the vicinity of wind farms describe the associated noise as 'thumping' (van den Berg, 2004) or 'rumbling' (Hansen et al., 2014) in character, indicating the presence of low-frequency, time-varying noise, which is far more annoying than steady traffic noise of the same level. This time-varying noise is almost always more apparent at night when high wind shear and low background noise levels occur at the same time. It is not surprising that many people who only experience daytime wind farm noise find it not to be at all annoying. AM is discussed in more detail in Section 2.2.7 and Section 6.2.10.

Listening tests have shown that for a given noise level, the presence of AM significantly contributes to perceived annoyance (Lee et al., 2011). In addition, the equal loudness contours at lower frequencies are closer together, meaning that low-frequency noise that is only moderately above the audibility threshold may be perceived as loud or even annoying (Møller and Pedersen, 2011).

In contrast to traffic noise, which consists of energy distributed over the mid-frequency range, wind farm noise at distances exceeding about 1–1.5 km from the nearest turbine is dominated by low-frequency energy. Although wind farms produce noise in both the low- and mid-frequency ranges, the dominance of low-frequency wind farm noise at distances greater than 1–1.5 km is because low-frequency noise is poorly absorbed by the atmosphere and ground, resulting in it being detected at much greater distances from the source than mid- to high-frequency noise (Leventhall, 2003). In addition, acoustic refraction arising from atmospheric wind and temperature gradients leads to reduced attenuation of low-frequency noise in the downwind direction but little change for mid- to high-frequencies (Hubbard and Shepherd, 1991). At a typical residence, noise in the mid- to high-frequency range is selectively attenuated by the walls and roof, resulting in the house structure behaving as a low-pass filter (Søndergaard et al., 2007). As a consequence, the spectrum of sound inside the house is even more heavily weighted towards lower frequencies, which is perceived as much more annoying than a well-balanced spectrum of equal loudness (Blazier Jr, 1997). Resonances in an average-sized room are well separated at low frequencies, causing a variation in sound pressure level of up to 20–30 dB as a function of location in the room (Pedersen et al., 2007). The perceived noise will therefore vary in amplitude as a resident walks around the room, increasing the probability of annoyance.

As discussed in Section 1.4.3, the WHO (Berglund et al., 1999; Hurtley, 2009) has stated that recommended limits of 30 dBA for bedrooms at night (corresponding to 45 dBA outside for traffic noise) do not apply when noise is dominated by low frequencies. So for noise dominated by low frequencies, the allowed levels should be much lower and an A-weighted noise level (see Section 2.2.11) is insufficient to properly characterise the noise. Yet in spite of this well-known fact, wind farm noise guidelines and regulations continue to be written in terms of A-weighted sound levels (see Section 2.2.11), so that annoying low-frequency noise produced by wind farms is essentially ignored even though it may be responsible for the adverse health effects that are suffered by some residents. The WHO states that the dose–response relationships for different types of transportation noise are different, so it follows that we would expect the dose–response relationship for wind farm noise to be different to that for road traffic noise, thus requiring a different metric for specifying allowable levels. A number of studies have confirmed the increase in annoyance of wind farm noise compared to traffic noise for the same L_{den} level (see Section 2.2.12) or A-weighted level (see for example, Pedersen et al. (2009) and Krogh et al. (2013)). Jabben et al. (2009) provided dose–response comparisons between wind farm noise and transportation noise as a function of L_{den}, as shown in Figure 7.1. The L_{den} descriptor (see Section 2.2.12) tends to penalise noise that continues at the same or higher level during the night. For example, if a noise source produced 35 dBA L_{Aeq} at a residence continuously, the corresponding L_{den} would be 41.7 dBA. Since 2008, wind farms have become larger in terms of number of turbines and rated power so the low-frequency noise output has increased without having much effect on the A-weighted level (see Section 2.2.11). It is therefore likely that there will be many situations where the noise is more annoying than indicated in Figure 7.1. The results shown in Figure 7.1 show a similar trend to more recent results obtained as part of the Health Canada study (Michaud et al., 2016) and shown in Figure 7.2.

Apart from the dominance of low frequencies in the noise spectrum, other aspects of wind farm noise that account for its more annoying nature are its amplitude variation, impulsiveness and beating character as well as its unpredictability.

It seems that there is no published literature that reports a dose–response relationship between adverse health effects or annoyance and levels of infrasound and low-frequency noise. Perhaps the prevalence of the unfortunate use of the A-weighting

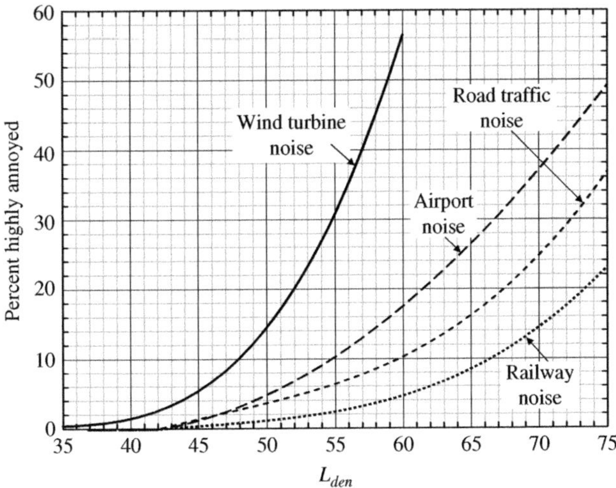

Figure 7.1 Dose–response functions for severe annoyance for wind turbine noise compared to transportation noise. Transportation noise curves from Miedema and Oudshoorn (2001) and wind farm curve from Janssen et al. (2008).

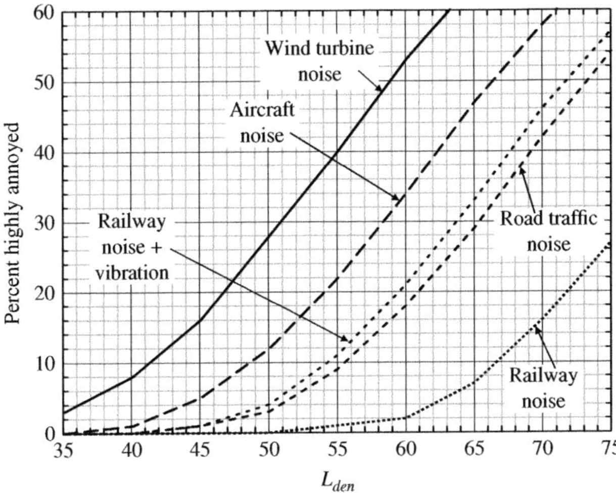

Figure 7.2 Dose response functions for high annoyance in communities found as part of the Health Canada study (Michaud et al., 2016)

to specify allowable wind farm noise levels, instead of a more appropriate measure that emphasises low-frequency noise and infrasound has arisen for two reasons.

1. Most current guidelines and regulations are based on ETSU-R-97 (see Section 1.4.1), which uses A-weighting criteria (see Section 2.2.11). Although this was in line with the current practice of environmental noise assessment, it did not take into account any annoying characteristics that made wind farm noise more annoying than indicated by its A-weighted level. Use of A-weighted criteria in isolation has resulted in

turbines being constructed closer to residences than may have been the case if the special characteristics of wind farm noise had been taken into account.
2. Environmental noise limits for other types of noise have been traditionally specified in terms of A-weighted noise levels and it is difficult for regulators to adopt an unfamiliar metric for wind farm noise, regardless of the justification for doing so.

In addition to the effects of audible wind farm noise on people, there is an ongoing debate and controversy on the physiological effects of infrasound. Nussbaum and Reinis (1985) published a report on human response to infrasound, reviewing a large amount of previously published work on the subject. The literature review provided many demonstrations of adverse effects of infrasound for short exposures at levels well above those produced by wind farms. Nussbaum and Reinis (1985) exposed 80 test subjects to two infrasound signals based on an 8-Hz tone at a total level of 130 dB for 30 min. The first signal had no higher-order harmonics while the second was rich in higher-order harmonics. The higher level of harmonics was found to be associated with headache and fatigue, while reduction in the harmonics resulted in symptoms of dizziness and nausea. They found that the reaction to infrasound varied considerably between individuals, most probably as a result of the inner ear structure or 'central adaptive mechanisms'. Some individuals exhibited symptoms that closely resembled motion sickness. With one exception, subjects reporting dizziness and nausea for the signal with no higher-order harmonics, experienced the symptoms hours later, after the experiment ended, but had raised respiratory rates during the experiment, which is consistent with vestibular system activation. Conversely, all subjects reporting only headache and fatigue noted these symptoms during the exposure, with headaches lasting up to 5 h beyond the exposure and fatigue lasting 15 h. No control subject reported any symptoms, thus validating the experiment.

Although there is no controversy on the adverse effects of short-duration exposures to relatively high levels of infrasound, there remains considerable controversy on the long-term effects of *periodic* infrasound at levels below the threshold of hearing. On the one hand some practitioners contend that infrasound should be ignored because wind farms produce infrasonic levels well below the hearing threshold so that noise levels are adequately documented using the A-weighting filter (see Section 2.2.11). On the other hand, a different perspective has been presented by Salt and Lichtenhan (2014) and co-workers. They have published a number of peer-reviewed articles on the effects of low-frequency noise and infrasound on the hearing mechanism and its physiology. They reported on a person suffering from Ménière's disease, who had her symptoms of vertigo, dizziness and nausea severely exacerbated when she was in the vicinity of wind turbines. People with this disease suffer from the condition called *endolymphatic hydrops*, in which the sensory organ can be displaced, thus obstructing the pressure relief opening at the apex of the cochlea (the helicotrema), resulting in a substantial increase in the low-frequency hearing response (by up to 20 dB) and symptoms of unsteadiness, vertigo, nausea (or seasickness), tinnitus and pressure fullness in the ear. Salt and Lichtenhan (2014) have suggested that hydrops can be developed in otherwise normal people when they are exposed to low-frequency noise or infrasound (based on experiments with short-term exposures to 50-Hz sound) and they have shown for these people the condition of hydrops resolves when they are no longer stimulated by low-frequency noise or infrasound. This is consistent with the relief from symptoms that wind farm noise sufferers experience some time after they leave the vicinity of a wind farm. No data are available on the effect of this condition on long-term exposure to low levels of infrasound, such as experienced by residents living near wind farms.

The ear is most sensitive to low-frequency noise and infrasound when audible noise in the 150–2000 Hz range is at a very low level or absent (Salt and Lichtenhan, 2012). This is because higher audible ambient background noise automatically causes the hearing threshold to be raised, thus protecting the ear against very high sound pressure levels. This is the reason why, in quiet environments such as in rural bedrooms, people can be maximally affected by infrasound: it is not attenuated much by the housing structure, whereas noise above 150 Hz is generally substantially attenuated. On the other hand, people in towns and cities have a raised low-frequency and infrasound hearing threshold as a result of other ambient noise, such as produced by traffic, and thus they are able to tolerate infrasound at levels that may cause problems to a rural resident. A study by Pedersen et al. (2010) showed that road traffic noise can also provide an additional effect of significant masking of wind farm noise, but only if the wind farm noise is between 35 and 40 dBA (L_{Aeq}) and not for higher or lower level wind farm noise. Masking only occurs when the road traffic noise is approximately 20 dBA or more louder than the wind farm noise.

Bell (2011), (2014) has also contributed some enlightened understanding to the debate, based on the physiological mechanism of how the ear hears (see Section 7.3). He explains how the middle ear acts continuously to position the eardrum in the middle of its operating range, while at the same time pushing or pulling on the oval window of the cochlea, thus increasing or decreasing the fluid pressure in the cochlea, which affects the gain of the hearing system. This gain control is an essential part of the ear having a dynamic range of over 120 dB (1 million to 1 in terms of pressure). The ear drum acts linearly to sound pressure, irrespective of the frequency, so that a 1-Hz sound at 70 dB has the same effect as a 1000-Hz sound in terms of affecting the middle ear muscle action in adjusting the ear drum position and cochlear pressure, even though the 1-Hz sound will be inaudible to the cochlea. Infrasonic sound in the range below 8 Hz is sensed as a series of impulses, corresponding to the peaks in sound pressure, rather than as a continuous stimulus, so in terms of the effect of infrasound on people, the peak of the signal is more important than the RMS value. Having the middle ear muscles acting every time they sense a peak pressure (which can be several times a second for infrasound) may lead to annoyance and a feeling of pressure in the ears of some people, and can result in serious sleep disturbance, even though the sound causing the problem is inaudible. As pointed out by Bell (2014), the level of infrasound needed to affect the operation of the middle ear is much lower than the level required for detection by the cochlea. In some people, it is possible that the action of the middle ear muscles in response to infrasound will increase the intralabyrinthine fluid pressure to a level sufficient to affect the balance organs of the inner ear and cause vertigo. People with Ménière's disease have episodes where they suffer vertigo and pressure fullness in their ears, and lose their balance. These effects may be attributed to a spasm in the middle ear muscles, increasing the intralabyrinthine fluid pressure beyond normal levels, in a similar way that exposure to low-level infrasound could affect some otherwise normal people.

In an interesting contribution to the debate, Swinbanks (2015a) pointed out that problems with low-frequency noise and infrasound from wind farms are most likely to manifest indoors rather than outdoors. This is because noise generated by pressure fluctuations as a result of wind blowing over our ears will mask low-level infrasound and low-frequency noise. However, indoors the effect is entirely different. There is no wind blowing over our ears and also pressure fluctuations due to wind noise do not penetrate a house as effectively as infrasound. This is because pressure fluctuations due to the wind at any particular location on the outside of the housing structure are only correlated with nearby locations, whereas infrasound pressure fluctuations are correlated over the entire housing structure. This results in wind-induced pressure fluctuations inside the house cancelling out, whereas the infrasound pressure fluctuations become more apparent. So perception of low-frequency sound and infrasound in a bedroom at a considerable distance from a wind farm can be considerably greater than what one would notice outside, even very close to a turbine.

As a representative of the alternative to the view that wind farm infrasound may affect the health of some people, Leventhall (2013) has stated categorically on a number of occasions that infrasound internally generated in our body is of the same order as wind turbine generated infrasound and that there is no evidence that wind turbine infrasound is harmful. He also writes:

> Internally generated infrasound from heartbeat and breathing, which enters the inner ear via the cochlear aqueduct, is greater than that received externally from wind turbines at similar frequencies, perhaps by 20dB or more. Levels of infrasound received from wind turbines at typical residential distances are well below hearing threshold and also mainly below the outer hair cell threshold, proposed by Salt and Hullar (See Section 7.3) as a possible onset level of adverse effects.

In response, Schomer (2013) writes:

> The signals impinging on the inner ear are 20 dB greater but this energy does not readily couple to the inner ear because its entry point is right next to the round window. The point is that the acoustic pressures (applied via the stapes) from the wind turbine may be as great, or greater than the body-generated pressures that reach the whole of the inner ear.

However, as Schomer points out, the brain knows when the heartbeat and breathing signals are occurring and can negate the associated response. Schomer also refers to the tonal nature of wind turbine infrasound being an important property that can cause similar seasickness-type symptoms as on a ship when signals to the brain from two different senses (sight and vestibular system) conflict.

In general, when people quote levels of environmental infrasound and state that wind turbine infrasound is at similar levels (see for example, Berger et al. (2015)), they have not taken into account three important aspects of wind turbine infrasound as listed below.

- The sound is regular and pulsating in amplitude. This rhythmic thumping stimulates the inner ear sensory functions, causing the symptoms mentioned above.
- Measurements of environmental infrasound are typically reported as average or RMS values. However, the peak values of the signal are what are likely to determine the detection threshold. The peak to RMS ratio for wind turbine infrasound is greater than the peak-to-RMS ratio for tonal sound used in infrasound threshold testing, so the RMS amplitude of wind turbine sound underestimates its ability to be detected.

- Some people are far more sensitive than others due to differences in physiology, so some may not be bothered at all but some may suffer severe symptoms when exposed to the same infrasound levels.

There are a sufficient number of case studies on small groups of people to indicate that wind farm noise does cause adverse health effects in some residents. It is not important to such people if the problem is infrasound or low-frequency noise or if their health problems are direct physiological effects or the effects of lack of sleep caused by stress and annoyance. The fact is that some people's health is adversely affected as a result of living in the vicinity of a wind farm.

Case studies on small groups of people are referred to as 'adverse event reports' in the medical literature and Phillips (2011) points out that the existence of thousands of these is a compelling case in support of the idea that wind farms are responsible for adverse health effects. Many in the medical profession do not believe that adverse event reports prove any cause–effect relationship, and that the only way to demonstrate a cause–effect relationship is to conduct a randomised clinical trial with all confounding factors accounted for. In the case of wind farm noise, this is clearly impractical and perhaps even unethical. In fact, randomised clinical trials have not been used in the past for testing the effects of noise exposure on people, simply because it is not possible to eliminate all confounding factors. Perhaps the next-best evidence could come from longitudinal field research. This would assess the health of a community before a turbine project is proposed and then continue to follow up during operation. However, it is difficult to find any community that has not been affected by the wind farm debate once a decision has been made to construct a wind farm in their community. The same is true of highways, railways and airports. This is why most of the studies undertaken to date on transportation noise and wind farm noise have been cross-sectional, which makes it impossible to determine whether the cause precedes the effect.

In the case of wind farm noise, it is unfortunate that because no randomised clinical trial data are available (or likely ever will be available), wind turbine manufacturers, wind farm operators and medical research bodies continue to declare that 'there is no evidence that wind farms cause adverse health effects in humans', while conveniently ignoring the extensive documentation of thousands of case studies. Of course, one or two case studies taken in isolation are insufficient evidence. However, when the number is in the thousands, with similar symptoms reported by residents living near wind farms and symptoms usually disappearing when the residents leave the wind farm vicinity for a week or two, it is difficult to reject such compelling evidence. In support of the findings from documented case studies, Krogh et al. (2011) surveyed 109 participants living in the vicinity of a wind farm and found a dose–response relationship between the logarithm of distance from the nearest turbine and the number of people suffering adverse health outcomes in the form of recurring sleep disturbance, excessive tiredness and headaches.

A research approach that could provide evidence of adverse health effects due to wind farms would be to remove the stimulus by removing residents from the vicinity of the wind farm for some weeks and then returning them and observing symptoms before, during and after removal.

McMurtry (2011) published a list of criteria to be satisfied for a diagnosis of probable adverse health effects caused by a wind farm. These are divided into first-order, second-order and third-order criteria. Some additional adverse health effects, which were found by Pierpont (2009), have been included in the following lists.

All four first-order criteria, which are listed below, must be satisfied.

1. Subjects must reside within 5 km of a wind farm.
2. Subjects must have altered health status within six months of wind farm start-up.
3. Symptoms must ameliorate when more than 5 km from a wind farm.
4. Symptoms must recur upon return to their residence.

Once people become sensitised to wind farm noise, it may be necessary for them to go further away than 5 km for their symptoms to subside; in some cases symptoms may not subside unless they are taken to a very quiet environment.

At least three of the following second-order criteria must be satisfied.

1. Subjects' quality of life must be compromised.
2. Subjects experience recurring sleep disruption and/or difficulty in falling asleep.
3. Subjects suffer annoyance, resulting in stress.
4. Subjects prefer to leave residence temporarily or permanently for sleep restoration or well-being.

At least three of the following third-order symptoms should occur or worsen following initial operation of the wind farm.

1. *Otological and vestibular*, including tinnitus (ringing in the ears), dizziness, vertigo (spinning dizziness), balance difficulties, ear ache, ear pressure and nausea.
2. *Cognitive*, including difficulty in concentrating and memory problems.
3. *Cardiovascular*, including hypertension (raised blood pressure), palpitations (irregular heart beat) and tachycardia (fast heart rate).
4. *Psychological*, including depression, anxiety, stress, frustration, distress, panic episodes and anger.
5. *Regulatory disorders*, including difficulty in controlling diabetes and the onset of thyroid problems.
6. *Systemic problems*, including fatigue, sleep disruption and insomnia.

The most commonly reported symptoms from the above list seem to be nausea, depression, anxiety, stress, frustration, recurring sleep disturbance, fatigue, insomnia, dizziness, elevated blood pressure, tachycardia and balance difficulties. It is not clear what health effects are a result of recurring sleep disturbance, poor sleep quality and annoyance rather than a direct physiological effect caused by wind farm noise. It is well known that sleep deprivation can lead to a range of illnesses, as sleep and associated periods of coordinated neuronal inactivity appear to be essential for the brain recuperation processes and reorganisation necessary for memory and learning, and also for good health. Sleep deprivation is well known to produce time-dependent reductions in the ability to concentrate, as well as reductions in work and learning performance. Recurring sleep disruption is much more common than sleep deprivation and also impairs cognition and performance, as well as glucose metabolism and immune function, which can lead to other physical symptoms. Poor sleep also causes traffic and work accidents, lost productivity and adverse health effects.

One should not neglect the possible effects of wind farm noise on children and their ability to do well at school. Bronzaft (2011) pointed out the adverse effects of train noise on students in a class room and also showed that high achievers more often came from families that respected quiet and provided quiet areas for their children to study and sleep. In addition, he asserted that if parents are stressed by wind farm noise to the extent that they are suffering sleep deprivation, then their children are likely to suffer from poorer parenting.

The effect of wind farm noise on people is highly variable, even between people living in the same house. The adverse health symptoms can occur immediately or within a few hours of exposure (Ambrose et al., 2012; Swinbanks, 2015a) or they can gradually develop over weeks and months. In many cases people become increasingly sensitised to wind farm noise as time goes on until they reach a point where they cannot bear to be near any low-frequency or infrasonic sound source. People with pre-existing migraines, motion sickness susceptibility or damage to inner ear structures (such as caused by noise-induced hearing loss) are more susceptible to the symptoms listed above.

Shain (2011) points out that wind farms represent a trade off of adverse health effects of a few to beneficial health effects of many in the form of reduced emissions. There is also an inherent argument that the government can interfere with the rights of a few for the benefit of many. However, in many other cases, the few disadvantaged people are usually fairly compensated by the corporation causing the disadvantage. Unfortunately the same is not true for people adversely affected by wind farms. The rapid multiplication of wind farms in recent times seems to be in conflict with the precautionary principle, that would require that a particular industrial activity not be undertaken unless there is no doubt that such an activity will not harm any people. In many countries there is also a common law requirement that there is a duty to ensure that one does not cause predictable harm to their neighbour's interests.

7.2.1 Amplitude Modulation, Amplitude Variation and Beating

The difference between AM, random amplitude variation and beating is illustrated in Figure 2.5. An amplitude modulated signal contains energy at the modulation frequency but a beating signal contains no energy at the beat frequency. Beating is explained in Section 2.2.6 and may be described as the interference between two signals that are close in frequency, which results in a single frequency that varies in amplitude. On the other hand, a modulated signal is a single-frequency, multi-frequency or random signal that has its amplitude varied in a periodic or regular, repetitive way. This phenomenon is described in detail in Section 2.2.7, where two different types of AM are defined and their generation mechanisms described: normal AM and enhanced AM (EAM). EAM is low-frequency in nature and is often experienced at night by residents living at large distances from the nearest wind farm.

It is of interest to associate the terms used to describe wind farm noise when it is disturbing, with the quantitative characteristics described above (see Renewable UK (2013)). Words like 'thumping' or 'whooping' seem to describe low-frequency EAM. Words like 'rumbling' seem to imply a much lower level of AM and again they apply to low-frequency noise perception. In some cases, rumbling may also be associated with the modulation of tonal noise arising from the gearbox.

Beating is the interference between two tonal noises that are close in frequency and its existence could be perceived as a slowly varying 'rumbling' noise when the two signals that are beating are low in frequency. Beating results in a regular variation in sound level and is thus also more annoying than randomly varying sound. Beating can be caused by sound pressure contributions arriving at the receiver from two or more turbines rotating at slightly different speeds and thus generating slightly different frequencies for any tonal noise that is associated with their operation. Yokoyama et al. (2015) carried out some experimental work, testing how people detect beating noise (although it is mistakenly defined as AM in the paper). They showed that 'noisiness' increases as the 'AM' strength increases and first becomes noticeable for an 'AM index' greater than 1.7 dBA for a 45-dBA wind turbine noise. The 'AM index', ΔL (dB) is defined as $\Delta L = 20 \log_{10}(A_{max}/A_{min})$, where A_{max} and A_{min} are defined in Figure 2.5.

An additional effect of the presence of more than one turbine is that the sound arriving at the receiver from each turbine will be characterised by a phase difference from the sound arriving from the other turbines, due mainly to different propagation differences. At low frequencies, atmospheric turbulence only has a small effect on the phase-angle differences, but as the frequency increases and the wavelength of the sound becomes smaller, atmospheric turbulence has a greater effect on the relative phase angles, resulting in only small and rapid changes in sound level at the receiver, which are not detectable. However, at low frequencies, the relative phase between the sound arriving from different turbines can vary slowly and there will be times and places where the sound from different turbines will reinforce one another, resulting in much greater levels than calculated by assuming random combination of the different contributions, as is done in standard noise propagation modelling. Huson (2015) reported on the amplitude variation of infrasound tones and showed that 3-min averaged spectra would often exceed the 1-h averaged spectra by 6 dB, indicating that a significant amount of amplitude variation is often present at frequencies below 20 Hz. These large variations can be attributed to interference and reinforcement of tonal noise from multiple turbines. The relative phase of the sound arriving from each turbine is affected by small changes in wind speed and atmospheric turbulence.

Our hearing mechanism responds to AM, amplitude variation and beating signals differently, even if they have the same maximum and minimum amplitudes, especially in the infrasound and low-frequency ranges. We find that many environmental sounds may be characterised as randomly varying, as there is no regularity to their variation. However, wind farm noise, especially infrasound and low-frequency noise, is often amplitude modulated so that it varies in a regular way. This type of sound is more intrusive and annoying than randomly varying sound, even if it is at the same sound pressure level.

When noise is amplitude modulated, the mean sound pressure levels are not very representative of the actual levels that occur as a function of time, as the peaks in sound pressure level are much higher. Therefore, if a low-frequency noise is amplitude modulated as well as being above the normal hearing threshold, it is likely to be annoying to many people, especially as the detection and annoyance thresholds are close together for low-frequency noise. Even if the noise is slightly below the hearing threshold, it will still be heard by some people due to the natural spread in hearing thresholds (Møller and Pedersen, 2011). In this case, the noise could manifest as inaudible or soft to some people, yet could be loud and annoying to others. In more recent work, Yoon et al. (2015) showed that generally the detection threshold for an individual's response to wind turbine noise decreases as the extent of variation of the noise increases.

The degree of AM that will be experienced varies considerably, depending on the weather conditions. It seems to be worse at night in stable, high wind shear conditions, which would produce a regularly varying load on, and possibly stall of, the turbine blades (and thus regularly varying noise radiation) as they pass from locations of low wind speed to locations of high wind speed. Amplitude modulation is also more often found when atmospheric turbulence intensity is low and there is an atmospheric temperature inversion (increasing air temperature with increasing altitude) in the first few hundred metres above the ground, which often occurs in the early hours of the morning, when people are likely to find wind farm noise most disturbing. In downwind conditions (wind blowing from the turbines to the receiver), wind turbine noise is louder, which makes any AM more annoying, especially if stable atmospheric conditions and an associated temperature inversion exists near the ground.

There are a number of methods that are in use for quantifying the extent of AM of wind farm noise and these are discussed in Section 6.2.10. Most are based on variations in the A-weighted sound pressure level (see Section 2.2.11). In terms of estimating the effect of the AM on humans, this is inappropriate because it will relegate to insignificance any modulation of the infrasonic component of the noise and possibly the low-frequency component as well. As Salt and Lichtenhan (2014) point out, using the A-weighting metric will underestimate the degree of modulation perceived by humans. The various metrics that have been investigated are discussed in detail in Section 6.2.10.

von Hünerbein and Piper (2015) tested the annoyance of subjects to wind turbine sound with varying degrees of modulation depth for L_{Aeq} values ranging from 25 to 45 dBA. Modulation depth is defined by von Hünerbein and Piper (2015) as the difference between the mean peak level and the mean trough level in the A-weighted RMS time series over the length of the test stimulus, where each data point is derived from 100-ms averages of L_{Aeq}. Modulation depths greater than 12 dBA were not studied because these are not expected to occur with wind farm noise. It was found that annoyance and perceived noisiness increased with both increase in level above 20–25 dBA and increase in modulation depth.

7.3 Hearing Mechanism

Before considering the effects of wind farm noise on people in detail, it is instructive to describe the principles underlying the operation of our hearing mechanism. As shown in Figure 7.3, our hearing mechanism consists of three major parts: the external ear (pinna), the middle ear and the inner ear (cochlea).

7.3.1 External Ear

The convolutions of the pinna give rise to multiple reflections and resonances within it, which are frequency and direction dependent. These effects and the location of the pinna on the side of the head make the response of the pinna directionally variable to incident sound in the frequency range of 3 kHz and above, so that a stimulus in the latter frequency range is best heard when incident from the side of the head.

7.3.2 Middle Ear

Sound enters the ear through the ear canal, which is an approximately straight tube between 23 and 30 mm in length, at the end of which is the eardrum. The eardrum is a diaphragm-like structure and is often referred to as the tympanic membrane. It moves in response to acoustic pressure fluctuations within the ear canal and its motion is transmitted through a mechanical linkage provided by three tiny bones, called ossicles, to a second membrane at the oval window of the middle ear. Sound is transmitted through the oval window to the inner ear (see Figure 7.3).

The middle ear cavity is kept at atmospheric pressure by occasional opening, during swallowing, of the eustachian tube, which is also shown in Figure 7.3. If an infection causes swelling or mucus to block the eustachian tube, preventing pressure equalisation, the air in the middle ear will be gradually absorbed, causing the air pressure to decrease below atmospheric pressure and the tympanic membrane will be unable to vibrate in response to incoming sound. The listener then will experience temporary deafness.

The three tiny bones located in the air-filled cavity of the middle ear are referred to as the malleus (hammer), incus (anvil) and stapes (stirrup). They provide a mechanical advantage of about 3:1, while the relative sizes of the larger eardrum and smaller oval window provide an additional mechanical advantage of about 5:1, resulting in an overall mechanical advantage of about 15:1. As the average length of the ear canal is about 25 mm, its first longitudinal resonance occurs at a frequency of about 4 kHz, giving rise to a further mechanical advantage about this frequency of the order of three.

The bones of the middle ear are attached to a muscular structure (see Figure 7.3), which allows some control of the motion of the linkage, and thus control of the mean position of the tympanic membrane as well as control of transmission of sound to the inner ear. For example, a moderately loud buzz introduced into the earphones of a gunner may be used to induce tensing of the muscles of the middle ear to stiffen the ossicle linkage, and to protect the inner ear from the loud percussive noise of firing. Some individuals suffer from what is most likely a hereditary disease, which takes the form of calcification of the joints of the middle ear, rendering them stiff and the victim deaf.

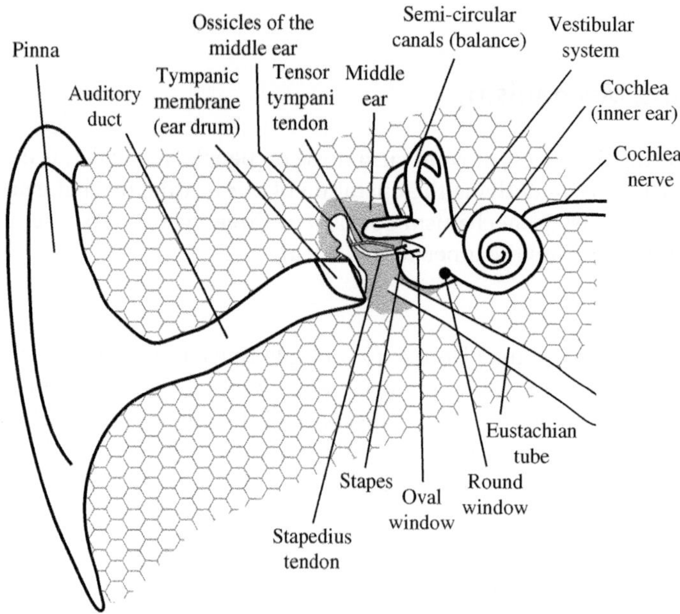

Figure 7.3 Illustration of the hearing mechanism.

In this case the cure may take the form of removal of part of the ossicles and replacement with a functional prosthesis.

7.3.3 Inner Ear

As can be seen by inspection of Figure 7.3, the oval window at the entrance to the liquid-filled inner ear is attached to the stapes, which in turn is connected to the middle ear ossicles. The inner ear consists of the cochlea (for hearing) and the semicircular canals (for balance), which is why an ear infection can sometimes induce dizziness or uncertainty of balance.

The human cochlea is a small tightly rolled spiral cavity, as illustrated Figure 7.3. Internally, the cochlea is divided into an upper gallery and a lower gallery by an organ called the cochlear duct, which runs the length of the cochlea (see Figure 7.4). The two galleries are connected at the apex of the cochlea by a small opening called the helicotrema. At the other end of the cochlea, the upper and lower galleries terminate at the oval and round windows respectively. The round window is a membrane-covered opening situated in the same general location of the lower gallery as the oval window in the upper gallery (see Figure 7.3). A schematic representation of the cochlea *unrolled* is shown in Figure 7.4 and a cross-sectional view is shown in Figure 7.5.

In humans, the average length of the cochlea from the basal end to the apical end is about 34 mm. The fluid that fills the upper and lower galleries within the cochlea is sodium rich, while the fluid that fills the cochlear partition is potassium rich and both fluids are essentially salt water.

The central partition acts as a mechanical shunt between the upper and lower galleries. Any displacement of the central partition, which tends to increase the volume of one gallery, will decrease the volume of the other gallery by exactly the same amount. Consequently, it may be concluded that the fluid velocity fields in the upper and lower galleries are essentially the same but of opposite phase.

Cochlear Duct

The cochlear duct (see Figure 7.5), which divides the cochlea into upper and lower galleries, is triangular in cross-section, being bounded on its upper side by Reissner's membrane, and on its lower side by the basilar membrane. The cochlear duct is anchored at its apical end (apex) to a bony ridge on the inner wall of the cochlear duct formed by the core of the cochlea, and the auditory nerve is connected to the central partition through this core. The closed three sides of the cochlear duct form a triangular partition between the upper and lower galleries.

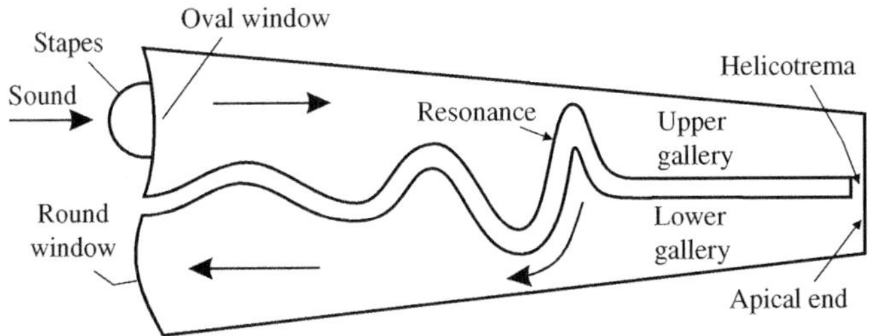

Figure 7.4 Schematic model of the cochlea (unrolled).

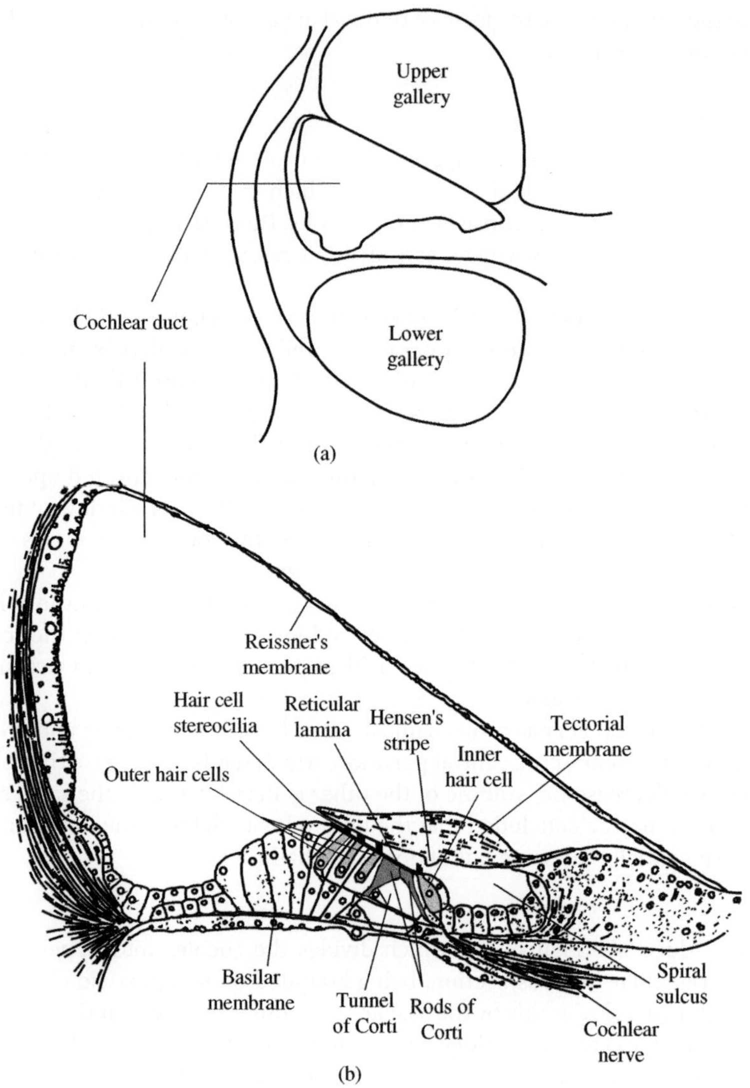

Figure 7.5 (a) Cross-section of the cochlea. (b) Cross section of the cochlear duct.

Most of the hearing sensation occurs via the inner hair cells, as shown in Figure 7.5: 95% of the nerves responsible for hearing are attached to these cells while the remaining 5% are attached to the outer hair cells.

The cochlear partition, the basilar membrane and upper and lower galleries form a coupled system in which sound transmitted into the cochlea through the oval window proceeds to travel along the cochlear duct as a travelling wave. The amplitude of this wave depends on the flexibility of the cochlear partition, which varies along its length. Depending on its frequency, this travelling wave will build to a maximum amplitude at a particular location along the cochlear duct, as shown in Figure 7.4. After that location, it will decay quite rapidly. Thus, a tonal sound incident upon the ear results in excitation along the cochlear partition that gradually increases up to a place of maximum response. The tone is sensed in the narrow region of the cochlear partition, where the velocity response is a maximum. The ability of the ear to detect the pitch of a sound appears to be dependent upon its ability to localise the region of maximum velocity response in the cochlear duct.

In summary, any sound incident upon the ear ultimately results in a disturbance of the cochlear duct, beginning at the stapes end, up to a length determined by the frequency of the incident sound. High-frequency sounds, for example 6 kHz and above, will result in motion restricted to about the first quarter of the cochlear duct nearest the oval window. The situation for an intermediate audio frequency is illustrated in Figure 7.4. All stimulus frequencies lower than the maximum response frequency result in some motion at all locations towards the basal end of the cochlear duct where high frequencies are sensed. For example, a heavy base note at 50 Hz drives the cochlear partition over its entire length to be heard finally at the apical end. This is perhaps why exposure to excessive noise results first in the loss of high-frequency hearing sensitivity.

Hair Cells

The sound-sensing hair cells are arranged in rows on either side of a rigid triangular construction formed by the rods of Corti, sometimes called the tunnel of Corti. As shown in Figure 7.5, the hair cells are arranged in a single row of inner hair cells and three rows of outer hair cells. The hair cells are each capped with a tiny hair bundle called hair cell stereocilia, which are of the order of 6 or 7 μm in length.

The stereocilia of the inner hair cells are free-standing in the space between the tectorial membrane above and the reticular lamina below, on which they are mounted. They are velocity sensors, responding to any slight motion of the fluid that surrounds them. Referring to Figure 7.5, it can be seen that motion of the basilar membrane upward in the figure results in rotation of the triangular tunnel of Corti and a decrease of the volume of the inner spiral sulcus and an outward motion of the fluid between the tectorial membrane and the reticular lamina.

By contrast with the inner hair cells, the outer hair cells are firmly attached to and sandwiched between the radially stiff tectorial membrane and the radially stiff reticular membrane, as shown in Figure 7.5. Thus they are displacement sensors and respond well to low-frequency sound and infrasound.

7.3.4 Frequency Response of the Human Ear

The frequency response of the ear is the loudness of noise as a function of frequency (or pitch) for the same sound pressure level at the entrance to the ear canal. Although the frequency range of human hearing is generally considered to be between 20 and 20 000 Hz, individuals can detect sound at frequencies well into the infrasonic region in the range 1–20 Hz. The term, *audio range* has traditionally been used to describe the frequency range from 20 to 20 kHz, which covers our most sensitive range of hearing. However, most people can hear noise at frequencies well below 20 Hz if it is sufficiently loud, and many can hear noise at frequencies above 20 kHz, so the term audio range can be confusing.

The mechanism by which humans detect infrasound at low levels is the hearing mechanism, as shown by Yeowart and Evans (1974), as similar threshold levels were determined with earphones and with whole-body immersion in the sound field. They also found that, as with higher frequency sound, the threshold levels when two ears are exposed are approximately 3 dB lower than if only one ear is exposed. The idea of the hearing mechanism being the most sensitive part of the body to infrasound was further demonstrated by Landström et al. (1983), who measured the tactile response for deaf people as well as hearing people, the latter being asked to ignore their hearing response, and showed that the infrasound-detection threshold was 15–25 dB higher than the hearing threshold. The vibro-tactile response was described as soft vibrations in the lumbar, buttock, thigh and calf regions, although Møller and Pedersen (2004) pointed out that some of their test subjects also experienced vibro-tactile sensations as a feeling of pressure in the upper part of the chest and in the throat region. Although the work of Landström et al. (1983) confirmed that the ear was the most sensitive receptor on the body for infrasound, it also showed that other parts of the body were only 10–25 dB less sensitive and that hearing subjects were able to distinguish between the hearing sensation and the vibro-tactile sensation.

As the annoyance threshold for low-frequency noise, such as experienced indoors by residents more than 1 km and less than 8 km from a wind farm, is very close to the hearing threshold for most people, it is of interest to examine the hearing threshold as a function of frequency. The international standard, ISO 226 (2003) provides generally agreed hearing-threshold curves for frequencies between 20 and 12 500 Hz. The standard is based on a range of data from studies between 1956 and 2002. Unfortunately, there have been only a few studies that tested hearing thresholds for frequencies below 20 Hz, and these have been compiled and analysed by Møller and Pedersen (2004) to produce recommended curves for frequencies in this range. These findings, together with some equal loudness curves for frequencies below 1000 Hz, are shown in Figure 7.6, alongside the standard curves for frequencies above 20 Hz. Curves showing threshold levels that exclude the hearing mechanism (vibro-tactile thresholds) are also included.

There are two characteristics of Figure 7.6 that are notable. The first is that the difference between the hearing threshold and an 80 Phon sound becomes very small at low

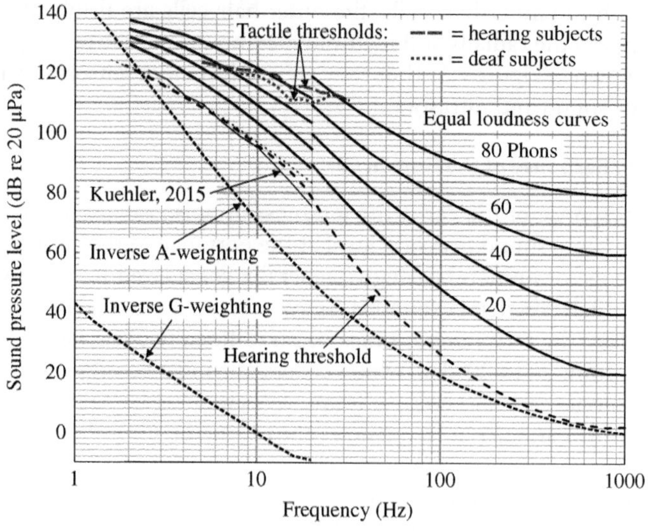

Figure 7.6 Hearing thresholds and equal loudness contours.

and infrasonic frequencies. The second characteristic is the discontinuity in the equal loudness curves at 20 Hz. This is because the infrasound curves are based on fewer and different studies. The original curve suggested by Møller and Pedersen (2004) for the hearing threshold below 20 Hz is shown as the faint dotted line and is the average of ten past studies. Unfortunately this curve does not meet the ISO 226 curve at 20 Hz, missing it by approximately 4 dB. On investigation of the data used by Møller and Pedersen (2004) to obtain the average, it is apparent that there are two studies for which the data match the ISO standard at 20 Hz and one of these is a study done by Watanabe and Møller (1990). Thus, when preparing Figure 7.6, it was decided to use the data from these latter two studies in the dashed curve shown in the figure for frequencies between 8 and 20 Hz so that the curve merged with the ISO standard at 20 Hz. The only significant differences between this curve and the one suggested by Møller and Pedersen (2004) is 4 dB at 20 Hz and 2 dB at 16 Hz. The hearing-threshold curve below 20 Hz shown in Figure 7.6 is further justified by more recent threshold tests (Kuehler et al., 2015), shown as the faint solid line in the figure.

An inverse A-weighting curve can be obtained by changing the sign of all of the A-weighting values so the curve can then be directly related to equal loudness curves. The inverse A-weighting curve is shown in Figure 7.6, demonstrating that it is not a very good approximation of the perceived loudness of noise below 20 Hz at any loudness level, nor is it a good approximation of the loudness of noise at a loudness level near our hearing threshold at any frequency. However, it is a reasonable approximation of the loudness of noise at a level of 60 Phons above 20 Hz (but not necessarily representative of the annoyance of the noise, which is known to increase as the frequency decreases). The inverse G-weighting curve is also shown in Figure 7.6, demonstrating that it is a good approximation of the slope of the hearing threshold curve for noise below 20 Hz, although it is not at all good for noise above 20 Hz (see Figure 2.6). The slope of the hearing threshold curve below 20 Hz is 6 dB per octave, which implies that the hearing mechanism is responsive to the rate of the rate of change of pressure (that is, the acceleration of the pressure or the rate of change of the rate of change in sound pressure). Thus a sharp impulse is readily detected by the hearing mechanism as a result of the rate-of-rate-of-change of its leading edge and this is often closely correlated with the peak amplitude of the impulse (Swinbanks, 2015b), indicating that the sensitivity of the ear in the infrasonic range is related to the signal peak rather than its RMS value.

The thresholds and equal loudness curves in Figure 7.6 represent the median of the population. One standard deviation is between 5 and 6 dB, so that if the hearing threshold in decibels is normally distributed, then 2.5% of the population (representing two standard deviations on one side) would have a hearing threshold that was 10–12 dB less than that shown in Figure 7.6. The threshold curve in the infrasonic region of Figure 7.6 represents a noise level of 97 dBG. If we subtract two standard deviations from this number, we arrive at 85 dBG, which is the recognised hearing threshold used to evaluate noise dominated by infrasound. In fact 2.5% of the population would have a lower hearing threshold than this and 0.15% would have a hearing threshold lower than 79 dBG.

It is important to note that the hearing thresholds in Figure 7.6 are for single-frequency tonal noise. Møller and Pedersen (2004) state that 1/3-octave broadband steady noise of a similar level is characterised by a similar hearing threshold. However, James (2012) states that the threshold of perception for a complex set of tones that are modulated in frequency and amplitude is likely to be much lower.

The mechanism by which the hearing system detects infrasound and very low-frequency sound is complex. von Gierke and Nixon (1976) suggest that detection results from non-linearities of conduction in the middle ear, which generate higher frequency harmonics in the more audible higher-frequency range. More recently, Salt and Lichtenhan (2014) point out that the outer hair cells (OHCs) in the cochlea are more sensitive than the inner hair cells (IHCs) to low-frequency sound and infrasound and respond at levels well below the threshold of hearing. The threshold of hearing is determined by the response of the IHCs (Salt and Hullar, 2010). The reason that OHCs are more sensitive to low-frequency sound and infrasound is that they are displacement sensors, as a result of the ends of the stereocilia being attached to the tectorial membrane. On the other hand, the IHCs stereocilia are not attached at one end and so respond to the vibrations of the fluid near the tectorial membrane, resulting in them responding to velocity rather than displacement. They are also innervated by type I afferent nerve fibres and so are responsible for our hearing response.

Salt and Hullar (2010) estimate that at 5 Hz, the OHCs are 40 dB more sensitive than the IHCs. As Salt and Lichtenhan (2014) point out, the human hearing response to sound, as the frequency decreases from 500 to 20 Hz, reduces at the rate of 18 dB per octave but below 10 Hz the rate of sensitivity loss is much smaller and more like 12 dB/octave (see Figure 7.6). Salt and Hullar (2010) suggest that this may be due to stimulation of the IHCs that are responsible for hearing by extracellular voltages generated by the OHCs. Stimulation of the OHCs at levels below the hearing threshold results in information transfer via pathways that do not involve conscious hearing, which may lead to various sensations, such as awakening from a deep sleep without having heard the noise that caused the awakening. According to Salt, this sort of stimulation can result in sleep disturbance and feelings of panic on awakening, with chronic sleep deprivation leading to blood pressure elevation and memory dysfunction.

Actual low-frequency and infrasonic noise levels measured in an unoccupied house 3.5 km from the Waterloo wind farm, which consists of 37 Vestas V90 3-MW wind turbines, are shown in Figure 7.7, along with the threshold of hearing and IHC and

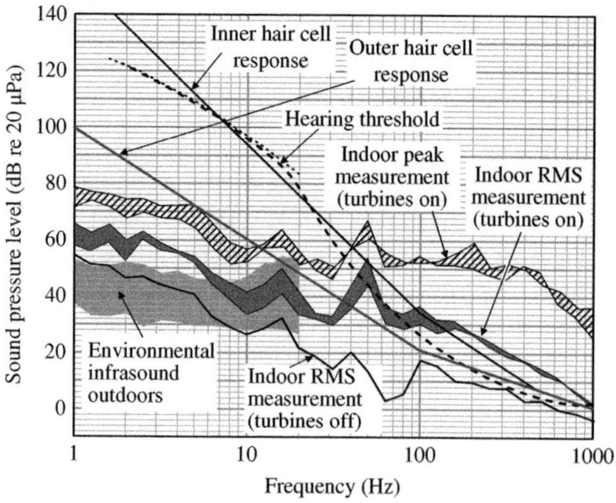

Figure 7.7 Inner and outer hair cell responses, hearing thresholds, environmental infrasound and wind farm infrasound.

OHC thresholds. Two representative shaded curves of wind farm noise, for downwind propagation conditions and measured inside the residence, are shown. The dark-shaded curve represents a range of measured RMS 1/3-octave levels and data represented by this curve only exceed the OHC threshold above about 35 Hz. Also plotted is a hatched curve, representing a range of measured peak 1/3-octave sound levels. Data represented by this hatched curve exceed the OHC threshold at frequencies above about 13 Hz, well into the infrasound range. The peak-level curve was determined by measuring the highest level that occurred in the 10-min sample when the noise was filtered with a standard 1/3-octave filter. At low frequencies, the rise time of 1/3-octave filters is much longer than the response time of the ear (see Table 2.4), so it is possible that the peak levels detected by the ear at frequencies below 50 Hz could be even higher. However, this possibility was investigated using a 100-Hz bandwidth filter centred at 50.5 Hz (with a rise time similar to the hearing mechanism) and it was found that the difference between RMS and peak levels obtained with this filter were similar to the differences obtained with the 1/3-octave filters in the same frequency range.

For comparison, a range of outdoor infrasound measurements provided by Turnbull et al. (2012) is represented by the light-shaded area on the graph. The higher limit of the shaded area represents measurements made during the day at a power station and a busy central business district of a city. The lower limit represents measurements made in a quiet suburb during the day. Measurements made with the wind farm shut down inside the house 3.5 km from the Waterloo wind farm are also shown. It can be seen that the wind farm results in a substantial increase in infrasound levels in the house over what would be experienced in its absence. Zajamšek et al. (2014) show that infrasound levels from the Waterloo wind farm are just detectable by the OHCs of a normal person at a distance of 8 km from the nearest turbine. If the same variation that occurs in the hearing threshold also occurs in the OHC sensitivity, then 2.5% of people would have an OHC sensitivity that is 12 dB less than shown in Figure 7.7, which would make the infrasound produced by the Waterloo wind farm easily detectable by the OHCs and also detectable by the IHCs of these people at a distance of up to 8 km from the nearest turbine, in downwind conditions.

As mentioned previously, the greatest response of the OHCs to low-frequency sound and infrasound occurs when stimulus from the mid-frequency range between 200 and 2000 Hz is absent. This helps to explain why people in quiet rural environments may be more susceptible to wind farm noise than people in urban and suburban environments, where the background noise levels in the mid-frequency range are considerably higher (as found by Pedersen et al. (2009)). This idea is further supported by the work of Krahé (2010), who showed that low-frequency noise (between 20 and 100 Hz) without the presence of significant noise above 100 Hz produced more stress than noise at the same level below 100 Hz but which also included frequencies up to 1000 Hz. Thus the noise with the higher A-weighted level was found to be less annoying, supporting the suggestion that the A-weighted level (see Section 2.2.11) is an inappropriate measure of the disturbance caused by low-frequency noise. However, in light of the effect that mid- and high-frequency noise levels have on the detectability of low-frequency noise, the practice of basing allowable turbine noise levels on existing ambient A-weighted noise levels seems to be at least partially justified, even though the use of A-weighted noise levels to define acceptable turbine noise levels is not. However, there are two problems, as follows, associated with the practice of basing regulations on existing A-weighted background noise levels as a function of wind speed at hub height.

1. It is assumed that the wind rustling leaves in trees will mask the low-frequency thumping noise from the wind farm, which is clearly not the case.
2. It is assumed that the noise level at a residence is correlated with the wind speed at hub height. This is also not the case, especially at night in conditions of high wind shear, as there can be times when the turbines are operational and the wind speed at the residence is very small or zero.

A particularly insidious effect of prolonged exposure to inaudible, periodic low-frequency noise and infrasound at levels of up to 10 dB below the normal hearing threshold is that the hearing threshold of the exposed person can be reduced by as much as 20–30 dB (Oud, 2013), making an otherwise inaudible sound audible. This means that such a sensitised person may not find relief on leaving the vicinity of the wind farm because there are many other environmental noises that they may now be able to hear that they could not hear previously (Oud, 2013).

Some researchers have suggested that wind turbine noise, particularly infrasound, can be perceived through physiological mechanisms other than hearing. According to Dooley (2013), infrasound can be described as fluctuations or cyclic changes in the local barometric pressure which are comparable to fluctuations in the surrounding barometric pressure experienced by an individual on a ship in high seas. The pressure fluctuations experienced by the individual on the ship occur due to changes in elevation as the ship moves between the crest and trough of ocean waves. Dooley (2013) proposed that this cyclic pressure variation may be the cause of motion sickness on ships as well as nausea in the vicinity of wind farms, as the amplitude of vertical motion in the 2–4 Hz range required to produce seasickness in sensitive individuals corresponds to an atmospheric pressure variation that is similar to the levels of acoustic pressure variation experienced by people living in the vicinity of wind farms. The regular, periodic nature of these variations may explain why similar levels of randomly varying environmental infrasound do not result in motion sickness symptoms in sensitive individuals. Dooley also suggests that the problems of nausea are exacerbated when the sensation that is detected by one of our senses is not reflected by another. For the case of seasickness, the feeling of nausea improves when the sufferer comes onto the deck and can see the horizon, so that the sight sense input matches that of the vestibular system. With wind farm infrasound, there is no corresponding visual reference to satisfy the sight sense. In this case, Bray (2012) suggests that a person's mental construct of comfort and safety can be weakened or even destroyed, leading to conscious distress. Thus it seems that regular pulsing infrasound, even at inaudible levels, is capable of affecting the vestibular system in some people, even though it may not be perceptible by our hearing mechanism.

Nevertheless, there are those who draw no distinction between randomly varying infrasound and periodic amplitude modulated infrasound at the same decibel level (see for example, Leventhall (2015a)). This then leads to the conclusion that infrasound from wind farms is below the threshold of hearing and therefore is not capable of generating adverse health effects in anyone, leaving us to ponder on the cause of the ill effects reported by Ambrose et al. (2012) and Swinbanks (2015a).

7.4 Reproduction of Wind Farm Noise for Adverse Effects Studies

Although it is relatively straight forward to reproduce sound above 30 Hz using a good hi-fi system, it is more difficult to reproduce wind farm infrasound and low-frequency noise in a laboratory for testing its effects on people. Researchers testing hearing thresholds or wind farm noise sensitivities have either used specially designed headphones or airtight enclosures that have been pressurised by several large loudspeakers driven by a DC amplifier.

An early headset system for testing hearing thresholds was used by Yeowart et al. (1967). They used a conventional ear muff, where each cup was connected to a 30-cm diameter loudspeaker via a 2.5-cm diameter port. The face of the loudspeaker was sealed with a 6-mm thick aluminium plate, with a 2.5-cm opening for connection to the ear muff cup. The headset cups were sealed to the head of the test subject by using a 3-cm thick surround of soft rubber. The sound pressure level in the ear cup was measured using a microphone that was mounted in a hole in the cup. More recently, Tonin et al. (2016) used specially designed headphones connected via tubing to a sound source driven by a DC amplifier, to test the sensitivity of subjects to wind farm noise. He found that for short exposure times, simulated wind farm infrasound had no statistically significant effect on symptoms reported by test subjects, but that prior concern about infrasound did have a significant effect on their symptoms.

Infrasonic test chambers were developed in the 1970s and used for testing threshold levels as well as effects of higher levels of infrasound on people (Yeowart, 1976). These chambers were quite small; just large enough to fit a seated person. One design used loudspeakers in the walls of the chamber (Yeowart and Evans, 1974) and another design used a heavy steel pipe to connect the loudspeaker enclosure to the chamber (Whittle et al., 1972).

Walker and Celano (2015) showed results for infrasound levels achieved in an ordinary residential room using loudspeakers in an airtight enclosure and driven by a DC amplifier. They managed to reproduce wind farm noise in the range 0.8–32 Hz using three 18-inch sub-woofer speakers, type JBL2245H, mounted in three faces of an airtight cubic enclosure with a volume of 0.44 m^3, with the three speakers producing a maximum volume displacement of 0.0036 m^3. Their enclosure is made from two layers of 19-mm thick MDF board, laminated with construction adhesive. The six sides are bolted together through a framework of 40 × 40 mm cleats so that they can be disassembled for transport. The three speakers are connected in series and driven with a Crown DC300 power amplifier. The free-air resonance of the speakers is approximately 23 Hz, which changes to approximately 55 Hz when the speakers are mounted on three of the faces of the speaker enclosure. Thus the speaker mechanical impedance is stiffness controlled below about 50 Hz, and the speaker displacement should then be frequency independent below 50 Hz. The sound level that can be generated in a room by the speakers mounted in their enclosure is dependent on the room size and the extent to which the room can be made airtight. If the speaker enclosure is placed in a room of dimensions 4 × 5 × 3 m and corresponding volume 60 m^3, the lowest room resonance frequency would be 37 Hz. For a room of this size, the maximum sound pressure level that could be produced if the room were airtight and the walls were infinitely stiff would be 106 dB, corresponding to a speaker displacement of 0.0036 m^3, and the frequency response should be flat below about 32 Hz. In practice it is difficult to control tiny leaks in the room as well as leaks through the speaker enclosure. For the test room used by Walker and Celano (2015), this caused the frequency response to begin rolling off at 4 Hz until it was 15 dB down at 0.8 Hz, and the maximum sound pressure level produced was 97 dB. Some of the issues that need to be addressed when installing such a system include:

- maximising output at the lowest frequencies
- sufficiently attenuating magnetostrictive noises from the amplifiers (by placing the amplifiers in a heavy, air-tight enclosure or by placing them in a separate room or both)
- sufficiently attenuating amplifier power supply hum so it is inaudible (see item above)
- removing protective covers on speakers to eliminate buzzing noise
- eliminating air leaks in the speaker enclosure and investigating tiny air leaks in the test room to reduce any air jet noise
- minimising the risk of the speakers overheating, which can result if there is excessive DC present or high amplitudes at very low-frequencies; it is usually a good idea to include a DC offset adjustment circuit in the signal path prior to the amplifier, to minimise the risk.

When using either headphones or an airtight enclosure, it is necessary to equalise the wind farm noise signal with the inverse of the frequency response of the sound system, so that the wind farm noise is accurately reproduced. This can be done by using importing the originally recorded wave file into MATLAB® and then convolving it (see Section 2.4.15) with the inverse of the sound-system frequency response.

There are two other systems that are commercially available and can be used for reproducing wind farm noise in a test room at frequencies ranging from just below 30 Hz to less than 1 Hz. One of these (the TR-17, manufactured by Eminent Technologies) is based on a rotating fan with variable-pitch blades. Varying the pitch of the blades while they are rotating produces sound that is directly related to the signal used to vary the pitch. It can produce relatively high levels of infrasound (115 dB between 1 and 20 Hz) with 3% harmonic distortion. The main part of the energy generating the infrasound comes from the electric motor that drives the fan. The energy used to vary the pitch of the blades is much less, so the 150W amplifier used in the system is required to produce much less power than one might expect. However, the system has to be installed by the manufacturer in a side room adjacent to the room in which the sound is to be generated.

A prototype system (RIS-10) that has been developed by Kevin Allan Dooley Inc (Dooley and Morris, 2015) is based on an electromagnetically controlled reversible axial compressor. It covers a frequency range from 0.2 to 40 Hz with an output of approximately 100 dB. The flow magnitude and direction (into the control volume or out of the control volume) as well as the frequency of variation are determined by the analogue input signal. A feedback pressure sensor included in the control loop can be used to minimise harmonic distortion.

Finally there is a traditional sub-woofer speaker called the Torus manufactured by Wilson-Benesch. Although this can produce 100 dB at 1 m, its low-end frequency is limited to 10 Hz (with a high end of 150 Hz). Perhaps the response can be extended below 10 Hz with some equalisation, a properly designed speaker enclosure and the use of a DC amplifier.

7.5 Vibration Effects

A number of residents living near wind farms complain that they can 'feel' vibrations caused by wind turbines when they lie down to sleep. Generally they feel the vibrations through their pillow. Whether or not the vibration levels generated by wind farms are sufficiently large to be able to be detected by residents has not yet been rigorously determined, but the general feeling among practitioners is that the 'feeling' of vibration is most likely linked to the presence of low-frequency sound and perhaps infrasound, although the latter is still the subject of considerable debate. Wind turbine vibration can be generated in multiple ways, as discussed in Section 5.14. In all cases the vibration is transmitted to the support tower, then into the foundation and finally into the ground, where the bulk of the energy is transmitted to the receiver as Rayleigh waves. This type of wave is the wave that causes most damage during an earthquake and involves elliptical motion (which includes both sideways and up-and-down motion) of the ground particles as the wave is transmitted. The motion is greatest at the surface and decreases with increasing depth with the sideways motion decreasing faster than the up-and-down motion. The Rayleigh wave speed is both frequency dependent and ground-type dependent and typical values may range from 120 to 200 m/s. The vibration 'felt' in a residence could be a result of vibration transmitted through the ground or it could be low-frequency sound and infrasound exciting the building structure.

More likely, the 'vibration' that some people experience when their head is resting on a pillow could be related to the fact that infrasound is more apparent to the receiver when the higher frequency noise is at a low level. When a person rests one ear on a pillow, higher frequency sound is attenuated by the pillow but low-frequency sound and infrasound are not, resulting in the low-frequency sound and infrasound being more noticeable in the ear against the pillow. This could be easily interpreted as pillow vibration.

Based on our own measurements and those of others, it is unlikely that the vibration transmitted through the ground from the wind farm will be sufficient to generate a level in a pillow that is sufficient to be detected by a human. The same can be said of airborne infrasound. It is unlikely that it will cause the pillow to vibrate, but it is likely that the infrasound would appear to be coming from the pillow due to the higher level of sensation felt in the ear that is against the pillow for the reason mentioned above.

7.6 Nocebo Effect

There are a few pro-wind-farm advocates who claim, via various web sites, that any adverse health effects due to wind farms are imagined, that wind farms do not make any audible noise and that complaints only began after anti-wind-farm activists began their campaigns to stop the construction of future wind farms. They attribute reported adverse health effects to mass hysteria and use pseudo-scientific terms such as 'mass psychogenic illness' or 'psychogenic nocebo phenomenon'. Others (for example, Knopper et al. (2013)) agree that wind farms produce sufficient noise to annoy some people, but suggest that the only adverse effect of wind farm noise is annoyance, which can manifest as physical symptoms in some people. Whether this is the case or whether low-frequency noise and infrasound generated by wind farms have a direct adverse physiological effect, as suggested by Salt and Lichtenhan (2014), is still unproven. However, all of the literature claiming to prove that all adverse health effects are a result of a nocebo effect is based on suspect data, such as records of complaints produced by the wind farm industry, with no acknowledgement that over the past few years, wind turbines and the number of them in a wind farm have become much greater and that it takes time for people to realise what is causing their symptoms. Literature in favour of a nocebo-effect explanation universally fails to acknowledge work published by researchers in the USA, New Zealand and Canada (Krogh et al., 2011; Møller and Pedersen, 2011; Shepherd et al., 2011), who found that many wind farm neighbours who were in favour of wind farm operations prior to their construction, changed their views after the turbines began operating and adverse health effects became noticeable.

In support of the argument denying the nocebo effect, there are at least two instances of acoustic engineering professionals succumbing to illness during and soon after undertaking measurements in homes of people complaining of wind farm noise (Ambrose et al., 2012; Swinbanks, 2015a). These symptoms came on entirely unexpectedly and the professionals involved did not have any prior expectations that they would suffer any effects. It is interesting to note that in the Swinbanks case, periodic infrasound was measured, with the highest peak level of 55 dB in the narrowband frequency spectrum and the cumulative effects of all the harmonics raising the peak level in the time domain to between 69 and 72 dB, indicating that spectral analysis can considerably underestimate the true peak of a noise. However, even this level is below what would normally be detected by a human subject, indicating that infrasound levels below the normal detection threshold may have ill effects on some people.

There are also a few reports of laboratory studies in which test subjects were exposed to wind farm noise for short periods of time, usually minutes (for example, Tonin (2015)), although at least one tested subjects for 3 h. Short-duration tests do not measure the effects that long-term chronic exposure has on some people. Much more extensive testing has to be done to determine a dose–response relationship that reflects real-world wind farm noise exposure and this will take a long time, even if funding becomes available to undertake the work. Before this work is complete we must make do with case studies and self-reporting incidents. These latter have now become so numerous that the construction of future wind farms should only proceed with care, ensuring that resulting low-frequency noise levels and infrasound at residences is below the thresholds discussed in Section 1.6. There should also be a commitment by a wind farm developer to pay appropriate compensation to, or purchase residential property from, anyone suffering adverse effects within the first six months of turbines beginning to operate.

7.7 Summary and Conclusion

There is a considerable amount of anecdotal evidence and published research that suggests that wind farm noise does seriously and adversely affect a small but significant proportion of the population. Nevertheless, many acoustic, social science and medical professionals ignore this evidence by defining 'scientific evidence' in a very narrow way and then stating that there is no 'scientific evidence' available that proves beyond doubt that wind farm noise can harm anyone. As pointed out by May and McMurtry (2015), there seems to be a tendency in the medical research discipline to interpret the lack of rigorous evidence of a causation link as proof that the link does not exist. There have been many hundreds of case studies that demonstrate that the health of a significant percentage of people has been affected by wind turbines. Whether or not these adverse effects are directly caused by wind turbine noise or are indirectly caused by such things as sleep disturbance as a result of a nearby wind farm is a moot point, as either case results in the same conclusion; that is, wind turbines result in some people living in their vicinity suffering adverse health effects.

It seems that the only type of scientific evidence that would be generally acceptable to those who do not believe that there are any adverse health effects resulting from wind farms, would be a longitudinal epidemiological study of thousands of people with all confounding factors removed, which is clearly an impossible task for all of the reasons mentioned earlier in this chapter. The situation is very similar to the debate on anthropogenic global warming. This too has not been proven beyond any reasonable doubt, but there is plenty of evidence that global warming is occurring and that it is a result of the activities of man. However, there are always those who for whatever reason, will deny the patently obvious and continue to do so regardless of any evidence with which they are confronted.

It is time to stop debating whether or not a problem exists. It is well known that wind farm noise does result in sleep disturbance and adverse health effects for some people and the time has come to decide what to do about it. However, further research is needed to determine whether the problem is inaudible infrasound or low-level, audible, low-frequency noise.

Although a number of researchers and acoustical consultants have stated that infrasound produced by wind farms is not a problem, it may be too early to completely discount the possibility that infrasound is at least part of the problem. If we assume a normal distribution of hearing sensitivity, 2.5% of the population would have hearing sensitivities up to 12 dB lower than indicated in Figure 7.7, indicating that infrasound generated by wind turbines would be detectable by them at considerable distances from the nearest turbine, especially considering that the level of infrasound does not decay with distance as rapidly as higher frequency sound. However, no-one has performed controlled laboratory and field tests on a sufficient number of subjects to demonstrate whether such low levels of infrasound are detectable. Nevertheless, there is an opinion shared by some researchers that some residents near wind farms become sensitised to wind turbine infrasound, so that they can detect much lower levels than are detectable by normal people. However, this idea has not been demonstrated in laboratory tests or controlled field tests either.

If the problem is low-frequency noise and not infrasound, then there is likely that there is no direct physiological effect but rather that adverse health effects are generated by stress as a result of the audible noise being annoying or depriving sensitive people of sleep. This low-frequency noise from wind farms is particularly prominent in rural residences where the ambient noise level is low, especially at night. The idea of low-frequency noise rather than infrasound being the cause of complaints of excessive wind farm noise was expressed by Bowdler (2015) in his summary of the 6th International Conference on Wind Turbine Noise, in Glasgow in April 2015, and by Leventhall (2015b) in his submission to the Australian Senate inquiry. Thus it is worthy of serious consideration.

Affected people find wind farm noise to be very annoying and as time goes on they become more and more sensitised to it, so they become more stressed and also suffer from insomnia and/or sleep disruption. Whether the problem is infrasound or low-frequency noise is not the most important issue. The fact remains that some people are so affected by wind farm noise that their health suffers and some are forced to leave their homes in order to achieve an acceptable quality of life. Thus it is important to begin work on new noise guidelines for new wind farm developments, based on maximum acceptable outdoor levels of A-weighted noise (see Section 2.2.11) as well as acceptable levels of low-frequency noise. Existing allowable levels (even the lowest of 35 dBA) are too high for a significant proportion of the population who live in quiet rural areas; the levels of 40 or 45 dBA allowed in some jurisdictions are well over what would be considered acceptable by most people. More research is needed on what can be done to ameliorate the noise problems caused by existing wind farms. In addition, all jurisdictions should determine what would be a fair offer of compensation for those who have suffered ill effects, which would at a minimum include the offer to purchase their residence at the price it would have realised prior to construction of the wind farm.

References

Ambrose S., Rand R and Krogh, C 2012 Wind turbine acoustic investigation: infrasound and low-frequency noise – a case study. *Bulletin of Science, Technology and Society*, **32**(2), 128–141.

Arra I, Lynn H, Barker K, Ogbuneke Cand Regalado S 2014 Systematic review 2013: association between wind turbines and human distress. *Cureus*, **6**(5), e183.

Bakker R, Pedersen E, van den Berg G, Stewart R, Lok W and Bouma J 2012 Impact of wind turbine sound on annoyance, self-reported sleep disturbance and psychological distress. *The Science of the Total Environment*, **425**, 42–51.

Bell A 2011 How do middle ear muscles protect the cochlea? Reconsideration of the intralabyrinthine pressure theory. *Journal of Hearing Science*, **1**, 9–23.

Bell A 2014 Annoyance from wind turbines: role of the middle ear muscles. *Acoustics Australia*, **42**(1), 57.

Bengtsson J, Persson-Waye Kand Kjellberg A 2004 Evaluations of effects due to low-frequency noise in a low demanding work situation. *Journal of Sound and Vibration*, **278**(1), 83–99.

Berger R, Ashtiani P, Ollson C, Aslund M, McCallum L, Leventhall H and Knopper L 2015 Health-based audible noise guidelines account for infrasound and low frequency noise produced by wind turbines. *6th International Meeting on Wind Turbine Noise*, Glasgow, UK.

Berglund B, Hassmén P and Job R 1996 Sources and effects of low-frequency noise. *Journal of the Acoustical Society of America*, **99**(5), 2985–3002.

Berglund B, Lindvall T and Schwela D 1999 *Guidelines for community noise*. WHO.

Blazier Jr W 1997 RC Mark II,: A refined procedure for rating the noise of heating, ventilating, and air-conditioning (HVAC) systems in buildings. *Noise Control Engineering Journal*, **45**(6), 243–150.

Bockstael A, Dekoninck L, Can A, Oldoni D, De Coensel B and Botteldooren D 2012 Reduction of wind turbine noise annoyance: an operational approach. *Acta Acustica United with Acustica*, **98**(3), 392–401.

Bowdler R 2015 Post conference report. *6th International Meeting on Wind Turbine Noise*, Glasgow, UK.

Bray W 2012 Relevance and applicability of the soundscape concept to physiological or behavioral effects caused by noise at very low frequencies which may not be audible. *Journal of the Acoustical Society of America*, **132**(3), 1925.

Broner N 1978 The effects of low frequency noise on people – a review. *Journal of Sound and Vibration*, **58**(4), 483–500.

Bronzaft A 2011 The noise from wind turbines: potential adverse impacts on children's well-being. *Bulletin of Science, Technology & Society*, **31**(4), 291–295.

Cooper S 2015 The results of an acoustic testing program – Cape Bridgewater wind farm. Technical report, The Acoustic Group.

Doolan C 2013 A review of wind turbine noise perception, annoyance and low frequency emission. *Wind Engineering*, **37**(1), 97–104.

Doolan C and Moreau D 2013 An on-demand simultaneous annoyance and indoor noise recording technique. *Acoustics Australia*, **41**(2), 144–148.

Dooley K 2013 Significant infrasound levels a previously unrecognized contaminant in landmark motion sickness studies. *Journal of the Acoustical Society of America*, **134**(5), 4097–4097.

Dooley K and Morris E 2015 Systems and methods for control of infrasound pressures. Patent number: CA 2887832. Canadian Intellectual Property Office.

Farboud A, Crunkhorn R and Trinidade A 2013 'Wind turbine syndrome': fact or fiction? *Journal of Laryngology & Otology*, **127**(03), 222–226.

Finlaysons 2014 Neighbour deed. Technical report, Trust Power Holdings.

Hanning C and Nissenbaum M 2011 Selection of outcome measures in assessing sleep disturbance from wind turbine noise. *Fourth International Meeting on Wind Turbine Noise*, Rome, Italy.

Hansen K, Zajamšek B and Hansen C 2014 Comparison of the noise levels measured in the vicinity of a wind farm for shutdown and operational conditions. *Internoise2014*, Melbourne, Australia.

Hubbard H and Shepherd K 1991 Aeroacoustics of large wind turbines. *Journal of the Acoustical Society of America*, **89**(6), 2495–2508.

Hurtley C 2009 *Night noise guidelines for Europe*. WHO Regional Office Europe.

Huson W 2015 Stationary wind turbine infrasound emissions and propagation loss measurements. *6th International Meeting on Wind Turbine Noise*, Glasgow, UK.

HWL Ebsworth Lawyers 2013 Participation agreement. Technical report, Moorabool Wind Farm Pty Ltd.

Ising H and Ising M 2002 Chronic cortisol increases in the first half of the night caused by road traffic noise. *Noise and Health*, **4**(16), 13.

ISO 226 2003 Acoustics: Normal equal-loudness-level contours. International Standards Organisation.

Jabben J, Verheijen E and Schreurs E 2009 Impact of wind turbine noise in the Netherlands. *Third International Conference on Wind Turbine Noise*, Aalborg, Denmark.

James R 2012 Wind turbine infra and low-frequency sound. warning signs that were not heard. *Bulletin of Science, Technology & Society*, **32**(2), 108–127.

Janssen S, Vos H and Eisses A 2008 Hinder door geluid van windturbines. Technical report number 2008-D-R1051/B, TNO (in Dutch).

Jeffery R, Krog C and Horner B 2014 Industrial wind turbines and adverse health effects. *Canadian Journal of Rural Medicine*, **19**(1), 21–26.

Karpova N, Alekseev S, Erokhin V, Kadyskina E and Reutov O 1970 Early response of the organism to low-frequency acoustic oscillations. *Noise and Vibration Bulletin*, **11**(65), 100–103.

Kelley N 1987 A proposed metric for assessing the potential of community annoyance from wind turbine low-frequency noise emissions. Technical report, Solar Energy Research Institute.

Knopper L, Whitfield Aslund M, McCallum L and Ollson C 2013 Research into aerodynamic modulation of wind turbine noise. *5th International Meeting on Wind Turbine Noise*, Denver, Colorado.

Krahé D 2010 Low frequency noise - strain on the brain. *14th International Meeting on Low Frequency Noise and Vibration and its Control*, Aalborg, Denmark.

Krogh C, Gillis L, Kouwen Nand Aramini J 2011 WindVOiCe, a self-reporting survey: adverse health effects, industrial wind turbines, and the need for vigilance monitoring. *Bulletin of Science, Technology & Society*, **31**(4), 334–345.

Krogh C, Horner B, Morris J, May M, Papadopoulos G and Watts A 2013 Trading off human health: Wind turbine noise and government. *5th International Meeting on Wind Turbine Noise*, Denver, Colorado.

Kuehler R, Fedtke T and Hensel J 2015 Infrasonic and low-frequency insert earphone hearing threshold. *Journal of the Acoustical Society of America*, **137**(4), EL347–EL353.

Landström U, Lundström R and Byström M 1983 Exposure to infrasound-perception and changes in wakefulness. *Journal of Low Frequency Noise and Vibration*, **2**(1), 1–11.

Lee S, Kim K, Choi W and Lee S 2011 Annoyance caused by amplitude modulation of wind turbine noise. *Noise Control Engineering Journal*, **59**(1), 38–46.

Leventhall G 2003 A review of published research on low frequency noise and its effects. Technical report, DEFRA.

Leventhall G 2013 Infrasound and the ear. *5th International Meeting on Wind Turbine Noise*, Denver, Colorado.

Leventhall G 2015a On the overlap region between wind turbine infrasound and infrasound from other sources and its relation to criteria. *6th International Meeting on Wind Turbine Noise*, Glasgow, UK.

Leventhall G 2015b Submission to the select committee on wind turbines: Application of regulatory governance and economic impact of wind turbines. Technical report, Australian Senate.

Magari S, Smith C, Schiff M and Rohr A 2014 Evaluation of community response to wind turbine-related noise in western new york state. *Noise and Health*, **16**(71), 228.

May M and McMurtry R 2015 Wind turbines and adverse health effects: a second opinion. *Journal of Occupational and Environmental Medicine*, **57**(10), e130–e132.

McBride D, Shepherd D and Thorne R 2014 Investigating the impacts of wind turbine noise on quality of life in the australian context: a case study approach. *Inter-Noise and Noise-Con Congress and Conference Proceedings*, Melbourne, Australia, pp. 4559–4562.

McMurtry R 2011 Toward a case definition of adverse health effects in the environs of industrial wind turbines: facilitating a clinical diagnosis. *Bulletin of Science, Technology & Society*, **31**(4), 316–320.

Michaud D, Feder K, Keith S, Voicescue S, Marro L, Than J, Guay M, Denning A, Bower T, Villeneuve P, Russell E, Koren G and van den Berg F 2016 Self-reported and measured stress related responses associated with exposure to wind turbine noise. *Journal of the Acoustical Society of America*, **139**(3), 1467–1479.

Miedema H and Oudshoorn C 2001 Annoyance from transportation noise: relationships with exposure metrics DNL and DENL and their confidence intervals. *Environmental Health Perspectives*, **109**(4), 409.

Møller H 1984 Physiological and psychological effects of infrasound on humans. *Journal of Low Frequency Noise and Vibration*, **3**(1), 1–16.

Møller H and Pedersen C 2004 Hearing at low and infrasonic frequencies. *Noise and Health*, **6**(23), 37.

Møller H and Pedersen CS 2011 Low-frequency noise from large wind turbines. *Journal of the Acoustical Society of America*, **129**(6), 3727–3744.

Morris M 2012 Waterloo wind farm survey, April 2012. Technical report (published by the author).

Nissenbaum M, Aramini Jand Hanning C 2012 Effects of industrial wind turbine noise on sleep and health. *Noise and Health*, **14**(60), 237–243.

Nussbaum D and Reinis S 1985 Some individual differences in human response to infrasound. Technical report No. 282, UTIAS, University of Toronto.

Onakpoya I, O'Sullivan J, Thompson M and Heneghan C 2015 The effect of wind turbine noise on sleep and quality of life: a systematic review and meta-analysis of observational studies. *Environment International*, **82**, 1–9.

Oud M 2013 Explanation for suffering from low-frequency sound. *Geluid (English translation)*, March 2013. URL: http://home.kpn.nl/oud/publications/2013_03_OudM_journalGeluid.pdf.

Pedersen E, Van den Berg F, Bakker R and Bouma J 2009 Response to noise from modern wind farms in the Netherlands. *Journal of the Acoustical Society of America*, **126**(2), 634–643.

Pedersen E, Van den Berg F, Bakker R and Bouma J 2010 Can road traffic mask sound from wind turbines? Response to wind turbine sound at different levels of road traffic sound. *Energy Policy*, **38**(5), 2520–2527.

Pedersen S, Møller H and Persson-Waye K 2007 Indoor measurements of noise at low frequencies-problems and solutions. *Journal of Low Frequency Noise, Vibration and Active Control*, **26**(4), 249–270.

Persson-Waye K 2004 Effects of low frequency noise on sleep. *Noise and Health*, **6**(23), 87–91.

Persson-Waye K, Rylander R, Benton S and Leventhall H 1997 Effects on performance and work quality due to low frequency ventilation noise. *Journal of Sound and Vibration*, **205**(4), 467–474.

Persson-Waye K, Clow A, Edwards S, Hucklebridge F and Rylander R 2003 Effects of nighttime low frequency noise on the cortisol response to awakening and subjective sleep quality. *Life Sciences*, **72**(8), 863–875.

Phillips C 2011 Properly interpreting the epidemiologic evidence about the health effects of industrial wind turbines on nearby residents. *Bulletin of Science, Technology & Society*, **31**, 303–315.

Pierpont N 2009 *Wind Turbine Syndrome. A Report on a Natural Experiment*. K-Selected Books.

Renewable UK 2013 Wind turbine amplitude modulation: research to improve understanding as to its cause & effect. Technical report, Renewable UK.

SAEPA 2013 Waterloo wind farm - environmental noise study. Technical report, South Australian Environmental Protection Authority.

Salt A and Hullar T 2010 Responses of the ear to low frequency sounds, infrasound and wind turbines. *Hearing Research*, **268**(1), 12–21.

Salt A and Lichtenhan J 2012 Perception-based protection from low-frequency sounds may not be enough. *InterNoise 2012*, New York, pp. 3999–4010.

Salt A and Lichtenhan J 2014 How does wind turbine noise affect people? *AcousticsToday*, **10**, 20–28.

Schmidt J and Klokker M 2014 Health effects related to wind turbine noise exposure: a systematic review. *PloS One*, **9**(12), e114183.

Schneider P 2012 Cullerin Range wind farm survey. Technical report (published by the author).

Schneider P 2013 Cullerin Range wind farm survey: Follow-up survey. Technical report (published by the author).

Schomer P 2013 Comments on recently published article, 'Concerns about infrasound from wind turbines'. *Acoustics Today*, **9**(4), 7–9.

Shain M 2011 Public health ethics, legitimacy, and the challenges of industrial wind turbines: the case of Ontario, Canada. *Bulletin of Science, Technology & Society*, **31**(4), 346–353.

Shepherd D, McBride D, Welch D, Dirks K and Hill E 2011 Evaluating the impact of wind turbine noise on health-related quality of life. *Noise and Health*, **13**(54), 333.

Søndergaard B, Hoffmeyer Dand Plovsing B 2007 Low frequency noise from large wind turbines. *Second International Meeting on Wind Turbine Noise*, Lyon, France.

Swinbanks M 2015a Direct experience of low frequency noise and infrasound within a windfarm community. *Sixth International Meeting on Wind Turbine Noise*, Glasgow, UK.

Swinbanks M 2015b Supplementary submission to the Senate Wind Turbine Select Committee. Technical report (published by the author).

Tempest W 1973 Loudness and annoyance due to low frequency sound. *Acta Acustica United with Acustica*, **29**(4), 205–209.

Thorne B 2012 The perception and effect of wind farm noise at two Victorian wind farms. Technical report, Noise Measurement Services.

Thorne B 2013 Wind farm noise and human perception: a review. Technical report, Noise Measurement Services.

Tonin R 2015 Response to simulated wind farm infrasound including effect of expectation. *Sixth International Meeting on Wind Turbine Noise*, Glasgow, UK.

Tonin R, Brett J and Colagiuri B 2016 The effect of infrasound and negative expectations to adverse pathological symptoms from wind farms. *Journal of Low Frequency Noise, Vibration and Active Control*, **35**(1), 77–90.

Turnbull C, Turner J and Walsh D 2012 Measurement and level of infrasound from wind farms and other sources. *Acoustics Australia*, **40**(1), 45–50.

van den Berg G 2004 Effects of the wind profile at night on wind turbine sound. *Journal of Sound and Vibration*, **277**(4), 955–970.

van den Berg G 2005 The beat is getting stronger: the effect of atmospheric stability on low frequency modulated sound of wind turbines. *Journal of Low Frequency Noise, Vibration and Active Control*, **24**(1), 1–24.

Verzini A, Frassoni C and Skarp A 1999 A field research about effects of low frequency noises on man. *Acta Acustica*, **85**, S16.

von Gierke, H.E. and Nixon, C. 1976 Effects of intense infrasound on man. In: W. Tempest (ed.), *Infrasound and Low Frequency Vibration*. Academic Press, New York.

von Hünerbein S and Piper B 2015 Affective response to amplitude modulated wind turbine noise. *6th International Meeting on Wind Turbine Noise*, Glasgow, UK.

Walker B and Celano J 2015 Progress report on synthesis of wind turbine noise and infrasound. *6th International Meeting on Wind Turbine Noise*, Glasgow, UK.

Watanabe T and Møller H 1990 Hearing thresholds and equal loudness contours in free field at frequencies below 1 kHz. *Journal of Low Frequency Noise Vibration and Active Control*, **9**(4), 135–148.

Whittle L, Collins S and Robinson D 1972 The audibility of low frequency sounds. *Journal of Sound and Vibration*, **21**, 431–448.

Yeowart N 1976 Thresholds of hearing and loudness for very low frequencies. In: W. Tempest (ed.) *Infrasound and Low Frequency Vibration*. Academic Press.

Yeowart N and Evans M 1974 Thresholds of audibility for very low-frequency pure tones. *Journal of the Acoustical Society of America*, **55**(4), 814–818.

Yeowart N, Bryan M and Tempest W 1967 The monaural m.a.p. threshold of hearing at frequencies from 1.5 to 100 c/s. *Journal of Sound and Vibration*, **6**, 335–342.

Yokoyama S, Kobayashi T, Sakamoto S and Tachibana H 2015 Subjective experiments on the auditory impression of the amplitude modulation sound contained in wind turbine noise. *6th International Meeting on Wind Turbine Noise*, Glasgow, UK.

Yoon K, Gwak D, Lee S and Lee S 2015 Experimental study of relationship between amplitude modulation and detection threshold of wind turbine noise. *6th International Meeting on Wind Turbine Noise*, Glasgow, UK.

Zajamšek B, Hansen K and Hansen C 2014 Investigation of the time dependent nature of infrasound measured near a wind farm. *Internoise2014*, Melbourne, Australia.

8

Wind Farm Noise Control

8.1 Introduction

There are many existing wind farms that result in annoyance and health problems for some residents living within 5 km or so of the nearest turbine. It is also likely that many future wind farms will result in problems for some residents in their near vicinity. It seems that the main problems are a feeling of nausea and sleep deprivation, which in turn can lead to other medical problems. Given that there are a significant number of people who are affected by wind farm noise and there seems to be a push by many governments and green groups to continue with new wind farm developments in the vicinity of rural communities, it is important that considerable research effort is devoted to finding ways to ameliorate the problems experienced by residents. It is now almost universally accepted that noise problems that affect residents are in the low-frequency range and possibly in the infrasonic range. Annoyance is also exacerbated by the periodically time-varying nature of the noise (known as 'amplitude modulation' (AM)). The main reasons for this are:

- Low-frequency noise is inherently more annoying than a balanced noise spectrum.
- Low-frequency noise is less attenuated by propagation through the atmosphere and by reflection from the ground and as a result it dominates the turbine noise spectrum at distances from the nearest turbine that exceed 1.5–2 km.
- Low-frequency noise is less attenuated by transmission through roofs, walls and windows, which results in the wind farm noise spectrum inside residences being heavily dominated by low-frequency noise and infrasound.
- Amplitude modulated noise is more annoying than relatively steady noise or randomly-varying noise.

Control of noise in the infrasonic and low-frequency ranges can be attempted by modifying the turbine design or by controlling the noise level in the affected residences. Both of these approaches are discussed in the following sections. Noise in the mid-frequency range (160 to 1000 Hz) can also be a problem for residents living within 1 km of a wind farm. Further, it is possible that the AM of these higher-frequency components may be responsible for some of the symptoms experienced by close-by residents and perceived as low-frequency noise.

Wind Farm Noise: Measurement, Assessment and Control, First Edition.
Colin H. Hansen, Con J. Doolan, Kristy L. Hansen.

8.2 Noise Control by Turbine Design Modification

The importance of a wind turbine's geometry gives some insight into the need to control aerodynamic noise on a wind turbine. The torque T produced by a turbine with rotor diameter D and rotor tip speed V_{tip} can be estimated from the power P extracted from the wind:

$$T = \frac{P \times D/2}{V_{\mathrm{tip}}} \tag{8.1}$$

There is increasing pressure to produce turbines with ever higher power ratings and size (Ceyhan 2012), which increases the torque on the gearbox, as indicated by Eq. (8.1). Increased torque means a more expensive and heavier gearbox, thus increasing the overall cost of the turbine. The increase in torque may be moderated by allowing the tip speed to increase but, as we have seen in Chapter 3, noise level is controlled by tip speed and in many cases the sound power associated with the blade trailing edge is proportional to V_{tip}^5. Thus we can immediately see that any strategy that can reduce aerodynamic noise from the rotor will not only help to protect residents living nearby wind farms, but will contribute directly to the design of larger turbines by reducing the torque on the gearbox and by extension, the overall cost.

Wind turbines generate noise in a number of ways:

- *Aerodynamic noise generated by the turbine blades.* This can be produced by a number of mechanisms:
  - trailing edge noise due to air flow over the blades
  - turbine blades passing through disturbed air flow caused by the presence of the tower, even though it is downwind from the blades – in the short time intervals between the blades passing the tower, the low-frequency noise generation will be at a lower amplitude, thus resulting in AM of the sound
  - stalling of the blades as their angle of attack changes with variations in wind speed as a function of altitude, the noise again being amplitude modulated as the blades pass through the varying-speed air flow
  - turbulent inflow to the blades caused by terrain effects or by the wake of upstream turbines.
- *Aeroacoustic excitation of the blades* as a result of the air flow over them and as a result of passing through the disturbed flow in front of the support tower. This in turn results in the blades vibrating and radiating noise.
- *Aeroacoustic excitation of the tower* as a result of wind blowing past it.
- *Mechanical excitation of the tower and blades* (and subsequent noise radiation) as a result of equipment mounted in the nacelle (gearbox and cooling fans).

Possible methods to control some of these noise sources are discussed in the following sections and sub-sections. They include:

- optimisation of the turbine blade design
- treatment of the blade trailing edge
- continuous blade pitch control using random fluctuations in pitch angle (de-synchronisation) (van den Berg, F 2013); (van den Berg, G 2005)
- non-random blade pitch control to avoid stall and to compensate for the change in air flow as the blade passes the tower (van den Berg, G 2005)

- blade phase control relative to other turbines
- control of gearbox noise by using tuned mass damper systems.

8.2.1 Optimisation of Blade Design

In order to produce turbines with quieter blade designs, it is first necessary to understand how the noise is generated. Brooks et al. (1989) shows that the mean-square acoustic pressure from trailing-edge noise $\langle p^2 \rangle$ (the major aerodynamic noise source on a turbine) is proportional to a variety of flow and geometrical parameters

$$\langle p^2 \rangle \propto \rho_0^2 \langle v^2 \rangle \frac{U_c^3}{c_0} \left(\frac{L\delta^*}{r^2} \right) \overline{D} \tag{8.2}$$

where ρ_0 is the air density, $\langle v^2 \rangle$ is the mean-square turbulence velocity, U_c is the convection velocity of the turbulence over the edge, L is the span, δ^* is the displacement thickness at the trailing edge, r is the distance from the trailing edge to a receiver and $\overline{D}$ is the directivity.

Hence Eq. (8.2) informs us that effective noise control may occur if a blade designer can reduce the level of turbulence in the boundary layer ($\langle v^2 \rangle$), the convection speed (U_c) or the length scale of turbulence (δ^*). These parameters can be controlled by optimising the sectional profile of the wind turbine blade.

Sørensen (2001) used a numerical blade design procedure to minimise the trailing edge noise of axial fans. While not a wind turbine, the study is significant as it is an early example of using numerical optimisation to design a quiet rotor. They were able to achieve an optimally quiet rotor design with only a limited reduction in fan efficiency. Marsden et al. (2007) used a computational aeroacoustics approach with derivative-free numerical shape optimisation to design a quiet hydrofoil shape. Again, while not a wind turbine blade, the novel use of numerical simulation and optimisation was able to minimise trailing-edge noise by manipulating the turbulence levels and length scales at the trailing edge. Marsden et al. (2007) were able to reduce the sound power by 9.5 dB, mainly in the lower frequency bands, through the control of vortex shedding.

Schepers et al. (2007) and Lutz et al. (2007) describe the numerical wind turbine blade optimisation process that was developed as part of the European Union's Fifth Framework research project 'SIROCCO'. Here, an efficient trailing-edge noise prediction procedure (Parchen 1998, known as the TNO method) is coupled with an aerodynamic solver (Drela 1989) and a numerical optimisation scheme. While specific details of the effect on the turbulent boundary layer properties are not given for proprietary reasons, the project participants were able to provide significant noise reductions (0.5–3 dB) and improvements in aerodynamic performance. Full-scale field tests of the optimised SIROCCO blades (Oerlemans et al. 2009) show, on average, a 0.5 dB noise reduction.

Another numerical optimisation study using the TNO method (Parchen 1998) was performed by Bertagnolio et al. (2010). Here the TNO method was coupled with an airfoil design code originally used for aerodynamic optimisation but extended to include trailing-edge noise. The numerical study found that noise could be reduced by up to 3.5 dB and this was achieved by a combination of a reduction of turbulent kinetic energy and boundary-layer height.

Lee et al. (2013) used the BPM model of Brooks et al. (1989; see also Appendix B) with a numerical optimisation procedure to design quieter blades for a 10-kW wind turbine.

The numerical optimisation procedure predicted that a 2.3-dB noise reduction would be possible whilst maintaining aerodynamic performance by optimising the shape of the blades. Wind tunnel testing in an anechoic facility achieved a 2.6-dB noise reduction, thus validating the methodology.

In Section 3.3.4, the effects of a trailing-edge bevel are summarised. It was found to reduce noise during the testing of new wind turbine blade-tip designs. Bevelling the trailing edge was shown to be an effective means of noise control, providing 3–5 dB of noise reduction.

8.2.2 Trailing-edge Treatments

While altering the blade profile allows some control over the turbulence level and length scales in the boundary layer, the effect on noise level has been shown to be of the order of 0.5–3 dB. Another, possibly more effective, method to control noise is to change the scattering properties of the trailing edge, thus making the edge less efficient at converting the turbulent flow energy into sound. In fact, many researchers are turning their attention to owl-wings in their search for quiet airfoil designs. Owls must fly silently in order to use bi-aural source localisation to target their prey. Acoustic stealth is a requirement in order to prevent prey from being alerted to their presence, as well as reducing background noise about the owl so as to allow accurate targeting. In fact, aeroacoustic measurements on live birds flying over an array (Sarradj et al. 2011), wing specimens in a wind tunnel (Geyer et al. 2013) and wing specimens in a reverberation room (Gruschka et al. 1971) show that owls are effectively quiet over a frequency range of 1.6–10 kHz. If the secrets of the owl wing can be determined, then they may be applicable to wind turbines for a very effective noise control method.

It is believed that it is the unique construction of an owl's wing that provides its superior acoustic performance. Figure 8.1 illustrates some of the major features of an owl wing's construction. These include serrated and poro-elastic (simultaneously porous and elastic) edges, with a fine fringe material at the leading edge. In addition to the

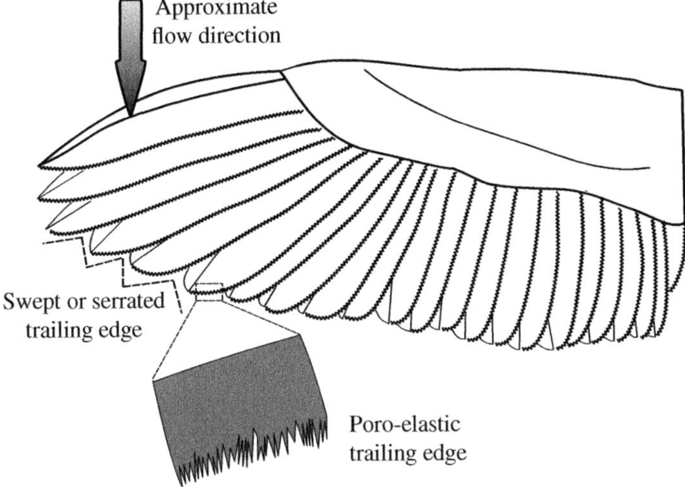

Figure 8.1 Illustration of the construction of an owl wing.

poro-elastic edge, the wing is covered by a soft downy coat (Lilley 1998). Finally, there is also a leading edge fringe. How exactly these components act together to reduce noise is an open question, but there are some indications in the literature to guide us.

The leading-edge fringe consists of a series of barb-like structures (Chen et al. 2012) that appear to act as a type of vortex generator. It has been speculated (Chen et al. 2012; Lilley 1998) that the leading-edge fringe creates a series of small counter-rotating stream-wise vortices over the wing that keep the boundary layer attached and suppress the formation of large turbulent eddies, thus controlling low-frequency noise (such as separation-stall noise).

Serrated trailing edges have long been considered an effective means of trailing-edge noise control (Howe 1991). However, recent experiments using rigid serrations in anechoic wind tunnel facilities (Chong et al. 2013; Gruber et al. 2010; Moreau and Doolan 2013) and on wind turbines in the field (Oerlemans et al. 2009) show only modest noise reductions in some frequency bands and an increase in others. The reason for this is not clear, but it is speculated that the serrations interact with the turbulent boundary layer in a way that was not anticipated when the theory of Howe (1991) was developed. More work is needed to understand the fluid dynamics of the serration/boundary-layer interaction to better understand why serrations do not appear to be as effective as theory suggests.

Poro-elastic trailing edges are also similarly thought to be effective for controlling trailing-edge noise. There are no comprehensive experimental data from poro-elastic edges, but the theoretical work of Jaworski and Peake (2013) provides an explanation. Jaworski and Peake (2013) show that elastic and porous edges are able to reduce the efficiency of the edge in converting turbulent energy into noise. In fact, acoustic power from porous edges were shown to have an M^6 dependence and elastic edges an M^7 dependence. The Mach number M of flight for owls and for rotating turbine blades is always less than 1 and is given by $M = U/c$, where c is the speed of sound and U is the speed of the wing (or the rotating blade) through the air. Note that as the turbine blade is rotating, the Mach number will be a function of the radial location on the blade, with the maximum value corresponding to the blade tip. Given that a rigid edge has an M^5 dependence, poro-elastic edges promise, at least theoretically, to provide effective noise control.

As with the serrated edges, new experimental and computational work is needed to understand poro-elastic edge noise generation in order to design new technology for wind turbine blades. In particular, investigation of the effect of poro-elasticity on the boundary layer must be done; as in the case of serrations, modification of the turbulence levels may increase noise in some frequency bands, which must be avoided if used on wind turbines.

Wagner et al. (1996) reported on a DELTA project that used porous edges on a wind turbine. In this case, sound power was reduced by only 0.5 dB. More recently, Geyer et al. (2010) performed aeroacoustic measurements on porous airfoils in a wind tunnel. Compared with a rigid, non-porous airfoil, noise reductions of up to 10 dB were measured. The level of noise reduction was found to be dependent on the flow resistivity level of the porous material used to construct the airfoil model. In a similar manner to serrations, an increase in noise was observed at high frequencies, and this was attributed to increased turbulence levels generated by the flow over the rough porous surface of the airfoil.

Kinzie et al. (2013) presented noise results for two porous (metallic foam) wind turbine trailing-edge designs and compared them with results using trailing-edge serrations. Results were presented for full-scale wind turbine tests as well as anechoic wind tunnel tests. The results show that the metallic foam edges are somewhat effective at lower frequencies, but produce more noise at higher frequencies. This is a similar result to what is observed with the use of trailing-edge serrations, but Kinzie et al. (2013) were able to show that if the blade profile is chosen correctly, then serrations can provide effective noise control over a wide frequency range. Proprietary restrictions prevents many details of these designs from being published.

The fine fringe at the downstream edge of the trailing edge of the owl wing most likely acts as a gradual impedance change, thus reducing the diffraction of pressure waves about the edge, thereby reducing trailing-edge noise radiation to the far-field. In some ways, this is similar to the use of trailing-edge brushes (Herr and Dobrzynski 2005), which have been shown to be effective in reducing trailing-edge noise over a wide frequency range, by 2–14 dB.

Finally, there is the soft downy coating on an owl's wing. Clark et al. (2014) show both experimentally and theoretically that the soft downy coat acts as a canopy layer that suppresses the hydrodynamic pressure fluctuations beneath it. They speculate that the reduced pressure fluctuations underneath the canopy have less energy available to scatter to the far field than when the canopy is absent.

In summary, there are many different facets of an owl's wing that contribute to noise reduction. Attempts to date have focussed upon individual aspects of the wing, such as serrations and porosity. While such detailed attention upon specific aspects is needed to isolate their effects on sound production, it is most likely that it is the combination of all of these aspects simultaneously that will enable the design of a much quieter wind turbine blade.

8.2.3 Blade-pitch Control

The blade pitch is adjusted by the turbine operational control algorithm to maintain a constant rotational speed and/or to maximise power production for a particular wind speed. The blade angle of attack relative to the incoming air flow is what determines how close to stall that the blade will operate and is related to the pitch angle of the blade (which is controllable) and the blade angle of twist (which is fixed for a particular blade design). Operating a blade with a pitch angle just below that which would produce stall, maximises turbine output in smooth and uniform flow conditions, but in the presence of wind shear, the blade may stall for a proportion of the time when the air flow speed exceeds the stall speed for the particular angle of attack. In addition, parts of the blade may stall due to turbulence causing localised areas of excessive flow as the blades rotate. Stalling causes a turbine to generate less power and so is to be avoided if possible. Stalling also results in excessive AM (Makarewicz and Golebiewski 2015), which results in the turbine noise being more annoying. The AM caused by atmospheric turbulence and wind shear is generally only noticeable at distances greater than 500 m from the turbine responsible for it.

Controlling the 'thumping' noise or AM caused by conditions of high wind shear can be approached by cyclic variation of the blade pitch as it rotates (Bertagnolio et al. 2014). This control methodology can result in reduced turbine output power, but it is better than having to shut down the turbines during high wind shear conditions (which could be a large percentage of nights in many cases). The control algorithm could be programmed so that it is only invoked during weather conditions that may result in generation of the 'thumping' noise. A very simple algorithm used by some operators is to decrease the blade pitch angle a few degrees in wind speeds that have been known by testing to cause AM problems (Cassidy and Bass 2015).

Controlling the amplitude variation of the blade noise as a result of turbulent inflow can be achieved by operating the turbine with the blade pitch far enough below the angle necessary for stall that the blade does not exhibit any stall behaviour for reasonable levels of turbulence. However, this may come at a cost of lower turbine power output in these conditions. This technique is different to that discussed in the previous paragraph in that the blade pitch is kept below a fixed level to account for random and unpredictable inflow turbulence. However, the two techniques may also be used together to control noise generated by both mechanisms.

As shown and discussed by Wagner et al. (1996), reducing the pitch of wind turbine blades also reduces the radiated sound power, but this will also reduce its electrical power output. Wagner et al. (1996) show that a 1° reduction in pitch reduces the sound power level by approximately 1 dB(A). This level of noise reduction comes with a cost of a 1–3% reduction in electrical power output.

Amplitude modulation of trailing-edge noise also results from the directivity of the sound radiation from the turbine blades changing (relative to a stationary reference frame) as the blades rotate, and is an inherent part of turbine operation, which cannot be changed. Although the difference between the maximum and minimum noise experienced at a nearby receiver as a result of this mechanism cannot be changed, the actual sound pressure levels can be minimised by efficient blade design, as discussed in Section 8.2.1. However, this form of AM is only noticeable at distances less than 500 m from a turbine and so it is not usually a problem for most residents (van den Berg, G 2004).

While most noise control methods focus on frequencies higher than about 300 Hz, van den Berg, G (2008) has proposed that AM (suspected to be the cause of low-frequency noise problems and 'thumping') may be controlled by random blade-pitch excitation. If the mean blade pitch remains the same during one revolution, then it may be possible to retain energy production and reduce noise. Further, the enhanced amplitude modulation mechanism (EAM, see Section 3.3.6 and Oerlemans (2011)) may also be controlled through blade-pitch control. However, as EAM and 'blade thump' are most likely due a combination of complex factors (see Section 3.3.6), pitch control is likely to be only part of a broader solution to this problem that takes into account operational and meteorological factors as well.

Another possible application of pitch control would be to control the so-called 'blade–tower-interaction' effect summarised in Chapter 3 (Section 3.3.7). Here, the blade experiences a rapid angle-of-attack change as it passes the tower. If this effect could be properly characterised and shown to be an important noise source, then it is theoretically possible to vary the blade pitch to compensate for the tower and reduce noise. However, care must be taken to ensure that the effort required to change blade pitch as it passes the tower does not create additional unwanted noise (mechanical or aerodynamic).

8.2.4 Phase Control

It is possible (theoretically) that a wind farm, especially one arranged in a line upon a ridge, can act as a form of acoustic array, even at very low frequencies. An array of acoustic sources can reinforce each other and create regions of high- and low-level noise. In addition, if each source is amplitude modulated, there will be areas where AM peaks from different turbines arrive at the same time, thus greatly increasing the effective AM over that for a single turbine. Figure 8.2 illustrates an array of two wind turbines. The reinforcement of trailing-edge and blade–tower-interaction (BTI) noise sources may create regions about the wind farm where noise amplitudes are high. Figure 8.2 shows noise propagating upwind of the turbines only (other directions are omitted for clarity) and regions where broadband swish (trailing-edge) noise and BTI noise may be reinforced. Of course, the sound will couple with atmospheric propagation effects, making the actual sound paths more complicated than are represented in the figure, but conceptually the idea is the same.

Mathematically, the directivity $D(\phi)$ about a linear array of N turbines radiating noise with frequency f separated by distance d can be written as (Blackstock 2000),

$$D(\phi) = \frac{\sin N\phi}{N \sin \phi} \tag{8.3}$$

where $\phi = (kd/2)\sin\theta$ is half the difference in phase between two turbines, where $k = 2\pi f/c$, f is frequency and c is the speed of sound. The angle θ represents the angle measured to a point in the far field (say a residence) from the centre of the line of turbines.[1] An example is presented in Figure 8.3 of an array of six turbines placed in a line. Each turbine (represented as a dot) is assumed to radiate noise at 1 Hz (corresponding to an

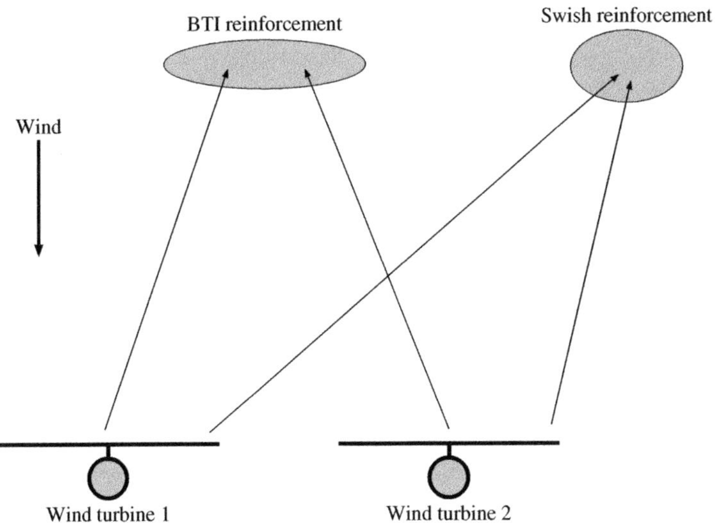

Figure 8.2 Plan view of two wind turbines with possible zones of noise reinforcement.

1 The wind farm need not be arranged in a line for such reinforcement to occur, but is more difficult to represent mathematically here.

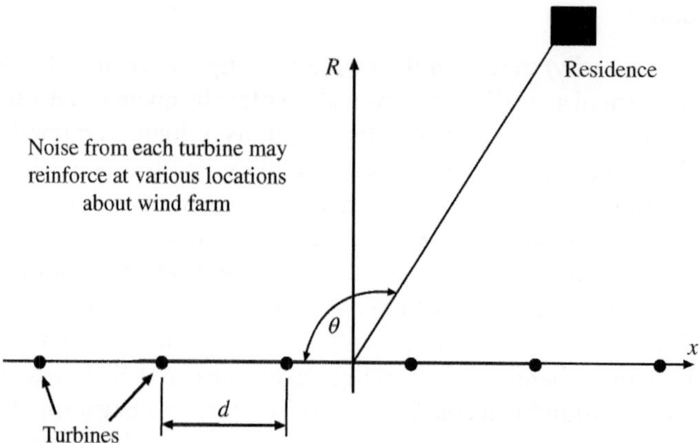

Figure 8.3 Depiction of a line of turbines as an acoustic line array. Noise from each turbine may reinforce (or not) at specific locations about the wind farm if the phasing between each turbine allows it.

infrasonic blade-pass noise source) with zero phase difference between each. Calculations of the directivity pattern are presented, in Figure 8.4 for four cases with different separation distances:

- $d = 200$ m ($kd/2 = 1.83$)
- $d = 400$ m ($kd/2 = 3.66$)
- $d = 800$ m ($kd/2 = 7.33$)
- $d = 1000$ m ($kd/2 = 9.16$)

The value $kd/2 = \pi d/\lambda$, which gives a relative measure of the turbine separation distance d to the acoustic wavelength λ.

The directivity patterns are quite different depending on the separation distances of the turbines. When the turbines are close, the turbines radiate in the forward and backward directions, similar to a dipole (or a *broadside* array). As the distance between the turbines increases, multiple lobes appear at different locations about the wind farm, indicating regions where the sound is reinforced and areas where it is not. As the widths of these lobes are quite narrow, this may explain the anecdotal observations of some residences experiencing wind turbine noise problems and others not. Also, these directivity patterns will be affected by many factors, including shifting wind conditions and phasing between turbines. Therefore it is not unlikely that a residence may find itself within a high-noise lobe during one operating condition and weather condition for the wind farm, but for it not to be so when these conditions change. The effect of increasing the number of turbines in the array is to narrow the width of the lobes, making the noise highly directional from the wind farm.

If this model is correct, it may explain why some residents become annoyed and some do not. While broadband swish noise may annoy people outside, its high-frequency components may be attenuated inside a home. However, if BTI reinforcement occurs at the same location, noise from BTI-excited structural vibration may also be apparent inside the home. While much more work is required to understand BTI and swish reinforcement, the model presented provides a framework for understanding and addressing public concerns about wind turbine noise.

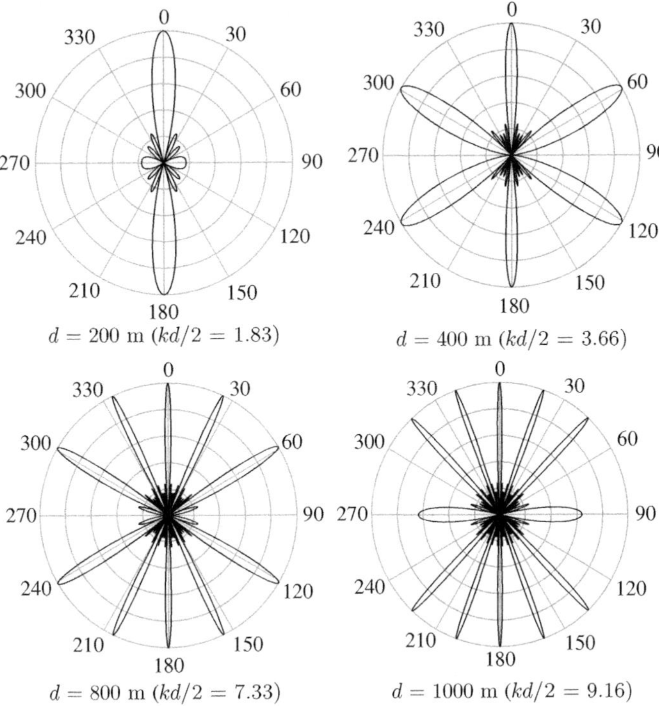

Figure 8.4 Theoretical directivity patterns about the line array of turbines depicted in Figure 8.3, for different separation distances and for each turbine emitting sound at $f = 1$ Hz with identical phase.

An obvious method of noise control would be to ensure that the phase between each turbine is regularly changed or randomised, but care needs to be taken to understand if and how the phase distribution amongst the turbines of a wind farm causes noise problems, and how such problems can be avoided for the various meteorological and operating conditions experienced by the wind farm. Buck et al. (2013) proposed a method for controlling the phase relationship between constant-speed turbines in a wind farm, as a means of directing incoherent interference maxima in directions where there are no residents.

8.2.5 Control of Noise Resulting from Aeroacoustic Excitation of the Blades

Although it has been established that wind turbine blades can radiate noise as a result of their vibration levels, the relative importance of this noise has not been quantified and neither have the relative contributions of aeroacoustic excitation and gearbox excitation. Gearbox excitation and control is discussed in Section 8.2.6 and aeroacoustic excitation of the turbine blades and subsequent noise radiation is a function of the design of the turbine blades, as discussed in Section 8.2.1. More efficient designs in terms of aerodynamics result in lower fluctuating loads on the blades and thus lower levels of aeroacoustic excitation.

8.2.6 Control of Noise Resulting from Mechanical Excitation of the Gearbox, Blades and Tower

Section 3.5 describes the noise created by wind turbine gearboxes and radiated by the gearbox, blades and tower. The most effective way to control gearbox noise is to treat the noise at the source. Hence the design of quiet gears is the best solution. Helical gears are often suggested to be the best type for quiet operation and these should used or specified as much as possible. Experience tells us that a helical design is not enough to prevent gear noise, so other treatments are necessary.

Gearbox vibration transmission into the tower can be reduced by vibration isolation, especially of the high-speed final stage. However, this approach cannot reduce low-frequency vibration due to the requirement for the vibration isolation system to be too 'soft', providing limited possibilities for success as a result of problems associated with securing the gearbox to the tower. Pinder (1992) provides an interesting discussion regarding the use of vibration isolation of wind turbine gearboxes, warning that care must be taken when using resilient mountings on heavy components.

Vibration isolation of wind turbine gearboxes does not address vibration transmission into the blades, which have been shown to be responsible for the radiation of most of the gearbox noise (Engelhardt et al. 2015). Thus the preferred treatment for controlling tonal noise is to use a tuned vibration absorber (mass-spring system) or tuned mass damper attached to the gearbox support base plate or other problematic parts of the drive train (Dawson and MacKenzie 2014). A tuned mass damper is the same as a tuned mass absorber, except that the former has damping added to it.

Illgen et al. (2007) described tuneable vibration absorbers that could be adjusted to be optimal for a fixed-speed wind turbine. They eliminated a gearbox tone at 160 Hz in the structure-borne noise and radiated noise spectra for a 1.5-MW turbine, using two absorber masses, each weighing 150 kg. They also suggested that variable speed turbines would benefit from active tuneable vibration absorbers, the tuning of which could be continuously adjusted using an appropriate control law.

Miyazaki et al. (2013) have shown that the use of two-degree-of-freedom active tuned mass dampers mounted on the gearbox platform can reduce the transmission of gearbox vibrational energy into the support tower of a model wind turbine. The reason that the dampers need to be actively controlled is that they have to track a variable planetary gear rotational speed. A feedback control system was used to adjust the stiffness of the damper so that its resonance matched that of the gearbox on its support structure. At the time of writing, scaling up to a full-size turbine has not been reported by these researchers.

Engelhardt et al. (2015), reported on noise reduction achieved using an active vibration absorber on a variable-speed turbine. They used multiple retrofittable absorbers, each of which was capable of addressing several gearbox harmonics, and achieved reductions of 8–12 dB for the first three harmonics. Each of their absorber systems weighed about 110 kg and generated a dynamic force of up to 5 kN. The purpose of the absorbers was to reduce the vibration transmission into the blades, which the authors suggested were the dominant gearbox noise radiators. However, the authors did not provide any details of the turbine type or size used in their testing.

Nacelle insulation is an effective means to reduce radiated noise from the gearbox, with approximately 15 dB reduction possible.

With the aim of resolving some of the problems associated with wind turbine gearboxes, there have been attempts to remove them completely from the design. The European company Enercon has developed the direct-drive wind turbine, explained in some detail by Ragheb and Ragheb (2010). In this model, the generator has been designed with 50–150 poles, thereby requiring only a slow rotational speed to create 50–60 Hz electrical power, thus eliminating the need for a gearbox. Use of more or fewer poles, depending on the rotational speed, can maintain the required line frequency. Such a design has been implemented in Europe and Japan, with good results, but is considered to be expensive by some parts of the wind power industry, as these models need special construction techniques and have additional weight and logistical issues (Cotrell 2002; Oyague 2009).

8.3 Optimisation of Turbine Layout

Turbines must be separated by appropriate distances to avoid excessive interaction with the wake from upstream turbines, which leads to reduced fatigue life and operating power output. The same unsteady forces acting on the turbine blades also lead to the generation of infrasound and low-frequency noise. In situations where there is a dominant prevailing wind, turbines are usually arranged in rows spaced 5 blade diameters apart, with each turbine in a row separated by 4 blade diameters (Brunskill and Hassan 2013). Larger offshore turbines are usually spaced 7 blade diameters apart and Meyers and Meneveau (2012) suggest that optimum power output can be achieved using a turbine spacing of 15 rotor diameters. However, many onshore farms have a limited area over which they must be installed and this leads to turbines being placed much closer together than is optimal, resulting in reduced life and more infrasonic and low-frequency noise generation, especially when the wind is blowing in a direction in which many turbines are in the wake of one or more upstream turbines.

Unfortunately, there is no work to date that relates the wakes from turbines with noise generation in any detail. While we understand the physics of noise production reasonably well (see Section 3.3), the specifics of the nature of the turbulence in the wake, how it develops over the 5–20 rotor diameters to the next turbine, how it extracts energy from the atmospheric boundary layer and how this turbulence interacts with downstream rotor blades are relatively open questions, and need further investigation.

While the details of interaction are not clear, there are some general principles that can be followed to optimise wind farm layout to avoid noise from wake interaction. The first of these is to maximise the distance between each turbine. This will allow the wakes to recover and mix with the surrounding air, reducing the turbulence intensity within them. Maximum spacing also allows greater power output and lower fatigue loads. Second, turbines should not be placed close to hills or ridges that could generate wakes when upwind of the turbines. Any large object that creates a large turbulent wake (even a building) should be avoided because the wake can interact with the rotor blades and create noise. Also, great care should be taken when placing turbines on a high ridge. The ridge may, depending on geography, create a large upwash, effectively changing the angle of attack of the blades. This could stall the blades in some operating conditions, creating low-frequency noise and contributing towards AM. The use of computational fluid dynamics and other analysis methods can assess the effects of ridges and should be used to determine if they contribute towards blade stall.

8.4 Options for Noise Control at the Residences

As the problematic aspect of wind farm noise is low-frequency in nature, it is difficult to achieve much noise reduction by treating a residence with acoustic materials. To reduce the transmission of low-frequency noise into a residence, the windows would need to be double glazed and remain shut. The ceiling would need to be of a double-panel construction with staggered studs and the roof would need to be heavy gauge steel. Most of these options are unacceptable to residents and there is no guarantee that a substantial reduction in very low-frequency noise and infrasound would be achieved even with the treatments outlined above. However, there are two other, less expensive and less intrusive possibilities for reducing the effect of low-frequency noise and infrasound on residents while they are in their houses. They are:

- use of an active noise-control system to generate anti-sound
- introduction of higher-frequency masking sound to produce a more balanced and less annoying noise spectrum in the residence, which also has the benefit of raising the threshold of hearing for low-frequency noise and infrasound.

Each will be discussed in the following sub-sections.

8.4.1 Active Noise Control

As the noise to be controlled is low-frequency in nature, typical passive methods of control, such as using sound-absorbing material in the roof space or double-glazed windows, are not very effective. For this reason, an active noise-cancellation system should be considered. The most appropriate type of system for this application is a feedforward system, for which a reference microphone is used outdoors between the wind farm and the residence to provide the control system with the noise signal to be controlled. The reference microphone should be at a distance from the residence sufficient to allow the control system to generate the cancelling sound by the time that the sound arrives at the residence. The other type of system is a feedback system, which requires no reference microphone. However, the amount of control that can be obtained in a large room using a feedback system is limited, and so this approach will not be considered further here.

There are two physical mechanisms whereby an active noise-cancellation system can reduce low-frequency noise and infrasound in a residence. The first is *cancellation*, whereby low-frequency turbine noise inside a residence is reduced at the expense of slightly increasing it outside. The second possible mechanism is that the control loudspeakers change the impedance that the walls, windows and ceiling experience when they vibrate and radiate noise, thus reducing the amount of acoustic energy that is radiated into the residence.

A typical active noise-control system is shown schematically in Figure 8.5 and consists of:

- One or more error microphones (and associated signal conditioning amplifiers) to sample the controlled sound field for the purpose of continually updating the control filter weights to minimise the signal experienced by the error microphones. These microphones would ideally be placed in the room corners near the ceiling. Twice as many error microphones as control loudspeakers should be used. Low-cost electret microphones may be used provided they are temperature compensated and that

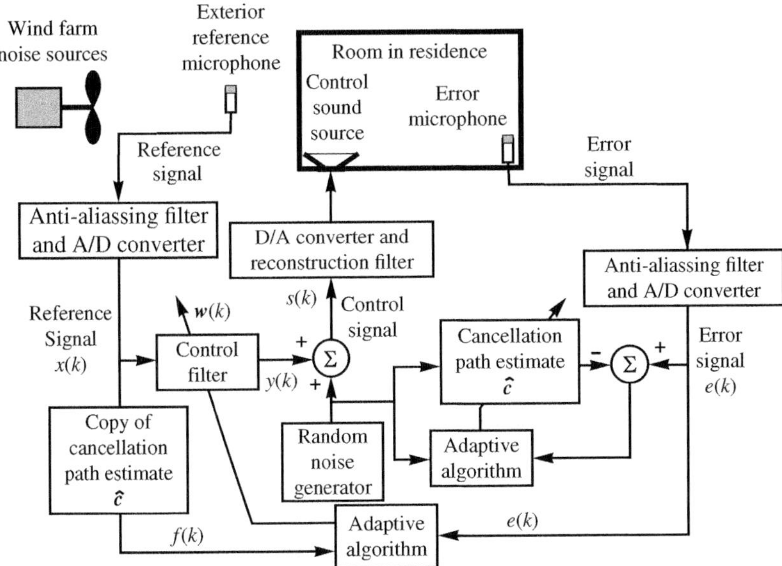

Figure 8.5 Schematic of a typical single-channel active noise-control system, where k represents the variable value at the kth time sample and $\hat{c}$ is the estimate of the cancellation-path impulse response.

they have the required dynamic range (maximum-to-minimum sound level that can be measured). The microphone signals will normally need to be passed through a signal-conditioning module prior to entering the electronic controller.

- One or more control loudspeakers (and associated power amplifiers) to provide the anti-noise signal. These would be placed on the floor in the corners of the rooms where the noise is to be reduced. One or two per room may be needed, or it may be possible to control noise in the entire house with one or two loudspeakers. For the control of noise below 30 Hz, it will be necessary to use a cube containing three loudspeakers: one in each side that is not against a wall or the floor. The loudspeakers must be low-frequency sub-woofers and driven with a DC amplifier so that they are all in-phase.
- One or more reference microphones (and associated signal-conditioning amplifiers) to sample the incoming noise in advance of it arriving at the location where it needs to be controlled. This gives the control system time to work out what signal it needs to send to the loudspeakers to cancel the noise at the error microphones. So in practice, the reference microphone(s) would be placed within 20 m of the residence and between the residence and the wind farm. If the wind farm has turbines in several different directions from the residence, then more than one reference microphone can be used and then all reference microphone signals would be used to determine the most appropriate signals to be sent to the loudspeakers to minimise the sound at the error microphones.
- A control system containing:
  - a transversal control filter (tapped delay line), which has weights that can be adjusted by a control algorithm to provide the signal to drive the control loudspeakers

- analogue-to-digital (A/D) converters to digitise the incoming microphone signals and to convert the outgoing signals to analogue signals for driving the loudspeakers
- anti-aliasing filters (see Section 2.4.5) and reconstruction filters on the controller outputs to smooth out the signal from the digital-to-analogue (D/A) converters
- a random noise source for determining the transfer function of the path between each output of the control system to each error microphone input to the control system.

Referring to Figure 8.5, the sound in the residence is controlled by the loudspeakers, and the success of the control is measured by the error microphones. The cancellation path (see next paragraph) for one control-loudspeaker/error-sensor combination is the path from the control-signal input to the D/A converter linked to the loudspeaker, to the output of the A/D converter linked to the error-sensor output; an estimate of the impulse response of this path is represented by the variable, $\hat{\mathbf{c}}$, in Figure 8.5. For our purposes here it is assumed that the cancellation-path impulse response (see Section 2.4.13) is constant with time. However, it is relatively simple to include a time-varying cancellation path in the analysis as explained in Hansen et al. (2012).

The signal provided to the control loudspeakers originates from the reference microphones and is passed through the control filter, then the D/A converter and smoothing filter, before being input to the loudspeaker amplifier. The adaptive algorithm uses the error-signal inputs and the reference signals modified by an estimate of the transfer function of each cancellation path (between the control sources and error sensors) to estimate the optimum control-filter weights to minimise the sum of the squares of the error signals. The schematic shown in Figure 8.5 is accurate for a single-channel active-control system with one reference sensor, one error sensor and one control source. In practice there will be more than one of each of the above items of hardware and a multi-channel active noise-control system and associated control algorithm will have to be used. The most likely system that will provide an adequate degree of control would be one with four error sensors, two control sources and between one and four reference sensors (depending on how many sides of the house face one or more turbines). A generalised active noise-control system arrangement with N_e error sensors, N_c control sources and four reference sensors is shown in Figure 8.6, where anti-aliasing filters, reconstruction filters and the random noise generator for determining the cancellation-path transfer functions have been omitted for clarity. The algorithm and controller outputs are only shown for the jth control loudspeaker, The Σ signs on the right-hand side of the figure represent the complex addition (taking relative phase into account) of wind farm and control loudspeaker sound pressures at the N_e error sensor locations. If there is more than one reference microphone, the reference microphone signals are added together with an electronic summing unit prior to being used by the active noise-control system, as indicated by the Σ sign on the left-hand side of the figure. The cancellation-path transfer function models are estimated by feeding a low-level random noise signal into each loudspeaker and measuring the result with the error sensors, as shown for the single-channel system in Figure 8.5.

The control filters shown in Figures 8.5 and 8.6 are finite impulse response filters, implemented as illustrated in Figure 8.7. From the figure it can be seen that the output of the control filter is a weighted sum of the current input sample and a number, N, of previous input samples, where N is the number of stages in the filter (or filter taps).

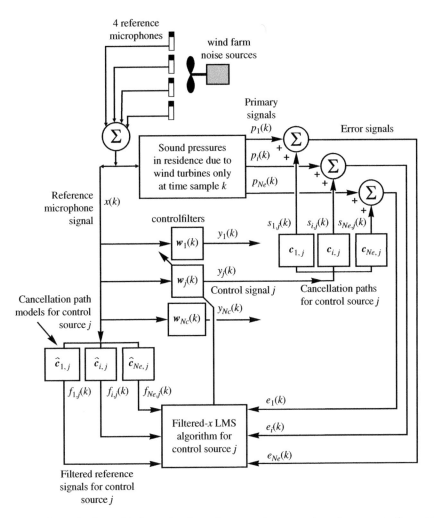

Figure 8.6 Schematic of a multi-channel active noise-control system, where k represents the variable value at the kth time sample and $\hat{c}_{i,j}$ is the estimate of the cancellation-path impulse response between the input to source, j and the output of error sensor, i. LMS=least mean square.

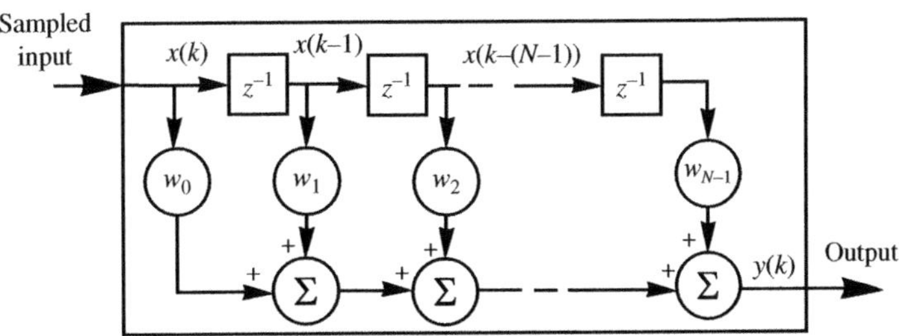

Figure 8.7 Schematic of a finite impulse response filter (tapped delay line) used in an active noise-control (ANC) controller.

The control algorithms for determining the filter weights for both single- and multi-channel systems are discussed in Appendix H.

8.4.2 Masking

Masking is the process of adding 'neutral' noise to an environment so that annoying noise is less easily heard. It is generally accepted that our hearing mechanism's threshold for low-frequency noise and infrasound would be raised in the presence of masking sound, so it is also possible that feelings of nausea caused by wind farm noise may be ameliorated by an appropriate level and spectrum shape of masking sound. It is possible that appropriate masking noise can positively influence the annoyance caused by wind farm noise, while to a lesser extent reducing its apparent loudness. Masking is likely to be particularly effective in rural areas where both ambient noise and wind farm noise are at relatively low levels. However, as stated by Chanaud (2007), it is easy to misuse masking and cause adverse reactions in people by using the wrong spectrum shape or the wrong level (see for example, Pedersen et al. (2010)).

The spectrum shape recommended by Chanaud (2007) as the least intrusive in an office environment is random noise with the 1/3-octave band levels listed in Table 8.1. The levels in the table are relative levels only and currently represent a level of 30 dBA. It is very likely that acceptable levels in a bedroom in a rural environment could be even less than this. However, it is relatively easy to adjust the overall level using the volume control on an amplifier once the correct spectrum shape is generated using a graphic equaliser.

8.5 Administrative Controls

The annoyance of a particular wind farm can be minimised by following the administrative guidelines listed below.

- Shut down turbines that are within 10 blade diameters downstream from an operating turbine. This will minimise rumbling noise resulting from turbulent inflow.
- Shut down turbines after 11pm and before 6am in conditions of high wind shear when there is little masking noise at residences due to negligible wind and when blades could be stalling as they travel into higher-speed air flows at the top of their path, possibly causing the thumping noise that is so annoying. Normally, there is not as much need for electricity generation in these nighttime hours as there is at other times, so one would hope this sort of control is possible.

Table 8.1 Recommended 1/3-octave white-noise masking spectra for an A-weighted level of 30 dBA

1/3-octave band centre frequency	160	200	250	315	400	500	630	800	1000
Level (dB)	29	28	27	26	24	23	22	20	19
1/3-octave band centre frequency	1250	1600	2000	2500	3150	4000	5000	6300	8000
Level (dB)	18	16	15	13	11	9	6	3	1

8.6 Summary

As has been demonstrated, there are a number of ways in which wind farm noise can be reduced. Aerodynamic noise may be controlled via the optimisation of the rotor blade shape or by treating the trailing edge of the rotor blade itself. Using the owl as inspiration, serrated and poro-elastic edges have recently been proposed to control trailing-edge noise.

Perhaps the most annoying aspects of wind farm noise are the low-frequency part of the spectrum and impulsive noise. Considerable reductions in these types of noise are possible: by optimising the blade-pitch control system to ensure that no part of the blade ever stalls, and optimising the blade design to minimise the production of impulsive noise as the blade stalls. However, this may come at the cost of reduced power output. Minimising the possibility of turbines being affected by the wake of others is another way of reducing excessive low-frequency and impulsive noise. Low-frequency noise and infrasound may also be reduced by randomising the phase between turbines, especially if there is a significant contribution to this noise from a blade–tower-interaction phenomenon. There is also the possibility of using active noise-control systems or masking systems inside people's homes, but these options are not very desirable. Future research should therefore focus on optimising wind farm layout to minimise low-frequency noise, optimising turbine control systems and optimising turbine blade design so that noise is minimised for a minimal sacrifice in turbine output power.

References

Bertagnolio F, Madsen H, Fischer A and Bak C 2014 Cyclic pitch for the control of wind turbine noise amplitude modulation. *Internoise 2014*, Melbourne, Australia.

Bertagnolio F, Madsen HA and Bak C 2010 Trailing edge noise model validation and application to airfoil optimization. *Journal of Solar Energy Engineering*, **132**(3), 031010.

Blackstock D 2000 *Fundamentals of Physical Acoustics*. Wiley-Interscience.

Brooks T, Pope D and Marcolini M 1989 Airfoil self-noise and prediction. Reference publication 1218, NASA.

Brunskill A and Hassan G 2013 Wind farm layout optimization in noise constrained areas. *5th International Meeting on Wind Turbine Noise*, Denver, Colorado.

Buck S, Scott S and Moriarty P 2013 Application of phased array techniques for amplitude modulation mitigation. *5th International Meeting on Wind Turbine Noise*, Denver, Colorado.

Cassidy M and Bass J 2015 Addressing the issue of amplitude modulation: a developer's perspective. *6th International Meeting on Wind Turbine Noise*, Glasgow, UK.

Ceyhan O 2012 Towards 20 MW wind turbine: high Reynolds number effects on rotor design. AIAA 2012-1157 in *50th AIAA Aerospace Sciences Meeting*.

Chanaud R 2007 *Sound Masking Manual*.

Chen K, Liu Q, Liao G, Yang Y, Ren L, Yang H and Chen X 2012 The sound suppression characteristics of wing feather of owl (*Bubo bubo*). *Journal of Bionic Engineering*, **9**(2), 192–199.

Chong TP, Vathylakis A, Joseph PF and Gruber M 2013 Self-noise produced by an airfoil with nonflat plate trailing-edge serrations. *AIAA Journal*, **51**(11), 2665–2677.

Clark IA, Devenport W, Jaworski JW, Daly C, Peake N and Glegg S 2014 The noise generating and suppressing characteristics of bio-inspired rough surfaces. AIAA 2014-2911 in *20th AIAA/CEAS Aeroacoustics meeting*.

Cotrell JR 2002 A preliminary evaluation of a multiple-generator drivetrain configuration for wind turbines. *ASME 2002 Wind Energy Symposium*, pp. 345–352.

Dawson B and MacKenzie N 2014 Tonal characteristics of wind turbines drive trains. *Inter-Noise and Noise-Con Congress and Conference Proceedings*, vol. **249**, pp. 2699–2708.

Drela M 1989 XFOIL: An analysis and design system for low Reynolds number airfoils. In: Mueller TJ, *Low Reynolds Number Aerodynamics*. Springer.

Engelhardt J, Katz S and Pankoke S 2015 Reduction of tonalities in wind turbines by means of active vibration absorbers. *6th International Meeting on Wind Turbine Noise*, Glasgow, UK.

Geyer T, Sarradj E and Fritzsche C 2010 Measurement of the noise generation at the trailing edge of porous airfoils. *Experiments in Fluids*, **48**(2), 291–308.

Geyer T, Sarradj E and Fritzsche C 2013 Silent owl flight: comparative acoustic wind tunnel measurements on prepared wings. *Acta Acustica United with Acustica*, **99**(1), 139–153.

Gruber M, Joseph P and Chong T 2010 Experimental investigation of airfoil self noise and turbulent wake reduction by the use of trailing edge serrations number. AIAA 2010-3803 in *16th AIAA/CEAS Aeroacoustics Conference*, Stockholm, Sweden.

Gruschka H, Borchers I and Coble J 1971 Aerodynamic noise produced by a gliding owl. *Nature*, **233**, 409–411.

Hansen C, Snyder S, Qiu X, Brooks L and Moreau D 2012 *Active Control of Noise and Vibration*. Spon Press.

Herr M and Dobrzynski W 2005 Experimental investigation in low-noise trailing edge design. *AIAA Journal*, **43**(6), 1167–1175.

Howe M 1991 Aerodynamic noise of a serrated trailing edge. *Journal of Fluids and Structures*, **5**(1), 33–45.

Illgen A, Drossel WG, Wittstock V, Neugebauer R, Hanus KH, Mitsch F and Wiedemann L 2007 Passive and active dynamic vibration absorbers for gearbox noise reduction in wind turbines. *Second International Meeting on Wind Turbine Noise*, Lyon, France.

Jaworski JW and Peake N 2013 Aerodynamic noise from a poroelastic edge with implications for the silent flight of owls. *Journal of Fluid Mechanics*, **723**, 456–479.

Kinzie K, Drobietz R, B. Petitjean and Honhoff S 2013 Concepts for wind turbine sound mitigation. *AWEA Windpower 2013*, Chicago, USA.

Lee S, Lee S, Ryi J and Choi JS 2013 Design optimization of wind turbine blades for reduction of airfoil self-noise. *Journal of Mechanical Science and Technology*, **27**(2), 413–420.

Lilley GM 1998 A study of the silent flight of the owl. AIAA 1998-2340 in *4th AIAA Aeroacoustics Conference*.

Lutz T, Herrig A, Wurz W, Kamruzzaman M and Kramer E 2007 Design and wind-tunnel verification of low-noise airfoils for wind turbines. *AIAA Journal*, **45**(4), 779–785.

Makarewicz R and Golebiewski R 2015 Influence of blade pitch on amplitude modulation of wind turbine noise. *Noise Control Engineering Journal*, **63**(2), 195–201.

Marsden AL, Wang M, Dennis JE and Moin P 2007 Trailing-edge noise reduction using derivative-free optimization and large-eddy simulation. *Journal of Fluid Mechanics*,**572**, 13–36.

Meyers J and Meneveau C 2012 Optimal turbine spacing in fully developed wind farm boundary layers. *Wind Energy*, **15**(2), 305–317.

Miyazaki T, Uchino K, Kuono K, Inoue Y and Yasuda M 2013 Suppression of the structure-borne sound from a wind turbine generator using active vibration control devices: model experiment. *5th International Meeting on Wind Turbine Noise*, Denver, Colorado.

Moreau DJ and Doolan CJ 2013 Noise-reduction mechanism of a flat-plate serrated trailing edge. *AIAA Journal*, **51**(10), 2513–2522.

Oerlemans S 2011 An explanation for enhanced amplitude modulation of wind turbine noise. Contractor report NLR-CR-2011-071, Nationaal Lucht- en Ruimtevaartlaboratorium.

Oerlemans S, Fisher M, Maeder T and Kögler K 2009 Reduction of wind turbine noise using optimized airfoils and trailing-edge serrations. *AIAA Journal*, **47**(6), 1470–1481.

Oyague F 2009 Gearbox modeling and load simulation of a baseline 750-kW wind turbine using state-of-the-art simulation codes. Technical report NREL/TP-500-41160, National Renewable Energy Laboratory, USA.

Parchen RR 1998 Progress report DRAW: a prediction scheme for trailing edge noise based on detailed boundary layer characteristics. TNO, Institute of Applied Physics.

Pedersen E, Van den Berg F, Bakker R and Bouma J 2010 Can road traffic mask sound from wind turbines? Response to wind turbine sound at different levels of road traffic sound. *Energy Policy*, **38**(5), 2520–2527.

Pinder JN 1992 Mechanical noise from wind turbines. *Wind Engineering*, **16**(2), 158–168.

Ragheb A and Ragheb M 2010 Wind turbine gearbox technologies. *1st International Nuclear & Renewable Energy Conference (INREC)*, Ammann, Jordan.

Sarradj E,Fritzsche C and Geyer T 2011 Silent owl flight: bird flyover noise measurements. *AIAA Journal*, **49**(4), 769–779.

Schepers J, Curvers A, Oerlemans S, Braun K, Lutz T, Herrig A, Wuerz W, Mantesanz A, Garcillan L, Fischer M *et al.* 2007 Sirocco: Silent rotors by acoustic optimisation. *Proceedings of the Second International Meeting on Wind Turbine Noise*, Lyon, France.

Sørensen DN 2001 Minimizing the trailing edge noise from rotor-only axial fans using design optimization. *Journal of Sound and Vibration*, **247**(2), 305–323.

van den Berg F 2013 Wind turbine noise: an overview of acoustical performance and effects on residents. *Acoustics 2013*, Victor Harbor, Australia.

van den Berg G 2004 Effects of the wind profile at night on wind turbine sound. *Journal of Sound and Vibration*, **277**(4-5), 955–970.

van den Berg G 2005 Mitigation measures for night-time wind turbine noise. *First International Meeting on Wind Turbine Noise: Perspectives for Control*, Berlin, Germany.

van den Berg G 2008 Wind turbine power and sound in relation to atmospheric stability. *Wind Energy*, **11**(2), 151–169.

Wagner S, Bareiss R and Guidati G 1996 *Wind Turbine Noise*. Springer Verlag. pp 67–143.

9

Recommendations for Future Research

9.1 Introduction

The effects of wind farm noise on people are poorly understood and many claims have been made based on anecdotal evidence and hearsay. Perhaps the most outrageous claim is that wind farms have no measurable effect on people and that people who are adversely affected have the anti-wind-farm activists to blame for telling them that wind farms should adversely affect them. The purpose of this chapter is to explore what needs to be done to identify the problems, understand them and then devise ways to ameliorate them. There are a number of aspects of wind farm noise and its effects on people that warrant further study, as listed below.

- The effect of wind farm noise on people.
  - Which part of the spectrum is most disturbing?
  - Does it awaken people who are already asleep?
  - Does it prevent people from going to sleep (insomnia)?
  - Does the infrasound and/or low-frequency part of the spectrum cause physiological problems, including feelings of nausea in some people?
- Improvements to regulations to properly reflect the annoyance characteristics of wind farm noise.
- Propagation model improvements, including:
  - the range of expected noise levels outside of each vulnerable residence as a result of varying atmospheric conditions
  - the uncertainty in the predictions of noise levels due to the wind farm, at each vulnerable residence
  - the calculation of long-term nighttime averages (between 11pm and 5am), rather than combining daytime and nighttime to calculate the long-term expected average sound pressure levels
  - the effect of sound rays arriving at the receiver via more than one ground reflection (the multiple ground reflection phenomenon).
- Measurement and data-analysis enhancements, including:
  - development of improved and universal AM detection algorithms
  - development of a universal AM metric for low-frequency noise and infrasound
  - specification of a standardised secondary wind screen design for immission measurements
  - inclusion of a greater number of operating conditions for tonality assessment.

Wind Farm Noise: Measurement, Assessment and Control, First Edition.
Colin H. Hansen, Con J. Doolan, Kristy L. Hansen.

- Identification of the problematic noise sources on wind turbines.
  - Which part of the turbine radiates the low-frequency noise?
  - Which part of the turbine radiates the mid- and high-frequency noise?
  - How is the energy transmitted from the noise source to the part of the structure that radiates the noise and where this is relevant?
  - What causes the amplitude of the noise heard at the receiver to vary?
  - What are the different mechanisms causing periodic amplitude variation (AM) and random amplitude variation.
- Possible turbine design and layout modifications to minimise noise generation.
- Possible methods for reducing the part of the noise spectrum inside residences that people find most disturbing or annoying.

9.2 Further Investigation of the Effects of Wind Farm Noise on People

In Australia, the National Health and Medical Research Council conducted two studies on this topic, with the results of the most recent study being released on 11 February 2015 (NHMRC 2015). This release stated:

> Examining whether wind farm emissions may affect human health is complex, as both the character of the emissions and individual perceptions of them are highly variable.
>
> After careful consideration and deliberation of the body of evidence, NHMRC concludes that there is currently no consistent evidence that wind farms cause adverse health effects in humans.
>
> Given the poor quality of current direct evidence and the concern expressed by some members of the community, high quality research into possible health effects of wind farms, particularly within 1500 metres (m), is warranted.

They further expand,

> There is consistent but poor quality direct evidence that wind farm noise is associated with annoyance. While the parallel evidence suggests that prolonged noise-related annoyance may result in stress, which may be a risk factor for cardiovascular disease, annoyance was not consistently defined in the studies and a range of other factors are possible explanations for the association observed.
>
> There is less consistent, poor quality direct evidence of an association between sleep disturbance and wind farm noise. However, sleep disturbance was not objectively measured in the studies and a range of other factors are possible explanations for the association observed. While chronic sleep disturbance is known to affect health, the parallel evidence suggests that wind farm noise is unlikely to disturb sleep at distances of more than 1500 m from wind farms.

The problem with this study and many other 'official' statements by government bodies is that all publications that do not meet the strict requirements of an epidemiological study are either rejected outright or the evidence is labelled 'poor quality', even though many of these studies have been published in reputable journals. However, it is virtually impossible to conduct a longitudinal study meeting all the required criteria. This would mean that the health of three random sample groups of the population would need to be studied before they had read anything at all about wind farms and the study would need to continue for all three groups, with one group now living in the vicinity of a wind farm and not exposed to anti- or pro-wind farm literature, another now living in the vicinity of a wind farm and exposed to anti- and pro-wind farm literature and the third group not living in the vicinity of a wind farm. It would be necessary for the only change in the noise environment for all three groups to be due to wind farms. Large numbers of people would need to be included in each group for the results to render the statistical analysis valid. Of course it is not possible to ever study such groups and so some research bodies such as the NHMRC will always find a way to discredit any study. It is noteworthy that although the NHMRC openly criticised and labelled as 'poor quality' all studies that have been done to date, they did not specify how a study should be done that would satisfy their requirements to be of 'excellent quality'. They just state that more research needs to be done knowing full well that an epidemiological study that would satisfy the requirements to be of 'excellent quality' is impossible to do.

It is interesting that critics of wind farm noise studies do not accept the validity of studies that focus on individuals who complain of symptoms and then test their symptoms by removing themselves from the wind farm vicinity for a period of weeks or months and then returning to the wind farm vicinity again. Even if the return of symptoms is psychological, one may ask why that is important, as the symptoms would not be there if the wind farm was not there. There seems to be an undue focus on proving whether or not there is a direct physiological link between wind farms and illness symptoms and any possible psychological link is blamed on anti-wind farm activists, with the implication that it would go away if the activists were shut down.

Perhaps there is one type of study that may produce evidence of good quality and which could demonstrate a physiological link between wind farm noise and health, if such a link exists. This would involve selecting a random sample of the population and undertaking a sleep study on them in a sleep laboratory for several nights for each person. They could be exposed to the full wind farm noise spectrum measured in a complainant's bedroom and then the infrasound part could be removed, followed by removal of all the noise. Throughout the process, their state of sleep could be measured continuously and linked to any changes in the loudness of the noise. The intention would be to determine whether or not wind farm noise causes sleep disturbance and, if so, what is the extent of such disturbance. Once this is known, other studies can be used to determine likely effects on health. In a parallel study, people could also be tested for the effects of different parts of the wind farm noise spectrum on feelings of nausea when exposure is for long periods, beginning when people arrive for the sleep study and continuing until they leave. Such studies as recommended above will be very expensive if they are to produce results that are acceptable by the medical community, and the population to be studied will need to be carefully selected to minimise as many 'confounding factors' as possible, such as strong anti- or pro-wind farm feelings, pre-existing medical conditions or sleep problems and many others that no doubt will be thought of at the beginning of any investigation. It is also possible that such studies could demonstrate that wind farm noise by itself does not affect the ability of the general population to sleep.

9.3 Improvements to Regulations and Guidelines

In studying the effects of wind farm noise on people, it has become apparent that the most annoying part of the noise spectrum is the low-frequency part, below frequencies of about 200 Hz. It is equally apparent that the A-weighting scale used for noise assessment does not adequately measure annoyance caused by low-frequency noise, for which the decibel level difference that separates the level corresponding to detection and that corresponding to annoyance is much smaller than it is for higher frequencies. As the frequency decreases towards the infrasound region, the detection and annoyance levels are very similar. This, coupled with the large variation in detection threshold levels for a particular population, means that some people can be annoyed by the noise when others cannot even detect that it exists. Thus a reasonable regulation would include upper allowable limits to octave band sound levels between 4 and 200 Hz in addition to an allowed A-weighted overall level. These permitted levels should be specified as indoor levels to minimise the effect of wind noise on the measurements. They also should be determined at night between 11pm and 5am, and they should be determined in conditions of high wind shear so that the noise of the wind blowing on the house structure and any nearby trees is minimised, while at the same time, there is sufficient wind to drive the turbines. Of course, the noise levels should be measured in a downward-refracting atmosphere when the wind is blowing from the wind farm towards the house being measured. There have been three attempts in the past to include the low-frequency contribution to annoyance in guidelines for the assessment of low-frequency noise. These are:

- Danish low-frequency calculated indoor limit of a maximum of a total of 20 dBA for the frequency range 10–160 Hz (see Section 5.4).
- DEFRA criteria of allowed indoor 1/3-octave band L_{eq} levels in the frequency range 10–160 Hz (see Section 1.4.4).
- Japan Ministry of Environment criteria of allowed outdoor 1/3-octave band L_{eq} levels in the frequency range 5–80 Hz (see Table 9.1).

It would be a significant improvement if regulators would consider audibility when preparing noise regulations for wind farms. It is well known that the difference between the threshold of audibility and the annoyance threshold reduces as the frequency reduces. Thus regulations are needed that accounts for this by producing the following.

Table 9.1 Japan Ministry of Environment criteria for low-frequency noise (JMEF 2004)

1/3-octave band centre frequency (Hz)	5	6.3	8	10	12.5	16	20	25	31.5	40	50	63	80
SPL for rattling complaints (L_{eq}, dB)	70	71	72	73	75	77	80	83	87	93	99	–	–
SPL for mental and physical complaints discomfort (L_{eq}, dB)	–	–	–	92	88	83	76	70	64	57	52	47	41

1. Allowed wind farm noise levels as a function of 1/3-octave band centre frequency for low frequency noise below 200 Hz. Suggestions for these levels between 10 and 160 Hz have already been published, as discussed above, but more research is needed on long-term exposures to determine if these levels are appropriate. We suspect that they may be too high for a significant proportion of the population living in rural areas. It is likely that appropriate levels would be close to thresholds of audibility at very low frequencies.
2. The percentage of time that wind-farm-only noise is allowed to exceed the target level in each 1/3-octave band.
3. The relationship between the percentage of time that the target level is exceeded and the number of 1/3-octave bands in which the target level is exceeded simultaneously.

However, whether or not the wind farm is heard will depend on the level of background noise, which will tend to mask the wind farm noise. Thus we need to resort to statistical analysis and calculate the probability that the wind farm noise is above a certain level at the same time that background noise is below a certain level. This could be done by slightly modifying the approach described by Ashtiani (2015). In this case all instances where the background noise level and wind turbine noise level are within a certain number of decibels (say 2 or 3 dB) would be discarded and the analysis of Ashtiani (2015) carried out on the remaining data. This would be repeated for each 1/3-octave band and each integer wind-speed bin to produce a probability of the wind farm noise exceeding the pre-determined limit in each 1/3-octave band for each integer wind speed. The overall probability of a target noise level being exceeded in each 1/3-octave band could be calculated using the following steps:

1. For each wind-speed bin, multiply the probability of the wind farm noise alone exceeding the target level by the probability that the wind speed will lie in that wind-speed bin.
2. Add all the combined probability results for each wind-speed bin calculated in Step 1 to obtain the overall probability of the target noise level being exceeded by the wind farm noise in the 1/3-octave band under consideration for all wind speeds.

It would also be necessary to take into account masking of the wind farm noise in the 1/3-octave band under consideration by noise in other 1/3-octave bands. However, the means to do this would require future research. It would also be necessary for regulators to determine how many 1/3-octave bands could have their target levels exceeded and for what percentage of the time this target level could be exceeded in each 1/3-octave band.

Another issue that is not properly addressed by current regulations is amplitude modulation (AM), which is a regular periodic variation of the noise that makes it more annoying. Random variation of noise is not amplitude modulation, although there seems to be some confusion on this point, as some published suggestions for quantifying amplitude modulation actually quantify random amplitude variation. However, random variation can also result in subjective annoyance and in the case of wind farms it may be caused when noise from individual turbines arriving at a receiver location varies in relative phase as a function of time due to varying atmospheric conditions. This phase variation results in random reinforcement and cancellation of sound of the same frequency from different turbines. This problem would be eliminated if variable-speed turbines were used in a wind farm. However, a new problem of 'beating' may arise, which is the interaction of tonal signals of slightly different frequencies, which results in a varying amplitude of the original tone, as described in Section 2.2.6.

Noise regulations are invariably written with little regard to the environment to which they are applied. For example, the same overall A-weighted noise levels are allowed for residents in suburban areas as are allowed for residents in rural townships where the background noise levels are much lower. In addition, rural residences outside of townships are sometimes zoned 'rural industry' as a result of the farming activities they support, and in these cases the allowable noise levels are usually 5 dBA higher for these residences. Future regulations should take into account existing background noise levels when setting the allowable average dBA level resulting from the introduction of a wind farm. It does not seem reasonable that turbines should be able to increase nighttime noise levels in quiet rural areas from less than 20 to 40 dBA, and expect that residents' amenity has been preserved. In fact, if wind farms are to be accepted in rural communities, their noise contribution should not be greater than the existing background noise, so that would mean that the total of wind farm noise and background noise should not be more than 3 dBA above existing background noise levels if measured using the same metric. However, if background noise is measured using the L_{A90} metric and the total of wind farm noise and background noise is measured using the L_{Aeq} metric, then it is reasonable that a 5-dB increase above background noise is allowed. If multiple wind farms are introduced, then the background noise used for comparison is that measured prior to the introduction of any wind farms. To reduce the influence of extraneous noise and to take into account the time of day or night when wind farm noise is most disturbing, all noise measurements should be conducted during the hours of midnight to 5am.

Another problem with most current regulations is that the allowed noise levels (as well as any background noise levels prior to construction of the wind farm) are based on 10-min or 15-min averages. Each 10-min average represents one data point that is plotted with up to 2000 other similar data points on a graph of 10-min average noise level versus 10-min average wind speed at hub height, where the wind speed at hub height may be measured at that height or inferred from a measurement made at a lower height and assuming a logarithmic wind speed profile. A regression analysis is then done (effectively drawing a line of best fit through the data) and the turbine noise levels are then deemed to be those corresponding to the regression line. This means that there can be a considerable number of data points (representing 10-min averages) that are above the regression line and the wind farm can thus be deemed compliant even though it exceeds the allowable level over a number of 10-min periods. Within these 10-min periods, the instantaneous noise levels can be even higher. Adding to the problem is the fact that noise levels measured during the daytime are averaged with those measured at night when the background noise levels are much lower and the contribution of the wind farm is more apparent. This combination of daytime and nighttime noise levels makes compliance checking difficult, as the daytime measurements are often contaminated by background noise. An improvement to all regulations would be to require that all compliance checking measurements be done between midnight and 5am during conditions of high wind shear so the wind speed at the noise measurement location is sufficiently small that the wind noise is negligible. A further improvement would be to use the statistical analysis method suggested in Section 6.2.14 and write regulations in terms of 'the wind farm only noise cannot exceed x dB more than y% of the time', where x and y are to be decided upon by the relevant jurisdiction.

The measurement results are further compromised in some cases by the use of an L_{A90} measure rather than an L_{Aeq} measure to quantify wind farm noise. The former measure is the level which is exceeded 90% of the time, or in other words the level that the noise is below for only 10% of the time and this is an appropriate measure for background noise. However, when applied to characterising an intrusive noise source, it is only valid if the noise due to the sound source being measured is steady and non-varying at the location of measurement, and this is certainly not true for wind farm noise when distances between the nearest turbine and the point of measurement exceed 500–1000 m. Even at closer distances, the noise level varies considerably due to the sound power output of the turbines varying as a result of turbulent inflow. The unfortunate use of L_{A90} for compliance testing in some jurisdictions, rather than L_{Aeq}, means that measured data considerably underestimate the true sound level experienced by the surrounding community. A number of people have measured both L_{Aeq} and L_{A90} near turbines and have found that L_{A90} data are consistently less than L_{Aeq} data; by 2–3 dBA. However, this is not an accurate estimate of the difference at larger distances from a wind farm where the noise levels vary considerably more than they do near a wind farm due to atmospheric wind, temperature and turbulence variations, which can be very substantial over a 10-min measurement period, making the likely difference even greater. Nevertheless, most regulations do not recommend that any correction be made to L_{A90} data for wind farm noise to compare with allowable and predicted L_{Aeq} levels, even though it is universally accepted that those data considerably underestimate the true average noise levels. The main reason for using L_{A90} data is to eliminate short-term extraneous noise events. However, it would be preferable to take L_{Aeq} measurements in the early hours of the morning when extraneous noise events are at a minimum, but this is a more expensive exercise.

Compliance assessment of wind farms is problematic at any distance greater than 500 m or so from the nearest turbine in a wind farm. This is because regulations rarely require that measurements are done in downward-refracting atmospheric conditions and after midnight but before 5am. Measurements made during the day often do not capture the worst-case turbine noise levels and data are invariably contaminated by other local noise sources, leading to the conclusion that 'there is no evidence that the wind farm is non-compliant'.

When measuring background noise levels prior to the construction of a wind farm, as specified in some regulations in order to determine allowed noise levels from the wind farm, some serious problems arise as listed below:

- The measured background noise level data are plotted as a function of wind speed estimated at hub height, and this wind speed may bear no relationship to the wind speed at the residence, especially in conditions of high wind shear.
- The wind speed at hub height is usually estimated from wind speed data from a 30-m high mast and this estimate can be in considerable error for some atmospheric conditions.
- Daytime and nighttime measurements are all plotted on the same graph, with each data point representing a 10-min or 15-min average. A regression analysis is then done to determine the average background noise level as a function of estimated wind

speed at hub height. If the average background noise level plus 5 dBA is more than the legislated allowed level, then the allowed level becomes background noise plus 5 dBA. There are three problems with this approach:

- Mixing up daytime and nighttime data results in a much higher background noise level than would occur if only the most important hours (from the annoyance and disturbance viewpoints) of midnight to 5am were used.
- The critical high-wind-shear times – when the wind speed at the residence is close to zero while it is sufficient at hub height to drive the turbines close to their maximum output – are not considered separately. This is important as these times represent a low background noise level and are times when the wind farm noise will be most disturbing to residents.
- Regulations do not contain information about the effect of the electronic noise floor of instrumentation when measuring low sound levels. Most instrumentation used by consultants to undertake the background noise measurements has an electronic noise floor of about 18 dBA. This means that they are only able to provide accurate data (with an error of less than 0.5 dBA) for sound levels above about 28 dBA. The error versus actual sound pressure level is shown in Figure 9.1.

It is possible to correct the measured data by mathematically subtracting the electronic noise floor of the instrumentation from the measured data. However, this method is usually only acceptable for measured data at least 3 dB above the electronic noise floor of the instrumentation and it is not appropriate for statistical noise measures such as L_{A90}, as discussed in Section 6.2.14. For L_{Aeq} data, the subtraction should be done ideally in 1/3-octave bands, as the instrumentation noise floor is frequency dependent. The result of reporting incorrect sound levels below about 28 dBA means that the regression line used to fit the data will be grossly distorted, and rather than compensate for the problem by correcting the data, the usual solution is to use higher-order cubic splines rather than just a straight line. The result is that the average background noise levels determined from the regression line and used to determine allowable turbine noise levels are inaccurate and generally overestimates.

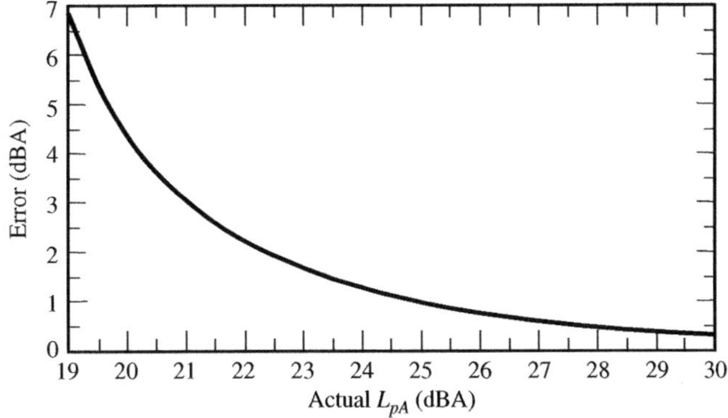

Figure 9.1 Error in sound pressure level for an instrument with an electronic noise floor of 18 dBA. The sound pressure level indicated by the instrument is the actual sound pressure level plus the error.

9.4 Propagation Model Improvements

Noise propagation models have become increasingly sophisticated in recent years, as can be seen from the discussion in Chapter 5. The European Union has spent considerable resources in determining the best model for predicting environmental noise and have concluded that the NMPB-2008 model is the best compromise between accuracy and practicality. However, the NMPB model does not account for multiple ground reflections in a downward-refracting atmosphere. It is true that this effect will only be noticed at a considerable distance from the wind farm and will be mostly apparent for low-frequency noise, as that noise is not absorbed by the atmosphere or the ground to the same extent as mid- and high-frequency noise. As discussed in Chapter 5, the distances from the source at which the first and subsequent ground reflections will occur are a function of the difference in source and receiver heights, as well as the sound-ray radius of curvature, which becomes smaller as the vertical atmospheric sonic gradient becomes larger. A transition distance is defined as the minimum distance of a receiver from a source for which it is possible for two sound rays leaving the sound source at different angles, to arrive at the receiver via one and two ground reflections, respectively. At this distance, the sound level will begin to decay at a rate less than the standard 6 dB per doubling of distance and measured data has shown that it quickly reaches 3 dB per doubling of distance, where it remains. One improvement for noise propagation models would be to include this effect and add contributions from more than one ground-reflected ray to the total sound pressure level.

Noise propagation model outputs typically provide contours of expected noise levels around a wind farm together with expected noise levels at sensitive receiver locations. Data are usually provided as single values, with no indication of what the expected range of levels may be as atmospheric conditions vary, nor are any estimates of uncertainty in the predictions provided. The expected range and uncertainty should be added to standard model outputs, so that people using the data for planning can make more informed decisions.

Many noise prediction models provide estimates of the long-term average sound pressure levels, with a weighting given to nighttime. However, for the purpose of wind farm noise annoyance, it is far more useful to provide long-term average estimates for the nighttime period of 11pm–5am.

9.5 Identification and Amelioration of the Problem Noise Sources on Wind Turbines

Identification of the contribution in terms of magnitude and frequency range of all noise sources on a wind turbine that contribute significantly to noise levels at residences in the near vicinity of wind farms is an important precursor to developing quieter turbines. The purpose of this section is to summarise what has been achieved and what remains to be done, as well as summarising progress in the development of quieter turbines.

9.5.1 Identification of Noise Sources

Chapter 6 provided an overview of methods used to measure and identify noise sources on a wind turbine. Aeroacoustic beamforming arrays, constructed using multiple

(20–200) phase-matched and synchronously acquired microphones with a unique spacing pattern, have recently become popular and will continue to be so. Used in the immediate vicinity of the turbine, they can localise the position and strength of noise sources, but are usually limited to the mid- and high-frequency range. Low-frequency source localisation could be performed using near-field acoustic holography, but this has not been attempted, due to the size of the turbines and the requirement to locate the arrays within one wavelength of the source. Still, one-off measurements using cranes holding long booms of microphones may be possible. Such measurements would clearly show the location of infrasonic and low-frequency sources.

Microphone arrays could also be used at a distance from a wind farm. Possibly, they could be used at a residence to identify the direction from which annoying sounds are coming. Such systems are already used by airports and mine sites to ensure noise compliance. Arrays such as these could conceivably be used to ensure compliance of wind farms.

9.5.2 Amelioration of Noise Sources

Chapter 8 summarised ways to control wind turbine noise at the source. The focus of noise reduction will continue to be optimisation of the blade shape and treatment of the trailing edge. As simulation methods and computing affordability improves, we can expect the development of new airfoil shapes that will increase aerodynamic efficiency and reduce noise emission. The development of trailing-edge devices will also continue. Trailing-edge serrations are being used to reduce noise and we can expect further innovations. Much research and development work has recently been performed based on the inspiration of the owl. New trailing-edge quietening devices that include porous and elastic properties will no doubt make their way onto operational turbines. Such advances could also be applied to the tip region of the blade.

Low-frequency noise generated by blade–tower interaction (BTI) will be more difficult to control. Passive methods of control may include redesign of the tower at the position where the blade passes. The use of absorptive materials and novel tower shapes may go some way towards controlling low-frequency noise of this type.

Active pitch control is one conceivable way of reducing the so-called enhanced amplitude modulation (EAM) noise, or 'blade-thump'. If the proposed mechanism of EAM, where the blade stalls at the top of the arc, proves to be valid, then once-per-revolution blade pitch control may go some way to reducing EAM noise by preventing stall at all angles of rotation. Active phase control, where the relative phases between each turbine are measured and controlled, may be another way of controlling infrasound and low-frequency noise generated by BTI if reinforcement of this noise by multiple turbines is the problem. Another suggested approach is to more carefully optimise the layout of the wind farm, ensuring propagation effects and wake ingestion from upstream turbines and geographical features do not cause additional noise.

Finally, gearbox noise can be controlled through vibration isolation, acoustic foam inside the nacelle and quiet gear design. Further reductions are possible by eliminating the gearbox altogether, using direct drive turbines.

In all of these cases, further progress cannot be achieved without investment by government and industry.

9.6 Reducing Low-frequency Noise Levels in Residences

There are two approaches that should be investigated for reducing the annoyance of wind farm noise in residences:

- masking by using pleasant higher frequency sounds to mask the low-frequency annoying noise
- active noise cancellation to reduce the level of low-frequency noise and also infrasound if that proves to be a significant issue.

As the annoying part of wind farm noise is the low-frequency part, it is difficult to reduce it significantly by structural modifications to residents' houses, especially if residents like to have their windows open. Thus, this option is not considered here. Both the masking and active noise cancellation options are discussed in Chapter 8, but they have not yet been implemented.

References

Ashtiani P 2015 Spectral discrete probability density function of measured wind turbine noise in the far field. *6th International Meeting on Wind Turbine Noise*, Glasgow, UK.

JMEF 2004 Evaluation guide to solve low frequency noise problems. Technical Report TNO report number 2008-D-R1051/B, Japan Ministry of Environment.

NHMRC 2015 NHMRC statement: Evidence on wind farms and human health. Technical report, National Health and Medical Research Council, Australia.

A

Basic Mathematics

A.1 Introduction

The purpose of this appendix is to explain some of the basic mathematical concepts that are necessary for understanding even the most simple noise measurement and analysis concepts described in this book. These concepts include logarithms (needed for understanding the meaning of the decibel, dB), complex numbers (needed for understanding frequency analysis of noise as well as the more advanced noise generation and propagation models) and the exponential function (needed for understanding noise generation as well as the moderately complex and very complex noise propagation models).

A.2 Logarithms

Logarithms have been used extensively throughout this book and, as they are essential to the understanding of the meaning of the word 'decibels' or 'dB', a brief explanation will be provided here.

The purpose of using logarithms is to compress the range of sound pressure levels that we can hear so that it is more manageable.

Perhaps logarithms are best explained by example. Taking the example:

$$2^5 = 2 \times 2 \times 2 \times 2 \times 2 = 32$$

Looking at the problem another way, if we know the answer of 2 raised to some power is 32, we may want to find what that power is, or the value of x in the following relation

$$2^x = 32 \Rightarrow x = 5$$

This problem can be rewritten as log to the base 2 of 32 is equal to 5, or

$$\log_2 32 = 5$$

Thus, logs are a way of finding out the power that a number (the base) is raised to. In this book we only use logs to the base 10. As an example, we may wish to find the logarithm to the base 10 of 100. We know that $100 = 10^2$ so the answer is 2. In acoustics we always divide a number by a reference quantity before taking logs and when we are calculating sound pressure we multiply the result by 20. The use of a reference level allows us to make sure that 0 dB corresponds to the threshold of hearing (for a pure tone at 1800 Hz) for people with normal hearing. Of course, some people can hear to lower

Wind Farm Noise: Measurement, Assessment and Control, First Edition.
Colin H. Hansen, Con J. Doolan, Kristy L. Hansen.

(negative) decibel levels while for others the threshold of hearing is above 0 dB. The use of the factor of 20 means that the decibel range is not too compressed, so that at a level of 124 dB we are experiencing considerable discomfort and even pain. A-weightings are discussed in Section 2.2.11.

A.3 Complex Numbers

Complex numbers are used to a large extent in Section 2.4 and Appendix C. A complex number, C, is defined as

$$C = a + jb \tag{A.1}$$

The complex conjugate of C is defined as

$$C^* = a - jb \tag{A.2}$$

The complex number, C, can also be written in trigonometric form as

$$C = A(\cos \beta + \mathrm{j} \sin \beta) = A\mathrm{e}^{\mathrm{j}\beta} \tag{A.3}$$

where

$$A = |C| = \sqrt{a^2 + b^2} \tag{A.4}$$

and

$$\beta = \arctan{(b/a)} \tag{A.5}$$

A.4 Exponential Function

The exponential function is denoted e and is equal to 2.718. We use it to express complex numbers, so that for example,

$$\mathrm{e}^{\mathrm{j}\theta} = \cos \beta + \mathrm{j} \sin \beta \tag{A.6}$$

Thus the complex number, c, can also be written as

$$c = A\mathrm{e}^{\mathrm{j}\beta} \tag{A.7}$$

The exponential function of argument, x is sometimes written as $\exp(x)$, which is the same as e^x.

B

The BPM model

This appendix presents formulae required to calculate the spectral functions A and D for the sound power methods in Chapter 4. Additionally, the empirical models for boundary layer height are included. Note that these models are described in full in Brooks et al. (1989); a concise summary is provided here. The formulation has been modified so be compatible with the sound power formulation of Chapter 4. The BPM model is named after its developers, Brooks, Pope and Marconi.

B.1 Boundary-layer Parameters

For the *tripped* case (the heavy trip used by Brooks et al. (1989)), the boundary layer heights at the trailing edge for a NACA 0012 airfoil at zero angle of attack can be calculated using

$$\log_{10}\left(\frac{\delta_0}{b}\right) = 1.892 - 0.9045\log_{10}(R_b) + 0.0596[\log_{10}(R_b)]^2 \tag{B.1}$$

$$\log_{10}\left(\frac{\delta_0^*}{b}\right) = \begin{cases} \log_{10}(0.0601) - 0.114\log_{10}(R_b) & (R_b \leq 0.3\times 10^6) \\ 3.411 - 1.5397\log_{10}(R_b) + 0.1059\,[\log_{10}(R_b)]^2 & (R_b > 0.3\times 10^6) \end{cases} \tag{B.2}$$

$$\log_{10}\left(\frac{\theta_0}{b}\right) = \begin{cases} \log_{10}(0.0723) - 0.1765\log_{10}(R_b) & (R_b \leq 0.3\times 10^6) \\ 0.5578 - 0.7079\log_{10}(R_b) + 0.0404\,[\log_{10}(R_b)]^2 & (R_b > 0.3\times 10^6) \end{cases} \tag{B.3}$$

where δ_0 is the boundary layer thickness (defined as the height from the trailing edge to the location where the velocity of the flow is 99% of the potential flow stream velocity), δ_0^* is the displacement thickness, θ_0 is the momentum thickness, b is chord and R_b is the Reynolds number based on chord (as defined in Chapter 3).

For the *untripped* case (natural boundary-layer transition), the boundary layer can be described using

$$\log_{10}\left(\frac{\delta_0}{b}\right) = 1.6569 - 0.9045\log_{10}(R_b) + 0.0596[\log_{10}(R_b)]^2 \tag{B.4}$$

$$\log_{10}\left(\frac{\delta_0^*}{b}\right) = 3.0187 - 1.5397\log_{10}(R_b) + 0.1059[\log_{10}(R_b)]^2 \tag{B.5}$$

$$\log_{10}\left(\frac{\theta_0}{b}\right) = 0.2021 - 0.7079\log_{10}(R_b) + 0.0404[\log_{10}(R_b)]^2 \tag{B.6}$$

Wind Farm Noise: Measurement, Assessment and Control, First Edition.
Colin H. Hansen, Con J. Doolan, Kristy L. Hansen.

At non-zero angles of attack, the zero-angle-of-attack thicknesses are corrected using expressions for the pressure and suction side. For the pressure side for both tripped and untripped boundary layers, the corrections are

$$\log_{10}\left(\frac{\delta_p}{\delta_0}\right) = -0.04175\alpha + 0.00106\alpha^2 \tag{B.7}$$

$$\log_{10}\left(\frac{\delta_p^*}{\delta_0^*}\right) = -0.0432\alpha + 0.00113\alpha^2 \tag{B.8}$$

$$\log_{10}\left(\frac{\theta_p}{\theta_0}\right) = -0.04508\alpha + 0.000873\alpha^2 \tag{B.9}$$

where α is the angle of attack of the airfoil. Note that in the original report by Brooks et al. (1989), a corrected angle of attack was used. As the tests were performed in a free-jet tunnel, the geometric angle of attack is only equivalent to a lower effective aerodynamic angle of attack, due to deflection of the free jet by the airfoil itself. In the case of wind turbines, the angle of attack of the blade element under consideration should be used in this model.

For the suction side for tripped boundary layers, the corrections are

$$\log_{10}\left(\frac{\delta_s}{\delta_0}\right) = \begin{cases} 0.0311\alpha & (0^\circ \leq \alpha \leq 5^\circ) \\ \log_{10}(0.3468) + 0.1231\alpha & (5^\circ < \alpha \leq 12.5^\circ) \\ \log_{10}(5.718) + 0.0258\alpha & (12.5^\circ < \alpha \leq 25^\circ) \end{cases} \tag{B.10}$$

$$\log_{10}\left(\frac{\delta_s^*}{\delta_0^*}\right) = \begin{cases} 0.0679\alpha & (0^\circ \leq \alpha \leq 5^\circ) \\ \log_{10}(0.381) + 0.1516\alpha & (5^\circ < \alpha \leq 12.5^\circ) \\ \log_{10}(14.296) + 0.0258\alpha & (12.5^\circ < \alpha \leq 25^\circ) \end{cases} \tag{B.11}$$

$$\log_{10}\left(\frac{\theta_s}{\theta_0}\right) = \begin{cases} 0.0559\alpha & (0^\circ \leq \alpha \leq 5^\circ) \\ \log_{10}(0.6984) + 0.0869\alpha & (5^\circ < \alpha \leq 12.5^\circ) \\ \log_{10}(4.0846) + 0.0258\alpha & (12.5^\circ < \alpha \leq 25^\circ) \end{cases} \tag{B.12}$$

For the suction side for untripped boundary layers, the corrections are

$$\log_{10}\left(\frac{\delta_s}{\delta_0}\right) = \begin{cases} 0.03114\alpha & (0^\circ \leq \alpha \leq 7.5^\circ) \\ \log_{10}(0.0303) + 0.2336\alpha & (7.5^\circ < \alpha \leq 12.5^\circ) \\ \log_{10}(12) + 0.0258\alpha & (12.5^\circ < \alpha \leq 25^\circ) \end{cases} \tag{B.13}$$

$$\log_{10}\left(\frac{\delta_s^*}{\delta_0^*}\right) = \begin{cases} 0.0679\alpha & (0^\circ \leq \alpha \leq 7.5^\circ) \\ \log_{10}(0.0162) + 0.3066\alpha & (7.5^\circ < \alpha \leq 12.5^\circ) \\ \log_{10}(52.42) + 0.0258\alpha & (12.5^\circ < \alpha \leq 25^\circ) \end{cases} \tag{B.14}$$

$$\log_{10}\left(\frac{\theta_s}{\theta_0}\right) = \begin{cases} 0.0559\alpha & (0^\circ \leq \alpha \leq 7.5^\circ) \\ \log_{10}(0.0633) + 0.2157\alpha & (7.5^\circ < \alpha \leq 12.5^\circ) \\ \log_{10}(14.977) + 0.0258\alpha & (12.5^\circ < \alpha \leq 25^\circ) \end{cases} \tag{B.15}$$

B.2 Turbulent Trailing-edge Noise Model

The BPM spectral shape and amplitude function, A, to be used in the model developed in Chapter 4 is detailed here. For angles of attack $0° \leq \alpha < \alpha_0$, the function A is defined as

$$A = 10\log_{10}(10^{A_\alpha/10} + 10^{A_s/10} + 10^{A_p/10}) \tag{B.16}$$

where

$$A_p = A_1\left(\frac{St_p}{St_1}\right) + (K_1 - 3) + \Delta K_1 \tag{B.17}$$

$$A_s = A_1\left(\frac{St_s}{\overline{St_1}}\right) + (K_1 - 3) \tag{B.18}$$

and

$$A_\alpha = B\left(\frac{St_s}{St_2}\right) + K_2 \tag{B.19}$$

For $\alpha > \alpha_0$

$$A_p = A_s = -\infty \tag{B.20}$$

and

$$A_\alpha = A_2\left(\frac{St_s}{St_2}\right) + K_2 \tag{B.21}$$

here, A_2 is determined using the curve for A_1, but using a value of R_b that is three times the actual value.

Strouhal numbers are given by

$$St_p = \frac{f\delta_p^*}{U}; \qquad St_s = \frac{f\delta_s^*}{U} \tag{B.22}$$

$$St_1 = 0.02\,M^{-0.6} \tag{B.23}$$

$$\overline{St_1} = \frac{St_1 + St_2}{2} \tag{B.24}$$

$$St_2 = St_1 \times \begin{cases} 1 & (\alpha < 1.33°) \\ 10^{0.0054(\alpha-1.33)^2} & (1.33° \leq \alpha \leq 12.5°) \\ 4.72 & (12.5 < \alpha) \end{cases} \tag{B.25}$$

where f is frequency in hertz. A_1 is presented as a function of a, where,

$$a = \left|\log_{10}\left(\frac{St}{St_{\text{peak}}}\right)\right| \tag{B.26}$$

and where $St = St_s, St_p$ and $St_{\text{peak}} = St_1, \overline{St_1}$ or St_2. The function A_1 is then

$$A_1(a) = A_{\min}(a) + A_R(a_0)[A_{\max}(a) - A_{\min}(a)] \tag{B.27}$$

where

$$A_{min}(a) = \begin{cases} \sqrt{67.552 - 886.788a^2} - 8.219 & (a < 0.204) \\ -32.665a + 3.981 & (0.204 \le a \le 0.244) \\ -142.795a^3 + 103.656a^2 - 57.757a + 6.006 & (0.244 < a) \end{cases} \tag{B.28}$$

$$A_{max}(a) = \begin{cases} \sqrt{67.552 - 886.788a^2} - 8.219 & (a < 0.13) \\ -15.901a + 1.098 & (0.13 \le a \le 0.321) \\ -4.669a^3 + 3.491a^2 - 16.699a + 1.149 & (0.321 < a) \end{cases} \tag{B.29}$$

$$a_0(R_b) = \begin{cases} 0.57 & (R_b < 9.52 \times 10^4) \\ (-9.57 \times 10^{-13})(R_b - 8.57 \times 10^5)^2 + 1.13 & (9.52 \times 10^4 \le R_b \le 8.57 \times 10^5) \\ 1.13 & (8.57 \times 10^5 < R_b) \end{cases} \tag{B.30}$$

$$A_R(a_0) = \frac{-20 - A_{min}(a_0)}{A_{max}(a_0) - A_{min}(a_0)} \tag{B.31}$$

Similarly, to calculate B, the value g is used

$$g = \left| \log_{10}\left(\frac{St_s}{St_2}\right) \right| \tag{B.32}$$

with the spectral shape

$$B(g) = B_{min}(g) + B_R(g_0)[B_{max}(g) - B_{min}(g)] \tag{B.33}$$

$$B_{min}(g) = \begin{cases} \sqrt{16.888 - 886.788g^2} - 4.109 & (g < 0.13) \\ -83.607g + 8.138 & (0.13 \le g \le 0.145) \\ -817.810g^3 + 355.210g^2 - 135.024g + 10.619 & (0.145 < g) \end{cases} \tag{B.34}$$

$$B_{max}(g) = \begin{cases} \sqrt{16.888 - 886.788g^2} - 4.109 & (g < 0.10) \\ -31.330g + 1.854 & (0.10 \le g \le 0.187) \\ -80.541g^3 + 44.17g^2 - 39.381g + 2.344 & (0.187 < g) \end{cases} \tag{B.35}$$

$$g_0(R_b) = \begin{cases} 0.30 & (R_b < 9.52 \times 10^4) \\ (-4.48 \times 10^{-13})(R_b - 8.57 \times 10^5)^2 + 0.56 & (9.52 \times 10^4 \le R_b \le 8.57 \times 10^5) \\ 0.56 & (8.57 \times 10^5 < R_b) \end{cases} \tag{B.36}$$

with

$$B_R(g_0) = \frac{-20 - B_{min}(g_0)}{B_{max}(g_0) - B_{min}(g_0)} \tag{B.37}$$

The amplitude function K_1 is

$$K_1 = \begin{cases} -4.31\log_{10}(R_b) + 156.3 & (R_b < 2.47\times 10^5) \\ -9.0\log_{10}(R_b) + 181.6 & (2.47\times 10^5 \le R_b \le 8.0\times 10^5) \\ 128.5 & (8.0\times 10^5 < R_b) \end{cases} \tag{B.38}$$

The level adjustment ΔK_1 is

$$\Delta K_1 = \begin{cases} -\alpha[1.43\log_{10}(R_{\delta_p^*}) - 5.29] & (R_{\delta_p^*} \le 5000) \\ 0 & (2.47\times 10^5 \le (R_{\delta_p^*} > 5000) \end{cases} \tag{B.39}$$

Here, $R_{\delta_p^*} = \frac{U_\infty \delta_p^*}{\nu}$ is the Reynolds number based on the pressure side trailing-edge displacement thickness, as defined in Chapter 3 (U_∞ is free-stream velocity and ν is the kinematic viscosity).

The amplitude function K_2 is given by

$$K_2 = K_1 \begin{cases} -1000 & (\alpha < \gamma_0 - \gamma) \\ \sqrt{\beta^2 - (\beta/\gamma)^2(\alpha - \gamma_0)^2} + \beta_0 & (\gamma_0 - \gamma \le \alpha \le \gamma_0 + \gamma) \\ -12 & (\gamma_0 + \gamma < \alpha) \end{cases} \tag{B.40}$$

where

$$\gamma = 27.094M + 3.31 \qquad \gamma_0 = 23.43M + 4.651 \tag{B.41}$$

$$\beta = 72.65M + 10.74 \qquad \beta_0 = -34.19M - 13.82 \tag{B.42}$$

B.3 Blunt Trailing-edge Noise Model

The spectral shape and amplitude function, D, to be used in the model presented in Chapter 4 is detailed here.

$$D = G_4\left(\frac{h}{\delta_{avg}^*}, \Psi\right) + G_5\left(\frac{h}{\delta_{avg}^*}, \Psi, \frac{St'''}{St'''_{peak}}\right) \tag{B.43}$$

where h is the thickness of the trailing edge. The Strouhal number is defined by

$$St''' = \frac{fh}{U} \tag{B.44}$$

where f is frequency in hertz and

$$St'''_{peak} = \begin{cases} \dfrac{0.212 - 0.0045\Psi}{1 + 0.235(h/\delta_{avg}^*)^{-1} - 0.0132(h/\delta_{avg}^*)^{-2}} & (0.2 \le h/\delta_{avg}^*) \\ 0.1(h/\delta_{avg}^*) + 0.095 - 0.00243\Psi & (h/\delta_{avg}^* > 0.2) \end{cases} \tag{B.45}$$

The average of the displacement thicknesses on either side of the airfoil trailing edge is

$$\delta_{avg}^* = \frac{\delta_s^* + \delta_p^*}{2} \tag{B.46}$$

Ψ is the angle between the sloping surfaces upstream of the trailing edge. This will vary from airfoil to airfoil. For a NACA 0012, $\Psi = 14°$.

The function G_4 is defined as

$$G_4\left(\frac{h}{\delta^*_{avg}}, \Psi\right) = \begin{cases} 17.5\log_{10}\left(\frac{h}{\delta^*_{avg}}\right) + 157.5 - 1.114\Psi & (h/\delta^*_{avg} \le 5) \\ 169.7 - 1.114\Psi & (h/\delta^*_{avg} > 5) \end{cases} \tag{B.47}$$

The function G_5 is calculated using the following.

$$G_5\left(\frac{h}{\delta^*_{avg}}, \Psi, \frac{St'''}{St'''_{peak}}\right) = (G_5)_{\Psi=0°} + 0.0714\Psi[(G_5)_{\Psi=14°} - (G_5)_{\Psi=0°}] \tag{B.48}$$

where

$$(G_5)_{\Psi=14°} = \begin{cases} m\eta + k & (\eta < \eta_0) \\ 2.5\sqrt{1-(\eta/\mu)^2} - 2.5 & (\eta_0 \le \eta < 0) \\ \sqrt{1.5625 - 1194.99\eta^2} - 1.25 & (0 \le \eta < 0.03616) \\ -155.543\eta + 4.375 & (0.03516 \le \eta) \end{cases} \tag{B.49}$$

$$\eta = \log_{10}\left(\frac{St'''}{St'''_{peak}}\right) \tag{B.50}$$

$$\mu = \begin{cases} 0.1221 & (h/\delta^*_{avg} \le 0.25) \\ -0.2175(h/\delta^*_{avg}) + 0.1755 & (0.25 \le h/\delta^*_{avg} < 0.62) \\ -0.0308(h/\delta^*_{avg}) + 0.0596 & (0.62 \le h/\delta^*_{avg} < 1.15) \\ 0.0242 & (1.15 \le h/\delta^*_{avg}) \end{cases} \tag{B.51}$$

$$m = \begin{cases} 0 & (h/\delta^*_{avg} \le 0.02) \\ 68.724(h/\delta^*_{avg}) - 1.35 & (0.02 < h/\delta^*_{avg} \le 0.5) \\ 308.475(h/\delta^*_{avg}) - 121.33 & (0.5 < h/\delta^*_{avg} \le 0.62) \\ 224.811(h/\delta^*_{avg}) - 69.35 & (0.62 < h/\delta^*_{avg} \le 1.15) \\ 1583.28(h/\delta^*_{avg}) - 1631.59 & (1.15 < h/\delta^*_{avg} < 1.2) \\ 268.344 & (1.2 < h/\delta^*_{avg}) \end{cases} \tag{B.52}$$

$$\eta_0 = -\sqrt{\frac{m^2\mu^4}{6.25 + m^2 + \mu^2}} \tag{B.53}$$

and

$$k = 2.5\sqrt{1 - \left(\frac{\eta_0}{\mu}\right)^2} - 2.5 - m\eta_0 \tag{B.54}$$

The method of calculating $(G_5)_{\Psi=0^\circ}$ is exactly the same as for $(G_5)_{\Psi=14^\circ}$, except that (h/δ^*_{avg}) is replaced with $(h/\delta^*_{avg})'$ in the calculation.

$$(h/\delta^*_{avg})' = 6.724\left(\frac{h}{\delta^*_{avg}}\right)^2 - 4.019\left(\frac{h}{\delta^*_{avg}}\right) + 1.107 \tag{B.55}$$

Reference

Brooks T, Pope D and Marcolini M 1989 Airfoil self-noise and prediction. Reference Publication 1218, NASA.

C

Ground Reflection Coefficient Calculations

C.1 Introduction

In Chapter 5, various sound propagation algorithms for estimating noise levels produced by wind farms in surrounding communities were discussed in detail. The more complex procedures relied on the calculation of various types of absorption coefficient, and means of calculating these coefficients are the subject of this appendix.

All the reflection coefficient types discussed here require an estimate of the characteristic impedance of the ground, which in turn requires an estimate of the flow resistivity of the ground surface. For this reason this appendix begins with a discussion of ground surface flow resistivity followed by an explanation of how this may be used to calculate the ground impedance. This is followed by separate sections for calculating the plane-wave reflection coefficient, the spherical-wave reflection coefficient and the incoherent reflection coefficient. The plane-wave reflection coefficient is calculated based on the assumption that the sound wave incident on the ground is a plane wave. This assumption can lead to considerable errors in the calculation of the phase shift as well as the amplitude of the reflected wave, which is why the more advanced propagation models prefer to use the spherical-wave reflection coefficient. This latter coefficient is derived from the plane-wave reflection coefficient, but provides more accurate results for the phase and amplitude of the reflected wave, which is important if the direct and reflected waves from the same source are to be combined coherently at the receiver location. However, the coherence of the waves at the receiver location is affected by atmospheric turbulence as well as by frequency averaging the results over a 1/3-octave or octave band. This effect becomes more pronounced at higher frequencies so that at some frequency there is effectively incoherent combination of the various waves from the same source at a particular receiver location. In this case, it is preferred to use the incoherent reflection coefficient because it is much simpler to calculate and uses much fewer computer resources.

Reflection coefficient calculations for a ground surface require a knowledge of the normal surface impedance of the earth, which is approximated as the characteristic impedance, Z_m. The normal surface impedance can be calculated from the flow resistivity of the earth for the particular ground cover of interest. The concepts of characteristic impedance and flow resistivity are discussed in the following sections.

Wind Farm Noise: Measurement, Assessment and Control, First Edition.
Colin H. Hansen, Con J. Doolan, Kristy L. Hansen.

C.2 Flow Resistivity

The flow resistivity of a ground surface is a measure of how resistant the ground is to a mean air flow through it. It can be determined using a method described in the standards, ANSI SI.18 (1996) and Nordtest (1999), which involves measuring the normal surface impedance using two microphones and a sound source. The method uses two fixed microphone positions, referred to as the bottom and the top microphone respectively, located 0.2 m and 0.5 m above the ground at a specified distance from the sound source, which is at a height of 0.5 m above the ground. The sound source should be located 1750 m from the microphones. At least four independent measurements should be used at different locations around the sound source. The sound source is excited by random noise over the frequency range that includes the 1/3-octave bands with centre frequencies from 200 to 2500 Hz. The measured sound pressure level difference, ΔL_{pm}, between the microphone outputs is used to determine the surface normal impedance, which is approximately equal to the ground characteristic impedance defined in Section C.3 below. The impedance is determined by finding the value that minimises the difference between the measured and calculated sound pressure level difference, $\Delta L_{pm} - \Delta L_{pc}$, between the two microphones. The calculated level difference, ΔL_{pc} is related to the impedance via the spherical-wave reflection coefficient described in Section C.5. It is given by the following equation:

$$\Delta L_{pc} = L_{pu} - L_{p\ell} = 20 \log_{10} \frac{p_u}{p_\ell}$$
$$= 20 \log_{10} \left[\left(\frac{\mathrm{e}^{-\mathrm{j}kr_{du}}}{r_{du}} + \frac{Q_u \mathrm{e}^{-\mathrm{j}kr_{ru}}}{r_{ru}} \right] \Big/ \left[\frac{\mathrm{e}^{-\mathrm{j}kr_{d\ell}}}{r_{d\ell}} + \frac{Q_\ell \mathrm{e}^{-\mathrm{j}kr_{r\ell}}}{r_{r\ell}} \right) \right] \tag{C.1}$$

where k is the wavenumber defined in Section 2.2.3, p_u is the sound pressure measured by the upper microphone and p_ℓ is the sound pressure measured by the lower microphone. The spherical-wave reflection coefficient (see Section C.5) for the wave travelling from the source via a ground reflection and then to the upper microphone is Q_u, while that for the wave travelling from the source via a ground reflection and then to the lower microphone is Q_ℓ. Similarly, the distance of the ground reflected wave to the upper microphone is r_{ru} and to the lower microphone it is $r_{r\ell}$, while the distance travelled by the direct (non-reflected) paths from the source to upper and lower microphones are r_{du} and $r_{d\ell}$ respectively.

Values for the flow resistivity of various ground surfaces are listed in Table C.1 below. The values were obtained from ANSI S1.18 (1996), Nordtest (1999), Embleton et al. (1983) and Attenborough et al. (2006).

As can be seen from Table C.1, there are great variations in measured flow resistivity data from the different sources. For this reason, the NORD2000 propagation model classifies ground surfaces into eight types, labelled A–H and listed in Table C.2

Type H also represents the upper limit set by thermal conduction and the viscous boundary layer.

Table C.1 Flow resistivities measured for some common ground surfaces

Ground surface type	**Flow resistivity, R_1 (kPa s/m^2)**
Dry snow, newly fallen 0.1 m over about 0.4 m older snow	10–30
Sugar snow	25–50
Soft forest floor with blueberry greens and moss	40
Forest floor covered by weeds	63–100
Pine or hemlock forest floor	20–80
Soft forest floor covered with pine needles	160
Sandy forest floor	630–2000
Dense shrubbery, 20 cm high	100
Earth and bark, sparse vegetation	100
Peat or turf area, homogeneous organic material	100
Earth covered with leaves and twigs	160–250
Earth mixed with sawdust	250
Relatively dense soil sparsely covered by grass and other low greens	630
Short grass, green moss and blueberry greens	40
Rough grassland and pasture	100–300
Grass, soccer field	630–2000
Lawn, moderately stepped on	160–250
Lawn, seldom stepped on	250
Lawn	250–400
Agricultural field	160–250
Hard soil	400–2000
Earth, exposed and rain packed	4000–8000
Wet, sandy loam	1500
Moistened sand	500
Bare sandy plain	250–500
Dry sand	60–140
Sandy silt, hard packed by vehicles	800–2500
Quarry dust, hard packed by vehicles	5000–20 000
Mixed paving stones and grass	630–2000
Old gravel field with sparse vegetation	2000
Gravel road, stones and dust	2000
Gravel parking lot	630–2000
Asphalt sealed by dust and light use	30 000
concrete	20 000

C.3 Characteristic Impedance

The complex characteristic impedance, Z_m, of the ground can be calculated from the flow resistivity, R_1, using the following relation (assuming $\exp^{+j\omega t}$ time dependence), which is valid up to a flow resistivity of 500 kPa s/m^2.

$$Z_m = \rho c \left(1 + 0.0571X^{-0.754} - j0.087X^{0.732}\right) \tag{C.2}$$

Table C.2 Flow resistivities for some common ground surface types (to be used with the ISO9613-2, NMPB-2008, NORD2000 and Harmonoise propagation models)

Ground surface class	Value of ISO9613-2 parameter, *G*	Value of NMPB-2008 parameter, *G*	Representative flow resistivity R_1 (kPa s/m^2) (Harmonoise)	Ground surface description
A	1	1	12.5	Very soft (snow or moss)
B	1	1	31.5	Soft forest floor
C	1	1	80	Uncompacted, loose ground
D	1	1	200	Normal uncompacted ground (pastures, forest floors)
E	0	0.7	500	Compacted fields, lawns and gravel
F	0	0.3	2000	Compacted dense ground (gravel road, parking lot)
G	0	0	20 000	Asphalt, concrete
H	0	0	200 000	Water

where

$$X = \rho f/R_1 \tag{C.3}$$

and where f is the frequency of sound (Hz), ρ is the density of air and c is the speed of sound at the ground.

For values of R_1 above 500 kPa s/m^2, it is necessary to use more accurate, although more complicated equations. The following equations hold for all values of R_1. The characteristic impedance is given by

$$Z_m = \rho c\sqrt{\rho_m \kappa} \tag{C.4}$$

where the normalised complex density, ρ_m, of the ground is written as

$$\rho_m = [1+\sigma]^{-1} \tag{C.5}$$

and where the normalised complex compressibility, κ, of the ground is written as

$$\kappa = [1+(1-\gamma)\tau]^{-1} \tag{C.6}$$

where γ is the ratio of specific heats of air (= 1.4) and

$$\sigma = a(X) + jb(X) \tag{C.7}$$

$$\tau = 0.592a(X_1) + jb(X_1) \tag{C.8}$$

$$a(X) = \frac{T_3(T_1-T_3)T_2^2 - T_4^2T_1^2}{T_3^2T_2^2 + T_4^2T_1^2} \tag{C.9}$$

$$b(X) = \frac{T_1^2T_2T_4}{T_3^2T_2^2 + T_4^2T_1^2} \tag{C.10}$$

$$T_1 = 1 + 9.66X; \qquad T_2 = X(1+0.0966X) \tag{C.11}$$

$$T_3 = 2.537 + 9.66X; \qquad T_4 = 0.159(1+0.7024X) \tag{C.12}$$

The quantities $a(X_1)$ and $b(X_1)$ are calculated by substituting $X_1 = 0.856X$ for the quantity X in Eqs. (C.9)–(C.12).

C.4 Plane-wave Reflection Coefficient

The plane-wave reflection coefficient is a complex number characterised by an amplitude and a phase. It is multiplied by the complex amplitude of the wave incident on the ground to determine the complex amplitude of the reflected wave arriving at the receiver, after taking into account the propagation distance. According to Attenborough (1988), the plane-wave reflection coefficient is valid when the source and receiver are well elevated from the ground or when the angle, θ, in Figure 5.2 is less than 85°.

The complex amplitude reflection coefficient, R_p for plane waves, may be written as follows (Bies and Hansen 2009):

$$R_p = \frac{Z_m \cos\theta - \rho c \cos\psi}{Z_m \cos\theta + \rho c \cos\psi} \tag{C.13}$$

where θ is defined in Figure 5.2 and

$$\cos\psi = \sqrt{1 - \left(\frac{k}{k_m}\right)^2 \sin^2\theta} \tag{C.14}$$

where

$$k_m = \frac{2\pi f}{c}\sqrt{\frac{\rho_m}{\kappa}} \tag{C.15}$$

and where ρ_m and κ are defined in Eqs. (C.5) and (C.6).

The complex reflection coefficient, R_p, may be written in terms of an amplitude and phase as

$$R_p = |R_p| \mathrm{e}^{\,j\alpha} \tag{C.16}$$

where $\alpha = \tan^{-1}[\mathrm{Im}\{R_p\}/\mathrm{Re}\{R_p\}]$ is the phase angle between reflected and incident rays.

Reference to Eq. (C.14) shows that when $k_m >> k$, the angle, ψ tends to zero and Eq. (C.13) reduces to the following form:

$$R_p = \frac{Z_m \cos\theta - \rho c}{Z_m \cos\theta + \rho c} \tag{C.17}$$

which is the equation for a locally reactive surface, for which propagation of sound in the lateral direction is negligible. A surface can be considered locally reactive (that is the displacement at a particular location on a surface is not affected by the displacement at any other location) if $k_m > 100k$, and this will occur if

$$\rho f < 10^{-3} R_1 \tag{C.18}$$

C.5 Spherical-wave Reflection Coefficient

According to Attenborough (1988), when the angle, θ, in Figure 5.2 is greater than 85°, the curvature of the wavefront striking the ground must be taken into account by using a complex spherical-wave coefficient in place of the plane-wave coefficient discussed in the previous section. In the following analysis, the equations are slightly different to the original references, as positive time dependence, $e^{j\omega t}$ has been used to be consistent with the rest of the book, rather than negative time dependence, $e^{-i\omega t}$, as used in the original references.

The complex amplitude reflection coefficient, Q, of a spherical wave incident upon a reflecting surface (Attenborough et al. 1980) may be written as follows:

$$Q = R_p + (1 - R_p)[BG(w)]^{n_G} = |Q|e^{j\alpha} \tag{C.19}$$

where $\alpha = \tan^{-1}[\mathrm{Im}\{Q\}/\mathrm{Re}\{Q\}]$ is the phase angle between reflected and incident rays, and where the exponent n_G has been added to the original spherical-wave reflection coefficient model to provide a more realistic solution for grazing incidence (Salomons and Janssen 2011). The value of n_G is given by

$$n_G = 1 - 0.7e^{-16(h'_S + h'_R)/\lambda} \tag{C.20}$$

where h'_S and h'_R are, respectively, the source and receiver heights above the ground segment under consideration or its extension (see Figure G.4). For high sources such as wind turbines, $n_G \approx 1$.

In almost all cases, the ground may be considered to be extensively reactive, which implies that wave propagation into the ground at a point on the ground is affected by the properties of the earth at nearby locations, and sound waves travelling horizontally in the earth have a significant amplitude for a short distance from their origin. In this case, B of Eq. (C.19) is defined as follows:

$$B = \frac{B_1 B_2}{B_3 B_4 B_5} \tag{C.21}$$

where

$$B_1 = \left[\cos\theta + \frac{\rho c}{Z_m}\left(1 - \frac{k^2}{k_m^2}\sin^2\theta\right)^{1/2}\right]\left[1 - \frac{k^2}{k_m^2}\right]^{1/2} \tag{C.22}$$

$$B_2 = \left\{\left[1 - \frac{\rho}{\rho_m^2}\right]^{1/2} + \frac{\rho c}{Z_m}\left[1 - \frac{k^2}{k_m^2}\right]^{1/2}\cos\theta + \left[1 - \left(\frac{\rho c}{Z_m}\right)^2\right]\sin\theta\right\}^{1/2} \tag{C.23}$$

$$B_3 = \cos\theta + \frac{\rho c}{Z_m}\left(1 - \frac{k^2}{k_m^2}\right)^{1/2}\left(1 - \frac{\rho}{\rho_m^2}\right)^{-1/2} \tag{C.24}$$

$$B_4 = \left(1 - \frac{k^2}{k_m^2}\sin\theta\right)^{1/2} \tag{C.25}$$

$$B_5 = \left[1 - \frac{\rho}{\rho_m^2}\right]^{3/2}\left[1 - \left(\frac{\rho c}{Z_m}\right)^2\right]^{1/2}[2\sin\theta]^{1/2} \tag{C.26}$$

The argument, w, of $G(w)$ in Eq. (C.19) is referred to as the numerical distance and is calculated using the following equation, where r_1 and r_2 are defined in Figure 5.2:

$$w = \frac{1}{2}(1 - \mathrm{j})[2k(r_S + r_R)]^{1/2}\frac{B_3}{B_6^{1/2}} \tag{C.27}$$

In Eq. (C.27), B_3 is defined as in Eq. (C.24) and B_6 is defined as

$$B_6 = 1 + \left[\frac{\rho c}{Z_m}\left(1 - \frac{k^2}{k_m^2}\right)^{1/2} \cos\theta + \left(1 - \frac{\rho c}{Z_m}\right)^{1/2} \sin\theta\right]\left[1 - \frac{\rho}{\rho_m^2}\right]^{-1/2} \tag{C.28}$$

The term $G(w)$ in Eq. (C.19) is defined as

$$G(w) = 1 - \mathrm{j}\sqrt{\pi}wg(w) \tag{C.29}$$

where

$$g(w) = \mathrm{e}^{-w^2}\operatorname{erfc}(-\mathrm{j}w) \tag{C.30}$$

where erfc() is the error function (Abramowitz and Stegun 1964) and w is a complex number so that $w = w_r + \mathrm{j}w_i$.

For small w, where the real part is less than or equal to 3.9 and the imaginary part is less than or equal to 3,

$$g(w) = -K_1(w_r, w_i) + \mathrm{j}K_2(w_r, w_i) \tag{C.31}$$

where

$$\begin{aligned} K_1(w_r, w_i) = {} & \frac{hw_i}{\pi(w_r^2 + w_i^2)} + \frac{2w_ih}{\pi}\sum_{n=1}^{\infty}\frac{\mathrm{e}^{-n^2h^2}(w_r^2 + w_i^2 + n^2h^2)}{(w_i^2 - w_r^2 + n^2h^2)^2 + 4w_r^2w_i^2} \\ & - \frac{w_i}{\pi}E(h) \\ & + P \quad \text{if} \quad w_i < \pi/h \\ & + 0.5P \quad \text{if} \quad w_i = \pi/h \\ & + 0 \quad \text{if} \quad w_i > \pi/h \end{aligned} \tag{C.32}$$

$$\begin{aligned} K_2(w_r, w_i) = {} & \frac{hw_r}{\pi(w_r^2 + w_i^2)} + \frac{2w_rh}{\pi}\sum_{n=1}^{\infty}\frac{\mathrm{e}^{-n^2h^2}(w_r^2 + w_i^2 - n^2h^2)}{(w_i^2 - w_r^2 + n^2h^2)^2 + 4w_r^2w_i^2} \\ & + \frac{w_r}{\pi}E(h) \\ & - F \quad \text{if} \quad w_i < \pi/h \\ & - 0.5F \quad \text{if} \quad w_i = \pi/h \\ & - 0 \quad \text{if} \quad w_i > \pi/h \end{aligned} \tag{C.33}$$

and where

$$P = 2e^{-[w_r^2+(2w_i\pi/h)-w_i^2]}\left[\frac{(A_1C_1 - B_1D_1)}{(C_1^2 + D_1^2)}\right] \tag{C.34}$$

$$F = 2e^{-[w_r^2+(2w_i\pi/h)-w_i^2]}\left[\frac{(A_1D_1 + B_1C_1)}{(C_1^2 + D_1^2)}\right] \tag{C.35}$$

$$\begin{cases} A_1 = \cos(2w_r w_i) \\ B_1 = \sin(2w_r w_i) \\ C_1 = e^{-2w_i\pi/h} - \cos(2w_r\pi/h) \\ D_1 = \sin(2w_r\pi/h) \end{cases} \tag{C.36}$$

The error bound can be estimated from (Attenborough et al. 2006):

$$E(h) \le \frac{2\sqrt{\pi}\ e^{-(\pi^2/h^2)}}{\left(1 - e^{-(\pi^2/h^2)}\right)} \tag{C.37}$$

Note that h is a constant selected by the user. If $h = 1$, then $E(h) \le 2 \times 10^{-4}$ and only three or four terms are needed in the infinite sums included in the expressions for K_1 and K_2. If h is reduced to 0.8, then $E(h) \le 10^{-6}$ and five terms will be needed in the infinite sums included in the expressions for K_1 and K_2. It is recommended that $h = 0.8$ be used (Plovsing 2006).

For values of w where the real part is greater than 3.9 or the imaginary part is greater than 3 and both are less than or equal to 6:

$$g(w) = jw\left[\frac{0.4613135}{w^2 - 0.1901635} + \frac{0.09999216}{w^2 - 1.7844927} + \frac{0.002883894}{w^2 - 5.5253437}\right] \tag{C.38}$$

For real or imaginary parts of w greater than 6:

$$g(w) = jw\left[\frac{0.5124242}{w^2 - 0.275255} + \frac{0.05176536}{w^2 - 2.724745}\right] \tag{C.39}$$

For the ground to be considered as essentially locally reactive, the following must be satisfied:

$$\frac{\rho f}{R_1} < 10^{-3} \tag{C.40}$$

In this case, the following simplifications are possible:

$$B_1 = B_3 = \cos\theta + \frac{\rho c}{Z_m}; \quad B_2 = (1 + \sin\theta)^{1/2} \tag{C.41}$$

$$B_4 = 1; \quad B_5 = (2\sin\theta)^{1/2}; \quad B_6 = B_2^2 \tag{C.42}$$

$$w = \frac{1}{2}(1 - j)[2k(r_S + r_R)]^{1/2}\left(\cos\theta + \frac{\rho c}{Z_m}\right)(1 + \sin\theta)^{-1/2} \tag{C.43}$$

Equation (C.19) consists of two terms, with the first being R_p. The entire second term acts as a correction for the fact that the wavefronts are spherical rather than plane. Although this contribution is referred to as a 'ground wave', it is actually a correction to the reflected wave that takes into account the diffusion of the image source position as a result of the spherical wavefront striking the ground (Attenborough et al. 2006).

At low frequencies a 'surface wave' may exist. This is an entirely separate wave that propagates along the ground surface, with exponentially decreasing amplitude above the ground. This wave is not affected by atmospheric refraction due to wind and temperature gradients and is more likely to exist at low frequencies. The pressure amplitude associated with surface waves only decays with the inverse of the square root of distance (3 dB per doubling of distance), compared to a decay with inverse distance of other waves. However, the resistive impedance components of most outdoor ground surfaces give rise to significant exponential attenuation along the surface, which tends to reduce the amplitude so that the resulting overall decay rate is larger than for the other waves. One exception to this is propagation in an upwind direction, in which case the other waves are severely attenuated as a result of upward refraction due to the negative vertical atmospheric wind gradient. In this case the sound levels in the shadow zone caused by upward refraction of the other waves may be dominated by the surface wave. The surface-wave contribution is already included in the calculation of the ground-wave term in the preceding equations, but in order for it to exist, the following condition has to be satisfied, and it is often satisfied for long propagation distances over porous ground surfaces such as grass:

$$\mathrm{Re}[1 + \beta \cos\theta - (1 - \beta^2)^{1/2} \sin\theta] > 0 \tag{C.44}$$

where Re[X] is the real part of the function, X, β is the ratio of the ground surface admittance to the air admittance. It is defined as the reciprocal of the ratio of the ground impedance of Eqs. (C.2) or (C.4) to the characteristic impedance in air, ρc. That is,

$$\beta = \frac{\rho c}{Z_m} = \beta_r - \mathrm{j}\beta_i \tag{C.45}$$

Equation (C.44) can be rewritten in terms of the real and imaginary parts of the admittance as Attenborough et al. (2006):

$$\frac{(\beta_r + \cos\theta)(1 + \beta_r \cos\theta)}{\sqrt{1 + 2\beta_r \cos\theta + \beta_r^2}} > \beta_i \sin\theta \tag{C.46}$$

indicating that surface waves will propagate when the acoustic compliance of the ground exceeds its acoustic resistance.

Surface waves become important in upwind propagation, when upward refraction of the other waves results in an acoustic shadow but the surface wave is unaffected. Surface waves can also be responsible for low-frequency sound levels in some cases being greater than expected purely from spherical spreading at large distances (over about 300 m) and for low frequencies. The surface wave has a speed of sound less than the speed of sound in air and it decays exponentially as the height of the receiver above the ground increases. Perhaps the most favourable ground surface for surface-wave propagation is fresh snow on frozen ground.

C.6 Incoherent Reflection Coefficient

The incoherent reflection coefficient, $\overline{\mathfrak{R}}$ is related to the statistical absorption coefficient, α_{st} of the ground surface, which can be calculated from the real, $R_i = \mathrm{Re}\{Z_m/(\rho c)\}$, and imaginary, $X_i = \text{Im}\{Z_m/(\rho c)\}$, parts of the normalised characteristic impedance, $Z_m/(\rho c)$ of the ground (Bies and Hansen 2009). Thus,

$$\frac{Z_m}{\rho c} = R_i + \mathrm{j}X_i = \xi \mathrm{e}^{\,\mathrm{j}\psi} \tag{C.47}$$

$$\xi = \sqrt{R_i^2 + X_i^2}; \quad \psi = \tan^{-1}(X_i/R_i) \tag{C.48}$$

and

$$\overline{\Re} = \sqrt{1 - \alpha_{st}} \tag{C.49}$$

where

$$\alpha_{st} = \left\{\frac{8\cos\psi}{\xi}\right\}\left\{1 - \left[\frac{\cos\psi}{\xi}\right]\log_{\mathrm{e}}(1 + 2\xi\cos\psi + \xi^2) + \left[\frac{\cos(2\psi)}{\xi\sin\psi}\right]\tan^{-1}\left[\frac{\xi\sin\psi}{1 + \xi\cos\psi}\right]\right\} \tag{C.50}$$

References

Abramowitz M and Stegun I 1964 *Handbook of mathematical Functions.* National Bureau of Standards, USA.

ANSI S1.18 1996 Template method for ground impedance. Technical report, ANSI.

Attenborough K 1988 Review of ground effects on outdoor sound propagation from continuous broadband sources. *Applied acoustics* **24**(4), 289–319.

Attenborough K, Hayek S and Lawther J 1980 Propagation of sound above a porous half-space. *Journal of the Acoustical Society of America* **68**(5), 1493–1501.

Attenborough K, Li K and Horoshenkov K 2006 *Predicting outdoor sound.* CRC Press.

Bies D and Hansen C 2009 *Engineering noise control: theory and practice* 4th edn. Spon press.

Embleton T, Piercy J and Daigle G 1983 Effective flow resistivity of ground surfaces determined by acoustical measurements. *Journal of the Acoustical Society of America* **74**(4), 1239–1244.

Nordtest 1999 NT-ACOU-104: ground surfaces: Determination of the acoustic impedance.

Plovsing B 2006 Comprehensive outdoor sound propagation model. part 1: Propagation in an atmosphere without significant refraction. Technical report AV 1849/00, Delta.

Salomons E and Janssen S 2011 Practical ranges of loudness levels of various types of environmental noise, including traffic noise, aircraft noise, and industrial noise. *International journal of environmental Research and public health* **8**(6), 1847–1864.

D

Calculation of Ray Path Distances and Propagation Times for the Nord2000 Model

D.1 Introduction

In Chapter 5, the Nord2000 model (Plovsing 2006a,b, 2014) was described and the idea of wind and temperature gradients affecting the curvature of the sound rays as they travelled from source to receiver was discussed. A method was also discussed that could be used to find the radius of curvature of the various rays for a log-lin variation of sound speed with height above the ground. The method used an expression for the maximum height of a non-circular ray calculated using Snell's law (see Eq. 4.9 in Salomons (2001)). The radius of a circular ray path that reached the same height was then determined, allowing propagation times and travel distances along this ray path to be calculated. The Nord2000 model uses a different approach in which the log-lin sound-speed profile is converted to an equivalent atmospheric vertical sound-speed profile that is linear, so that the curved rays will travel along the arc of a circle.

In this appendix, means are provided for calculating the radius of the circular arc as well as circular arc distances and propagation times for direct and reflected rays arriving at the receiver from the sound source. The circular path radius is a function of local vertical sonic gradients, which are a function of local wind and temperature gradients. Here, only cases that result in downward-diffracting waves (corresponding to a positive sound-speed gradient, which is represented by increasing sound speed with increasing altitude) are considered, as this represents the case of least attenuation of the propagating sound. This sound-speed gradient can be a result of downwind propagation (as the wind speed increases with altitude) and/or a temperature inversion in which the air temperature increases with altitude within a few hundred metres of the ground.

The methods for calculating the ray path length and propagation time are also applicable to paths involving diffraction over a barrier, in which case the top of the barrier replaces the receiver in order to calculate the path from the source to the barrier top, and the top of the barrier replaces the source in order to calculate the path from the barrier top to the receiver. The Nord2000 model recommends adjusting the barrier heights when the source or receiver is placed on the barrier top, replacing the barrier with one of an equivalent height that depends on the propagation distance. The equivalent height, h''_b, of the screen or barrier is calculated as follows (Plovsing 2014):

$$h''_b = \begin{cases} \dfrac{4(d''_{SbR} - 75)}{135} & d''_{SbR} > 75 \\ 0 & d''_{SbR} \leq 75 \end{cases} \tag{D.1}$$

Wind Farm Noise: Measurement, Assessment and Control, First Edition.
Colin H. Hansen, Con J. Doolan, Kristy L. Hansen.

where

$$d''_{SbR} = d_{SR} + \min(d_{Sb}, d_{Rb}) \tag{D.2}$$

and d_{SR} is the source receiver distance, d_{Sb} is the distance from the source to the actual barrier top and d_{Rb} is the distance from the receiver to the actual barrier top.

The parameters and variables that need to be calculated to find the relative phases of the rays arriving at the receiver location from a single sound source, for the Nord2000 model, are:

- equivalent linear atmospheric vertical sound-speed profile.
- ray path variables: length and propagation times of all ray paths from the source to the receiver.

D.2 Equivalent Linear Atmospheric Vertical Sound-speed Profile

The equivalent linear vertical sound-speed profile is needed for the calculation of ray travel times and distances because the log-lin profile of Eq. (5.26) does not produce circular arcs, making travel times and distances very difficult to calculate.

The equivalent linear sound-speed profile is defined as (Plovsing 2006b):

$$c_e(h) = c_0 + \frac{\Delta c}{\Delta h} h = c_0(1 + \xi_n h) \tag{D.3}$$

where c_0 is the speed of sound at height $h = 0$ in stationary air. The quantity, $\Delta c / \Delta h$ is the average gradient between source and receiver heights, h_S and h_R, respectively and xi_n is this gradient normalised with the speed of sound, c_0. That is,

$$\frac{\Delta c}{\Delta h} = \frac{c(h_R) - c(h_S)}{h_R - h_S} = c_0 \xi_n \tag{D.4}$$

where the sound speeds, $c(h_R)$ and $c(h_S)$, are calculated using Eq. (5.26), repeated below for convenience.

$$c(h) = B_m \log_e \left(\frac{h}{z_0} \right) + A_m h + c_0 \tag{D.5}$$

A_m is calculated using Eq. (5.22), B_m is calculated using Eq. (5.30) and c_0 is the speed of sound corresponding to the temperature at ground level, calculated using Eq. (5.21).

If $h_R = h_S$, then the two quantities are slightly modified by adding 0.005 m to h_R and subtracting 0.005 m from h_S. The quantities h_R and h_S are not permitted to be less than $5z_0$, where z_0 is the roughness length of Table 5.3.

If $h_R > h_S$, the mean sound speed is given by

$$\bar{c} = \frac{1}{h_R - h_S} \int_{h_S}^{h_R} c(h) \mathrm{d}h \tag{D.6}$$

and the sound speed, c, at height, $h = 0$ is given by

$$c_0 = \bar{c} - \frac{\Delta c}{\Delta h} \frac{h_R + h_S}{2} = \frac{\bar{c}}{1 + \xi_n (h_R + h_S)/2} \tag{D.7}$$

The sound speed at $h = 0$, calculated using Eq. (D.7), is not the same as the actual sound speed at $h = 0$. Also, if $|\xi_n| < 10^{-6}$, ξ_n is set equal to 0 and both $\bar{c}$ and c are set equal to the actual speed of sound at height, h_0. If the source is higher than the receiver, the subscripts R and S in Eq. (D.6) are interchanged.

For the case of non-hard surfaces, such as grassland, the equivalent linear sound-speed gradient that is used for calculation of ray path lengths and propagation times is a function of frequency. In this case, the above equivalent linear sound-speed gradient should be replaced with a modified equivalent linear sound-speed gradient, $(\Delta c/\Delta h)'$. This defined as

$$\left(\frac{\Delta c}{\Delta h}\right)' = \begin{cases} \Delta c/\Delta h & f > f_H \\ \dfrac{\log_{10} f - \log_{10} f_L}{\log_{10} f_H - \log_{10} f_L} \dfrac{\Delta c}{\Delta h} & f_L \le f \le f_H \\ 0 & f < f_L \end{cases} \tag{D.8}$$

and

$$c'_0 = \bar{c} - \left(\frac{\Delta c}{\Delta h}\right)' \frac{h_R + h_S}{2} \tag{D.9}$$

f_L and f_H are defined as follows in terms of the sound-speed difference, Δc_{10}, at height 0 m and height 10 m, $(\Delta c_{10} = c(0) - c(10))$ calculated using Eq. (D.3):

$$f_L = \begin{cases} f_\pi & \Delta c_{10} \le 1 \\ f_\pi \dfrac{43 - 3\Delta c_{10}}{40} & 1 < \Delta c_{10} < 5 \\ 0.7 f_\pi & \Delta c_{10} \ge 5 \end{cases} \tag{D.10}$$

and

$$f_H = \max(\sqrt{f_L f_{2\pi}}, 1.25 f_L) \tag{D.11}$$

where f_π and $f_{2\pi}$ correspond to the frequencies that are found iteratively such that they produce a phase shift, $\Delta\alpha$, between direct and ground-reflected sound rays of π and 2π, respectively. $\Delta\alpha$ is given approximately by

$$\Delta\alpha = \frac{2\pi f}{c_0} \Delta r + \tan^{-1} \frac{\mathrm{Im}(R_p)}{\mathrm{Re}(R_p)} \tag{D.12}$$

where Δr is the difference in straight-line distance between the direct path and the ground-reflected path for rays travelling from the source to the receiver, and is given by the equation below. The quantities, $\mathrm{Im}(R_p)$ and $\mathrm{Re}(R_p)$ are the imaginary and real parts, respectively, of the plane-wave reflection coefficient, which is defined in Appendix C.

$$\Delta r = \sqrt{d^2 + (h_S + h_R)^2} - \sqrt{d^2 + (h_S - h_R)^2} \tag{D.13}$$

where d is the horizontal distance between source at height, h_S and receiver at height, h_R.

D.3 Calculation of Ray Path Lengths and Propagation Times

The calculation of ray path lengths and propagation times is done separately for direct and reflected rays. The difference in propagation time is used, together with the phase shift due to ground reflection, to determine the phase difference between the direct and ground-reflected rays arriving at the receiver. This is then used together with partial coherence analysis discussed in Section 5.8.1 to determine the total sound pressure at the receiver.

D.3.1 Direct Ray

The variables used in the calculation of ray path lengths and propagation times are defined in Figure D.1 for various relative source and receiver positions.

In the following analysis, the lower of the source and receiver heights will be denoted h_L and the higher of the source and receiver heights will be denoted h_U. To be able to use the Nord2000 analysis outlined by Hidaka et al. (1985) and later by L'Espérance et al. (1992), it is necessary to replace the sound speed c_0 at height $h = 0$ in Eq. (D.3) with the sound speed c_L at height $h = h_L$, so that,

$$c_e(h) = c_L \left[1 + \xi_n(h - h_L)\right] \tag{D.14}$$

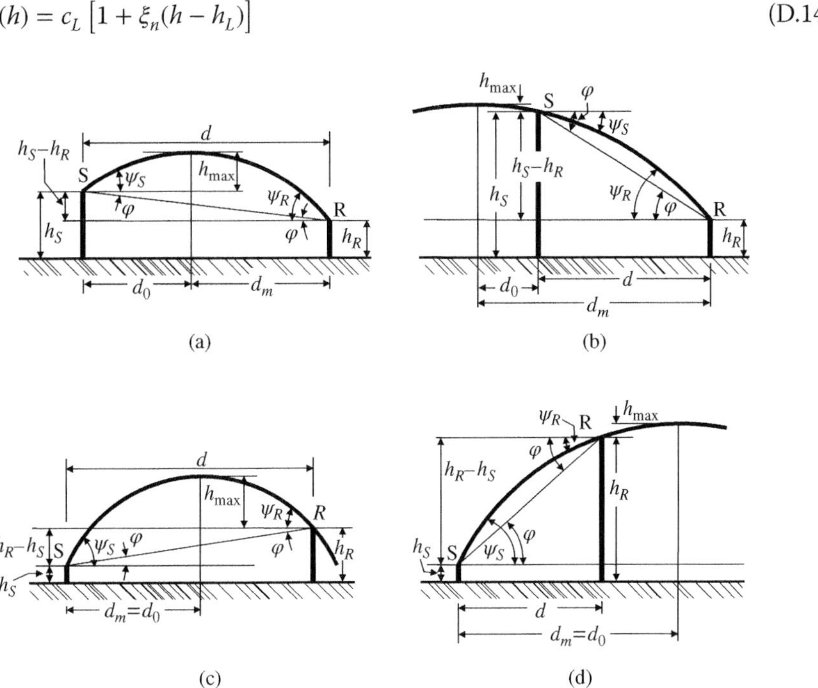

Figure D.1 Geometrical parameters for a reflected ray from a source to a receiver: (a) source higher than the receiver and the maximum height of the ray path lying between the source and receiver; (b) source higher than the receiver with the maximum height of the ray path lying on the opposite side of the source from the receiver; (c) receiver higher than the source with the maximum height of the ray path lying between the source and receiver; (d) receiver higher than the source with the maximum height of the ray path lying on the opposite side of the receiver from the source.

where the normalised sound-speed gradient, ξ_n, is defined as

$$\xi_n = \frac{\Delta c/\Delta h}{c_L} \tag{D.15}$$

$\Delta c/\Delta h$ is defined by Eq. (D.4), with the sound speeds calculated using Eq. (D.5), and c_L is calculated using,

$$c_L = c_0(1 + \xi_n h_L) \tag{D.16}$$

The travel distance r_{SR}, from the source to the receiver along the curved ray path is calculated by modifying the equation given by Hidaka et al. (1985) to account for $h_L \neq 0$ to give the following for $d \leq d_m$ and for a downward-diffracting atmosphere (Plovsing 2006b):

$$r_{SR} = \frac{1}{\xi_n \cos\psi_L}\{\arcsin[(1 + \xi_n|h_S - h_R|)\cos\psi_L] - (\pi/2) + \psi_L\} \tag{D.17}$$

where

$$\tan\psi_L = \frac{\xi_n d}{2} + \frac{|h_S - h_R|(2 + \xi_n|h_S - h_R|)}{2d} \tag{D.18}$$

and the horizontal distance, d_m, between the lower of the source and receiver and the highest point reached by the circular arc is,

$$d_m = \frac{\tan\psi_L}{\xi_n} = \tan\psi_L(R_c - h'_{\max}) \tag{D.19}$$

where R_c is the radius of curvature of the ray, $h'_{\max}$ is the maximum height reached by the ray above the lower of the source and receiver and $\psi_L = \psi_S$ if the source is lower than the receiver and $\psi_L = \psi_R$ if the source is higher than the receiver (see Figure D.1).

The time taken for the ray to travel over the curved path from the source to the receiver (for $d \leq d_m$) is given by (Plovsing 2006b),

$$\tau_{SR} = \frac{1}{2\xi_n c_L}\log_e\left(\frac{f(h_L)}{f(|h_S - h_R|)}\right) \tag{D.20}$$

where

$$f(h_L) = \frac{1 + \sin\psi_L}{1 - \sin\psi_L} \tag{D.21}$$

and

$$f(|h_S - h_R|) = \frac{1 + \sqrt{1 - (1 + \xi_n|h_S - h_R|)^2\cos^2\psi_L}}{1 - \sqrt{1 - (1 + \xi_n|h_S - h_R|)^2\cos^2\psi_L}} \tag{D.22}$$

When $d > d_m$, the ray path distance and travel time are calculated as follows,

$$\begin{aligned} r_{SR} =& \frac{2}{\xi_n \cos\psi_L}\left\{\arcsin\left[(1 + \xi_n(h_{\max} - h_L))\cos\psi_L\right] - (\pi/2) + \psi_L\right\} \\ &- \frac{1}{\xi_n \cos\psi_L}\left\{\arcsin\left[(1 + \xi_n|h_S - h_R|)\cos\psi_L\right] - (\pi/2) + \psi_L\right\} \end{aligned} \tag{D.23}$$

and

$$\tau_{SR} = \frac{1}{2\xi_n c_L}\left[2\log_e\left(\frac{f(h_L)}{f(|h_{\max} - h_L|)}\right) - \log_e\left(\frac{f(h_L)}{f(|h_S - h_R|)}\right)\right] \tag{D.24}$$

where $f(|h_{max} - h_L|)$ is found using Equation D.22 with the appropriate substitutions for h_S and h_R. For a homogeneous atmosphere, when $\xi_n = 0$, the same equations cannot be used, so in this case ξ_n is set equal to 10^{-10} and then the same equations are used (Plovsing 2006a).

D.3.2 Reflected Ray

The reflected ray is made up of contributions from each ground segment, as discussed in Appendices E and F. Here, the calculation of the reflected path length and propagation time for a single ground segment (or its extension to include the reflection point P), is outlined. In the following discussion, reference will be made to Figure D.2, which shows four different possibilities for the path of a ground-reflected ray travelling from the source to the receiver where the source is higher than the receiver. A similar set of figures can be drawn for the case of the receiver being higher than the source.

First, the point of reflection, P, must be found and then the same analysis as used in the previous section for the direct ray can be used. For the ray on the source side of the ground-reflection point, the analysis of the previous section is used, with the reflection point, P, used as the receiver. For the ray on the receiver side of the ground reflection, the analysis of the previous section is used, with the reflection point, P, used as the source.

The horizontal distance d_G to the reflection point from the source is calculated by solving the following cubic equation (Plovsing 2006b).

$$2d_G^3 - 3dd_G^2 + (b_R^2 + b_S^2 + d^2)d_G - b_S^2 d = 0 \tag{D.25}$$

where d is the horizontal source–receiver separation distance, and

$$b_S^2 = \frac{h_S}{\xi_n}(2 + \xi_n h_S) \tag{D.26}$$

$$b_R^2 = \frac{h_R}{\xi_n}(2 + \xi_n h_R) \tag{D.27}$$

If Eq. (D.25) has more than one solution for d_G, the reflection point closest to the source ($\min d_G$) is chosen if the source is lower than the receiver, while the reflection point closest to the receiver ($\max d_G$) is chosen if the source is higher than the receiver. If the source and receiver are at the same height, then the source height should be increased slightly and the receiver height should be reduced by the same amount to avoid discontinuities in the reflection point location when the number of solutions switches between one and three. Equation (D.25) has a maximum of three solutions, yet Figure D.2 shows four possibilities for the location of the reflection point, P. This is because any particular geometry will not allow both solutions represented in Figures D.2, parts (a) and (d) to occur.

The travel distance and times for the reflected ray are calculated using the same method as for the direct ray described in Section D.3.1. When undertaking calculations for the reflected ray path between the source and reflection point, P, $h_L = h_R = 0$ in the analysis in Section D.3.1 and for the calculations for the reflected ray path between the reflection point, P, and the receiver, $h_L = h_S = 0$. The total path length is the sum of the path lengths between the source and reflection point and between the reflection point and the receiver. Likewise for the propagation times.

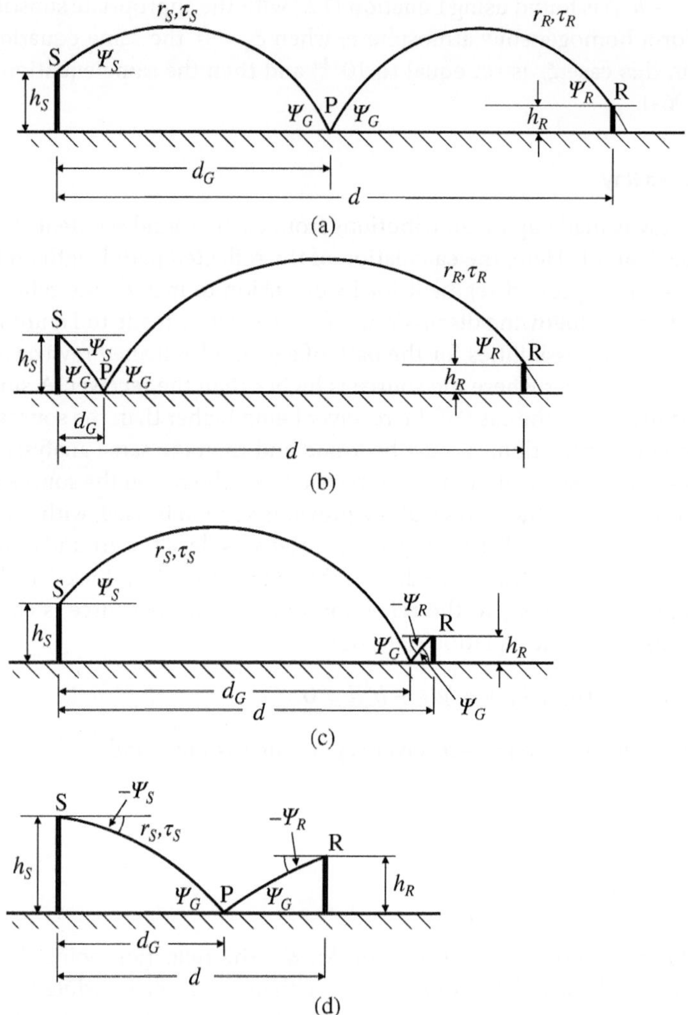

Figure D.2 Geometrical parameters for a reflected ray from a source to a receiver, with the source higher than the receiver: (a) maximum height of the ray path, both prior to and after reflection, lying between the source and receiver; (b) maximum height of the ray path prior to reflection lying on the opposite side of the source from the receiver and after reflection lying between the source and receiver; (c) maximum height of the ray path prior to reflection lying between the source and receiver and after reflection, lying on the opposite side of the receiver to the source; (d) maximum height of the ray path before reflection lying on the opposite side of the source from the receiver and after reflection, lying on the opposite side of the receiver to the source.

References

Hidaka T, Kageyama K and Masuda S 1985 Sound propagation in the rest atmosphere with linear sound velocity profile. *Journal of the Acoustical Society of Japan (E)*, **6**(2), 117–125.

L'Espérance A, Nicolas J, Herzog P and Daigle G 1992 Heuristic model for outdoor sound propagation based on an extension of the geometrical ray theory in the case of a linear sound speed profile. *Applied Acoustics*, **37**(2), 111–139.

Plovsing B 2006a Comprehensive outdoor sound propagation model. Part 1: Propagation in an atmosphere without significant refraction. Technical report AV 1849/00, Delta.

Plovsing B 2006b Comprehensive outdoor sound propagation model. Part 2: Propagation in an atmosphere with refraction. Technical report AV1851/00, Delta.

Plovsing B 2014 Proposal for Nordtest Method: Nord2000–prediction of outdoor sound propagation. Technical report, Delta.

Salomons E 2001 *Computational Atmospheric Acoustics*. Springer Science & Business Media.

E

Calculation of Terrain Parameters for the Nord2000 Sound Propagation Model

E.1 Introduction

In the Nord2000 propagation model, the terrain and barrier effects are combined with the ground effect. This is because any diffraction edge (barrier or screen) as a result of terrain or manmade construction produces additional ground reflections between the source and receiver and additional sound rays that need to be combined at the receiver as described in Section 5.8.1. The purpose of this appendix is to provide the calculations necessary to estimate the additional attenuation produced for any ray or wave as a result of terrain or manmade barriers (Plovsing 2006, 2014). The effect of ground reflection on the amplitude and phase of reflected rays is discussed in detail in Sections C.4 and C.5.

The calculations of the attenuation due to ground and barrier effects is combined with the effects of downward refraction caused by positive vertical wind and temperature gradients, as well as the partial coherence effects, as discussed in Section 5.8.1.

E.2 Terrain Effects

The Nord2000 model provides procedures for the following terrain situations.

1. Flat terrain with one type of surface.
2. Flat terrain with more than one type of surface.
3. Non-flat terrain without screening effects (valley-shaped terrain).
4. Terrain with one diffraction edge having one edge.
5. Terrain with one diffraction edge having two edges.
6. Terrain with two diffraction edges.

Flat terrain with one type of surface is the simplest case and the excess attenuation due to the terrain is calculated using the following equations and reference to Figure E.1. For flat terrain, the ground effect is taken into account here by finding the amplitude reduction and phase delay due to reflection using the spherical-wave reflection coefficient equation described in Section C.5. The phase change, ϕ, due to reflection, is converted to an additional time delay for the reflected path, given by,

$$\tau_R = \frac{\phi}{2\pi f} \tag{E.1}$$

Wind Farm Noise: Measurement, Assessment and Control, First Edition.
Colin H. Hansen, Con J. Doolan, Kristy L. Hansen.

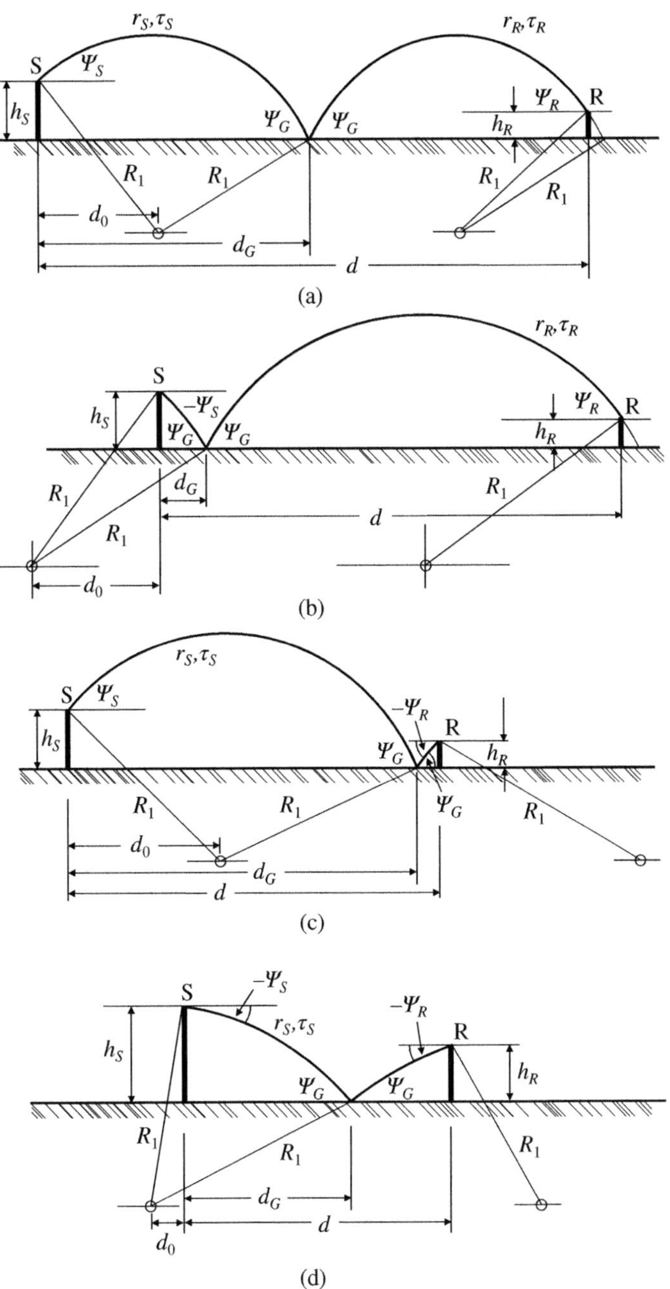

Figure E.1 Geometrical parameters for a reflected ray from source to receiver

Flat terrain with more than one type of surface above is treated in a similar way, except that Fresnel coefficients corresponding to each surface type are calculated, and the contribution from each surface type to the total reflected ray is determined by its Fresnel coefficient.

Non-flat terrain without screening effects (valley-shaped terrain) is treated in the Nord2000 model as a very complex situation (Plovsing 2006). In that model, the cross section of the terrain between the source and receiver is divided into segments, as explained in Section E.3. Each segment is then classified as concave, convex or transition (Plovsing 2014) and the terrain effect is calculated for each segment as follows. A typical valley-shaped terrain where there is line of sight between the sound source and the receiver is shown in Figure E.2, where the source and receiver heights are the shortest perpendicular distances from the local terrain and the terrain baseline is the line (not necessarily horizontal) joining the two points where vertical lines from the source and receiver, respectively, intersect the ground.

It is necessary to distinguish between three different types of segment, based on the source height, h'_S, and receiver height, h'_R, measured relative to the segment or its extension segment, as shown in Figure E.3 and described below. The physical line segment is extended to form an imaginary extension segment when it is not possible for a line drawn from the source or receiver to intersect the physical segment at right angles, as illustrated in Figure E.3. The three segment types are:

- concave segment; $h_{S,\text{ rel}} = 1$ and $h_{R,\text{ rel}} = 1$
- convex segment; $h_{S,\text{ rel}} = 0$ or $h_{R,\text{ rel}} = 0$
- transition segment; neither concave nor convex.

h'_S and h'_R are defined in Figure E.3 and

$$h_{S,\text{ rel}} = \begin{cases} 1 & h'_S \geq h''_S \\ \dfrac{h'_S}{h''_S} & 0 < h'_S < h''_S \\ 0 & h'_S \leq 0 \end{cases} \tag{E.2}$$

$$h_{S,\text{ rel}} = \begin{cases} 1 & h'_R \geq h''_R \\ \dfrac{h'_R}{h''_R} & 0 < h'_R < h''_R \\ 0 & h'_R \leq 0 \end{cases} \tag{E.3}$$

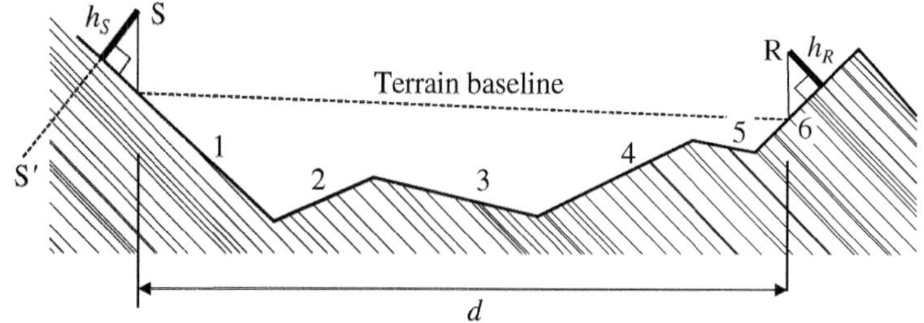

Figure E.2 Typical valley-shaped ground profile, where the individual ground segments are numbered 1–6

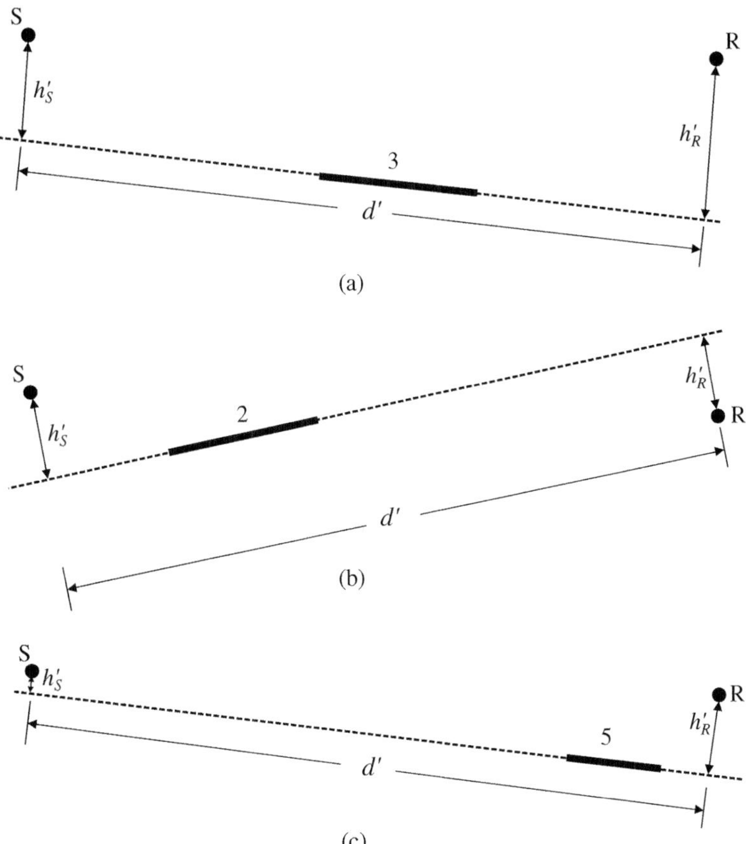

Figure E.3 Various types of line segment: (a) concave, where $h'_S > h_S$ and $h'_R > h_R$; (b) convex, where $h'_S > h_S$ and $h'_R < 0$; (c) transition, (neither concave nor convex). The line segments are taken from Figure E.2 and the identifying numbers correspond in each figure.

The quantities, h''_S and h''_R are defined as

$$h''_S = \min(h_S, h_{S,F}) \tag{E.4}$$

$$h''_R = \min(h_R, h_{R,F}) \tag{E.5}$$

The quantities, $h_{S,F}$ and $h_{R,F}$ are defined using Figure E.4, where $h_{S,F}$ is the height where the following condition is satisfied (see Figure E.4a).

$$|S_F P_1| + |P_1 R'| - |S_F R'| = \lambda / n_F \tag{E.6}$$

where $|S_F P_1|$ is the straight-line distance between point S_F and point P_1, and n_F is set equal to 16 for terrain effects calculations, 8 for reflection from vertical surfaces and 4 for reflection from a variable-impedance surface (Plovsing 2006). The quantity, $h_{S,F}$ can be calculated using

$$h_{S,F} = \frac{-B - \sqrt{B^2 - 4AC}}{2A} \tag{E.7}$$

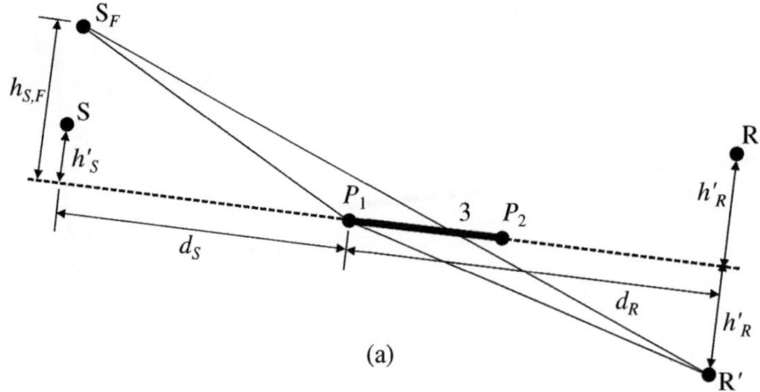

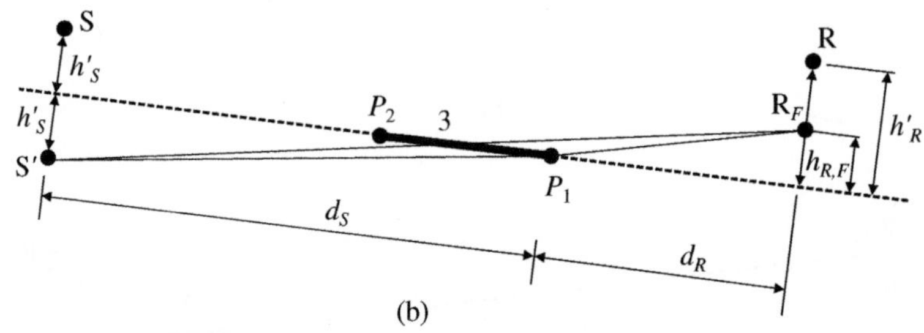

Figure E.4 Construction for the calculation of (a) $h_{S,F}$ and (b) $h_{R,F}$.

where

$$A = 4h_R'^2 - 4x_1 \tag{E.8}$$

$$B = 4h_R' x_2 \tag{E.9}$$

$$C = x_2^2 - 4d_S^{\ 2} x_1 \tag{E.10}$$

$$x_1 = (\sqrt{h_R'^2 + d_R^{\ 2}} - \lambda/n_F)^2 \tag{E.11}$$

$$x_2 = h_R'^2 + (d_S + d_R)^2 - d_S^{\ 2} - x_1 \tag{E.12}$$

$h_{R,F}$ can be calculated the same way as described above for $h_{S,F}$, by replacing S_F with R_F and h_R' with h_S', and interchanging P_1 and P_2, and d_S and d_R, as shown in Figure E.4b.

The last three terrain situations – items 4, 5 and 6 in the list on p. 547 – involve segmentation of the cross-section between the source and receiver and including the effects of diffraction over any diffraction edges as well as reflection from the ground on either side of one or two diffraction edges. The model can only include a maximum of two diffraction edges.

To find the excess attenuation due to the combination of barrier and terrain effects, characterised by the six terrain situations in the list, the following steps are needed.

1. Approximation of the terrain profile by straight-line segments.
2. Identification of the two most efficient diffraction edges (if any).

3. Calculation of the lengths and propagation times for each ray path and hence the difference between the direct path length and each reflected or diffracted path length.
4. Calculation of the effect of ground reflections on the phase and amplitude of any ground-reflected rays.
5. Use of the calculated path differences and phase and amplitude effects resulting from ground reflections to calculate the contribution at the receiver due to each diffracted or reflected path.
6. Converting to decibels, the total sound pressure at the receiver due to the direct and all reflected and diffracted rays, and then arithmetically subtracting this total from the contribution due to the direct ray (also in dB), to evaluate the ground and terrain effect, $A_{\text{gr+bar}}$.

The above steps are discussed in detail in the following sections.

E.3 Approximating Terrain Profiles by Straight-line Segments

To calculate the ground effect for a propagating wave, it is necessary to divide the terrain over which it propagates into segments. Typical segmented profiles are shown in Figure E.5.

According to the Nord2000 method (Plovsing 2006), any terrain profile can be approximated using straight-line segments by following the procedure outlined below and by referring to Figure E.6. In the figure, S_g and R_g are points on the ground below the source

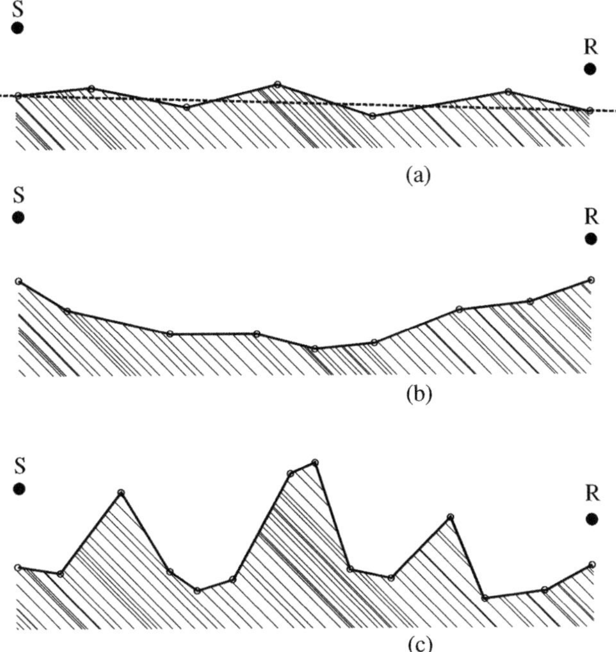

Figure E.5 Examples of segmented terrain profiles: (a) approximately flat terrain; (b) valley-shaped terrain; (c) hilly terrain, where the start and end points of each segment are indicated by open circles.

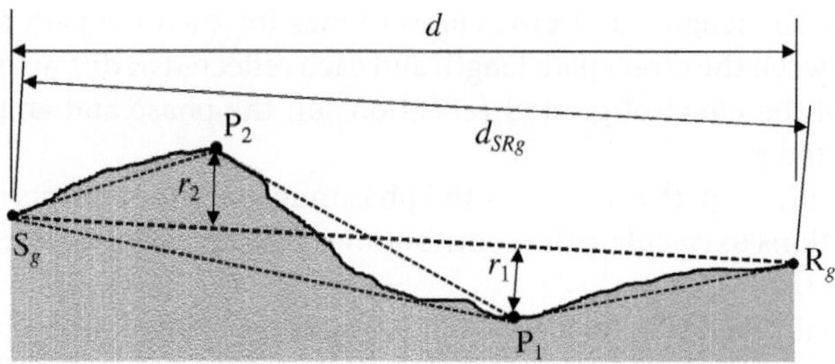

Figure E.6 Ground profile cross section used for the segmentation example

and receiver respectively, such that lines drawn from the source and receiver locations intersect the ground perpendicularly at ground points S_g and R_g, respectively and d_{SRg} is the straight-line distance between these two points.

Segmentation of the ground profile shown in Figure E.6 begins by finding the point, P_1, of maximum deviation of the ground profile from the straight line between points S_g and R_g. Connecting S_g and R_g via point P_1 with straight lines produces the first two segments approximating the terrain profile. The point of maximum deviation between these two lines and the terrain profile is found to define another point, P_2. This point is then used to form two more line segments by joining points S_g and P_2 and points P_2 and R_g for the particular case shown in Figure E.6. The process is repeated until one of the following three conditions is satisfied:

1. The number of segments is 10
2. r_i satisfies the following equation

$$r_i \leq \begin{cases} 0.1; & d \leq 50\text{m} \\ 0.002d; & 50\text{m} < d < 500\text{m} \\ 1; & d \geq 500\text{m} \end{cases} \tag{E.13}$$

3. The smallest segment length, $d_{\min}$ satisfies the following equation

$$d_{\min} \leq \begin{cases} 1; & d \leq 20\text{m} \\ 0.05d; & 20\text{m} < d < 200\text{m} \\ 10; & d \geq 200\text{m} \end{cases} \tag{E.14}$$

E.4 Calculation of the Excess Attenuation due to the Ground Effect for Relatively Flat Terrain with no Diffraction Edges

This case is illustrated in Figure E.5a, where the terrain is replaced by the flat dashed line. There are only two rays arriving at the receiver that need to be combined coherently: a ground-reflected ray and a direct ray or a direct ray that has been reflected by a vertical surface. For these cases, the sound pressure amplitude at the receiver is given by

$$|p_t|^2 = |p_1|^2 \left[\left|1 + F_2 \frac{p_2}{p_1}\right|^2 + (1 - F_2^2)\left|\frac{p_2}{p_1}\right|^2 \right] \tag{E.15}$$

where the subscript 1 applies to the direct ray (which is denoted the primary ray and which may or may not have been reflected by a vertical surface), and the subscript 2 applies to the ground-reflected ray. The coefficient of coherence, F_2, for the reflected ray (ray 2), is defined in Eq. (5.155). The sound pressures, p_1 and p_2, corresponding to rays 1 and 2, are defined as

$$p_1 = \frac{e^{-jkr}}{kr} \tag{E.16}$$

$$p_2 = Q\frac{e^{-jk(r_S + r_R)}}{k(r_S + r_R)} \tag{E.17}$$

where $k = 2\pi/\lambda$ is the wave number and Q is the frequency-dependent complex spherical-wave reflection coefficient of the ground. For a non-diffracting atmosphere, $r = \sqrt{d^2 + (h_S - h_R)^2}$ and r, r_S, r_R, d, h_S and h_R are as defined in Figure 5.2. When there is a downward-refracting atmosphere, the distances, r, r_S and r_R are the distances along curved paths, which can be calculated following the methods outlined in Appendix D.

The sound pressures given by Eqs. (E.16) and (E.17) are not actual pressures because their amplitude is not properly accounted for. However, their relative amplitudes are correct and as we only require a ratio of amplitudes to obtain the excess attenuation, the relative amplitudes and phases between p_1, p_2 and p_t, for each frequency considered, are all that are important.

The excess attenuation due to the ground effect only (as there are no diffraction edges) is then given by

$$A_{\text{gr}} = A_{\text{gr+bar}} = -20\log_{10}\frac{|p_t|}{|p_1|} \quad \text{(dB)} \tag{E.18}$$

If the ray being considered has been reflected from a vertical surface, then the contribution from that reflected ray is added incoherently to the contributions from non-reflected rays. This also applies to rays that have been reflected from the ground as well as from a vertical surface. The sound pressure level arriving at the receiver due to the reflected ray is calculated in the same way as the sound pressure level arriving due to the direct ray with spherical spreading, scattering and atmospheric absorption already taken into account. The reflection loss, A_r (see Eq. (5.142)) must be subtracted from the decibel level of the reflected ray prior to it being logarithmically added to the sound pressure level at the receiver arising from contributions from other rays.

E.5 Calculation of the Excess Attenuation due to the Ground Effect for Relatively Flat Terrain with a Variable Impedance Surface and no Diffraction Edges

This case uses the same approach as in the previous section, including the same Figure E.5a, except that an allowance is made for different parts of the surface having different impedances. The ground between the source and receiver is first approximated by the straight dashed line shown in Figure E.5a. Then the Fresnel diffraction zone is defined and separated into a source region and a receiver region as shown in Figure F.1b. For each combination of ground impedance, i, and surface roughness, j, a frequency-dependent Fresnel weighting, w_{ijL}, is calculated, which for low frequencies is the ratio of the total surface area, S_{ij}, in the Fresnel zone associated with surface, i, j, to the total Fresnel zone area, S_F. Thus,

$$w_{ijL} = \frac{S_{ij}}{S_F} \tag{E.19}$$

where $S_{ij} = S_{ijS} + S_{ijR}$, S_{ijS} is the Fresnel area associated with surface, i, j, in the source region and S_{ijR} is the Fresnel area associated with surface, i, j, in the receiver region (see Figure F.1b). The calculation of the ratios of the Fresnel areas for this case is described in Appendix F, where it is shown that for the two-dimensional modelling case, the ratio of Fresnel areas is equivalent to the ratio of Fresnel lengths in the terrain cross-section.

In the high-frequency range, the calculation of w_{ijH} is more complex and follows Plovsing (2006). Thus,

$$w_{ijH} = (r_{ij} - r'_{ij})r_H + r'_{ij} \tag{E.20}$$

where

$$r_H = \begin{cases} 1; & \tan\psi_G \geq 0.04 \\ \dfrac{\log_{10}(200\tan\psi_G)}{\log_{10}(8)}; & 0.005 < \tan\psi_G < 0.04 \\ 0; & \tan\psi_G \leq 0.005 \end{cases} \tag{E.21}$$

and

$$r'_{ij} = \frac{r''_i}{\sum_{i=1}^{N} r''_i} \frac{r_{ij}}{r_i} \tag{E.22}$$

$$r''_i = 3.1r_i - 10.69r_i^2 + 21.76r_i^3 - 21.95r_i^4 + 8.78r_i^5 \tag{E.23}$$

$$r_i = \sum_{j=1}^{M} r_{ij} \tag{E.24}$$

$$r_{ij} = 0.5\left(\frac{S_{ijS}}{S_{FS}} + \frac{S_{ijR}}{S_{FR}}\right) \tag{E.25}$$

For a terrain profile with N ground impedance values and M surface roughness values, the frequency-dependent excess attenuation due to the ground effect is given by

$$A_{\text{gr}} = A_{\text{gr+bar}} = -\left[\sum_{i=1}^{N}\sum_{j=1}^{M} w_{ij}\Delta L_{ij}\right] \quad \text{(dB)} \tag{E.26}$$

where ΔL_{ij} is calculated for reflected path, i, j, using Eq. (E.27), rearranged to give

$$\Delta L_{ij} = 20\log_{10}\frac{|p_{tij}|}{|p_1|} \tag{E.27}$$

where p_1 is the sound pressure at the receiver due to the direct ray and p_{tij} is the total sound pressure at the receiver due to the combination of the direct ray and the ray reflected from segment i, j (or its extension), calculated using Eq. (E.15) with the subscript 2 replaced with the subscript i, j.

The Fresnel weighting, w_{ij}, corresponding to the 1/3-octave band with centre frequency f, is calculated as follows,

$$w_{ij} = \begin{cases} w_{ijL}; & f \le f_L \\ \dfrac{\log_{10} f_H - \log_{10} f}{\log_{10} f_H - \log_{10} f_L}(w_{ijL} - w_{ijH}) + w_{ijH}; & f_L < f < f_H \\ w_{ijH}; & f \ge f_H \end{cases} \tag{E.28}$$

The frequencies, f_H and f_L are based on the phase difference, $\Delta\alpha$, between the direct and ground-reflected rays. Sufficient accuracy is obtained (Plovsing 2006) by using for the calculation the plane-wave reflection coefficient, R_p, rather than the spherical-wave reflection coefficient (see Appendix C). Thus,

$$\Delta\alpha = k(r_S + r_R - r) + Arg\{R_p\} \tag{E.29}$$

where $Arg\{R_p\}$ is the phase difference between the reflected and incident rays, calculated as described in Section C.4.

The frequency, f_H is the smallest frequency where $\Delta\alpha = \pi$. The frequency, f_L is the smallest frequency where $\Delta\alpha = \Delta\alpha_L$, where

$$\Delta\alpha_L = \pi - (1.9483 \log_e(h_{\min}) + 18.052) \tan\psi_G \tag{E.30}$$

where ψ_G is the angle of incidence and reflection for the ray and

$$h_{\min} = \begin{cases} \text{Min}(h_S,\ h_R) & h_S > 0.01 \text{ and } h_R > 0.01 \\ 0.01 & h_S \le 0.01 \text{ or } h_R \le 0.01 \end{cases} \tag{E.31}$$

E.6 Calculation of the Excess Attenuation due to the Ground Effect for Valley-shaped Terrain

This case is illustrated in Figure E.5b, where there are a number of ground segments between the source and receiver and the contribution of each is included and weighted by its frequency-dependent Fresnel weighting $w_i(f)$ for segment, i, calculated as described in Sections F.3 and F.4 (see Eq. (F.21)). For a terrain profile that has been divided into N segments, the frequency-dependent excess attenuation due to the ground effect is given by

$$A_{\text{gr}} = A_{\text{gr+bar}} = -\left[\sum_{i=1}^{N} w_i \Delta L_i\right] \quad \text{(dB)} \tag{E.32}$$

where ΔL_i is calculated for reflected path, i, using Eq. (E.18), rearranged to give

$$\Delta L_i = 20 \log_{10} \frac{|p_{ti}|}{|p_1|} \quad \text{(dB)} \tag{E.33}$$

where p_1 is the sound pressure at the receiver due to the direct ray and p_{ti} is the total sound pressure at the receiver due to the combination of the direct ray and the ray reflected from segment i (or its extension), calculated using Eq. (E.15) with the subscript 2 replaced with the subscript i.

E.7 Identification of the Two Most Efficient Diffraction Edges

The hilly ground profile illustrated in Figure E.5c can be sub-divided into several sub-categories, as illustrated in Figure E.7. As the Nord2000 model only allows for two (most efficient) diffraction edges to be considered, it is necessary to first identify these for each of the cases illustrated in Figure E.7. Then the excess attenuation effects for these various ground profiles can be calculated, as shown in following sections.

The first case illustrated in Figure E.7a is trivial because it only contains one diffraction edge. In the second case (part (b)), there is a single hill with a number of convex diffraction edges and in this case, the two most efficient edges must be found. An example of this process is shown in Figure E.8, where a single hill is first represented by a number of segments, joined by points P_3–P_8. In the first step, the path-length difference between the direct path and the path over the diffracting edge, $|SP| + |PR| - |SR|$, is calculated for each point P_3–P_8. If the point P is below the line of sight between S and R, the path-length difference is multiplied by −1. The most efficient diffraction edge is the point P that results in the largest path-length difference and this is P_6. The second most efficient diffracting edge on the source side of P_6 is found by replacing P_6 with the receiver point R, and then finding the largest path-length difference $|SP| + |PR| - |SR|$ for all points on the source side of P_6 and multiplying the result by −1 for points below the line of sight between S and the new location of R. The second most efficient diffracting edge

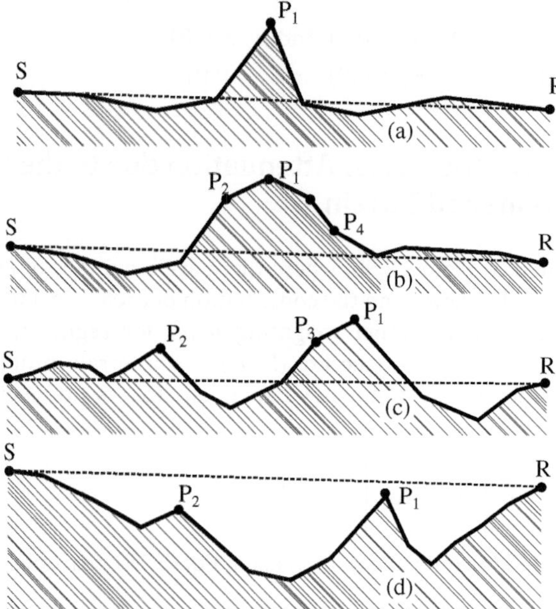

Figure E.7 Illustration of the two most efficient diffraction edges (P_1 and P_2): (a) terrain with a single diffraction edge above the line of sight between the source and receiver; (b) terrain with multiple diffraction edges, which are all part of the same terrain feature, above the line of sight between the source and receiver; (c) terrain with multiple diffraction edges above the line of sight between the source and receiver, including a double diffraction edge, P_1 and P_3; (d) terrain with diffraction edges below the line of sight between the source and receiver.

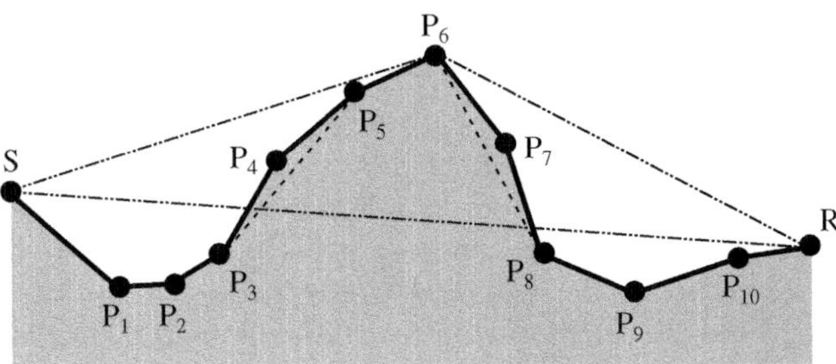

Figure E.8 Simplification of a segmented hill with multiple diffraction edges to a hill with two diffraction edges.

on the receiver side of P_6 is found by replacing P_6 with S and then finding the largest path-length difference |SP| + |PR| − |SR| for all points on the receiver side of P_6 and multiplying the result by −1 for points below the line of sight between the new location of R and S. The largest path-length difference value takes into account the sign, so that −1 is larger than −2 and 0.5 is larger than −1. The largest of the two path differences calculated above is then defined as the second most efficient diffraction edge. For the example shown in Figure E.8, the second most efficient diffraction edge is P_5, which has a negative path-length difference because it is below the line of sight between S and P_6. The new segmented hill shape is then defined by the dashed line joining points P_3, P_5, P_6 and P_8 and this is the shape used in the following sections for calculating the attenuation due to diffraction.

When there is more than a single hill or screen blocking the line of sight between the source and receiver (as illustrated in Figure E.7c), the identification of the two most efficient diffraction edges follows a similar procedure to that described above for a single hill.

The most efficient diffraction edge is the edge for which the difference between the direct-ray path length d and the length of the path, $(d_S + d_R)$ over the top of the edge is the greatest (if the diffraction edge is above the line of sight between the source and receiver).

The second most efficient diffraction edge on the source side of the first diffraction edge is found by replacing the receiver point with the top of the first diffraction edge and on the receiver side of the first diffraction edge it is found by replacing the source point with the top of the first diffraction edge. The second most efficient diffraction edge is then the most efficient of the two edges identified above and is labelled as P_2 in Figure E.7. Diffraction edge P_3, in Figure E.7c is ignored because it is the third most efficient diffraction edge.

If no diffraction edge is above the line of sight, then the most efficient diffraction edge is the one below the line of sight and closest to it (point P_1 in Figure E.7d). If the highest diffraction edge is well below the line of sight, the terrain may be considered to be a 'valley'. Four cases, illustrated in Figure E.7, where the most efficient diffraction edge is labelled P_1 and the second most efficient labelled P_2, will be considered in the following sections. These cases are:

- single wedge-shaped diffraction edge (Figure E.7a)
- two closely spaced diffraction edges that are part of the same terrain feature (Figures E.7b and E.8)

- two well-separated diffraction edges (Figure E.7c)
- most efficient diffraction edge below the line of sight between the source and receiver (Figure E.7(d)).

Artificial screens can also be approximated by straight-line segments, as explained in Plovsing (2006), but this is rarely worthwhile in practice. It is usually sufficient to consider the highest point of an artificial barrier as the *diffraction point*. The diffraction efficiency of the edges of artificial screens and barriers should be compared with the two found for the terrain and the two most efficient of all of them is then chosen for further analysis. If an artificial screen is relatively short, so that diffraction around its ends is important, then those paths also need to be taken into account when considering all the sound rays of importance arriving at the receiver. In this case, it is possible that the top edge of a screen (or barrier), identified as the most efficient or second most efficient diffraction edge, may no longer be so identified when diffraction around its ends is taken into account. A ray arriving at the receiver as a result of diffracting around the vertical edge of a screen is treated the same as one that is diffracted over the top of a screen or barrier, except that there is only one ground-reflected wave associated with diffraction around a screen end.

In wind farm noise propagation calculations, most screening is a result of terrain such as hills between the turbines and receiver location. In this case, for implementation of the Nord2000 model, it is necessary to take into account reflections from the finite-impedance ground that makes up the wedge. For manmade barriers close to the receiver, which can be considered as thin, vertical surfaces, it is still possible to use the same wedge equations to calculate the diffraction effect, except that the β is set equal to 2π (see Figure E.9) and the reflection coefficients of the barrier walls are set equal to 1.

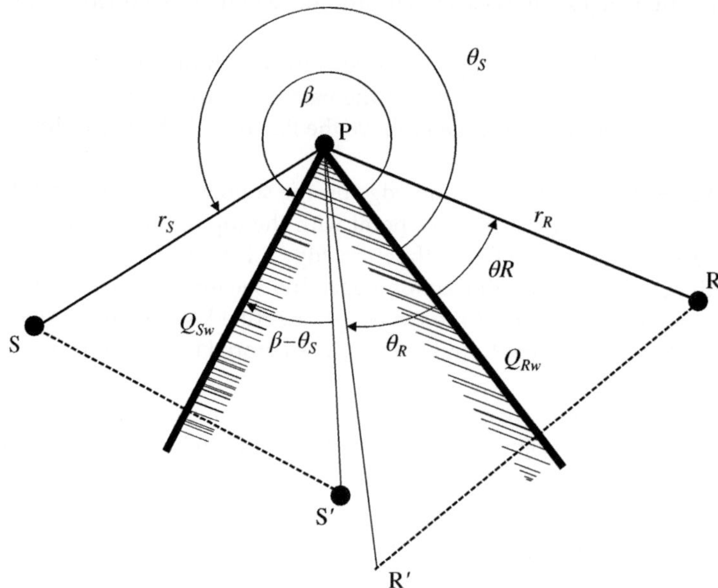

Figure E.9 Definition of variables used to calculate the sound pressure at the receiver due to diffraction by a finite-impedance wedge in free space.

E.8 Calculation of the Sound Pressure at the Receiver for each Diffracted Path in Hilly Terrain

The noise reduction of a ray as a result of diffraction over a diffraction edge (barrier or screen) is a function of the wavelength of sound and the path difference between a direct ray from source to receiver and a ray that travels from the source to the diffraction edge and then on to the receiver. The phase of the diffracted wave relative to the direct wave is derived from the path difference and travel-time difference of the two waves (originating from the same source and arriving at the same receiver) for a manmade barrier in the form of a thin, hard screen such as a brick wall or a gypsum board wall. However, if the barrier is a hill that has been approximated by a wedge of finite impedance, the solution for the diffracted pressure arriving at the receiver is more complex, and is also a function of the complex spherical-wave reflection coefficients of each side of the wedge. The calculation of the effect of diffraction over a wedge in free space (in the absence of ground) is needed as a basis for the calculation of the diffraction effect with finite-impedance ground on either side of the wedge, which is the construction used to represent shielding by terrain. The Nord2000 model does not discriminate between terrain and manmade screens; both types of barrier are treated as wedge-shaped screens, as outlined in the analysis in the next section (Plovsing 2014).

In the following sections, the analysis is carried out for straight rays, which characterise a non-refracting (or homogeneous) atmosphere. For a downward-refracting atmosphere, the straight-ray paths must be replaced with curved paths having path lengths and propagation times that are calculated as described in Appendix D.

E.8.1 Diffraction over a Single Finite-impedance Wedge-shaped Screen

The geometry for the calculation of the sound pressure at the receiver, R, for sound diffracted by a finite-impedance wedge (see Figure E.7a), is shown in Figure E.9, for a non-refracting atmosphere. Note that in this book, equations for sound pressure use positive time dependence, $e^{j\omega t}$, rather than the less common negative time dependence, $e^{-i\omega t}$, used by Hadden Jr and Pierce (1981) and Plovsing (2014).

All angles are referenced to the wedge surface on the side of the receiver. The complex sound pressure at the receiver is made up of a combination of four diffracted rays: SPR, S′PR, SPR′ and S′PR′. Image source (S′) and receiver (R′) locations are shown for reflection from the wedge surface on either side of the apex, P. As can be seen, the path lengths of the reflected and diffracted rays are the same length as the non-diffracted rays. Although this result fits well with the concept of image source and receivers representing the source and receipt respectively of reflected rays and it is also mathematically valid, it is difficult to justify physically. However, it is part of the Nord2000 model and thus is included here.

The complex sound pressure at the receiver may be calculated as follows (Hadden Jr and Pierce 1981; Plovsing 2014).

$$p_{\text{diff}} = -\frac{1}{\pi}\sum_{n=1}^{4} Q_n A(\theta_n) E_v(A(\theta_n)) \frac{e^{-j\omega\tau}}{k\ell} \tag{E.34}$$

where $\omega\tau = k\ell$, Q_n and $E_\upsilon(A(\theta_n))$ are complex numbers (characterised by an amplitude and a phase) and

$$\begin{aligned}\ell &= r_S + r_R;\ \tau = \tau_S + \tau_R;\ \theta_1 = \theta_S - \theta_R \\ \theta_2 &= \theta_S + \theta_R;\ \theta_3 = 2\beta - (\theta_S + \theta_R);\ \theta_4 = 2\beta - (\theta_S - \theta_R)\end{aligned} \tag{E.35}$$

Complex numbers are defined in Section A.3.

Again, the expression for p_{diff} does not represent its true amplitude, but as mentioned earlier we are only interested in relative amplitudes in order to determine the excess attenuation, $A_{\text{gr+bar}}$.

The quantity $\omega\tau$ is used in the exponent of e in Eq. (E.34) rather than $k\ell$, as that allows the equation to be used for curved rays that arise in a downward-refracting atmosphere in which the speed of sound increases with altitude (positive sonic gradient). In that case, the travel time is not linearly related to the path length. This is important for the phase calculation of the diffracted ray but not for the amplitude calculation, because slight errors in path length do not significantly affect the amplitude of the ray. Hence it is acceptable to use $k\ell$ in the denominator of Eq. (E.34).

The spherical-wave reflection coefficients (see Section C.5) are defined as follows.:

$$Q_1 = 1;\ Q_2 = Q_{Rw};\ Q_3 = Q_{Sw};\ Q_4 = Q_{Sw} + Q_{Rw} \tag{E.36}$$

As the direct ray path over the top of the wedge does not involve any reflections, $Q_1 = 1$. Reflection coefficients Q_{Sw} and Q_{Rw} represent the spherical-wave reflection coefficients of the wedge sides on the source side and receiver side respectively. The second and third equations in Eq. (E.36) have been interchanged to correct what is written in Plovsing (2014).

The real-valued function, $A(\theta_n)$ of Eq. (E.34) is defined as

$$A(\theta_n) = \frac{\nu}{2}(-\beta - \pi + \theta_n) + \pi H(\pi - \theta_n) \tag{E.37}$$

where β is defined in Figure E.9, $\nu = \pi/\beta$ is the wedge index and the Heaviside step function $H(x) = 1$ if $x \geq 0$ and $H(x) = 0$ if $x < 0$.

The complex function, $E_\upsilon(A(\theta_n))$, is defined as

$$E_\upsilon(A(\theta_n)) = \frac{\pi \sin|A(\theta_n)|}{\sqrt{2}\,|A(\theta_n)|} \frac{\mathrm{e}^{-\mathrm{j}\pi/4}}{\sqrt{1 + \left(\frac{2\tau_S\tau_R}{\tau^2} + \frac{1}{2}\right)\frac{\cos^2|A(\theta_n)|}{\nu^2}}} \frac{B}{|B|}[f(|B|) + \mathrm{j}g(|B|)] \tag{E.38}$$

where τ_S is the travel time (in seconds) from the source, S, to point P at the top of the wedge and τ_R is the travel time from point P to the receiver, R and $\mathrm{j} = \sqrt{-1}$.

The real-valued function, B, is defined as,

$$B = \sqrt{\frac{4\omega\tau_S\tau_R}{\pi\tau}} \frac{\cos|A(\theta_n)|}{\sqrt{\nu^2 + \left(\frac{2\tau_S\tau_R}{\tau^2} + \frac{1}{2}\right)\cos^2|A(\theta_n)|}} \tag{E.39}$$

and the functions, $f(|B|)$ and $g(|B|)$ are defined as

$$f(|B|) = \begin{cases} \dfrac{1}{\pi|B|} & |B| \geq 5 \\ \sum_{n=0}^{12} a_n |B|^n & |B| < 5 \end{cases} \tag{E.40}$$

and

$$g(|B|) = \begin{cases} \dfrac{1}{\pi^2|B|^3} & |B| \geq 5 \\ \sum_{n=0}^{12} b_n |B|^n & |B| < 5 \end{cases} \tag{E.41}$$

Values of the coefficients, a_n and b_n are given in Table E.1

If the top of the wedge is above the line of sight from the source to the receiver, as shown in Figure E.7a, $(\theta_1 > \pi)$, then the frequency-dependent complex sound pressure, p_1 at the receiver is

$$p_1 = p_{\text{diff}} \tag{E.42}$$

If the top of the wedge is below the line of sight from the source to the receiver as shown in Figure E.7d without point P_2, $(\theta_1 < \pi)$, the sound pressure, p_F, at the receiver due to the direct ray needs to be added to the diffracted ray. Thus,

$$p_1 = p_{\text{diff}} + p_F = p_{\text{diff}} + \frac{e^{-j\omega\tau_0}}{kr_0} \tag{E.43}$$

Table E.1 Coefficients a_n and b_n as a function of n, to be used in Eqs (E.40) and (E.41)

n	a_n	b_n
0	0.49997531354311	0.50002414586702
1	0.00185249867385	−1.00151717179967
2	−0.80731059547652	0.80070190014386
3	1.15348730691625	−0.06004025873978
4	−0.89550049255859	−0.50298686904881
5	0.44933436012454	0.55984929401694
6	−0.15130803310630	−0.33675804584105
7	0.03357197760359	0.13198388204736
8	0.00447236493671	−0.03513592318103
9	0.00023357512010	0.00631958394266
10	0.00002262763737	−0.00073624261723
11	−0.00000418231569	0.00005018358067
12	0.00000019048125	−0.00000151974284

where $\omega\tau_0 = kr_0$ for a non-refracting atmosphere and r_0 is the straight-line distance between the source and receiver, given by

$$r_0 = \sqrt{r_S^2 + r_R^2 - 2r_S r_R \cos\theta_1} \tag{E.44}$$

and

$$\tau_0 = \sqrt{\tau_S^2 + \tau_R^2 - 2\tau_S \tau_R \cos\theta_1} \tag{E.45}$$

The frequency-dependent, complex diffraction coefficient for the wedge is defined as

$$D = p_1 k \ell \, \mathrm{e}^{-\mathrm{j}\omega\tau} \tag{E.46}$$

The sound pressure at the receiver due to a ground reflected wave on either side of the wedge can be calculated for three cases:

- Ray reflected from the ground on the source side of the wedge. In this case, the above calculations are done for an image source location that is as far below the ground as the source is above the ground (in a perpendicular direction to the ground surface).
- Ray reflected from the ground on the receiver side of the wedge. In this case, the above calculations are done for an image receiver location that is as far below the ground as the receiver is above the ground.
- Ray reflected from the ground on the source side of the wedge and also on the receiver side of the wedge. In this case, the above calculations are done for an image source location and an image receiver location.

Effect of a Downward-refracting Atmosphere

When propagation over the wedge occurs in a downward-refracting atmosphere, the above analysis has to be done for curved sound rays, with the path lengths and travel times calculated as described in Appendix D.

E.8.2 Diffraction over a Finite-impedance Thick screen with Two Diffraction Edges

A thick screen is one with two diffraction edges, as shown in Figures E.10 and E.7b. The primary diffraction edge is the one for which the difference between the diffracted path and the direct path is the largest. For the case shown in Figure E.10a, the primary diffraction edge is P_2 and this path is $SP_2 + P2R - SR$. The secondary diffraction path is obtained by replacing the primary diffraction edge with R if it is nearer R than S (the case in the figure) and with S if it is nearer to S than *R*. For the case shown in Figure E.10b the relevant path difference is $SP_1 + P_1P_2 - SP_2$. If the diffraction edges are below the line of sight between the source and receiver, as shown in Figure E.10c, the primary diffraction edge is the one for which difference between the diffracted path and the direct path is the smallest. For the case shown in Figure E.10c, this path is $SP_1 + P_1P_2 - SP_2$.

The frequency-dependent complex sound pressure, p_{diff2} at the receiver due to diffraction over the thick barrier is

$$p_{\mathrm{diff2}} = 0.5 D_1 D_2 \frac{\mathrm{e}^{-\mathrm{j}\omega(\tau_S + \tau_R + \tau_m)}}{k(r_S + r_R + r_m)} \tag{E.47}$$

where r_m and τ_m represent the distance and ray travel time, respectively, for a ray travelling between diffraction edges P_2 and P_1. The diffraction coefficients, D_1 and D_2, correspond to the primary and secondary diffraction paths respectively and each is calculated

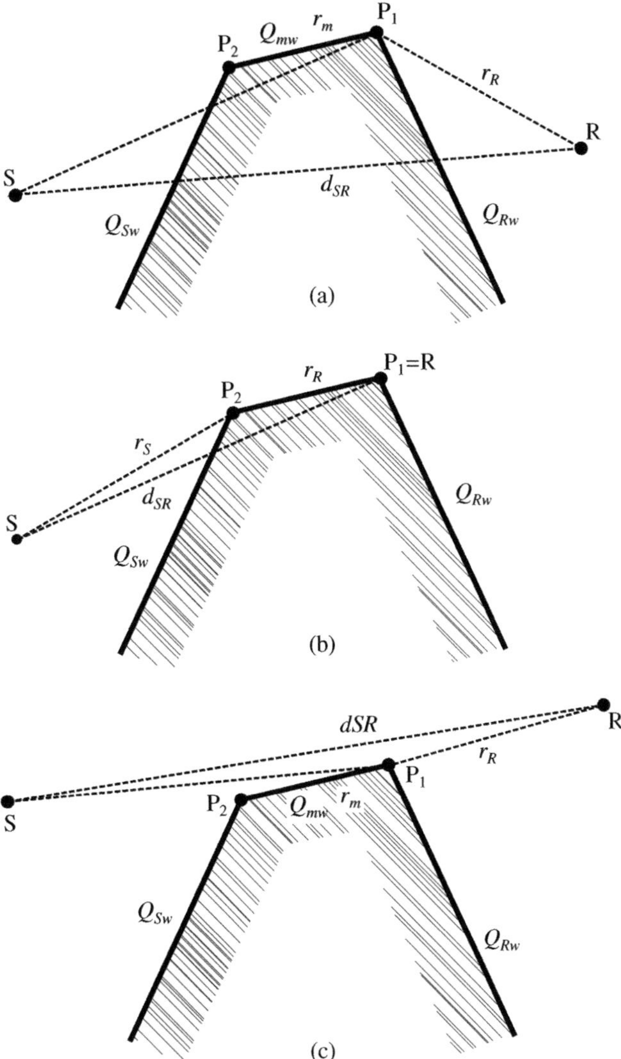

Figure E.10 Primary and secondary diffraction edges for a thick barrier with two diffraction edges: (a) diffraction over primary diffraction edge, P_2; (b) diffraction over secondary diffraction edge, P_1; (c) diffraction over primary edge, P_2, for a receiver with a direct line of sight to the source.

in the same way as for the wedge in the previous section, using Eq. (E.46). This calculation will be detailed as follows and will refer to Figure E.11.

Diffraction over Edge P_1, Referred to as the First Edge

For the purposes of the following analysis, P_1 is the edge closest to the source and is referred to as the first edge, while P_2 is closest to the receiver and is referred to as the second edge. If the first edge, P_1, is the primary diffraction edge, then the diffracted pressure, p_1, at the receiver is calculated using Eqs. (E.34) to (E.45) if edge P_1 is above the line of sight between S and R. The same equations may be used if edge P_1 is below

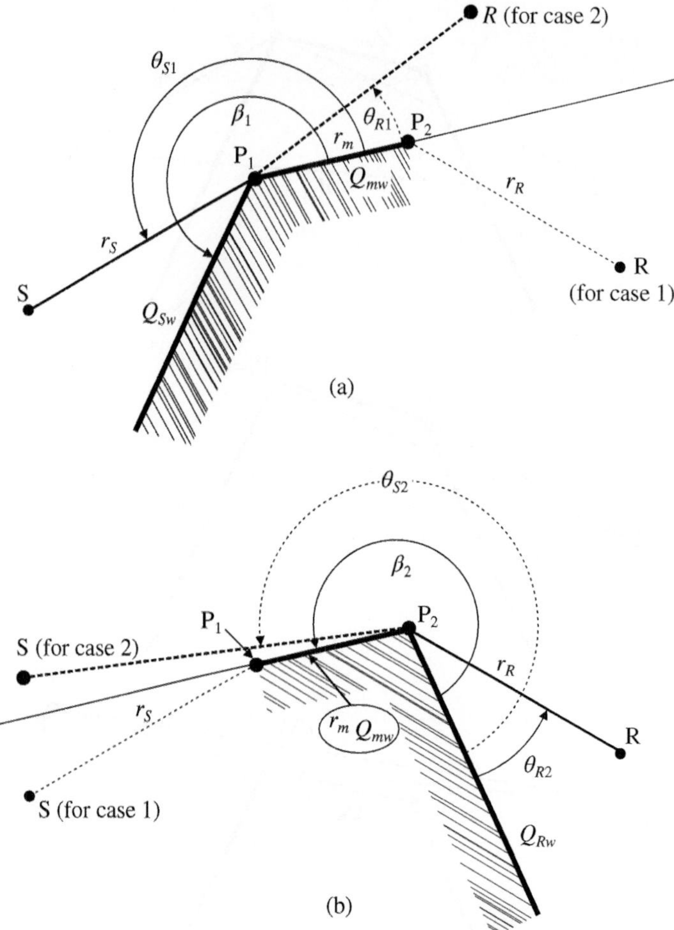

Figure E.11 Definition of variables used to calculate the sound pressure at the receiver due to diffraction by a thick barrier, in free space, with two diffraction edges, with the primary diffraction edge closest to the receiver: (a) diffraction over the first edge, P_1; (b) diffraction over the second edge, P_2.

the line of sight between S and R and P_1 is still the primary diffraction edge. However, in this case, the variables in those equations must be replaced as follows:

$$\begin{aligned} r_S &= r_S;\ r_R = r_m + r_R;\ \theta_S = \theta_{S1};\ \beta = \beta_1;\ \ell = r_S + r_m + r_R \\ Q_{Sw} &= Q_{Sw};\ Q_{Rw} = Q_{Rw};\ \tau = \tau_S + \tau_m + \tau_R;\ \tau_R = \tau_m + \tau_R \\ \tau_S &= \tau_S \end{aligned} \tag{E.48}$$

When calculating the diffraction coefficient D_1 of edge P_1, if point P_2 is above the line P_1R (case 1), the receiver is moved to point P_2, with diffraction angle θ_{R1} determined by the direction P_1P_2 and equal to zero. However, if point P_2 is below the line P_1R (case 2), the diffraction angle θ_{R1} is determined by the direction P_1R instead (dashed line in Figure E.11a)

$$\theta_R = \theta_{R1} \begin{cases} = 0; & \text{if } \mathrm{P_2} \text{ is at or above } \mathrm{P_1R} \text{ (case 1)} \\ \text{angle from } \mathrm{P_1P_2} \text{ to } \mathrm{P_1R}; & \text{if } \mathrm{P_2} \text{ is below } \mathrm{P_1R} \text{ (case 2)} \end{cases} \quad \text{(E.49)}$$

If instead P_1 is the secondary diffraction edge (as shown in Figure E.11), then the following replacements are made in Eqs. (E.34)–(E.45). Equation (E.49) applies for this case as well.

$$\begin{aligned} r_S &= r_S;\ r_R = r_m;\ \theta_S = \theta_{S1};\ \beta = \beta_1;\ \ell = r_S + r_m \\ Q_{Sw} &= Q_{Sw};\ Q_{Rw} = Q_{mw};\ \tau = \tau_S + \tau_m;\ \tau_R = \tau_m \\ \tau_S &= \tau_S \end{aligned} \quad \text{(E.50)}$$

The diffraction coefficient, D_1, due to diffraction over edge P_1 is then

$$D_1 = p_1 k \ell_1 \mathrm{e}^{-\mathrm{j}\omega\tau_1} \quad \text{(E.51)}$$

where $\ell_1 = \ell$ and $\tau_1 = \tau$, both of which are defined in Eq. (E.48) or Eq. (E.50). The quantity p_1 is calculated using Eq. (E.34) and either Eq. (E.42) or (E.43), depending on whether the diffraction edge interrupts the line of sight between the source and receiver.

Diffraction over Edge P_2, Referred to as the Second Edge

For the purposes of the following analysis, P_1 is the edge closest to the source and is referred to as the first edge, while P_2 is the edge closest to the receiver and is referred to as the second edge. The diffracted pressure, p_2, is calculated using Eqs. (E.34)–(E.42) if edge P_2 is above the line of sight between S and R and Eqs. (E.34)–(E.41) and (E.43) (but excluding Eq. (E.42)) if edge P_2 is below the line of sight between S and R.

If P_2 is the primary diffraction edge, then the following replacements in Eqs. (E.34)–(E.45) are made:

$$\begin{aligned} r_S &= r_S + r_m;\ r_R = r_R;\ \theta_R = \theta_{R2};\ \beta = \beta_2;\ \ell = r_S + r_m + r_R \\ Q_{Sw} &= Q_{Sw};\ Q_{Rw} = Q_{Rw};\ \tau = \tau_S + \tau_m + \tau_R;\ \tau_R = \tau_R \\ \tau_S &= \tau_S + \tau_m;\ p_1 = p_2 \end{aligned} \quad \text{(E.52)}$$

When calculating the effect of diffraction edge P_2, if point P_1 is above the line SP_2, the source is moved to edge P_1, as shown in Figure E.11b, with diffraction angle θ_{S2} determined by the direction P_2P_1 and equal to β_2. However, if point P_1 is below the line SP_2, the diffraction angle θ_{S2} is determined by the direction P_2S instead (dashed line in Figure E.11b).

$$\theta_S = \theta_{S2} \begin{cases} \text{Angle to } \mathrm{P_1P_2} = \beta_2; & \text{if } \mathrm{P_1} \text{ is at or above } \mathrm{P_2S} \text{ (case 1)} \\ \text{Angle to } \mathrm{SP_2}; & \text{if } \mathrm{P_1} \text{ is below } \mathrm{P_2S} \text{ (case 2)} \end{cases} \quad \text{(E.53)}$$

If instead P_2 is the secondary diffraction edge, then the following replacements are made in Eqs. (E.34)–(E.45). Equation (E.53) applies for this case as well.

$$\begin{aligned} r_S &= r_m;\ r_R = r_R;\ \theta_R = \theta_{R2};\ \beta = \beta_2;\ \ell = r_m + r_R \\ Q_{Sw} &= Q_{Sw};\ Q_{Rw} = Q_{Rw};\ \tau = \tau_m + \tau_R;\ \tau_R = \tau_R \\ \tau_S &= \tau_m;\ p_1 = p_2 \end{aligned} \quad \text{(E.54)}$$

The diffraction coefficient D_2, due to diffraction over edge P_2, is then

$$D_2 = p_2 k \ell_2 \mathrm{e}^{-\mathrm{j}\omega\tau_2} \quad \text{(E.55)}$$

where $\ell_2 = \ell$ and $\tau_2 = \tau$, both of which are defined in Eq. (E.52) or Eq. (E.54).

The quantity p_2 in Eq. (E.55) is calculated using Eq. (E.34) and either Eq. (E.42) or (E.43), depending on whether the diffraction edge interrupts the line of sight between the source and receiver.

The total diffracted pressure arriving at the receiver due to the two diffraction edges is then calculated using Eq. (E.47), where D_1 is defined by Eq. (E.51) and D_2 is defined by Eq. (E.55).

E.8.3 Diffraction over Two Finite-impedance Wedges

The sound pressure at the receiver due to sound travelling over two separate wedges, as illustrated in Figures E.7c,d and E.12, is calculated using Eq. (E.47) in a similar way to the case of a thick finite-impedance barrier discussed in the previous section. The only difference is that the '0.5' factor in Eq. (E.47) is excluded for this case. The calculation for the sound pressure arriving at the receiver after diffraction over two finite-impedance wedges will be detailed in the following paragraphs and will refer to Figure E.12.

Diffraction over Edge P_1

If P_1 in Figure E.12a is the primary diffraction edge, then the diffracted pressure, p_1, at the receiver is calculated using Eqs. (E.34)–(E.42) if edge P_1 is above the line of sight between S and R, and Eqs. (E.34)–(E.41) and (E.43) if edge P_1 is below the line of sight between S and R. However, the variables in Eqs. (E.34)–(E.45) must be replaced as follows.

$$\begin{aligned} &r_S = r_{S1};\ r_R = r_m + r_{R2};\ \theta_S = \theta_{S1};\ \beta = \beta_1 \\ &\ell = r_{S1} + r_m + r_{R2};\ Q_{Sw} = Q_{Sw1};\ Q_{Rw} = Q_{Rw1} \\ &\tau = \tau_{S1} + \tau_m + \tau_{R2};\ \tau_R = \tau_m + \tau_{R2};\ \tau_S = \tau_{S1};\ p = p_1 \end{aligned} \tag{E.56}$$

The propagation times, τ_{S1}, τ_m and τ_{R2} are calculated for curved ray paths corresponding to a downward-refracting atmosphere as described in Section D.3.

When calculating the diffraction coefficient D_1 of wedge P_1, the receiver is moved to point P_2, with diffraction angle θ_{R1} determined by the direction P_1P_2. However, if point P_2 is below the line P_1R, the diffraction angle θ_{R1} is determined by the direction P_1R instead (dashed line in Figure E.12a).

$$\theta_R = \theta_{R1} \begin{cases} \text{to } \mathrm{P_1P_2}; & \text{if } \mathrm{P_2} \text{ is at or above } \mathrm{P_1R} \text{ (case 1)} \\ \text{to } \mathrm{P_1R}; & \text{if } \mathrm{P_2} \text{ is below } \mathrm{P_1R} \text{ (case 2)} \end{cases} \tag{E.57}$$

If P_1 is the secondary diffraction edge (as shown in Figure E.11), then the following replacements are made in Eqs. (E.34)–(E.45). Equation (E.57) applies for this case as well.

$$\begin{aligned} &r_S = r_{S1};\ r_R = r_m;\ \theta_S = \theta_{S1};\ \beta = \beta_1;\ \ell = r_{S1} + r_m \\ &Q_{Sw} = Q_{Sw1};\ Q_{Rw} = Q_{Rw1};\ \tau = \tau_{S1} + \tau_m;\ \tau_R = \tau_m \\ &\tau_S = \tau_{S1};\ p = p_1 \end{aligned} \tag{E.58}$$

The diffraction coefficient D_1 due to diffraction over edge P_1 is then

$$D_1 = p_1 k \ell_1 \mathrm{e}^{-\mathrm{j}\omega\tau_1} \tag{E.59}$$

where $\ell_1 = \ell$ and $\tau_1 = \tau$, both of which are defined in Eq. (E.56) or (E.58).

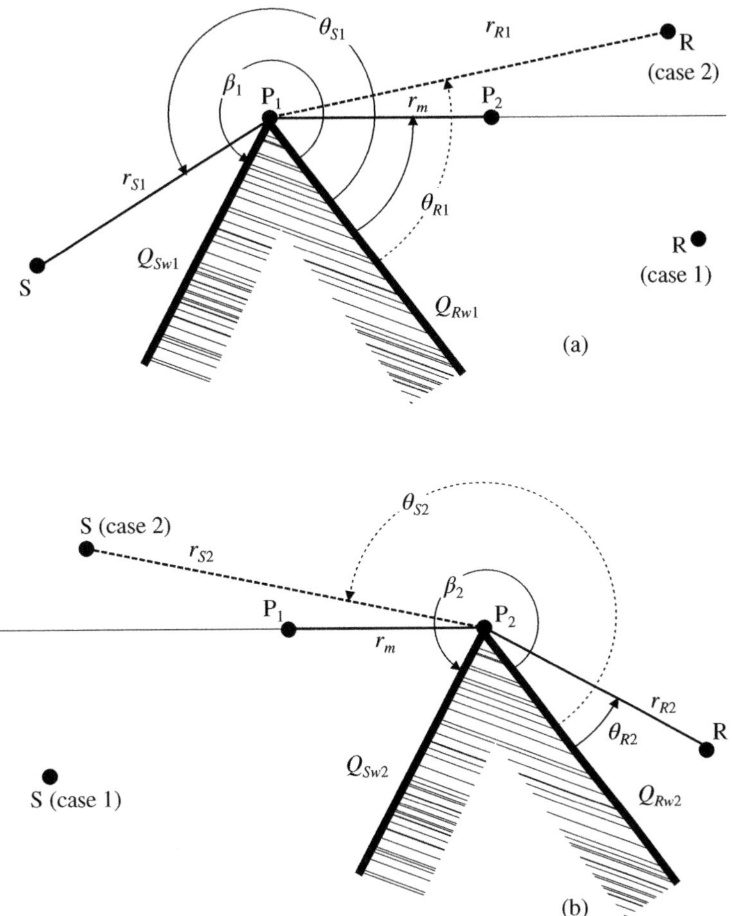

Figure E.12 Definition of variables used to calculate the sound pressure at the receiver due to diffraction over two finite-impedance wedges, in free space, with the primary diffraction edge closest to the receiver: (a) diffraction over the primary wedge; (b) diffraction over the secondary wedge.

The quantity p_1 in Eq. (E.59) is calculated using Eq. (E.34) and either Eq. (E.42) or (E.43), depending on whether the diffraction edge interrupts the line of sight between the source and receiver, as shown in Figure E.7c for line-of-sight interruption and Figure E.7d for no interruption.

Diffraction over Edge P_2

When calculating the effect of diffraction edge P_2, the source is moved to edge P_1, as shown in Figure E.12b, with diffraction angle θ_{S2} determined by the direction P_2P_1. However, if point P_1 is below the line SP_2, the diffraction angle θ_{S2} is determined by the direction P_2S instead (dashed line in Figure E.12b).

The diffracted pressure p_2 is calculated using Eqs. (E.34) and (E.42) if edge P_2 is above the line of sight between S and R and Eqs. (E.34) and (E.43) if edge P_2 is below the line of sight between S and R, as shown in Figure E.7d.

If P_2 is the primary diffraction edge, then the following replacements are made in Eqs. (E.34)–(E.45).

$$\begin{aligned} &r_S = r_{S1} + r_m;\ r_R = r_{R2};\ \theta_R = \theta_{R2};\ \beta = \beta_2 \\ &\ell = r_{S1} + r_m + r_{R2};\ Q_{Sw} = Q_{Sw2};\ Q_{Rw} = Q_{Rw2} \\ &\tau = \tau_{S1} + \tau_m + \tau_{R2};\ \tau_R = \tau_{R2};\ \tau_S = \tau_{S1} + \tau_m;\ p = p_2 \end{aligned} \tag{E.60}$$

$$\theta_S = \theta_{S2} \begin{cases} \text{to } P_1P_2; & \text{if } P_1 \text{ is at or above } P_2S \text{ (case 1)} \\ \text{to } SP_2; & \text{if } P_1 \text{ is below } P_2S \text{ (case 2)} \end{cases} \tag{E.61}$$

If instead P_2 is the secondary diffraction edge, then the following replacements are made in Eqs. (E.34)–(E.45). Equation (E.61) applies for this case as well.

$$\begin{aligned} &r_S = r_m;\ r_R = r_{R2};\ \theta_R = \theta_{R2};\ \beta = \beta_2;\ \ell = r_m + r_{R2} \\ &Q_{Sw} = Q_{Sw2};\ Q_{Rw} = Q_{Rw2};\ \tau = \tau_m + \tau_{R2};\ \tau_R = \tau_{R2} \\ &\tau_S = \tau_m;\ p = p_2 \end{aligned} \tag{E.62}$$

The diffraction coefficient D_2, due to diffraction over edge P_2, is then

$$D_2 = p_2 k \ell_2 e^{-j\omega\tau_2} \tag{E.63}$$

where $\ell_2 = \ell$ and $\tau_2 = \tau$, both of which are defined in Eq. (E.60) or Eq. (E.62).

The quantity p_2 in Eq. (E.63) is calculated using Eq. (E.34) and either Eq. (E.42) or Eq. (E.43), with p_1 replaced with p_2, depending on whether the diffraction edge interrupts the line of sight between the source and receiver. Figure E.7c shows the diffraction edge interrupting the line of sight and Figure E.7d shows no interruption.

The total diffracted pressure, p_{diff2}, arriving at the receiver due to the two diffraction edges is then calculated as follows.

$$p_{\text{diff2}} = D_1 D_2 \frac{e^{-j\omega(\tau_S + \tau_R + \tau_m)}}{k(r_S + r_R + r_m)} \tag{E.64}$$

where D_1 is defined by Eq. (E.59) and D_2 is defined by Eq. (E.63).

In the preceding analyses, if the barriers have hard sides, the reflection coefficients, Q_S, Q_R and Q_m, can be set equal to one, and if the barriers are absorptive, the reflection coefficients can be set equal to 0.

E.9 Calculation of the Combined Ground and Barrier Excess-attenuation Effects

In the previous section, it was outlined how to calculate the diffraction coefficient for the four types of situation shown in Figures E.7a–d, involving one or more diffraction edges between the source and receiver. For the purpose of the calculation procedure, it was assumed that the diffraction edge was in free space with no ground-reflected waves. In this section we show how to combine the effects of ground reflections on the source and receiver sides of the diffraction edge(s) as well as the ground reflection between the diffraction edges in the case of two diffraction edges. Every ground segment on either side of, and between the diffraction edges if there are two, generates a reflected wave, and the interaction of these waves on the source and receiver sides of the diffraction edge must be taken into account when calculating the resulting sound pressure at the receiver due to all ray paths. Details of the procedures for doing this for each type of diffraction edge will be provided in separate subsections to follow. In each section, the increase, ΔL_D, in sound pressure level due to the screening effect of any terrain (usually negative) and the increase, ΔL_g, due to the ground reflections are calculated separately and the excess attenuation, $A_{\text{gr+bar}}$, is then calculated as

$$A_{\text{gr+bar}} = -\Delta L = -(\Delta L_D + \Delta L_g) \qquad \text{(dB)} \tag{E.65}$$

E.9.1 Terrain Involving a Single Diffraction Wedge

The calculation of the sound pressure at the receiver p_1 due to rays travelling over a finite-impedance wedge in the absence of ground was discussed in a previous section, where the pressure p_1 at the receiver is defined in Eq. (E.42) or Eq. (E.43). Here the effect of the ground will be included. When sound travels from source to receiver over a barrier or wedge, there are four diffracted rays (p_1–p_4) that have to be included if there is only one ground segment on each side of the wedge, or if the wedge is on flat ground as shown in Figure E.13a, where the four paths are labelled from 1 to 4. Ray p_1 is the ray that was considered in the previous sections for diffraction over a wedge in free space in the absence of any ground effects.

As we are interested in the excess attenuation (in dB) due to the combined effect of the ground and diffraction edge, it is convenient to write the sound pressure, p_t, at the receiver as a ratio of the free field sound pressure, p_F (defined in (E.43)), that would exist

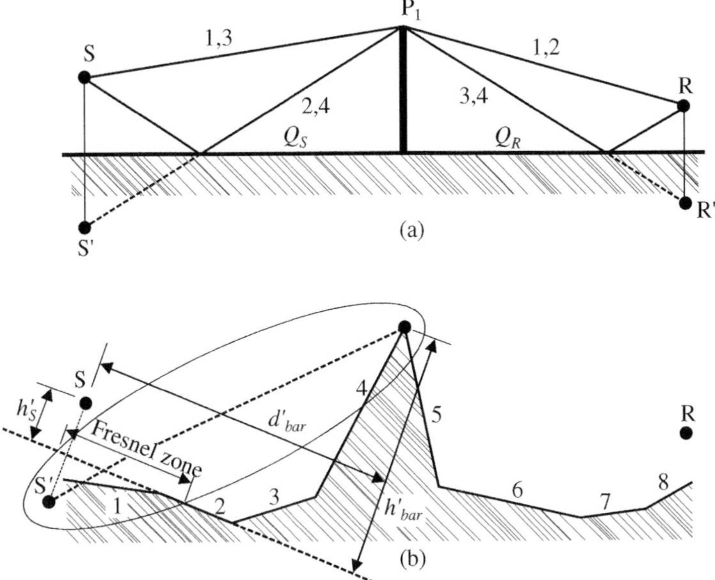

Figure E.13 Diffraction over a finite-impedance wedge including the effect of the ground: (a) flat ground; (b) segmented ground profile.

in the absence of any diffraction edge, ground and atmospheric refraction.

$$\frac{p_t}{p_F} = \frac{p_1}{p_F}\left(1 + Q_S\frac{p_2}{p_1} + Q_R\frac{p_3}{p_1} + Q_SQ_R\frac{p_4}{p_1}\right) \tag{E.66}$$

where Q_S and Q_R are the complex spherical-wave reflection coefficients on the source and receiver sides of the diffraction edge, respectively. The term outside of the brackets on the right-hand side of the equation represents the excess attenuation due to the diffraction edge and the second term represents the ground effect. Thus,

$$\Delta L = \Delta L_D + \Delta L_g = 20\log_{10}\left|\frac{p_1}{p_F}\right| + 20\log_{10}\left|1 + Q_S\frac{p_2}{p_1} + Q_R\frac{p_3}{p_1} + Q_SQ_R\frac{p_4}{p_1}\right| \quad \text{(dB)} \tag{E.67}$$

The diffracted sound pressure, p_1, is calculated using Eq. (E.42) or (E.43) for diffraction over a diffraction edge in free space. The diffracted sound pressure, p_2, is obtained in the same way as p_1, except that the source is replaced by the image source, which is as far below the ground plane as the source is above it. The sound pressure, p_3, is obtained in the same way as p_1, except that the receiver is replaced by the image receiver; the sound pressure, p_4, is obtained in the same way as p_1, except that the source is replaced by the image source and the receiver is replaced by the image receiver. To calculate the excess attenuation due to the ground effect, ΔL_g, Eq. (E.67) must be modified to account for the effects of reduced coherence between the rays that make up the ground-reflected contribution at the receiver, in accordance with the coherence function (Eq. (5.152)). The modified form of the second modulus term on the right-hand side of Eq. (E.67), for a single ground segment on each side of the wedge, is

$$\begin{aligned}\Delta L_g = 10\log_{10}\Bigg[&\left|1 + F_2\frac{w_{St}^2 Q_S p_2}{p_1} + F_3\frac{w_{Rt}^2 Q_R p_3}{p_1} + F_4\frac{w_{St}^2 Q_S w_{Rt}^2 Q_R p_4}{p_1}\right|^2 \\ &+ (1 - F_2^2)\left|\frac{w_{St}^2 \overline{\mathfrak{R}}_S p_2}{p_1}\right|^2 + (1 - F_3^2)\left|\frac{w_{Rt}^2 \overline{\mathfrak{R}}_R p_3}{p_1}\right|^2 \\ &+ (1 - F_4^2)\left|\frac{w_{St}^2 \overline{\mathfrak{R}}_S w_{Rt}^2 \overline{\mathfrak{R}}_R p_4}{p_1}\right|^2\Bigg] \quad \text{(dB)}\end{aligned} \tag{E.68}$$

where $\overline{\mathfrak{R}}_S$ and $\overline{\mathfrak{R}}_R$ are the incoherent ground-reflection coefficients (see Section C.6) on the source side of the diffraction edge and receiver side, respectively, and w_{St} and w_{Rt} are the total Fresnel weights on the source side of the diffraction edge and receiver side, respectively. Of course for flat ground on both sides of the diffraction edge, these latter two values are each 1. However, for non-flat terrain it is possible for these values to be considerably less than 1. The Fresnel zone weights need to be modified (to account for small segment sizes) and normalised (to account for the total on the source or receiver side of the diffraction edge exceeding 1), as discussed in the following paragraphs.

For a segmented terrain, the ground-effect term in brackets of Eq. (E.66), or the right hand modulus term of Eq. (E.67), is expanded to include all combinations of reflections on the source and receiver sides of the diffraction edge (Plovsing 2006).

Each reflection is associated with a Fresnel zone, as discussed in Appendix F. The Fresnel zone for each reflection is associated with a Fresnel weight to represent the relative contribution of the reflected ray from a particular ground segment, as explained in Appendix F. The Fresnel zone on the source side of the diffraction edge is determined using the image source position and the receiver placed on top of the diffraction edge, as illustrated for reflection from ground segment 2 in Figure E.13b. Similarly, the Fresnel zone on the receiver side is found using the source placed on top of the diffraction edge and the image receiver position.

The Fresnel weights on the receiver side of the diffraction edge with N_i ground segments, are denoted w_{Si} and on the receiver side, with N_j ground segments, they are denoted w_{Rj}. If the source (or receiver, for reflection from a ground segment on the receiver side) or the top of the diffraction edge is below the segment, the Fresnel weight for that reflection is set equal to 0.

If the size of the segments are not sufficiently large, the ground effect has to be corrected to account for the limited segment size. This is done by modifying the Fresnel weights as follows, to obtain the modified weights w_{Si} (Plovsing 2014). The modified weight, w_{Si}, is calculated using as a basis the original weight, $w_{Si,or}$, which, in turn, is calculated as discussed in Appendix F. Thus,

$$w_{Si} = w_{Si,or} G_1 G_2 \tag{E.69}$$

where

$$G_1 = \begin{cases} 1 & h'_S \geq h''_S \\ \dfrac{h'_S}{h''_S} & 0 < h'_S < h''_S \\ 0 & h'_S \leq 0 \end{cases} \tag{E.70}$$

$$h''_S = \begin{cases} h_S & h_S < h''_{\text{bar}} \\ h''_{\text{bar}} & h'_S \geq h''_{\text{bar}} \end{cases} \tag{E.71}$$

where h_S is the shortest distance from the source to the terrain below.

$$G_2 = \begin{cases} 1 & h'_{\text{bar}} \geq h''_{\text{bar}} \\ \dfrac{h'_{\text{bar}}}{h''_{\text{bar}}} & 0 < h'_{\text{bar}} < h''_{\text{bar}} \\ 0 & h'_{\text{bar}} \leq 0 \end{cases} \tag{E.72}$$

$$h''_{\text{bar}} = \begin{cases} 0.0005 d'_{\text{bar}} & d'_{\text{bar}} < 400 \\ 0.2 & d'_{\text{bar}} \geq 400 \end{cases} \tag{E.73}$$

where the variables, h'_S, d'_{bar} and h'_{bar} are defined in Figure E.13b for ground segment 2 on the source side of the diffraction edge.

Although Eq. (E.73) reflects what is presented in Plovsing (2014), this is different by a factor of 10 to the original document (Plovsing 2006).

Equations similar to Eqs. (E.69)–(E.73) are used to define the modified Fresnel zone weights for the receiver side of the diffraction edge.

The frequency-dependent Fresnel weights, w_{Si}, on the source side of the diffraction edge, as calculated using the procedure in Appendix F and Eq. (E.69), are further normalised with w_{St} to obtain the weights, w'_{Si}, as follows (Plovsing 2014).

$$w'_{Si} = \begin{cases} \dfrac{w_{Si}}{w_{St}}\left(\dfrac{\Delta w_{St}}{\Delta(w_{St}+w_{Rt})}+1\right) & w_{St} > 1 \\ \dfrac{w_{Si}}{w_{St}} & 0 < w_{St} \leq 1 \\ 0 & w_{St} = 0 \end{cases} \tag{E.74}$$

where Δw_{St} is the amount by which the total value, w_{St}, of the weights on the source side of the diffraction edge (calculated in Eq. (E.69)) exceed 1 and is equal to 0 if the sum of the weights is less than 1. A similar definition holds for Δw_{Rt}, which represents the sum of the modified weights on the receiver side of the diffraction edge. The frequency-dependent modified and normalised Fresnel weights, w'_{Rj}, on the receiver side of the diffraction edge are calculated in a similar way to those on the source side. The sum of weights, w_{St}, for N ground segments on the source side of the wedge, is given by,

$$w_{St} = \sum_{i=1}^{N} w_{Si} \tag{E.75}$$

The reflection coefficients for each segment are also modified by multiplying each one by the square of the sum of the modified (not normalised) Fresnel weights, w^2_{St}, on the source side of the diffraction edge, provided that the multiplier is less than or equal to 1. A similar operation is performed for the Fresnel weights and reflection coefficients for the segments on the receiver side of the diffraction edge, using w_{Rt} as the sum of the modified (not normalised) Fresnel weights on the receiver side of the diffraction edge.

The resulting ground effect for propagation from the source to receiver in the presence of a single barrier or diffraction wedge for the case of reflection from ground segment i on the source side and ground segment j on the receiver side is,

$$\begin{aligned} \Delta L_{g,ij} = 10\log_{10}\Bigg[& \left|1 + F_{2i}\frac{w^2_{St}Q_{Si}p_{2i}}{p_1} + F_{3j}\frac{w^2_{Rt}Q_{Rj}p_{3j}}{p_1} \right. \\ & \left. + F_{4ij}\frac{w^2_{St}Q_{Si}w^2_{Rt}Q_{Rj}p_{4ij}}{p_1}\right|^2 \\ & + (1-F^2_{2i})\left|\frac{w^2_{St}\overline{\Re}_{Si}p_{2i}}{p_1}\right|^2 + (1-F^2_{3j})\left|\frac{w^2_{Rt}\overline{\Re}_{Rj}p_{3j}}{p_1}\right|^2 \\ & + (1-F^2_{4ij})\left|\frac{w^2_{St}\overline{\Re}_{Si}w^2_{Rt}\overline{\Re}_{Rj}p_{4ij}}{p_1}\right|^2\Bigg] w'_{Si}w'_{Rj} \end{aligned} \tag{E.76}$$

where:

- Q_{Si} and Q_{Rj} are the spherical-wave reflection coefficients for segment i on the source and segment j on the receiver side of the wedge, respectively
- $\overline{\Re}_{Si}$ and $\overline{\Re}_{Rj}$ are the incoherent reflection coefficients for segment i on the source and segment j on the receiver side of the wedge, respectively
- F_{2i} and p_{2i} refer to path 2 in Figure E.13a, involving reflection from the ith segment on the source side of the diffraction edge
- F_{3j} and p_{3j} refer to path 3 in Figure E.13a, involving reflection from the jth segment on the receiver side of the diffraction edge
- F_{4ij} and p_{4ij} refer to path 4 in Figure E.13a, involving reflection from the ith segment on the source side of the diffraction edge and the jth segment on the receiver side of the diffraction edge.

The image sources and receivers are unique to each ground segment, as shown in Figure E.13b for segment 2 on the source side of the diffraction edge. If any Fresnel weight, w'_{Si} or w'_{Rj} is greater than 1, it is set equal to 1.

The diffracted sound pressures, p_{2i}, p_{3j} and p_{4ij}, corresponding to rays reflected from ground segment i on the source side of the wedge and segment j on the receiver side, are obtained using Eq. (E.42) or (E.43) with the appropriate substitutions, as explained previously for the calculation of p_2, p_3 and p_4. The sound pressure p_1 is the diffracted sound pressure of Eq. (E.42) or (E.43).

The total ground attenuation is obtained by summing arithmetically all of the possible i, j combinations, bearing in mind that each combination is made up of four waves, as can be seen by inspection of Eq. (E.76)

$$\Delta L_g = \sum_{i=1}^{N_i} \sum_{j=1}^{N_j} \Delta L_{g,ij} \quad \text{(dB)} \tag{E.77}$$

E.9.2 Terrain involving a Double Diffraction Wedge

A double diffraction wedge is a single wedge with two significant diffraction edges, as shown in Figures E.6c and E.9b. The main difference to the calculations for the single wedge is that the spherical-wave ground-reflection coefficients Q_S and Q_R on the source and receiver side of the diffraction edge, respectively, are calculated as if the receiver is located at P_2 when calculating Q_S or the source is located at P_1 when calculating Q_R, as shown in Figure E.9b. The equations used for calculating the diffracted sound pressure due to each path are those derived in Section E.8.2 for a double diffraction wedge. Ray p_1 is the diffracted ray with no ground reflections and it is equal to p_{diff2}, as defined in Eq. (E.47).

E.9.3 Terrain involving Two Single Diffraction Wedges

With two diffraction edges, there are eight sound ray paths that need to be considered for each combination of ground segment on the source side, ground segment between the diffraction edges and ground segment on the receiver side, as shown in Figure E.14.

Referring to Figure E.14, Eq. (E.66) for a single wedge becomes, for two wedges,

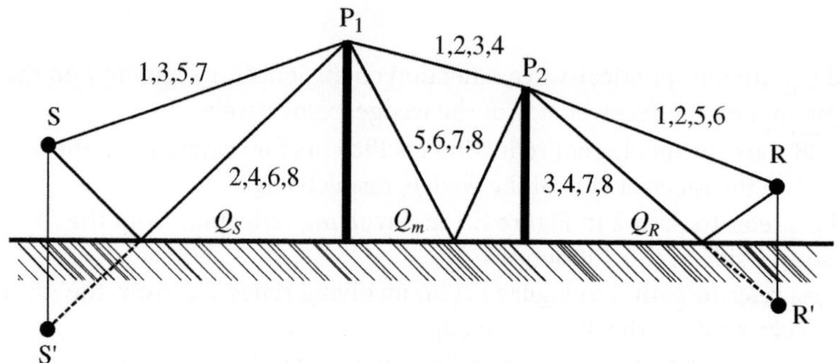

Figure E.14 Diffraction paths (eight) for a double wedge.

$$\frac{p}{p_F} = \frac{p_1}{p_F}\left(1 + Q_S\frac{p_2}{p_1} + Q_R\frac{p_3}{p_1} + Q_SQ_R\frac{p_4}{p_1} + Q_m\frac{p_5}{p_1} + Q_SQ_m\frac{p_6}{p_1} + Q_mQ_R\frac{p_7}{p_1} + Q_SQ_mQ_R\frac{p_8}{p_1}\right) \tag{E.78}$$

and Eq. (E.68) for a single wedge becomes for two wedges,

$$\begin{aligned}\Delta L_{g,ij\ell} = 10\log_{10} \Bigg[\Bigg| & 1 + F_{2i}\frac{w_{St}^3 Q_{Si} p_{2i}}{p_1} + F_{3j}\frac{w_{Rt}^3 Q_{Rj} p_{3j}}{p_1} \\ & + F_{4ij}\frac{w_{St}^3 Q_{Si} w_{Rt}^3 Q_{Rj} p_{4ij}}{p_1} + F_{5\ell}\frac{w_{mt}^3 Q_{m\ell} p_{5\ell}}{p_1} \\ & + F_{6i\ell}\frac{w_{St}^3 Q_{Si} w_{mt}^3 Q_{m\ell} p_{6i\ell}}{p_1} + F_{7j\ell}\frac{w_{mt}^3 Q_{m\ell} w_{Rt}^3 Q_{Rj} p_{7j\ell}}{p_1} \\ & + F_{8ij\ell}\frac{w_{St}^3 Q_{Si} w_{mt}^3 Q_{m\ell} w_{Rt}^3 Q_{Rj} p_{8ij\ell}}{p_1}\Bigg|^2 \\ & + (1 - F_{2i}^2)\left|\frac{w_{St}^3 \overline{\mathfrak{R}}_{Si} p_{2i}}{p_1}\right|^2 + (1 - F_{3j}^2)\left|\frac{w_{Rt}^3 \overline{\mathfrak{R}}_{Rj} p_{3j}}{p_1}\right|^2 \\ & + (1 - F_{4ij}^2)\left|\frac{w_{St}^3 \overline{\mathfrak{R}}_{Si} w_{Rt}^3 \overline{\mathfrak{R}}_{Rj} p_{4ij}}{p_1}\right|^2 + (1 - F_{5\ell}^2)\left|\frac{w_{mt}^3 \overline{\mathfrak{R}}_{m\ell} p_{5\ell}}{p_1}\right|^2 \\ & + (1 - F_{6i\ell}^2)\left|\frac{w_{St}^3 \overline{\mathfrak{R}}_{Si} w_{mt}^3 \overline{\mathfrak{R}}_{m\ell} p_{6i\ell}}{p_1}\right|^2 \\ & + (1 - F_{7j\ell}^2)\left|\frac{w_{mt}^3 \overline{\mathfrak{R}}_{m\ell} w_{Rt}^3 \overline{\mathfrak{R}}_{Rj} p_{7j\ell}}{p_1}\right|^2 \\ & + (1 - F_{8ij\ell}^2)\left|\frac{w_{St}^3 \overline{\mathfrak{R}}_{Si} w_{mt}^3 \overline{\mathfrak{R}}_{m\ell} w_{Rt}^3 \overline{\mathfrak{R}}_{Rj} p_{8ij\ell}}{p_1}\right|^2 \Bigg] w'_{Si} w'_{Rj} w'_{m\ell}\end{aligned} \tag{E.79}$$

The sound pressure, p_1, is the diffracted sound pressure, p_{diff2}, of Eq. (E.64) for diffraction over two wedges in free space. The sound pressure, p_2, is obtained in the same way as p_1, except that the source is replaced by the image source, which is as far below the ground plane as the source is above it. The sound pressure, p_3, is obtained in the same way as p_1, except that the receiver is replaced by the image receiver. The sound pressure, p_4, is obtained in the same way as p_1, except that the source is replaced by the image source and the receiver is replaced by the image receiver. The sound pressure, p_5, is obtained in the same way as p_1, except that P_1 is replaced by its image when diffraction over P_2 is considered and P_2 is replaced by its image when diffraction over P_1 is considered. The sound pressure, p_6, is obtained in the same way as p_1, except that the source is replaced by the image source and P_1 is replaced by its image when diffraction over P_2 is considered and P_2 is replaced by its image when diffraction over P_1 is considered. The sound pressure, p_7, is obtained in the same way as p_1, except that P_1 is replaced by its image when diffraction over P_2 is considered and P_2 is replaced by its image when diffraction over P_1 is considered and the receiver is replaced by the image receiver. The sound pressure, p_8, is obtained in the same way as p_1, except that the source is replaced by the image source, P_1 is replaced by its image when diffraction over P_2 is considered and P_2 is replaced by its image when diffraction over P_1 is considered and the receiver is replaced by the image receiver.

The total ground attenuation is obtained by summing arithmetically all of the possible i, j, ℓ combinations, bearing in mind that each combination is made up of eight waves, as can be seen by inspection of Eq. (E.79).

$$\Delta L_g = \sum_{i=1}^{N_i} \sum_{j=1}^{N_j} \sum_{\ell=1}^{N_\ell} \Delta L_{g,ij\ell} \quad \text{(dB)} \tag{E.80}$$

Effect of a Downward-refracting Atmosphere

When propagation over the wedge occurs in a downward-refracting atmosphere, the preceding analysis has to be done for curved sound rays, with the propagation distances and times calculated as described in Appendix D.

References

Hadden Jr, WJ and Pierce A 1981 Sound diffraction around screens and wedges for arbitrary point source locations. *Journal of the Acoustical Society of America* **69**(5), 1266–1276.

Plovsing B 2006 Comprehensive outdoor sound propagation model. part 1: Propagation in an atmosphere without significant refraction. Technical Report AV 1849/00, Delta.

Plovsing B 2014 Proposal for Nordtest Method: Nord2000–prediction of outdoor sound propagation. Technical report, DELTA Acoustics,.

F

Calculation of Fresnel Zone Sizes and Weights

F.1 Introduction

The concept of a Fresnel zone is used to take into account the fact that reflection of a sound ray from the ground does not just involve the ground surface at the point of specular reflection. The finite size of the wavelength of sound means that the ground on either side of the specular reflection point will affect the amplitude and phase of the reflected ray. The ground surface involved in the reflection is referred to as the 'Fresnel zone', its extent is wavelength- (and hence frequency-) dependent and it is taken into account in the Nord2000 propagation model. In this appendix, we explain how to calculate the size of the Fresnel zone to be used in the calculations outlined in Section E.5 and E.9. The concept of a Fresnel zone also applies to reflection from any object and is used in the calculation of the excess attenuation due to reflection, A_r.

For 3-D terrain modelling, it is necessary to calculate Fresnel areas based on the intersection of a Fresnel ellipsoid with the reflecting surface and the method for doing this is described in Plovsing (2006). The Nord2000 model is restricted to 2-D terrain modelling, so the 2-D Fresnel areas needed for 3-D terrain modelling become 1-D Fresnel lengths for 2-D modelling. However, reflection from vertical surfaces requires the use of the Fresnel ellipsoid and the corresponding Fresnel ellipse, which represents the area of intersection of the Fresnel ellipsoid with the reflecting surface. The Nord2000 model uses a rectangular area as an approximation to the area of this ellipse and the calculation of this area will also be discussed in this appendix. As terrain effects are only represented as 2-D in the Nord2000 model, calculation of the required Fresnel lengths to represent the length of ground involved in the reflection of a ray will be discussed first.

F.2 Fresnel Zone for Reflection from Flat Ground

The frequency-dependent Fresnel zone length for reflection from a plane ground between the source and receiver is the length of the ground surface that is intersected by an ellipse that has its foci at the image source and receiver, as shown in Figure F.1.

The Fresnel zone length, $a(f) = a_1(f) + a_2(f)$, is calculated as follows.

$$a_i(f) = \frac{-B_i + \sqrt{B_i^2 - 4A_iC_i}}{2A_i} \qquad i = 1, 2 \tag{F.1}$$

Wind Farm Noise: Measurement, Assessment and Control, First Edition.
Colin H. Hansen, Con J. Doolan, Kristy L. Hansen.

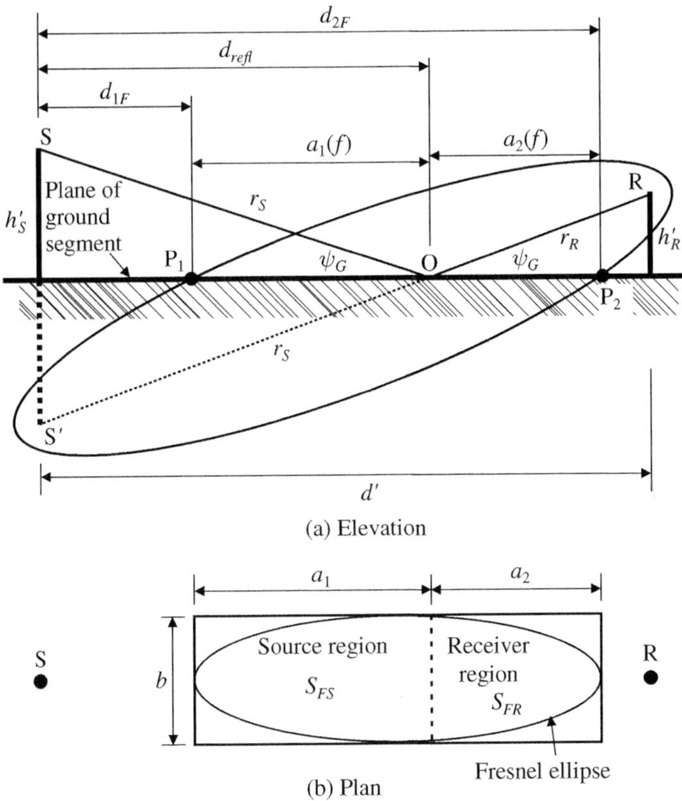

Figure F.1 Definition of the 1-D Fresnel zone for reflection of a ray travelling from the source S to the receiver R, showing the image source S′ as far below the ground segment (or its extension) as the actual source is above the ground.

where

$$A_1 = 4\{\ell^2 - [(r_S + r_R)\cos(\pi - \psi_G)]^2\} \quad \text{(F.2)}$$
$$A_2 = 4\{\ell^2 - [(r_S + r_R)\cos\psi_G]^2\} \quad \text{(F.3)}$$
$$B_1 = 4(r_S + r_R)\cos(\pi - \psi_G)(r_R^2 - r_S^2)) + 4\ell^2\cos(\pi - \psi_G)(r_S - r_R) \quad \text{(F.4)}$$
$$B_2 = 4(r_S + r_R)\cos\psi_G(r_R^2 - r_S^2)) + 4\ell^2\cos\psi_G(r_S - r_R) \quad \text{(F.5)}$$
$$C_1 = C_2 = -\ell^4 + 2(r_S^2 + r_R^2)\ell^2 - (r_S^2 - r_R^2) \quad \text{(F.6)}$$

and where

$$\ell = r_S + r_R + \lambda/n_F \quad \text{(F.7)}$$
$$\psi_G = \arctan\left(\frac{h'_S + h'_R}{d'}\right) \quad \text{(F.8)}$$
$$r_S = \frac{h_S\sqrt{(h'_S + h'_R)^2 + d'^2}}{h'_S + h'_R} \quad \text{(F.9)}$$

$$r_R = \frac{h'_R\sqrt{(h'_S + h'_R)^2 + d'^2}}{h'_S + h'_R} \tag{F.10}$$

where the prime sign on the source and receiver heights indicates that the heights are above the ground segment (or its extension) being considered and d' represents the source–receiver distance along a line parallel to the ground segment.

In Eq. (F.7), the Fresnel parameter, $n_F = 16$ for terrain-effects calculations, $n_F = 8$ for vertically reflecting obstacles and $n_F = 4$ for flat terrain with non-uniform surface properties.

For vertically reflecting obstacles it is necessary to calculate an approximate Fresnel zone area. The Nord2000 model approximates the elliptical Fresnel zone (intersection of the Fresnel ellipsoid with the reflecting surface, as shown in Figure G.6) with an equivalent rectangular area. The rectangular area is determined by calculating the frequency-dependent rectangle width $b(f)$ perpendicular to the plane of Figure F.1a (in a direction into the page) and shown in Figure F.1b. This is given by

$$b(f) = \frac{2b_0(f)}{\sqrt{1 - \left(\frac{a_2(f)-a_1(f)}{a_2(f)+a_1(f)}\right)^2}} \tag{F.11}$$

The approximate area of the Fresnel zone is then given by

$$S_F \approx b(f)[a_1(f) + a_2(f)] \approx S_{FS} + S_{FR} \tag{F.12}$$

where $a_1(f)$ and $a_2(f)$ are given by Eq. (F.1) and $b_0(f)$ is calculated using

$$b_0(f) = \frac{\sqrt{C}}{2\ell} \tag{F.13}$$

where

$$C = \ell^4 - 2(r_S^2 + r_R^2)\ell^2 + (r_S^2 - r_R^2) \tag{F.14}$$

and ℓ is defined by Eq. (F.7).

In Figure F.1, the quantities a_1, a_2 and b refer to quantities $a_1(f)$, $a_2(f)$ and $b(f)$, respectively, in the above equations. The term (f) has been added to indicate that the quantities are frequency dependent.

When there is a sonic gradient resulting in downward-curved rays, the curvature of the rays has to be taken into account when determining the Fresnel zone size. The curved rays are transformed into equivalent straight rays as shown in Figure F.2. In this case the new angle of reflection $\overline{\psi}_G$ will be different to the angle ψ_G shown in Figure F.1, and can be calculated as described in Section D.3.1. In addition, the distance between the source and receiver, d', along a line parallel to the ground segment under consideration, the source height h'_S above the ground segment and the receiver height h'_R in Eqs. (F.1)–(F.10) are replaced with $\overline{d}'$, $\overline{h}'_S$ and $\overline{h}'_R$, which are defined in terms of the new angle, $\overline{\psi}_G$ as follows (see Figure F.2). The new Fresnel ellipse has foci at points $\overline{S}''$ and $\overline{R}'$, where $\overline{S}''$ is the image of the new source in the plane of the ground segment under consideration.

$$\overline{d}' = (r_S + r_R)\cos\overline{\psi}_G \tag{F.15}$$

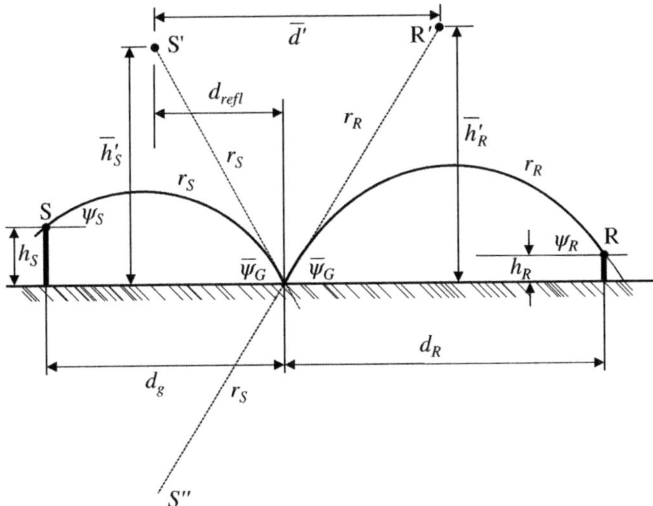

Figure F.2 Definition of Fresnel variables for curved rays resulting from a downward-refracting atmosphere.

where r_S and r_R are defined in Figure F.1 and are the distances to the reflection point from the source and receiver, respectively.

$$\overline{h}'_S = r_S \sin \overline{\psi}_G \tag{F.16}$$

$$\overline{h}'_R = r_R \sin \overline{\psi}_G \tag{F.17}$$

Thus the size of the Fresnel zone for a refracting atmosphere can be obtained by substituting $\overline{d}'$, $\overline{h}'_S$, $\overline{h}'_R$ and $\overline{\psi}_G$ for the corresponding quantities (without the $^-$) in Eqs. (F.1)–(F.10).

F.3 Fresnel Weights for Reflection from a Concave or Transition Ground Segment

Fresnel weights for reflection from a segmented terrain are used to weight the relative importance of reflection from each segment. A particular Fresnel weight applies to a particular terrain segment and represents the ratio of the segment area inside the source/receiver Fresnel zone to the total area of the source/receiver Fresnel zone. However, as mentioned previously, the Nord2000 model is a 2-D model so the Fresnel model is a 1-D model. Thus the Fresnel weight is the ratio of the segment length inside the source/receiver Fresnel zone to the total length of the source/receiver Fresnel zone.

The Fresnel weights to be used in Section E.2 are calculated as follows. In addition to Eqs. (F.15)–(F.17), the following equations will be needed.

$$d_{refl} = r_S \cos \overline{\psi}_G \tag{F.18}$$

$$d_{1,F} = d_{refl} - a_1(f) \tag{F.19}$$

$$d_{2,F} = d_{refl} + a_2(f) \tag{F.20}$$

where r_S is the distance from the source to the specular reflection point and r_R is the distance from the receiver to the specular reflection point. The specular reflection point is where the line from the image source (in ground segment 1 or its extension) to the receiver intersects ground segment 1 or its extension.

The frequency-dependent Fresnel weights, $w(f)$, for a terrain segment, i, (see Section E.3 for calculation of segment sizes and locations), can be calculated using

$$w_i(f) = 0.5[w_S(f) + w_R(f)] \tag{F.21}$$

where the (f) indication of frequency dependence has been dropped in the equations in Section E.9 for simplicity.

$$w_S(f) = \begin{cases} 0 & d_1 \geq d_{\text{refl}} \text{ or } d_2 \leq d_{1,F} \\ 1 & d_1 \leq d_{1,F} \text{ and } d_2 \geq d_{\text{refl}} \\ (d_2 - d_1)/a_1(f) & d_1 > d_{1,F} \text{ and } d_2 < d_{\text{refl}} \\ (d_2 - d_{1,F})/a_1(f) & d_1 \leq d_{1,F} \text{ and } d_2 < d_{\text{refl}} \\ (d_{2,F} - d_1)/a_1(f) & d_1 > d_{1,F} \text{ and } d_2 \geq d_{\text{refl}} \end{cases} \tag{F.22}$$

$$w_R(f) = \begin{cases} 0 & d_1 \geq d_{2,F} \text{ or } d_2 \leq d_{\text{refl}} \\ 1 & d_1 \leq d_{\text{refl}} \text{ and } d_2 \geq d_{2,F} \\ (d_2 - d_1)/a_2(f) & d_1 > d_{\text{refl}} \text{ and } d_2 < d_{2,F} \\ (d_2 - d_{\text{refl}})/a_2(f) & d_1 \leq d_{\text{refl}} \text{ and } d_2 < d_{2,F} \\ (d_{2,F} - d_1)/a_2(f) & d_1 > d_{\text{refl}} \text{ and } d_2 \geq d_{2,F} \end{cases} \tag{F.23}$$

where $w_S(f)$ is the frequency dependent Fresnel weight that represents the reflection area of the ray on the source side of the reflection point, $w_R(f)$ is the frequency-dependent Fresnel weight that represents the reflection area of the ray on the receiver side of the reflection point, d_1 is the horizontal distance from the source S to the end point of the ground segment closest to S, as measured along the extended ground segment, and d_2 is the horizontal distance measured from the source S to the end point of the ground segment closest to R, as measured along the extended ground segment, as illustrated in Figures F.3 (for a non-refracting atmosphere) and F.4 (for a refracting atmosphere) for ground segment 1. Each segment requires calculation of the reflection point, reflection angle and a new Fresnel ellipse, based on new values of h'_S and h'_R, shown in Figures F.3 and F.4 for ground segment 1. So values of a_1 and a_2 will also be different for each ground segment.

If a barrier exists between the source and the ground segment under consideration, the receiver is moved to the top of the barrier and if the barrier exists between the receiver and ground segment under consideration, the source is moved to the top of the barrier and the calculations proceed in the same way as they do in the absence of the barrier but with the new source or receiver position.

If the total of the Fresnel weights exceeds 2, they are normalised to a sum of 2. In cases involving diffraction over one or two barriers or screens, the weights on the source and receiver side of the barrier(s) (and in between the barriers if there are two barriers) are normalised to a sum of unity on the source side, the receiver side and also the zone between the barriers, resulting in a possible maximum total of 3. Fresnel zone weights

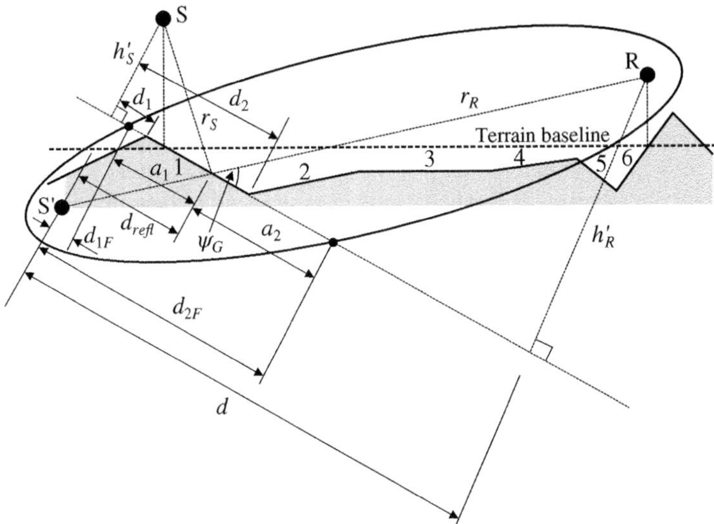

Figure F.3 Definition of distances d_1 and d_2 for a specified ground segment under a non-diffracting atmosphere.

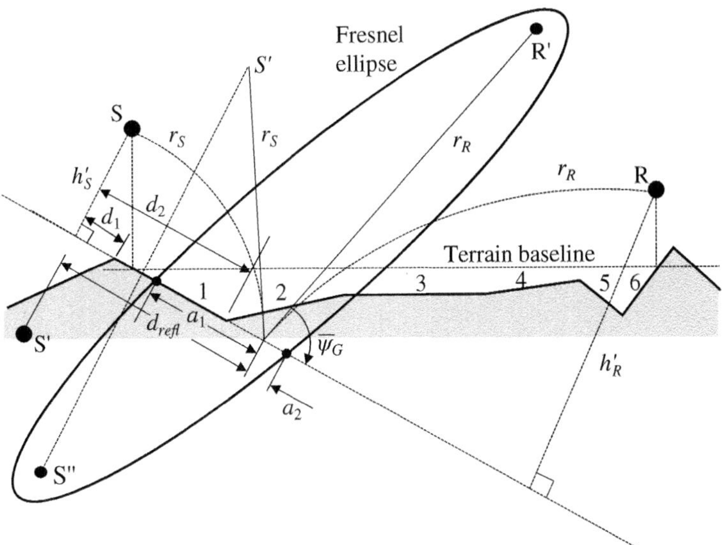

Figure F.4 Definition of distances, d_1 and d_2 for a specified ground segment under a diffracting atmosphere

calculated as described above are used only for concave and transition terrain segments (but not convex segments). The calculation of a Fresnel weight for convex segments requires a different approach, as discussed in Section F.4. See Section E.2 for definitions of element types.

When screens or terrain barriers are involved, the Fresnel zone length is calculated for each side of the barrier by replacing the source with the barrier top for calculations on the receiver side and by replacing the receiver with the barrier top for calculations on the source side. If two barriers are involved, with a ground reflection in between, the Fresnel zone length for segments between the barriers is calculated by replacing the source with the top of the barrier nearest the source and the receiver with the top of the barrier nearest the receiver. The calculations then proceed as described in Section F.2.

F.4 Fresnel Weights for Reflection from a Convex Ground Segment

A terrain segment will be convex if the source or receiver is below the extended segment at the closest point. In this case it is not possible to calculate the Fresnel zone weight for the segment as described in Section F.3, and a modified segment must be used. First the segment must be extended to form an equivalent wedge as shown for segment 2 (with secondary segment 5) in Figure F.5a. If the top of the equivalent wedge is on or above the terrain base line, then the projection, P_1P_2, of the segment onto the terrain base line is the new segment that is used to determine the Fresnel zone size and Fresnel zone weights, as shown in Figure F.5a for segment 2. If the top of the equivalent wedge is

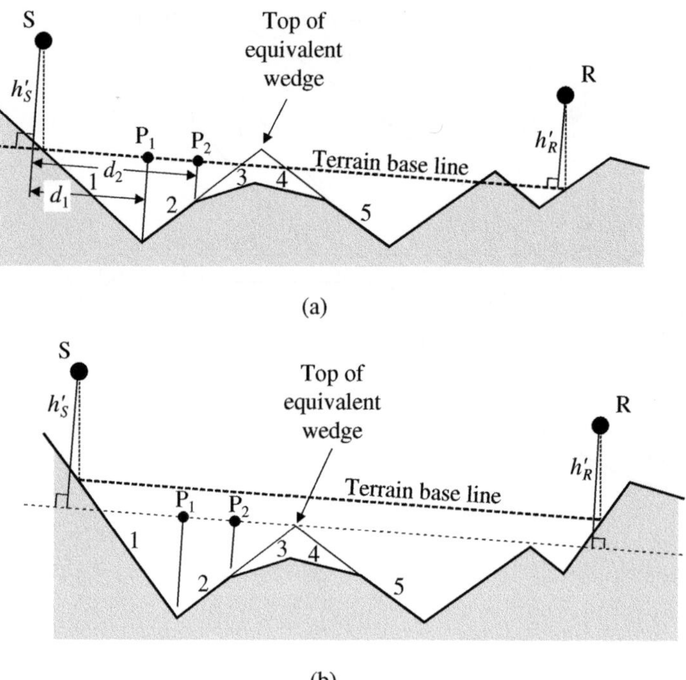

Figure F.5 Definition of equivalent segment size and orientation to replace a convex terrain segment, showing that the equivalent segment is the projection of the concave segment on the terrain base line or on a line parallel to it: (a) top of equivalent wedge above the terrain base line; (b) top of equivalent wedge below the terrain base line.

below the terrain base line, then the projection, P_1P_2, of the segment on to a line parallel with the terrain base line, and passing through the top of the equivalent wedge, is the new segment that is used to determine the Fresnel zone size and weights, as shown in Figure F.5b for segment 2. The terrain base line is the line connecting the points on the ground surface that are vertically beneath the source and receiver, as shown in Figure F.5.

Reference

Plovsing B 2006 Comprehensive outdoor sound propagation model. part 1: Propagation in an atmosphere without significant refraction. Technical Report AV 1849/00, Delta.

G

Calculation of Diffraction and Ground Effects for the Harmonoise Model

G.1 Introduction

In Chapter 5, the excess attenuation (in dB) for the Harmonoise propagation model, as a function of the 1/3-octave band frequency, was given by Eq. (5.207) as a function of the quantities, ΔL_D (diffraction effect) and ΔL_g (ground effect). That is,

$$A_{\text{gr+bar}} = -(\Delta L_D + \Delta L_g) \qquad \text{(dB)} \tag{G.1}$$

The diffraction-effect term is usually negative, indicating a reduction in sound level at the receiver. The ground-effect term is usually positive (indicating an increase in sound level at the receiver) but it can also be negative at certain frequencies where destructive interference with the direct sound ray occurs. The purpose of this appendix is to outline how to calculate ΔL_D and ΔL_g.

When there are no diffraction edges (sometimes called barriers or screens) between the source and receiver, ΔL_D will be zero. For the case of a single diffraction edge, the diffraction effect is calculated using the actual source and receiver locations. When there are two diffraction edges, as illustrated in Figure G.1, there will be two diffraction effects to be calculated. The first calculation will be based on using actual source and receiver positions together with the most efficient diffracting edge. The diffraction efficiency is a function of the path difference between the direct path from source to receiver and the path from the source to the top of the diffraction edge and then to the receiver. If the diffraction edge blocks the line of sight between source and receiver, the path difference is positive and if there is direct line of sight from the source past the diffraction edge to the receiver, then the path difference is negative. The most efficient diffraction edge is the one that is characterised by the largest path-difference number, where all positive numbers are larger than all negative numbers. This path-length difference is used to determine the diffraction effect for the most efficient diffraction edge.

In Figure G.1, the most efficient diffraction edge is P_1 and the corresponding path length difference for calculating the diffraction effect of edge P_1 is the difference between paths SP_1R (solid line) and SR (dashed line), as shown in Figure G.1a. The next most efficient diffraction edge is P_2 and the corresponding path-length difference for calculating the diffraction effect of edge P_2 is the difference between paths P_1P_2R (solid line) and P_1R (dashed line), as shown in Figure G.1b.

Continuing with the case of two diffraction edges between the source and receiver, if the second most efficient diffracting edge is between the source and the most efficient diffracting edge, then for the calculation of the path difference, the receiver position used

Wind Farm Noise: Measurement, Assessment and Control, First Edition.
Colin H. Hansen, Con J. Doolan, Kristy L. Hansen.

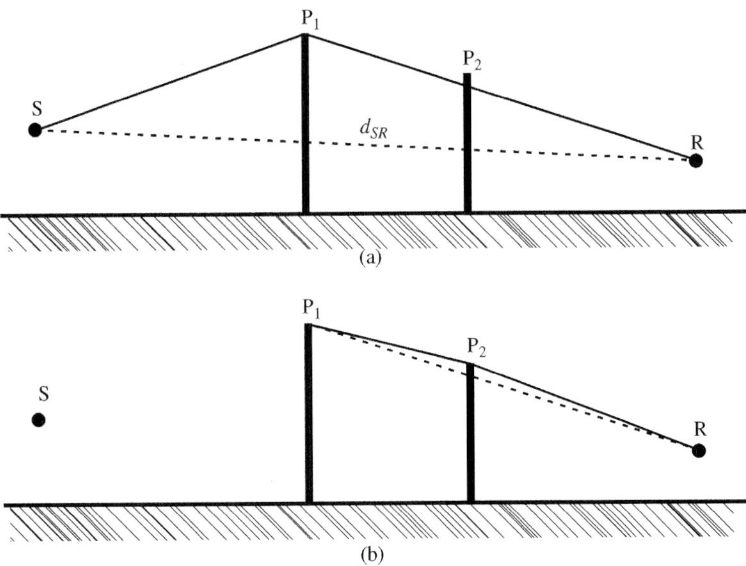

Figure G.1 Double diffraction for the purpose of illustrating the calculation of the diffraction effect: (a) diffraction over the most efficient diffraction edge; (b) diffraction over the second most efficient diffraction edge.

will be the top of the most efficient diffracting edge (referred to as a secondary receiver position). If the second most efficient diffracting edge is between the receiver and the most efficient diffracting edge (as is the case in Figure G.1), then for the calculation of the path difference, the source position used will be the top of the most efficient diffracting edge (referred to as a secondary source position).

When there are three or more diffraction edges between the source and receiver, a similar process to that described above is used to rank order the diffraction edges and to calculate, for each diffracting edge, the corresponding path differences that are used to calculate the diffraction effects. The iterative process to find the path-length differences corresponding to each diffraction edge is as follows. All diffraction edges labelled before each iteration will be referred to as *identified* diffraction edges. First, the most efficient diffraction edge between the actual source and receiver is identified, as discussed above, and labelled as an identified diffraction edge. The general procedure for finding the next most efficient diffraction edge and the next one after that, is as follows. The first iteration finds the second most efficient diffraction edge and in this case the identified diffraction edge in items 1 and 2 below is the most efficient diffraction edge, which is found as described above. The second iteration uses the first and second most efficient diffraction edges as identified edges and finds the third most efficient diffraction edge. The process continues until no more unidentified diffraction edges remain.

1. Find the most efficient diffraction edge between the source and the identified diffraction edge nearest to the source. This is done by placing the receiver at a secondary receiver position that is on the identified diffraction edge nearest the source.
2. Find the most efficient diffraction edge between the receiver and the identified diffraction edge nearest to the receiver. This is done by placing the source at a

secondary source position that is on the identified diffraction edge nearest the receiver.

3. Place the source on the identified diffraction edge nearest to the source. This is the secondary source position.
4. Place the receiver on the nearest identified diffraction edge (secondary receiver position) on the receiver side of the secondary source position determined in item 3 above.
5. Find the most efficient diffraction edge between the secondary source and secondary receiver position. Only diffraction edges that block the line of sight between the secondary receiver and secondary source are considered when searching for the most efficient edge, so there may be no significant diffraction edges between the secondary source and secondary receiver position.
6. Move the secondary source position to the secondary receiver position used in item 5 above, and move the secondary receiver position to the next identified diffraction edge in the direction of the receiver.
7. Repeat items 5 and 6 until there are no more identified diffraction edges between the secondary source and secondary receiver.
8. Find the most efficient new diffraction edge from all the diffraction edges found using steps 1 to 7 above and call this an *identified diffraction edge.*
9. Repeat steps 1 to 8 above until no new diffraction edges are identified.

The diffraction edges found using the iterative procedure just described are numbered in the same order in which they are identified, so the most efficient is labelled 1 and the next most efficient labelled 2, and so on.

When there are no diffraction edges between the source and receiver, $\Delta L_D = 0$ and ΔL_g will be obtained using a single calculation for a single sound ray. When there is a single diffraction edge between the source and receiver, there will be a ground reflection on the source side and one on the receiver side. When there are multiple diffraction edges, there will be ground reflections between each diffraction edge.

G.2 Diffraction Effect, ΔL_D

The overall diffraction effect is the sum of the diffraction effects attributed to all of the diffraction edges between the source and receiver. For M significant diffraction edges,

$$\Delta L_D = \sum_{i=0}^{M} \Delta L_{Di} \tag{G.2}$$

The diffraction effect does not take into account the effect of finite impedance of wedge-shaped diffraction edges, as this is included in the ground-effect calculation described in the next section. All diffraction-effect calculations are based on the thin-screen model, for which the noise reduction due to diffraction over the top of the diffraction edge (ignoring ground reflected paths), is calculated using Eq. (G.3) for the ith diffraction edge (Salomons and Janssen 2011). The Harmonoise model is not limited to two diffraction edges as required for the Nord2000 model. For the Harmonoise model, the diffraction edge numbering ($i = 1$) begins with the most efficient diffraction edge and then moves to the next most efficient diffraction edge,

$i = 2$, continuing with diffraction edges in decreasing order of importance until all have been numbered. For each diffraction edge, the diffraction effect is calculated by placing the source and receiver at the secondary locations actually used in the itemised procedure above to identify the diffraction edge under consideration. For the most efficient diffraction edge, the source and receiver positions are their actual positions.

$$\Delta L_{Di} = \begin{cases} 0 & N_{Fi} < -0.25 \\ -6 + 12\sqrt{-N_{Fi}} & -0.25 \leq N_{Fi} < 0 \\ -6 - 12\sqrt{N_{Fi}} & 0 \leq N_{Fi} < 0.25 \\ -8 - 8\sqrt{N_{Fi}} & 0.25 \leq N_{Fi} < 1 \\ -16 - 10\log_{10} N_{Fi} & N_{Fi} \geq 1 \end{cases} \tag{G.3}$$

where N_{Fi} is defined by

$$N_{Fi} = \pm(2/\lambda)(r_{Si} + r_{Ri} - d_{SR,i}) \tag{G.4}$$

where r_{Si} is the distance from the source (or the secondary source position) to the ith diffraction edge, r_{Ri} is the distance from the receiver (or the secondary receiver position) to the ith diffraction edge and $d_{SR,i}$ is the distance between source and receiver (or between the secondary source and receiver positions as the case may be). For $\theta_i \leq \pi$

$$d_{SR,i} = \sqrt{r_{Si}^2 + r_{Ri}^2 - 2r_{Si}r_{Ri}\cos\theta_i} \tag{G.5}$$

where $\theta_i = \theta_{Si} - \theta_{Ri}$, and r_{Si}, r_{Ri}, θ_{Si} and θ_{Ri} are defined in Figure G.2. To retain consistency with the Nord2000 model, the definitions of θ_{Si} and θ_{Ri} are slightly different to the definitions used by Salomons and Janssen (2011). Although Eq. (G.5) holds for $d_{SR,i}$ when $\theta_i > \pi$, it is not used in that case. Instead, when $\theta_i > \pi$, $d_{SR,i}$ is set equal to $r_{Si} + r_{Ri}$.

The positive sign is used in Eq. (G.4) when the diffraction edge blocks the line of sight between the source (or the secondary source position) and the receiver (or the secondary receiver position). The negative sign is used when there is no blockage to the line of sight.

Only path 1 in Figure E.13a is included in the diffraction-effect calculation. No paths involving a ground reflection are included here.

To enable the calculation of N_F to continue to be valid as θ exceeds 1.5π, Salomons and Janssen (2011) suggest that for the case where the line of sight is blocked ($\theta_i = (\theta_{Si} - \theta_{Ri}) > \pi$), Eq. (G.4) should be replaced with

$$N_{Fi} = (2/\lambda)d_{SR,i}\left(\frac{\epsilon^2}{2} + \frac{\epsilon^4}{3}\right) \tag{G.6}$$

where $d_{SR,i} = r_{Si} + r_{Ri}$ and

$$\epsilon = \frac{\sqrt{r_{Si}r_{Ri}}}{d_{SR,i}}(\theta_i - \pi) \tag{G.7}$$

Once ΔL_{Di} has been calculated, the complex diffracted sound pressure amplitude for use in the ground-effect calculations is given by,

$$p_{Di} = \frac{e^{-jkd_{SR,i}}}{kd_{SR,i}} 10^{\Delta L_{Di}/20} \tag{G.8}$$

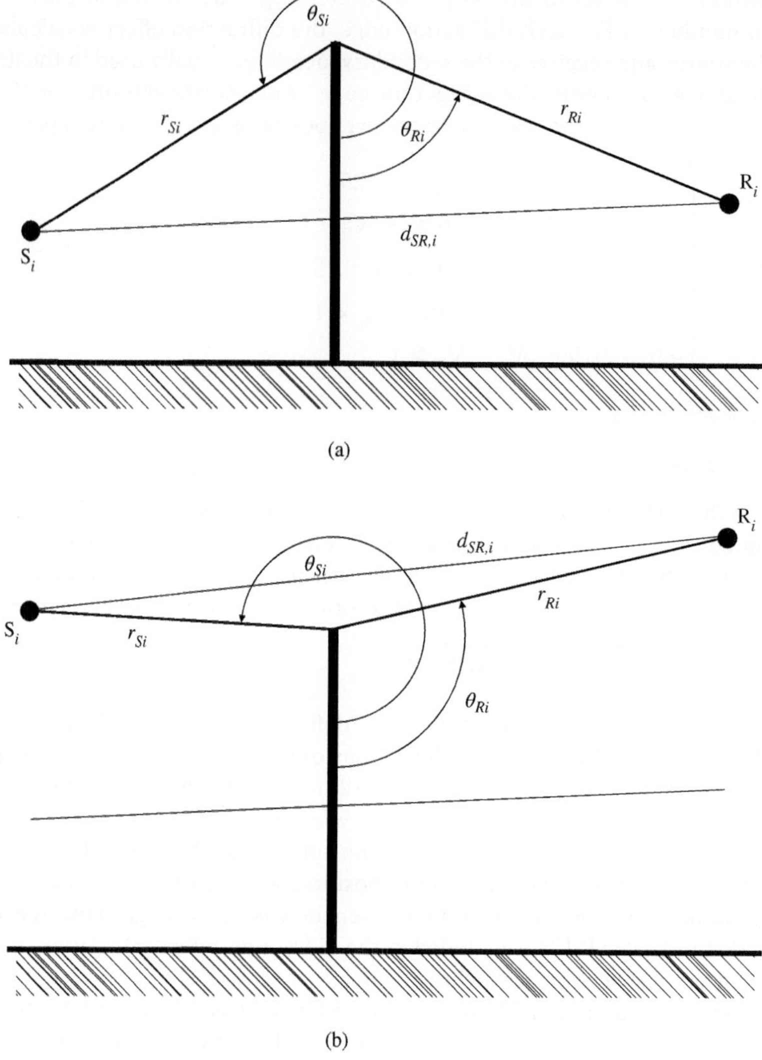

Figure G.2 Definition of parameters used in diffraction effect calculations: (a) screen blocks the line-of-sight between the source and receiver; (b) screen does not block the line-of-sight between the source and receiver.

where $d_{SR,i}$ is defined above differently for $\theta_i \leq \pi$ and $\theta_i > \pi$. Of course p_{Di} is not the actual sound pressure amplitude. It is a relative quantity that can be used to calculate excess attenuation quantities but it cannot be used directly to calculate the sound pressure level at the receiver.

If $N_{Fi} < 0.25$, the diffraction effect, ΔL_{Di}, for the ith diffraction edge is set equal to 0 dB and no further diffraction edges are considered. For cases where there are no diffraction edges for which $N_{Fi} > -0.25$, the total diffraction effect, $\Delta L_D = 0$.

G.3 Ground Effect

The overall ground effect is the sum of the ground effect between the source and its nearest diffraction edge, between the receiver and its nearest diffraction edge and then between each pair of diffraction edges. For the case of two diffraction edges, as shown in Figure G.1a, there will be a total of three ground effects: between S and P_1, between P_1 and P_2 and between P_2 and R. For M ground effects,

$$\Delta L_g = \sum_{i=0}^{M} \Delta L_{gi} \tag{G.9}$$

As for the Nord2000 model (see Section E.2 and Figure E.3), a ground effect is calculated for the ground between the source and its nearest diffraction edge, for the ground between the receiver and its nearest diffraction edge and for the ground between each diffraction edge. Each of the preceding three types of ground section is usually divided into a number of segments using the procedure outlined in Section E.3. The segments may be classified into several types as follows:

$$\text{Segment type} = \begin{cases} \text{concave} & h'_S > 0 \text{ and } h'_R > 0 \\ \text{convex} & h'_S < 0 \text{ or } h'_R < 0 \\ \text{hull} & h'_S = 0 \text{ and } h'_R = 0 \end{cases}$$

where h'_S and h'_R refer to source and receiver heights above ground segments (or their extension) as shown in Figure E.3, and will vary in magnitude from one segment to another.

The ground effect, ΔL_g, is calculated differently, depending on the type of ground segment under consideration. For a concave ground segment or hull segment, the ground effect is a combination of the ground effect for flat ground, ΔL_{gF}, and the ground effect for valley-shaped terrain, ΔL_{gV}, as discussed in Section G.3.1. For a convex ground segment, the ground effect, $\Delta L_g = \Delta L_{gT}$, as discussed in Section G.3.2.

A hull segment is a ground segment that forms a part of the convex hull and for which both the source and the receiver and all of the identified diffraction edges are below its extension. Thus there will only be one ground segment between the two adjacent diffraction edges that bound the hull segment, as shown in Figure G.3. In this figure, the convex hull is defined by the dotted line joining $SP_1P_2P_3P_4R$, and the ground segment

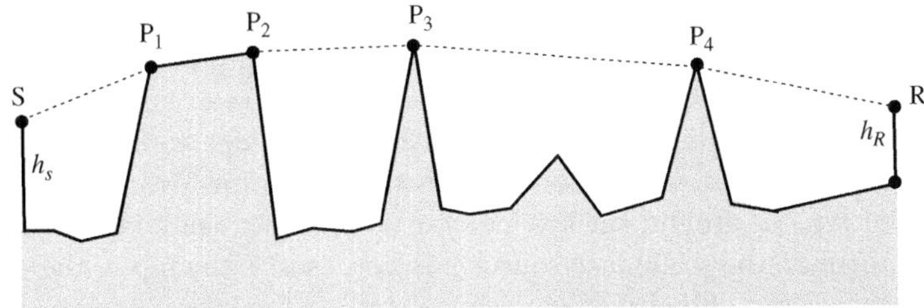

Figure G.3 Construction of a convex hull for the ground profile between the source, *S* and receiver, *R*. The convex hull is defined by the dotted line passing through $SP_1P_2P_3P_4R$ and a hull ground segment is that between points P_1 and P_2.

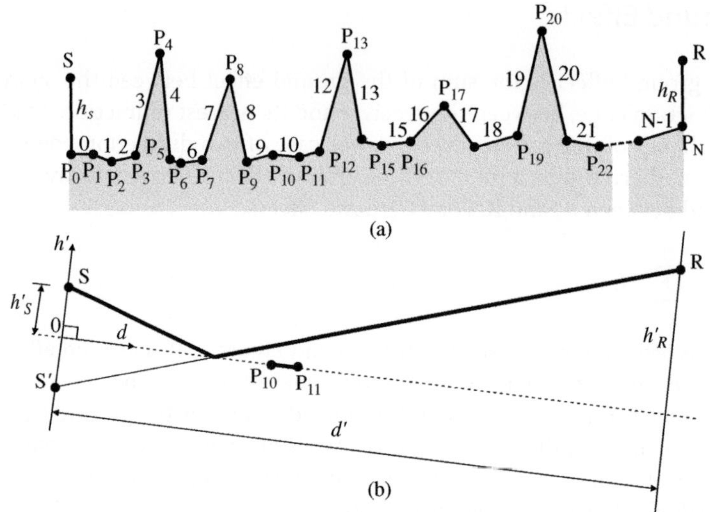

Figure G.4 Ground profile with diffraction edges: (a) aegmentation example with a number of diffraction edges, P_4, P_8, P_{13} and P_{20}; (b) illustration of local (h', d) coordinate system for segment P_{10} to P_{11}.

between P_1 and P_2 is a hull segment. When calculating the weighting coefficient, w_ℓ of Eq. (G.21) for use in the hull segment P_1 to P_2 ground-effect calculation, the source position is moved to P_1 and the receiver position is moved to P_2 and so for the hull ground segment, $h'_S = 0$ and $h'_R = 0$. However, for the calculation of the geometry coefficient, D_ℓ of Eq. (G.22) for hull segment P_1 to P_2, the source position remains at the actual source position and the receiver position is moved to P_3.

The reflection coefficient for hull segments is set equal to 1.0 to avoid excessive attenuation effects obtained at some frequencies. The procedure for calculating the ground effect for a hull ground segment is the same as for calculating the ground effect for a concave ground segment.

One ground-effect model (the concave model) is used when the ground section between two adjacent diffraction edges contains no convex segments and another (the transition model) is used when the ground section between two adjacent diffraction edges contains one or more convex segments. An example of a segmented ground profile that has a number of diffracting edges is shown in Figure G.4a. In this case, the convex hull is defined by points $SP_4P_{13}P_{20}R$.

When the ground effect is being calculated for a section of ground between the source and a diffracting edge, between two diffraction edges or between a diffracting edge and the receiver, the ground segments making up the diffracting edge are included. However, one calculation aspect that can be confusing is that when the Fresnel weighting (see Section G.4) or type of ground surface (convex or other) is being identified between two diffraction edges, the secondary source position used is the top of the diffraction edge that is nearest (out of the two being considered) to the source and the secondary receiver position used is the top of the diffraction edge that is nearest (out of the two being considered) to the receiver. However, when the geometry factor, D_ℓ is being considered, the secondary source is moved one diffraction edge back towards the source and the secondary receiver is moved one diffraction edge towards the receiver. If no diffraction edges are available between the diffraction edge of interest and the source, then the secondary source position becomes the actual source position and similarly for the secondary receiver position.

G.3.1 Concave Model

The concave ground model only considers ground-profile sections that contain concave or no convex ground segments. The ground sections of interest are:

- between significant diffraction edges
- between the source and the nearest significant diffraction edge
- between the receiver and its nearest significant diffraction edge.

Sections that contain one or more convex segments are discussed in the next section. The following analysis calculates the ground effect, $\Delta L_{Gc}(\mathrm{P}_i, \mathrm{P}_j)$ for propagation from a secondary source position P_i, to a secondary receiver position P_j. If $i = 0$, the secondary source position is the actual source position. If $i > 0$, the secondary source position for the calculation of the geometric weighting factor, D_ℓ is determined as follows.

$$\begin{array}{l}\text{Secondary}\\ \text{source}\\ \text{position}\end{array} = \begin{cases} \text{actual source position} & \text{if no diffraction edges between S and } \mathrm{P}_i \\ \begin{array}{l}\text{diffraction edge nearest } \mathrm{P}_i\\ \text{on the source side of } \mathrm{P}_i\end{array} & \begin{array}{l}\text{if there are one or more diffraction}\\ \text{edges between S and } \mathrm{P}_i\end{array} \end{cases}$$

If $j = N$, where N is the number of terrain points, the secondary receiver position is the actual receiver position. If $j < N$, the secondary receiver position for the calculation of the geometric weighting factor D_ℓ is determined as follows (see Figure G.4):

$$\begin{array}{l}\text{Secondary}\\ \text{receiver}\\ \text{position}\end{array} = \begin{cases} \text{actual receiver position} & \text{if no diffraction edges between R and } \mathrm{P}_j \\ \begin{array}{l}\text{diffraction edge nearest } \mathrm{P}_j\\ \text{on the receiver side of } \mathrm{P}_j\end{array} & \begin{array}{l}\text{if there are one or more diffraction}\\ \text{edges between R and } \mathrm{P}_j\end{array} \end{cases}$$

As an example, if we are calculating the ground effect for ground segments between P_8 and P_{13}, there is a diffraction edge (P_4) between P_8 and S and another (P_{20}) between P_{13} and R. Thus the secondary source position used in the following equations for this example would be P_4 and the secondary receiver position would be P_{20}. Calculation of the geometrical weighting factor, D_ℓ is discussed later on in this section.

The concave model for calculating the ground effect, ΔL_g, uses a weighted average of the ground-attenuation equation for flat ground ΔL_{gF}, and the equation for valley-shaped terrain ΔL_{gV}. That is,

$$\Delta L_g = F_g \Delta L_{gF} + (1 - F_g)\Delta L_{gV} \tag{G.10}$$

where

$$F_g = 1 - e^{-1/x_g^2} \tag{G.11}$$

$$x_g = \frac{w_t}{\sqrt{1+(f/f_c)^2}} \tag{G.12}$$

$$w_t = \sum_{\ell=i}^{j-1} w_\ell \tag{G.13}$$

$$f_c = \sqrt{f_{\min} f_{\max}} \tag{G.14}$$

and w_ℓ is the modified Fresnel weighting for ground segment ℓ (defined in Section G.4) and the frequencies $f_{\max}$ and $f_{\min}$ are the frequencies for which the following relationships hold.

$$\phi_{\max}(f_{\min}) = \pi/2 \tag{G.15}$$

$$\phi_{\max}(f_{\max}) = \pi \tag{G.16}$$

$$\phi_{\max}(f) = \max_{i \leq \ell < j}\{\phi_\ell(f)\} \tag{G.17}$$

and

$$\phi_\ell(f) = \mathrm{Arg}(Q_\ell) + k(d_{S\ell} + d_{R\ell} - d_\ell) \tag{G.18}$$

where k is the wave number, Q_ℓ is the complex spherical wave reflection coefficient (see Section C.5) for segment ℓ, and the function $Arg(x) = \tan^{-1}[\mathrm{Im}(x)/\mathrm{Re}(x)]$ radians.

However, the Harmonoise model only does calculations at 1/3-octave band centre frequencies, so it is necessary to interpolate to obtain $f_{\max}$ and $f_{\min}$. To calculate $f_{\max}$, the lowest 1/3-octave band centre frequency f_n for which $\phi(f_n) \geq \pi$ is found and then $f_{\max}$ is found by linear interpolation,

$$f_{\max} = f_{n-1} + (f_n - f_{n-1})\frac{\pi - \phi(f_{n-1})}{\phi(f_n) - \phi(f_{n-1})} \tag{G.19}$$

A similar calculation is used to find $f_{\min}$, with π replaced with $\pi/2$, but in this case, if $n = 1$, then $f_{\min} = f_n$.

The Fresnel parameter, n_F, used to calculate modified Fresnel weights (which are used later in ground-effect calculations) is given by

$$n_F = 32[1 - e^{f_c^2/f^2}] \tag{G.20}$$

The calculation procedure for the Fresnel weights is slightly different to that used for the Nord2000 model and is discussed in Section G.4.

The ground effect, ΔL_{gF}, is given by

$$\Delta L_{gF} = \sum_{\ell=i}^{j-1} w_\ell \Delta L_{gF,\ell} \tag{G.21}$$

where

$$\Delta L_{gF,\ell} = 10\log_{10}\left[\,|1 + F_\ell D_\ell Q_\ell|^2 + (1 - F_\ell^2)|D_\ell Q_\ell|^2\,\right] \tag{G.22}$$

The ground effect, ΔL_{gV}, is given by

$$\Delta L_{gV} = 10\log_{10}\left[\left|1+\sum_{\ell=i}^{j-1} w_\ell F_\ell D_\ell Q_\ell\right|^2 + \sum_{\ell=i}^{j-1} w_\ell (1-F_\ell^2)|D_\ell Q_\ell|^2\right] \quad \text{(G.23)}$$

- F_ℓ is the coherence factor to account for the phase difference between the direct sound and the sound reflected from ground segment ℓ, (or its extension) given by Eq. (5.195).
- Q_ℓ is the complex spherical wave reflection coefficient for ground segment ℓ, calculated as described in Section C.5.
- D_ℓ is a geometrical weighting factor calculated as described below.
- w_ℓ is a modified Fresnel weight, calculated as described in Section G.4.

For calculating the geometrical weighting factor, four cases are considered.

1. $i = 0, j = N$ (no diffraction)

$$D_\ell = \frac{p_F(\mathrm{S}'_\ell \mathrm{R})}{p_F(\mathrm{SR})} \quad \text{(G.24)}$$

where the subscript, F, represents non-diffracted sound pressure, S'_ℓ is an image source location corresponding to ground segment ℓ or its extension, $p_F(\mathrm{S}'_\ell\mathrm{R})$ represents the sound pressure arriving at the receiver along the path $(\mathrm{S}'_\ell\mathrm{R})$ and $p_F(\mathrm{SR})$ represents the sound pressure arriving at the receiver along the path (SR), given by,

$$p_F(\mathrm{SR}) = \frac{\mathrm{e}^{-jkr}}{kr} \quad \text{(G.25)}$$

where r is the distance between the source and receiver, given by $r = \sqrt{d'^2 + (h'_S - h'_R)^2}$ and d', h'_S and h'_R are defined for ground segment 10 in Figure G.4b. A similar equation applies for $p_F(\mathrm{S}'_\ell\mathrm{R})$, where in this case, r is the path length between the image source (in the ℓth ground segment or its extension) and the receiver.

2. $i = 0, j < N$ (ground reflection between the rightmost diffraction edge and the receiver)

$$D_\ell = \frac{p_D(\mathrm{S}'_\ell \mathrm{P}_j \mathrm{R})}{p_D(\mathrm{SP}_j\mathrm{R})} \quad \text{(G.26)}$$

where $p_D(\mathrm{SP}_j\mathrm{R})$ represents the diffracted pressure for path $(\mathrm{SP}_j\mathrm{R})$ and is given by Eq. (G.8), with the path length $d_{SR,i}$ equal to the diffracted path length.

3. $i > 0, j = N$ (ground reflection between the leftmost diffraction edge and the source)

$$D_\ell = \frac{p_D(\mathrm{SP}_i \mathrm{R}'_\ell)}{p_D(\mathrm{SP}_i\mathrm{R})} \quad \text{(G.27)}$$

4. $i > 0, j < N$ (ground reflection between two diffraction edges)

$$D_\ell = \frac{p_D(\mathrm{SP}_i \mathrm{P}'_{j,\ell})}{p_D(\mathrm{SP}_i\mathrm{P}_j)} \frac{p_D(\mathrm{P}'_{i,\ell} \mathrm{P}_j\mathrm{R})}{p_D(\mathrm{P}_i\mathrm{P}_j\mathrm{R})} \quad \text{(G.28)}$$

where the prime represents a source, diffraction edge or receiver image in the ground segment, ℓ, or its extension, so that the image is as far below the ground segment (or its extension) as the source, diffraction edge or receiver, respectively, is above. In all cases the diffracted sound pressure, p_D is calculated using Eq. (G.8).

As an example, if we consider the ground between diffraction edges P_8 and P_{13} in Figure G.4a, then $i = 8, j = 13$, and the equations are evaluated for all ℓ between 8 and 12 inclusive. The source position to be used in Eq. (G.28) is position P_4 and the receiver position is P_{20}. As another example, if we consider the ground between diffraction edges P_{13} and P_{20}, then $i = 13, j = 20$, and the equations are evaluated for all ℓ between 13 and 19 inclusive. The source position to be used in Eq. (G.28) is position P_8 and the receiver position is the actual receiver position. Note, however, that the ground between P_{13} and P_{20} contains a convex peak P_{17} that is below the line of sight between $S = P_{13}$ and $R = P_{20}$. In this case, the ground attenuation between P_{13} and P_{20} must be calculated using the transition model described in the next section.

It is stated by van Maercke and Defrance (2007), and Salomons and Janssen (2011) that the ground segments representing the sides of diffraction wedges (for example, segment 3 in Figure G.4) should be included in the calculation of the ground effect.

The ground effect between two significant diffraction edges is calculated by using a secondary source location on the diffraction edge nearer the source and a secondary receiver location on the diffraction edge nearer the receiver. For the hull section shown as P_1P_2 in Figure G.3, both h'_S and h'_R are zero as a result of the source and receiver being located on the two diffraction edges. Thus the ground between P_1 and P_2 can be treated as a concave segment and, for best results, the absorption coefficient of this segment should be set equal to 1.

G.3.2 Transition Model

Here, the calculation of the ground effect for a section containing a convex ground segment for which $h'_S < 0$ and $h'_R > 0$ (Salomons and Janssen 2011) is explained. The calculation for a segment for which $h'_S > 0$ and $h'_R < 0$ follows a similar procedure. The convex segment used as an example here is segment 16 in Figure G.4a.

The quantities, Q_ℓ, D_ℓ and F_ℓ for a convex ground segment are calculated in the same way as for a concave segment as outlined in Section G.3.1, but with two modifications:

1. The receiver is replaced with the image receiver so h'_R is replaced with $|h'_R|$.
2. The geometrical weighting factor D_ℓ, calculated for a concave segment, is multiplied by the factor, F_T.

$$F_T = \frac{p_D(S'XR)}{p_D(SXR)} \tag{G.29}$$

where S is the relevant secondary source position ($=P_{13}$ in our example) and R is the relevant secondary receiver position ($=P_{20}$ in our example; see Figure G.5a). Location S′ is the image of the secondary source position in the ground segment under consideration (or its extension). The diffracted pressures, p_D, are calculated using Eq. (G.8), with the angle $\theta = \pi$, defined in Figure G.5b, for use in calculating p_D in the denominator of Eq. (G.29) and the angle $\theta' > \pi$, defined in Figure G.5b for use in calculating p_D in the numerator of Eq. (G.29).

The location, X_{16}, is the point of specular reflection in the ground segment under consideration (segment 16 in Figure G.5b).

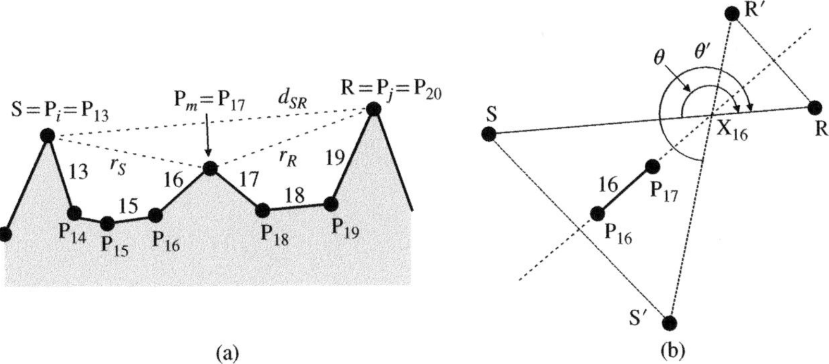

Figure G.5 (a) Geometry for the transition model example showing secondary source and receiver positions at the top of diffraction edges between which the ground effect is to be calculated; (b) location of the specular reflection point, X_{16}, for segment 16.

The procedure will now be outlined for calculating the ground effect, ΔL_{Gt}, between points P_i and P_j (shown as secondary source position S and secondary receiver position R in Figure G.5), with one diffraction edge, $P_\ell = P_m = P_{17}$ below the line of sight between S and R. If there is more than one diffraction edge below the line of sight between S and R, the one closest to the line joining S and R is the one considered in the following analysis.

$$\Delta L_{gt} = \chi \Delta L_1 + (1 - \chi)\Delta L_2 \tag{G.30}$$

where

$$\Delta L_1 = \Delta L_D(SP_mR) + \Delta L_{gc}(SP_m) + \Delta L_{gc}(P_mR) \tag{G.31}$$

and

$$\Delta L_2 = \Delta L_{gc}(SR) \tag{G.32}$$

where $\Delta L_D(\mathrm{SP}_m\mathrm{R})$ is the diffraction attenuation for the path $(\mathrm{SP}_m\mathrm{R})$ and $\Delta L_{gc}(\mathrm{SR})$ is the ground attenuation for path (SR), with the geometrical weighting factor D_ℓ multiplied by the factor F_T of Eq. (G.29), and any negative source and receiver heights multiplied by −1.

The factor χ in Eq. (G.30) is defined as

$$\chi = \chi_2 + (1 - \chi_1)(1 - \chi_2) \tag{G.33}$$

where

$$\chi_1 = \begin{cases} 1 - e^{-1/\tau_1^2} & \tau_1 > 0 \\ 1 & \tau_1 \le 0 \end{cases} \tag{G.34}$$

with

$$\tau_1 = \frac{\delta_{s,av} - \delta_d}{\lambda/8} \tag{G.35}$$

and

$$\chi_2 = \begin{cases} 1 - e^{-1/\tau_2^2} & \tau_2 > 0 \\ 1 & \tau_2 \le 0 \end{cases} \tag{G.36}$$

with

$$\tau_2 = \frac{\delta_d}{\lambda/64} \tag{G.37}$$

where δ_d is the path-length difference between the diffracted path (SP_mR) and the direct path (SR), and where in the example of Figure G.5, $P_m = P_{17}$ and $\delta_d = r_S + r_R - d_{SR}$. The quantity, $\delta_{s,\,av}$ is defined as

$$\delta_{s,av} = \frac{\sum_{\ell=i}^{j-1} w_\ell \delta_{s,\ell}}{\sum_{\ell=i}^{j-1} w_\ell} \tag{G.38}$$

where w_ℓ is the Fresnel weighting for ground segment ℓ, which is calculated as described in Section G.4.

For concave segments, $\delta_{s,\ell}$ is the path-length difference between the reflected path ($SX_\ell R$) and the direct path (SR), where X_ℓ is the specular reflection point for a ray travelling from S to R, and reflecting from ground segment ℓ or its extension, as the case may be. For convex segments, $\delta_{s,\ell}$ is negative, as the replacement of h'_S with $|h'_S|$ or h'_R with $|h'_R|$ is not done for this calculation.

G.4 Fresnel Zone for Reflection from a Ground Segment

A Fresnel zone is used to take into account the fact that reflection of a sound ray from the ground does not just involve the ground surface at the point of specular reflection. The finite size of the wavelength of sound means that the ground on either side of the specular reflection point will affect the amplitude and phase of the reflected ray. The extent of the ground surface involved in the reflection is wavelength- (and hence frequency-) dependent. In this section, we explain the Harmonoise method used to calculate the size of the Fresnel zone to be used in the calculations outlined in Sections G.3.1 and G.3.2. The concept of a Fresnel zone also applies to reflection from any object and is used in the calculation of the excess attenuation due to reflection from a vertical surface, A_r.

The Harmonoise terrain model usually used is 2-D (as for the Nord2000 model), so the 2-D Fresnel areas needed for 3-D terrain modelling become 1-D Fresnel lengths for 2-D modelling. As terrain effects are represented as 2-D, calculation of the required Fresnel lengths to represent the length of ground involved in the reflection of a ray will be discussed first.

Reflection from vertical surfaces is done with a 3-D model and thus requires a 2-D Fresnel model. This requires the use of the Fresnel ellipsoid and the corresponding Fresnel ellipse that represents the area of intersection of the Fresnel ellipsoid with the reflecting surface. A rectangular area is used as an approximation to the area of the ellipse (see the white, unshaded area in Figure G.6a) and the calculation of this area is discussed in Section F.2.

The intersection of the Fresnel ellipsoid with a ground segment (or its extension) creates a Fresnel ellipse, as shown in Figure G.6a. The Fresnel ellipsoid for a reflected ray path is a 3-D surface defined by a set of points P that satisfy the following relation:

$$|S', P| + |P, R| = |S', P| + \lambda/n_F \tag{G.39}$$

where $|S', P|$ is the distance between points S′ and P.

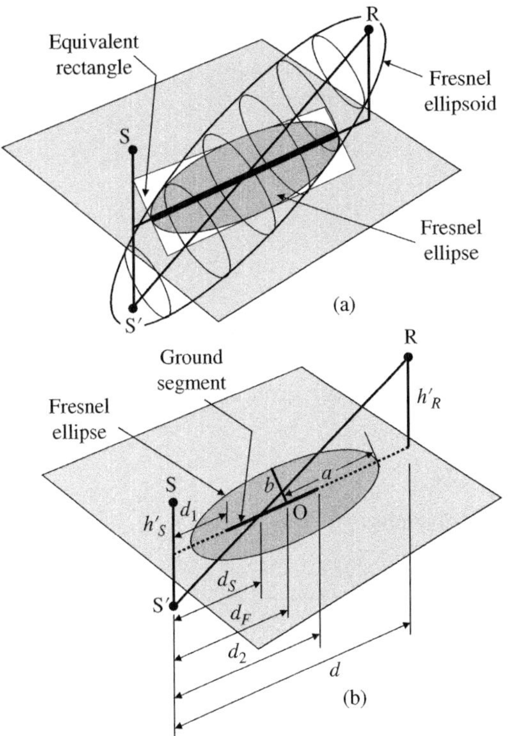

Figure G.6 Intersection with a ground segment of the Fresnel ellipsoid with foci at the image source, S′, and receiver, R: (a) 2-D Fresnel length used for terrain modelling is shown as the intersection of the Fresnel ellipse with the line representing the ground segment (or its extension), shown in the figure as the thick line; (b) Fresnel ellipse in the plane of the ground segment is shown, with centre at O, major axis length $2a$ and minor axis length $2b$. The ground segment is shown as a thick line extending from d_1 to d_2.

In Figure G.6, S and R represent the source and receiver locations, respectively. However, when diffraction edges exist between the source and receiver, S and R may represent a diffraction edge, depending on the part of the terrain for which the ground effect is being calculated.

In Figure G.6b, the source and receiver are at heights, h'_S and h'_R above the ground segment (or its extension in cases where the source or receiver are not directly above it). Although it is possible to calculate Fresnel weights based on the area of the ground segment inside the Fresnel ellipse, most implementations of Harmonoise use the 2-D version, which is based on the length of the ground segment within the length defined by the Fresnel ellipse intersecting the ground segment or its extension, as shown in Figure G.6b. The following calculation procedure for Fresnel weights is slightly different to that used in the Nord2000 model.

Referring to Figure G.6b, the following relations are used to calculate the indicated variables.

$$d_F = \frac{d}{2}\left(1 + \frac{h'^2_S - h'^2_R}{D^2 - d^2}\right) \tag{G.40}$$

and a, which is half the length of the major axis of the ellipse, is given by:

$$a = \frac{1}{2}\sqrt{\frac{D^4 + (D_S^2 - D_R^2)^2 - 2D^2(D_S^2 + D_R^2)}{D^2 - d^2}} \tag{G.41}$$

where

$$D = \frac{\lambda}{n_F} + \sqrt{(h'_S + h'_R)^2 + d^2} \tag{G.42}$$

$$D_S^2 = d_F^2 + h'^2_S \tag{G.43}$$

$$D_R^2 = (d - d_F)^2 + h'^2_R \tag{G.44}$$

The Fresnel weight is calculated in 2-D so each ground segment is represented by a 1-D line. The Fresnel weight, $w_{F,\ell}$, for ground segment ℓ extending from d_1 to d_2, is given by,

$$w_{F,\ell} = F_w(\xi_2) - F_w(\xi_1) \tag{G.45}$$

with the function, $F_w(x)$, given by

$$F_w(x) = \begin{cases} 0 & x \le -1 \\ 1 - \frac{1}{\pi}(\cos^{-1}(x) - x\sqrt{1 - x^2}\,) & -1 < x < 1 \\ 1 & x \ge 1 \end{cases} \tag{G.46}$$

and

$$\xi_m = \frac{d_m - d_F}{a}; \qquad m = 1, 2 \tag{G.47}$$

Since the model uses no ground segments to the left of the source or to the right of the receiver, the function $F(w) = 0$ for the leftmost point of the first segment (beneath the source or secondary source) and $F(w) = 1$ for the rightmost point of the last segment (beneath the receiver or secondary receiver).

Salomons and Janssen (2011) suggest that the Fresnel weights, $w_{F,\ell}$, which are calculated as just described, do not give accurate results at higher frequencies and so suggest that a modified Fresnel weighting, w_ℓ, be used. This is defined as,

$$w_\ell = F_w(\xi'_2) - F_w(\xi'_1) \tag{G.48}$$

with

$$\xi'_m = \frac{\xi_m - \xi_C}{1 - \xi_m \xi_C} \qquad m = 1, 2 \tag{G.49}$$

$$\xi_C = \frac{d_C - d_F}{a} \tag{G.50}$$

$$d_C = \alpha(f)d_F + [1 - \alpha(f)]d_{SP} \tag{G.51}$$

$$d_{SP} = d\frac{h'_S}{h'_S + h'_R} \tag{G.52}$$

$$\alpha(f) = \left[1 + \left(\frac{f}{f_c}\right)^2\right]^{-1} \tag{G.53}$$

where f is the 1/3-octave band centre frequency and f_c is defined by Eq. (G.14). The quantities, h'_S, h'_R, d_F and d are defined in Figure G.6b. For hull segments where $h'_S = 0$ and $h'_R = 0$, Eq. (G.52) is replaced with $d_{SP} = 0.5d$.

References

Salomons E and Janssen S 2011 Practical ranges of loudness levels of various types of environmental noise, including traffic noise, aircraft noise, and industrial noise. *International Journal of Environmental Research and Public Health*, **8**(6), 1847–1864.

van Maercke D and Defrance J 2007 Development of an analytical model for outdoor sound propagation within the harmonoise project. *Acta Acustica United with Acustica*, **93**(2), 201–212.

H

Active Noise-control System Algorithms

H.1 Introduction

Algorithms are needed to calculate values for filter coefficients used to generate the cancelling signal in an active noise-control system. Control algorithms for determining the filter weights are derived in (Hansen et al. 2012, Ch. 6), so only the results will be presented here, first for the single-channel system of Figure 8.5 and second for the multi-channel system of Figure 8.6.

H.2 Single-input, Single-output (SISO) Weight Update Algorithm

The weights of the FIR control filter in the single-input, single-output (SISO) system of Figures 8.5 and 8.7 are updated each time the system obtains a new error-signal sample, k. The error signal at time sample k is given by the sum of the primary-source-generated disturbance $p(k)$ and the control-source-generated disturbance, $s(k)$.

$$e(k) = p(k) + s(k) \tag{H.1}$$

Referring back to Figure 8.7, the output, $y(k)$, of the N-stage FIR control filter at time k is equal to the convolution operation,

$$y(k) = \sum_{i=0}^{N-1} w_i(k)x(k-i) = \boldsymbol{w}^{\mathrm{T}}(k)\boldsymbol{x}(k) = \boldsymbol{x}^{\mathrm{T}}(k)\boldsymbol{w}(k) \tag{H.2}$$

where $\boldsymbol{x}(k)$ is the $(N \times 1)$ vector of reference samples, x in the filter delay chain at time k, given by,

$$\boldsymbol{x}(k) = \begin{bmatrix} x(k) & x(k-1) & x(k-2) & \dots & x(k-(N-1)) \end{bmatrix}^{\mathrm{T}} \tag{H.3}$$

and $\boldsymbol{w}(k)$ is the $(N \times 1)$ vector of filter-weight coefficients at time, k, given by

$$\boldsymbol{w}(k) = \begin{bmatrix} w_0(k) & w_1(k) & w_2(k) & \dots & w_{(N-1)}(k) \end{bmatrix}^{\mathrm{T}} \tag{H.4}$$

In an active noise or vibration-control system, the feedforward control signal $y(k)$, derived by the FIR control filter, will not be equal to the control-source component $s(k)$ of the error signal. This is because active noise-control systems target 'physical' signals (sound disturbances), as opposed to the electrical signals directly utilised by the

Wind Farm Noise: Measurement, Assessment and Control, First Edition.
Colin H. Hansen, Con J. Doolan, Kristy L. Hansen.

adaptive electronic components, and so loudspeakers and microphones are required to convert between the two regimes. These have characteristic frequency responses, or transfer functions. Also, there will be an acoustic,transfer function between the loudspeaker location and the error-sensor location, which will include a propagation delay due to the separation distance between the loudspeaker and the error microphone. In addition, there is an inherent time delay in analogue anti-aliasing filters, as well as a transfer-function component due to the characteristic response of the system being controlled (in this case the sound field in the residence).

Another peculiarity of implementing the filtered-x LMS (or FXLMS) algorithm in an active noise-control system is that microphones can only add signals arriving at their location; they cannot subtract one signal from another as can be done in an electronic system. All of the transfer functions between the control filter output and the error sensor output can be lumped into a single 'cancellation path transfer function', which can be modelled in the time domain as the m-order FIR function estimate (vector), $\hat{\boldsymbol{c}}$.

$$\hat{\boldsymbol{c}} = \begin{bmatrix} c_0 & c_1 & \dots & c_{m-1} \end{bmatrix}^{\mathrm{T}} \tag{H.5}$$

where T denotes the vector transpose.

The control-source-generated component $s(k)$ of the error signal $e(k)$ at time sample k, is then equal to the convolution of the filter output $\boldsymbol{y}(k)$ and this FIR function, $\hat{\boldsymbol{c}}$.

$$s(k) = \boldsymbol{y}(k) * \hat{\boldsymbol{c}} = \sum_{i=0}^{m-1} y(k-i)c_i = \boldsymbol{y}^{\mathrm{T}}(k)\hat{\boldsymbol{c}} \tag{H.6}$$

where $\boldsymbol{y}(k)$ is an $(m \times 1)$ vector of present and past control filter outputs, given by,

$$\boldsymbol{y}(k) = \begin{bmatrix} y(k) & y(k-1) & \dots & y(k-m+1) \end{bmatrix}^{\mathrm{T}} \tag{H.7}$$

The quantity, $\boldsymbol{y}(k)$ can be obtained by matrix multiplication as follows,

$$\boldsymbol{y}(k) = \boldsymbol{X}^{\mathrm{T}}(k)\boldsymbol{w} \tag{H.8}$$

where the T denotes the matrix transpose and $\boldsymbol{X}(k)$ is an $(N \times m)$ matrix of present and past reference signal vectors, the columns of which are the m most recent reference signal vectors, given by,

$$\begin{aligned} \boldsymbol{X}(k) &= \begin{bmatrix} \boldsymbol{x}(k) \; \boldsymbol{x}(k-1) \dots \boldsymbol{x}(k-m+1) \end{bmatrix} \\ &= \begin{bmatrix} x(k) & x(k-1) & \dots & x(k-m+1) \\ x(k-1) & x(k-2) & \dots & x(k-m) \\ & & \vdots & \\ x(k-N+1) & x(k-N) & \dots & x(k-m-N+2) \end{bmatrix} \end{aligned} \tag{H.9}$$

The control source component of the error signal can now be re-expressed by substituting Eq. (H.8) into Eq. (H.6).

$$\begin{aligned} s(k) &= [\boldsymbol{X}^{\mathrm{T}}(k)\boldsymbol{w}]^{\mathrm{T}}\boldsymbol{c} \\ &= \boldsymbol{w}^{\mathrm{T}}\boldsymbol{X}^{\mathrm{T}}(k)\boldsymbol{c} \\ &= \boldsymbol{w}^{\mathrm{T}}\boldsymbol{f}(k) \\ &= \boldsymbol{f}^{\mathrm{T}}(k)\boldsymbol{w} \end{aligned} \tag{H.10}$$

where $\boldsymbol{f}(k)$ is the 'filtered' reference signal vector (see Figure 8.5).

$$\boldsymbol{f}(k) = \boldsymbol{X}^{\mathrm{T}}(k)\boldsymbol{c} = \left[f(k) \quad f(k-1) \quad \dots \quad f(k-N+1)\right]^{\mathrm{T}} \tag{H.11}$$

The ith term in $\boldsymbol{f}(k)$, $f(k-i)$, is equal to the m most recent reference signal samples, $x(k-i)$ through $x(k-i-m+1)$, at the ith control filter stage (the ith position in the input delay chain of Figure 8.7) used in the generation of the control filter output, convolved with the impulse-response function model of the cancellation path, represented as $\hat{\boldsymbol{c}}$ in Figure 8.5.

The gradient descent algorithm used to adjust the control-filter weights has the form,

$$\boldsymbol{w}(k+1) = \boldsymbol{w}(k) - \mu \boldsymbol{\Delta w}(k) \tag{H.12}$$

where $\boldsymbol{\Delta w}(k)$ is the gradient of the error criterion with respect to the weights in the filter, and μ is the portion of the negative gradient to be added to the current weight coefficients with the aim of improving the performance of the system, known as the convergence coefficient.

Squaring Eq. (H.1) gives,

$$e^2(k) = [p(k) + s(k)]^2 = \left[p(k) + \boldsymbol{w}^{\mathrm{T}}\boldsymbol{f}(k)\right]^2 \tag{H.13}$$

Differentiating this with respect to the weight coefficient vector gives,

$$\boldsymbol{\Delta w}(k) \approx \frac{\partial e^2(k)}{\partial \boldsymbol{w}} = 2e(k)\boldsymbol{f}(k) \tag{H.14}$$

Substituting Eq. (H.13) into Eq. (H.11), the gradient descent algorithm used for adapting the weights in the control source FIR filter is,

$$\boldsymbol{w}(k+1) = \boldsymbol{w}(k) - 2\mu e(k)\boldsymbol{f}(k) \tag{H.15}$$

This is known as the (SISO) filtered-x LMS algorithm, which is the standard algorithm that has been used in active noise-control systems for many years. The adaptation of the weight coefficients involves use of the 'filtered' reference signal vector, $\boldsymbol{f}$, as opposed to just the reference signal vector $\boldsymbol{x}$. Thus the characteristics of the cancellation path and the accuracy of its estimate will have an influence upon the stability of the algorithm.

H.3 Multiple-input, Multiple-output Weight Update Algorithm

In Figure 8.6, it can be seen that the signal supplied to each control source is generated by a separate FIR filter, and it will be assumed, without loss of generality, that each control source FIR filter has N stages. The output, $e_i(k)$, of the ith error sensor at time sample k, can be thought of as comprising two parts: that due to primary excitation, $p_i(k)$, and that due to the sum of contributions from each of the N_c control sources, $s_{i,j}(k)$.

$$e_i(k) = p_i(k) + \sum_{j=1}^{N_c} s_{i,j}(k) \tag{H.16}$$

As discussed in the previous section for the SISO case, the jth control source component of the ith error signal is not, in general, equal to the output of the jth control filter, $y_j(k)$, but is rather equal to a version of the control signal that has been modified by the cancellation path transfer function, c_{ij}, of Figure 8.6 between the output of the jth control filter and the output of the ith error sensor. Modelling this transfer function as an m-stage FIR function (vector), $\hat{\boldsymbol{c}}_{i,j}$, and assuming that the system is time-invariant:

$$s_{i,j}(k) = y_j(k) * \hat{\boldsymbol{c}}_{i,j} = \boldsymbol{y}_j^{\mathrm{T}}(k)\hat{\boldsymbol{c}}_{i,j} \tag{H.17}$$

where $\boldsymbol{y}_j(k)$ is an $(m \times 1)$ vector of most recent outputs from the jth control filter.

$$\boldsymbol{y}_j(k) = \begin{bmatrix} y_j(k) & y_j(k-1) & \dots & y_j(k-m+1) \end{bmatrix}^{\mathrm{T}} \tag{H.18}$$

If the output of the jth control filter is expanded as,

$$\boldsymbol{y}_j(k) = \boldsymbol{x}^{\mathrm{T}}(k)\boldsymbol{w}_j \tag{H.19}$$

then $s_{i,j}(k)$ can be expressed in terms of the 'filtered reference signal' used in the previous section as,

$$s_{i,j}(k) = [\boldsymbol{X}^{\mathrm{T}}(k)\boldsymbol{w}_j]^{\mathrm{T}}\boldsymbol{c}_{i,j} \;\; = \boldsymbol{w}_j^{\mathrm{T}}[\boldsymbol{X}(k)\boldsymbol{c}_{i,j}] \;\; = \boldsymbol{w}_j^{\mathrm{T}}\boldsymbol{f}_{i,j}(k) \tag{H.20}$$

where $\boldsymbol{X}(k)$ is an $(N \times m)$ matrix of the m most recent reference signal vectors (the mth row of $\boldsymbol{X}(k)$ is the reference signal used in deriving the control filter output at time $(k - m + 1)$) and $\boldsymbol{f}_{i,j}$ is the i, jth filtered reference signal vector, 'filtered' by the cancellation path transfer function between the jth control filter output and ith error sensor output. Each element, $f_{i,j}(k)$, at time k in this vector is equal to the reference signal $x(k)$ convolved with the FIR model of the cancellation path transfer function.

$$f_{i,j}(k) = \boldsymbol{x}(k) * \boldsymbol{c}_{i,j} \;\; = \boldsymbol{x}^{\mathrm{T}}(k)\boldsymbol{c}_{i,j} \tag{H.21}$$

In the multiple-input, multiple-output (MIMO) control arrangement, the aim of the exercise is to minimise the sum of the mean square values of the signal from each of the error sensors. Thus the quantity to be minimised is,

$$\sum_{i=1}^{N_e} e_i^2(k) = \sum_{i=1}^{N_e} \left[p_i(k) + \sum_{j=1}^{N_c} \boldsymbol{w}_j^{\mathrm{T}}\boldsymbol{f}_{i,j}(k) \right]^2 \tag{H.22}$$

Differentiating this with respect to the jth weight coefficient vector produces the following expression for the gradient estimate.

$$\boldsymbol{\Delta w}_j(k) \approx \sum_{i=1}^{N_e} \frac{\partial e_i^2(k)}{\partial \boldsymbol{w}_j} \;\; = 2\sum_{i=1}^{N_e} \boldsymbol{f}_{i,j}(k)e_i(k) \tag{H.23}$$

Using this gradient estimate in the standard gradient descent format produces the MIMO filtered-x LMS algorithm. For the jth control source this is expressed as (Elliott and Nelson 1985; Elliott et al. 1987),

$$\boldsymbol{w}_j(k+1) = \boldsymbol{w}_j(k) - 2\mu \sum_{i=1}^{N_e} \boldsymbol{f}_{i,j}(k)e_i(k) \tag{H.24}$$

Of course, there are many other possible control algorithms that can be used, and which may perform even better than the multi-channel **FXLMS** just outlined. These are discussed in considerable detail in Hansen et al. (2012, Ch. 6).

References

Elliott S and Nelson P 1985 Algorithm for multichannel LMS adaptive filtering. *Electronics Letters* **21**, 979–981.

Elliott S, Stothers I and Nelson P 1987 A multiple error lms algorithm and its application to the active control of sound and vibration. *IEEE Transactions on Acoustics, Speech, and Signal Processing* **ASSP-35**, 1423–1434.

Hansen C, Snyder S, Qiu X, Brooks L and Moreau D 2012 *Active control of noise and vibration*. Spon Press.

Index

Wind Farm Noise: Measurement, Assessment and Control, First Edition.
Colin H. Hansen, Con J. Doolan, Kristy L. Hansen.

Printed and bound by CPI Group (UK) Ltd, Croydon, CR0 4YY
16/04/2025
14658394-0001